Cold-Adapted Yeasts

Pietro Buzzini · Rosa Margesin
Editors

Cold-Adapted Yeasts

Biodiversity, Adaptation Strategies and Biotechnological Significance

Springer

Editors
Pietro Buzzini
Department of Agricultural, Environmental and Food Sciences,
Industrial Yeasts Collection DBVPG
University of Perugia
Perugia
Italy

Rosa Margesin
Institute of Microbiology
University of Innsbruck
Innsbruck
Austria

ISBN 978-3-642-39680-9 ISBN 978-3-642-39681-6 (eBook)
DOI 10.1007/978-3-642-39681-6
Springer Heidelberg New York Dordrecht London

Library of Congress Control Number: 2013948372

Printed on acid-free paper

Springer is part of Springer Science+Business Media (www.springer.com)

Preface

Why a book on *Cold-Adapted Yeasts*? What is the scientific background that encouraged us to edit a book on the biodiversity, adaptation strategies and biotechnological implications of these attractive microorganisms that are apparently able to survive and even to grow at low temperatures in really inhospitable habitats?

Over 80 % of the Earth's environments are permanently or periodically exposed to temperatures below 5 °C. Cold environments include deep seas, cold deserts and glacial habitats (i.e. glaciers and related habitats, icy seas, ice caps and continental ice sheets and frozen ground, which are characterised by the presence of ice in extensive masses and cover about 10 % of the Earth's surface at the present time). In many natural ecosystems, cold conditions are frequently associated with other limiting environmental factors (e.g. low water activity and nutrient availability, high hydrostatic pressure and oxidative stress, high solar irradiation, etc.) which make such extreme habitats very inhospitable (or even life-limiting) ecosystems. A few cold habitats are also associated with human activities (e.g. refrigeration technology in food industry).

Current knowledge on microbial biodiversity and ecology has shown that cold habitats harbour a wide diversity of psychrophilic prokaryotic and eukaryotic microbial life, including archaea, bacteria, cyanobacteria, yeasts, filamentous fungi, algae and protozoa. Low temperatures have a strong influence on microbial life, both indirectly (change of the physical state of water) and directly (low metabolic rate due to the reduced enzymatic activity). The key feature of both prokaryotic and eukaryotic organisms adapted to cold is to successfully overcome the negative effects of low temperatures through the development of structural and functional adaptations.

Yeasts are a group of eukaryotic organisms belonging to the Kingdom of Fungi which are widely distributed in worldwide microbiomes. Their manifest ubiquity in the Earth's biosphere is however balanced by their great diversity and specificity for the different habitats. Yeasts are probably one of the most relevant microbial groups in both traditional fermentation technologies and biotechnological applications. Most people associate yeasts with the ascomycetous species *Saccharomyces cerevisiae* (the so-called "baker yeast"), traditionally involved in the production of alcoholic beverages and in leavened bread, although this technologically domesticated species represents only a infinitesimal bit of the vast biodiversity occurring inside the yeast world. Since the 1950s, the study of cold-adapted yeasts has attracted an increasing number of scientists: consequently the number of papers published and the number of psychrophilic and psychrotolerant yeast species described in literature spectacularly increased.

In this book, prominent authors from universities and research centres present an up-to-date state-of-the-art on the biodiversity, adaptation strategies and biotechnological significance of cold-adapted yeasts in order to provide an additional source of information to all those scientists who are interested in the microbiology of these microorganisms. The book is subdivided into four main Parts:

i. an introductive part devoted to conceptual and methodological aspects related to the study of cold-adapted yeasts in natural ecosystems and the role of culture collections in handling the ever-increasing number of preserved strains and relevant data
ii. a second part reporting an overview on the diversity and ecology of cold-adapted yeasts (including the so-called "black yeasts") in worldwide cold habitats (i.e. Arctic and Antarctic regions, Alpine, Apennine and Patagonian cold areas)
iii. a third part describing the different physiological, biochemical and molecular adaptation strategies used by cold-adapted yeasts to survive or even thrive successfully in cold habitats
iv. a final part devoted to the biotechnological impact of cold-adapted yeasts as biocatalysts in traditional and advanced (actual and potential) biotechnologies or as food spoilers.

As studies of cold-adapted yeasts have been carried out over more than 60 years, original strain identification was performed using taxonomic criteria of current use at the time of isolation: hence, many species names cited in the early literature are not adjourned. Accordingly, all original taxonomic designations reported in the cited references were checked and, if necessary, updated according to the latest taxonomic guidelines published in Kurtzman et al. (2011), or more recent literature.

It is obvious that the book does not claim to be comprehensive of all aspects concerning the life of these fascinating organisms. Some topics not covered here are dealt with elsewhere, in particular the general introduction to the climate of snow and ice as boundary condition for microbial life, published in Margesin et al. (2008). Other topics are still characterized by a general insufficiency of studies (e.g. diversity and ecology of cold-adapted yeast populations in Himalayan areas) and are therefore not covered in this book.

The editors of this book want to thank all the authors for their excellent contributions and hope that this book will provide a useful tool to increase the interest in cold-adapted yeasts which, hopefully, will stimulate increasing efforts in supporting research on this attractive field. P. Buzzini would like to thank his retired colleague Prof. Ann Vaughan-Martini for the trust she demonstrated over the years. He would also like to dedicate this book in memory of his teacher (and friend) Prof. Alessandro Martini, who supported him since early 2000s in the study of this fascinating matter. Finally, the editors would like to thank the Springer team, especially Dr. Jutta Lindenborn, for valuable and continuous support during the preparation of this book.

Perugia, Italy, May 2013 Pietro Buzzini
Innsbruck, Austria, May 2013 Rosa Margesin

References

Kurtzman CP, Fell JW, Boekhout T (eds) (2011) The yeasts, a taxonomic study. Elsevier, Amsterdam

Margesin R, Schinner F, Marx JC, Gerday C (eds) (2008) Psychrophiles: from biodiversity to biotechnology. Springer, Berlin

Contents

Part I Introduction

1 **Cold-Adapted Yeasts: A Lesson from the Cold and a Challenge for the XXI Century** 3
Pietro Buzzini and Rosa Margesin

2 **Methods for the Isolation and Investigation of the Diversity of Cold-Adapted Yeasts and Their *Ex Situ* Preservation in Worldwide Collections** 23
Kyria L. Boundy-Mills

Part II Diversity and Ecology of Cold-Adapted Yeasts in Worldwide Cold Habitats

3 **Cold-Adapted Yeasts in Arctic Habitats** 49
Polona Zalar and Nina Gunde-Cimerman

4 **Cold-Adapted Yeasts in Antarctic Deserts** 75
Laurie B. Connell, Russell R. Rodriguez, Regina S. Redman and Joseph J. Dalluge

5 **Cold-Adapted Yeasts in Alpine and Apennine Glaciers** 99
Benedetta Turchetti, Marta Goretti, Pietro Buzzini and Rosa Margesin

6 **Cold-Adapted Yeasts in Patagonian Habitats** 123
Virginia de Garcia, Diego Libkind, Martín Moliné, Carlos A. Rosa and Maria Rosa Giraudo

7 **Cold-Adapted Yeasts in Deep-Sea Environments** 149
Yuriko Nagano, Takahiko Nagahama and Fumiyoshi Abe

8 **Black Yeasts in Cold Habitats** 173
Laura Selbmann, G. Sybren de Hoog, Laura Zucconi, Daniela Isola and Silvano Onofri

Part III Adaptation Strategies of Cold-Adapted Yeasts to Low Temperatures

9 **Production of Pigments and Photo-Protective Compounds by Cold-Adapted Yeasts** 193
Martín Moliné, Diego Libkind, Virginia de Garcia and María Rosa Giraudo

10 **Changes in Lipids Composition and Fluidity of Yeast Plasma Membrane as Response to Cold** 225
Nina Gunde-Cimerman, Ana Plemenitaš and Pietro Buzzini

11 **Cold-Shock Response and Adaptation to Near-Freezing Temperature in Cold-Adapted Yeasts** 243
Masayori Inouye and Sangita Phadtare

12 **Production of Antifreeze Proteins by Cold-Adapted Yeasts** 259
Hak Jun Kim, Jun Hyuck Lee, Hackwon Do and Woongsic Jung

13 **Role of Sterol Metabolism and Endoplasmic Reticulum-Associated Degradation of Proteins in Cold Adaptation of Yeasts** 281
Kelaine Haas

14 **Subzero Activity of Cold-Adapted Yeasts** 295
Nicolai S. Panikov

15 **Fundamentals of Cold-Active Enzymes** 325
Charles Gerday

Part IV Biotechnological Significance of Cold-Adapted Yeasts

16 **Cold-Active Yeast Lipases: Recent Issues and Future Prospects** 353
Mirosława Szczęsna-Antczak, Janina Kamińska, Tomasz Florczak and Marianna Turkiewicz

17 Miscellaneous Cold-Active Yeast Enzymes of Industrial Importance 377
Aneta Białkowska and Marianna Turkiewicz

18 Production of Polymers and Other Compounds of Industrial Importance by Cold-Adapted Yeasts 397
Kostantsa I. Pavlova

19 Low-Temperature Production of Wine, Beer, and Distillates Using Cold-Adapted Yeasts 417
Maria Kanellaki, Argyro Bekatorou and Athanasios A. Koutinas

20 Cold-Adapted Yeasts as Biocontrol Agents: Biodiversity, Adaptation Strategies and Biocontrol Potential 441
Marcela P. Sangorrín, Chistian Ariel Lopes, Silvana Vero and Michael Wisniewski

21 Bioremediation and Biodegradation of Hydrocarbons by Cold-Adapted Yeasts 465
Rosa Margesin

22 Heterologous Expression of Proteins from Cold-Adapted Yeasts in Suitable Hosts: Methods and Applications 481
Rosli Md. Illias, Aizi Nor Mazila Ramli, Kheng Oon Low, Nor Muhammad Mahadi, Abdul Munir Abdul Murad and Amir Rabu

23 Food Spoilage by Cold-Adapted Yeasts 497
Anna Maráz and Mónika Kovács

Subject Index 533

Microorganism Index 543

Part I
Introduction

Chapter 1
Cold-Adapted Yeasts: A Lesson from the Cold and a Challenge for the XXI Century

Pietro Buzzini and Rosa Margesin

Abstract Yeasts are a versatile group of eukaryotic organisms and widely distributed in worldwide microbiomes. The literature on cold-adapted yeasts considerably increased in the last two decades. For some decades, Antarctica has been the geographical area preferred by microbiologists for studying the diversity of cold-adapted microorganisms including yeasts. However, in more recent years, the biodiversity and the ecology of cold-adapted yeast populations colonizing worldwide non-Antarctic cold habitats have attracted an increasing number of scientists. Like other microbial taxa, yeasts inhabiting cold ecosystems have evolved a set of structural and functional adaptation strategies to overcome the negative effects of cold (sometimes associated with other limiting conditions) that make such extreme habitats very inhospitable for microbial life. Due to their singular phenotypic traits, cold-adapted yeasts have also been studied in recent years for their biotechnological potential.

Keywords Psychrophilic yeasts · Psychrotolerant yeasts · Cold ecosystems · Cold adaptation

---

P. Buzzini (✉)
Department of Agricultural, Environmental and Food Sciences and Industrial Yeasts Collection DBVPG, University of Perugia, Borgo XX Giugno 74, 06121 Perugia, Italy
e-mail: pietro.buzzini@unipg.it

R. Margesin
Institute of Microbiology, University of Innsbruck, Technikerstrasse 25, 6020 Innsbruck, Austria
e-mail: rosa.margesin@uibk.ac.at

P. Buzzini and R. Margesin (eds.), *Cold-Adapted Yeasts*,
DOI: 10.1007/978-3-642-39681-6_1, © Springer-Verlag Berlin Heidelberg 2014

Contents

1.1 Introduction 4
1.2 Cold-Adapted Organisms: Psychrophilic and Psychrotolerant Life Forms 5
1.3 Fundamentals of Cold-Adapted Yeasts 7
1.4 Ecology of Yeasts in Cold Ecosystems 10
1.5 Adaptation of Yeasts to Cold: The Basis of Successful Colonization and Survival 12
1.6 Cold-Adapted Yeasts: Trends and Perspectives in Biotechnology 15
1.7 Future Outcomes and Challenges for the XXI Century 16
References 18

1.1 Introduction

More than 80 % of the Earth's environments exhibit temperatures below 5 °C (Cavicchioli and Tortsen 2000). The major fraction of cold habitats is represented by deep oceans, Arctic and Antarctic snow- and ice caps, permafrost soils, sea ice and high mountain glaciers. Antarctica (about 14 million km^2) is the most representative example of a terrestrial cold habitat, of which almost the totality of surface is covered by ice and snow (Holdgate 1977). In many ecosystems, cold conditions are frequently associated with other limiting factors (e.g. low water availability (A_w) and nutrient availability, high hydrostatic pressure and oxidative stress, high solar irradiation) which make such extreme habitats very inhospitable (or even life-limiting) ecosystems (Claridge and Campbell 1977; Vincent 1988; Campbell and Claridge 2000).

Current knowledge about the microbial biodiversity and ecology has shown that even the most extreme cold habitat (e.g. Antarctica) harbours enormously diverse, viable and metabolically active cold-adapted micro- and mesobiomes comprising representatives of all three domains of life (*Bacteria, Archaea* and *Eukarya* including yeasts, filamentous fungi, protists, lichens and small invertebrates) (Cameron et al. 1970, 1971; Vishniac and Mainzer 1972; Vincent 1988; Wynn-Williams 1990; Staley and Gosink 1999; Bej et al. 2010; Miller and Whyte 2012).

Of course, some cold habitats are also associated with human activities, principally thanks to refrigeration technology in food industry: the study of cold-adapted spoilage microbial populations surviving (and even actively growing) under these conditions makes the efforts to rid these habitats of microbial contamination more effective (Stratford 2006).

Low temperatures have a strong influence on whether a certain organism can survive or even successfully thrive in a given habitat. This effect may be indirect (through its influence on the physical state of water) and direct (through the degree of metabolic rate due to the activity of enzymes) (Poindexter 2009). Despite the influence of other environmental factors beside temperature on both growth and metabolic efficiency, the key attribute of both prokaryotic and eukaryotic organisms adapted to cold is to successfully overcome the negative effects of low temperatures through the development of a range of structural and functional adaptations (Deming 2002; Siddiqui and Cavicchioli 2006; Feller 2007; Margesin et al. 2008; Deming 2009; Bej and Mojib 2010; Gostincar et al. 2010; Shivaji and Prakash 2010; Margesin and Miteva 2011).

Yeasts are a group of eukaryotic organisms belonging to the kingdom of fungi. Since the pioneer investigations carried out in the 1860s by Pasteur, the study of the yeast world has experienced a very important advance in terms of its understanding, phenotypic and taxonomic characterization and commercial exploitation (Walker 1998; Johnson and Echavarri-Erasun 2011; Kurtzman et al. 2011a). Most people associate yeasts with the ascomycetous species *Saccharomyces cerevisiae*, with the production of alcoholic beverages and with leavened bread. In fact, it is not uncommon in some areas of molecular biology to treat the words "yeast" and "*Saccharomyces*" as synonyms. Besides, owing to advances in functional genomics and systems biology, *S. cerevisiae* is presently assumed to represent the primary model eukaryotic organism for biological studies. This is in spite of the fact that this domesticated species represents only a tiny fragment of the vast biodiversity occurring in the yeast world. In fact, yeasts are distributed between the Basidiomycota and Ascomycota phyla due to their paraphyletic nature, like filamentous fungi (Kurtzman et al. 2011a). Moreover, in recent decades, an increasing body of the academic and industrial research has paid its attention to the so-called non-conventional yeasts (NCYs) (Wolf et al. 2003; Buzzini and Vaughan-Martini 2006), of which cold-adapted yeasts can be considered a not irrelevant portion.

Some fundamental aspects on the ecology, adaptation strategies to low temperatures and biotechnology significance of cold-adapted yeasts are treated in this chapter.

1.2 Cold-Adapted Organisms: Psychrophilic and Psychrotolerant Life Forms

The classical definitions based on thermal dependence of growth kinetic parameters (Eddy 1960; Morita 1975; Cavicchioli and Tortsen 2000) are still used to distinguish psychrophilic from mesophilic and thermophilic organisms. However, although all organisms that successfully thrive on cold environments are usually referred to as psychrophiles, a clear dichotomy should be highlighted. Psychrophilic ("cold-loving") organisms are differentiated from psychrotolerant ("cold-tolerant") organisms on the basis of their limit of temperature for growth. Accordingly, psychrophiles (often labelled as obligate psychrophiles) have an optimum growth temperature of about 15 °C or lower, an upper growth temperature of about 20 °C and a minimum growth temperature of 0 °C. In contrast, organisms that grow at 0 °C, but have growth optima of 20–30 °C, are called psychrotolerant or facultatively psychrophilic (van Uden 1984; Cavicchioli and Tortsen 2000; Raspor and Zupan 2006; Vishniac 2006a). Moyer and Morita (2007) also termed such organisms psychrotrophic; however, Canganella and Wiegel

(2011) considered this term ("cold-eaters") obsolete. Nonetheless, the questionableness of using a maximum growth temperature as an indicator of how well an organism is adapted to the cold has been emphasized (Cavicchioli 2006).

Psychrophiles are commonly prevalent in permanently cold habitats (e.g. polar regions, high altitudes or deep sea). In contrast, psychrotolerant organisms, which are able to grow over a wider range of temperature and show higher growth rates above 20 °C, are predominant in environments characterized by fluctuating low temperatures (Russell 1990). Based on the above definitions, psychrotolerants have evolved to tolerate cold, but they are not apparently as physiologically specialized as psychrophiles (Gounot 1986). The ability of psychrophiles to grow slowly at low temperatures may be in reality an advantage in oligotrophic environments, where a rapid exhaustion of available nutrients would lead to cell starvation (Russell 1997). Besides, psychrophiles can be considered as true extremophiles as they are adapted not only to low temperatures, but frequently also to further environmental constrains, such as extremely high pressures (Yayanos 1995), low A_w and nutrient availability (Friedmann 1982; Staley and Gosink 1999) and exposition to strong UV radiation (Carpenter et al. 2000).

van Uden (1984) investigated in his monumental studies growth temperature profiles of psychrophilic, mesophilic and thermophilic yeasts. Recent studies focussed on the relationships between growth kinetic parameters and temperatures and demonstrated that the temperature at which yeasts grow fastest is generally different from that allowing the highest production of biomass (growth yield). A relationship between the temperatures allowing maximum growth rate and the source of the strain was also observed (Sabri et al. 2000; Margesin 2009a; Rossi et al. 2009).

It has been emphasized that the temperature resulting in the maximum growth rate (often erroneously termed "optimal growth temperature") in both prokaryotes and eukaryotes could reflect only kinetic effects and occur above the linear part of the Arrhenius curve, where the physiological conditions could not be ideal, thus inducing cellular stress (Gounot and Russell 1999; Glansdorff and Xu 2002; Feller and Gerday 2003; D'Amico et al. 2006; Feller 2007). Accordingly, Feller and Gerday (2003) considered the use of the classical definitions of Morita (1975) based on thermal dependence of growth kinetic parameters for three main reasons to be ambiguous: (1) the limits of temperature result apparently arbitrarily selected, because they do not correspond to any clear separation of biological processes or environmental conditions; (2) the classical definition cannot be successfully applied to most eukaryotes; (3) microorganisms work both as thermodynamic (any increase in temperature increases in parallel reaction and growth rate) and as biological units (at a given temperature, some key heat-labile metabolic steps can impair the performance of some pathways). In other words, although in some cases a temperature shift (e.g. from 5 to 25 °C) can increase the growth rate of some cold-adapted organisms, their physiological state can appear seriously altered (i.e. reduced viable counts, enzyme synthesis and membrane functionality) (Margesin 2009a). Accordingly, few authors seriously reconsider the use of growth rates as an exclusive arbiter to define the optimum growth temperature.

The terms stenothermal and eurythermal have also been proposed for organisms growing over narrow and wide temperature ranges, respectively. Accordingly, psychrophiles could be labelled as stenothermal psychrophiles, whereas psychrotolerant would be eurythermal psychrophiles. Such a definition takes into account that psychrotolerants (which grow within a wide range of temperature) are much more abundant in cold environments, probably because they can tolerate larger temperature fluctuations (Feller and Gerday 2003). Nevertheless, due to the facts that some evidences suggest that there is a continuum in temperature adaptation for life and that a wide or narrow growth temperature range depends on the organism, in this chapter we use the general term "cold-adapted" to designate both psychrophilic and psychrotolerant organisms.

1.3 Fundamentals of Cold-Adapted Yeasts

Yeasts can be defined as those fungi whose asexual growth predominantly results from budding or fission and which do not form their sexual states within or upon a fruiting body (Kurtzman et al. 2011b). For ascomycetous yeasts, this distinction has been substantiated by molecular comparisons, which demonstrate that budding and fission yeasts are phylogenetically distinct from one another and from Euascomycetes (Kurtzman 2011). A similar distinction can be made for basidiomycetous yeasts, which are phylogenetically distinct from filamentous fungi that form complex fruiting bodies (Kurtzman et al. 2011c).

Yeasts are probably one of the most relevant microbial groups in biotechnology. Mankind has benefited from yeasts for millennia in traditional fermentation technologies. Studies in recent decades aimed at exploring the metabolic diversity of yeasts have revealed innumerable promising properties relevant to biotechnological applications, such as interesting activities of cells and metabolites, the use of yeast cells as hosts for heterologous expression (e.g. for the synthesis of fine chemicals), as well as their role in important agricultural and environmental interactions. Some of these processes and products have reached commercial utility, while others are still confined at the laboratory scale. The impact of yeast biotechnology has been extensively documented in a number of reviews (Walker 1998; Buzzini and Vaughan-Martini 2006; Johnson and Echavarri-Erasun 2011; Schisler et al. 2011).

The significance of yeasts in biotechnology is not exclusively restricted to their role as "metabolic factories". Like bacteria and filamentous fungi, they also make a significant contribution as food spoilers. Food spoilage catalysed by yeasts is often predictable, principally occurring in those products where competition from faster-growing bacteria is either retarded or prevented by the properties of the food and the conditions of processing and storage (e.g. low temperatures) (Stratford 2006; Fleet 2011).

With regard to their ecological relevance, yeasts are widely distributed in worldwide microbiomes (Starmer and Lachance 2011). The evident ubiquity of yeasts in the Earth's biosphere is balanced by their diversity, specificity for habitats and relationships with other organisms. As recently underlined by Shivaji and Prasad (2009), yeasts are a versatile group of eukaryotic microorganisms exhibiting heterogeneous nutritional profiles and a surprising ability to survive in a wide range of natural and altered habitats (Starmer and Lachance 2011). They colonize a variety of ecosystems and are well adapted to a wide range of climates, altitudes, substrates and geographical locations. It is possible to find yeasts in glaciers, high-salinity lakes, water, soil, air, intestines of vertebrates and invertebrates (Starmer and Lachance 2011), and even in acid waters (Russo et al. 2008) and marine deep-sea environments (Nagahama 2006).

Following a classical food chain model, free-living yeasts are primarily decomposers due to their heterotrophic status. Typically, they are among the earlier colonizers of nutrient-rich substrates, where they are followed by a succession of organisms that degrade dead organic matter. However, yeasts are not just decomposers but can assume a diversity of forms and functions in the natural world. Along with their role in the biocycling of nutrients, they can engage in intimate relationships with other organisms as mutualists, competitors, parasites or pathogens (Starmer and Lachance 2011). Nevertheless, despite a plethora of studies, it is estimated that only 1 % of all existing yeast species is currently known. From 1820 to 2011, the number of described yeasts has increased spectacularly. Currently, there are approx. 1,500 recognized yeast species, which means the expected number of yeast species on Earth would be around 150,000 (Hawksworth 2004; Lachance 2006; Starmer and Lachance 2011; Kurtzman et al. 2011a).

It has been suggested that yeasts might be better adapted to low temperatures than bacteria (Turkiewicz et al. 2003; Shivaji and Prasad 2009). For many years, Antarctica has been the geographical area preferred by microbiologists for studying the diversity of cold-adapted microorganisms including yeasts (Vishniac 2006a; Onofri et al. 2007; Shivaji and Prasad 2009). However, in more recent years, the role of cold-adapted yeast populations in non-Antarctic cold habitats has attracted an increasing number of scientists (Table 1.1). The diversity of yeasts in worldwide cold ecosystems is reported in Chaps. 3, 4, 5, 6, 7, 8.

The chronological sequence of the literature so far published on the taxonomy, ecology, physiology and biotechnology of cold-adapted yeasts, the number of yeast species (including their synonyms) cited in those studies and the ratio between the number of synonyms and that of recognized yeast species is reported in Fig. 1.1. Since the 1950s, the number of papers published and the number of species of cold-adapted yeasts cited in literature dramatically increased (Buzzini et al. 2012). In contrast, the percentage of species misidentified (i.e. use of synonymous) decreased from over 50 % (observed until the 1980s) to slightly more than 10 % (in the 2010s). This is certainly due to the refinement of the molecular taxonomy techniques, which reduced the margin of error in the taxonomic description of yeast species.

As early ecological studies of cold-adapted yeasts have been carried out about 60 years ago, original strain identification was performed using taxonomic criteria

Table 1.1 Occurrence of cold-adapted yeast species in cold ecosystems

Sampling area	Isolation source	No. of yeast species identified
Antarctica	Decaying wood and algae	23
	Ice cores	5
	Lake water and sediments	8
	Lichen and moss	4
	Moulting feathers of penguins	6
	Rock cryptoendolithic habitat	1
	Sea sponge	1
	Sea water	3
	Snow	3
	Soil	54
Arctic	Ice cores	27
	Permafrost	7
	Puddles on snow	1
	Sand and sediment	3
	Sea water	4
	Puddles	7
	Soil	15
Alps	Glacier cryoconites	4
	Ice cores	4
	Mud	2
	Glacier rock debris	19
	Soil	8
	Meltwater	4
Apennines	Ice cores	2
	Meltwater	3
	Glacier rock debris	30
Andes	Lagoon of glacial origin	3
	Meltwater	44

of current use at the time of isolation; hence, many designations resulted not updated or even incorrect. The traditional phenotypic methods used to identify yeasts are rapidly being replaced by sequence-based approaches. Accordingly, all original taxonomic designations of cold-adapted yeasts reported in the literature before the 1990s should be checked (and, if necessary, updated) according to the latest taxonomic guidelines reported by Kurtzman et al. (2011c). Molecular techniques used since the 2000s (sequencing of variable D1/D2 regions of the 26S rDNA gene, 18S, 5.8S and mitochondrial small sub-unit rDNAs gene, as well as ITS and RFLP-ITS) have greatly contributed to the increase in the number of novel species identified and, at the same time, to the decrease in the number of those misidentified (Kurtzman and Robnett 1998; Fell et al. 2000; Kurtzman et al. 2011c).

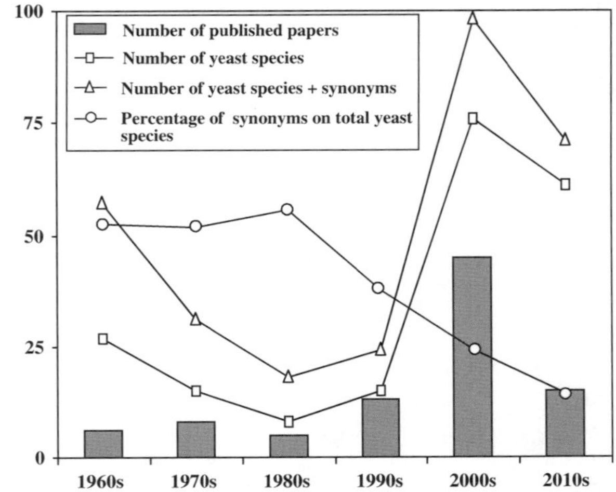

Fig. 1.1 Development of the literature published on cold-adapted yeasts and of cited yeast species from the 1960s to the 2010s

However, cold-adapted yeasts are in all probability more abundant than assumed, because some are difficult to culture under standardized laboratory conditions. With no doubt, recent studies of worldwide cold habitats have resulted in the description of numerous culturable yeast species exhibiting temperature optima for growth (fastest growth rates) between 4 and 15 °C. There is also evidence for novel, not yet described taxa. Interestingly, on average, a percentage of about 10–20 % of the yeast strains isolated during the last 60 years was attributed to species still not described (later identified as novel species), thus confirming that cold ecosystems represent a well-established reservoir of so far unknown yeast diversity (Buzzini et al. 2012). Recently, also some non-culturable taxa were observed (Arenz et al. 2006; Butinar et al. 2007; Bridge and Newsham 2009).

1.4 Ecology of Yeasts in Cold Ecosystems

Although some authors (Price and Sowers 2004) underlined that the physiological characters of microorganisms cultured under controlled laboratory conditions do not necessarily reflect their *in situ* performances, the impact of metabolic activity of cold-adapted microbial populations in biogeochemistry of cold ecosystems is one of the most stimulating environmental questions. The functional diversity (including interspecific ecological interactions) of cold-adapted yeasts in worldwide cold habitats may also be characterized on the basis of their physiological traits. However, although most of them are belonging to species acknowledged for their versatile metabolic traits and high resistance to stress associated with cold environment (e.g. long-term freezing, freeze–thaw cycles, low A_w and nutrient availability, high solar irradiation), little is known to date about the ecology of yeasts in cold habitats, because most of the literature so far published was merely explorative (Shivaji and Prasad 2009; Buzzini et al. 2012).

Antarctica was considered so far the privileged terrestrial model to explore the possible relationships between geographical (e.g. latitude, altitude), physical and chemical parameters and yeast diversity (Shivaji and Prasad 2009). Pioneer studies carried out by di Menna (1960, 1966a, b) found a general difficulty in isolating cold-adapted yeasts from some Antarctic samples. Viable yeasts were preferably found in acidic substrates, although inspection of the results showed that high pH values were not in themselves inhibitory. It was also observed that the isolation of yeasts from Antarctic soils was apparently dependent on the presence of plant residues (di Menna 1966a). Furthermore, Babyeva and Golubev (1969) isolated more yeasts at 5 °C than at higher temperature and showed that about 40 % of total isolates were psychrophiles, failing to grow above 20 °C.

Some years later, Vishniac (1996, 2006a, b) questioned the role of temperature as the exclusive parameter affecting yeast distribution in Antarctica and concluded that the biodiversity of fungi (including yeasts) in terrestrial ecosystems increases with the availability of water and energy. On the contrary, Connell et al. (2008) identified pH and electrical conductivity as significant predictors of yeast biodiversity in Antarctic soils. Obviously, the large metabolic variability existing between different species (and even between different strains) can affect their survival and growth in different cold ecosystems, thereby influencing yeast biodiversity.

A number of studies described viable cells in ancient Antarctic ice cores or permafrost soil (Abyzov 1993; Margesin 2009b). So, one question has recently drawn significant interest: Do microbial cells (including cold-adapted yeasts) simply survive entrapped frozen for hundreds of thousands of years or are they metabolically active and responsible for certain natural processes? It has been postulated that microbial activity at sub-zero temperatures could be restricted to unfrozen water inside the ice cores (possibly adhering to the internal surface of inner gas bubbles) and permafrost soil, and to brine channels (D'Amico et al. 2006; Deming 2009). These microamounts of water contain high concentrations of salts, polymeric compounds and particulate matter, and fluid flow may be maintained by concentration and temperature gradients. Hence, although this question is still unresolved (and probably is dependent on the sites and the studied organisms), two possible mechanisms could be postulated (Psenner et al. 2002; Price and Sowers 2004): (1) survival with low metabolic activity (with or without reproduction); (2) long-period dormancy: in this case, traces of metabolic activity are exclusively devoted to support macromolecular damage repair.

The genus *Cryptococcus* has been described as the most important yeast taxon in worldwide cold ecosystems (Vishniac 1996, 2006a; Shivaji and Prasad 2009; Buzzini et al. 2012). This dominance is generally ascribed to its ability to produce a polysaccharide capsule (Vishniac 2006a). However, Connell et al. (2008) questioned the above hypothesis, supposing that some additional adaptation mechanisms (e.g. a wider carbon and nitrogen assimilation pattern) could support its superior ability to colonize cold natural habitats.

Another important point to be underlined is that some species of cold-adapted yeasts are apparently able to colonize exclusively some well-defined geographical areas: (1) *Cryptococcus antarcticus*, *Cryptococcus friedmannii*, *Cryptococcus nyarrowii*, *Cryptococcus vishniacii*, *Dioszegia antarctica*, *Dioszegia cryoxerica*, *Dioszegia statzelliae* and *Glaciozyma antarctica* have been so far exclusively found in Antarctica (Vishniac 1996, 2006a; Shivaji and Prasad 2009; Buzzini et al. 2012); (2) *Rhodotorula psychrophila* has been isolated so far exclusively in cryoconites of Alpine glaciers (Margesin et al. 2007); (3) *Rhodotorula himalayensis* has been found only in cold Himalayan soil (Shivaji et al. 2008). So, the ecological question is, are such species endemic or else? Vishniac (1999, 2006a) underlined that Antarctica (due to its remoteness for millions of years) should be the favourite Earth place that could support the presence of endemic organisms, probably as the consequence of evolutionary processes of microbial speciation. This hypothesis could be satisfactory also for explaining the presence of endemic yeast species in the Himalayas, but certainly it could be much less convincing for Alpine ecosystems. However, the same author also underlined that the isolation of certain species exclusively from a well-defined geographical area does not constitute sufficient confirmation supporting its endemism, because the possibility that such species may in the future be found in other habitats cannot be excluded (Vishniac 1999). For instance, *Pseudozyma antarctica* (formerly *Candida antarctica*) and *Cryptococcus victoriae* were first isolated from Antarctica (Goto et al. 1969; Montes et al. 1999), but later these species were found in other temperate non-Antarctic habitats. In other words, some cold-adapted organisms (including yeasts) are extremely versatile in their adaptive abilities, and therefore, they would change their apparent endemic status to attain a more ubiquitous distribution (Vishniac 1999).

On the other hand, some ubiquitous species has also been recently isolated from cold habitats (Vishniac 2006a; Shivaji and Prasad 2009; Buzzini et al. 2012). So, another ecological question is, have such species become cold-adapted or else? This is hard to establish. Connell et al. (2008) underlined that the simple isolation of a given yeast taxa from a cold site does not a priori indicate that such organism has developed some form of adaptation to low temperatures. Thus, the significance of their presence in cold habitats is at present unknown.

1.5 Adaptation of Yeasts to Cold: The Basis of Successful Colonization and Survival

Since permanently cold habitats exert a highly selective pressure on microbial populations, the evolution of a number of adaptation mechanisms with regard to metabolic activities and protection strategies enables these organisms to be active at low (even at sub-zero) temperatures. Although organisms in cold habitats are subjected to temperature fluctuations and frequent freeze–thaw events, the lowest temperature limit for life, reported for microorganisms living in sea ice and in permafrost soil, seems to be around −20 °C (Russell 2002; Price and Sowers 2004; D'Amico et al. 2006; Bakermans 2008). However, substantial growth and metabolic activity (respiration and biosynthesis) of permafrost bacteria and fungi at temperatures even down to −35 °C have been demonstrated (Panikov and Sizova 2007). Sub-zero activity of cold-adapted yeasts is described in Chap. 14.

As above stated, other factors such as low A_w and nutrient availability, high hydrostatic pressure, osmotic and oxidative stress, and solar irradiation also strongly affect living conditions. Therefore, adaptation to cold is often combined with adaptation to other limiting conditions (Vishniac 1996, 2006a; Shivaji and Prasad 2009; Buzzini et al. 2012). An exhaustive description of pigments and photoprotective compounds synthesized by cold-adapted yeasts in highly irradiated cold ecosystems is reported in Chap. 9.

To successfully colonize cold environments, both prokaryotes and eukaryotes have evolved a number of genetically based adaptive and acclimatory responses. The timescale exposure is essential to interpret the effects of sub-optimal temperature on microbial physiology. A rapid decrease in temperature is expected to generate fast, highly dynamic stress response on living cells. If the exposure to sub-optimal temperature is quite prolonged, this leads to acclimation, which implicates regulatory mechanisms resulting in the full adjustment of both genomic expression and physiological traits during a lifetime. Examples of acclimatory responses to fluctuating cold conditions are dependent upon sensor/signal pathways and involve modulations of enzyme transcription or translation rates (Morgan-Kiss et al. 2006; Rossi et al. 2009). In contrast, when the evolutionary selection of the gene alleles (which increases the fitness for a given environmental niche, e.g. cold habitat) occurs over a timescale of several generations, the response is termed adaptation. Examples of adaptive mechanisms to cold include the evolution towards an increased membrane fluidity through the accumulation of polyunsaturated fatty acids, the ability to synthesize cold-shock and anti-freeze proteins, the modulation of kinetics of key enzymes and the synthesis of other protecting compounds (Vishniac 1999; Gerday et al. 1997, 2000; Morgan-Kiss et al. 2006; Margesin et al. 2008; Rossi et al. 2009; Buzzini et al. 2012).

To increase membrane fluidity, cold-adapted yeasts apply various strategies. When growth temperature is lowered, the most frequently noted change in fatty acid composition is an increased extent of unsaturation (Vishniac 2006a; Rossi et al. 2009). This determines an improved membrane fluidity causing both the maintenance of the appropriate physical state of the lipid bilayer and the good functionality of membranes in perception of environmental signals (Los and Murata 2004; Morgan-Kiss et al. 2006). Differences in fatty acid composition have been described in yeasts growing at diverse temperatures (McMurrough and Rose 1973; Arthur and Watson 1976; Watson and Arthur 1976; Watson 1978; Watson et al. 1978). Changes in fatty acid composition and fluidity of cell membranes of cold-adapted yeasts are reported in detail in Chap. 10.

Compared to mesophilic and thermophilic counterparts, cold-adapted fungi (including yeasts) permanently display a superior ability to synthesize particular types of proteins (cold-shock and heat-shock) as a response to abrupt temperature changes. The number of proteins synthesized increases with the severity of the cold-shock (Julseth and Inniss 1990a, b; Deegenaars and Watson 1998; Phadtare et al. 1999; Phadtare and Inouye 2008). An in-depth overview of cold-shock response and adaptation to near-freezing temperature of cold-adapted yeasts is reported in Chap. 11.

Osmotic protection of microbial cells against cold conditions (Robinson 2001) is also achieved by the accumulation of anti-freeze proteins. They are ice-binding proteins that have the ability to inhibit the growth of ice crystal structure through their modification (Gilbert et al. 2004, 2005). These proteins lower the freezing point of water without altering the melting point (thermal hysteresis) and, at the same time, inhibit the growth of large cytoplasm crystals at the expense of small crystals at sub-zero temperatures, thus protecting the cell against intracellular freezing or minimizing the deleterious effects of ice crystal formation. The ability of cold-adapted yeasts to synthesize anti-freeze proteins is reviewed in Chap. 12.

Besides, cold-adapted yeasts have developed the ability to adapt their metabolism to low temperatures through the synthesis of cold-active enzymes. The conformation and 3D structures of cold-active enzymes so far studied are not markedly different from their mesophilic homologues. On the contrary, they exhibit some important differences if compared with their mesophilic homologues (Gerday et al. 1997; Feller and Gerday 2003; D'Amico et al. 2006): (1) up to ten times higher activity at low and moderate temperatures; (2) heat lability and possible inactivation at moderate temperatures that are not detrimental for mesophilic enzymes; and (3) maintenance of an appropriate flexibility and dynamics of the active site at low temperatures at which their mesophilic and thermophilic counterparts have severely restricted molecular motions. The fundamentals of cold-active enzymes are discussed in Chap. 15.

Finally, recent studies reported additional metabolic adaptation strategies used by cold-adapted yeasts to overcome cold conditions. The roles of sterol metabolism and endoplasmic-reticulum-associated protein degradation (ERAD) in cold adaptation and the sub-zero activity of cold-adapted yeasts are in depth described in Chaps. 13 and 14, respectively.

Despite the above plethora of studies, modern genomic approaches are needed for a better understanding of the molecular basis of cold adaptation in yeasts. While genome sequence is the first step in understanding the adaptation of a given organism to life under extreme conditions, a more powerful tool should be the integration of genomics with metabolic function through physiological and biochemical investigations. As of May 2013, at least 80 ascomycetous and basidiomycetous yeast genomes have been sequenced (www.ncbi.nlm.nih.gov/genome). However, to the authors' knowledge, only the genomic sequence of the species *P. antarctica* is available online as a representative of cold-adapted yeast species. Only a considerable increase in the number of sequenced genomes of cold-adapted microorganisms (including yeasts) and wider application of novel metagenomic and metaproteomic technologies will provide additional valuable information on the mechanisms of cold adaptation. In fact, more detailed genomic approaches are needed to advance our knowledge of the dominant life forms at low temperature as well as to identify new eukaryotic models of low-temperature adaptation. The emerging genomic and proteomic analyses may provide further new insights into cold-adapted lifestyle. In the future, genome-wide analysis of gene expression of cold-adapted yeasts could enable the efficient monitoring of the genes involved in the physiological state that governs their cold adaptation.

1.6 Cold-Adapted Yeasts: Trends and Perspectives in Biotechnology

Biodiversity of yeasts is being studied not only to catalogue life on Earth. Yeasts are probably one of the first organisms domesticated by humankind, for example they are used in traditional foods for immemorial time. Production of wine, beer and bread are three examples of the importance of yeasts in human nutrition and culture (Walker 1998). Apart from the importance in traditional technologies, yeast biotechnology is a growing field where physiological abilities of a number of yeast species (both wild-type strains and strains improved by means of metabolic engineering) have been recently studied as a source of novel chemicals of commercial value (Buzzini and Vaughan-Martini 2006; Johnson and Echavarri-Erasun 2011).

Yeasts are potent metabolic factories for biotechnology and their use as source of enzymes represents undoubtedly one of the most fascinating yeast-associated biotechnological aspects. Compared with certain fungi (e.g. *Aspergillus niger*) and bacteria (e.g. *Bacillus* spp.), yeasts are not particularly rich sources of industrially useful enzymes. Nevertheless, some species have been tested since the 1980s as enzyme producers for potential industrial exploitations (Johnson and Echavarri-Erasun 2011). However, the enzymes obtained from mesophilic yeasts exhibit sometimes poor performances in some mild technologies, such as the biocatalysis of heat-labile molecules (Walker 1998; Buzzini and Vaughan-Martini 2006; Johnson and Echavarri-Erasun 2011). The biotechnological relevance of cold-active enzymes synthesized by cold-adapted yeasts is a well-established trend (Gerday et al. 1997; Feller and Gerday 2003). Lipases, α-amylases, glucoamylases, proteases and xylanases represent the most famous examples of cold-active enzymes produced by cold-adapted yeasts (Shivaji and Prasad 2009; Johnson and Echavarri-Erasun 2011; Buzzini et al. 2012). These enzymes find optimal application in the case of (food and non-food) low-temperature processing (Gerday et al. 1997, 2000; Feller and Gerday 2003). Besides, their high activity at mild temperatures offers potential economic benefits in biocatalysis through substantial energy savings in large-scale processes. The industrial importance of cold-active enzymes produced by yeasts is extensively reported in Chaps. 16 and 17.

The biotechnology of cold-adapted yeasts is not merely confined to the production of cold-active enzymes: a number of recent studies carried out at the laboratory scale have demonstrated that their future perspectives could also be directed towards the production of polymeric compounds (Alchihab et al. 2009; Pavlova et al. 2009). The heterologous expression of proteins from cold-adapted yeasts in suitable hosts is an additional possible venture (Ramli et al. 2011). These topics are in depth treated in Chaps. 18, 19, 20, 21, 22.

The use of cold-adapted yeasts for agricultural, food and environmental biotechnologies could represent another attracting way to ensure their possible economical exploitation: low-temperature wine-making and beer-making, control of post-harvest diseases of fruits and seeds, and bioremediation of hydrocarbon-polluted cold ecosystems represent undoubtedly the most relevant challenges (Margesin et al. 2003; Bergauer et al. 2005; Krallish et al. 2006; Margesin and Feller 2010; Robiglio et al. 2011; Lainioti et al. 2012) An in-depth discussion of these interesting topics is reported in Chaps. 19, 20, 21.

However, it would be too easy to believe that the biotechnological significance of cold-adapted yeasts is exclusively limited to their role as pro-biotech microorganisms. Because of their ability to successfully thrive in cold environments, some yeasts can act as contaminants (and potential spoilers) of foods stored at low temperatures. Cold-adapted spoilage yeasts can potentially cause physical and chemical deterioration of foods, thus determining a potential risk to consumers and a relevant economic burden for food companies (Stratford 2006; Fleet 2011). This topic is treated in Chap. 23.

1.7 Future Outcomes and Challenges for the XXI Century

Although cold ecosystems constitute one of the largest biospheres on Earth, prokaryotic and eukaryotic organisms that not only survive, but also thrive in these extreme habitats are still poorly understood. Accordingly, cold habitats can be considered one of the last unexplored frontiers of the planet. The emerging field of biogeography has demonstrated that some dominant species appear to be endemic to their ecosystem. Polar cold environments provide ideal model systems of similar ecological characteristics separated geographically by climatic barriers. Research on the biodiversity of cold-adapted yeasts in worldwide cold ecosystems is essential to establish yeast species richness, to learn more about their adaptation strategies and to decrypt the molecular basis of their adaptation to cold and to examine new strains for the production of useful compounds. The comparison of genomes of cold-adapted yeasts (when they will become available) with those of their phylogenetically related mesophilic counterparts could give new insights into the genetic basis for the phenotypes that support growth at low temperatures, a feature which has ecological and biotechnological significance. Accordingly, their study is currently being boosted by more and more scientific teams in academia and industry.

The development of improved methods to collect samples and to isolate and culture cold-adapted yeasts may lead to an increase in the recovery of viable cells and novel taxa. Culture-dependent and culture-independent methods should be applied to better understand their biodiversity and ecology in cold ecosystems. One critical point that may possibly enhance the chances for successful studies on the taxonomy, physiology and biotechnology of cold-adapted yeasts could be the availability of a large number of strains for large, multi-laboratory comparative studies. Unfortunately, many researchers underestimated in the past years the importance of depositing their cultures in known culture collections. As a result, some yeast cultures isolated from cold habitats during the last decades are lost forever. Yeast collections play a fundamental role in the preservation, identification and characterization of these microorganisms and are essential for biotechnological research, because they represent safe repositories where biodiversity is preserved for the future. Only in more recent years, a number of cold-adapted yeast strains became available in the worldwide culture collections (Shivaji and Prasad 2009). An in-depth discussion on the methods for the isolation and investigation of the diversity of cold-adapted yeasts and their *ex-situ* preservation is reported in Chap. 2.

Due to their distinctive ability to grow and metabolize at low temperatures, cold-adapted yeasts have been extensively studied for their huge biotechnological potential. However, only a few products from these organisms are currently on the market, whereas a number of other possible chemicals are still confined to the laboratory (Shivaji and Prasad 2009; Margesin and Feller 2010; Buzzini et al. 2012). A few cold-adapted enzymes, namely lipases A and B (produced by *P. antarctica*), are probably the most famous example of patented cold-active enzymes from cold-adapted yeasts. Lipase B is at present sold as Novozym 435 by Novozymes (Denmark) and is characterized by a moderate heat stability, whereas Lipase A exhibits a higher thermal stability (about 90 °C) than other lipases (Russell 1998; Margesin and Schinner 1999; Cavicchioli et al. 2002; De Maria et al. 2005; Joseph et al. 2008).

As underlined by a few authors (Joseph et al. 2008), the lower number of patented cold-active enzymes from yeasts may be the consequence of the high risk and cost concerned in pursuing this largely unexplored field. So, it is not surprising that the number of companies (largely related to pharmaceutical, chemical and food industries) that fund large-scale screening surveys and apply for cold-adapted yeast-based patents is rather limited. At present, most patents involving the use of both lipases from *P. antarctica* reported in the WIPO website (www.wipo.int/patentscope) are process-based for the production of polymeric compounds, whereas a few are product-based describing the production of enzyme variants or intermediate derivatives (Joseph et al. 2008).

The use of novel recombinant DNA technologies (e.g. metagenomics and site-directed mutagenesis) will probably have a profound and positive effect on the expression and production of greater and greater amounts of recombinant enzymes from cold-adapted yeasts in suitable hosts. Further investigations should also consider modelling of cold-active enzymes for various industrial applications. This would imply more competitive prices and the introduction of new or tailored catalytic activities of cold-adapted enzymes at low temperature.

References

Abyzov SS (1993) Microorganisms in the Antarctic ice. In: Friedman EI (ed) Antarctic microbiology. Wiley, New York, pp 265–295

Alchihab M, Destain J, Aguedo M, Majad L, Ghalfi H, Wathelet JP, Thonart P (2009) Production of γ-decalactone by a psychrophilic and a mesophilic strain of the yeast *Rhodotorula aurantiaca*. Appl Biochem Biotechnol 158:41–50

Arenz BE, Held BW, Jurgens JA, Farrell RL, Blanchette RA (2006) Fungal diversity in soils and historic wood from the Ross Sea Region of Antarctica. Soil Biol Biochem 38:3057–3064

Arthur H, Watson K (1976) Thermal adaptation in yeast: growth temperatures, membrane lipid, and cytochrome composition of psychrophilic, mesophilic, and thermophilic yeasts. J Bacteriol 128:56–68

Babyeva IP, Golubev WI (1969) Psychrophilic yeasts in the Antarctic oases. Microbiology 38:436–440

Bakermans C (2008) Limits for microbial life at subzero temperatures. In: Margesin R, Schinner F, Marx JC, Gerday C (eds) Psychrophiles: from biodiversity to biotechnology. Springer, Berlin Heidelberg, pp 17–28

Bej AK, Mojib N (2010) Cold adaptation in Antarctic biodegradative microorganisms. In: Bej AK, Aislabie J, Atlas RM (eds) Polar microbiology: the ecology, biodiversity and bioremediation potential of microorganisms in extremely cold environments. CRC Press, Boca Raton, FL, pp 157–178

Bej AK, Aislabie J, Atlas RM (eds) (2010) Polar microbiology: the ecology, biodiversity and bioremediation potential of microorganisms in extremely cold environments. CRC Press, Boca Raton, FL

Bergauer P, Fonteyne PA, Nolard N, Schinner F, Margesin R (2005) Biodegradation of phenol and phenol-related compounds by psychrophilic and cold-tolerant alpine yeasts. Chemosphere 59:909–918

Bridge PD, Newsham KK (2009) Soil fungal community composition at Mars Oasis, a southern maritime Antarctic site, assessed by PCR amplification and cloning. Fungal Ecol 2:66–74

Butinar L, Spencer-Martins I, Gunde-Cimerman N (2007) Yeasts in high Arctic glaciers: the discovery of a new habitat for eukaryotic microorganisms. A van Leeuwenhoek 91:277–289

Buzzini P, Vaughan-Martini A (2006) Yeast biodiversity and biotechnology. In: Rosa CA, Peter G (eds) Biodiversity and ecophysiology of yeasts. Springer, Berlin, pp 533–559

Buzzini P, Branda E, Goretti M, Turchetti B (2012) Psychrophilic yeasts from worldwide glacial habitats: diversity, adaptation strategies and biotechnological potential. FEMS Microbiol Ecol 82:217–241

Cameron RE (1971) Antarctic soil microbial and ecological investigations. In: Quam LO, Porter HD (eds) Research in the Antarctic. AAAS, Washington, DC, pp 137–189

Cameron RE, King J, David CN (1970) Microbiology, ecology, and microclimatology of soil sites in dry valleys of Southern Victoria Land, Antarctica. In: Holdgate M (ed) Antarctic ecology, vol 2. Academic Press, New York, pp 702–716

Campbell IB, Claridge GGC (2000) Soil temperature, moisture and salinity patterns in Transantarctic Mountain cold desert ecosystems. In: Davidson W, Howard-Williams C, Broady P (eds) Antarctic ecosystems: models for wider understanding. Caxton Press, Christchurch, pp 2332–2340

Canganella F, Wiegel J (2011) Extremophiles: from abyssal to terrestrial ecosystems and possibly beyond. Naturwissenschaften 98:253–279

Carpenter EJ, Lin S, Capone DG (2000) Bacterial activity in South Pole snow. Appl Environ Microbiol 66:4514–4517

Cavicchioli R (2006) Cold-adapted archaea. Nat Rev 4:331–343

Cavicchioli R, Tortsen T (2000) Extremophiles. Inmederberg J (ed) Encyclopedia of microbiology. Academic Press, San Diego, pp 317–337

Cavicchioli R, Siddiqui KS, Andrews D, Sowers KR (2002) Low-temperature extremophiles and their applications. Curr Opin Biotechnol 13:253–261

Claridge GG, Campbell IB (1977) The salts in Antarctic soils, their distribution and relationship to soil processes. Soil Sci 123:337–384

Connell LB, Redman R, Craig S, Scorzetti G, Iszard M, Rodriguez R (2008) Diversity of soil yeasts isolated from South Victoria Land, Antarctica. Microb Ecol 56:448–459

D'Amico S, Collins T, Marx JC, Feller G, Gerday C (2006) Psychrophilic microorganisms: challenges for life. EMBO Rep 7:385–389

de Maria PD, Carboni-Oerlemans C, Tuin B, Bargeman G, van der Meer AB, van Gemert R (2005) Biotechnological applications of *C. antarctica* lipase A: state of the art. J Mol Catal B Enzym 37:36–46

Deegenaars ML, Watson K (1998) Heat-shock response in psychrophilic and psychrotrophic yeast from Antarctica. Extremophiles 2:41–49

Deming JW (2002) Psychrophiles and polar regions. Curr Opin Microbiol 5:301–309

Deming JW (2009) Extremophiles: cold environments. In: Schaechter M (ed) Encyclopedia of microbiology. Elsevier, Oxford, pp 147–158

di Menna ME (1960) Yeasts from Antarctica. J Gen Microbiol 23:295–300

di Menna ME (1966a) Three new species from Antarctic soil: *Candida nivalis*, *Candida gelida* and *Candida frigida* spp. nov. A van Leeuwenhoek 32:25–28

di Menna ME (1966b) Yeasts in Antarctic soil. A van Leeuwenhoek 32:29–38

Eddy BP (1960) The use and meaning of the term psychrophilic. J Appl Bacteriol 23:189–190

Fell JW, Boekhout T, Fonseca A, Scorzetti G, Statzell-Tallman A (2000) Biodiversity and systematics of basidiomycetous yeasts as determined by large-subunits rDNA D1/D2 domain sequence analysis. Int J Syst Evol Microbiol 50:1351–1371

Feller G (2007) Life at low temperatures: is disorder the driving force? Extremophiles 11:211–216

Feller G, Gerday C (2003) Psychrophilic enzymes: hot topics in cold adaptation. Nat Rev Microbiol 1:200–208

Fleet GH (2011) Yeast spoilage of foods and beverages. In: Kurtzman CP, Fell JW, Boekhout T (eds) The yeasts. A taxonomy study, vol 1. Elsevier, New York, pp 53–63

Friedmann EI (1982) Endolithic microorganisms in the Antarctic cold desert. Science 215:1045–1053

Gerday C, Aittaleb M, Arpigny JL, Baise E, Chessa JP, Garsoux G, Petrescu I, Feller G (1997) Psychrophilic enzymes: a thermodynamic challenge. Biochim Biophys Acta 1342:119–131

Gerday C, Aittaleb M, Bentahir M, Chessa JP, Claverie P, Collins T, D'Amico S, Dumont J, Garsoux G, Georlette D, Hoyoux A, Lonhienne T, Meuwis MA, Feller G (2000) Cold-adapted enzymes: from fundamentals to biotechnology. Trend Biotechnol 18:103–107

Gilbert JA, Hill PJ, Dodd CER, Laybourn-Parry J (2004) Demonstration of antifreeze protein activity in Antarctic lake bacteria. Microbiology 150:171–180

Gilbert JA, Davies PL, Laybourn-Parry J (2005) A hyperactive, Ca^{2+} dependent antifreeze protein in an Antarctic bacterium. FEMS Microbiol Lett 245:67–72

Glansdorff N, Xu Y (2002) Microbial life at low temperatures: mechanisms of adaptation and extreme biotopes. Implications for exobiology and the origin of life. Recent Res Dev Microbiol 6:1–21

Gostincar C, Grube M, de Hoog S, Zalar P, Gunde-Cimerman N (2010) Extremotolerance in fungi: evolution on the edge. FEMS Microbiol Ecol 71:2–11

Goto S, Sugiyama J, Iizuka H (1969) A taxonomic study of Antarctic yeasts. Mycologia 61:748–774

Gounot AM (1986) Psychrophilic and psychrotrophic microorganisms. Experientia 42:1192–1197

Gounot AM, Russell NJ (1999) Physiology of cold-adapted microorganisms. In: Margesin R, Schinner F (eds) Cold-adapted organisms: ecology, physiology, enzymology and molecular biology. Springer, Berlin, pp 33–55

Hawksworth DL (2004) Fungal diversity and its implications for genetic resource collections. Stud Mycol 50:9–18

Holdgate MV (1977) Terrestrial ecosystems in the Antarctic. Philos Trans Royal Soc B 279:5–25

Johnson EA, Echavarri-Erasun C (2011) Yeast biotechnology. In: Kurtzman CP, Fell JW, Boekhout T (eds) The yeasts. A taxonomy study, vol 1. Elsevier, New York, pp 21–44

Joseph B, Ramteke PW, Thomas G (2008) Cold active microbial lipases: some hot issues and recent developments. Biotechnol Adv 26:457–470

Julseth CR, Inniss WE (1990a) Heat shock protein induction and the acquisition of thermotolerance in the psychrotrophic yeast *Trichosporon pullulans*. Curr Microbiol 20:391–396

Julseth CR, Inniss WE (1990b) Induction of protein synthesis in response to cold shock in the psychrotrophic yeast *Trichosporon pullulans*. Can J Microbiol 36:519–524

Krallish I, Gonta S, Savenkova L, Bergauer P, Margesin R (2006) Phenol degradation by immobilized cold-adapted yeast strains of *Cryptococcus terreus* and *Rhodotorula creatinivora*. Extremophiles 10:441–449

Kurtzman CP (2011) Discussion of teleomorphic and anamorphic ascomycetous yeasts and yeast-like taxa. In: Kurtzman CP, Fell JW, Boekhout T (eds) The yeasts. A taxonomy study, vol 1. Elsevier, New York, pp 292–307

Kurtzman CP, Robnett CJ (1998) Identification and phylogeny of ascomycetous yeasts from analysis of nuclear large subunit (26S) ribosomal DNA partial sequences. A van Leeuwenhoek 73:331–371

Kurtzman CP, Fell JW, Boekhout T (2011a) The yeasts. A taxonomy study, vol 1–3. Elsevier, New York

Kurtzman CP, Fell JW, Boekhout T (2011b). Definition, classification and nomenclature of the yeasts In: Kurtzman CP, Fell JW, Boekhout T (eds) The yeasts. A taxonomy study, vol 1. Elsevier, New York, pp 3–5

Kurtzman CP, Fell JW, Boekhout T (2011c) Gene sequence analyses and other DNA-based methods for yeast species recognition. In: Kurtzman CP, Fell JW, Boekhout T (eds) The yeasts. A taxonomy study, vol 1. Elsevier, New York, pp 137–144

Lachance MA (2006) Yeast biodiversity: how many and how much? In: Rosa CA, Peter G (eds) Biodiversity and ecophysiology of yeasts. Springer, Berlin, pp 1–9

Lainioti GC, Kapolos J, Koliadima A, Karaiskakis G (2012) The study of the influence of temperature and initial glucose concentration on the fermentation process in the presence of *S. cerevisiae* yeast strain immobilized on starch gels by reversed-flow gas chromatography. Prep Biochem Biotechnol 42:489–506

Los DA, Murata N (2004) Membrane fluidity and its roles in the perception of environmental signals. Biochim Biophys Acta 1666:142–157

Margesin R (2009a) Effect of temperature on growth parameters of psychrophilic bacteria and yeasts. Extremophiles 13:257–262

Margesin R (ed) (2009b) Permafrost soils. Springer, Berlin

Margesin R, Feller G (2010) Biotechnological applications of psychrophiles. Environ Technol 31:835–844

Margesin R, Miteva V (2011) Diversity and ecology of psychrophilic microorganisms. Res Microbiol 162:346–361

Margesin R, Schinner F (eds) (1999) Biotechnological applications of cold-adapted organisms. Springer, Berlin

Margesin R, Gander S, Zacke G, Gounot AM, Schinner F (2003) Hydrocarbon degradation and enzyme activities of cold-adapted bacteria and yeasts. Extremophiles 7:451–458

Margesin R, Fonteyne PA, Schinner F, Sampaio JP (2007) *Rhodotorula psychrophila* sp. nov., *Rhodotorula psychrophenolica* sp. nov. and *Rhodotorula glacialis* sp. nov., novel psychrophilic basidiomycetous yeast species isolated from alpine environments. Int J Syst Evol Microbiol 57:2179–2184

Margesin R, Schinner F, Marx JC, Gerday C (eds) (2008) Psychrophiles. From biodiversity to biotechnology. Springer, Berlin

McMurrough I, Rose AH (1973) Effects of temperature variation on the fatty acid composition of a psychrophilic *Candida* species. J Bacteriol 114:451–452

Miller RV, Whyte LG (eds) (2012) Polar microbiology. Life in a deep freeze. ASM Press, Washington, DC

Montes MJ, Belloch C, Galiana M, Garcia MD, Andrés C, Ferrer S, Torres-Rodriguez JM, Guinea J (1999) Polyphasic taxonomy of a novel yeast isolated from Antarctic environment: description of *C. victoriae* sp. nov. Syst Appl Microbiol 22:97–105

Morgan-Kiss RM, Priscu JC, Pocock T, Gudynaite-Savitch L, Huner NP (2006) Adaptation and acclimation of photosynthetic microorganisms to permanently cold environments. Microbiol Mol Biol R 70:222–252

Morita RY (1975) Psychrophilic bacteria. Bacteriol Rev 39:144–167

Moyer CL, Morita RY (2007) Psychrophiles and psychrotrophs. Encyclopedia of life sciences. Wiley, New York, pp 1–6

Nagahama T (2006) Yeast biodiversity in freshwater, marine and deep-sea environments. In: Rosa CA, Peter G (eds) Biodiversity and ecophysiology of yeasts. Springer, Berlin, pp 241–262

Onofri S, Zucconi L, Tosi S (2007) Continental Antarctic fungi. IHW Verlag, Munchen

Panikov NS, Sizova MV (2007) Growth kinetics of microorganisms isolated from Alaskan soil and permafrost in solid media frozen down to −5 degrees C. FEMS Microbiol Ecol 59:500–512

Pavlova K, Panchev I, Krachanova M, Gocheva M (2009) Production of an exopolysaccharide by Antarctic yeast. Folia Microbiol 54:343–348

Phadtare S, Inouye M (2008) Cold-shock proteins. In: Margesin R, Schinner F, Marx JC, Gerday C (eds) Psychrophiles: from biodiversity to biotechnology. Springer, Berlin, pp 191–209

Phadtare S, Alsina J, Inouye M (1999) Cold-shock response and cold-shock proteins. Curr Opin Microbiol 2:175–180

Poindexter JS (2009) Encyclopedia of microbiology. Elsevier, Amsterdam

Price PB, Sowers T (2004) Temperature dependence of metabolic rates for microbial growth, maintenance, and survival. PNAS 101:4631–4636

Psenner R, Wille A, Sattler B, Priscu JC, Felip M, Wagenbach D (2002) Ice ecosystems and biodiversity. In: Encyclopedia of life support system. Extremophiles, vol II. UNESCO-EOLSS Publisher

Ramli AN, Mahadi NM, Rabu A, Murad AM, Bakar FD, Illias RM (2011) Molecular cloning, expression and biochemical characterisation of a cold-adapted novel recombinant chitinase from *G. antarctica* PI12. Microb Cell Fact 10:94

Raspor P, Zupan J (2006) Yeasts in extreme environments. In: Rosa CA, Peter G (eds) Biodiversity and ecophysiology of yeasts. Springer, Berlin, pp 371–417

Robiglio A, Sosa MC, Lutz MC, Lopes CA, Sangorrín MP (2011) Yeast biocontrol of fungal spoilage of pears stored at low temperature. Int J Food Microbiol 147:211–216

Robinson CH (2001) Cold adaptation in Arctic and Antarctic fungi. New Phytol 151:341–353

Rossi M, Buzzini P, Cordisco L, Amaretti A, Sala M, Raimondi S, Ponzoni C, Pagnoni UM, Matteuzzi D (2009) Growth, lipid accumulation and fatty acid composition in obligate psychrophilic, facultative psychrophilic, and mesophilic yeasts. FEMS Microbiol Ecol 69:363–372

Russell NJ (1990) Cold adaptation of microorganisms. Philos Trans R Soc Lond B 326:595–611

Russell NJ (1997) Psychrophilic bacteria. Molecular adaptations of membrane lipids. Comp Biochem Physiol A 118:489–493

Russell NJ (1998) Molecular adaptations in psychrophilic bacteria: potential for biotechnological applications. Adv Biochem Eng Biotechnol 61:1–21

Russell NJ (2002) Psychrophily and resistance to low temperature. In: Encyclopedia of life support system. Extremophiles, vol 2. UNESCO-EOLSS Publisher

Russo G, Libkind D, Sampaio JP, van Broock M (2008) Yeast diversity at the Volcanic acidic environment of the Lake Caviahue and Rio Agrio (Patagonia, Argentina). FEMS Microbiol Ecol 65:415–424

Sabri A, Jacques P, Wekkers F, Bare G, Hiligsmann S, Moussaïf M, Thonart DP (2000) Effect of temperature on growth of psychrophilic and psychrotrophic members of *Rhodotorula aurantiaca*. Appl Biochem Biotechnol 84–86:391–399
Schisler DA, Janisiewicz WJ, Boekhout T, Kurtzman CP (2011) Agriculturally important yeasts: biological control of field and postharvest diseases using yeast antagonists, and yeasts as pathogens of plants. In: Kurtzman CP, Fell JW, Boekhout T (eds) The yeasts. A taxonomy study, vol 1. Elsevier, New York, pp 45–52
Shivaji S, Prakash JSS (2010) How do bacteria sense and respond to low temperature? Arch Microbiol 192:85–95
Shivaji S, Prasad GS (2009) Antarctic yeasts: biodiversity and potential applications. In: Satyanarayana T, Kunze G (eds) Yeast biotechnology: diversity and applications. Springer, Berlin, pp 3–16
Shivaji S, Bhadra B, Rao RS, Pradhan S (2008) *Rhodotorula himalayensis* sp. nov., a novel psychrophilic yeast isolated from Roopkund Lake of the Himalayan mountain ranges. India. Extremophiles 12:375–381
Siddiqui KS, Cavicchioli R (2006) Cold-adapted enzymes. Annu Rev Biochem 75:403–433
Staley JT, Gosink JJ (1999) Poles apart: biodiversity and biogeography of sea ice bacteria. Annu Rev Microbiol 53:189–215
Starmer WT, Lachance MA (2011) Yeast ecology. In: Kurtzman CP, Fell JW, Boekhout T (eds) The yeasts. A taxonomy study, vol 1. Elsevier, New York, pp 65–83
Stratford M (2006) Food and beverage spoilage yeasts. In: Querol A, Fleet G (eds) Yeasts in food and beverages. Springer, Berlin, pp 335–379
Turkiewicz M, Pazgier M, Kalinowska H, Bielecki S (2003) A cold-adapted extracellular serine proteinase of the yeast *Leucosporidium antarcticum*. Extremophiles 7:435–442
van Uden N (1984) Temperature profiles of yeasts. Advances Microb Physiol 25:195–251
Vincent CF (1988) Microbial ecosystems of Antarctica. Cambridge University Press, Cambridge
Vishniac HS (1996) Biodiversity of yeasts and filamentous microfungi in terrestrial Antarctic ecosystems. Biodivers Conserv 5:1365–1378
Vishniac HS (1999) Psychrophilic yeasts. In: Seckbach J (ed) Enigmatic microorganisms and life in extreme environments. Kluwer Academic Publishers, Amsterdam, pp 317–324
Vishniac HS (2006a) Yeast biodiversity in the Antarctic. In: Rosa CA, Peter G (eds) Biodiversity and ecophysiology of yeasts. Springer, Berlin, pp 420–440
Vishniac HS (2006b) A multivariate analysis of soil yeasts isolated from a latitudinal gradient. Microb Ecol 52:90103
Vishniac VW, Mainzer SE (1972) Soil microbiology studied *in situ* in the dry valleys of Antarctica. Antarct J US 7:88–89
Walker GM (1998) Yeast physiology and biotechnology. Wiley, New York
Watson K (1978) Thermal adaptation in yeasts: correlation of substrate transport with membrane lipid composition in psychrophilic and thermotolerant yeasts. Biochem Soc T 6:293–296
Watson K, Arthur H (1976) *Leucosporidium* yeasts: obligate psychrophiles which alter membrane-lipid and cytochrome composition with temperature. J Gen Microbiol 97:11–18
Watson K, Morton H, Arthur H, Streamer M (1978) Membrane lipid composition: a determinant of anaerobic growth and petite formation in psychrophilic and psychrophobic yeasts. Biochem Soc T 6:380–381
Wolf K, Breunig K, Barth G (2003) Non conventional yeasts in genetics, biochemistry and biotechnology. Springer, Berlin
Wynn-Williams DD (1990) Ecological aspects of Antarctic microbiology. Adv Microbial Ecol 11:71–146
Yayanos AA (1995) Microbiology to 10,500 m in the deep sea. Annu Rev Microbiol 49:777–805

Chapter 2
Methods for the Isolation and Investigation of the Diversity of Cold-Adapted Yeasts and Their *Ex Situ* Preservation in Worldwide Collections

Kyria L. Boundy-Mills

Abstract Cold-adapted (psychrophilic and psychrotolerant) yeasts have been isolated from a variety of substrates, using a variety of cultivation methods. Yeasts able to grow at as low as 0 °C have been isolated from cold substrates such as glaciers, snow, and deep-sea sediment, but also from temperate and tropical climates. A broad diversity of media and culture conditions have been used to isolate and cultivate these yeasts. Low-temperature incubation is used to select for psychrophiles, thus depending on the strains relatively long incubation time (up to 14 weeks) may be required. Cold-adapted yeast strains belong to many species in many clades of Ascomycota and Basidiomycota. Numerous strains have been deposited in public culture collections. Online strain catalogs of some public yeast culture collections include searchable fields for growth temperatures, allowing selection of yeasts able to grow at desired temperatures. Culture-independent methods for profiling yeast diversity in mixed communities can be used to profile populations, allowing detection of yeasts whose DNA is present in a specimen but that were not cultivated.

Keywords Psychrophilic and psychrotolerant yeasts • Yeast cultivation methods • Yeast culture collections

Contents

2.1 General Methods of Isolation of Yeasts from the Environment 24
2.2 Niches that Harbor Cold-Adapted Yeasts 26
2.3 Aseptic Collection of Specimens 27
2.4 Surface Sterilization of Specimens 30

K. L. Boundy-Mills (✉)
Phaff Yeast Culture Collection, Department of Food Science and Technology, University of California Davis, Davis, CA 95616, USA
e-mail: klbmills@ucdavis.edu

P. Buzzini and R. Margesin (eds.), *Cold-Adapted Yeasts*,
DOI: 10.1007/978-3-642-39681-6_2, © Springer-Verlag Berlin Heidelberg 2014

2.5 Homogenization, Dilution, or Concentration of Specimens 30
2.6 Cultivation Methods for Cold-Adapted Yeasts 32
2.7 Cold-Adapted Yeast Species 35
2.8 Preservation of Cold-Tolerant Yeasts in Public Culture Collections 36
2.9 Methods for Detecting Uncultured Yeasts 40
2.10 Conclusions 42
References 42

2.1 General Methods of Isolation of Yeasts from the Environment

Several considerations must be taken into account when designing a protocol for isolation of cold-adapted yeasts. As many of these considerations are identical for all yeasts, a brief discussion of general yeast isolation and cultivation protocols is presented. It is recommended that a combination of protocols be utilized when sampling a new substrate, to ensure that the desired results can be obtained.

Culture-based methods for isolation and cultivation of yeasts in natural substrates are desirable when living cultures are needed for subsequent studies. The major limitation of culture-based methods is that some yeast species and strains may not be culturable under the growth conditions used. Culture-independent methods (described in Sect. 2.9) have an advantage in that a broader cross section of taxa can be detected, minority species can be detected, and relative quantities may be determined. However, examination of the physiological properties of yeasts, and subsequent biotechnological developments, requires pure, living cultures. Therefore, this chapter focuses primarily on methods used for isolation of pure cultures.

Culture-based methods for enumeration and cultivation of yeasts from foods and environmental substrates have been described in detail in many recent reviews and book chapters (Martini et al. 1996; Welthagen and Viljoen 1997; Beuchat 1998; Fung 1999; Boundy-Mills 2006; Solís et al. 2009).

Three general types of culture-based methods are used for enumeration of microbes: plate count (PC) procedures (either pour plates or spread plates), membrane filtration (MF) followed by growth of colonies on the membrane placed on an agar plate, and most probable number (MPN) procedures. PC and MF methods utilize agar plates and are discussed in this chapter.

Prior to plating, a substrate is often processed to make it more amenable to growth of a statistically valid number of single colonies on or in an agar plate (Boundy-Mills 2006). Processing may include some of the following steps:

(1) Aseptic collection of the specimen;
(2) Storage of the specimen under appropriate conditions, such as refrigeration;
(3) If needed, surface sterilization to avoid cultivation of microbes on the surface of the substrate;

(4) Homogenization, disruption, or mixing of the sample to release and evenly distribute microbes;
(5) Enrichment;
(6) Dilution of specimens with high yeast density in an appropriate diluent;
(7) Concentration of specimens with low yeast density, such as by filtration or precipitation.

After appropriate processing, agar plate methods can involve either spread plates or pour plates. Many researchers prefer spread plates as they are simpler to perform, and it is easier to recover a yeast from a colony on the surface of the agar, rather than embedded in the agar. Furthermore, the warm temperature of the molten agar used in pour plates may be particularly injurious to cold-adapted yeasts.

There are numerous nutrient agars available for cultivation of various types of yeasts (Beuchat 1998; Boundy-Mills 2006). Liquid or agar-based complex nutritionally rich media generally include an energy source such as glucose, a nitrogen source such as ammonium or hydrolyzed protein (peptone, tryptone, etc.), and a vitamin source such as yeast extract or malt extract. Other supplements often include compounds such as rose Bengal or dichloran to inhibit or slow growth of filamentous fungi (Beuchat 1998) and antibiotics such as tetracycline or chloramphenicol to prevent growth of bacteria. Alternatively, the pH of the medium can be reduced by addition of acid to discourage growth of bacteria. A broad range of yeasts and filamentous fungi can grow on these media. Selective and differential media containing inhibitors, selective nutrients, and indicator dyes are used to differentiate or select for growth of specific types of yeasts. For example, lysine agar is used to select for growth of non-*Saccharomyces* yeasts in brewing, as *Saccharomyces* cannot utilize lysine as a sole carbon source but unwanted spoilage yeasts can (Heard and Fleet 1986).

When selecting a media formulation, it is important to consider the composition of the substrate of origin and the types of yeast to be cultivated. For example, osmotolerant yeasts that inhabit high-sugar environments like as fruit juice, nectar, and dried fruit, such as several species of *Metschnikowia, Zygosaccharomyces*, and *Starmerella* species, grow well in media containing 50 % (w/v) glucose, but most other yeasts cannot. In contrast, microbes such as soil oligotrophs may fail to grow on full-strength media and can only be cultivated on diluted media such as 1/10 strength rich media (Hattori 1980).

Following enumeration of yeasts by plate counts or membrane filtration, yeasts can be selected and purified for further characterization. It is common to select one or more colonies of each morphology for purification. However, multiple species may have similar or identical colony morphologies, so some species may be overlooked. Culture-independent methods (discussed in Sect. 2.9) should be used if a more complete view of the yeast diversity is needed.

2.2 Niches that Harbor Cold-Adapted Yeasts

The term “cold-adapted” is used in this chapter to refer to yeasts able to grow at low temperatures in the laboratory. An in-depth overview of the concept of psychrophily and psychrotolerance in yeasts is reported in Chap. 1.

In the food industry, the term “psychrotroph” is sometimes used to describe microbes that are able to grow at refrigeration temperatures (0–7 °C), but have optimal growth temperatures at or above 20 °C (Cousin et al. 1992). Most food-associated cold-adapted microbes are psychrotolerant (Tomkin 1963), as discussed in more detail in Chap. 23.

When searching for a yeast strain with specific properties, one should search in natural habitats that have those properties. For example, acid-tolerant yeasts thrive in grape juice and other acidic fruit juices. Osmotolerant yeasts can be found in dried fruit and honey. Ethanol-tolerant yeasts can be found in wine fermentations. Thus, it should not be surprising that cold-adapted yeasts are prevalent in cold habitats. However, it is important to note that yeasts able to grow as low as 4 °C have also been isolated from plant surfaces, water, soil, and insects in temperate and even tropical climates (see Sect. 2.8).

Cold-adapted yeasts have been reliably found in numerous cold habitats, including permafrost, snow, cold deserts, and glacial ice, meltwater and sediment (Chaps. 3, 4, 5, 6), the deep sea (Chap. 7), and frozen and refrigerated foods (Chap. 23). In addition to cold tolerance, yeasts able to grow in ice, permafrost, and frozen foods must tolerate low water activity, as very little liquid water is present in these substrates. Yeasts from glacial ice therefore must be osmotolerant.

It is unclear whether cold-adapted yeasts are endemic to cold habitats. With the limited number of yeast strains isolated to date, and the limited number of yeast ecologists, it is difficult to establish where a given yeast species is *not* found. For example, the yeast species *Cryptococcus victoriae,* first isolated from South Victoria Land, Antarctica (Montes et al. 1999), was subsequently found in many other habitats and geographic locations, such as cherries in California (Hamby et al. 2012). In contrast, yeast species *Cryptococcus vishniacii* is currently unknown outside of Antarctica. Yeasts belonging to the genera *Mrakia* and *Mrakiella* have been isolated exclusively from cold habitats such as Antarctica (Turchetti et al. 2008; Singh and Singh 2012), the European Alps (Margesin et al. 2002), and subglacial waters of northwest Patagonia, Argentina (Brizzio et al. 2007).

Endemic microbes, as well as exotics blown in by the wind from temperate and tropical regions, can be preserved for thousands of years deep in glacial, Arctic and Antarctic ice (Catranis and Starmer 1991; Abyzov 1993; Ma et al. 1999, 2000). Microbial species thought to be extinct, such as pathogens, could theoretically be revived re-appear as glaciers and polar ice melt due to global warming or other conditions (Ma et al. 1999).

2.3 Aseptic Collection of Specimens

For reasons of intellectual property, biosafety, and adherence to inter-institutional and international agreements and treaties such as the Convention on Biological Diversity, descriptive data should be collected including site, habitat, and other descriptive data. Projects involving isolation of large numbers of microbes, especially those performed by multi-institutional or international teams, often utilize standardized data sheets to ensure that all relevant data are collected. A very thorough data collection plan and data sheet template for collection of microbes has been recently compiled (Tadych 2008). Some types of data that are gathered in general microbiology surveys, and specific data related to cold-adapted yeasts, are listed in Table 2.1.

Some types of data are specific for the type of habitat. For example, a recent publication of yeasts isolated from a glacial lake includes descriptions of the elevation, slope, sun exposure, surface area, maximum depth, pH, dissolved solids (conductivity), Secchi disk transparency, summer air and water temperatures, and how many months per year the lake is frozen (Libkind et al. 2004). A Secchi disk is a black-and-white disk lowered into the water on a rope to determine water clarity, a property that may reflect nutrient levels in the water.

Replicate samples are recommended if possible. For example, Butinar et al. (2011) filtered and plated between four and eleven replicates per sample of melted glacial ice.

Although collection of water from shallow lakes and rivers may be relatively easy, many of the habitats that harbor cold-adapted yeasts are quite inaccessible, such as deep-sea water and sediment, and the interior of glaciers. Aseptic sampling techniques have been developed for these substrates.

Deepwater sediment: Sediment core samplers have been developed for aseptic sampling of deep-sea sediment (Ikemoto and Kyo 1993) that can be used in either a manually controlled or remote-controlled submersible.

Deepwater, lake and river water sampling: Water sampling is performed with specially designed sampling devices such as the Nansen bottle or Niskin sampler. The Nansen bottle was designed in 1910 by oceanographer Fridtjof Nansen, and later improved by Shale Niskin. It is a metal or plastic cylinder, open at both ends, that is lowered in a vertical position on a cable to the desired depth. The spring-loaded valves are activated by releasing a metal weight ("messenger") down the cable, by a pressure switch, or by remote control, sealing the ends of the tube. The tube is then raised to the surface of the water. Similarly, the van Dorn sampler is a sample tube that is lowered into the water in a horizontal position with both ends open, and at the desired depth, the ends are sealed with spring-loaded valves by releasing a messenger down the cable. Note that in both types of sampler, the interior of the sampling tube is exposed to surface water as it is lowered. Microbes from the higher zones are diluted but not excluded. A variant, the Niskin bio-sampler, allows aseptic sampling. Another sampling apparatus, the J-Z sampler, has been modified to reduce contamination (Kimball et al. 1963).

Table 2.1 Categories and types of data to be collected in general microbial surveys, and specific data related to cold-adapted yeasts

Category	Data collected in general microbial surveys	Data collected for cold-adapted yeasts
Collector documentation	Names and institutions of primary and secondary personnel	Name and institution of experiment station facility, including cold storage facility
	Date specimen was collected	
	Date specimen was processed and plated	
	Photograph filenames, name of photographer, type of camera	
	Collecting permit documentation	
Locality	Site description: public or private land, national park, city, province, state, country	Name of mountain, glacier, fjord, bay, river, or deepwater trench
		Depth of river, lake, or ocean
		Surface area of lake
	GPS coordinates	Average salinity and pH of river, lake, or ocean
	Elevation	Average summer air and water temperatures
	Forest, soil, or water type	How many months per year a lake is frozen
	Slope (such as north-facing mountain); sun exposure	Average ice thickness of a glacier
	Type of ecosystem	
Ambient conditions	Water and air temperatures, humidity	Depth of ice core specimen
		Secchi disk transparency of water
	Relevant weather conditions such as recent rain, flood, drought, heat wave	Underground temperature at various depths after collecting ice or permafrost core samples
	Sun exposure	Temperature of sediment, water, and ice
	Season	
Substrate	Plant or insect host, and name and institution of person who identified it	Depth of ice core specimen
		Whether sample was from glacier superficial meltwater, supraglacial sediments, deep piping sediments, subglacial sediments, etc.
		Whether ice core or meltwater specimens are sediment rich or clear
		Physicochemical parameters including dry mass, pH, salinity, Na^+, Mg^{2+}, K^+, total C, N, P; organic C, N, P

(continued)

Table 2.1 (continued)

Category	Data collected in general microbial surveys	Data collected for cold-adapted yeasts
Sampling regimen	Number of sites samples and number of replicate specimens collected per site	Temperature and length of storage of ice core samples
	Aseptic sampling method	Type of sampling device used, such as Nansen bottle or Niskin sampler, spatula, rotary drill
	Size of samples	
	How samples were stored after collection and before microbial plating	
	Amount of time between collecting specimen and plating	
	Processing methods: surface sterilization, homogenization, concentration, or dilution	
	Volume plated	

Glacial ice cores: Viable yeasts have been cultivated from ice samples up to 3250 years old (Abyzov 1993). Ice cores are collected using a hollow drill which cuts in a cylindrical pathway. The tube surrounds the core as it proceeds. The length of the core is limited by the length of the drill barrel, such as the 6-m drill at Vostok Station in Antarctica (Abyzov 1993). Deep cores are collected by cutting to the length of the drill assembly, raising it to the surface, emptying the barrel, and repeating. Deep cores more than 300 m may require use of a fluid to prevent the hole from deforming and closing due to pressure. Cores are sealed in plastic bags, stored frozen, and analyzed in clean rooms. Precautions such as storing the core below −15 °C and transporting in foam shippers are taken to avoid introduction of microfractures that could allow microbes to enter the interior of the ice core. Ice cores are examined visually on a light table to detect fractures (Ma et al. 1999). More extensive measures to detect microbial contamination into the interior of ice core samples have been developed, including fluorescent microspheres, a chemical tracer, dissolved organic carbon (DOC) signatures, and comparison of bacterial 16S clone libraries from drilling mud to bacterial species recovered from core samples (Gronstal et al. 2009). When collecting core samples, prevention of contamination from the borehole can be difficult. Gilichinsky et al. (2005) devised a method to investigate the contribution of contaminating bacteria from surface and borehole microbes in cryopeg samples, by measuring incorporation of ^{14}C from glucose to determine relative metabolic activity under hypersaline conditions. They determined that metabolic activity in sterilized brine inoculated with surface contaminants was low, but high in native brine; thus, the contaminants were relatively metabolically inactive in brine.

Permafrost: The microbial diversity of permafrost was recently reviewed (Wagner 2008; Margesin 2009). It has been suggested that microbes, especially bacteria, may survive in a hypometabolic state deep in permafrost for millions of years (Vorobyova et al. 1997). Although less prevalent than bacteria, yeasts are present in deep permafrost and were the dominant culturable aerobic eukaryotes in samples of ancient Oleorian permafrost, 2–3 million years old (Vorobyova et al. 1997). Yeasts were more frequently found in young permafrost with high plant residue content such as peat.

2.4 Surface Sterilization of Specimens

Ice and permafrost cores have been collected from depths up to 4 km. Several researchers have examined these ice cores for viable microbes or microbial DNA, as described by Rogers et al. (2004). It can be difficult to confirm that viable microbes were indeed isolated from the interior of the ice cores, as contamination from external sources such as drilling equipment is hard to exclude. Ma et al. (2000) surface sterilized the ends of Greenland glacier ice cores by UV irradiation, then used a sterile drill and saw to remove a subcore from the center of the ice sample. Rogers et al. (2004) compared a variety of protocols for decontaminating ice core samples, including exposure to bleach, ethanol, UV radiation, acid and base, and found that bleach was the most effective method in killing microbial contaminants on the surface of the ice. They confirmed that microorganisms that they spread on the surface of the ice core were not recovered in the melted material. Abyzov (1993) devised a method to aseptically sample the interior of Antarctic ice cores, using a circular knife blade to crack off the end of an ice core sample, followed by melting and collecting water from an interior section using a cone-shaped heater applied to the freshly revealed interior surface.

Ice samples collected from surface environments such as glacier surfaces and sea ice do not require such extensive surface sterilization. For example, Butinar et al. (2011) simply melted and discarded the surface layer from ice samples and then melted the remaining ice sample.

2.5 Homogenization, Dilution, or Concentration of Specimens

Homogenization: Some samples must be homogenized, disrupted, or mixed to release and disperse microbes. Depending on the composition and texture of the substrate, a sample may be aseptically ground, shaken, pounded, pummeled, stirred, swirled, or pureed (Boundy-Mills 2006). For example, samples of giant white clams and tube worms collected at over 1,000 m depth in Sagami Bay, Japan, were surface sterilized with ethanol and then pulverized in artificial seawater (Nagahama et al. 2001, 2003). After surface sterilization, ice core and subcore samples are simply melted to release microbes from the ice matrix and

Table 2.2 Yeast abundance in select substrates known to harbor cold-adapted yeasts

Substrate	Yeast abundance	Reference
Cryoconite	7×10^3 to 1.4×10^4 CFU g^{-1} dry mass	Singh and Singh (2012)
Cryoconite	500–1.5×10^5 CFU g^{-1} dry mass	Margesin et al. (2002)
Antarctic soil	5–1×10^5 cells g^{-1} soil	Di Menna (1966)
Surface layers of glacier ice	Up to 25 CFU ml^{-1}	Butinar et al. (2007)
Glacial ice cores and meltwater	0.11–0.56 CFU ml^{-1}	Turchetti et al. (2008)
Supra- and subglacial sediment	12–21 CFU g^{-1} dry mass	Turchetti et al. (2008)
Cryokarst formations in glacier	Up to 400 CFU ml^{-1}	Butinar et al. (2007)
Basal glacier ice	Up to 4,000 CFU ml^{-1}	Butinar et al. (2007)
Permafrost	200–2,000 CFU g^{-1} dry mass	Vorobyova et al. (1997)
Cryopegs (hypersaline underground water surrounded by permafrost)	3–400 CFU ml^{-1}	Gilichinsky et al. (2005)
Arctic subglacial environments	Up to 4×10^3 CFU ml^{-1}	Butinar et al. (2007)
Supraglacial samples	25 CFU ml^{-1}	Butinar et al. (2005)
Subglacial ice	Up to 10 CFU ml^{-1} (ascomycetes)	Butinar et al. (2011)
Sea ice	84 CFU L^{-1} (ascomycetes) and 2×10^3 CFU L^{-1} (basidiomycetes)	Butinar et al. (2011)
Seawater in a fjord	Up to 500 CFU L^{-1} (ascomycetes) and 400 CFU L^{-1} (basidiomycetes)	Butinar et al. (2011)
Snow/ice in the tidal zone bordering the fjord	6.7×10^3 CFU L^{-1} ascomycetes	Butinar et al. (2011)

evenly distribute the microbes (Abyzov 1993; Rogers et al. 2004). Yeast abundance varies significantly in different substrates, or in the same type of substrate sampled in different locations and seasons. Examples of yeast abundance reported for several different types of substrates that harbor cold-adapted yeasts are listed in Table 2.2. For example, yeast populations were reported to be denser in freshwater than marine water in Brazil (Hagler and Ahearn 1987).

Dilution: Some substrates such as cryoconite (organic and inorganic dust containing microbes found on the surface of glaciers), soil or glacial or marine sediment require suspension or washing to release suspend microbes. They may also contain sufficiently high concentrations of viable yeasts to require dilution prior to plating. Diluents used include liquid media, saline, water, or various phosphate buffers (Boundy-Mills 2006). A surfactant such as 0.01–0.05 % (v/v) Tween 80 can be added to aid in separation of cells from soil particles, cell clumps, and filamentous structures (Deak 2003). For example, cryoconite was shaken with phosphate buffer and Tween 80 at 170 rpm, and then, appropriate dilutions in saline were surface plated onto agar plates (Margesin et al. 2002). The water activity of the diluent should resemble that of the substrate, or yeast viability may be affected (Hocking et al. 1992; Hernandez and Beuchat 1995). Because frozen substrates such as ice and permafrost have very little available liquid water, media with quite low water activity have been successfully used to isolate yeasts from melted snow and ice (Butinar et al. 2011). Yeasts in diluent must be plated promptly, particularly in saline solutions, which may have adverse effects on yeast viability (Andrews et al. 1997).

Concentration: Specimens with low yeast density can be concentrated using filtration or centrifugation. Filters such as nitrocellulose membranes, 0.2- or 0.45-μm pore size, are used in a sterile filtration device to concentrate the microbes in the water. To exclude particles larger than yeasts, the material can be pre-filtered with a larger pore size membrane such as 5 μm (López-García et al. 2001). The filter is then aseptically removed from the filtration device and placed filtrate side up on the surface of an agar plate. The volume of water to be filtered depends on the concentration of resident microbes. Table 2.2 lists yeast abundance reported in selected substrates from which cold-adapted yeasts have been isolated. As it may be difficult to predict the yeast density in advance, multiple concentrations should be plated in order to obtain a statistically significant number of yeasts.

Organic contents including yeast cells in dilute samples such as lake water have been concentrated using a coagulant (Sláviková et al. 1992). Water samples can also be concentrated using a Foerst-type continuous flow centrifuge (Kimball et al. 1963).

2.6 Cultivation Methods for Cold-Adapted Yeasts

Many different media and growth conditions have been used for isolation and cultivation of cold-adapted yeasts. Table 2.3 describes a variety of media, incubation temperatures, and incubation times used in selected published descriptions of isolation and cultivation of cold-adapted yeasts from many of the cold substrates discussed in this chapter.

Media composition: The composition of media including nutrient concentration and water activity can dramatically affect recovery of viable microbes. Recovery of microbes from oligotrophic environments such as water may be more successful

Table 2.3 Media and growth conditions used to isolate and cultivate cold-adapted yeasts from various cold substrates in selected recent publications

Substrate, location	Isolation medium and growth conditions	Reference
Glacier, Patagonia, Argentina	Meltwater filtered on 0.45-μm nitrocellulose filter placed on MYP agar with chloramphenicol. Incubated at 4–15 °C for up to 1 month	de Garcia et al. (2007)
Glacier, Ny-Ålesund, Norway	Cryoconite diluted in unnamed diluent, plated on six media by pour plate and spread plate: YPD, MYP, MEA, PDA, SDA, PCA, rose Bengal plus tetracycline. Incubated at 5, 10, 15, 20, and 25 °C for 2–4 weeks	Singh and Singh (2012)
Glacier, Kongsfjorden, Svalbard, Norway	Basal ice, subglacial meltwater, subsurface ice from cryokarst formations, snow/ice mixtures, seasonal meltwaters, seawater, and sea ice. Up to 100 mL of liquid filtered onto 0.22- and 0.45-μm membranes, and filters placed on DRBCA, DG18, MY10-12, MY20G, MY35G, MY50G, MEA, MEA5NaCl, MEA10NaCl, MEA15NaCl, MEA17NaCl, MEA24NaCl, MEA30NaCl (all with chloramphenicol, 50 mg L^{-1}). Incubated 4, 10, and 24 °C for up to 14 weeks	Butinar et al. (2011)
Glacier, Tyrolean Alps, Austria	Cryoconite suspended in diluent and then plated on MP plus chloramphenicol and lactic acid. Incubated at four temperatures: 2 °C for 10 days, 20 °C for 3–5 days, 37 °C for 2 days or 55 °C for 2–4 days	Margesin et al. (2002)
Glacial cores and meltwater, Forni and Szforzellina glaciers, Italy	50–100 ml of melted glacial cores or meltwater filtered through 0.22-μm filters, and filters placed on RB + tetracycline, DG18, and DRB + chloramphenicol. Incubated at 4 °C for 12 weeks and 20 °C for 3 weeks	Turchetti et al. (2008)
Supra- and subglacial sediment, Forni and Szforzellina glaciers, Italy	Solid samples diluted with 0.1 % sodium pyrophosphate, serial dilutions plated in triplicate on RB + tetracycline, DG18, and DRB + chloramphenicol. Incubated at 4 °C for 12 weeks and 20 °C for 3 weeks	Turchetti et al. (2008)
Ancient ice cores, Greenland	200 uL of melted subcores plated on eight media: SAB, YMA, acidified YMA, NA, MEA, PDA, MA, OMA. Incubated at 8 °C for 6 weeks, then 15 °C for 2 weeks	Ma et al. (1999, 2000)

(continued)

Table 2.3 (continued)

Substrate, location	Isolation medium and growth conditions	Reference
Polythermal glacier ice from edge of glacier, Kongsfjorden, Norway	Ice melted, then either diluted or filtered and filter plated on agar; plated on MEA, MEA5NaCl, and MEA20G. Incubated at 4, 10, and 24 °C, number of days not stated	Butinar et al. (2007)
Cryopegs, East Siberian Sea coast	Cryopeg brine plated directly, or filtered and filter placed on agar, on MA with or without sodium puryvate, and on Czapek media, with added sucrose or NaCl. Incubated at 4 or 26 °C for 1 month	Gilichinsky et al. (2005)
Permafrost, Dry Valley, Antarctica	Core sample surface sterilized, 1 ml of melted microcore spread plated on RB with tetracycline. Incubated at 4 °C for 12 weeks or 20 °C for 3 weeks	Zucconi et al. (2012)
Deep-sea sediments, Suruga Bay, Japan	Sediment plated on YMA, 1/5 YMA, PDA, and CMA, all made with artificial seawater and supplemented with 0.01 % chloramphenicol. Incubated 5–10 °C for 2 weeks then 20 °C for 1 month	Nagahama et al. (2003)
Deep-sea sediments, giant white clams and tubeworm from Suruga Bay, Iheya Ridge and Mariana Trench	Sediments plated directly on agar; animals surface sterilize and then pulverized in artificial water. Specimens plated on YMA, PDA, NA with 0.5 % glucose, YNB without amino acids containing 0.5 % glucose and 2 % agar; all dissolved in artificial seawater with 0.01 % chloramphenicol, 0.002 % streptomycin. Plates incubated 5–10 °C for 2 weeks, then 20 °C for one month	Nagahama et al. (2001)

YPD agar yeast extract peptone dextrose agar; *MYP agar* malt extract yeast extract soytone agar (Bandoni 1972); *MP agar* malt extract mycological peptone agar (3 % malt extract, 0.5 % mycological peptone, 1.5 % agar, pH 5.4) (Margesin et al. 2002); *SAB* Sabouraud's agar; *YMA* yeast–malt extract agar; *NA* nutrient agar; *MEA* malt extract agar; *MEA5NaCl* malt extract agar containing 5 % (w/v) NaCl; *MEA20G* malt extract agar containing 20 % glucose; *PDA* potato dextrose agar; *MA* mycobiotic agar; *OMA* oatmeal agar; *SDA* Sabouraud dextrose agar; *PCA* potato carrot agar; *RB* rose Bengal agar; *CMA* corn meal agar; *DRB* dichloran rose Bengal agar (Difco); *DRBCA* dichloran rose Bengal chloramphenicol agar (King et al. 1979); *DG18* dichloran 18 % glycerol agar (Hocking and Pitt 1980); *MY* 10–12 malt yeast 10 % glucose and 12 % NaCl agar (Samson et al. 2004); *MY20G* malt yeast 20 % glucose agar (Gunde-Cimerman et al. 2003); *MY35G* malt yeast 35 % glucose agar (Gunde-Cimerman et al. 2003); *MY50G*, malt yeast 50 % glucose agar (Samson et al. 2004); *MEA5NaCl* malt extract 5 % NaCl (Gunde-Cimerman et al. 2003); *MEA10NaCl* malt extract 10 % NaCl (Gunde-Cimerman et al. 2003); *MEA15NaCl* malt extract 15 % NaCl (Gunde-Cimerman et al. 2003); *MEA17NaCl* malt extract 17 % NaCl (Gunde-Cimerman et al. 2003); *MEA24NaC* malt extract 24 % NaCl (Gunde-Cimerman et al. 2003); *MEA30NaCl* malt extract 30 % NaCl (Gunde-Cimerman et al. 2003); *YNB*, yeast nitrogen base

with low nutrient media. Vishnivetskaya et al. (2000) observed that recovery of viable bacteria from permafrost samples up to 2 million years old was more successful using 1/10 strength TSA (tryptone soy agar) than with full-strength TSA, but recovery of bacteria from surface tundra soil was higher on full-strength TSA. Water activity is also an important consideration. Although frozen substrates such as ice and permafrost have very little dissolved solutes, they also have very little available liquid water and thus have low water activity. Media with low water activity have been successfully used to isolate yeasts from melted snow and ice (Butinar et al. 2011).

Temperature: Cultivation temperatures must be carefully selected for cold-adapted yeasts. These yeasts can be either psychrophilic or psychrotolerant. For example, three to ten times more yeasts were isolated from cryoconite on the surface of a glacier at 2 °C than at 20 °C, and none could be isolated at 30 °C (Margesin et al. 2002). Kutty and Philip (2008) recommend that after inoculation onto agar plates, yeasts from polar and deep-sea habitats should be incubated at 5 °C. However, the growth temperature range of yeasts isolated from cold substrates can be surprisingly high: Butinar et al. (2011) found that most ascomycetous yeasts they isolated from glacier ice and meltwater could grow up to 25–30 °C, in addition to growing at 4 °C.

2.7 Cold-Adapted Yeast Species

Cold-adapted yeasts belong to multiple clades of Ascomycota and Basidiomycota. While ascomycetous yeasts are the predominant agent of spoilage of chilled or frozen foods (Davenport 1980; Schmidt-Lorenz 1983), they are less abundant than basidiomycetous yeasts in polar soil (Vishniac and Onofri 2003), cold seawater, and subglacial ice (Butinar et al. 2011). The medium used for isolation of yeasts can impact the species recovered: Butinar et al. (2011) found that ascomycetous yeasts dominated on media containing elevated NaCl, but basidiomycetous yeasts dominated on other media with lower NaCl concentrations. In samples from the northwest Pacific Ocean, Nagahama et al. (2001, 2003) found that ascomycetes dominated in deep-sea sediments sampled from less than 2,000 m depth, but basidiomycetes dominated in deeper sediments.

Several recent publications detail the range of yeast species isolated from numerous cold habitats and locations. A review on yeasts isolated from Antarctica recently published by Buzzini et al. (2012) includes a list of species found in various Antarctic habitats. Shivaji and Prasad (2009) reviewed the biodiversity of yeasts from Antarctica, including a list of strains available from the Centraalbureau voor Schimmelcultures, CBS). Recent reports listing numerous yeast species include yeasts from four glaciers in Argentina (de García et al. 2007) and Italy (Turchetti et al. 2008), deep-sea sediment in Japan (Nagahama et al. 2001, 2003), and permafrost (Zucconi et al. 2012).

Some of these species are ubiquitous, with strains found in cold as well as warm habitats, such as *Cryptococcus laurentii, Cryptococcus macerans* (anamorph of *Cystofilobasidium macerans*), *Rhodotorula mucilaginosa* (de García et al. 2007). These species are also able to grow at temperatures of 20 °C and higher. However, some species isolated from Antarctica are psychrophiles. Species in the genus *Mrakia* and its anamorphic genus *Mrakiella,* by definition of the genera, have a maximum growth temperature of 20 °C or lower. Some species names reflect the cold-adapted nature of this genus, including *Mrakia frigida, Mrakia gelida*, and *Mrakia psychrophila*. Many *Mrakia* strains were isolated from cold climate soil and glaciers, such as the many *M. gelida* and *M. frigida* strains isolated from glaciers in Italy (Turchetti et al. 2008).

2.8 Preservation of Cold-Tolerant Yeasts in Public Culture Collections

Many scientific journals require that microbial strains cited in scientific publications be made available to the scientific public for research. Some are maintained in research collections at the institution of the researcher that performed the study, and others are deposited in public collections to allow broader availability and long-term preservation. For example, yeasts from glaciers and seawater described in a recent publication (Butinar et al. 2011) are maintained in the EX-F Culture Collection of Extremophilic Fungi of the Department of Biology, Biotechnical Faculty, University of Ljubljana, Slovenia.

The mission of public microbial culture collections is to preserve and distribute pure, authenticated microbial strains. Culture collections acquire yeasts through deposit by researchers from other institutions, through in-house research, or a combination of these (Boundy-Mills 2012). Many of these collections have online strain catalogs that can be searched by genus, species, geographic origin, and habitat of origin such as Antarctica, Norway, or Alaska, or cold habitats such as glaciers, deep-sea sediment, frozen foods, or permafrost. Researchers searching for a specific strain or species can search the species or strain fields in the online catalogs of these collections to locate strains. Representative culture collections, their websites, and examples of habitats sampled to obtain cold-adapted yeasts are listed in Table 2.4.

In addition to this basic descriptive data, some public culture collections have made physiological data including growth temperatures available in a searchable format in their online catalogs. For instance, the Industrial Yeasts Collection (DBVPG) at the University of Perugia, Italy, has a strong emphasis on environmental isolates. Historically, this collection focused on wine yeasts and has expanded in recent decades to include yeasts useful for other biotechnological applications such as cold-active enzymes. To fill this need, DBVPG personnel have built a sizeable collection of yeasts isolated from glaciers and snow. The

Table 2.4 Cold-adapted yeast species and strains available from public yeast culture collections

Yeast culture collection	Website	Website searchable for growth temperatures	Examples of cold substrates and geographic origins	Genera of yeasts able to grow at 4 °C
Industrial yeasts collection, University of Perugia, Italy (DBVPG)	http://www.dbvpg.unipg.it	Yes	Glacial ice, glacial meltwater, subglacial sediment	*Aureobasidium, Barnettozyma, Bensingtonia, Bullera, Bulleromyces, Candida, Cryptococcus, Cuniculitrema, Curvibasidium, Cystofilobasidium, Debaryomyces, Dioszegia, Glaciozyma, Guehomyces, Hannaella, Hanseniaspora, Holtermanniella, Kazachstania, Kluyveromyces, Lachancea, Leucosporidiella, Leucosporidium, Metschnikowia, Mrakia, Mrakiella, Ogataea, Pichia, Rhodosporidium, Rhodotorula, Saccharomyces, Sporidiobolus, Sporobolomyces, Torulaspora, Trichosporon, Udeniomyces, Wickerhamomyces, Williopsis*, and *Xanthophyllomyces*
Centraalbureau voor Schimmelcultures, Utrecht, The Netherlands (CBS)	http://www.cbs.knaw.nl/Collections/DefaultInfo.aspx?Page=Home	Yes (12 temperatures between 4 and 45 °C)	Snow, sea ice, glacial ice, glacial meltwater, subglacial sediment, glacier cryoconite, deep-sea sediment, refrigerated and frozen foods	*Barnettozyma, Bensingtonia, Botryozyma, Bullera*, 36 species of *Candida*, 20 species of *Cryptococcus, Cuniculitrema, Curvibasidium, Cystofilobasidium, Debaryomyces, Dioszegia, Guehomyces, Hannaella, Hanseniaspora, Holtermanniella, Kazachstania, Kluyveromyces, Lachancea, Leucosporidiella, Metschnikowia, Mrakia, Ogataea, Phytophthora, Pichia, Rhodosporidium, Rhodotorula, Saccharomyces, Sporidiobolus, Sporobolomyces, Torulaspora, Trichosporon, Udeniomyces, Wickerhamomyces, Williopsis,* and *Xanthophyllomyces*
National collection of yeast cultures (NCYC)	http://www.ncyc.co.uk	Yes (opt., max and min temperatures)	Frozen foods, frozen soil, Antarctica	*Cystofilobasidium, Cryptococcus*
Phaff yeast culture collection, University of California Davis, USA (UCDFST)	http://www.phaffcollection.org	No	Glaciers, Antarctic sea, Alaska, frozen foods	n/a
Colección Española de Cultivos Tipo (CECT), Universidad de Valencia, Spain	http://www.cect.org	No	Antarctic seawater, soil, and lake sediment, ice cream	n/a
Belgian coordinated collections of microorganisms (BCCM/MUCL), Belgium	http://bccm.belspo.be/db/mucl_search_form.php	No	Snow, Antarctica, fjord	n/a

(continued)

Table 2.4 (continued)

Yeast culture collection	Website	Website searchable for growth temperatures	Examples of cold substrates and geographic origins	Genera of yeasts able to grow at 4 °C
Culture collection of extremophilic fungi (EX-F), University of Ljubljana, Slovenia	http://www.ex-genebank.com	No	Sea ice, seawater, subglacial ice, surface ice, surface glacier ice, glacier meltwater	*Cryptococcus, Rhodotorula, Mrakia*
NITE biological resource center (NBRC), Tokyo, Japan	http://www.nbrc.nite.go.jp/NBRC2/NBRCDispSearchServlet	No	Antarctica, snow, frozen foods	*M. frigida*
USDA-ARS culture collection (US department of agriculture–agricultural research service), USA (NRRL), Peoria, Illinois, USA	http://nrrl.ncaur.usda.gov	No	n/a	n/a
Culture collection of the National Institute of Chemistry, Slovenia (MZKI)	http://www.mbc.ki.si/slo_katalog.html	No	Glacial basal ice, surface ice	*Bulleromyces, C. Rhodotorula, Rhodosporidium, Trichosporon*

n/a not available

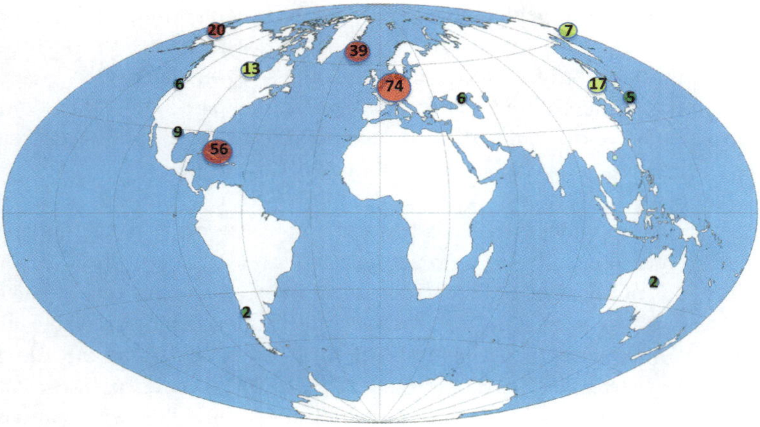

Fig. 2.1 Geographic origin of yeast strains in the Centraalbureau voor Schimmelcultures (CBS) collection (Utrecht, The Netherlands) that tested positive for growth at 4 °C. The number of CBS strains from each region is indicated. Data taken from the CBS online strain database (www.cbs.knaw.nl/Collections/Biolomics.aspx)

online DBVPG strain database is searchable for many fields including growth at low temperatures and "enzymatic and industrial surveys." A search for "growth at 4 °C" in this field generated a list of over 500 strains, most of which were isolated from snow and glacial ice, meltwater, and sediments from Italy and Antarctica.

The Centraalbureau voor Schimmelcultures (CBS) in Utrecht, the Netherlands, has posted quite extensive physiological data for the strains in the public CBS catalog. Data are posted for growth of CBS yeasts across a broad range of growth temperatures ranging from 4 to 45 °C. Over 300 strains able to grow at 4 °C were found in a search of the CBS online strain database (www.cbs.knaw.nl/Collections/Biolomics.aspx). Some interesting trends appear in this set of strains.

The geographic origin of CBS psychrotolerant strains is depicted in Fig. 2.1. Psychrotolerant yeasts able to grow at 4 °C were isolated from chilly geographic locations such as Antarctica (1 strain) and Iceland (39 strains), but also from unexpectedly warm temperate and tropical climates such as Egypt (2 strains) and the Bahamas (4 strains). For example, *Lachancea meyersii* CBS 9925 was isolated in 1999 by J. W. Fell and colleagues from seawater in a mangrove creek on Andros Island in the Bahamas (Fell et al. 2004) and can grow at 4 °C through 30 °C, but not at 35 °C or higher (CBS online database for growth temperatures). The bias toward temperate and tropical origins indicated in Fig. 2.1 most likely reflects the travel preferences of the field microbiologists who collect yeasts, rather than the actual biogeography of psychrotolerant yeasts.

Psychrotolerant yeasts in the CBS catalog were isolated from cold habitats such as glaciers, snow, and chilled beef, but also from various mid-temperature habitats such as fruits, flowers, trees, seawater, soil, and insects. These yeasts could be useful for many applications as discussed in Part IV of this book. A broad variety of psychrotolerant yeast genera and species are represented in the CBS catalog, as listed in Table 2.4. Both basidiomycetous and ascomycetous yeasts are represented, though basidiomycetes are more prevalent, as has been seen in an evaluation of yeasts isolated from Antarctica (Buzzini et al. 2012). Basidiomycetes made up 85 % of yeasts isolated from subglacial ice, glacial meltwater, seawater, sea ice, and melted snow in Norway (Butinar et al. 2011).

The National Collection of Yeast Cultures (NCYC), located at the Institute of Food Research, Norwich, UK, has a historic emphasis on brewing strains and has expanded in recent decades to embrace yeasts with biotechnology or food value. Like CBS, NCYC has an online strain database (www.ncyc.co.uk) that can be searched for strains with specific growth temperatures. Eighty-six strains had a reported minimum growth temperature of 4 °C. Notable strains include *Phaffia rhodozyma* type strain NCYC 874 (UCDFST 67-210), a species used for industrial production of the carotenoid pigment astaxanthin, and multiple wine, brewing and baking strains of *Saccharomyces cerevisiae*. Use of low-temperature-adapted yeasts for wine making is addressed further in Chap. 19.

2.9 Methods for Detecting Uncultured Yeasts

As discussed in Sect. 2.6, cultivation methods can bias recovery of different types of yeasts. Many yeast species may not be cultivable under the conditions selected. Molecular methods of profiling yeast biodiversity do not generate living cultures for further analysis, but can provide more detailed profiles of mixed populations. Molecular methods for the analysis of mixed communities of bacteria or yeasts were reviewed recently (Bokulich 2012).

Targeted methods are aimed at specific taxonomically defined groups of microbes, such as all bacteria, or a specific genus, species, or strain. These methods include fluorescence *in situ* hybridization (FISH) and quantitative PCR (qPCR). Broader profiling methods utilize universal primers aimed at larger taxonomic groups, such as kingdom or phylum level. The PCR products can be analyzed and profiled using several methods.

Amplified fragments can be ligated into a cloning vector, transformed into *E. coli*, and sequenced. This method is rather low-throughput, as each clone must be individually sequenced.

Denaturing gradient gel electrophoresis (DGGE) and its close relative temperature gradient gel electrophoresis (TGGE) have been used for the past 20 years to profile mixed microbial communities. These methods separate short PCR products along a chemical or temperature gradient. DGGE was used by Cocolin et al. (2000) to profile the succession of yeasts in wine fermentations. Yeasts are identified by comparison of band lengths to those of standards, or by extracting from the gel, re-amplifying, and sequencing. This method is limited by the low-throughput, and difficulty in performing the procedure. More information about DGGE can be found in a review by Ercolini (2004).

Terminal restriction fragment length polymorphism (TRFLP) differentiates microbial members of a community based on differences in length of a fluorescently labeled PCR fragment, cut with restriction enzymes, and separated by capillary electrophoresis on a DNA sequence analyzer. The fragment lengths are compared to those in databases, such as that compiled by Bokulich et al. (2012) for analysis of yeasts associated with wine fermentations. This method can be used in high-throughput, but is limited to fragment sizes for the species found in relevant databases.

Next-generation sequencing (NGS) methods include several recently developed technologies for massively parallel sequencing of a mixture of diverse DNA sequences. PCR is used to amplify marker genes, primarily 16S rRNA for prokaryotes and ITS genes for fungi. The two methods currently used for profiling microbial communities are 454 Life Sciences pyrosequencing (Margulies et al. 2005) and Illumina sequencing platforms (Bennett 2004). These two methods both generate large numbers of short segments of DNA sequence, with the fragments being physically partitioned and the growing DNA strand detected using different technologies. They differ in error rates, sequence lengths, and cost. Pyrosequencing generates longer reads (600 bp) than Illumina (150 bp), but Illumina generates greater sequence coverage (10^9 reads versus 10^6).

Users of these methods are cautioned to carefully select the reference database for identification of species from the raw DNA sequence data. Public DNA sequence databases contain sequences that are too short, misannotated, or contain too many ambiguous base calls to be useful. Fungal databases that are curated include the UNITE database of hand-curated fungal ITS sequences, primarily ectomycorrhizal fungi, and the Silva reference database of fungal large subunit (LSU) rDNA sequences. Different reference databases will yield different taxonomic assignments (Bokulich 2012). Illumina sequencing was recently used to profile both bacterial and yeast communities in a winery (Bokulich et al. 2013).

Some molecular methods have been used to profile fungal diversity in cold habitats. Bass et al. (2007) amplified an 18S rDNA segment from genomic DNA isolated from filtered seawater collected at several marine sites between 250 and 4,000 m depth, cloned the PCR fragments into the TOPO TA vector, and sequenced the inserts; 115 sequences were fungal, and included 19 fungal 18S sequence types, with 3–11 sequence types per site. Eleven of these sequences have only been detected in deep-sea samples. Yeasts predominated, particularly basidiomycetous yeasts. DNA sequences were similar to those of know cold-adapted yeast species including *C. vishniacii, Rhodosporidium diobovatum, Filobasidium globisporum, R. mucilaginosa,* and other basidiomycetous and ascomycetous yeast species.

Bellemain et al. (2013) used next-generation sequencing of the ITS ribosomal region to analyze the fungal diversity in 26 samples of Siberian permafrost aged 16,000–32,000 years. Seventy-five fungal operational taxonomic units (OTUs) belonging to 21 orders in three phyla were detected, 75 % of which were ascomycetous. Species detected from this study include many filamentous fungi, as well as yeasts that have been cultivated from other cold substrates including

Mrakiella aquatica (99.5 % identical), *Rhodotorula minuta* (100 % identical), and *Cryptococcus albidus* (100 % identical).

These two examples demonstrate that molecular methods, including next-generation sequencing methods, can be useful to detect, identify, and estimate the relative proportions of yeasts in substrates such as deep-sea water and may be used more extensively in coming years.

2.10 Conclusions

General methods used to isolate and cultivate yeasts have been modified to allow cultivation of cold-adapted yeasts. These modifications include selection of appropriate cold incubation temperature, osmolarity, and nutrient levels. Cold-adapted yeasts can be isolated from rather inaccessible habitats, such as deep-sea sediment and glacial ice cores, which requires particular aseptic sampling and processing methods. Some species isolated from these cold habitats only grow at cold temperatures, while others can tolerate mesophilic temperatures as well. Because these yeasts are particularly difficult to sample, isolate, cultivate, and preserve, it is important to archive them in properly equipped institutions with personnel knowledgeable about their cultivation, such as public culture collections. As culture-independent methods of microbe community profiling gain prominence, it may be important to archive the hard-won environmental samples and associated environmental DNA samples in public archives as well. Culture-independent methods, although powerful, still rely on data acquired using representative pure cultures to identify the species of microbes present. Isolation of cold-adapted yeasts is helping to expand our knowledge of the ecology of cold habitats and is leading to taxonomic discoveries and biotechnology advances.

References

Abyzov S (1993) Microorganisms in the Antarctic ice. In: Friedmann E (ed) Antarctic microbiology. Wile-Liss, New York, pp 265–295

Andrews S, DeGraaf H, Stamatin H (1997) Optimization of methodology for enumeration of xerophilic yeasts from foods. Int J Food Microbiol 35:109–116

Bandoni RJ (1972) Terrestrial occurrence of some aquatic hyphomycetes. Can J Bot 50:2283–2288

Bass D, Howe A, Brown N, Barton H, Demidova M, Michelle H, Li L, Sanders H, Watkinson SC, Willcock S (2007) Yeast forms dominate fungal diversity in the deep oceans. Proc Royal Soc B: Biol Sci 274:3069–3077

Bellemai E, Davey ML, Kauserud H, Epp LS, Boessenkool S, Coissac E, Geml J, Edwards M, Willerslev E, Gussarova G (2013) Fungal paleodiversity revealed using high- throughput metabarcoding of ancient DNA from arctic permafrost. Environ Microbiol 15:1176–1189

Bennett S (2004) Solexa ltd. Pharmacogenomics 5:433–438

Beuchat LR (1998) Progress in conventional methods for detection and enumeration of foodborne yeasts. Food Technol Biotechnol 36:267–272

Bokulich NA (2012) Next-generation approaches to the microbial ecology of food fermentations. Biochem Mol Biol Rep 45:377–389

Bokulich N, Hwang C, Liu S, Boundy-Mills K, Mills D (2012) Profiling the yeast communities of wine fermentations using terminal restriction fragment length polymorphism analysis. Am J Enol Vitic 63:185–194

Bokulich N, Ohta M, Richardson P, Mills D (2013) Monitoring seasonal changes in winery-resident microbiota. PLoS ONE, (in press)

Boundy-Mills K (2006) Methods for investigating yeast biodiversity. In: Rosa CA, Péter G (eds) Biodiversity and ecophysiology of yeasts. Springer, pp 67–100

Boundy-Mills K (2012) Yeast culture collections of the world: meeting the needs of industrial researchers. J Ind Microbiol Biotechnol 39:673–680

Brizzio S, Turchetti B, de Garcia V, Libkind D, Buzzini P, van Broock M (2007) Extracellular enzymatic activities of basidiomycetous yeasts isolated from glacial and subglacial waters of northwest Patagonia (Argentina). Can J Microbiol 53:519–525

Butinar L, Santos S, Spencer Martins I, Oren A, Gunde-Cimerman N (2005) Yeast diversity in hypersaline habitats. FEMS Microbiol Lett 244:229–234

Butinar L, Spencer-Martins I, Gunde-Cimerman N (2007) Yeasts in high Arctic glaciers: the discovery of a new habitat for eukaryotic microorganisms. A van Leeuwenhoek 91:277–289

Butinar L, Strmole T, Gunde-Cimerman N (2011) Relative incidence of ascomycetous yeasts in Arctic coastal environments. Microb Ecol 61:832–843

Buzzini P, Branda E, Goretti M, Turchetti B (2012) Psychrophilic yeasts from worldwide glacial habitats: diversity, adaptation strategies and biotechnological potential. FEMS Microbiol Ecol 82:217–241

Catranis C, Starmer W (1991) Microorganisms entrapped in glacial ice. Antarct J US 26:234–236

Cocolin L, Bisson L, Mills D (2000) Direct profiling of the yeast dynamics in wine fermentations. FEMS Microbiol Lett 189:81–87

Cousin M, Jay J, Vasavada P (1992) Psychrotrophic microorganisms. Compendium Methods Microbiol Exam Food 3:153–165

Davenport R (1980) Cold-tolerant yeasts and yeast-like organisms. In: Skinner F, Passmore S, Davenport R (eds) Biology and activities of yeasts. Academic Press, London, pp 215–230

de García V, Brizzio S, Libkind D, Buzzini P, Van Broock M (2007) Biodiversity of cold-adapted yeasts from glacial meltwater rivers in Patagonia, Argentina. FEMS Microbiol Ecol 59:331–341

Deak T (2003) Detection, enumeration and isolation of yeasts. In: Boekhout T, Robert V (eds) Yeasts in food: beneficial and detrimental aspects. B. Behr's Verlag GmbH & Co, Hamburg, pp 39–68

Di Menna ME (1966) Yeasts in Antarctic soils. A van Leeuwenhoek 32:29–38

Ercolini D (2004) PCR-DGGE fingerprinting: novel strategies for detection of microbes in food. J Microbiol Meth 56:297–314

Fell JW, Statzell-Tallman A, Kurtzman CP (2004) *L. meyersii* sp. nov., an ascosporogenous yeast from mangrove regions in the Bahama Islands. Stud Mycol 50:359–363

Fung DY (1999) Overview of rapid methods of microbiological analysis. Food Microbiol Analytical Methods: New Technol 12:1

Gilichinsky D, Rivkina E, Bakermans C, Shcherbakova V, Petrovskaya L, Ozerskaya S, Ivanushkina N, Kochkina G, Laurinavichuis K, Pecheritsina S (2005) Biodiversity of cryopegs in permafrost. FEMS Microbiol Ecol 53:117–128

Gronstal AL, Voytek MA, Kirshtein JD, Nicole M, Lowit MD, Cockell CS (2009) Contamination assessment in microbiological sampling of the Eyreville core, Chesapeake Bay impact structure. Geol Soc Am Special Papers 458:951–964

Gunde-Cimerman N, Zalar P, Hoog S, Plemenitaš A (2000) Hypersaline waters in salterns–natural ecological niches for halophilic black yeasts. FEMS Microbiol Ecol 32:235–240

Gunde-Cimerman N, Sonjak S, Zalar P, Frisvad JC, Diderichsen B, Plemenitaš A (2003) Extremophilic fungi in arctic ice: a relationship between adaptation to low temperature and water activity. Phys Chem Earth A/B/C 28:1273–1278
Hagler A, Ahearn D (1987) Ecology of aquatic yeasts. In: Rose A, Harrison J (eds) The yeasts, vol 2., Yeasts and the environmentAcademic Press, London, pp 181–205
Hamby KA, Hernández A, Boundy-Mills K, Zalom FG (2012) Associations of yeasts with spotted-wing Drosophila (*Drosophila suzukii*; *Diptera: Drosophilidae*) in cherries and raspberries. Appl Environ Microbiol 78:4869–4873
Hattori T (1980) A note on the effect of different types of agar on plate count of oligotrophic bacteria in soil. J Gen Appl Microbiol 26:373–374
Heard G, Fleet G (1986) Evaluation of selective media for enumeration of yeasts during wine fermentation. J Appl Bacteriol 60:477–481
Hernandez P, Beuchat L (1995) Evaluation of diluents and media for enumerating *Zygosaccharomyces rouxii* in blueberry syrup. Int J Food Microbiol 25:11–18
Hocking AD, Pitt JI (1980) Dichloran-glycerol medium for enumeration of Xerophilic fungi from low-moisture foods. Appl Environ Microbiol 39:488–492
Hocking A, Pitt J, Samson R, King A (1992) Recommendations from the closing session of SMMEF II. In: Samson R, Hocking A, Pitt J, King A (eds) Modern methods in food mycology. Elsevier, Amsterdam, pp 359–368
Ikemoto E, Kyo M (1993) Development of microbiological compact mud sampler. JAMSTEC Res 30:1–16
Kimball J, Corcoran EF, Wood FE (1963) Chlorophyll-containing microorganisms in the aphotic zone of the oceans. Bull Marine Sci 13:574–577
King A, Hocking A, Pitt J (1979) Dichloran-rose bengal medium for enumeration and isolation of molds from foods. Appl Environ Microbiol 37:959–964
Kutty SN, Philip R (2008) Marine yeasts: a review. Yeast 25:465–483
Libkind D, Brizzio S, Van Broock M (2004) *R. mucilaginosa*, a carotenoid producing yeast strain from a Patagonian high-altitude lake. Folia Microbiol 49:19–25
López-García P, Rodríguez-Valera F, Pedrós-Alió C, Moreira D (2001) Unexpected diversity of small eukaryotes in deep-sea Antarctic plankton. Nature 409:603–607
Ma L, Catranis CM, Starmer WT, Rogers SO (1999) Revival and characterization of fungi from ancient polar ice. Mycologist 13:70–73
Ma L-J, Rogers SO, Catranis CM, Starmer WT (2000) Detection and characterization of ancient fungi entrapped in glacial ice. Mycologia 92:286–295
Margesin R (ed) (2009) Permafrost soils. Springer, Berlin
Margesin R, Zacke G, Schinner F (2002) Characterization of heterotrophic microorganisms in Alpine glacier cryoconite. Arctic Antarct Alp Res 34:88–93
Margulies M, Egholm M, Altman WE, Attiya S, Bader JS, Bemben LA, Berka J, Braverman MS, Chen Y-J, Chen Z (2005) Genome sequencing in microfabricated high-density picolitre reactors. Nature 437:376–380
Martini A, Ciani M, Scorzetti G (1996) Direct enumeration and isolation of wine yeasts from grape surfaces. American J Enol Viticult 47:435–440
Montes MJ, Belloch C, Galiana M, Garcia MD, Andrés C, Ferrer S, Torres-Rodriguez JM, Guinea J (1999) Polyphasic taxonomy of a novel yeast isolated from Antarctic environment; description of *Cryptococcus victoriae* sp. nov. Syst Applied Microbiol 22:97–105
Nagahama T, Hamamoto M, Nakase T, Takami H, Horikoshi K (2001) Distribution and identification of red yeasts in deep-sea environments around the northwest Pacific Ocean. A van Leeuwenhoek 80:101–110
Nagahama T, Hamamoto M, Nakase T, Takaki Y, Horikoshi K (2003) *Cryptococcus surugaensis* sp. nov., a novel yeast species from sediment collected on the deep-sea floor of Suruga Bay. Int J Syst Evol Microbiol 53:2095–2098
Rogers S, Theraisnathan V, Ma L, Zhao Y, Zhang G, Shin S-G, Castello J, Starmer WT (2004) Comparisons of protocols for decontamination of environmental ice samples for biological and molecular examinations. Appl Environ Microbiol 70:2540–2544

Samson RA, Hoekstra ES, Frisvad JC (2004) Introduction to food-and airborne fungi, 7th edn. Centraalbureau voor Schimmelcultures (CBS), Utrecht

Schmidt-Lorenz W (1983) Indicator organisms in frozen foods in relation to spoilage. A van Leeuwenhoek 48:625–633

Shivaji S, Prasad G (2009) Antarctic yeasts: biodiversity and potential applications. In: Satyanarayana T, Kunze G (eds) Yeast biotechnology: diversity and applications. Springer, Berlin, pp 3–18

Singh P, Singh SM (2012) Characterization of yeast and filamentous fungi isolated from cryoconite holes of Svalbard, Arctic. Polar Biol 35:575–583

Slávikovä E, Vadkertiová R, Kocková-Kratochvílová A (1992) Yeasts isolated from artificial lake waters. Can J Microbiol 38:1206–1209

Solís L, Sánchez E, García S, Heredia N (2009) Traditional methods for detection of foodborne pathogens. In: Heredia N, Wesley I (eds) Microbiologically safe foods. Wiley, Hoboken, pp 525–546

Tadych M (2008) International cooperative biodiversity groups central Asia project. Rutgers, The State University of New Jersey, USA

Tomkin R (1963) Refrigeration temperature as an environmental factor influencing the microbial quality of food: a review. Food Technol 27:54–58

Turchetti B, Buzzini P, Goretti M, Branda E, Diolaiuti G, D'Agata C, Smiraglia C, Vaughan-Martini A (2008) Psychrophilic yeasts in glacial environments of Alpine glaciers. FEMS Microbiol Ecol 63:73–83

Vishniac HS, Onofri S (2003) *Cryptococcus antarcticus* var. *circumpolaris* var. nov., a basidiomycetous yeast from Antarctica. A van Leeuwenhoek 83:231–233

Vishnivetskaya T, Kathariou S, McGrath J, Gilichinsky D, Tiedje JM (2000) Low-temperature recovery strategies for the isolation of bacteria from ancient permafrost sediments. Extremophiles 4:165–173

Vorobyova E, Soina V, Gorlenko M, Minkovskaya N, Zalinova N, Mamukelashvili A, Gilichinsky D, Rivkina E, Vishnivetskaya T (1997) The deep cold biosphere: facts and hypothesis. FEMS Microbiol Rev 20:277–290

Wagner D (2008) Microbial communities and processes in Arctic permafrost environments. In: Dion P, Nautiyal C (eds) Microbiology of extreme soils. Springer, Berlin, pp 133–154

Welthagen J, Viljoen B (1997) Comparison of ten media for the enumeration of yeasts in dairy products. Food Res Int 30:207–211

Zucconi L, Selbmann L, Buzzini P, Turchetti B, Guglielmin M, Frisvad J, Onofri S (2012) Searching for eukaryotic life preserved in Antarctic permafrost. Polar Biol 35:749–757

Part II
Diversity and Ecology of Cold-Adapted Yeasts in Worldwide Cold Habitats

Chapter 3
Cold-Adapted Yeasts in Arctic Habitats

Polona Zalar and Nina Gunde-Cimerman

Abstract The Arctic area is a polar region at the northernmost part of the Earth, north of the Arctic Circle (66° 33′N). It is a region of an ice-covered Arctic ocean, surrounded by treeless permafrost. The climate within the Circle is extremely cold, and much of the area is permanently covered with ice. Unlike in the temperate zones, where filamentous fungi prevail, extremophilic yeasts are particularly well adapted to these conditions and thus thrive in different polar environments. The range of Arctic environments inhabited by yeasts includes glacial ice with brine networks and inclusions, subglacial environments, soil, permafrost, Arctic sea and sea ice. Yeast isolates from cryoconite holes and cryptoendolithic environments are rare. The physiological activity of yeasts has been demonstrated at temperatures below −20 °C. Studies of the abundance, distribution and taxonomy of Arctic yeasts are very much limited in comparison with the studies of the Antarctic yeasts. Interestingly, the dominant yeasts in the Arctic are also of basidiomycetous affinity and belong to the same genera that occur frequently in the Antarctic soils and polar offshore sea waters. These are different species of the genera *Aureobasidium, Candida, Debaryomyces, Metschnikowia, Pichia, Protomyces, Bensingtonia, Bulleromyces, Cryptococcus, Cystofilobasidium, Filobasidium, Leucosporidiella, Malassezia, Kondoa, Mastigobasidium, Mrakia, Mrakiella, Rhodosporidium, Rhodotorula, Sporidiobolus, Sporobolomyces, Trichosporon* and *Udeiomyces*. Amongst these, the polyphyletic genera *Cryptococcus* and *Rhodotorula* appear to be the predominant and are also represented by the highest numbers of species. Classical culture-based methods, molecular techniques and more recently also metagenomic analyses have revealed the existence of yet unknown species, genera and even new phylogenetic lineages.

P. Zalar · N. Gunde-Cimerman (✉)
Department of Biology, University of Ljubljana, Večna pot 111, 1000 Ljubljana, Slovenia
e-mail: nina-gunde-cimerman@bf.uni-lj.si

N. Gunde-Cimerman
Centre of Excellence for Integrated Approaches in Chemistry and Biology of Proteins (CIPKeBiP), Jamova 39, 1000 Ljubljana, Slovenia

P. Buzzini and R. Margesin (eds.), *Cold-Adapted Yeasts*,
DOI: 10.1007/978-3-642-39681-6_3,

Keywords Arctic • Permafrost • Glaciers • Ice • *Cryptococcus* • *Rhodotorula* • Black yeasts

Contents

3.1 Introduction ... 50
3.2 Yeasts on Arctic Vegetation, in Soil and Permafrost ... 53
3.3 Yeasts and Yeast-Like Fungi in/on Rocks ... 60
3.4 Yeasts in Sea Water and Sea Ice ... 62
3.5 Yeasts in Glaciers and Subglacial Environments ... 63
3.6 Yeasts in Cryoconite Holes ... 66
3.7 Conclusions ... 67
References ... 69

3.1 Introduction

The Arctic area is a polar region located at the northernmost part of the Earth, north of the Arctic Circle (66° 33′N). It is a region of an ice-covered Arctic ocean, surrounded by treeless permafrost. Geographically, it is spread over Greenland, parts of Iceland, Alaska and Norway (Svalbard; Fig. 3.1a), Sweden, Finland, Canada and Russia. The climate within the Circle is very cold, and much of the area is permanently covered with ice. In the mid-winter months, the temperatures can reach −50 °C in the higher latitudes. In the summer months, sunlight melts the seas and the topsoil. The average temperature of the warmest month (July) does not exceed 10 °C.

The main stress factors that influence microbial and therefore also fungal biodiversity in extremely cold polar areas are the low temperature, closely related to desiccation and low water availability (low A_w) due to freezing, the relatively high concentrations of ions and the generally low levels of nutrients in the liquid water and sometimes high UV irradiation and hypoxia. It appears that unlike in the temperate zones, where filamentous fungi prevail, yeasts are particularly well adapted to these conditions and thus thrive in different polar environments.

In spite of hard conditions and logistic problems, the initial studies on the presence of yeasts in the Earth's polar areas started more than a century ago with a study on Antarctic fungi performed in the early 1870s (Bridge et al. 2013). Thirty-seven research stations were established in the Arctic region, which prompted the research even in the most remote places (Fig. 3.1b). Initial studies of yeasts in polar areas were culture-based and concentrated primarily on the scarce vegetation, wooden remains, soil and permafrost, followed by studies of the aerosphere and rock-inhabiting yeast-like fungi. More recent studies have focused on water-based environments, such as sea water, glacial melt-water, snow, glacial and lake

Fig. 3.1 Arctic landscapes. **a** Glaciers in Kongsfjorden, (Spitsbergen, Svalbard); **b** Ablation zone of a glacier, icebergs on the sea surface; **c** characteristic buildings of Ny-Ålesund settlement (78° 55′ N, 11° 56′ E, Spitsbergen, Svalbard), where research stations maintained by several countries are located

ice. The methods used for yeast detection have been time appropriate, from the classical early microscopical visualisation to molecular DNA-based techniques (Fell et al. 2006), which have been complemented lately by metagenomic studies (Simon et al. 2009; Kennedy et al. 2010; Ziolkowski et al. 2013), in which fungi were most often assessed only as a part of the overall analysis, mainly concentrated on the prokaryotic diversity.

Detailed taxonomic analyses based on physiological profiles and relevant molecular markers have allowed the description of numerous novel yeast species originating predominantly from the Antarctic (Montes et al. 1999; Scorzetti et al. 2000; Thomas-Hall and Watson 2002; Thomas-Hall et al. 2002; Vishniac 2002;

Libkind et al. 2005; Xin and Zhou 2007; Thomas-Hall et al. 2010), but also from other cold environments (Margesin et al. 2007; Margesin and Fell 2008; de García et al. 2010a, b; Thomas-Hall et al. 2010; Vishniac and Takashima 2010). Both in Arctic and Antarctic regions, basidiomycetous yeasts prevail (Branda et al. 2010). The majority of data on the occurrence of yeasts in the Arctic region can be retrieved as scattered reports (Gilichinsky et al. 2007; Frisvad 2008; Ludley and Robinson 2008; Libkind et al. 2009; Ozerskaya et al. 2009; Branda et al. 2010), or as information in public fungal databases, such as the CBS Fungal Biodiversity Centre or the American Type Culture Collection (ATCC), which can also serve as valuable sources for data on yeasts from cold regions. An overview of the public services collections preserving cold-adapted yeasts isolated from worldwide cold habitats is reported in Chap. 2.

The species with the broadest adaptive potential and the widest distributions across various polar niches are in particular different species of the genera *Bulleromyces, Cryptococcus, Cystofilobasidium, Dioszegia, Erythrobasidium, Filobasidium, Guehomyces, Leucosporidiella, Leucosporidium, Malassezia, Kondoa* (unpublished, from the Arctic), *Mastigobasidium, Mrakia, Mrakiella, Rhodosporidium, Rhodotorula, Sporidiobolus, Sporobolomyces, Trichosporon* and *Udeiomyces*. Among these, *Cryptococcus* and *Rhodotorula*, which are both of polyphyletic origins that occur in several phylogenetic lineages of the Agaricomycotina, appear to be the predominant genera and are also represented by the highest numbers of species (Branda et al. 2010; Bellemain et al. 2013).

A special group of yeasts, extremely adapted to different types of extreme environments (Selbmann et al. 2008; Zalar et al. 2008), are melanised, black, yeast-like fungi (Onofri et al. 1999). These study organisms can inhabit environments as different and extreme as Antarctic rocks (Selbmann et al. 2005; Gorbushina 2007) and Arctic glacial ice with high salt concentrations (Butinar et al. 2011). The black yeasts can colonise environments on Earth that resemble most environments on the planet Mars, and have therefore been used as the closest eukaryotic models for exobiological speculations. Their astonishing viability after freezing and thawing, as well as after UV exposure, and their tolerance to osmotic imbalances have shown their uncommon ability to survive under harsh external pressures (Selbmann et al. 2005; Onofri et al. 2008).

In comparison with the studies of the Antarctic yeasts, the studies of the Arctic yeasts with respect to their abundance, distribution and taxonomy are very much limited. Taxonomic characterisations have shown that the dominant yeasts in the Arctic are also of basidiomycetous affinity and belong to the same genera that occur frequently in the Antarctic soils and polar offshore sea waters (Jones 1976; Abyzov 1993; Abyzov et al. 2004; Onofri et al. 2004; Vishniac 2006). The species of the genera *Cryptococcus, Cystofilobasidium, Rhodosporidium, Trichosporon* and *Rhodotorula* are predominant yeasts in the various habitats of the Arctic (Starmer et al. 2005; Raspor and Zupan 2006; Butinar et al. 2007, 2011; Vishniac and Takashima 2010; de García et al. 2012). The distribution of the discovered Arctic yeasts ranges from circumpolar to endemic and cosmopolitan species. Whereas some yeasts that are endemic to the Arctic region show psychrophilic behaviour (Margesin and Fell 2008; de García et al. 2012), the majority are psychrotolerant and globally distributed. The global warming phenomena, resulting in the accelerated melting of polar ice, facilitate their transportation via global water systems (Grabińska-Łoniewska et al. 2007) and air currents. Some Arctic yeasts were previously known only from human proximity; as inhabitants of freezers, cold storage rooms, refrigerated and even frozen food. In Table 3.1 to the best of our knowledge, all known yeast species from diverse Arctic environments are listed.

Many recovered Arctic yeast species represent reservoirs of rare or still unknown microbial species, and therefore also of novel biological processes. Although most investigations of yeasts in the Arctic have been limited to their biodiversity, an increasing number of studies have been oriented towards extracellular enzymatic activities of yeasts, antifreeze proteins, the biosynthesis of secondary metabolites and bioremediation at low temperatures (Margesin 2009; Simon et al. 2009; Pathan et al. 2010; Lee et al. 2012), providing us with a better understanding of the complexity of these processes at 'the edge of life' (Robinson 2001; Ruisi et al. 2007; Xiao et al. 2010). Recently, adaptations to cold and consequentially low water activity at the molecular level have also received more attention, expanding the realm of extremophilic model organisms from prokaryotes to yeasts (Gostinčar et al. 2008; Turk et al. 2011).

3.2 Yeasts on Arctic Vegetation, in Soil and Permafrost

Limiting factors for microbial growth and activity in the Arctic soils are extremely low temperatures, cycles of freezing and thawing, low annual precipitation and low nutrient content (Margesin and Miteva 2011). Annual soil temperatures recorded at 10 cm depth can range between −27 and 14 °C in the High Arctic, and between −7 and 11 °C in the Low Arctic (Timling and Taylor 2012). Because of non uniform snow distribution, wide temperature differences can be recorded in the underlying soil. A considerable microbial activity occurs in the region between the snow packs and above permafrost region, even at the temperatures below freezing point (Sturm et al. 2005). The activity is prompted during the short growing season, lasting up to 6 weeks in the High Arctic up to 4 months in the Low Arctic, when the upper 30–60 cm of soil is thawing, which can result in anaerobic condition due to the underlying permafrost preventing drainage. On the other hand, freezing can cause desiccation and salt accumulation in salt crusts. Due to diverse cryogenic processes, Arctic soils are shaped into extremely heterogeneous environment, with pH values ranging from 4 to 9, and relatively variable amount of carbon (Timling and Taylor 2012).

Yeasts have been reported in connection with the subarctic vegetation and soil. Mainly, basidiomycetous yeasts have been isolated from soil and adjacent berries, flowers, vegetation of the littoral zone, forest trees and grasses, while ascomycetous yeasts have been recovered more in connection with vegetation (Babjeva and

Table 3.1 Ascomycetous and basidiomycetous yeasts reported from different Arctic habitats. Letters indicate references, listed in the legend

	Plants	Soil	Glaciers, subglacial habitats	Permafrost	Sea water, sea ice	Others
Ascomycota						
Aureobasidium pullulans	a	a	f	e, j	j	
Candida apicola	a					
Pichia kudriavzevii (formerly *Candida krusei*)	a					
Candida molischiana (anamorph of *Kuraishia molischiana*)	a					
Candida norvegica		g				
Candida parapsilosis					m	
Candida pseudointermedia	a					
Candida quercitrusa	a					
Candida sake	a					
Candida santamariae	a					
Debaryomyces hansenii	a			e*	m	
Debaryomyces maramus					m	
Metschnikowia bicuspidata					m	
Metschnikowia pulcherrima	a					
Metschnikowia reukaufii	a					
Metschnikowia zobellii					m	
Meyerozyma guilliermondii (formerly *Pichia guilliermondii*)				e*	m	
Scheffersomyces spartinae (formerly *Pichia spartinae*)	a					
Protomyces inouyei					m	
Basidiomycota						
Bensingtonia yamatoana		g*				
Bulleromyces albus			h			
Cryptococcus adeliensis			h			
Cryptococcus aerius		g				

(continued)

Table 3.1 (continued)

	Plants	Soil	Glaciers, subglacial habitats	Permafrost	Sea water, sea ice	Others
Cryptococcus albidosimilis	a		h			
Cryptococcus albidus		g	f, h			
Cryptococcus antarcticus		g				
Cryptococcus aquaticus		g				d(DFS)
Cryptococcus carnescens			n, h			
Cryptococcus flavus	a					
Cryptococcus fonsecae			n			
Cryptococcus gastricus	a					k(S)*
Cryptococcus gilvescens			h			o(CH)
Cryptococcus heimaeyensis		c				
Cryptococcus heveanensis	a					
Cryptococcus humicola			f			
Dioszegia hungarica (formerly *Cryptococcus hungaricus*)	a					
Cryptococcus laurentii	a	a	h	e*		
Cryptococcus liquefaciens		g	h			
Cryptococcus luteolus	a					
Cryptococcus macerans			h			d(DFS)
Cryptococcus magnus	a		h			
Cryptococcus oeirensis			h			
Cryptococcus podzolicus		a				
Cryptococcus psychrotolerans			n			
Cryptococcus saitoi			h			
Cryptococcus skinnerii	a					
Cryptococcus tephrensis		c				
Cryptococcus terricola	a	a, g				k(S)*

(continued)

Table 3.1 (continued)

	Plants	Soil	Glaciers, subglacial habitats	Permafrost	Sea water, sea ice	Others
Cryptococcus victoriae		g	n, h	e*		
Cryptococcus watticus		g				
Cystofilobasidium capitatum (formerly *Cystofilobasidium lari-marini*)	a	a				d(DFS)
Cystofilobasidium inirmo-miniatum	a	a				
Filobasidium uniguttulatum			h			
Leucosporidiella fragaria			h			
Leucosporidiella creatinivora		g				
Leucosporidium scottii	a	g*				
Leucosporidium antarcticum	a					
Mrakia frigida		g				
Mrakia gelida						k(S)*
Mrakia psychrophyla						k(S)*
Mrakiella cryoconiti						j(CR)
Rhodosporidium diobovatum			h			
Rhodosporidium kratochvilovae		g				
Rhodotorula arctica		l				
Rhodotorula aurantiaca	a					
Rhodotorula buffonii	a					
Leucosporidiella creatinivora (formerly *Rhodotorula creatinivora*)				b		
Rhodotorula fujianensis (anamorph of *Curvibasidium cygneicollum*)	a					
Rhodotorula glacialis	a					k(S)*
Rhodotorula hylophila	a					
Rhodotorula laryngis			f, h			
Rhodotorula minuta	a		h			
Rhodotorula mucilaginosa	a		f, h			

(continued)

Table 3.1 (continued)

	Plants	Soil	Glaciers, subglacial habitats	Permafrost	Sea water, sea ice	Others
Leucosporidiella muscorum (formerly *Rhodotorula muscorum*)	a					k(S)*
Leucosporidiella yakutica (formerly *Rhodotorula yakutica*)				b		
Sporidiobolus salmonicolor (teleomorph of *Sporobolomyces salmonicolor*)	a		f			
Sporobolomyces roseus (anamorph of *Sporidiobolus metaroseus*)	a					
Trichosporon cutaneum	a					
Trichosporon mucoides			h			
Guehomyces pullulans (formerly *Trichosporon pullulans*)	a	b				

S sediments; *CH* cryoconite holes; *DFS* diverse frozen samples
*nearest phylogenetic neighbour, related to
a Babjeva and Reshetova (1998), *b* Golubev (1998), *c* Vishniac (2002), *d* Birgisson et al. (2003), *e* Gilichinsky et al. (2005), *f* Starmer et al. (2005), *g* Vishniac (2006), *h* Butinar et al. (2007), *i* Margesin and Fell (2008), *j* Zalar et al. (2008), *k* Pathan et al. (2010), *l* Vishniac and Takashima (2010), *m* Butinar et al. (2011), *n* de García et al. (2012), *o* Singh and Singh (2012)

Reshetova 1998). The multivariant analysis of soil samples collected all over the globe over the latitudinal gradient has shown the temperature among other factors (precipitation, electrical conductivity, pH, vegetation types) as the major factor for yeast distribution. The success of the soil-inhabiting polyphyletic genus *Cryptococcus* is mainly attributed to their capsules, which enable them to grow in mineral soil types distributed all over the globe (Vishniac 2006). A list of yeast species inhabiting polar soils of Arctic is reported in Table 3.1.

Permafrost, ground that is comprised of soil, sediment or rock, and includes ice and organic material, which remains at or below 0 °C for at least two consecutive years, presents even harsher conditions for microbial activity. Permafrost is a strictly thermal phenomenon and does not depend on the composition of the ground. Permafrost in polar regions covers more than 25 % of the land surface and significant parts of the coastal sea shelves (Wagner 2008). Together with seasonally frozen soils, permafrost represents a large part (approximately 50 %) of the terrestrial Earth (Panikov 2009). Permafrost reaches depths of 600–800 m in the Eurasian tundra and a depth of 1,450 m in the Antarctic deserts (Steven et al. 2006). Usually, it is divided into three temperature-depth layers. The surface, active layer is influenced by air temperature fluctuations, with a thickness from a few cm to 10 m and temperatures fluctuating from +15 to −35 °C. The second layer is composed of perennially frozen permafrost sediments, with a thickness from 10 to 20 m and temperatures from 0 to −15 °C. The third and deepest layer is composed of stable permafrost sediments, with temperatures from −5 to −10 °C (Wagner 2008). In the active layer and in the middle layer of permafrost, intensive physicochemical processes take place, strongly influencing microbial diversity, whereas in the deeper permafrost sediments, the conditions can remain stable for long periods of time, limiting the microbial processes (Wagner 2008). Thermal conditions influence the formation of different cryogenic structures, such as 'ice wedges' (a crack in the ground formed by a narrow or thin piece of ice), connected with the atmosphere, unfrozen layers of ground in permafrost, called 'taliks', and lenses of brine in permafrost, called 'cryopegs'.

Different types of permafrost habitats are characterised by extremely low temperatures, freeze–thaw cycles, low water and nutrient availability, and low gas permeability (Morozova et al. 2007; Wagner 2008; Panikov 2009). For many years, permafrost was considered a depository of ancient microbial life, but the recently discovered intensive winter gas fluxes of methane and carbon dioxide from the tundra to the atmosphere have led to an intensified search for organisms able to metabolise below the freezing point. Now we know that the total microbial biomass in permafrost is comparable to that of the communities of temperate soil ecosystems (Wagner 2008; Wagner et al. 2005). Microbial communities in permafrost are represented by dormant propagules of mesophilic cosmopolitan species that have been deposited from the surrounding areas by wind and animals, but also by active indigenous culturable or non-culturable species of Archaea, Bacteria and Eukarya, particularly fungi (Ruisi et al. 2007; Panikov 2009). It has been demonstrated that yeast species not only survive under permafrost conditions, but also can sustain an active metabolism (Rivkina et al. 2004; Wagner 2008). It appears that yeasts are more resistant to hostile permafrost environments and show more vigorous growth in frozen habitats than bacteria (Steven et al. 2006). Depending on the depth and age of the samples, the numbers of culturable yeasts can vary from 10 to almost 100,000 CFU g^{-1} material. The peaks of fungal populations in permafrost are microfocal and, importantly, they show a parallel decrease in species numbers (Ozerskaya et al. 2008).

Permafrosts in Arctic and Antarctic areas differ considerably. Permafrost in the Antarctic regions is typically dry, with low intertidal water, while the Arctic permafrost is, in contrast, impregnated with ice. During the thaw, the underlying permafrost can prevent drainage of soils, leading to temporary anoxic conditions, while subsequent freezing can lead to desiccation and an increased salinity, especially in the High Arctic, where salt crusts can form on the soil surface due to the high rates of evaporation. The Arctic soils are shaped also by cryogenic processes, such as repeated freeze–thaw cycles, cryoturbation, frost heaving, thermal cracking, and the formation of needle ice and ice lenses. As a result, the Arctic soils are extremely heterogeneous at small scales. Soil pH values can vary between 4 and 9, nutrient contents (N, P, K) are generally low, while carbon can vary substantially (Timling and Taylor 2012).

Fungal diversity in the Arctic permafrost has been studied mainly over the last decade. The first studies concentrated on the longevity of frozen cells, while in later studies mainly biodiversity was investigated. Since DNA may be preserved in permafrost soils either as an intracellular DNA within the intact, cryopreserved cells, or as an extracellular DNA adsorbed to sediment particles, the recent studies also include DNA metabarcoding (Bellemain et al. 2013).

Viable yeasts and fungi have been isolated from Siberian permafrost sediments that have remained in a frozen state for up to 3 million years, although they were preserved in considerably lower numbers than the prokaryotic cells (Golubev 1998; Faizutdinova et al. 2005). Upon thawing, these microorganisms were able to resume their metabolic activities (Takano et al. 2004). In some permafrost regions, yeasts represented an important, or even the major (up to 100 %), part of all of the fungi isolated, and 20–25 % of the total aerobic heterotrophs (Vorobyova et al. 1997; Steven et al. 2006). Researchers have even managed to isolate yeasts below freezing point on solid media (cellulose powder or plastic film) with ethanol as the sole carbon source without using artificial antifreezes. Enrichment from soil and permafrost obtained on such frozen solid media contained mainly fungi, and further purification resulted in an isolation of basidiomycetous yeasts of the genera *Mrakia* and *Leucosporidium,* as well as of ascomycetous fungi of the genus *Geomyces* (Panikov and Sizova 2007).

Most of the yeasts isolated from the Arctic permafrost were cosmopolitan mesophilic species, which are easily dispersed via air, water and animals (Samson et al. 2002). These yeasts might represent a recent contamination of the permafrost, and this thus makes the identification of the indigenous species more difficult (Lydolph et al. 2005; Ruisi et al. 2007). However, it has been shown that permafrost contains a considerable taxonomic diversity with significant numbers of new taxa (Golubev 1998; Vishniac and Takashima 2010). The taxonomic diversity and ecological diversity of the ancient fungal communities of the Arctic permafrost fungi were assessed by combining next generation sequencing and metabarcoding of the DNA, preserved in permafrost. Twenty-six sediment samples from two localities in Siberia dated 16,000–32,000 radiocarbon years were analysed for fungal ITS. Among the operational taxonomic units (OTUs), there were sequences that belonged to almost all of the known yeast groups, and the following genera were detected: (1) *Malassezia*, a lipophilous basidiomycete yeast genus and (2) *Cryptococcus*, *Mrakiella* and *Rhodotorula*, putative psychrophilic fungi, which might represent a modern DNA from metabolically active individuals. Saccharomycetales were not identified to the genus level (Bellemain et al. 2013).

In culture, the most frequently isolated yeasts from these permafrost sediments belong to the genera *Cryotococcus*, *Rhodotorula* and *Saccharomyces*. By the use of standard culture techniques, yeast species of the genera *Bulleromyces*, *Candida*, *Clavispora*, *Debaryomyces*, *Dioszegia*, *Guehomyces*, *Leucosporidium*, *Mrakia*, *Pichia*, *Pseudozyma*, *Rhodosporidium*, *Sporidiobolus*, *Sporobolomyces*, *Tilletiopsis*, *Torulaspora*, *Trichosporon* and *Leucosporidium* have also been found in permafrost (Vishniac 1993; Thomas-Hall and Watson 2002; Vishniac 2006; Margesin et al. 2007; Xin and Zhou 2007).

The microbial biodiversity in cryopegs in permafrost originating from the Arctic Ocean regression in Siberia were studied by Gilichinsky et al. (2005). From the brine in cryopegs, aerobic and anaerobic, halotolerant and halophilic, psychrophilic and psychrotolerant bacteria, mycelial fungi and yeasts of the genera *Cryptococcus*, *Debaryomyces* and *Pichia* have been identified, as well as black yeast *Aureobasidium pullulans* var. *pullulans*. Activities were detected below 0 °C, and the isolated microbes were considered as survivors of 43,000 years at −10 °C (Gilichinsky et al. 2005).

3.3 Yeasts and Yeast-Like Fungi in/on Rocks

The Arctic region is rich in stony and rocky habitats. The rock represents a dwelling place as well as a substrate for endolithic microorganisms, including yeasts. The conditions on the surface of the rock are the harshest ones seen, and they can only be sustained by lichens (Friedmann 1982; Nienow and Friedmann 1993; Selbmann et al. 2005), assigned as epiliths. The interiors of the rock protect—to some extent—life from environmental extremes, which allows for the growth of different kinds of endolithic microorganisms (Friedmann and Koriem 1989; de los Ríos et al. 2003, 2005; McLoughlin et al. 2007). Non-lichenised, rock-inhabiting fungi in polar regions are cryptoendoliths that establish themselves under an abiotic crust on the successive layers of the rock surface. The darkly pigmented non-lichenised yeasts inhabit the black layer, and *Rhodotorula* and *Sporobolomyces*, together with hyaline filamentous fungi, inhabit the white layer (Burford et al. 2003).

Yeasts living within the rock are exposed to different kinds of stress factors, which can include a high UV irradiation, low temperatures, desiccation, repeated freeze-thawing and a lack of nutrients. The general response of the cryptoendolithic fungi to these conditions is a high level of simplification. They primarily grow as multicellular clumps and have the ability to shift to a simpler life cycle and meristematic growth, which also includes the loss of sexual, and in some fungi even asexual, reproduction (Selbmann et al. 2005). Through such simplification, they can conclude their life cycle in a shorter time, with less energy needed and hence at a lower metabolic cost (Ruisi et al. 2007).

While the Antarctic cryptoendolithic fungi, including yeasts, such as *Cryptococcus friedmannii* (Vishniac 1985), have been long investigated (Friedmann 1982; Onofri et al. 1999, 2004; Selbmann et al. 2005, 2008), studies on the Arctic cryptoendolithic fungi have been initiated only recently (Omelon et al. 2006, 2007; Selbmann 2013, personal communication). The studies performed on cryptoendoliths in the sandstone outcrops of the Canadian High Arctic have revealed the presence of black-pigmented yeast-like fungi (Omelon et al. 2006, 2007). Although these isolates have not been identified, it appears that the general microbial diversity in the Arctic rock is higher in comparison with similar habitats in Antarctica. The main reasons for this are probably the higher temperatures and moisture and to a lesser extent the lower pH and higher concentrations of iron, aluminium and silicon in the overlying surface (Omelon et al. 2007). The total time that is available for ideal metabolic activity within the cryptoendolithic environment in the Canadian High Arctic has been estimated as 2,500 h per year (Omelon et al. 2006), whereas for the Antarctic Dry Valleys it has been estimated as between 50 and 500 h per year (Omelon et al. 2006). These and other studies performed over the last few years have shown that the cryptoendolithic yeasts are much more widespread and common than previously thought. It appears that the conditions in the rocks are prohibitive enough to prevent the settlement and the growth of competitive cosmopolitan yeasts, yet they enable a slow growth of only the most specialised black yeasts. To our current knowledge, only the genera of black yeasts *Friedmanniomyces* and *Cryomyces* are endemic to Antarctica (Selbmann et al. 2005), whereas other recently described genera and species of black yeasts *Elasticomyces*, *Recurvomyces* and *Acidomyces* have also been found in diverse, geographically remote, mountainous regions, e.g. the Andes of Argentina and the Italian Alps (Selbmann et al. 2008). An in-depth review of the distribution of black yeasts in worldwide cold habitats is reported in Chap. 8.

The most recent study performed in a gypsum-based endolithic community in the polar desert of the Canadian High Arctic characterised the microbial community diversity, turnover and microbe–mineral interactions. Pyrotag sequencing of 16S/18S rRNA demonstrated the presence of a diverse community of phototrophic and heterotrophic bacteria, algae and fungi. The diversity observed by molecular techniques was confirmed by stable carbon isotope analysis of the lipids in viable microbial membranes, which indicated that the atmospheric carbon is assimilated into the microbial community biomass. Among fungi, the sequences of Hypocreales, Verrucariales, Chaetothyriales, Helotiales, Capnodiales, Erysiphales,

Lecanoreales and Dothideales were detected, which excludes true yeasts (Ziolkowski et al. 2013).

3.4 Yeasts in Sea Water and Sea Ice

Yeast communities in oceanic environments appear to be composed of ubiquitous and endemic species (Kohlmeyer and Kohlmeyer 1979). Typical ubiquitous marine yeasts are the ascomycetous yeast *Debaryomyces hansenii* and the basidiomycetous yeast genera *Cryptococcus* and *Rhodotorula*. Although the isolation frequency of yeasts falls with depth, ascomycetous yeasts (e.g. *Candida, Debaryomyces, Kluyveromyces, Pichia* and *Saccharomyces*) are more common in shallow waters, while yeasts belonging to the Basidiomycetes (*Cryptococcus, Rhodosporidium, Rhodotorula, Sporobolomyces*) are more common in deep waters. Mainly basidiomycetous yeasts have also been isolated from polar offshore sea waters (Jones 1976). The genera *Rhodotorula, Cryptococcus, Debaryomyces* and *Candida,* which predominate among yeasts isolated from the deep-sea waters, also prevail in cold polar areas and in the vicinity of sea ice (Butinar et al. 2011). In particular, *Rhodotorula* spp. yeasts show a strong ubiquity, since they were isolated from deep-sea vents, from a depth of 11,000 m (Gadanho and Sampaio 2005), from deep-sea sediments (Nagahama et al. 2001), from coastal waters (Gadanho et al. 2003; Gadanho and Sampaio 2004), and from the Arctic ocean (Gunde-Cimerman et al. 2003). Recent data have shown that animals living in deep-sea hydrothermal vents are a new niche for colonisation by yeasts, which frequently occur either in the Arctic water-based environments or in the hypersaline environments (Kutty and Philip 2008; Burgaud et al. 2010). For example, the psychrophilic marine yeast *Leucosporidium scottii*, previously known as autochthonous in the oceanic regions close to Antarctica, was found recently in association with deep-sea hydrothermal animals. The authors hypothesised that this could be an evidence of global exchanges from polar environments to deep-sea vents (Burgaud et al. 2010). An overview of the occurrence of cold-adapted yeasts in the deep sea is reported in Chap. 7.

When seasonal temperatures drop and frozen seawater forms a semisolid matrix, seasonally transient sea ice is formed. It is constantly broken up and reshaped by the wind and the ocean currents, and it has a highly variable temperature, both temporally and spatially, with temperatures ranging from −1 °C to as low as −50 °C in winter. The seasonally transient sea ice is permeated by a network of channels and pores, filled with brine. These sea-ice brine channels remain liquid up to −35 °C and can reach 20 % (w/v) NaCl concentration. Changes in the salinity and temperature are thus the dominant factors that influence the biological communities within the sea ice (Brown and Bowman 2001; Thomas and Dieckmann 2002). At very low external temperatures, most of the microbial biomass finds refuge in the central mass of the ice within brine pockets, whereas in spring, dense microbial communities can develop at the bottom of the ice (Gosink et al. 1993). Although autotrophic algal communities are relatively well described, sea-ice-dwelling fungal communities have been less documented (Brown and Bowman 2001; Gunde-Cimerman et al. 2003, 2005). For example, when DNA was extracted from an Arctic sea-ice sample, some of the eukaryotic clones detected were described as fungal (Thomas and Dieckmann 2002). From Svalbard (Arctic), conventional culture-based methods with sea-ice samples have yielded up to 7,000 CFU fungi L^{-1} (Gunde-Cimerman et al. 2003). Most (85 %) of the isolates were basidiomycetous yeasts, which were mainly represented by *Cryptococcus adeliensis* and *Rhodotorula mucilaginosa.* The neighbouring sea water contained *D. hansenii, D. maramus, Meyerozyma guilliermondii* (formerly *Pichia guilliermondii*), a novel species resembling *Candida galli* and *Metschnikowia bicuspidata.* Of all these, only *M. guilliermondii* was isolated exclusively from sea ice, while snow/ice in the fjord tidal zone included *Candida parapsilosis, D. hansenii, M. guilliermondii* and *Metschnikowia zobellii* (Butinar et al. 2011). All of these isolated strains were characterised as psychrotolerant and xero/halotolerant. When sea ice melts, the entrapped yeasts and other microbes are released into the oceans; however, up to our knowledge, no studies report on the dynamics of these processes. Only a few studies have reported on the occurrence of yeasts in the Arctic aquatic habitats and their global distribution, particularly in deep-sea waters.

3.5 Yeasts in Glaciers and Subglacial Environments

Glacial ice was long considered as an extremely stable and static environment. Ice-core analyses were used to document and date past climate changes, geological events and human activities. Isolation of viable microorganisms from ice cores has unravelled ancient, 'living microbial fossils' in the older layers, which originated from northern latitudes, and temperate as well as tropical regions (Abyzov 1993; Ma et al. 1999, 2000). Low numbers of filamentous fungi, but no yeasts, were isolated from 10,000–13,000-year-old Greenland ice (Ma et al. 1999, 2000). PCR amplification of fragments of the eukaryotic 18S rRNA genes that were extracted from 2,000-year-old to 4,000-year-old ice-core samples from north Greenland led to the identification of a diversity of fungi, plants, algae and protists (Price 2000). All of these findings of fungi in glacier ice cores have been interpreted as the result of coincidental aeolian deposits of spores or mycelium into the ice during its geological history.

Recent investigations have shown that ice in glaciers is a much more dynamic habitat than previously thought, on the microscale as well as at the level of geomorphology. Active microbes have been found in all types of glaciers, but their abundance and activity are greatest in polythermal and temperate types. Their presence is connected to the availability of water, which is in turn related to the heat budget and the distribution of drainage channels in the ice.

Four key glacial ecosystems have so far been identified: the proglacial environment, seasonally flooded by glacial melt-water; the glacier surface, i.e. the supraglacial system; the ice–bed interface, i.e. the subglacial system; and the englacial system, which exists *inside* glacial ice.

Ice in temperate glaciers is permeated by a continuous network of aqueous veins that are formed due to sea salts deposited as aerosols, as they are essentially insoluble in ice crystals. Due to the percolation of the salts from the top of the glacier to its bottom, salts can accumulate to relatively high concentrations in the bottom parts of glaciers (Price 2000). Although these liquid veins, located between the ice crystals in bulk polycrystalline ice, can have high ionic strength (Price 2000; Rohde and Price 2007), they contain living microorganisms of different physiological groups. To the authors' knowledge, the presence of yeasts within these veins has not been demonstrated yet. These microorganisms can serve as inocula when they reach the subglacial environment that contains more liquid water.

The heat budget of melting ice is controlled not only by physical phenomena such as solar radiation and air temperature, but also by quick seismic shifts (Ekstrom et al. 2003; Fahnestock 2003), which affect the formation of cryokarst phenomena and distribution of ice at the pressure melting point. Water in a glacier can thus exist on the surface of glaciers as channels, films, veins and pockets at grain interstices or boundaries and also as caves, interglacial lakes or moulins within the glaciers (Christner et al. 2000). If these waters reach the glacier bed and mix with the groundwater and glacial basal melt-water, rock and sediment, they create a dynamic subglacial environment.

Initially these environments were considered abiotic; however, large microbial communities dominated by aerobic, heterotrophic Betaproteobacteria were discovered in 2000 (Skidmore et al. 2000; Foght et al. 2004) and yeasts and other fungi in 2003 (Gunde-Cimerman et al. 2003, 2005; Sonjak et al. 2006; Butinar et al. 2007; Zalar et al. 2008). The yeast counts detected in the subglacial samples were two orders of magnitude greater when compared with those in supraglacial samples (with up to 4×10^6 CFU L^{-1}). Yeasts were primarily associated with the clear glacier ice, while filamentous fungi were detected with the highest counts in debris-rich subglacial ice (Butinar et al. 2007; Sonjak et al. 2006). In another study by Singh et al. (2013), the viable counts of yeast cells in ice-melted water from glacier samples were lower and ranged between 3×10^3 and 1×10^4 CFU L^{-1}.

According to the species diversity and the abundance, the majority of the species were assigned to the hymenomycetous yeasts (*Filobasidium*/*Cryptococcus albidus* taxa of the Tremellales). The stable core of the subglacial yeast communities was represented by *Cryptococcus liquefaciens*, *R. mucilaginosa*, *D. hansenii* and *M. guillermondii* (Butinar et al. 2007, 2011; de García et al. 2012), while in the study by Singh et al. (2013), *C. adeliensis, C. albidosimilis, C. saitoi, Rhodosporidium lusitaniae* and *R. mucilaginosa* prevailed.

Ascomycetous yeasts have been found in glaciers of the Southern and Northern Hemisphere only occasionally, the main reason probably being isolation media with water activity close to 1.0. When media with lowered water activity were

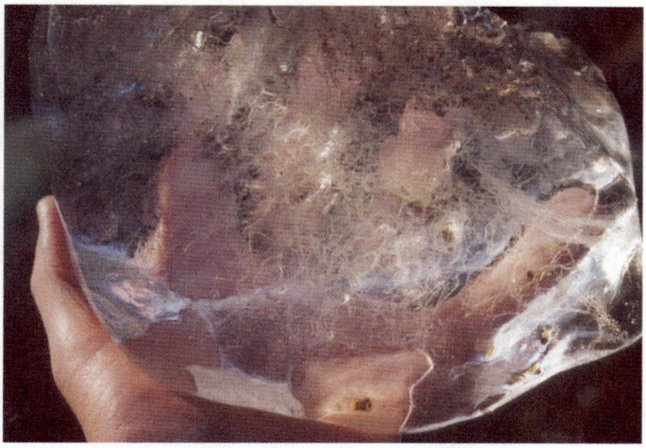

Fig. 3.2 Glacial ice with gypsum inclusions

used, a much higher number of ascomycetous yeasts were retrieved (Gunde-Cimerman et al. 2003). The dominant ascomycetous yeast species was *D. hansenii*, with different species of the genera *Candida*, *Metschnikowia* and *Pichia* (de García et al. 2010a). A surprising discovery was made in ice rich in gypsum precipitations (Fig. 3.2). Although such samples of ice contained some basidiomycetous yeasts, they were dominated by osmotolerant ascomycetous black yeast *A. pullulans*. This species is of great biotechnological importance due to its production of the ex-trapolysaccharide pullulan, as well as numerous extracellular hydrolytic enzymes (Chi et al. 2009). Recently, *A. pullulans* has been described to comprise four varieties, of which three have been detected in polar environments. The most well known of these, *A. pullulans* var. *pullulans*, is a cosmopolitan taxon that has been found at both poles in glaciers (Zalar et al. 2008) and in Greenland ice sheets (Starmer et al. 2005). The *A. pullulans* var. *melanogenum* has been isolated mainly from oligotrophic watery habitats, such as melted glacial water from the Italian Alps (Branda et al. 2010), and melted subglacial water of the Arctic Svalbard glaciers (Zalar et al. 2008). The sequence data of five molecular markers have revealed an additional variety, named *A. pullulans* var. *subglaciale*, which has so far been isolated only from the subglacial ice of Svalbard glaciers and the immediate surroundings (Zalar et al. 2008). This variety probably represents an example of geographic isolation and initiation of speciation of this panglobal species (Gostinčar et al. 2010), indicating its high genetic variability. It appears likely that more varieties of *A. pullulans* will be discovered and described in polar environments in the future.

In the area of Svalbard (Arctic Archipelago), yeasts have also been isolated from ice and melt-water and from puddles in the vicinity of glaciers (Pathan et al. 2010; Butinar et al. 2011). The isolated yeasts were phylogenetically affiliated to *Cryptococcus gastricus*, *C. terricolus*, *Leucosporidiella muscorum* (formerly *Rhodotorula muscorum*), *Mrakia psychrophila*, *Mrakia gelida* and *Rhodotorula glacialis*. Strains belonging to the genera *Mrakia* and *Mrakiella* have been found in many cold climates throughout the world, such as Antarctica, Greenland,

Siberia, Alaska and Patagonia. *Mrakia* is the dominant yeast genus in Antarctic soil, representing 24 % of the yeast species. Since it is not able to grow at temperatures above 20 °C, it can be defined as an obligate psychrophile. There was a major overlap in the detected species diversity with yeasts detected in Italian Alpine glaciers (Turchetti et al. 2008; Branda et al. 2010; Thomas-Hall et al. 2010) and in Patagonian glaciers (de García et al. 2007). Three novel species—*Mrakia robertii, Mrakia blollopis* and a related anamorphic species *Mrakiella niccombsii*—were isolated from Antarctica and the Italian Alps, as well as the novel genus *Glaciozyma* with two newly described species, namely *Glaciozyma martinii* and *Glaciozyma watsonii* and the reclassified *Leucosporidium antarcticum*, now *Glaciozyma antarctica* (Thomas-Hall et al. 2010). In the Italian Alpine glaciers, *Cr. gastricus* and *Rhodotorula psychrophenolica* (Branda et al. 2010) prevailed. Surprisingly, the opportunistic pathogenic black yeast *Exophiala dermatitidis* has been found in the supraglacial environment of an Italian glacier (Branda et al. 2010). A complete overview of the distribution of cold-adapted yeast species (including black yeasts) in worldwide cold habitats is reported in Chaps. 3, 4, 5, 6, 7 and 8.

3.6 Yeasts in Cryoconite Holes

Cryoconite holes are formed when solar-heated dark organic and inorganic wind-blown debris melts into the underlying ice. These melt-water-filled depressions in the surface ice are formed in the ablation zones of glaciers and ice sheets of the Arctic, Antarctic and Alpine regions, as a result of the low albedo of localised debris accumulations (Edwards et al. 2013). They consist mostly of both organic and inorganic matter, 90 % immersed in water, with the remaining space filled with air. The holes can range from less than 1 cm to 1 m in width. They seldom grow deeper than 60 cm, although they can coalesce into bigger holes or become interconnected by melt-water channels. The cryoconite holes in the Arctic seem to be more interconnected than the holes in the Antarctica, which is the most probable explanation for the difference in the community structures and dynamics between the two environments. As the solar radiation decreases, an ice cover can form on the water surface and grow downwards. With faster freezing rates, solutes can become trapped in brine pockets with salinities up to 10 % NaCl. These ponds can freeze only after reaching brine temperatures of −12 °C. In relation to the melt regime of the glacier surface, the cryoconite holes can last from several days to weeks, or to entire seasons. On some stable glaciers, they are thought to have remained for at least a 100 years, thus serving as biological refuges during periods of extreme cold (Mueller et al. 2001; Vincent et al. 2004).

The cryoconite holes provide niches within a permanently cold and hostile environment, assigned as ice-cold hot spots of microbial diversity (Edwards et al. 2013). The organic matter, also presenting the important driver in the process of cryoconite hole formation, mainly includes algae, cyanobacteria, bacteria, fungi and rotifers, while the inorganic matter is a mixture of minerals and trace elements (Singh and Singh 2012). During the polar summer, photosynthesis by algae and cyanobacteria within the cryoconite hole can provide sufficient nutrients for complex community development.

The dominant organisms are typically nitrogen-fixing, filamentous, mat-forming cyanobacteria. Light microscopy has also documented the presence of pollen grains, bacteria, algae, diatoms, occasional microinvertebrates, filamentous fungi and yeasts. The latter two groups drive the process of organic macromolecule degradation through cold-adapted enzyme secretion, thereby assisting in nutrient cycling (Singh and Singh 2012). Psychrotolerants prevail, while psychrophiles have been detected only rarely (Mueller et al. 2001; Vincent et al. 2004).

The best-known culturable components of the sampled cryoconite holes in Svalbard are the psychrophilic basidiomycete yeasts of the classes Tremellomycetes (*Mrakia* and *Cryptococcus*) and Microbotryomycetes (*Glaciozyma* and *Rhodotorula*), with several species known only from glacial habitats. These yeasts are often dominant in culture-based studies depending on the plating method used. For yeasts and filamentous fungi, the number of CFU g^{-1} of the sediment sample was calculated to be about 7×10^3 to 1.4×10^4 and 4×10^3 to 1.2×10^4, respectively. In the cryoconite holes at higher latitudes, more yeast species than fungi were recovered, while filamentous fungi were more abundant at the lowermost point (Singh and Singh 2012). However, the isolated species comprise only a minor component of the fungal communities, as assessed by a T-RFLP study of rRNA ITS amplicons (Edwards et al. 2013), since the most dominant T-RFLP peaks did not match any of the cultured isolates.

3.7 Conclusions

The extremely cold Arctic environment is inhabited by extremophilic yeasts that even today represent a largely unknown biodiversity. Classical culture-based methods, molecular techniques and more recently also metagenomic analyses (Simon et al. 2009; Kennedy et al. 2010; Bellemain et al. 2013) have revealed the existence of novel species and genera, and even novel phylogenetic lineages. Over the past decades, studies of yeasts were focused on the one hand on living fossils entrapped in the ice, providing us with an insight into the biological past, while on the other hand we discovered unexpectedly large metabolically active communities. The range of Arctic environments that are inhabited by yeasts and were previously considered abiotic or dominated exclusively by Bacteria has been expanded to include rocks, sea ice, microbial mats on sea-ice shelves, glacial brine networks and inclusions, subglacial environments, cryoconite holes and permafrost. The physiological activity of the yeasts has been demonstrated at temperatures below −20 °C (Rivkina et al. 2000; Deming 2002; Price and Sowers 2004). The adaptation of psychrophilic yeast species to cold environments is reflected in their structural and biochemical characteristics, e.g. the production of cold-adapted enzymes, cryoprotectant wall carbohydrates, pigments, as well as higher amounts of polyunsaturated fatty acids in cytoplasmic membranes and a high amount of intracellular lipids (Thomas-Hall et al. 2010). The complex adaptation strategies of cold-adapted yeasts to low temperatures are reviewed in Chaps. 9, 10, 11, 12, 13, 14 and 15.

In spite of the adaptations studied at the cellular level, the ecological role of the yeasts in cold environments remains mainly unknown. Their heterotrophic metabolism and ability to degrade organic macromolecules through the secretion of extracellular hydrolytic cold-adapted enzymes suggest their potential auxiliary role as biogeochemical nutrient recyclers; however, more studies are needed to clarify complex connections between yeasts and other microbial communities, particularly in the cryosphere. Also, more focus should be given to the complex interplay of interactions amongst not only biological, but also chemical and physical factors. Particularly processes occurring at water–ice interfaces, e.g. solute concentration, pressure and the physical state of the water and ice, rather than the effect of temperature alone should be emphasised. For example in permafrost, 2–7 % of the water persists as briny liquid films and lenses (cryopegs) that form as a result of the salting out of soils as the *in situ* temperature drops and remains at about −10 °C. Therefore, our understanding should not be limited only to psychrophilic and psychrotolerant yeasts, but should also include the eco-physiological group of yeasts, which live at the critical interface inherent to the phase change of (saline) water to ice.

It is well known that the polar environments are increasingly threatened by the global overheating; particularly the glaciers are retreating most rapidly. Since the glacier habitats (cryosphere) include some of the largest unexplored and extreme biospheres of Earth, this phenomenon determines the potential complete loss of an as yet unexplored reservoir of unknown microbial species. The influence of climate change on fungi is just beginning to be evaluated in the Arctic. Evidence from palaeobotanical studies and from contemporary warming experiments alike indicates that the Arctic soil fungal communities have responded to, and are likely to respond to, climate warming. The analyses of the DNA preserved in the ancient permafrost in northeastern Siberia have revealed that fungal communities changed in concert with plant communities after the last Ice Age. As the environment altered, the tundra became dominated by shrubs and trees, which expanded into the previous tundra steppe and fungal communities changed from yeast-like and parasitic fungi to communities with root-associated macrofungi (Timling and Taylor 2012). In this context, the isolation and *ex situ* conservation of the psychrophilic microorganisms can be seen as a means of avoiding their possible extinction. Yeasts inhabiting the Arctic environment efficiently spread via air, soil, water, marine animals, including those that inhabit deep sea vents (Burgaud et al. 2010), and polar birds (Dynowska et al. 2013), and seed their immediate environment, affecting the local geobiochemistry. It remains mainly unclear to what extent the air and water currents can transport the yeasts globally to temperate zones, where they can find refuge either in the refrigerated food or in other cold man-made environments. This is certainly an issue that is gaining importance with the accelerated disappearance of the polar environments and the increased release of microbes in non-polar areas.

The importance of psychrophilic and psychrotolerant yeasts has also increased due to the mounting evidence that the biosphere has experienced several extremely low temperature periods, perhaps even during the earliest stages in the emergence and evolution of life. The Earth has been covered in ice four times through the Snowball Earth periods. These episodes ranged from 3 to 30 million years each (Bodisetisch et al. 2005). The entire world ocean was frozen near the time of origin of the eukaryotic cell and the adaptive radiation of Metazoa. It has been hypothesised that the endemic descendants of the primitive cold-adapted microorganisms can live today in rock, lakes, sediments and in subglacial ice. Since 'microbial endemism' is currently a much debated issue in microbial ecology, the polar regions and other extreme environments should be amongst the first places to be examined for the evolutionary processes that could give rise to microbial speciation (Vincent 2000; Gostinčar et al. 2010). Since yeasts have reaffirmed themselves as one of the ecologically most successful eukaryotic lineages, they represent important model organisms, which have developed little-investigated adaptive strategies that are crucial for successful survival in some of the harshest and most extreme environments on our planet.

References

Abyzov SS (1993) Microorganisms in the Antarctic ice. In: Friemann EI (ed) Antarctic microbiology. Wiley, New York, pp 265–295

Abyzov SS, Hoover RB, Imura S, Mitskevich IN, Naganuma T, Poglazova MN, Ivanov MV (2004) Use of different methods for discovery of ice-entrapped microorganisms in ancient layers of the Antarctic glacier. Adv Space Res 33:1222–1230

Babjeva I, Reshetova I (1998) Yeast resources in natural habitats at polar circle latitude. Food Technol Biotechnol 36:1–5

Bellemain E, Davey ML, Kauserud H, Epp LS, Boessenkool S, Coissac E, Geml J, Edwards M, Willerslev E, Gussarova G, Taberlet P, Brochmann C (2013) Fungal palaeodiversity revealed using high-throughput metabarcoding of ancient DNA from arctic permafrost. Environ Microbiol 15:1176–1189

Birgisson HK, Delgado O, Arroyo LG, Hatti-Kaul R, Mattiasson B (2003) Cold-adapted yeasts as producers of cold-active polygalacturonases. Extremophiles 7:185–193

Bodisetisch B, Koeberl C, Master S, Reimold W (2005) Estimating duration and intensity of neoproterozoic snowball glaciation from Ir anomalies. Science 308:239–242

Branda E, Turchetti B, Diolaiuti G, Pecci M, Smiraglia C, Buzzini P (2010) Yeast and yeast-like diversity in the southernmost glacier of Europe (Calderone Glacier, Apennines, Italy). FEMS Microbiol Ecol 72:354–369

Bridge P, Spooner B, Roberts P (2013) List of non-lichenized fungi from the Antarctic region. Available online: http://www.antarctica.ac.uk/bas_research/data/access/fungi/. Accessed 3 Apr 2013

Brown MV, Bowman JP (2001) A molecular phylogenetic survey of sea-ice microbial communities (SIMCO). FEMS Microbiol Ecol 35:267–275

Burford EP, Fomina M, Gadd GM (2003) Fungal involvement in bioweathering and biotransformation of rocks and minerals. Mineral Mag 67:1127–1155

Burgaud G, Arzur D, Durand L, Cambon-Bonavita M-A, Barbier G (2010) Marine culturable yeasts in deep-sea hydrothermal vents: species richness and association with fauna. FEMS Microbiol Ecol 73:121–133

Butinar L, Spencer-Martins I, Gunde-Cimerman N (2007) Yeasts in high Arctic glaciers: the discovery of a new habitat for eukaryotic microorganisms. A van Leeuwenhoek 91:277–289

Butinar L, Strmole T, Gunde-Cimerman N (2011) Relative incidence of ascomycetous yeasts in Arctic coastal environments. Microb Ecol 61:832–843

Chi ZM, Wang F, Chi Z, Yue LX, Liu GL, Zhang T (2009) Bioproducts from *Aureobasidium pullulans*, a biotechnologically important yeast. Appl Microbiol Biot 82:793–804

Christner BC, Mosley-Thompson E, Thompson LG, Zagorodnov V, Sandman K, Reeve JN (2000) Recovery and identification of viable bacteria immured in glacial ice. Icarus 144:479–485

de García V, Brizzio S, Libkind D, Buzzini P, van Broock M (2007) Biodiversity of coldadapted yeasts from glacial meltwater rivers in Patagonia, Argentina. FEMS Microb Ecol 59:331–341

de García V, Brizzio S, Libkind D, Rosa CA, van Broock M (2010a) *Wickerhamomyces patagonicus* sp. nov., an ascomycetous yeast species from Patagonia. Argentina. Int J Syst Evol Microbiol 60:1693–1696

de García V, Brizzio S, Russo G, Rosa CA, Boekhout T, Theelen B, van Libkind D, Broock M (2010b) *Cryptococcus spencermartinsiae* sp. nov., a basidiomycetous yeast isolated from glacial waters and apple fruits. Int J Syst Evol Microbiol 60:707–711

de García V, Zalar P, Brizzio S, Gunde-Cimerman N, van Broock M (2012) *Cryptococcus* species (Tremellales) from glacial biomes in the southern (Patagonia) and northern (Svalbard) hemispheres. FEMS Microbiol Ecol 82:523–539

de los Ríos A, Sancho LG, Grube M, Wierzchos J, Ascaso C (2005) Endolithic growth of two lecidea lichens in granite from Continental Antarctica detected by molecular and microscopy techniques. New Phytol 165:181–189

de los A, Wierzchos J, Sancho LG, Ascaso C (2003) Acid microenvironments in microbial biofilms of antarctic endolithic microecosystems. Environ Microbiol 5:231–237

Deming JW (2002) Psychrophiles and polar regions. Curr Opin Microbiol 5:301–309

Dynowska M, Wojczulanis-Jakubas K, Pacynska JA, Jakubas D, Ejdys E (2013) Potentially pathogenic yeast isolated from the throat and cloaca of an Arctic colonial seabird: the little auk (Alle alle). Polar Biol 36:343–348

Edwards A, Douglas B, Anesio AM, Rassner SM, Irvine-Fynn TDL, Sattler B, Griffith GW (2013) A distinctive fungal community inhabiting cryoconite holes on glaciers in Svalbard. Fungal Ecol 6:168–176

Ekstrom G, Nettles M, Abers GA (2003) Glacial earthquakes. Science 302:622–624

Fahnestock M (2003) Geophysics: glacial flow goes seismic. Science 302:578–579

Faizutdinova RN, Suzina NE, Duda VI, Petrovskaya LE, Gilichinsky DA (2005) Yeasts isolated from ancient permafrost. In: Castello JD, Rogers SO (eds) Life in ancient ice. Princeton University Press, Princeton, pp 118–126

Fell JW, Scorzetti G, Connell L, Craig S (2006) Biodiversity of micro-eukaryotes in Antarctic dry valley soils with <5 % soil moisture. Soil Biol Biochem 38:3107–3119

Foght J, Aislabie J, Turner S, Brown CE, Ryburn J, Saul DJ, Lawson W (2004) Culturable bacteria in subglacial sediments and ice from two southern hemisphere glaciers. Microb Ecol 47:329–340

Friedmann EI (1982) Endolithic microorganisms in the Antarctic cold desert. Science 215:1045–1053

Friedmann EI, Koriem AM (1989) Life on Mars: how it disappeared (if it was ever there). Adv Space Res 9:167–172

Frisvad JC (2008) Fungi in cold ecosystems. In: Margesin R, Schinner F, Marx J-C, Gerday C (eds) Psychrophiles: from biodiversity to biotechnology. Springer, Berlin, pp 137–156

Gadanho M, Sampaio JP (2004) Application of temperature gradient gel electrophoresis to the study of yeast diversity in the estuary of the Tagus river, Portugal. FEMS Yeast Res 5:253–261

Gadanho M, Sampaio J (2005) Occurrence and diversity of yeasts in the Mid-Atlantic ridge hydrothermal fields near the azores archipelago. Microb Ecol 50:408–417

Gadanho M, Almeida JMGCF, Sampaio JP (2003) Assessment of yeast diversity in a marine environment in the south of Portugal by microsatellite-primed PCR. A van Leeuwenhoek 84:217–227

Gilichinsky DA, Rivkina E, Bakermans C, Shcherbakova V, Petrovskaya L, Ozerskaya S, Ivanushkina N, Kochkina G, Laurinavichuis K, Pecheritsina S, Fattakhova R, Tiedje JM (2005) Biodiversity of cryopegs in permafrost. FEMS Microbiol Ecol 53:117–128

Gilichinsky DA, Wilson GS, Friedmann EI, McKay CP, Sletten RS, Rivkina EM, Vishnivetskaya TA, Erokhina LG, Ivanushkina NE, Kochkina GA, Shcherbakova VA, Soina VS, Spirina EV, Vorobyova EA, Fyodorov-Davydov DG, Hallet B, Ozerskaya SM, Sorokovikov VA, Laurinavichyus KS, Shatilovich AV, Chanton JP, Ostroumov VE, Tiedje JM (2007) Microbial populations in Antarctic permafrost: biodiversity, state, age, and implication for astrobiology. Astrobiology 7:275–311

Golubev WI (1998) New species of basidiomycetous yeasts, *Rhodotorula creatinovora* and *R. yakutica*, isolated from permafrost soils of Eastern-Siberian Arctic. Mykologiya I Phytopathologiya 32:8–13

Gorbushina AA (2007) Life on the rocks. Environ Microbiol 9:1613–1631

Gosink JJ, Irgens RL, Staley JT (1993) Vertical distribution of bacteria in arctic sea ice. FEMS Microbiol Lett 102:85–90

Gostinčar C, Turk M, Trbuha T, Vaupotič T, Plemenitaš A, Gunde-Cimerman N (2008) Expression of fatty-acid-modifying enzymes in halotolerant black yeast *Aureobasidium pullulans* (de Bary) G. Arnaud under salt stress. Stud Mycol 61:51–59

Gostinčar C, Grube M, de Hoog GS, Zalar P, Gunde-Cimerman N (2010) Extremotolerance in fungi: evolution on the edge. FEMS Microbiol Ecol 71:2–11

Grabińska-Łoniewska A, Konińowicz-Kowalska T, Wardzyńska G, Boryn K (2007) Occurrence of fungi in water distribution system. Pol J Environ Stud 16:539–547

Gunde-Cimerman N, Sonjak S, Zalar P, Frisvad JC, Diderichsen B, Plemenitaš A (2003) Extremophilic fungi in arctic ice: a relationship between adaptation to low temperature and water activity. Phys Chem Earth 28:1273–1278

Gunde-Cimerman N, Butinar L, Sonjak S, Turk M, Uršič V, Zalar P, Plemenitaš A (2005) Halotolerant and halophilic fungi from coastal environments in the Arctics. In: Gunde-Cimerman N, Oren A, Plemenitaš A (eds) Adaptation to life at high salt concentrations in Archaea, Bacteria, and Eukarya. Springer, Dordrecht, pp 397–423

Jones GEB (1976) Recent advances in aquatic mycology. The Gresham Press, Old Woking, Surrey

Kennedy J, Flemer B, Jackson SA, Lejon DPH, Morrissey JP, O'Gara F, Dobson ADW (2010) Marine metagenomics: new tools for the study and exploitation of marine microbial metabolism. Mar Drugs 8:608–628

Kohlmeyer J, Kohlmeyer E (1979) Marine mycology: the higher fungi. Academic Press, New York

Kutty SN, Philip R (2008) Marine yeasts: a review. Yeast 25:465–483

Lee JH, Park AK, Do H, Sun Park KS, Moh SH, Chi YM, Kim HJ (2012) Structural basis for antifreeze activity of ice-binding protein from Arctic yeast. J Biol Chem 287:11460–11468

Libkind D, Gadanho M, van Broock M, Sampaio JP (2005) *Sporidiobolus longiusculus* sp. nov. and *Sporobolomyces patagonicus* sp. nov., novel yeasts of the *Sporidiobolales* isolated from aquatic environments in Patagonia, Argentina. Int J Syst Evol Microbiol 55:503–509

Libkind D, Moline M, Sampaio JP, van Broock M (2009) Yeasts from high-altitude lakes: influence of UV radiation. FEMS Microbiol Ecol 69:353–362

Ludley K, Robinson CH (2008) 'Decomposer' Basidiomycota in Arctic and Antarctic ecosystems. Soil Biol Biochem 40:11–29

Lydolph MC, Jacobsen J, Arctander P, Gilbert MT, Gilichinsky DA, Hansen AJ, Willerslev E, Lange L (2005) Beringian paleoecology inferred from permafrost-preserved fungal DNA. Appl Environ Microbiol 71:1012–1017

Ma L, Catranis CM, Starmer WT, Rogers SO (1999) Revival and characterization of fungi from ancient polar ice. Mycologist 13:70–73

Ma L, Rogers SO, Catranis CM, Starmer WT (2000) Detection and characterization of ancient fungi entrapped in glacial ice. Mycologia 92:286–295

Margesin R (2009) Effect of temperature on growth parameters of psychrophilic bacteria and yeasts. Extremophiles 13:257–262

Margesin R, Fell JW (2008) *Mrakiella cryoconiti* gen. nov., sp. nov., a psychrophilic, anamorphic, basidiomycetous yeast from alpine and arctic habitats. Int J Syst Evol Microbiol 58:2977–2982

Margesin R, Miteva V (2011) Diversity and ecology of psychrophilic microorganisms. Res Microbiol 162:346–361

Margesin R, Fonteyne PA, Schinner F, Sampaio JP (2007) *Rhodotorula psychrophila* sp. nov., *Rhodotorula psychrophenolica* sp. nov. and *Rhodotorula glacialis* sp. nov., novel psychrophilic basidiomycetous yeast species isolated from alpine environments. Int J Syst Evol Microbiol 57:2179–2184

McLoughlin N, Brasier MD, Wacey D, Green OR, Perry RS (2007) On biogenicity criteria for endolithic microborings on early Earth and beyond. Astrobiology 7:10–26

Montes MJ, Belloch C, Galiana M, Garcia MD, Andres C, Ferrer S, Torres-Rodriguez JM, Guinea J (1999) Polyphasic taxonomy of a novel yeast isolated from antarctic environment; description of *Cryptococcus victoriae* sp. nov. Syst Appl Microbiol 22:97–105

Morozova D, Mohlmann D, Wagner D (2007) Survival of methanogenic archaea from Siberian permafrost under simulated Martian thermal conditions. Orig Life Evol Biosph 37:189–200

Mueller DR, Vincent WF, Pollard WH, Fritsen CH (2001) Glacial cryoconite ecosystems: a bipolar comparison of algal communities and habitats. Nova Hedwig Beih 123:173–197

Nagahama T, Hamamoto M, Nakase T, Takami H, Horikoshi K (2001) Distribution and identification of red yeasts in deep-sea environments around the northwest Pacific Ocean. A van Leeuwenhoek 80:101–110

Nienow JA, Friedmann EI (1993) Terrestrial liophytic (rock) communities. In: Friedmann EI (ed) Antarctic microbiology. Wiley-Liss, New York, pp 343–412

Omelon C, Pollard W, Ferris F (2006) Environmental controls on microbial colonization of high Arctic cryptoendolithic habitats. Polar Biol 30:19–29

Omelon C, Pollard W, Ferris F (2007) Inorganic species distribution and microbial diversity within high Arctic cryptoendolithic habitats. Microb Ecol 54:740–752

Onofri S, Pagano S, Zucconi L, Tosi S (1999) *Friedmanniomyces endolithicus* (Fungi, Hyphomycetes), anam.-gen. and sp. nov., from continental Antarctica. Nova Hedwigia 68:175–181

Onofri S, Selbmann L, Zucconi L, Pagano S (2004) Antarctic microfungi as models for exobiology. Planet Space Sci 52:229–237

Onofri S, Barreca D, Selbmann L, Isola D, Rabbow E, Horneck G, de Vera JPP, Hatton J, Zucconi L (2008) Resistance of Antarctic black fungi and cryptoendolithic communities to simulated space and Martian conditions. Stud Mycol 61:99–109

Ozerskaya SM, Kochkina GA, Ivanushkina NE, Knyazeva EV, Gilichinskii DA (2008) The structure of micromycete complexes in permafrost and cryopegs of the Arctic. Microbiology 77:482–489

Ozerskaya S, Kochkina G, Ivanushkina N, Gilichinsky DA (2009) Fungi in permafrost. In: Margesin R (ed.) Permafrost soils. Springer-Verlag, Berlin Heidelberg, pp 85–95

Panikov NS (2009) Microbial activity in frozen soils. In: Margesin R (ed) Permafrost soils. Springer, Berlin Heidelberg, pp 119–147

Panikov NS, Sizova MV (2007) Growth kinetics of microorganisms isolated from Alaskan soil and permafrost in solid media frozen down to −35 °C. FEMS Microbiol Ecol 59:500–512

Pathan AAK, Bhadra B, Begum Z, Shivaji S (2010) Diversity of yeasts from puddles in the vicinity of Midre Lovenbreen Glacier, Arctic and Bioprospecting for enzymes and fatty acids. Curr Microbiol 60:307–314

Price PB (2000) A habitat for psychrophiles in deep Antarctic ice. Proc Natl Acad Sci USA 97:1247–1251

Price PB, Sowers T (2004) Temperature dependence of metabolic rates for microbial growth, maintenance, and survival. Proc Natl Acad Sci USA 101:4631–4636

Raspor P, Zupan J (2006) Yeasts in extreme environments. In: Rosa CA, Peter G (eds) Biodiversity and ecophysiology of yeasts. Springer, Berlin, pp 371–417

Rivkina EM, Friedmann EI, McKay CP, Gilichinsky DA (2000) Metabolic activity of permafrost bacteria below the freezing point. Appl Environ Microbiol 66:3230–3233

Rivkina E, Laurinavichius K, McGrath J, Tiedje J, Shcherbakova V, Gilichinsky DA (2004) Microbial life in permafrost. Adv Space Res 33:1215–1221

Robinson CH (2001) Cold adaptation in Arctic and Antarctic fungi. New Phytol 151:341–353

Rohde RA, Price PB (2007) Diffusion-controlled metabolism for long-term survival of single isolated microorganisms trapped within ice crystals. Proc Natl Acad Sci USA 104:16592–16597

Ruisi S, Barreca D, Selbmann L, Zucconi L, Onofri S (2007) Fungi in Antarctica. Rev Environ Sci Biotechnol 6:127–141

Samson RA, Hoekstra ES, Frisvad JC, Filtenborg O (2002) Introduction to food- and airborne fungi. Central Bureau for Schimmelcultures, Utrecht

Scorzetti G, Petrescu I, Yarrow D, Fell JW (2000) *Cryptococcus adeliensis* sp nov, a xylanase producing basidiomycetous yeast from Antarctica. A van Leeuwenhoek 77:153–157

Selbmann L, de Hoog GS, Mazzaglia A, Friedmann EI, Onofri S (2005) Fungi at the edge of life: cryptoendolithic black fungi from Antarctic desert. Stud Mycol 51:1–32

Selbmann L, de Hoog GS, Zucconi L, Isola D, Ruisi S, van den Ende AH, Ruibal C, De Leo F, Urzi C, Onofri S (2008) Drought meets acid: three new genera in a dothidealean clade of extremotolerant fungi. Stud Mycol 61:1–20

Simon C, Wiezer A, Strittmatter AW, Daniel R (2009) Phylogenetic diversity and metabolic potential revealed in a glacier ice metagenome. Appl Environ Microbiol 75:7519–7526

Singh P, Singh SM (2012) Characterization of yeast and filamentous fungi isolated from cryoconite holes of Svalbard, Arctic. Polar Biol 35:575–583

Singh P, Tsuji M, Singh SM, Roy U, Hoshino T (2013) Taxonomic characterization, adaptation strategies and biotechnological potential of cryophilic yeasts from ice cores of Midre Lovenbreen glacier, Svalbard, Arctic. Cryobiology 66:167–175

Skidmore ML, Foght JM, Sharp MJ (2000) Microbial life beneath a high Arctic glacier. Appl Environ Microbiol 66:3214–3220

Sonjak S, Frisvad JC, Gunde-Cimerman N (2006) *Penicillium* mycobiota in Arctic subglacial ice. Microb Ecol 52:207–216

Starmer W, Fell J, Catranis C, Aberdeen V, Ma L, Zhou S, Rogers S (2005) Yeasts in the genus *Rhodotorula* recovered from the Greenland ice sheet. In: Castello JD, Rogers SO (eds) Life in ancient ice. Princeton Univ Press, New Jersey, pp 181–195

Steven B, Leveille R, Pollard WH, Whyte LG (2006) Microbial ecology and biodiversity in permafrost. Extremophiles 10:259–267

Sturm M, Schimel J, Michaelson G, Welker JM, Oberbauer SF, Liston GE, Fahnestock J, Romanovsky VE (2005) Winter biological processes could help convert Arctic tundra to shrubland. Bioscience 55:17–26

Takano Y, Kobayashi K, Marumo K, Ishikawa Y (2004) Biochemical indicators and enzymatic activity below permafrost environment. In: Extremophiles 5th international conference on extremophiles, Abstract Book, Cambridge, Maryland, p 84

Thomas DN, Dieckmann GS (2002) Antarctic sea ice: a habitat for extremophiles. Science 295:641–644

Thomas-Hall S, Watson K (2002) *Cryptococcus nyarrowii* sp nov, a basidiomycetous yeast from Antarctica. Int J Syst Evol Microbiol 52:1033–1038

Thomas-Hall S, Watson K, Scorzetti G (2002) *Cryptococcus statzelliae* sp nov and three novel strains of *Cryptococcus victoriae,* yeasts isolated from Antarctic soils. Int J Syst Evol Microbiol 52:2303–2308

Thomas-Hall SR, Turchetti B, Buzzini P, Branda E, Boekhout T, Theelen B, Watson K (2010) Cold-adapted yeasts from Antarctica and the Italian Alps—description of three novel species: *mrakia robertii* sp nov, *Mrakia blollopis* sp nov and *Mrakiella niccombsii* sp nov. Extremophiles 14:47–59

Timling I, Taylor DL (2012) Peeking through a frosty window: molecular insights into the ecology of Arctic soil fungi. Fungal Ecol 5:419–429

Turchetti B, Buzzini P, Goretti M, Branda E, Diolaiuti G, D'Agata C, Smiraglia C, Vaughan-Martini A (2008) Psychrophilic yeasts in glacial environments of Alpine glaciers. FEMS Microbiol Ecol 63:73–83

Turk M, Plemenitaš A, Gunde-Cimerman N (2011) Extremophilic yeasts: plasma-membrane fluidity as determinant of stress tolerance. Fungal Biol 115:950–958

Vincent WF (2000) Evolutionary origins of Antarctic microbiota: invasion, selection and endemism. Antarctic Sci 12:374–385

Vincent WF, Mueller DR, Bonilla S (2004) Ecosystems on ice: the microbial ecology of Markham Ice Shelf in the high Arctic. Cryobiology 48:103–112

Vishniac HS (1985) *Cryptococcus friedmannii*, a new species of yeast from the Antarctic. Mycologia 77:149–153

Vishniac HS (1993) The microbiology of Antarctic soils. In: Friedman EI (ed) Antarctic microbiology. Wiley-Liss, Inc., New York, pp 297–341

Vishniac HS (2002) *Cryptococcus tephrensis*, sp. nov, and *Cryptococcus heimaeyensis*, sp. nov; new anamorphic basidiomycetous yeast species from Iceland. Can J Microbiol 48:463–467

Vishniac HS (2006) Yeast biodiversity in the Antarctic. In: Rosa CA, Péter G (eds) Biodiversity and ecophysiology of yeasts. Springer, Berlin, pp 419–440

Vishniac HS, Takashima M (2010) *Rhodotorula arctica* sp nov, a basidiomycetous yeast from Arctic soil. Int J Syst Evol Micr 60:1215–1218

Vorobyova E, Soina V, Gorlenko M, Minkovskaya N, Zalinova N, Mamukelashvili A, Gilichinsky DA, Rivkina E, Vishnivetskaya T (1997) The deep cold biosphere: facts and hypothesis. FEMS Microbiol Rev 20:277–290

Wagner D (2008) Microbial communities and processes in Arctic permafrost environments. In: Dionand P, Nautiyal CS (eds) Microbiology of extreme soils. Springer, Berlin, pp 133–154

Wagner D, Lipski A, Embacher A, Gattinger A (2005) Methane fluxes in permafrost habitats of the Lena Delta: effects of microbial community structure and organic matter quality. Environ Microbiol 7:1582–1592

Xiao N, Suzuki K, Nishimiya Y, Kondo H, Miura A, Tsuda S, Hoshino T (2010) Comparison of functional properties of two fungal antifreeze proteins from *Antarctomyces psychrotrophicus* and *Typhula ishikariensis*. FEBS J 277:394–403

Xin MX, Zhou PJ (2007) *Mrakia psychrophila* sp nov, a new species isolated from Antarctic soil. J Zhejiang Univ Sci B 8:260–265

Zalar P, Gostinčar C, de Hoog GS, Ursic V, Sudhadham M, Gunde-Cimerman N (2008) Redefinition of *Aureobasidium pullulans* and its varieties. Stud Mycol 61:21–38

Ziolkowski LA, Mykytczuk NCS, Omelon CR, Johnson H, Whyte LG, Slater GF (2013) Arctic gypsum endoliths: a biogeochemical characterization of a viable and active microbial community. Biogeosciences Discuss 10:2269–2304

Chapter 4
Cold-Adapted Yeasts in Antarctic Deserts

Laurie B. Connell, Russell R. Rodriguez, Regina S. Redman and Joseph J. Dalluge

Abstract Antarctica is a large continent and as such has a variety of soil habitats ranging from relatively warm, moist, and high in organic carbon content, found on the Antarctic Peninsula, through the cold arid oligotrophic dry valleys. Efforts to identify yeasts from Antarctica were spurred by the development of research stations initiated during the International Geophysical Year (IGY) (1957–1958). The combination of cold, dry, oligotrophic, and high UV conditions makes the Antarctic deserts a challenging place to live. The majority of yeast species found in the Antarctic deserts are from the genera of *Cryptococcus* and *Rhodotorula*. Adaptations of yeasts to the Antarctic soil habitat include psychrophily, alteration of sterols, ability to withstand desiccation as well as the ability to successfully scavenge minerals in an oligotrophic habitat. New techniques such as high throughput sequencing and advances in mass spectrometry-based metabolomics research offer the opportunity to further explore how yeasts at the edges of life function.

Keywords Antarctica · Sterol · Siderophore · Metabolomics

L. B. Connell (✉)
School of Marine Sciences, University of Maine, 5735 Hitchner Hall,
Orono, ME 04469, USA
e-mail: Laurie_Connell@umit.maine.edu

R. R. Rodriguez
Symbiogenics, 10738 Durland Ave NE, Seattle, WA 98125, USA

R. S. Redman
Adaptive Symbiotic Technologies, 10738 Durland Ave NE, Seattle, WA 98125, USA

J. J. Dalluge
Department of Chemistry, University of Minnesota, 207 Pleasant St SE, Minneapolis,
MN 55455-0431, USA

P. Buzzini and R. Margesin (eds.), *Cold-Adapted Yeasts*,
DOI: 10.1007/978-3-642-39681-6_4,

Contents

4.1 Historical Perspective ... 76
4.2 The Antarctic Soil Habitats ... 77
4.3 Dominant Yeasts in Antarctic Desert Habitats ... 80
4.4 Adaptations Required for Life in Antarctic Deserts ... 81
4.4.1 Mass Spectrometry-Based Metabolite Profiling Approaches to the Study of Biochemical Adaptation in Antarctic Psychrophilic Yeasts ... 82
4.4.2 Membrane Responses by Antarctic Yeast to Growth Temperatures ... 85
4.4.3 Desiccation Resistance in Antarctic Yeasts ... 86
4.4.4 Iron and Manganese Acquisition of Antarctic Yeasts in Oligotrophic Environments ... 88
4.5 Anthropogenic Influence on Yeast Diversity ... 89
4.6 Yeast Diversity in Other Antarctic Habitats ... 89
4.7 Conclusions ... 90
References ... 91

4.1 Historical Perspective

The study of yeasts from Antarctic soil is a relatively recent development compared with other soil microbiological studies worldwide. Most of the early work resulted from the first International Geophysical Year (IGY) 1957–1958 that was modeled after the International Polar Years (IPY) of 1882–1883 and 1932–1933. The first IGY was developed to coordinate scientific inquiry worldwide and twelve nations participated in Antarctic exploration during this effort. The Antarctic treaty was developed in the first IGY and resulted in building of several permanently established research stations around the continent, enhancing accessibility to the continent for scientific research (Fig. 4.1). The earliest research efforts by New Zealand (di Menna 1960, 1966a, b) and Japanese (Soneda 1961; Goto et al. 1969) expeditions revealed a number of psychrophilic yeasts isolated from soil and lake sediment. The New Zealand effort was focused around the McMurdo Dry Valleys of South Victoria Land, while the Japanese work reflected samples from soil collected during the Third and Fourth Japanese Antarctic Research Expeditions on Syowa Base and the Ongul Islands (Soneda 1961; Tubaki 1961) as well as the McMurdo Dry Valleys (Goto et al. 1969). There were intermittent reports of yeasts isolated from Antarctica throughout the next several decades as reviewed by Vishniac (2006a) where it is pointed out that many of the early cultures no longer exist and therefore cannot be re-examined. Much of this work was the result of interest in the Antarctic as an analog for exobiology, in particular for Mars exploration (Bruch 1966; Onofri et al. 2004; Tamppari et al. 2012). Increasingly, over the last decade descriptions of species new to science (Xin and Zhou 2007; Connell et al. 2010; Hao et al. 2010; Thomas-Hall et al. 2010; Turchetti et al. 2011; Wuczkowski et al. 2011) and additional species found (Onofri et al. 2007;

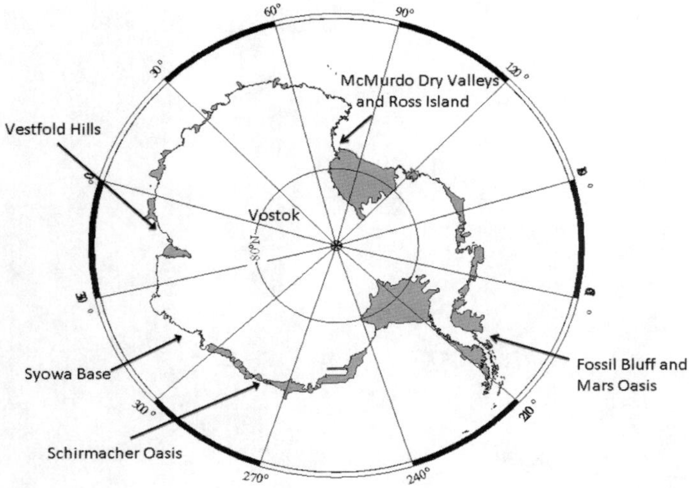

Fig. 4.1 Continental Antarctica showing important sites for terrestrial research

Bridge and Spooner 2012; Buzzini et al. 2012) have been added to the expanding list of microorganisms identified from Antarctica. Many of these species can be found in other parts of the world; however, most display adaptations for life in cold habitats, such as psychrophily, changes in membrane components, synthesis of pigments, and ability to resist desiccation, all traits that may be preadaptation for life 'at the ecological edge' (Gostincar et al. 2010).

4.2 The Antarctic Soil Habitats

Long-term isolation resulting from climatic and geological events may be among the most important historical factors influencing biogeographic patterns in the Antarctic. Continental Antarctica presents some of the most extreme conditions on the Earth's surface for life to exist. The exposed land area is less than 2 % (Ugolini and Bockheim 2008) and spans a variety of nutrient inputs especially between the polar deserts and high nunataks of South Victoria Land, including the McMurdo Dry Valley region near Ross Island, and the relatively moist and warmer Antarctic Peninsula. South Victoria Land soils are typically polar deserts (Bockheim 1997) (Fig. 4.2) although there are some areas near the Ross Sea that have higher soil moisture and areas with increased nutrient input associated with vertebrate animal (e.g., bird or seal colonies) and human activity. Most of the soils are defined as Anhyorthels because of the lack of cryoturbation (Bockheim 2002). The region is typical of other deserts with salt pans and kettle ponds and is among some of the oldest exposed surfaces on earth (Bockheim 1997). Relatively stable alpine glaciers with frozen basement cement line many of the valleys and supply water

Fig. 4.2 Soils of South Victoria Land. Rocky outcropping above 1,600 m a.s.l. on Sponsors Peak near the mouth of the Victoria Upper Glacier (*left*); Beacon sandstone in the Wright Valley (*upper right*) and soil from the Beardmore Glacier area (*lower right*)

Fig. 4.3 Glaciers on the sides of Taylor Valley, McMurdo Dry Valleys. Solas and Hughes glaciers in the Lake Bonney Basin (*left*); Commonwealth Glacier in the Lake Fryxell Basin (*right*)

during the short melt season for ponds and lakes (Fountain et al. 2006), many of which are perpetually ice covered. These glaciers are mostly free from debris, unlike temperate latitude glaciers, and have a sharp face of approximately 20 m high (Taylor 1916) (Fig. 4.3). The climate within the McMurdo Dry Valleys is much more extreme than the maritime or sub-Antarctic. Low temperatures dominate with the average annual air temperature between −17.7 °C and −20.8 °C (Doran et al. 2002) ranging from −60 °C in the winter to +10 °C during the short austral summer. Strong winds (Nylen et al. 2004), alkaline soils with pH values raging from pH 7 to pH 10 (di Menna 1966a; Cameron 1971; Connell et al. 2008), and low moisture are also dominant features of the landscape (for reviews see Cameron 1972; Campbell et al. 1998). Surface temperatures can vary from −15 to +27 °C in 3 h (Cameron and Morelli 1974), and many freeze–thaw cycles can be experienced in a matter of minutes when clouds pass over (Friedmann et al. 1987). Daylight ranges from 4 months of complete darkness in the winter to 4 months of complete exposure in the summer further reducing potential for primary productivity (Nylen et al. 2004).

The McMurdo Dry Valleys region is Antarctica's largest ice-free area with three large valleys (Taylor, Victoria, and Wright), a number of smaller valleys as well as uplands comprising nearly 15 % ($\sim$4,800 km^2) of the exposed land of the area below 60° S (Ugolini and Bockheim 2008). The entire region has been protected as an Antarctic Specially Managed Area (ASMA) since 2004 under the Antarctic Treaty and these protocols were substantially revised in 2011 (Antarctica-New-Zealand 2012). This landscape can be divided into three basic zones: coastal, intermediate, and interior (Marchant and Denton 1996). Steep gradients exist for carbon sources, salts, moisture availability, and temperature from the coastal through the interior regions as well as with increasing elevation up the valley sides (Cameron 1971, 1972; Campbell et al. 1998). The soils are low in organic matter content and are very weakly developed. A paucity of clay and organic material may lead to a low soil buffering capacity; therefore, the salt concentration directly effects the soil pH (for a review see Campbell et al. 1998). These habitats are described by persistent sub-zero temperatures, limited water availability (McKnight et al. 1999), intense winds (Nylen et al. 2004), high UV light inputs (Madronich et al. 1998; Tosi et al. 2005), and patchily distributed nutrients (Tiao et al. 2012). Nutrient availability and organic carbon levels are also very low in the Antarctic deserts (Burkins et al. 2000, 2001), especially when compared with the large amount of organic carbon contained in Arctic regions (Bockheim et al. 1999). The structure and organic content of desert soils in South Victoria Land are highly dependent on proximity to lacustrine sources of primary production (Elberling et al. 2006; Hopkins et al. 2006b) and on past geological events (Fountain et al. 1999; Moorhead et al. 1999). These soils have some of the lowest organic carbon levels on earth along with the high elevations of the Atacama and Patagonian desert regions of Chile and Argentina (Burkins et al. 2000; Barrett et al. 2006a; Hopkins et al. 2006a; Costello et al. 2009; Lynch et al. 2012). Legacy nutrient sources, left behind when the shoreline of paleolakes receded from $\sim$300 m above the current valley floor (Hall et al. 2000), can help to fuel soil communities functionally linking the ancient and modern (Moorhead et al. 1999). Areas at lake edges with an abundance of primary production do not need to be linked to legacy carbon sources; however, more distant communities dominated by soil heterotrophs may require closer coupling along with supplementation derived from aerial distribution of recent nutrient inputs (e.g., decadal pond size fluxes) (Moorhead et al. 2003; Elberling et al. 2006; Hopkins et al. 2006b). Glacial stream runoff during warming events supplements primary production in the McMurdo Dry Valley lakes (Foreman et al. 2004), further moving nutrients from lakes to soil through subsequent aerial distribution. An additional supplemental source of nutrients is flushing from cryoconites during glacial ablation (Foreman et al. 2007). The Antarctic deserts are among the driest in the world with an average of 45 mm or less precipitation per year (Schwerdtfeger 1984). Liquid water is intermittently available to soil communities for a short period of time (maximum of 10 weeks) during the austral summer season (Conovitz et al. 1998; McKnight et al. 1999) and ranges from a few hundred hours per year in coastal locations to once every several years in interior sections (Wall et al. 2004). Most soil moisture is rapidly lost through sublimation and evaporation, yielding an average soil moisture below 5 % (Freckman and Virginia 1998) and ranging from 14 to below 0.5 % (Fell et al. 2006; Connell et al. 2008). The dominant sources of soil moisture are marine influenced humidity and precipitation, glacial inputs, as well as lakes and streams, and permafrost (Conovitz et al. 1998). However, lakes and streams are only significant moisture sources within 8 m of the water source due to soil porosity (Campbell et al. 1998).

Despite the environmental challenges, habitable niches, such as soils often near moisture sources (Vishniac and Klinger 1986), and lithic substrates (Onofri et al. 2000; Pointing et al. 2009), can harbor relatively diverse fungal assemblages. Black fungi, many of which are cryptoendolithic, have been frequently found in the Antarctic (Selbman et al. 2005, 2011) and are discussed in Chap. 8. The deserts of Antarctica are unique in that they have no vascular plants, therefore no rhizosphere, and the soil communities are less complex than those found in lower latitudes and the Arctic (Robinson 2001). Therefore, most of the biological activity is restricted to below ground (Hogg et al. 2006). Hot spots where longer-term snow melt is available, such as troughs or cracks (Ayres et al. 2010), sites within a few meters of a lake or stream (McKnight et al. 1999) and endolithic conditions with little snow coverage (Sun 2013) exist. The diversity and community composition of modern-day Antarctic microbiota are known to be influenced by historical processes such as glacier and ice sheet evolution, continental drift, and geological activity that influence local geochemistry and landscape features (Barrett et al. 2006b). Thus, historical processes can act as primary constraints on modern microbial biodiversity and ecosystem development (Zeglin et al. 2009; Smith et al. 2010).

4.3 Dominant Yeasts in Antarctic Desert Habitats

Yeasts appear to be the dominant fungi in some locations and have been the only fungi identified in many of the locations sampled (Atlas et al. 1978; Thomas-Hall 2004; Connell et al. 2008). Beyond their role as decomposers, yeast may play a function, along with cyanobacteria, in early site colonization and soil structure development. Further, yeast may be one of the principle taxa available in more arid sites to synthesize sterols required by the soil invertebrates (Nes and McKean 1977; Weete 1974). Therefore, because of the relatively simple community structures in these extreme environments, yeasts may occupy a trophic level more important in Antarctic desert soils than in temperate or sub-Antarctic soils, functioning at several levels within the food web. Yeasts in the McMurdo Dry Valleys are dominated by basidiomycetous species (89 %), most particularly those from the genus *Cryptococcus* (33 %) (di Menna 1960; Connell et al. 2008). Habitat gradients from higher relative moisture (6 %) to very dry conditions (0.5 %) showed that yeasts dominated filamentous species in culture (Connell et al. 2006) and in culture-independent studies from the most xeric habitats (Fell et al. 2006). In a study of soil associated with Antarctic historic huts and a limited number of McMurdo Dry Valleys soil samples, Arenz et al. (2006) found that *Cryptococcus* species accounted for 67 % (4 out of 6) of the yeast species identified in McMurdo Dry Valley soil, and 72 % (13 of 18) in soils surrounding the historic huts. A later study (Arenz and Blanchette 2011) found basidiomycetous yeasts of the genera *Rhodotorula* and *Cryptococcus* made up 21 % of the yeasts in the Ross Sea region, including McKelvey Valley, and only 9.8 % from the more moist Antarctic Peninsula region. The dominance of *Cryptococcus* species in soil, particularly arid soil, has been ascribed to their ability to produce polysaccharide capsules (Vishniac 2006b). However, although yeasts, in general, are considered to utilize only simple sugars, that is not necessarily the case for basidiomycetous species (Kurtzman and Fell 1997). *Cryptococcus* species may be able to utilize nutrients available while most of the ascomycetous yeast species cannot, and their nutrient utilization profile can vary from isolate to isolate (Kurtzman and Fell 1997). Further, the ability of some taxa, such as *Cryptococcus*, to survive may be related to their ability to produce compounds that allow them to resist desiccation, scavenge nutrients from oligotrophic conditions, and maintain membrane fluidity at low temperatures.

A genomic-based study at the Mars Oasis found that black yeasts dominated in the most arid sites (Bridge and Newsham 2009). Because of their slow growth rate, black yeasts can require additional culture conditions for isolation from those that have been used in most Antarctic culture-based work to date (Sterflinger 2006). Therefore, although black yeasts have not been isolated as often in culture-based studies as other genera, they may form one of the dominant groups based on their ability to survive the extreme conditions found in these habitats. An additional yeast species that has been found several times in Antarctic culture-independent studies is *Malassezia*, in particular *Malassezia restricta* and *Malassezia globosa* (Arenz et al. 2006; Fell et al. 2006; Bridge and Spooner 2012; Connell and Staudigel 2013). These *Malassezia* species are typically associated with vertebrate animals and have a strict requirement of oil for growth (Batra et al. 2005). They have thus far only been reported in culture-independent studies, possibly because of their culture requirements. Alternatively, *Malassezia* species may be an indication of human or animal activity and may not be active members of the soil fungal community.

4.4 Adaptations Required for Life in Antarctic Deserts

Most investigations on yeasts in the Antarctic polar deserts have focused on descriptions of their biodiversity and distribution. Evidence for adaptation of yeasts to the soils of the Antarctic polar desert conditions means having the ability to grow at low temperatures, as well as the ability to survive multiple freeze–thaw cycles, desiccation, high UV exposure and eke an existence out of oligotrophic habitats. Experiments investigating properties related to adaptation in Antarctica with cultured isolates focused on growth at low temperatures (Sinclair and Stokes 1965; Stokes 1971; Vishniac 1987) later expanded to exploration of other products such as membrane lipids (Watson et al. 1976; Kates and Baxter 1982; Zlatanov et al. 2001), DNA synthesis (Silver et al. 1977), extracellular proteases and enzyme production (Ray et al. 1992; Gomes et al. 2000; Scorzetti et al. 2000; Sabri et al. 2001; Turkiewicz et al. 2001; Pavlova et al. 2002; Turkiewicz et al. 2003) including photoreactive compounds (Vaz et al. 2011), heat- and cold-shock proteins (Deegenaars and Watson 1998; Boo et al. 2013), growth under limited moisture conditions (Vishniac 1985), nutrient utilization (Katayama-Hirayama et al. 2003), and generation of odor and flavor compounds (Alchihab et al. 2009). Recently, several yeasts isolated from the McMurdo Dry Valleys have had both their genome and transcriptome sequenced by the Joint Genome Institute (Walnut Creek CA, USA), *Cryptococcus vishniacii* CBS 10616 and *Dioszegia cryoxerica* CBS 10919. These two isolates were also subjected to a metabolomics study at two temperatures to identify compounds associated with psychrophily (Dalluge and Connell 2013, unpublished data). With the increasing ability of sequencing technologies to rapidly generate genome data on which to build gene scaffolds more information from metagenome projects should be able to be applied to yeast species in the near future to further genomics of psychrophily (Casanueva et al. 2010) and expand it to psychrophilic yeasts, perhaps elucidating their role in communities. In addition, advances in mass spectrometry-based metabolic profiling methods should dovetail well with data garnered from genomic studies.

4.4.1 Mass Spectrometry-Based Metabolite Profiling Approaches to the Study of Biochemical Adaptation in Antarctic Psychrophilic Yeasts

The high sensitivity, selectivity, and mass accuracy of LC/MS employing high-resolution mass analyzers, such as time-of-flight (TOF), quadrupole-time-of-flight (QTOF), Orbitraps, and Fourier transform ion cyclotron resonance (FTICR), have made it a method of choice for the generation of metabolic signatures in biological systems and for generation of metabolic databases (Theodoridis et al. 2011). Technological advances in both chromatography and mass spectrometry including the advent of ultraperformance liquid chromatography (UPLC) that allows significantly shorter run times without sacrificing chromatographic fidelity, and MS^e that allows QTOF instrumentation to provide simultaneous acquisition of low- and high-collision energy mass spectra for every analyte in a single experiment without sacrificing exact mass information for precursor or product ions, have established UPLC/MS^e as both a rapid and robust method for metabolite profiling studies (Castro-Perez et al. 2010; Theodoridis et al. 2011).

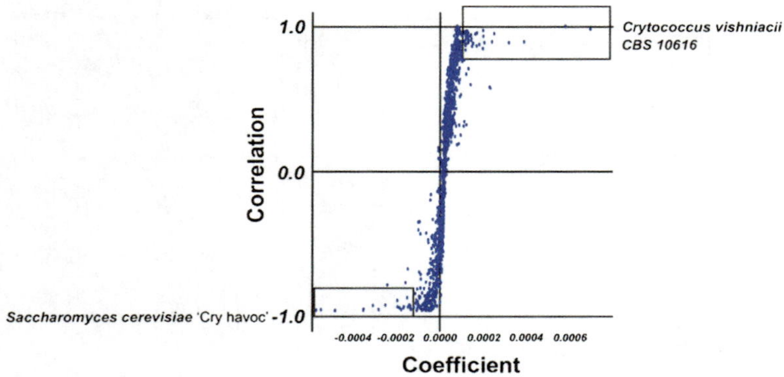

Fig. 4.4 Chemometric S plot resulting from orthogonal partial least squares-discriminant analysis (OPLS-DA) of UPLC/MS^e analysis of cell extracts (positive electrospray ionization) from *Saccharomyces cerevisiae* 'Cry havoc' (37 °C) versus *Cryptococcus vishniacii* CBS 10616 (4 °C). Similar S plots were generated from OPLS-DA analysis of *Dioszegia cryoxerica* versus *S. cerevisiae*. The coefficient axis represents the specificity of a given feature to a sample (positive = *D. cryoxerica*-specific features; negative = *S. cerevisiae*-specific features). The correlation axis represents the confidence that a given feature is species specific (1.0 = highest confidence *C. vishniacii*-specific features; −1.0 = highest confidence *S. cerevisiae*-specific features). The boxes surround features that are highly specific to each yeast extract

To move beyond more traditional targeted approaches to the biochemical characterization of yeast psychrophily and provide a more detailed holistic understanding of the chemistry of physiological adaptation of psychrophiles at the molecular level, Dalluge and Connell (2013) describe the use of UPLC/MS^e for a preliminary comparative analysis of cell extracts from two psychrophilic Antarctic yeasts (*C. vishniacii* CBS 10616 and *D. cryoxerica* CBS 10919) versus the mesophilic yeast *Saccharomyces cerevisiae* 'Cry havoc.'

Total ion chromatograms from UPLC/MS^e analysis of cell extracts of *C. vishniacii* were visually indistinguishable from those of *D. cryoxerica*. Visualization of the results of orthogonal partial least squares-discriminant analysis (OPLS-DA) chemometric analysis of detected features in psychrophilic versus mesophilic cell extract analysis can be accomplished via an S plot, as illustrated in Fig. 4.4. In such a plot, features most unique and statistically relevant to a given biological system reside at opposite ends of the S plot, respectively. The boxes in Fig. 4.4 surround individual features specific to either *C. vishniacii* (1.0) or *S. cerevisiae* 'Cry havoc' (−1.0). A trend plot for each feature illustrating the relative abundance of a specific feature in one species versus the other can also be used to corroborate determinant features of a given organism (Fig. 4.5a), together with manual interrogation of extracted ion chromatograms XICs (Fig. 4.5b), for each feature identified from the S plots. Of more than 50 determinant psychrophilic yeast-specific features corroborated in this way (from a pool of approximately 2,000 potential features detected), greater than 70 % are present in both *D. cryoxerica* and *C. vishniacii*.

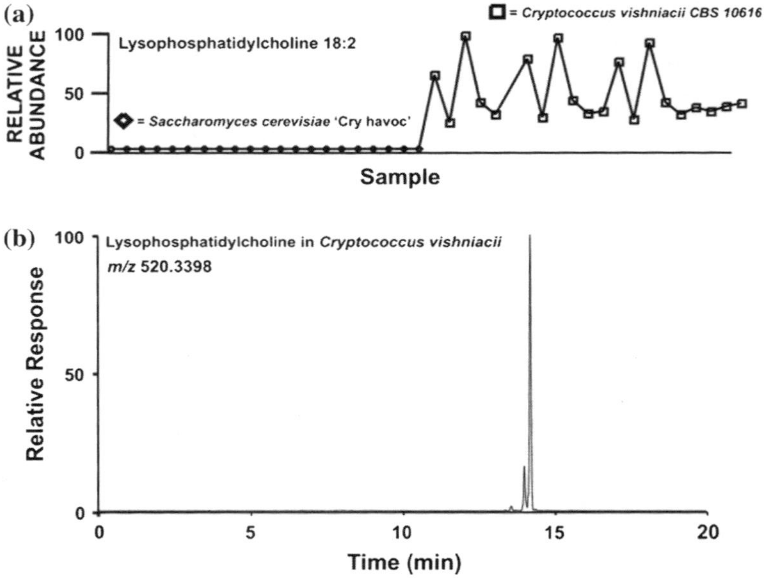

Fig. 4.5 **a** Chemometric trend plot illustrating the relative abundance of the psychrophile-specific compound lysophosphatidylcholine 18:2 in *Cryptococcus vishniacii* CBS 10616 cell extracts versus cell extracts from the mesophilic yeast *Saccharomyces cerevisiae* 'Cry havoc'; **b** extracted ion chromatogram (XIC) for *m/z* 520.3398 (LysoPC 18:2) in *C. vishniacii*. Separation and detection of two peaks indicates possible presence of distinct isomeric species of LysoPC in these organisms. Identical trend plots and extracted ion chromatograms were generated for this compound in extracts of *Dioszegia cryoxerica*

Preliminary identifications of psychrophilic yeast-specific chemical entities included lysophosphatidylethanolamine 18:2 and 18:3, as well as lysophosphatidylcholine 18:2 and 18:3, identified in both *D. cryoxerica* and *C. vishniacii*. Guanosine diphosphate (GDP) was detected in *D. cryoxerica* and *C. vishniacii* at a $\geq$5X level over that found in the mesophilic yeast *S. cerevisiae*. In addition to the chemical species above, the preliminary investigation yielded at least 10 additional non-lipid psychrophile-specific features unidentified to date (Table 4.1).

Preliminary identification of the unsaturated bioactive glycerophospholipids as specific to the psychrophilic yeasts is certainly interesting from the standpoint of general lipid metabolism, as well as membrane structure, composition, and dynamics of these psychrophiles. It should also be of interest, however, to investigate the potential roles these bioactive lysophospholipids may play in modulating activities of regulatory and signaling proteins (Resnick and Tomáska 1994; Yuan et al. 1996; Slessareva and Dohlman 2006). Of note, structural isomers of all of these glycerophospholipids were separated chromatographically, providing the additional opportunity to evaluate *cis–trans* lipid ratios in these organisms, thought to play a role in modulation of membrane fluidity (Loffeld and Keweloh 1996).

Table 4.1 Psychrophilic yeast-specific peptides discovered to date from *Cryptococcus vishniacii* CBS 10616 and *Dioszegia cryoxerica* CBS 10919

Peptides	Putative amino acid sequence
Peptide PG1097[a]	(GE)AGS(LS)PAGPR
Peptide PG1104[a]	GVQGPPGSPGPR
Peptide PG1302[a]	(GT)GGP(GS)PTGFGSPQ
Peptide PG1562[a]	(QAG)LAPTAGAWPG(LGVP)
Peptide PG1137[b]	SVPGPMGPSGPR

[a] Indicates peptides detected in psychrophilic yeast cell extracts and not detected in mesophilic yeast cell extracts

[b] Indicates peptide with relative abundance in psychrophilic yeast extracts versus mesophilic yeast extracts of ≥5X; peptide regions of known amino acid composition but unknown sequence are in parentheses

Elevated levels of GDP in psychrophilic cell extracts may also be of interest in the context of G-protein-dependent signal transduction and of sensing (Slessareva and Dohlman 2006), as well as purine energy metabolism. Finally, five psychrophilic yeast-specific peptides were discovered in *C. vishniacii* and *D. cryoxerica* (Table 4.1). Following the determination of these peptides as specific to psychrophilic yeast, subsequent precursor-selected MS/MS analysis revealed the amino acid compositions and sequence information for these peptides. Interestingly, this analysis demonstrated that these peptides were both proline and glycine rich. Bioactive natural peptides are ubiquitous in all life kingdoms and can be found either free or encrypted in proteins (Daffre et al. 2008). A high molar abundance of proline and/or glycine is characteristic of a variety of these bioactive peptides (Otvos 2000; Daffre et al. 2008; Rozgonyi et al. 2009; Zhang and Zhu 2012; Ilić et al. 2013) that can exhibit both intracellular and extracellular functions (Lazazzera 2001) and is reported to influence among other things, membrane assembly and dynamics (Rozgonyi et al. 2009; Ilić et al. 2013). Proline is also known to be a powerful antioxidant that could be very important in stress adaption and reduction in apoptosis under stressful conditions in fungi (Chen and Dickman 2005).

4.4.2 Membrane Responses by Antarctic Yeast to Growth Temperatures

Antarctic yeasts are adapted to cold temperature extremes and have temperature growth limits ranging from just above freezing to 20 °C. Although the mechanism of cold adaptation in Antarctic yeast is poorly defined, these organisms have been shown to modulate membrane fatty acids in response to temperature (Weinstein et al. 2000). To determine Antarctic yeasts modulated membrane sterols in

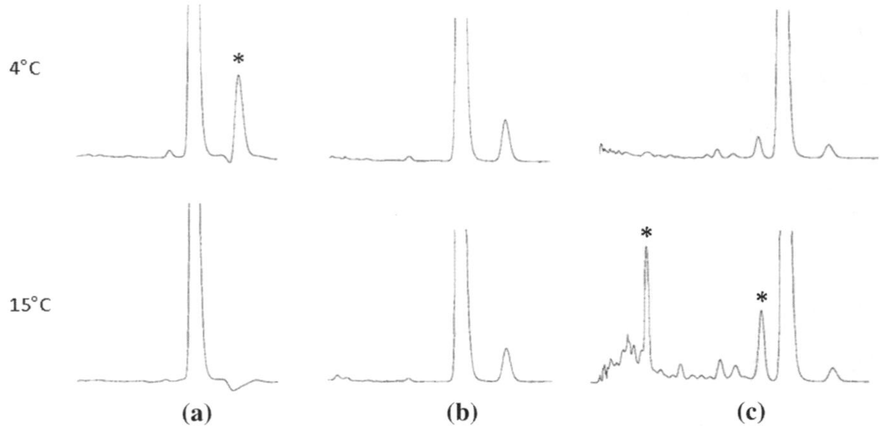

Fig. 4.6 Free sterol changes in response to growth temperature. **a** *Dioszegia antarctica* CBS 10769; **b** *Cryptococcus saitoi* CBS 10631; **c** *Glaciozyma antarctica* CBS 10640. The *asterisks* indicate the sterols that are significantly different between cells grown at 4 and 15 °C

response to temperatures, cultures were grown at 4 and 15 °C and membrane sterol composition was analyzed (Fig. 4.6). Interestingly, three patterns of sterol modulation were observed with regard to minor sterols depending on the yeast species. *Dioszegia antarctica* produced a minor sterol at 4 °C but not at 15 °C. *Cryptococcus saitoi* produced multiple minor sterols at 15 °C that were absent at 4 °C, and *Glaciozyma antarctica* showed no difference in minor sterols produced at the two temperatures. An in-depth review of the changes in lipid composition and fluidity of yeast plasma membrane as response to cold is reported in Chap. 10.

4.4.3 Desiccation Resistance in Antarctic Yeasts

Desiccation tolerance in hyper-arid regions of Antarctic deserts is another adaptation required of yeasts active in these communities. Desiccation tolerance is the ability of an organism to have as much as 95 % of its water removed and be able to return to normal metabolism after rehydration (Crowe et al. 1992). Desiccation pathways have been found in *S. cerevisiae* but this yeast can only survive desiccation if it enters the state while cultures are in stationary phase (Welch et al. 2013). Therefore, it is possible that Antarctic yeast species may utilize a different mechanism to resist desiccation and remain active in hyper-xeric soils.

Early suggestions about how yeasts survive desiccation considered that internal increases in trehalose concentrations may provide protection; however, trehalose is not sufficient or needed for desiccation resistance in yeasts (Ratnakumar and Tunnacliffe 2006). Mechanisms for desiccation survival in bacteria and *Ustilago* yeasts and mechanisms for resistance to exposure to radiation in the extremophilic

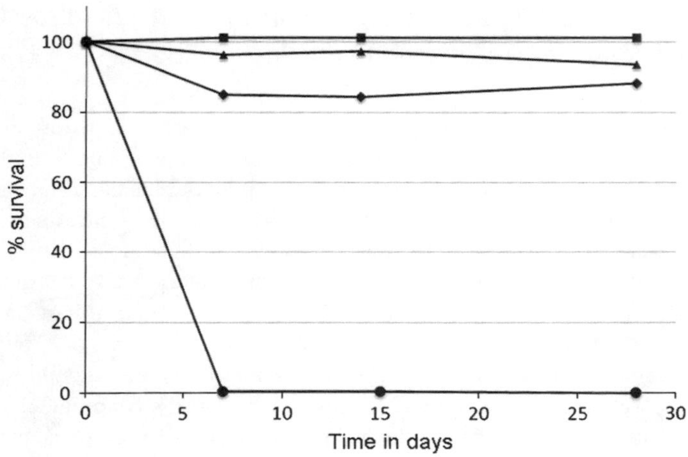

Fig. 4.7 Percent survival after desiccation (time in days) of the Antarctic yeasts *Cryptococcus vishniacii* (*square*), *Rhodotorula mucilaginosa* (*triangle*) and *Glaciozyma martinii* (*diamond*) and of the control *Saccharomyces cerevisiae* (*circle*). The relative humidity in all chambers fell and remained below 1 % for the remainder of the experiment within 12 h

bacterium *Deinococcus radiodurans* appear to be related to protein damage and repair, not DNA damage as first thought, and this repair is mediated through Mn-peptide-protein complexes (for a review see Daly 2012).

In a small study of three yeasts isolated from the McMurdo Dry Valleys from sites of low moisture (Connell et al. 2008), resistance to desiccation was determined by filtering cells grown at 10 °C to logarithmic phase onto filters and then placing those filters in desiccation chambers for 7, 15, or 27 days. The filters were then placed on petri dishes containing growth medium, and CFU were counted after 1 and 2 weeks (Fig. 4.7). The yeast isolates from the Wright Valley (*C. vishniacii*) and from above 16,000 m.a.s.l. on Sponsors Peak (*Rhodotorula mucilaginosa*) were very resistant to desiccation and showed almost 100 % survival after 27 days at a relative humidity below 1 %. *Glaciozyma martinii* had only a slightly lower survival rate.

It is not yet clear which mechanism Antarctic yeasts utilize to resist desiccation and UV radiation exposure but it is possible that similar mechanisms to those used by *D. radiodurans* may exist. If Antarctic yeasts use this mechanism, then acquisition of Mn for cellular construction into Mn-peptide-protein complexes could be enhanced by production of siderophores with high binding affinity. Thus, many of the adaptations for survival in Antarctic deserts may in fact be linked.

4.4.4 Iron and Manganese Acquisition of Antarctic Yeasts in Oligotrophic Environments

Much of the research on cold-adapted yeasts from Antarctica has thus far focused on determining how they function under cold or desiccating conditions. However, many of these yeasts are found in highly oligotrophic habitats and may play a role in biogeochemical cycling of metals such as iron (Fe) and manganese (Mn). In terrestrial weathering systems, microeukaryotes, dominated by fungi, are quite common, contributing to the weathering of silicate materials (Barker and Banfield 1998), including volcanic rocks that are commonly found in much of South Victoria Land (Etienne and Dupont 2002). Although Fe is one of the most abundant elements in the Earth's crust much of that Fe is biologically unavailable because the oxidized form, Fe(III), is insoluble under physiological conditions (Neilands 1995). Many microbes have developed mechanisms to acquire Fe and Mn using low molecular weight chelators, such as siderophores, that can then transport the Fe into the cell (Guerinot 1994). Siderophores are well known in filamentous fungi but are also produced by some yeasts (Renshaw et al. 2002). Siderophore production has been confirmed in five Antarctic yeast species (*D. antarctica*, *D. cryoxerica*, *G. antarctica*, *Glaciozyma watsonii*, and an undescribed *Glaciozyma* sp.) that were isolated from the McMurdo Dry Valleys (Fig. 4.8). Yeast isolates were grown on Chrom-Azurol Siderophore (CAS) agar and a color change of the medium from blue to purple or orange indicated siderophore production (Pérez-Miranda et al. 2007; Connell et al. 2009). The Antarctic yeasts investigated produced siderophores of potentially the catechol type, based on the purple color, not typical of the structures found in many filamentous fungi (Neilands 1995). This type of siderophore has the highest known binding

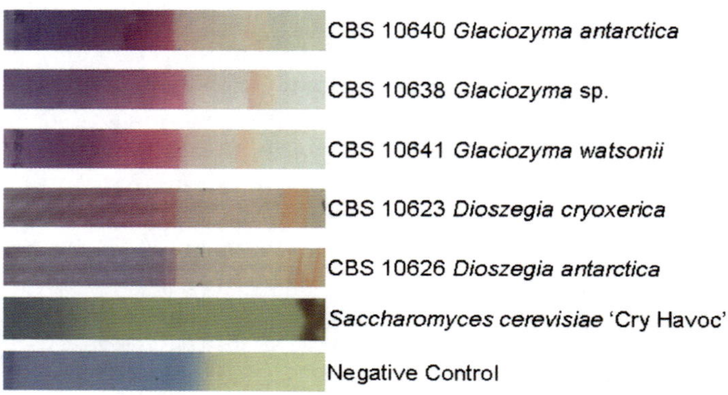

Fig. 4.8 Siderophore production of Antarctic yeasts. A *Saccharomyces cerevisiae* negative control, a no-isolate negative control, and isolates of yeasts collected from the McMurdo Dry Valleys of Antarctica were grown on CAS medium. A color change in the medium on the *left* from *light blue* to *pink* or *purple* indicates siderophore production

constant for ferric ion and may reflect these organisms need to acquire Fe and Mn from a highly oligotrophic environment (Raymond and Carrano 1979). It is interesting that Mn levels may be related to desiccation resistance; therefore, these siderophores may play a role in adaptation beyond mere nutritional requirements in oligotrophic environments.

4.5 Anthropogenic Influence on Yeast Diversity

Throughout the last 100 years, humans have ventured to Antarctica with increasing frequency over the last two decades. Now both tourists and researchers visit a few sites often. It is clear that humans have influenced the yeast diversity in some locations, for example local contamination associated with the building of stations and research camps, including the huts remaining from the heroic exploration era (Blanchette et al. 2004, 2010; Arenz et al. 2006; Arenz and Blanchette 2011; Farrell et al. 2011). *S. cerevisiae* has been found from 50 years old cashes near the Terra Nova hut at Cape Evans (Meyer et al. 1962), and the high percent of *Malassezia* clones found in a Mt. Erebus cave clone library likely represent contamination by visiting humans (Connell and Staudigel 2013). Clearly, a number of species that have been cultured from Antarctica are also found elsewhere; for example, *Glaciozyma* strains are also found in alpine settings (Turchetti et al. 2011) and in undersea volcanoes (Connell et al. 2009), and *Rhodotorula* are cosmopolitan.

Suggestions for organized monitoring (Rudolph and Benninghoff 1977) resulted in 1991 in the Protocol on Environmental Protection to the Antarctic Treaty (Madrid Protocol). Through the Madrid Protocol, the Parties committed themselves to a comprehensive protection of the Antarctic environment and dependent and associated ecosystems and designated Antarctica as a natural reserve, devoted to peace and science. Nonetheless, concerns have been raised about the introduction of organisms into the environment (Cowan and Tow 2004; Cowan et al. 2011). Increasingly strict protocols are being put into place to help reduce both habitat destruction and inadvertent contamination of sites, even cross-contamination between sites within Antarctica.

4.6 Yeast Diversity in Other Antarctic Habitats

Although we have focused on cold-adapted yeasts of the Antarctic desert, especially the McMurdo Dry Valleys region, other studies on Antarctic yeast show that they occur in many, if not all other habitats. Marine yeasts were first investigated in 1970s (Fell and Statzell 1971; Fell 1974; Fell and Hunter 1974), and later both marine waters and marine sponges were explored (Connell 1994; Zhang et al. 2012; Vaca et al. 2013). The Australian Antarctic Division maintains several bases in the Antarctic, mostly in the Vestfold Hills area, and have reported a number of yeast species including those of the genera *Cryptococcus*, *Mrakia*, *Rhodotorula*, and *Glaciozyma* (Thomas-Hall and Watson 2002; Thomas-Hall et al. 2002, 2010; Guffogg et al. 2004; Thomas-Hall 2004; Turchetti et al. 2011). Both *Cryptococcus* and *Rhodotorula* have been isolated from Adélie Land (Scorzetti et al. 2000; Alchihab et al. 2009). A culture-independent study of a Mt. Erebus dark fumarole cave on Ross Island in South Victoria Land has revealed a number of yeast genera dominated by *Malassezia* but also including *Rhodotorula*, *Saccharomyces*, *Clavispora*, *Sporobolomyces*, *Glaciozyma*, *Cryptococcus*, and *Candida* as well as the yeast-like dimorphic fungus *Aureobasidium pullulans* (Connell and Staudigel 2013). German, Indian, and Russian stations in and near the Schirmacher Oasis (Queen Maud Land) have all had yeasts identified (Ray et al. 1989). One of the most remote habitats that yeasts have been isolated from is Vostok Lake Ice-core drilling project where both *Cryptococcus* and *Rhodotorula* have been isolated and shown to be active at −5 °C (Amato and Christner 2009). *Cystofilobasidium* sp. and *A. pullulans* have also been isolated from accretion ice of that core (D'Elia et al. 2009), and *Cryptococcus* has been identified in a metagenome (Rogers et al. 2013).

In contrast, the Antarctic Peninsula is relatively warmer and moister than other Antarctic regions and has yielded a diversity of yeast genera in both culture and culture-independent studies (Baker 1970; Ellis-Evans 1985; Pavlova et al. 2001; Zlatanov et al. 2001; Turkiewicz et al. 2003; Lawley et al. 2004; Arenz et al. 2006; Xin and Zhou 2007; Yergeau et al. 2007; Bridge and Newsham 2009; Bridge and Spooner 2012; Yergeau et al. 2012). Yeasts associated with mosses and microbial mats (Goto et al. 1969; Baublis et al. 1991; Tosi et al. 2002), human activity (Baublis et al. 1991; Blanchette et al. 2004, 2010; Arenz et al. 2006; Held et al. 2006; Arenz and Blanchette 2011), and animals (Del Frate and Caretta 1990; Baublis et al. 1991) have also been described.

4.7 Conclusions

It is clear that the diversity of habitats on the Antarctic continent allows for a similar level of diversity among yeast species. Although the adaptive potential of yeasts to environmental stresses is easily inferred by survival in various habitats, adaptive mechanisms are poorly defined. However, the adaptive potential of yeasts from Antarctic deserts is likely to include membrane modifications, protein stability, antioxidation systems, and rapid responses to extreme changes in soil temperatures and dehydration/rehydration conditions. Now that distribution of yeast species in Antarctica is beginning to be understood, it is time to begin focusing more on adaptive mechanisms to develop predictive tools to understand how these unique communities will respond to climate change during this century. In addition, care must be given to ensuring that these unique sets of habitats remain uncontaminated for future research.

References

Alchihab M, Destain J, Aguedo M, Majad L, Ghalfi H, Wathelet J-P, Thonart P (2009) Production of gamma-decalactone by a psychrophilic and a mesophilic strain of the yeast *Rhodotorula aurantiaca*. Appl Biochem Biotechnol 158:41–50

Amato P, Christner B (2009) Macromolecular synthesis by yeasts under frozen conditions. Environ Microbiol 11:589–596

Antarctica-New-Zealand (2012) McMurdo Dry Valleys ASMA manual, Christchurch

Arenz BE, Held BW, Jurgens JA, Farrell RL, Blanchette RA (2006) Fungal diversity in soils and historic wood from the Ross Sea region of Antarctica. Soil Biol Biochem 38:3057–3064

Arenz BE, Blanchette RA (2011) Distribution and abundance of soil fungi in Antarctica at sites on the Peninsula, Ross Sea Region and McMurdo Dry Valleys. Soil Biol Biochem 43:308–315

Atlas RM, di Menna ME, Cameron RE (1978) Ecological investigations of yeasts in Antarctic soils. Antarct Res Ser Washington 30:27–34

Ayres E, Nkem JN, Wall DH, Adams BJ, Barrett JE, Simmons BL, Virginia RA, Fountain AG (2010) Experimentally increased snow accumulation alters soil moisture and animal community structure in a polar desert. Polar Biol 33:897–907

Baker JH (1970) Quantitative study of yeast and bacteria in a Signy Island peat. Brit Antarct Surv Bull 23:51–55

Barker WW, Banfield JF (1998) Zones of chemical and physical interaction at interfaces between microbial communities and minerals. Geomicrobiol J 15:223–244

Barrett JE, Virginia RA, Parsons AN, Wall D (2006a) Soil carbon turnover in McMurdo Dry Valleys, Antarctica. Soil Biol Biochem 38:3065–3082

Barrett JE, Virginia RA, Hopkins DW, Aislabie J, Bargagli R, Bockheim JG, Campbell IB, Lyons WB, Moorhead DL, Nkemh JN, Sletteni RS, Steltzerh H, Wall D, Wallenstein MD (2006b) Terrestrial ecosystem processes of Victoria Land, Antarctica. Soil Biol Biochem 38:3019–3034

Batra R, Boekhout T, Gueho E, Cabanes FJ, Dawson TL Jr, Gupta AK (2005) *Malassezia* Baillon, emerging clinical yeasts. FEMS Yeast Res 5:1101–1113

Baublis JA, Wharton RA Jr, Volz PA (1991) Diversity of micro-fungi in an Antarctic dry valley. J Basic Microbiol 31:3–12

Blanchette RA, Held BW, Jurgens JA, Mcnew DL, Harrington TC, Duncan SM, Farrell RL (2004) Wood-destroying soft rot fungi in historic expedition huts of Antarctica. Appl Environ Microbiol 70:1328–1335

Blanchette RA, Held BW, Arenz BE, Jurgens JA, Baltes NJ, Duncan SM, Farrell RL (2010) An Antarctic hot spot for fungi at Shackleton's historic hut on Cape Royds. Microb Ecol 60:29–38

Bockheim J (1997) Properties and classification of cold desert soils from Antarctica. Soil Soc Amer J 61:224–231

Bockheim JG, Everett LR, Hinkel KM, Nelson FE, Brown J (1999) Soil organic carbon storage and distribution in Arctic Tundra, Barrow Alaska. Soil Soc Amer J 63:934

Bockheim JG (2002) Landform and soil development in the McMurdo Dry Valleys, Antarctica: a regional synthesis. Arct Antarct Alp Res 34:308–317

Boo SY, Wong CMVL, Rodrigues KF, Najimudin N, Murad AMA, Mahadi NM (2013) Thermal stress responses in Antarctic yeast, *Glaciozyma antarctica* PI12, characterized by real-time quantitative PCR. Polar Biol 36:381–389

Bridge PD, Newsham KK (2009) Soil fungal community composition at Mars Oasis, a southern maritime Antarctic site, assessed by PCR amplification and cloning. Fungal Ecol 2:66–74

Bridge PD, Spooner BM (2012) Non-lichenized Antarctic fungi: transient visitors or members of a cryptic ecosystem? Fungal Ecol 5:381–394

Bruch CW (1966) Instruments for the detection of extraterrestrial life. In: Pittendrigh CS, Vishniac W, Pearman JPT (eds) Biology and exploration of Mars. National Academy of Sciences National Research Council, Washington DC, pp 487–502
Burkins MB, Virginia RA, Chamberlain CP, Wall DH (2000) Origin and distribution of soil organic matter in Taylor Valley, Antarctica. Ecology 81:2377–2391
Burkins MB, Virginia RA, Wall DH (2001) Organic carbon cycling in Taylor Valley, Antarctica: quantifying soil reservoirs and soil respiration. Global Change Biol 7:113–125
Buzzini P, Branda E, Goretti M, Turchetti B (2012) Psychrophilic yeasts from worldwide glacial habitats: diversity, adaptation strategies and biotechnological potential. FEMS Microbiol Ecol 82:217–241
Cameron RE (1971) Antarctic soil microbial and ecological investigations. In: Quam LO, Porter HD (eds) Research in Antarctic. American Association for the Advancement of Science, Washington DC, pp 137–189
Cameron RE (1972) Microbial and ecological investigations in Victoria Dry Valley, Southern Victoria Land, Antarctica. Antarct Res Ser Washington 20:195–260
Cameron RE, Morelli FA (1974) Viable microorganisms from Antarctic Ross Island and Taylor Valley Drill cores. Antarct J US 9:113–115
Campbell DI, Claridge GGC, Campbell DI, Balks MR (1998) The soil environment of the McMurdo Dry Valleys, Antarctica. In: Priscu JC (ed) Ecosystems dynamics in a polar desert, vol 72. American Geophysical Union, Washington DC, pp 297–322
Casanueva A, Tuffin M, Cary SC, Cowan DA (2010) Molecular adaptations to psychrophily: the impact of 'omic' technologies. Trends Microbiol 18:374–381
Castro-Perez JM, Kamphorst J, DeGroot J, Lafeber F, Goshawk J, Yu K, Shockcor JP, Vreeken RJ, Hankemeier T (2010) Comprehensive LC-MS E lipidomic analysis using a shotgun approach and its application to biomarker detection and identification in osteoarthritis patients. J Proteome Res 9:2377–2389
Chen C, Dickman MB (2005) Proline suppresses apoptosis in the fungal pathogen *Colletotrichum trifolii*. PNAS 102:3459–3464
Connell LB (1994) Biogeographic observations on South Georgia marine yeasts. Antarct J USA 29:143
Connell LB, Redman RS, Craig SD, Rodriguez RJ (2006) Distribution and abundance of fungi in the soils of Taylor Valley, Antarctica. Soil Biol Biochem 38:3083–3094
Connell LB, Redman RS, Craig SD, Scorzetti G, Iszard M, Rodriguez RJ (2008) Diversity of soil yeasts isolated from South Victoria Land, Antarctica. Microb Ecol 56:448–459
Connell LB, Templeton AS, Barrett A, Staudigel H (2009) Fungal diversity associated with an active deep sea volcano: Vailulu'u Seamount, Samoa. Geomicrobiol J 26:597–605
Connell LB, Redman RS, Rodriguez RJ, Barrett A, Iszard M, Fonesca Á (2010) *Dioszegia antarctica* and *D. cryoxerica* spp. nov., two novel psychrophilic basidiomycetous yeasts from polar desert soils in Antarctica. Int J Syst Evol Micr 60:1466–1472
Connell LB, Staudigel H (2013) Fungal diversity in a dark oligotrophic volcanic ecosystem (DOVE) on Mount Erebus, Antarctica. Biol 2013 2
Conovitz PA, McKnight DM, MacDonald LH, Fountain AG, House HR (1998) Hydrological processes influencing streamflow variation in Fryxell basin, Antarctica. Antarct Res Ser Washington 72:93–108
Costello EK, Halloy SRP, Reed SC, Preston S, Schmidt SK (2009) Fumerole-supported islands of biodiversity within a hyperarid, high-elevation landscape on Socompa Volcano, Puna de Atacama, Andes. App Environ Microbiol 75:735–747
Cowan DA, Tow LA (2004) Endangered Antarctic environments. Ann Rev Microbiol 58:649–690
Cowan DA, Chown SL, Convey P, Tuffin M, Hughes KA, Pointing S, Vincent WF (2011) Non-indigenous microorganisms in the Antarctic: assessing the risks. Trends Microbiol 19:540–548
Crowe JH, Hoekstra FA, Crowe LM (1992) Anhydrobiosis. Ann Rev Physiol 54:579–599

Dalluge JJ, Connell LB (2013) On the potential of mass spectrometry-based metabolite profiling approaches to the study of biochemical adaptation in psychrophilic yeast. Extremophiles [In Press]

D'Elia T, Veerapaneni R, Theraisnathan V, Rogers SO (2009) Isolation of fungi from Lake Vostok accretion ice. Mycologia 101:751–763

Daffre S, Bulet P, Spisni A, Ehret-Sabatier L, Rodrigues EG, Travassos LR (2008) Bioactive natural peptides. Stud Nat Prod Chem 35:597–691

Daly MJ (2012) Death by protein damage in irradiated cells. DNA Repair 11:12–21

Deegenaars ML, Watson K (1998) Heat shock response in psychrophilic and psychrotrophic yeast from Antarctica. Extremophiles 2:41–49

Del Frate G, Caretta G (1990) Fungi isolated from Antarctic material. Polar Biol 11:1–7

di Menna ME (1960) Yeasts from Antarctica. J Gen Microbiol 23:295–300

di Menna ME (1966a) Yeasts from Antarctic soils. A van Leeuwenhoek 32:29–38

di Menna ME (1966b) Three new yeasts from Antarctic soils: *Candida nivalis*, *Candida gelida* and *Candida frigida* spp.n. A van Leeuwenhoek 32:25–28

Doran PT, McKay CP, Clow GD, Dana GL, Fountain AG, Nylen TH, Lyons WB (2002) Valley floor climate observations from the McMurdo dry valleys, Antarctica, 1986–2000. J Geophys Res 107:ACL13.1–ACL13.12

Elberling B, Gregorich EG, Hopkins DW, Sparrow AD, Novis PM, Greenfield LG (2006) Distribution and dynamics of soil organic matter in an Antarctic dry valley. Soil Biol Biochem 38:3095–3106

Ellis-Evans JC (1985) Fungi from maritime Antarctic freshwater environments. Brit Ant Sur Bull 68:37–45

Etienne S, Dupont J (2002) Fungal weathering of basaltic rocks in a cold oceanic environment (Iceland): comparison between experimental and field observations. Earth Surf Proc 27:737–748

Farrell RL, Arenz BE, Duncan SM, Held BW, Jurgens JA, Blanchette RA (2011) Introduced and indigenous fungi of the Ross Island historic huts and pristine areas of Antarctica. Polar Biol 34:1669–1677

Fell JW, Statzell AC (1971) *Sympodiomyces* gen. n., a yeast-like organism from southern marine waters. A van Leeuwenhoek 37:359–367

Fell JW (1974) Yeasts in oceanic regions. In: Jones EBG (ed) Recent advances in amuatic microbiology. Paul Elek Ltd, London, pp 93–124

Fell JW, Hunter IL (1974) *Torulopsis austromarina* sp. nov. A yeast isolated from the Antarctic Ocean. A van Leeuwenhoek 40:297–306

Fell JW, Scorzetti G, Connell LB, Craig SD (2006) Biodiversity of micro-eukaryotes in Antarctic Dry Valley soil with <5 % soil moisture. Soil Biol Biochem 38:3107–3119

Foreman CM, Wolf CF, Priscu JC (2004) Impact of episodic warming events on the physical, chemical and biological relationships of lakes in the McMurdo Dry Valleys, Antarctica. Aquat Geochem 10:239–268

Foreman CM, Sattler B, Mikucki JA, Porazinska DL, Priscu JC (2007) Metabolic activity and diversity of cryoconites in the Taylor Valley, Antarctica. J Geophys Res 112:G04S32

Fountain AG, Lyons WB, Burkins MB, Dana GL, Doran PT, Lewis KJ, McKnight DM, Moorhead DL, Parsons AN, Priscu JC, Wall DH, Wharton RA Jr, Virginia RA (1999) Physical controls on the Taylor Valley ecosystem, Antarctica. Bioscience 49:961–971

Fountain AG, Nylen TH, MacClune KL, Dana GL (2006) Glacier mass balances (1993–2001), Taylor Valley, McMurdo Dry Valleys, Antarctica. J Glaciol 52:451–462

Freckman DHW, Virginia RA (1998) Soil biodiversity and community structure in the McMurdo Dry Valleys, Antarctica. In: Priscu JC (ed) Ecosystems dynamics in a polar desert, vol 72. American Geophysical Union, Washington DC, pp 323–335

Friedmann EI, McKay CP, Nienow J (1987) The cryptoendolithic microbial environment in the Ross Desert of Antarctica: continuous nanoclimate data. Polar Biol 7:273–287

Gomes J, Gomes I, Steiner W (2000) *Thermolabile xylanase* of the Antarctic yeast *Cryptococcus adeliae*: production and properties. Extremophiles 4:227–235

Gostincar C, Grube M, de Hoog S, Zalar P, Gunde-Cimerman N (2010) Extremotolerance in fungi: evolution on the edge. FEMS Microbol Ecol 71:2–11

Goto S, Sugiyama J, Iizuka H (1969) A taxonomic study of Antarctic yeasts. Mycologia 61:748–774

Guerinot ML (1994) Microbial iron transport. Annu Rev Microbiol 48:743–772

Guffogg SP, Thomas-Hall S, Holloway P, Watson K (2004) A novel psychrotolerant member of the hymenomycetous yeasts from Antarctica: *Cryptoccus watticus* sp. nov. Int J Syst Evol Microbiol 54:275–277

Hall BL, Denton GH, Hendy CH (2000) Evidence from Taylor Valley for a grounded ice sheet in the Ross Sea, Antarctica. Geogr Antarct 82A:275–303

Hao Y, S-y Chen, Blanchette RA, Liu X-Z (2010) *Sistotrema brinkmannii*, a psychrotolerant fungus from Antarctic soil. Mycosystema 29:864–868

Held BW, Jurgens JA, Duncan SM, Farrell RL, Blanchette RA (2006) Assessment of fungal diversity and deterioration in a wooden structure at New Harbor, Antarctica. Polar Biol 29:526–531

Hogg ID, Cary SC, Convey P, Newsham KK, O'Donnell AG, Adams BJ, Aislabie J, Fratif F, Stevens MI, Wall DH (2006) Biotic interactions in Antarctic terrestrial ecosystems: are they a factor? Soil Biol Biochem 38:3035–3040

Hopkins DW, Sparrow AD, Elberling B, Gregorich EG, Novis PM, Greenfield LG, Tilston EL (2006a) Carbon, nitrogen and temperature controls on microbial activity in soils from Antarctic dry valleys. Soil Biol Biochem 38:3130–3140

Hopkins DW, Sparrow AD, Novis P, Gregorich EG, Elberling B, Greenfield LG (2006b) Controls on the distribution of productivity and organic resources in Antarctic Dry Valley soils. Proc Royal Soc Biol Sci Ser B 273:2687–2695

Ilić N, Novković M, Guida F, Xhindoli D, Benincasa M, Tossi A, Juretić D (2013) Selective antimicrobial activity and mode of action of adepantins, glycine-rich peptide antibiotics based on anuran antimicrobial peptide sequences. Biochim Biophys Acta 1828:1004–1012

Katayama-Hirayama K, Koike Y, Kaneko H, Kobayashi K, Hirayama K (2003) Removal of nitrogen by Antarctic yeast cells at low temperature. Polar Biosci 16:43–48

Kates M, Baxter RM (1982) Lipid composition of mesophilic and psychrophilic yeasts (*Candida* species) as influenced by environmental temperature. Can J Biochem Phys 40:1213–1218

Kurtzman C, Fell JW (1997) The yeasts: a taxonomic study, 4th edn. Elsevier, New York

Lawley B, Ripley S, Bridge P, Convey P (2004) Molecular analysis of geographic patterns of eukaryotic diversity in Antarctic soils. Appl Environ Microbiol 7:5963–5972

Lazazzera BA (2001) The intracellular function of extracellular peptides. Peptides 22:1519–1527

Loffeld B, Keweloh H (1996) Cis/trans isomerization of unsaturated fatty acids as a possible control mechanism of membrane fluidity in *Pseudomonas putida* P8. Lipids 31:811–815

Lynch RC, King AJ, Farías ME, Sowell P, Vitry C, Schmidt SK (2012) The potential for microbial life in the highest-elevation (>6000 m.a.s.l.) mineral soils of the Atacama region. J Geophys Res 117:G02028

Madronich S, McKenzie RL, Björn LO, Caldwell MM (1998) Changes in biologically active ultraviolet radiation reaching the Earth's surface. J Photochem Photobiol B 46:5–19

Marchant DR, Denton GH (1996) Miocene and pliocene paleoclimate of the Dry Valley region, Antarctica, southern Victoria Land; A geomorphological approach. Mar Micropaleontol 27:253–271

McKnight DM, Niyogi DK, Alger AS, Bomblies A, Conovitz PA, Tate CM (1999) Dry valley streams in Antarctica: ecosystems waiting for water. BioSci 49:985–995

Meyer GH, Morrow MB, Wyss O (1962) Viable micro-organisms in a fifty-year-old yeast preparation in Antarctica. Nature 196:598

Moorhead DL, Doran PT, Fountain AG, Lyons WB, McKnight DM, Priscu JC, Virginia RA, Wall DH (1999) Ecological legacies: Impacts on the ecosystems of the McMurdo Dry Valleys. BioSci 49:1009–1019

Moorhead DL, Barrett JE, Virginia RA, Wall DH, Porazinska DL (2003) Organic matter and soil biota of upland wetlands in Taylor Valley, Antarctica. Polar Biol 26:567–576

Neilands JB (1995) Siderophores: structure and function of microbial iron transport compounds. J Biol Chem 276:26723–26726

Nes WR, McKean ML (1977) Biochemistry of steroids and other isopentenoids. Baltimore, MD, University Park Press

Nylen TH, Fountain AG, Doran PT (2004) Climatology of katabatic winds in the McMurdo Dry Valleys, southern Victoria Land, Antarctica. J Geophys Res 109:D03114

Onofri S, Fenice M, Cicalini AR, Tosi S, Magrino A, Pagano S, Selbmann L, Laure Z, Vishniac HS, Ocampo-Friedmann R, Friedmann EI (2000) Ecology and biology of microfungi from Antarctic rocks and soils. Ital J Zool 67:163–167

Onofri S, Selbmann L, Zucconi L, Pagano S (2004) Antarctic microfungi as models for exobiology. Planet Space Sci 52:229–237

Onofri S, Zucconi L, Tosi S (2007) Continental Antarctic fungi. IHW-Verlag, München

Otvos L Jr (2000) Antibacterial peptides isolated from insects. J Pep Sci 6:497–511

Pavlova K, Grigorova D, Hristozova T, Angelov A (2001) Yeast strains from Livingston Island, Antarctica. Folia Microbiol 46:397–401

Pavlova K, Angelova G, Savova I, Grigorova D, Kupenov L (2002) Studies of Antarctic yeast for β-glucosidase production. World J Microbiol Biotech 18:569–573

Pérez-Miranda S, Cabirol N, George-Téllez R, Zamudio-Rivera LS, Fernández FJ (2007) O-CAS, a fast and universal method for siderophore detection. J Microbiol Meth 70:127–131

Pointing SB, Chan Y, Lacap DC, Lau MC, Jurgens JA, Farrell RL (2009) Highly specialized microbial diversity in hyper-arid polar desert. Proc Natl Acad Sci USA 106:19964–19969

Ratnakumar S, Tunnacliffe A (2006) Intracellular trehalose is neither necessary nor sufficient for desiccation tolerance in yeast. FEMS Yeast Res 6:902–913

Ray MK, Shivajil S, Shyamala Rao N, Bhargaval PM (1989) Yeast strains from the Schirmacher Oasis, Antarctica. Polar Biol 9:305–309

Ray MK, Devi KU, Kumar GS, Shivaji S (1992) Extracellular protease from the Antarctic yeast *Candida humicola*. Appl Environ Microbiol 58:1918–1923

Raymond KN, Carrano CJ (1979) Coordination chemistry and microbial iron transport. Accounts Chem Res 12:183–190

Renshaw JC, Robson GD, Trinci APJ, Wiebe MG, Livens FR, Collison D, Taylor RJ (2002) Fungal siderophores: structures, functions and applications. Mycol Res 106:1123–1142

Resnick RJ, Tomáska L (1994) Stimulation of yeast adenylyl cyclase activity by lysophospholipids and fatty acids: implications for the regulation of Ras/effector function by lipids. J Biol Chem 269:32336–32341

Robinson CH (2001) Cold adaptation in Arctic and Antarctic fungi. New Phytol 151:341–353

Rogers S, Shtarkman Y, Koçer Z, Edgar R, Veerapaneni R, D'Elia T (2013) Ecology of subglacial Lake Vostok (Antarctica), based on metagenomic/metatranscriptomic analyses of accretion ice. Biol 2:629–650

Rozgonyi F, Szabo D, Kocsis B, Ostorházi E, Abbadessa G, Cassone M, Wade JD, Otvos L Jr (2009) The antibacterial effect of a proline-rich antibacterial peptide A3-APO. Cur Med Chem 16:3996–4002

Rudolph ED, Benninghoff WS (1977) Competitive and adaptive responses if invading versus indigenous biotas in Antarctica. In: Llano GA (ed) Adaptations within Antarctic ecosystems. Smithsonian Institution and Gulf Publishing Company, Houston, pp 1211–1225

Sabri A, Bare G, Jacques P, Jabrane A, Ongena M, Heugen JC, Devreese B, Thonart P (2001) Influence of moderate temperatures on myristoyl-CoA metabolism and Acyl-CoA thioesterase activity in the psychrophilic Antarctic yeast *Rhodotorula aurantiaca*. J Biol Chem 276:12691–12696

Schwerdtfeger W (1984) Weather and climate in Antarctica. Developments in atmospheric sciences, vol 15. Elsevier, Amsterdam

Scorzetti G, Petrescu I, Yarrow D, W FJ (2000) *Cryptococcus adeliensis* sp. nov., a xylanase producing basidiomycetous yeast from Antarctica. A van Leeuwenhoek 77:153–157

Selbman L, de Hogg GS, Mazzaglia A, Friedmann EI, Onofri S (2005) Fungi at the edge of life: cryptoendolithic black fungi from Antarctic desert. Stud Mycol 51:1–32

Selbmann L, Isola D, Zucconi L, Onofri S (2011) Resistance to UV-B induced DNA damage in extreme-tolerant cryptoendolithic Antarctic fungi: detection by PCR assays. Fungal Biol 115:937–944

Silver SA, Yall I, Sinclair NA (1977) Molecular basis for the maximum growth temperature of an obligatly psychrophilic yeast, *Leucosporidium stokesii*. J Bacteriol 132:676–680

Sinclair NA, Stokes JL (1965) Obligately psychrophilic yeasts from the Polar Regions. Can J Microbiol 11:259–269

Slessareva JE, Dohlman HG (2006) G protein signaling in yeast: new components, new connections, new compartments. Science 314:1412–1413

Smith JL, Barrett JE, Tusnády G, Rejtö L, Cary SC (2010) Resolving environmental drivers of microbial comunity structure in Antarctic soils. Antarct Sci 22:673–680

Soneda M (1961) Biological results of the Japanese Antarctic research expedition 15. On some yeasts from the Antarctic region. Jap Antarct Res Exp 1956–62 Sci Rep Ser E:1–10

Sterflinger K (2006) Black yeasts and meristematic fungi: ecology, diversity and identification. In: Rosa CA, Peter G (eds) Biodiversity and ecophysiology of yeasts. Springer, Berlin, pp 501–514

Stokes JL (1971) Influence of temperature on the growth and metabolism of yeasts. In: Rose AH, Harrison JS (eds) The yeasts, vol 2., Physiology and biochemistry of yeasts. Academic Press, London, pp 119–134

Sun HJ (2013) Endolithic microbial life in extreme cold climate: snow is required, but perhaps less is more. Biol 2013(2):693–701

Tamppari LK, Anderson RM, Archer PD, Douglas S, Kounaves SP, Mckay CP, Ming DW, Moore Q, Quinn JE, Smith PH, Stroble S, Zent AP (2012) Effects of extreme cold and aridity on soils and habitability: McMurdo Dry Valleys as an analogue for the Mars Phoenix landing site. Antarct Sci 24:211–228

Taylor G (1916) With Scott: the silver lining. Smith, Elder & Co, London

Theodoridis G, Gika HG, Wilson ID (2011) Mass spectrometry-based holistic analytical approaches for metabolite profiling in systems biology studies. Mass Spectrom Rev 30:884–906

Thomas-Hall SR, Watson K (2002) *Cryptococcus nyarrowii* sp. nov., a basidiomycetous yeast from Antarctica. Int J Syst Evol Microbiol 52:1033–1038

Thomas-Hall SR, Watson K, Scorzetti G (2002) *Cryptococcus statzelliae* sp. nov. and three novel strains of *Cryptococcus victoriae*, yeasts isolated from Antarctic soils. Int J Syst Evol Microbiol 52:2303–2308

Thomas-Hall SR (2004) Phylogenetic studies of fungi: part A. Physiological and biochemical analysis of novel yeast species from Antarctica. PhD Thesis, University of New England, Armidale, New South Wales

Thomas-Hall SR, Turchetti B, Buzzini P, Branda E, Boekhout T, Theelen B, Watson K (2010) Cold-adapted yeasts from Antarctica and the Italian Alps-description of three novel species: *Mrakia robertii* sp. nov., *Mrakia blollopis* sp. nov. and *Mrakiella niccombsii* sp. nov. Extremophiles 14:47–59

Tiao G, Lee CK, McDonald IR, Cowan DA, Cary SC (2012) Rapid microbial response to the presence of an ancient relic in the Antarctic Dry Valleys. Nature Com 3:660

Tosi S, Casado B, Gerdol R, Caretta G (2002) Fungi isolated from Antarctic mosses. Polar Biol 25:262–268

Tosi S, Onofri S, Brusoni M, Zucconi L, Vishniac HS (2005) Response of Antarctic soil fungi assemblages to experimental warming and reduction of UV radiation. Polar Biol 28:470–482

Tubaki K (1961) Notes on some fungi and yeasts from Antarctica. Antarct Rec (Tokyo) 11:161–162

Turchetti B, Thomas-Hall SR, Connell LB, Branda E, Buzzini P, Theelen B, Müller WH, Boekhou T (2011) Psychrophilic yeasts from Antarctica and European glaciers. Description of *Glaciozyma* gen. nov., *Glaciozyma martinii* sp. nov and *Glaciozyma watsonii* sp. nov. Extremophiles 15:573–586

Turkiewicz M, Pazgier M, Kalinowska H, Bielecki S (2001) Properties of cold-adapted subtilisine-like proteinase of endemic yeast *Leucosporidium antarcticum*. Med Fac Landbouww Univ Gent 66:329–332

Turkiewicz M, Pazgier M, Kalinowska H, Bielecki S (2003) A cold-adapted extracellular serine proteinase of the yeast *Leucosporidium antarcticum*. Extremophiles 7:435–442

Ugolini FC, Bockheim JG (2008) Antarctic soils and soil formation in a changing environment: a review. Geoderma 144:1–8

Vaca I, Faúndez C, Maza F, Paillavil B, Hernández V, Acosta F, Levicán G, Martínez C, Chávez R (2013) Cultivable psychrotolerant yeasts associated with Antarctic marine sponges. World J Microbiol Biotech 29:183–189

Vaz ABM, Rosa LH, Vieria MLA, de Garcia V, Brandão LR, Teixeira LCRS, Moliné M, Libkind D, van Broock M, Rosa CA (2011) The diversity, extracellular enzymatic activities and photoprotective compounds of yeasts isolated in Antarctica. Brazilian J Microbiol 42:937–947

Vishniac HS (1985) Yeast biomass in Ross Desert (dry valley) soils: evaluation of quantitation methods and sample transport. Antarct J US 19:174–176

Vishniac HS, Klinger JM (1986) Yeasts in the Antarctic deserts. In: Megusar F, Ganter M (eds) Perspectives in microbial ecology. Slovene Society for Microbiology, Ljubljana, pp 46–51

Vishniac HS (1987) Psychrophily and the systematics of yeast-like fungi. In: De Hoog GS, Smith MT, Weijman ACM (eds) Expanding realm of yeast-like fungi, vol 30., Studies in mycology. Elsevier, Amsterdam, pp 389–402

Vishniac HS (2006a) Yeast biodiversity in the Antarctic. In: Rosa CA, Péter G (eds) Biodiversity and ecophysiology of yeasts. Springer, Berlin, pp 419–440

Vishniac HS (2006b) A multivariate analysis of soil yeasts isolated from a latitudinal gradient. Microb Ecol 52:90–103

Wall A, Balks MR, Campbell DI, Paetzold RF (2004) Soil moisture measurement in the Ross Sea region of Antarctica using hydra soil moisture probes. In: Singh B (ed) Supersoil 2004: proceedings of 3rd Australian New Zealand soils conference, University of Sydney, Australia, 5–9 December 2004. Aust and New Zeal Soc Soil Sci, Sydney

Watson K, Arthur H, Shipton WA (1976) *Leucosporidium* yeasts: obligate psychrophiles which alter membrane-lipid and cytochrome composition with temperature. J Gen Microbiol 97:11–18

Weete JD (1974) Fungal lipid biochemistry: distribution and metabolism. New York, NY, Plenum Press

Weinstein RN, Montiel PO, Johnstone K (2000) Influence of growth temperature on lipid and soluble carbohydrate synthesis by fungi isolated from fellfield soil in the maritime Antarctic. Mycologia 92:222–229

Welch AZ, Gibney PA, Botstein D, Koshland DE (2013) TOR and RAS pathways regulate desiccation tolerance in *Saccharomyces cerevisiae*. Mol Biol Cell 24:115–128

Wuczkowski M, Passoth V, Turchetti B, Andersson A-C, Olstorpe M, Laitila A, Theelen B, van Broock MR, Buzzini P, Prillinger H, Sterflinger K, Schnurer J, Boekhout T, Libkind D (2011) Description of *Holtermanniella* gen. nov., including *Holtermanniella takashimae* sp. nov. and four new combinations, and proposal of the order Holtermanniales to accommodate tremellomycetous yeasts of the Holtermannia clade. Int J Syst Evol Microbiol 61:680–689

Xin M-x, Zhou P-j (2007) *Mrakia psychrophila* sp. nov., a new species isolated from Antarctic soil. J Zhejiang Univ Sci B 8:260–265

Yergeau E, Bokhorst S, Huiskes AHL, Boschker HTS, Aerts R, Kowalchuk GA (2007) Size and structure of bacterial, fungal and nematode communities along an Antarctic environmental gradient. FEMS Microbiol Ecolol 59:436–451

Yergeau E, Bokhorst S, Kang S, Zhou J, Greer CW, Aerts R, Kowalchuk GA (2012) Shifts in soil microorganisms in response to warming are consistent across a range of Antarctic environments. ISME J 6:692–702

Yuan Y, Schoenwaslder SM, Salem HH, Jackson SP (1996) The bioactive phospholipid lysophosphatidylcholine induces cellular effects via G-protein-dependent activation of adenylyl cyclase. J Biol Chem 271:27090–27098

Zeglin LH, R L Sinsabaugh RL, Barrett JE, Gooseff MN, Takacs-Vesbach CD (2009) Landscape distribution of microbial activity in the McMurdo Dry Valleys: linked biotic processes, hydrology, and geochemistry in a cold desert ecosystem. Ecosystems 12:562–573

Zhang X, Hua M, Song C, Chi Z (2012) Occurance and diversity of marine yeasts in Antarctica environments. J Ocean Univ China 11:70–74

Zhang Z, Zhu S (2012) Comparative genomics analysis of five families of antimicrobial peptide-like genes in seven ant species. Dev Comp Immunol 38:262–274

Zlatanov M, Pavlova K, Grigorova D (2001) Lipid composition of some yeast strains from Livingston Island, Antarctica. Folia Microbiol 46:402–406

Chapter 5
Cold-Adapted Yeasts in Alpine and Apennine Glaciers

Benedetta Turchetti, Marta Goretti, Pietro Buzzini and Rosa Margesin

Abstract Alpine and Apennine glaciers contain only a small part of the total ice mass existing on Earth and it is well known that these ice masses probably are going to disappear in the next future. The study of biodiversity and ecology of cold-adapted yeasts living in cold habitats of Alpine and Apennine glaciers (supraglacial and subglacial sediments, cryoconite, snow, ice, and melt water) represents a contribution to obtain a better defined picture of the microbial ecology of these peculiar ecosystems. Most frequently isolated yeast species belonged to the genera *Cryptococcus* (mainly of the order Filobasidiales) and *Rhodotorula* (mainly of the class Cystobasidiomycetes). The majority of strains exhibited psychrotolerant characteristics and some of them were able to secrete extracellular hydrolytic cold-active enzymes. From an ecological point of view, some yeasts species could be subjected to an *in situ* selective enrichment in both supraglacial and subglacial sediments and cryoconite. Due to their heterotrophic nature, the possible role of cold-adapted yeasts in biogeochemical cycling through the mineralization of organic matter in cold environments is discussed.

Keywords Alpine glaciers · Apennine glacier · Cryoconites · Subglacial - Supraglacial

B. Turchetti (✉) · M. Goretti · P. Buzzini
Department of Agricultural, Environmental and Food Sciences and Industrial Yeasts Collection DBVPG, University of Perugia, Borgo XX Giugno 74, 06121 Perugia, Italy
e-mail: benedetta.turchetti@unipg.it

R. Margesin
Institute of Microbiology, University of Innsbruck, Technikerstrasse 25, 6020 Innsbruck, Austria

P. Buzzini and R. Margesin (eds.), *Cold-Adapted Yeasts*,
DOI: 10.1007/978-3-642-39681-6_5, © Springer-Verlag Berlin Heidelberg 2014

Contents

5.1 Introduction 100
5.2 Cold-Adapted Yeasts in Alpine and Apennine Supraglacial and Subglacial Sediments and Soils 103
5.2.1 General Properties of Supraglacial and Subglacial Sediments 103
5.2.2 Abundance of Cold-Adapted Yeasts in Supraglacial and Subglacial Sediments of Alps and Apennines 104
5.2.3 Diversity and Ecology of Cold-Adapted Yeasts in Supraglacial and Subglacial Sediments of Alps and Apennines 105
5.2.4 Diversity of Cold-Adapted Yeasts in Alpine Soils 109
5.3 Cold-Adapted Yeasts in Alpine and Apennine Ice Cores, Snow, and Melt Water 109
5.3.1 General Properties of Ice Cores, Snow, and Melt Water 109
5.3.2 Abundance of Cold-Adapted Yeasts in Ice Cores, Snow, and Melt Water of Alps and Apennines 112
5.3.3 Diversity and Ecology of Cold-Adapted Yeasts in Ice Cores, Snow, and Melt Water of Alps and Apennines 112
5.4 Cold-Adapted Yeasts in Alpine Glacier Cryoconite 113
5.4.1 Properties of Alpine Cryoconite 113
5.4.2 Diversity and Ecology of Cold-Adapted Yeasts in Alpine Cryoconite 114
5.5 Contribution of Yeast Populations to the Ecology of Alpine and Apennine Cold Habitats 116
5.6 Conclusions 118
References 119

5.1 Introduction

Arctic and Antarctic regions have been studied since the 1950s as reservoir of psychrophilic and psychrotolerant microorganisms (Abyzov 1993; Skidmore et al. 2000; Deming 2002). Bacteria were firstly considered but eukaryotic populations were also investigated in depth more recently (Buzzini et al. 2012). Some decades later, Alpine cold regions attracted the early interest of a few research groups (Margesin et al. 2002; Margesin 2009; Lazzaro et al. 2012). The limited extension of the Alpine chain and its characteristic geo-morphological conformation confer to this area some peculiar ecological traits in comparison with the complex Arctic and Antarctic surface.

The Alps are one of the great mountain range systems of Europe. Their length is approximately 1,200 km, the highest mountain is Mont Blanc (4,810 m a.s.l.) which is placed on the Italian-French border, whereas the whole Alpine area contains many peaks higher than 4,000 m. Alpine glaciers contain only a small part of the total ice mass existing on Earth. A few authors calculated that they lost about 35 % of their total area from 1850 until the 1970s, and almost 50 % by 2000. A total glacier volume of more than 200 km^3 was estimated around 1850, whereas now only about one-third of this value is left. It was predicted that in the Alps most small glaciers (80 % of total glacial coverage, which represents an important contribution to local water resources) could disappear in the next decades (Cannone et al. 2008). In this context, global warming has a strong impact on Alpine mountains and a dramatic situation for future glacier developments including complete deglaciation of entire European mountain ranges can be depicted (Zemp et al. 2006).

Two different geo-morphological typologies of glaciers can be found in the European Alps: (1) the first is characterized by two or more different accumulation basins generating two or more ice flows usually ending in a common terminal plateau; (2) the second is represented by circular-shape glaciers (occasionally covered totally or in part by debris), which are usually smaller than the former and are representative of the majority of European Alpine glaciers.

The Apennines are a mountain chain consisting of parallel smaller chains extending around 1,200 km along the length of peninsular Italy. Their highest peak, Corno Grande (2,912 m a.s.l.), is partially covered by the Calderone glacier. With the disappearance of the Corral de la Veleta Glacier (Sierra Nevada, Spain) in 1913, the Calderone glacier became the southernmost one of Europe. Although still classified as a glacier, Calderone can be actually labeled as a debris-covered glacieret: a small ice mass of indefinite shape, placed over a protected slope that originated from snow drifting and/or avalanching, and exhibiting no marked ice flow pattern (Pecci et al. 2008). Global warming could particularly impact for the weak geo-morphological conditions of this small glacier, which is subjected to harsh fluctuations of temperature, and can be considered the most reliable witness of global warming (Pecci et al. 2008).

Because of this catastrophic scenario, all the studies on the biodiversity and ecology of microbial populations living in cold ecosystems of the Alps and the Apennines assume an increasing relevance: a temperature increase may determine an abundant loss of macro- and microbial diversity, thus influencing biogeochemical cycles. Additionally, *ex situ* conservation of microbial biodiversity collected in those ecosystems may be particularly interesting and represents a way to preserve species that are going to disappear, for future studies and utilization. An interesting description of the *ex situ* preservation of cold-adapted yeasts in worldwide collections is reported in Chap. 2.

Yeast diversity of Alpine and Apennine habitats was studied only in a few sites (Table 5.1, Fig. 5.1). Within this general framework, the ecology of cold-adapted (psychrophilic and psychrotolerant) yeasts inhabiting various ecological niches, including supraglacial and subglacial sediments, ice, melting water, snow and cryoconite, has been considered. Additionally, some studies described yeast diversity and physiology of polluted cold soils, also in view of their ability for bioremediation purposes. In this Chapter, an overview of yeast biodiversity of Alpine and Apennine cold habitats is presented.

Table 5.1 Sample sites studied for yeast diversity in the Alps and Apennines

Sample site	Geographical position	Altitude (m a.s.l.)	Type of samples	References
Stubaier glacier	Tyrolean Alps, Austria	2,000	Mud and sediments	Bergauer et al. (2005), Margesin et al. (2005b)
Etendard glacier	Grenoble, France	2,900	Mud and sediments	Bergauer et al. (2005)
Brenner pass	Border of Austria and Italy	1,372	Soil and sediments (oil-shale mine and railway area)	Bergauer et al. (2005)
Forni glacier	Ortles-Cevedale group, Italy	2,600–3,670	Supra and subglacial sediments, melt water, ice	Turchetti et al. (2008)
Sforzellina glacier	Ortles-Cevedale group, Italy	2,850–3,050	Supra and subglacial sediments, melt water, ice	Turchetti et al. (2008)
Calderone glacier	Gran Sasso d'Italia group, Italy	630–2,830	Supraglacial and deep-piping sediments, melt water, ice	Branda et al. (2010)
Glacier du Geant	Mont Blanc Massif, Italy	3,430	Snow, air	Turchetti et al. (2013)
Miage glacier	Mont Blanc Massif, Italy	1,720–2,400	Subglacial sediments, melt water, ice	Turchetti et al. (2013)

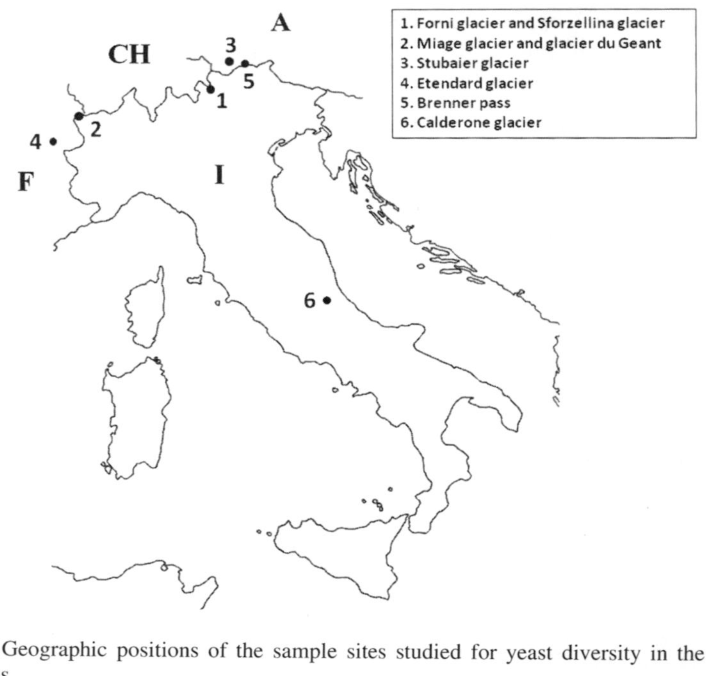

Fig. 5.1 Geographic positions of the sample sites studied for yeast diversity in the Alps and Apennines

5.2 Cold-Adapted Yeasts in Alpine and Apennine Supraglacial and Subglacial Sediments and Soils

5.2.1 General Properties of Supraglacial and Subglacial Sediments

Glacial sediments are generally associated with both glacial surface and ice-bed interface (Hodson et al. 2008). Some glaciers exhibit a continuous debris mantle covering the surface of the lower part of their ablation zone (supraglacial sediments). Such glaciers are common in high-relief mountain environments where mass-wasting processes deliver large volumes of debris to glacial areas. Supraglacial sediments are generally composed of poorly sorted debris sized from both fine grain matrix to large rock boulders. The presence of thick supraglacial sediment accumulation may retard glacier ablation (Benn and Evans 1998). In contrast, subglacial sediments include rocks, sand, silt, and clay derived from glacier grinding up and mixing of rock and soil debris beneath its base. They are poorly sorted sediments exhibiting a little structure and often containing water micro-flow channels in pores within the matrix of the fine-grained material. Supraglacial and subglacial ecosystems may differ vastly in terms of their water activity (A_w), nutrient abundance, redox potential, rock-water contact, pressure, solar irradiation, and pH (Paterson 2002). Among them, the presence of a metabolically active microbial community in subglacial sediments is strictly dependent on the availability of water. Liquid water is vital for any ecosystem and the presence of water coming from rainfall and snowfall and ice melting may represent the source that provides adequate water activity for microbial metabolisms in glacial habitats (Hodson et al. 2008; Anesio et al. 2010; Anesio and Laybourn-Parry 2012). Liquid water can occur at the bed of temperate glaciers (characterized by ice at the pressure and temperature close to melting point, which is typical of the Italian Alps) that additionally show abundant rock debris derived from ice ablation (Anesio and Laybourn-Parry 2012).

Recent studies have shown that supraglacial and subglacial sediments can act as suitable substrate for microbial growth and metabolism; this is apparently confirmed also by the positive correlation found between microbial abundance and concentration of sediments in ice and melt water (Sharp et al. 1999). However, the role of the microbial community in these habitats is still partially unclear.

Microbial communities inhabiting surface and deep habitats of worldwide glaciers have been extensively studied: the presence of algae and cyanobacteria performing photosynthetic metabolism and of aerobic heterotrophic prokaryotic and eukaryotic populations contributing to the cycle of organic matter was observed in supraglacial sediments (Hodson et al. 2008, 2010). Microorganisms in supraglacial sediments could derive mainly from air movement, snowfall, and rainfall, although only a small percentage could be able to survive under such extreme conditions (Simon et al. 2009). In contrast, in subglacial ecosystems, the predominance of chemoautotrophic prokaryotes and heterotrophic prokaryotes and eukaryotes, that even survive under aerobic conditions, has been observed (Sharp et al. 1999; Tranter et al. 2002; Skidmore et al. 2000, 2005; Foght et al. 2004; Priscu et al. 1999; Christner et al. 2001, 2006; Miteva et al. 2004). Some interesting information on forefields of receding Alpine glaciers (i.e., new terrain that becomes exposed due to the glacial retreat) has been recently reported by Lazzaro et al. (2012) and Zumsteg et al. (2012). According to both authors, microorganisms colonizing such forefields represented an exclusive biome, and it was suggested that these microorganisms could be involved in the formation of new soil, with a definite microbial succession from archaea to prokaryotes to eukaryotes. Their ability to interact with minerals and organic compounds through physical and metabolic processes makes them contributors to rock weathering (Zumsteg et al. 2012).

In recent years, the yeast community of unfrozen subglacial sediments of the glaciers in the Ortles-Cevedale complex and the one of the supraglacial sediments of both the Alps and Apennines (Miage and Calderone glaciers, respectively) have been investigated (Turchetti et al. 2008; Branda et al. 2010; Turchetti et al. 2013). Besides, dirt cones from Calderone glacier (Apennines) were also studied: these peculiar sediments originate from the melt out of either endoglacial or subglacial debris bands (Branda et al. 2010). These phenomena are due to differential ablation which produces such upstanding morphologies on the glacial surface. Dirt cones can also derive from deep-piping sediments originating from intra-glacier high water pressure. In both cases, dirt cones are composed predominantly of ice, water, and debris (Benn and Evans 1998).

5.2.2 Abundance of Cold-Adapted Yeasts in Supraglacial and Subglacial Sediments of Alps and Apennines

Overall, the numbers of culturable yeasts in Alpine and Apennine supra- and subglacial sediments were lower by one to several orders of magnitude than those of bacteria and filamentous fungi. In particular, numbers of yeast cells ranging from 10^2 to 10^3 CFU g^{-1} dry weight (DW) were observed in supraglacial sediments and dirt cones of the Calderone glacier, whereas the presence of culturable yeasts in supraglacial sediments of the Ortles-Cevedale complex was less abundant (about a few tens CFU g^{-1} DW), thus two orders of magnitude lower than in the corresponding subglacial habitat (10^2–10^3 CFU g^{-1} DW). Interestingly, similar amounts of total organic carbon, nitrogen, and phosphorous were observed in supraglacial and subglacial sediments of the Ortles-Cevedale complex, while the content of organic compounds in supraglacial sediments and dirt cones of the Calderone glacier was 10-fold (carbon and nitrogen) or 1000-fold (phosphorous) more concentrated (Turchetti et al. 2008; Branda et al. 2010).

5.2.3 *Diversity and Ecology of Cold-Adapted Yeasts in Supraglacial and Subglacial Sediments of Alps and Apennines*

A total of fifty different yeast species were found in Alpine and Apennine sediments: almost all were basidiomycetes. *Cryptococcus* was the prevalent genus with species belonging to both Filobasidiales and Tremellales orders (Table 5.2) (Turchetti et al. 2008; Branda et al. 2010; Turchetti et al. 2013). The prevalence of *Cryptococcus* species is consistent with previous studies carried out in worldwide glacial areas (Butinar et al. 2007; de Garcia et al. 2007; Connell et al. 2008). The high percentage of *Cryptococcus* strains could be justified by their ability to produce an extracellular capsule, which could result in a protection from external extreme conditions (Vishniac 2006). Other adaptation mechanisms, which could confer to this genus a superior ability to thrive and even to grow in such ecosystems (e.g., a wider carbon and nitrogen assimilation pattern), were also recently hypothesized (Connell et al. 2008).

The most frequently isolated species were *Cryptococcus gilvescens*, *Cryptococcus gastricus*, *Cryptococcus victoriae*, *Rhodotorula psychrophenolica* (Fig. 5.2), *Mrakia robertii*, *Cryptococcus terricola,* and *Cryptococcus antarcticus*. Around 15–20 % of the species resulted psychrophilic, whereas the remaining ones were psychrotolerant (Turchetti et al. 2008, 2013; Branda et al. 2010). The large part of the species observed in sediments of Alps and Apennines have been previously regarded as exclusive inhabitants of cold environments (Butinar et al. 2007; de Garcia et al. 2007; Connell et al. 2008). Interestingly, a few cosmopolitan species, namely *Cryptococcus fuscescens*, *Rhodotorula hordea,* and *Holtermanniella takashimae*, never been isolated so far from glacial ecosystems, were found in Alpine and Apennine sediments.

A heterogeneous distribution of the different yeast species was observed in sediments collected in different geographical areas. *C. gilvescens* was the dominant species (58 % of the total isolated strains) in sediments sampled in the Ortles-Cevedale complex, but this species was absent in sediments of the Calderone glacier. On the contrary, *C. gastricus* was the dominant species in sediments of the Calderone glacier (49 %), but it was lacking in sediments sampled in the glaciers of the Ortles-Cevedale complex (Turchetti et al. 2008; Branda et al. 2010). Furthermore, both species were found in sediments of Mont Blanc glaciers (Turchetti et al. 2013). *C. gilvescens* is phenotypically undistinguishable from *C. gastricus*: the analysis of the D1/D2 domain of 26S rRNA and ITS1&2 sequences showed that both species occupy distinct positions within the Filobasidiales lineage, but with a strong phylogenetic relation (they differ by only three nucleotide substitutions in both regions) (Fonseca et al. 2011). Both species appear to be regular inhabitants of cold environments (Fonseca et al. 2011). Their psychrotolerance, their ability to produce extracellular hydrolytic cold-active enzymes and their frequent occurrence in cold habitats could suggest their possible role in geochemical cycling of organic matter in supraglacial and subglacial sediments (Turchetti et al. 2008).

Table 5.2 Yeast species isolated from different glaciers and cold habitats of the Alps and Apennines

Species	Alps											Apennines		
	Ortles Cevedale complex (Turchetti et al. 2008)			**Miage glacier** (Turchetti et al., 2013)			**Glacier du Geant** (Turchetti et al. 2013)		**Stubaier glacier, Etendard glacier, Brenner pass** (Bergauer et al. 2005; Margesin et al. 2005, 2007)			**Calderone glacier** (Branda et al. 2010)		
	S	M	I	S	M	I	SN	A	C	So	I	S	M	I
Ascomycetous yeasts														
Candida santamariae				■								■		
Candida sp.				■			■							
Basidiomycetous yeasts														
Bulleromyces albus				■			■							
Cryptococcus adeliensis				■	■		■					■		
Cryptococcus aerius				■			■					■		
Cryptococcus albidosimilis					■		■					■		
Cryptococcus albidus				■			■							
Cryptococcus antarcticus				■	■									
Cryptococcus carnescens				■										
Cryptococcus chernovii							■							
Cryptococcus diffluens							■							
Cryptococcus dimennae													■	
Cryptococcus friedmanii					■									
Cryptococcus fuscescens				■										
Cryptococcus gastricus				■	■		■					■	■	■
Cryptococcus gilvescens	■	■	■	■	■	■	■							
Cryptococcus laurentii							■							
Cryptococcus macerans					■		■						■	
Cryptococcus magnus							■							
Cryptococcus oeirensis							■					■	■	
Cryptococcus saitoi	■		■									■		
Cryptococcus stepposus							■						■	
Cryptococcus tephrensis				■			■					■		
Cryptococcus terreus				■						■				
Cryptococcus terricola	■			■						■				
Cryptococcus victoriae				■	■	■	■					■	■	■
Cryptococcus wieringae							■					■		
Cryptococcus sp.	■		■	■	■	■	■					■	■	■
Cystofilobasidium capitatum					■								■	
Cystofilobasidium infirmominiatum							■							
Cystofilobasidium macerans												■	■	

(continued)

Table 5.2 (continued)

Dioszegia crocea							■					■		
Dioszegia fristingensis							■							
Dioszegia hungarica					■		■							
Dioszegia sp.					■	■	■							
Erythrobasidium hasegawianum												■		
Glaciozyma martinii												■		
Glaciozyma watsonii	■													
Guehomyces pullulans				■	■							■	■	
Holtermanniella festucosa				■			■						■	
Holtermanniella takashimae				■	■		■					■		
Holtermanniella wattica				■		■						■		
Leucosporidiella creatinivora	■			■						■				
Leucosporidium sp.				■										
Mastigobasidium intermedium										■			■	
Mrakia blollopis													■	
Mrakia gelida				■	■	■						■	■	
Mrakia robertii	■	■		■	■									
Mrakia psychrophila					■									
Mrakiella aquatica					■								■	
Mrakiella cryoconiti					■		■		■			■		
Mrakiella niccombsii					■									
Rhodosporidium lusitaniae										■				
Rhodotorula aurantiaca								■						
Rhodotorula bacarum	■													
Rhodotorula colostri					■								■	
Rhodotorula glacialis	■	■		■	■	■			■	■				
Rhodotorula hordea				■										
Rhodotorula ingeniosa										■				
Rhodotorula laryngis	■			■			■					■		
Rhodotorula psychrophenolica	■	■		■			■		■	■		■		
Rhodotorula psychrophila										■				
Rhodotorula sp.				■										
Sporobolomyces roseus				■			■			■	■	■		
Trichosporon dulcitum										■				
Udeniomyces pannonicus							■							
Yeast-like organisms														
Aureobasidium pullulans				■			■	■				■	■	
Aureobasidium sp.				■										
Exophiala dermatitidis												■	■	

S sediments, *M* melt water, *I* ice cores, *SN* snow, *A* air, *C* cryoconite, *So* soil

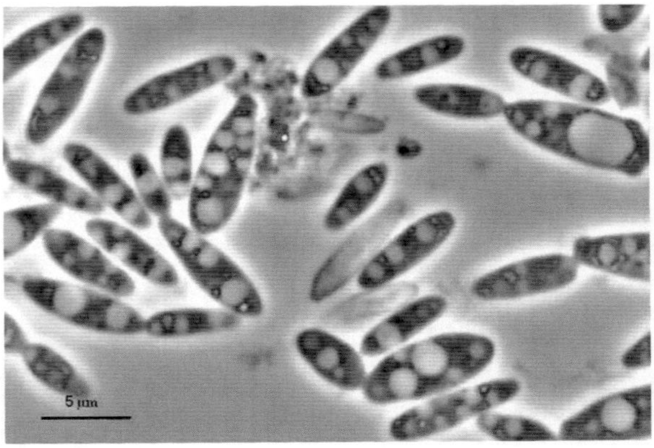

Fig. 5.2 Cells of *Rhodotorula psychrophenolica* DBVPG 4792 after 2 weeks of incubation at 10 °C on malt extract agar (*MEA*), showing budding cells and pseudohyphae. *Bar*, 5 µm

The species *C. victoriae* was isolated in the past from cold Arctic and Antarctic soils, but also from other temperate habitats (Herzberg et al. 2002; Gildemacher et al. 2004, 2006). Neighbor-joining analysis of D1/D2 sequences placed this species into the order Tremellales, particularly into the *victoriae* clade together with the species *Cryptococcus carnescens*, *Cryptococcus heimaeyensis,* and *Cryptococcus tephrensis* (Fonseca et al. 2011), which have been all frequently isolated from cold ecosystems.

A total of twenty-eight percentage of all strains isolated from Alpine and A-pennine glaciers were found to belong to novel species: *R. psychrophenolica* and *Rhodotorula glacialis* (Margesin et al. 2007), *Mrakiella cryoconiti* (Margesin and Fell 2008), *M. robertii*, *Mrakia blollopis* and *Mrakiella niccombsii* (Thomas-Hall et al. 2010), *Glaciozyma martinii* and *Glaciozyma watsonii* (Fig. 5.3) (Turchetti et al. 2011).

Interestingly, the species *M. niccombsii* was recently found in supraglacial sediments of the Miage glacier (Turchetti et al. 2013). Since this species has so far been considered endemic of Antarctic soils (Vishniac 1985; Vishniac and

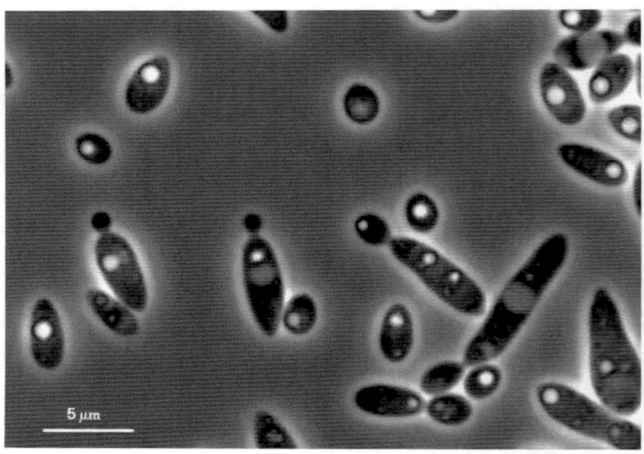

Fig. 5.3 Optical microscopy image of *Glaciozyma watsonii* DBVPG 4760 after 1 week of incubation at 10 °C on MEA, showing budding cells. *Bar*, 5 µm

Kurtzman 1992; Vishniac and Onofri 2002; Arenz et al. 2006; Vishniac 2006; Thomas-Hall et al. 2010), this evidence could lead to reconsider its status as endemic species.

Most of the yeasts isolated from supraglacial and subglacial sediments of Alpine and Apennine glaciers were common inhabitants of worldwide cold areas, such as Antarctica and the Arctic (Butinar et al. 2007; de Garcia et al. 2007; Connell et al. 2008; Libkind et al. 2009). Atmospheric circulation, via wind and dust particles, rainfall, and snowfall vectors, could be considered the most plausible way of their global dissemination (Xiang et al. 2009). Of course, the sole dissemination cannot be considered sufficient to justify the differential diversity of cold-adapted yeasts observed in geographically distinct cold habitats, because only some habitats could support their *in situ* survival and growth.

5.2.4 Diversity of Cold-Adapted Yeasts in Alpine Soils

Cold-adapted yeasts strains isolated from contaminated soils in the Kitzbühler Alps, Stubaier Alps, and at the Brenner pass were studied by Bergauer et al. (2005) and Margesin et al. (2005b). Eleven basidiomycetous species were identified; dominant genera were *Cryptococcus* and *Rhodotorula*. Some of them, namely *R. glacialis*, *R. psychrophenolica*, *Cryptococcus terreus*, and *C. terricola*, have been frequently found in Alpine environments. On the contrary, five other species were isolated from cold habitats for the first time, namely *Leucosporidiella creatinivora* (formerly *Rhodotorula creatinivora*), *Rhodosporidium lusitaniae*, *Rhodotorula ingeniosa*, *Rhodotorula psychrophila*, and *Trichosporon dulcitum* (Table 5.2). All the strains isolated from Alpine cold soils showed the ability to degrade phenolic compounds at low temperature (for more information, see Chap. 21).

5.3 Cold-Adapted Yeasts in Alpine and Apennine Ice Cores, Snow, and Melt Water

5.3.1 General Properties of Ice Cores, Snow, and Melt Water

Ice constituting glaciers is a substrate which can act as long-term, chronological repository of viable microorganisms. Due to harsh conditions (high pressure, darkness, acidity and low A_w), an englacial environment can be considered an extreme habitat (Price 2000, 2007). Some peculiar situations occurring within englacial ice can apparently simplify microbial life in this environment. For example, the part of englacial ice lying at the pressure of the melting point can determine the increase in the temperature and the A_w, thus increasing the

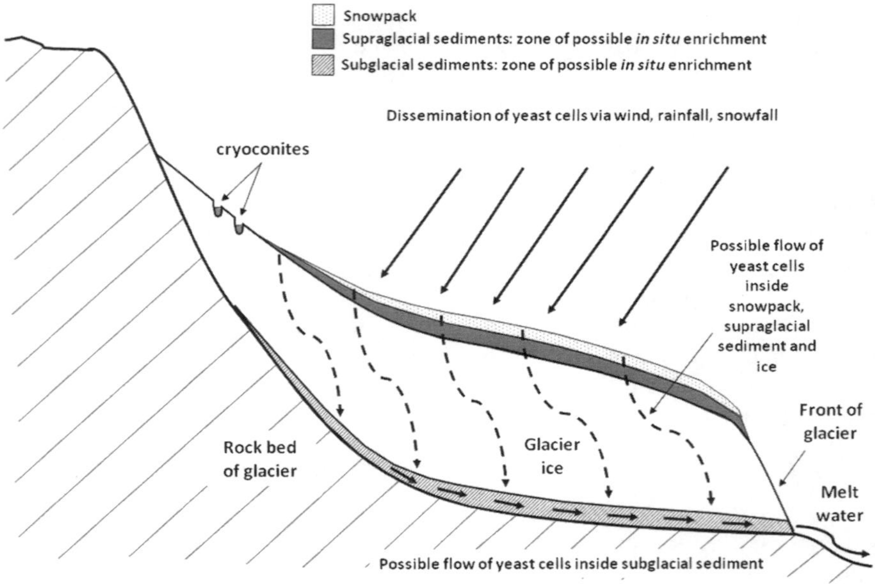

Fig. 5.4 Representation of a potential ecological model for the distribution of cold-adapted yeasts inside the ecosystem of a temperate glacier in the Alps and Apennines

possibility of microbial life and metabolic activity (Price 2007). Besides, the phenomenon labeled as "hydratation distance" at the surface of solid impurity or hydrophilic mineral substrates could determine the presence of an unfrozen layer of temperature-dependent thickness around particles, providing nutrients and water for microorganisms (Price 2007). In temperate glaciers, englacial habitats include crevasses and water channels (Fig. 5.4), which can act themselves as growing substrates for microbial communities and at the same time convey water, nutrients, atmospheric gases, and microbiota into the glacier (Price 2007).

Microbial biodiversity of worldwide icy habitats has attracted the interest of an increasing number of biologists. Early studies, conducted by Abyzov et al. (1982, 2001), found diverse microorganisms including cold-adapted yeasts (belonging to the species *Cryptococcus albidus* and *Rhodotorula mucilaginosa*) in the deep ice cores drilled from Lake Vostok, Antarctica. Likewise, Skidmore et al. (2000) demonstrated the presence of aerobic chemoheterotrophs, anaerobic nitrate- and sulfate-reducing bacteria and methanogens in the debris-rich basal ice layers and the surface of high Arctic glaciers. Later, englacial ecosystems were also studied in glacier ice cores extracted from different geographic areas: the presence of diverse bacterial communities including representatives of Proteobacteria, Bacteroidetes (formerly Cytophaga–Flavobacterium–Bacteroides) and Gram-positive bacteria was observed (Christner et al. 2003b). The increasing awareness that a large number of viable microorganisms, including yeasts, are entombed within glacial ice enables the hypothesis that englacial microbial populations could play an important role in nutrient cycling inside glaciers and ice sheets. However, despite the presence of viable microorganisms in ice cores (Castello and Rogers 2005; Price 2007), their *in situ* activity remains to be fully elucidated.

Snowpack covering ice surface is another interesting niche characteristic of glacial environments that could act as reservoir of different types of microorganisms. Snow accumulates on the surface of icy surface and contains an extremely low concentration of organic matter. The ecological role of snow is significant since it provides an important inoculation of microbial cells, nutrient, and water that enriches the glacial ecosystems beneath the snow superficial layer (Fig. 5.4). Microorganisms occurring on surface snow are often exposed to light, nutrients, and temperatures of around 0 °C in summer, which facilitates the growth of phototrophic cyanobacteria and algae and can result in the presence of colorful snow coverage (Stibal et al. 2007). Hodson (2006) found a high concentration of inorganic nitrogen and phosphate in such habitat due to the phototrophic activity of algae and cyanobacteria. By contrast, microbial communities in the deep snow are exposed to abrupt temperature drop and low light intensity, which reduces the growth rate of light-dependent microorganisms (Xiang et al. 2009). Besides, complex bacterial consortia were found in snowpack covering the surface of a few glaciers located in the Alps, Andes, and Svalbard archipelago (De Angelis and Gaudichet 1991; Amato et al. 2007).

Sun irradiation provokes snow and ice melting due to an increase in temperature, albedo effect, and strong pressure on the glacier bed. The different origin of melt water makes it difficult to define a general idea about the characteristics of this peculiar substrate (Paterson 2002):

1. streaming of melt water on the surface of the glacier derived from ice and snow ablation during summer time;
2. percolation of water from the surface into ice;
3. ice melting induced by pressure change and streaming in subglacial sediments or creating channels within ice;
4. water trapped inside pockets originated from ice formation;
5. lakes located close to the margins of the glacier, or on the glacier, originating from the melt water above described.

The abundance of microorganisms in melt water of worldwide temperate glaciers has been sporadically described (Sharp et al. 1999; Butinar et al. 2007; de Garcia et al. 2007) because more emphasis has been devoted to sediments and ice. The currently accepted hypothesis considers melt water as an important connection between different niches within the same glacier (or even as a natural link between glacial habitat and glacier forefield). Variable microbial abundance has been detected in melt water collected in glaciers of the Svalbard archipelago: bacterial numbers were about 10^5 cells ml^{-1} (Sharp et al. 1999), whereas yeast numbers ranged from 10 to 10^3 CFU ml^{-1} (Butinar et al. 2007).

5.3.2 Abundance of Cold-Adapted Yeasts in Ice Cores, Snow, and Melt Water of Alps and Apennines

According to recent studies, ice, snow, and melt water can be considered as extremely oligotrophic habitats in comparison with both supraglacial and subglacial sediments. This could apparently justify the lower yeast cells abundance ($10–10^2$ CFU L^{-1}), if compared with the corresponding sediments (Turchetti et al. 2008; Branda et al. 2010).

These results are consistent with the data observed in melt water sampled from Patagonian glaciers (de Garcia et al. 2007; Libkind et al. 2009). Surprisingly, ice cores and snow sampled from Alpine and Apennine glaciers showed a yeast abundance that was 10-fold to 100-fold lower than that observed in corresponding sediments, in contrast with the results reported for Svalbard ice cores ($10–10^3$ CFU ml^{-1}) (Butinar et al. 2007).

5.3.3 Diversity and Ecology of Cold-Adapted Yeasts in Ice Cores, Snow, and Melt Water of Alps and Apennines

The yeast diversity observed in ice cores collected from Alpine and Apennine glaciers partially overlapped with that observed in the corresponding supraglacial and subglacial sediments, i.e., *C. gilvescens* and *C. gastricus*, respectively, were the dominant species, although their relative abundance was 10-fold to 100-fold lower (Turchetti et al. 2008; Branda et al. 2010).

In contrast, snowpack covering the surface of the glacier du Geant (Mont Blanc) showed a wider yeast biodiversity spectrum: 38 different species were identified (Turchetti et al. 2013). *C. victoriae* (28 % of all isolated strains), *R. psychrophenolica* (24 %), the yeast-like species *Aureobasidium pullulans* (7 %), and *C. gilvescens* (6 %) were the most frequently found species. Some of them are regularly observed in worldwide cold environments, while some other are regarded as ubiquitous species (e.g., *A. pullulans*, *Cryptococcus chernovii*, *Cryptococcus diffluens*, *and R. laryngis*) (Turchetti et al. 2013). Their deposition on snow surface by air current, snowfalls, and rainfalls and/or their possible variable ability to survive (or even to grow) under these conditions could be the basis justifying their differential occurrence.

A large number of yeast species, already described as regular inhabitants of both supraglacial and sub-glacial sediments, were found in melt water sampled in Alpine and Apennine glaciers. The species *R. glacialis* and *C. gilvescens* represented over 50 and 20 %, respectively, of the yeasts isolated from meltwater collected in the Ortles-Cevedale complex (Turchetti et al. 2008). In contrast, *Cystofilobasidium macerans* and *Cystofilobasidium capitatum*, which were also found to be ubiquitous in water streams or glacial lakes (Sampaio 2011), were the predominant species (19 % and 17 %, respectively) in meltwater collected from the Apennine Calderone glacier (Branda et al. 2010).

5.4 Cold-Adapted Yeasts in Alpine Glacier Cryoconite

The term "cryoconite" (ice dust) was used for the first time by Nordenskjöld during his 1870 Greenland expedition and refers to small (0.1–3 mm in diameter) dark-colored wind-borne particles at the bottom of cryoconite holes. These particles consist of inorganic (e.g., mineral components of soils and sediments) and organic (e.g., litter, pollen, Bacteria, Eukarya) materials. Cryoconite holes are water-filled depressions (up to 50 cm deep and from a few centimeters to >1 m in diameter) on the glacial surface and form when the dark particles deposited on the glacier surface are warmed by solar radiation (black body effect) and melt into the underlying ice because of the lower albedo of cryoconite compared to the surrounding ice (Wharton et al. 1985; Takeuchi et al. 2001; Cameron et al. 2012). The reduced surface albedo may play an important role in the acceleration of glacier melting. Cryoconite holes are found worldwide in high-latitude and high-alpine glaciers (MacDonell and Fitzsimons 2008) as well in glaciers of temperate areas; they can cover 0.1–10 % of the glacier surface (Anesio et al. 2009). Most cryoconite holes are open to the environment, while those in Antarctic Dry Valley glaciers are covered by an ice lid and thus isolated from the atmosphere (Miteva 2008).

5.4.1 Properties of Alpine Cryoconite

The temperature in Alpine cryoconite ranges from about ≤ 0 °C to about +3 °C in summer (July to September), as it is permanently saturated with cold meltwater from ablating snow patches. During the residual time of the year, cryoconite is snow-covered and permanently frozen. Freezing also can occur every night during summer (Margesin et al. 2002). Cryoconite holes could serve as biological refuges during extended periods of subzero temperatures; members of cryoconite communities that survive these conditions might ensure the reseeding of the surrounding environment during warmer periods (Christner et al. 2003a).

Cryoconite contains substantial amounts of nutrients, which supports microbial growth and activity. Three Alpine cryoconite samples collected from an Austrian glacier (Margesin et al. 2002) were characterized by a pH value of 5, a temperature of 1.5–3 °C at the time of sampling (September), and a water content of 50 %, and contained 0.8–1.8 % (related to cryoconite dry mass) total carbon and 0.02–0.09 % total nitrogen. The C:N ratio ranged from 40:1 to 20:1. The $CaCO_3$ content was in the range of 1.9–2.1 % (corresponding to 0.2–0.4 % inorganic carbon). Alpine cryoconites of an Austrian glacier contained high amounts of anthropogenic radionuclides from nuclear weapons and Chernobyl fallouts (Tieber et al. 2009). In comparison, Arctic cryoconite contained 1.1–1.9 % organic C and had a water content of 67–79 % (Singh and Singh 2012). Most of the organic carbon in Arctic cryoconite holes originates from allochthonous rather than indigenous primary production (Stibal et al. 2008).

In some areas (e.g., glacier-skiing areas), the composition of cryoconite is strongly influenced by human presence. Studies on the influence of human impact on microbial community composition of Alpine cryoconite samples collected from six Austrian glaciers indicated the introduction of allochthonous bacterial species, as shown by the presence of isolates with higher maximum growth temperatures on sites with strong human impact compared to sites without or with low human impact (Lee et al. 2011). Such data have not yet been gathered for yeast population, however, it is likely that a similar result can be obtained.

5.4.2 Diversity and Ecology of Cold-Adapted Yeasts in Alpine Cryoconite

Since cryoconite holes contain abundant populations of actively living organisms, Wharton et al. (1981, 1985) pointed already 30 years ago to the important role of cryoconite holes in glacier ecosystems. During the last decade, cryoconite holes have been recognized as an important microbial ecosystem (Hodson et al. 2008; Anesio and Laybourn-Parry 2012). Every cryoconite hole is unique and therefore may support a novel and discrete ecosystem (Christner et al. 2003a). These supraglacial mini-ecosystems are considered "hot spots" of microbial metabolic activity, which explains the strong interest in their biogeochemical processes, chemical gradients, and nutrient cycling (Margesin and Miteva 2011; Singh and Singh 2012). Diversity studies of microbial communities in cryoconite holes demonstrated the presence of photosynthetic cyanobacteria and algae, heterotrophic bacteria, yeasts and filamentous fungi, metazoa and viruses (Wharton et al. 1985; Christner et al. 2003a; Anesio et al. 2009; Stibal et al. 2012). Archaeal communities have so far only been detected in Antarctic samples (Cameron et al. 2012).

Members of individual cryoconite communities were often found to be specific to individual locations, suggesting that they may be seeded primarily via localized wind transportation and/or during glacial melt from adjacent surrounding environments (Christner et al. 2003a; Cameron et al. 2012). Long-distance wind transport may provide a mechanism for immigration into cryoconite holes. Cryoconite microorganisms may also play a role in the colonizing of newly exposed areas after glacier retreats (Wharton et al. 1985).

There is little information on the abundance and role of yeasts in cryoconite (Margesin et al. 2002; Singh and Singh 2012; Uetake et al. 2012). Three Alpine cryoconite samples collected from an Austrian glacier contained predominantly culturable aerobic heterotrophic bacteria, but culturable yeasts were also present and were identified as representatives of the genera *Cryptococcus*, *Candida,* and *Rhodotorula* (Margesin et al. 2002). *Cryptococcus gilvescens* dominated among *Cryptococcus* species (Margesin, unpublished data). The description of novel cold-adapted bacterial and yeast (*Rhodotorula glacialis*, *R. psychrophenolica*, and *Mrakiella cryoconiti*) taxa from Alpine glacier cryoconite (Margesin et al. 2007; Margesin and Fell 2008) indicates the typical community composition of cryoconite (Margesin and Miteva 2011). The difference between the abundance of bacteria and yeasts was influenced by temperature: significantly higher numbers of yeasts were able to grow at 2 °C (2×10^3 to 2×10^5 CFU g^{-1} cryoconite dry mass) than at 20 °C (5×10^2 to 3×10^4 CFU g^{-1} dry mass) in all three cryoconite samples. Yeast populations were about 5–10-fold lower than bacterial numbers at 2 °C, but more than 100-fold lower at 20 °C (Margesin et al. 2002).

Many cryoconite microorganisms show remarkably good cold adaptation with regard to growth characteristics and metabolic activities (Margesin and Miteva 2011). The maximum temperatures for growth of yeast strains from Alpine cryoconite is usually 15 °C or 20 °C, while only few can grow at 25 °C (Margesin et al. 2002, 2007; Margesin and Fell 2008). These data and further studies led to the assumption that cold-adapted yeasts have a more restricted growth temperature range than cold-adapted bacteria (Margesin 2009).

The ability of cold-adapted yeasts to grow even below the freezing point of water (Panikov and Sizova 2007) indicates their capability to degrade organic compounds at low temperatures (Buzzini et al. 2012). Cryoconite microorganisms convert organic matter into dark-colored humic substances, which results in increased light absorbance (Takeuchi et al. 2001). Yeasts from Alpine glacier cryoconite utilize various organic compounds (cellulose, starch, tributyrin, polygalacturonic acid, diesel oil, and phenol) as the sole carbon and energy source (Margesin et al. 2002, 2005a; Bergauer et al. 2005).

Substrate utilization was more influenced by the cultivation temperature than by the composition of the substrate. All of 14 yeast strains investigated utilized the tested substrates when cultured at 5, 10, or 15 °C. The relative substrate utilization was higher at 0 °C than at temperatures above 15 °C, at which significantly reduced substrate utilization was noticed. The optimum temperature for biomass production and substrate hydrolysis is usually lower than the optimum growth temperature of cold-adapted bacteria and yeasts (Feller et al. 1996; Margesin 2009). This could also be demonstrated with a cold-active alkaline pectinase (pectate lyase) produced by *Mrakia frigida* from Alpine glacier cryoconite (Margesin et al. 2005a), later identified as a representative of a novel genus (*Mrakiella cryoconiti*; Margesin and Fell 2008).

5.5 Contribution of Yeast Populations to the Ecology of Alpine and Apennine Cold Habitats

Cold-adapted microorganisms colonizing glacier ecosystems are subjected to extreme low temperatures and A_w, and often to nutrient limitation. Such organisms are adapted to these extreme environments through some modifications of their metabolic processes, such as production of cold-active enzymes, anti-freezing proteins, and exopolymeric substances that provide cell protection against the potentially damaging effects of low temperatures. Besides, they maintain the fluidity of cell membranes until sub-zero temperatures by an increase in the proportion of unsaturated fatty acids and a decrease in the average fatty acid chain length (Siddiqui and Cavicchioli 2006; Buzzini et al. 2012). An exhaustive overview of the strategies adopted by cold-adapted yeasts to overcome the negative effect of cold is reported in Chaps. 9, 10, 11, 12, 13, 14, 15.

In glacier ecosystems, the nutrient input (especially in terms of carbon and nitrogen) derives principally from wind and alluvial deposition of allochtonous particulate material, which can release both inorganic and organic compounds, the last due to decomposing processes of organic materials catalyzed by microbial biomass (Anesio and Laybourn-Parry 2012). Both supraglacial and subglacial sediments are actively colonized by chemoautotrophic and heterotrophic microorgansims including cold-adapted yeasts that are able to make use of recalcitrant substrates more efficiently than bacteria. There is additional evidence that organic compounds in glacial sediments can be easily hydrolyzed aerobically bringing subglacial sediment to anoxia and guiding to the development of methanogens (Wadham et al. 2008). Microbial communities sharing glacial habitats could be responsible of some biogeochemical transformations, with some important implications at both local and global scales (Anesio and Laybourn-Parry 2012).

Chemical analysis of sediments and cryoconite (and to a lesser extent of ice, snow, and melt water) collected from Alpine and Apennine glaciers demonstrated the presence of organic carbon, nitrogen, and phosphorus in amounts that support survival and even growth of psychrophilic and psychrotolerant heterotrophic microorganisms including yeasts (Margesin et al. 2002; Turchetti et al. 2008; Branda et al. 2010). The majority of yeasts isolated from Alpine and Apennine glaciers exhibited the ability to secrete cold-active hydrolytic enzymes (Margesin et al. 2005a; Turchetti et al. 2008; Branda et al. 2010). These enzymes could be involved in the hydrolysis of organic macromolecules naturally present in these harsh environments (Anesio and Laybourn-Parry 2012).

Considering the above general findings and under the light of the studies on cold-adapted yeasts in Alpine and Apennine ecosystems (Sharp et al. 1999; Turchetti et al. 2008; Branda et al. 2010), it is possible to postulate that glacial habitats constitute an important reservoir of metabolically active heterotrophic prokaryotic and eukaryotic microbial communities. Although some important differences on abundance and biodiversity of yeast populations between eutrophic (i.e., sediments and cryoconite) and oligotrophic (ice, snow, melt water) habitats have been observed (Turchetti et al. 2008; Branda et al. 2010), it can be postulated that cold-adapted yeasts are able to survive (and even to grow) under such extreme conditions. In this framework, studies reporting their ability to perform some important bio-geochemical activities (such as organic matter decomposition and nutrient cycling) at low temperature could be considered an important milestone in the knowledge of the *in situ* ecological role of cold-adapted yeasts, with special emphasis on more eutrophic habitats, specifically supraglacial and subglacial sediments, and cryoconites (Turchetti et al. 2008; Branda et al. 2010; Buzzini et al. 2012).

Accordingly, the following models of the ecological significance of cold-adapted yeasts in Alpine and Apennine ecosystems could be postulated:

1. A few species (i.e., those occurring at a higher frequency, namely *C. gilvescens* and *C. gastricus*) could have probably a superior fitness with regard to the eutrophic conditions present in sediment and cryoconite ecosystems, which allows them not only to survive, but even to grow under such extreme conditions. Accordingly, an *in situ* selective enrichment of yeast populations could be postulated for these species. This hypothesis could be apparently supported by the results reported by Turchetti et al. (2008), who demonstrated that *C. gilvescens* is able to exponentially grow under laboratory-simulated *in situ* conditions.
2. Due to their substantial oligotrophy, ice, snow, and melt water associated with glacier ecosystems could result lesser suitable to support active growth of cold-adapted yeasts. This could suggest that only limited (if any) *in situ* enrichment could occur in these niches. In particular, ice could represent only a reservoir of surviving cells. Price (2000, 2007) demonstrated that biological activity may occur inside ice, but microorganisms inhabiting this environment are characterized by very low metabolic rates because most of the nutrients and energy that become available could be likely expected to serve just for the repair of cellular and molecular damages. Accordingly, it is possible to deduce that even if this strategy could permit cold-adapted yeasts to remain viable for months and years (thousands or even millions of years, considering Antarctic ice) (Abyzov et al. 1982, 2001), these microorganisms could play only a minor significant role inside the ice mass of Alpine and Apennine glaciers.
3. In this context, melt water could only act by transferring viable yeast cells from supraglacial to subglacial sediments. In fact, temperate glaciers of the Alps and Apennines exhibit crevassing and water channels (Fig. 5.4), which enhance ice melting and water movement from supraglacial to subglacial sides (passing by englacial ecosystems). Accordingly, yeast cells could be supposed to move from superficial snow, supraglacial sediments, water, and englacial ice to reach subglacial sediments.

Considering all above hypotheses, a possible pathway of cold-adapted yeasts inside the glacier ecosystem could be postulated (Fig. 5.4):

1. Yeasts may initially fall on supraglacial debris, cryoconite, snow, and ice surface by wind, rainfall, and snowfall dissemination. The combination of these components could determine the amount and composition of microbial species

in the glacier surface (Xiang et al. 2009). After this first step, an *in situ* post-depositional selection could occur, thus resulting in varying structures of the yeast community. Cryoconite and supraglacial sediments with sufficient concentration of organic compounds may support the selective *in situ* enrichment of yeasts in favor of the species more adaptable to the cold habitat and able to use complex organic matter (e.g., *C. gastricus* in supraglacial sediment of the Calderone glacier) (Branda et al. 2010). Other species exhibiting a lower fitness for such habitat could remain in very low cell concentration (or even not quantifiable by using classical, i.e., culture-dependent, methods of analysis).

2. Temperate glaciers are characterized by hydrological transfers from surface to glacier bed due to firn burial and glacier ice flow, but also due to crevassing enhancing flow of melt water, nutrients, and organisms inside the ice body from surface to the glacier bed, where the presence of pressure-melting-point ice at the glacier bed provides not only a basis for nutrient achievement following rock–water contact (Tranter 2005), but also an increased likelihood of crevassing due to enhanced basal sliding and subglacial till deformation (Paterson 1994). All these effects could result in the coupling of supraglacial and subglacial ecosystems allowing yeast cells to percolate from snow and supraglacial sediment coverage through ice, arriving to unfrozen subglacial sediments beneath the glacier bed or remaining within melt water, which could be regarded (together with ice) only a transition step (Fig. 5.4).
3. In subglacial sediments, a second *in situ* enrichment of yeast populations could be supposed (e.g., *C. gilvescens* in subglacial sediment in the Ortles-Cevedale complex) (Turchetti et al. 2008). Yeast cells growing in such an ecosystem could also be washed-out by the melt water flow running off from the glacier bed (Fig. 5.4).

5.6 Conclusions

Investigations on the ecology of glacial habitats have revealed that microbial community is active in biogeochemical transformations such as carbon fixation and respiration, iron cycling and methanogenesis (Anesio and Laybourn-Parry 2012). Cold-adapted yeasts can be considered as an active part of this community, because they have developed a complex suite of physiological and metabolic adaptations, which enhance their possibility to survive and, in some cases, to grow in glacial habitats. Therefore, they play a fundamental role in cold ecosystems, especially in relation to the hydrolysis of complex organic macromolecules connected with the mineralization of organic matter. Regarding their possible role in weathering of surfaces in forefields, their effective *in situ* quantitative contribution is a still open question, although a recent study hypothesized that cold-adapted yeasts may be the active part of this biogeochemical process (Brunner et al. 2011).

References

Abyzov SS (1993) Microorganisms in the Antarctic ice. In: Friedmann (ed) Antarctic microbiology. Wiley, New York, pp 265–295

Abyzov SS, Bobin NE, Koudryashov BB (1982) Quantitative assessment of microorganisms in microbiological studies of Antarctic glaciers. Biol Bull Acad Sci USSR 9:558–564

Abyzov SS, Mitskevich IN, Poglazova MN, Barkov NI, Lipenkov VY, Bobin NE, Koudryashov BB, Pashkevich VM, Ivanov MV (2001) Microflora of the basal strata at Antarctic ice core above the Vostok Lake. Adv Space Sci 28:701–706

Amato P, Hennebelle R, Magand O, Sancelme M, Delort AM, Barbante C, Boutron C, Ferrari C (2007) Bacterial characterisation of the snow cover at Spitsbergen, Svalbard. FEMS Microbiol Ecol 59:255–264

Anesio AM, Laybourn-Parry J (2012) Glaciers and ice sheets as a biome. Trends Ecol Evol 27:219–225

Anesio AM, Hodson AJ, Fritz A, Psenner R, Sattler B (2009) High microbial activity on glaciers: importance to the global carbon cycle. Global Change Biol 15:955–960

Anesio AM, Sattler B, Foreman C, Telling J, Hodson A, Trantner M, Psenner R (2010) Carbon fluxes through bacterial communities on glacier surface. Ann Glaciol 51:32–40

Arenz BE, Held BW, Jurgens JA, Farrell RL, Blanchette RA (2006) Fungal diversity in soils and historic wood from the Ross Sea Region of Antarctica. Soil Biol Biochem 38:3057–3064

Benn DI, Evans DJA (1998) Glacier and glaciation. Arnold, London, pp 212–239

Bergauer P, Fonteyne PA, Nolard N, Schinner F, Margesin R (2005) Biodegradation of phenol and phenol-related compounds by psychrophilic and cold-tolerant alpine yeasts. Chemosphere 59:909–918

Branda E, Turchetti B, Diolaiuti G, Pecci M, Smiraglia C, Buzzini P (2010) Yeast and yeast-like diversity in the southernmost glacier of Europe (Calderone glacier, Apennines, Italy). FEMS Microbiol Ecol 72:354–369

Brunner I, Plotze M, Rieder S, Zumsteg A, Furrer G, Frey B (2011) Pioneering fungi from the Damma glacier forefield in the Swiss Alps can promote granite weathering. Geobiology 9:266–279

Butinar L, Spencer-Martins I, Gunde-Cimerman N (2007) Yeasts in high Arctic glaciers: the discovery of a new habitat for eukaryotic microorganisms. A van Leeuw J Microbiol 91:277–289

Buzzini P, Branda E, Goretti M, Turchetti B (2012) Psychrophilic yeasts from worldwide glacial habitats: diversity, daptation strategies and biotechnological potential. FEMS Microbiol Ecol 82:217–241

Cameron KA, Hodson AJ, Osborn AM (2012) Structure and diversity of bacterial, eukaryotic and archaeal communities in glacial cryoconite holes from the Arctic and the Antarctic. FEMS Microbiol Lett 82:254–267

Cannone N, Diolaiuti G, Guglielmin M, Smiraglia C (2008) Accelerating climate change impacts on alpine glacier forefield ecosystems in the European Alps. Ecol Appl 18:637–648

Castello JD, Rogers SO (2005) Life in ancient ice. Princeton University Press, Princeton

Christner BC, Mosley-Thompson E, Thompson LG, Reeve JN (2001) Isolation of bacteria and 16 S rDNAs from Lake Vostok accretion ice. Environ Microbiol 3:570–577

Christner BC, Krivko BH II, Reeve JN (2003a) Molecular identification of bacteria and eukarya inhabiting an Antarctic cryoconite hole. Extremophiles 7:177–183

Christner BC, Mosley-Thompson E, Thompson LG, Reeve JN (2003b) Bacterial recovery from ancient glacial ice. Environ Microbiol 5:433–436

Christner BC, Royston-Bishop G, Foreman CM, Arnold BR, Tranter M, Welch K, Lyons WB, Tsapin AI, Priscu J (2006) Limnological conditions in subglacial Lake Vostok. Limnol Oceanogr 51:2485–2501

Connell LB, Redman R, Craig S, Scorzetti G, Iszard M, Rodriguez R (2008) Diversity of soil yeasts isolated from South Victoria Land, Antarctica. Microb Ecol 56:448–459

De Angelis M, Gaudichet A (1991) Saharan dust deposition over Mont Blanc (French Alps) during the last 30 years. Tellus 43B:61–75

de Garcia V, Brizzio S, Libkind D, Buzzini P, van Broock M (2007) Biodiversity of cold-adapted yeasts from glacial meltwater rivers in Patagonia. Argentina FEMS Microbiol Ecol 59:331–341

Deming WJ (2002) Psychrophiles and Polar regions. Curr Opin Microbiol 5:301–309

Feller G, Narinx E, Arpigny JL, Aittaleb M, Baise E, Genicot S, Gerday C (1996) Enzymes from psychrophilic organisms. FEMS Microbiol Rev 18:189–202

Foght J, Aislabie J, Turner S, Brown CE, Ryburn J, Saul DJ, Lawson W (2004) Culturable bacteria in subglacial sediments and ice from two southern hemisphere glaciers. Microbial Ecol 47:329–340

Fonseca A, Boekhout T, Fell JW (2011) Cryptococcus Vuillemin (1901). In: Kurtzman CP, Fell JW, Boekhout T (eds) The yeasts, a taxonomic study. Elsevier, London, pp 1661–1738

Gildemacher P, Heijne B, Silvestri M, Houbraken J, Hoekstra E, Boekhout T (2006) Interactions between yeasts, fungicides and apple fruit russeting. FEMS Yeast Res 6:1149–1156

Gildemacher PR, Heijne B, Houbraken J, Vromans T, Hoekstra ES, Boekhout T (2004) Can phyllosphere yeasts explain the effect of scab fungicides on russeting of Elstar apples? Eur J Plant Pathol 110:929–937

Herzberg M, Fischer R, Titze A (2002) Conflicting results obtained by RAPD-PCR and large-subunit rDNA sequences in determining and comparing yeast strains isolated from flowers: a comparison of two methods. Int J Syst Evol Microbiol 52:1423–1433

Hodson A (2006) Biogeochemistry of snowmelt in an Antarctic glacial ecosystem. Water Resources Res 42 (11) doi: 10.1029/2005WR004311

Hodson A, Anesio AM, Trantner M, Fountain A, Osborn M, Priscu J, Laybourn-Parry J, Sattler B (2008) Glacial ecosystems. Ecol Monograph 78:41–67

Hodson A, Cameron K, Bøggild C, Irvine-Fynn T, Langford H, Pearce D, Banwart S (2010) The structure, biological activity and biogeochemistry of cryoconite aggregates upon an Arctic valley glacier: Longyearbreen, Svalbard. J Glaciol 56:349–362

Lazzaro A, Brankatschk R, Zeyer J (2012) Seasonal dynamics of nutrients and bacterial communities in unvegetated alpine glacier forefields. Appl Soil Ecol 53:10–22

Lee YM, Kim S-Y, Jung J, Kim EH, Cho KH, Schinner F, Margesin R, Hong SG, Lee HK (2011) Cultured bacterial diversity and human impact on alpine glacier cryoconite. J Microbiol 49:355–362

Libkind D, Moline M, Sampaio JP, van Broock M (2009) Yeasts from high-altitude lakes: infuence of UV radiation. FEMS Microbiol Ecol 69:353–362

MacDonell S, Fitzsimons S (2008) The formation and hydrological significance of cryoconite holes. Prog Phys Geogr 32:595–610

Margesin R (2009) Effect of temperature on growth parameters of psychrophilic bacteria and yeasts. Extremophiles 13:257–262

Margesin R, Miteva V (2011) Diversity and ecology of psychrophilic microorganisms. Res Microbiol 162:346–361

Margesin R, Fauster V, Fonteyne PA (2005a) Characterization of cold-active pectate lyases from psychrophilic *Mrakia frigida*. Lett Appl Microbiol 40:453–459

Margesin R, Fonteyne PA, Redl B (2005b) Low-temperature biodegradation oh high amounts of phenol by *Rhodococcus* spp. and basidiomycetous yeasts. Res Microbiol 156:68–75

Margesin R, Fonteyne PA, Schinner F, Sampaio JP (2007) *Rhodotorula psychrophila* sp. nov., *Rhodotorula psychrophenolica* sp. nov. and *Rhodotorula glacialis* sp. nov., novel psychrophilic basidiomycetous yeast species from alpine environments. Int J Syst Evol Microbiol 57:2179–2184

Margesin R, Fell JF (2008) *Mrakiella cryoconiti* gen. nov., sp. nov., a psychrophilic, anamorphic, basidiomycetous yeast from alpine and arctic habitats. Int J Syst Evol Microbiol 58:2977–2982

Margesin R, Zacke G, Schinner F (2002) Characterization of heterotrophic microorganisms in alpine glacier cryoconite. Arct Antarct Alp Res 34:88–93

Miteva V (2008) Bacteria in snow and glacier ice. In: Margesin R, Schinner F, Marx JC, Gerday C (eds) Psychrophiles: from biodiversity to biotechnology. Springer, Berlin, pp 31–50

Miteva VI, Sheridan PP, Brenchley JE (2004) Phylogenetic and physiological diversity of microorganisms isolated from a deep Greenland glacier ice core. Appl Environ Microbiol 70:202–213

Panikov NS, Sizova MV (2007) Growth kinetics of microorganisms isolated from Alaskan soil and permafrost in solid media frozen down to −35°C. FEMS Microbiol Ecol 59:500–512

Paterson WSB (1994) The physics of glaciers, 3rd edn. Pergamon, Oxford

Paterson WSB (2002) The physics of glaciers. Butterworth-Heinemann, Elsevier Science, Burlington, pp 159–172

Pecci M, D'Agata C, Smiraglia C (2008) Ghiacciaio del Calderone (Apennines, Italy): the mass balance of a shrinking Mediterranean glacier. Geogr Fis Din Quat 31:55–62

Price PB (2000) A habitat for psychrophiles in deep, Antarctic ice. Proc Natl Acad Sci USA 97:1247–1251

Price PB (2007) Microbial life in glacial ice and implications for a cold origin of life. FEMS Microbiol Ecol 59:217–231

Priscu JC, Adams EE, Lyons WB, Voytek MA, Mogk DW, Brown RL, McKay CP, Takacs CD, Welch KA, Wolf CF, Kirshtein CD, Avci R (1999) Geomicrobiology of subglacial ice above Lake Vostok, Antarctica. Science 286:2141–2144

Sampaio JP (2011) *Cystofilobasidium* Oberwinkler & Bandoni (1983). In: Kurtzman CP, Fell JW, Boekhout T (eds) The yeasts, a taxonomic study. Elsevier, London, pp 1423–1432

Sharp M, Parkes J, Cragg B, Fairchild IJ, Lamb H, Tranter M (1999) Widespread bacterial populations at glacier beds and their relationship to rock weathering and carbon cycling. Geology 27:107–110

Siddiqui KS, Cavicchioli R (2006) Cold-adapted enzymes. Annu Rev Biochem 75:403–433

Simon C, Wiezer A, Strittmatter AW, Daniel R (2009) Phylogenetic diversity and metabolic potential revealed in a glacier ice metagenome. Appl Environ Microb 75:7519–7526

Singh P, Singh SM (2012) Characterization of yeast and filamentous fungi isolated from cryoconite holes of Svalbard, Arctic. Polar Biol 35:575–583

Skidmore M, Foght JM, Sharp MJ (2000) Microbial life beneath a high arctic glacier. Appl Environ Microbiol 66:3214–3220

Skidmore ML, Anderson SP, Sharp M, Foght J, Lanoil BD (2005) Comparison of microbial community compositions of two subglacial environments reveals a possible role for microorganisms in chemical weathering processes. Appl Environ Microbiol 71:6986–6997

Stibal M, Trantner M, Benning LG, Rehak J (2008) Microbial primary production on an Arctic glacier is insignificant in comparison with allochtonous organic carbon input. Environ Microbiol 10:2172–2178

Stibal M, Elster J, Sabacka M, Kastovska K (2007) Seasonal and diel changes in photosynthetic activity of the snow alga *Chlamydomonas nivalis* (Chlorophyceae) from Svalbard determined by pulse amplitude modulation fluorometry. FEMS Microbiol Ecol 59:265–273

Stibal M, Sabacka M, Zarsky J (2012) Biological processes on glacier and ice sheet surfaces. Nat Geosci 5:771–774

Takeuchi N, Kohshima S, Seko K (2001) Structure, formation and darkening process of albedo-reducing material (cryoconite) on a Himalayan glacier: a granular algal mat growing on the glacier. Arct Antarct Alp Res 33:115–122

Thomas-Hall SR, Turchetti B, Buzzini P, Branda E, Boekhout T, Theelen B, Watson K (2010) Cold adapted yeasts from Antarctica and the Italian Alps. Description of three novel species: *Mrakia robertii* sp. nov., *Mrakia blollopis* sp. nov. and *Mrakiella niccombsii* sp. nov. Extremophiles 14:47–59

Tieber A, Lettner H, Bossew P, Hubmer A, Sattler B, Hofmann W (2009) Accumulation of anthropogenic radionuclides in cryoconite on alpine glaciers. J Environ Radioact 100:590–598

Tranter M (2005) Geochemical weathering in glacial and proglacial environments. In: Holland HD, Turekian KK (eds) Treatise on geochemistry, vol 5. Elsevier, London, pp 189–205

Tranter M, Sharp MJ, Lamb H, Brown GH, Hubbard BP, Willis IC (2002) Geochemical weathering at the bed of Haut Glacier d'Arolla, Switzerland: a new model. Hydrol Process 16:959–993

Turchetti B, Buzzini P, Goretti M, Branda E, Diolaiuti G, D'Agata C, Smiraglia C, Vaughan-Martini A (2008) Psychrophilic yeasts in glacial environments of Alpine glaciers. FEMS Microbiol Ecol 63:73–83

Turchetti B, Thomas-Hall SR, Connell LB, Branda E, Buzzini P, Theelen B, Muller WH, Boekhout T (2011) Psychrophilic yeasts from Antarctica and European glaciers: description of *Glaciozyma* gen. nov., *Glaciozyma martinii* sp. nov. and *Glaciozyma watsonii* sp. nov. Extremophiles 15:573–586

Turchetti B, Goretti M, Branda E, Diolaiuti G, D'Agata C, Smiraglia C, Onofri A, Buzzini P (2013) Influence of abiotic variables on culturable yeast diversity in two distinct Alpine glaciers. FEMS Microbiol Ecol (in press) doi: 10.1111/1574-6941.12164

Uetake J, Yoshimura Y, Nagatsuka N, Kanda H (2012) Isolation of oligotrophic yeasts from supraglacial environments of different altitude on the Gulkana Glacier (Alaska). FEMS Microbiol Ecol 82:1–8

Vishniac HS (1985) *Cryptococcus friedmannii*, a new species of yeast from the Antarctic. Mycologia 77:149–153

Vishniac HS (2006) A multivariate analysis of soil yeasts isolated from a latitudinal gradient. Microb Ecol 52:90–103

Vishniac HS, Kurtzman CP (1992) *Cryptococcus antarcticus* sp. nov, and *Cryptococcus albidosimilis* sp. nov., basidioblastomycetes from antarctic soils. Int J Syst Bacteriol 42:547–553

Vishniac HS, Onofri S (2002) *Cryptococcus antarcticus* var. *circumpolaris* var. nov., a basidiomycetous yeast from Antarctica. A van Leeuwenhoek 83:231–233

Wadham JL, Tranter M, Tulaczyk S, Sharp M (2008) Subglacial methanogenesis: a potential climatic amplifier? Global Biogeochem Cycles 22, GB2021

Wharton RA Jr, Vinyard WC, Parker BC, Simmons GM Jr, Seaburg KG (1981) Algae in cryoconit holes on Canada glacier in southern Victoria Land, Antarctica. Phycologia 20:208–211

Wharton JRA, McKay CP, Simmons GM, Parker BC (1985) Cryoconite holes on glaciers. Bioscience 8:499–503

Xiang SR, Shang TC, Chen Y, Yao TD (2009) Deposition and post deposition mechanisms as possible drivers of microbial population variability in glacier ice. FEMS Microbiol Ecol 70:165–176

Zemp M, Haeberli W, Hoelzle M, Paul F (2006) Alpine glaciers to disappear within decades? Geophys Res Lett 33:L13504

Zumsteg A, Luster K, Göransson H, Smittenberg RH, Brunner I, Bernasconi SM, Zeyer J, Frey B (2012) Bacterial, archaeal and fungal succession in the forefield of a receding glacier. Microb Ecol 63:552–564

Chapter 6
Cold-Adapted Yeasts in Patagonian Habitats

Virginia de Garcia, Diego Libkind, Martín Moliné, Carlos A. Rosa and Maria Rosa Giraudo

Abstract The Patagonian Andes possess unique physical and environmental characteristics, such as a great variety of glaciers, large temperate ice masses, glacially formed water bodies and forests dominated principally by endemic tree species. Most of these environments are included in protected areas characterized by low anthropogenic impact and minimal atmospheric pollution. Extensive studies on the occurrence of both psychrotolerant and psychrophilic yeasts in these natural environments have been performed. Patagonian natural environments to date surveyed include freshwaters, meltwaters, glacial ice, sea water, flowers, phylloplane, sap exudates, bark, soil, rotten wood, fungal stromata and rhizosphere. This research work allowed the characterization of native cold-adapted yeast communities, the description of novel species and was a contribution to yeast taxonomy, biogeography and also to biotechnology. The occurrence of a broad yeast biodiversity in cold Patagonian ecosystems of high scientific and technological value is here presented.

Keywords Patagonia · Psychrophilic and psychrotolerant yeasts · Glaciers · Nothofagus

Contents

6.1 Andes and Patagonia ... 124
6.2 Cold-Adapted Yeast Species Described in Patagonia ... 126

V. de Garcia (✉) · D. Libkind · M. Moliné · M. R. Giraudo
Laboratorio de Microbiología Aplicada y Biotecnología, Centro Regional Universitario Bariloche, Universidad Nacional del Comahue, INIBIOMA (CONICET-UNCo), Quintral 1250, San Carlos de Bariloche C.P. 8400 Río Negro, Argentina
e-mail: vdegarcia@comahue-conicet.gob.ar

C. A. Rosa
Departamento de Microbiologia, ICB, C. P. 486, Universidade Federal de Minas Gerais, Belo Horizonte 31270-901, Minas Gerais, Brazil

P. Buzzini and R. Margesin (eds.), *Cold-Adapted Yeasts*,
DOI: 10.1007/978-3-642-39681-6_6,

6.3 Diversity, Taxonomy and Ecology of Cold-Adapted Yeasts in Patagonia 130
6.3.1 Aquatic Environments: Freshwater, Meltwater and Glacial Ice 130
6.3.2 Terrestial Environments: Cold-Adapted Yeasts in Phyllosphere 137
6.3.3 Occurrence of Cold-Adapted Yeasts in Soil 138
6.4 A Few "Cool Cases" from Cold Habitats of Patagonia 139
6.4.1 *Phaffia rhodozyma* 139
6.4.2 Wild *Saccharomyces* and the Discovery of *Saccharomyces eubayanus* 141
6.5 Conclusions 142
References 142

6.1 Andes and Patagonia

The Andes, located along the western coast of South America, is the longest continental continual mountain range in the world. They are about 7,000 km long, 200–700 km wide and have an average height of about 4,000 m; they extend from north to south through seven South American countries: Venezuela, Colombia, Ecuador, Peru, Bolivia, Chile and Argentina (Lliboutry 1998). The Andes are the world's highest mountain range outside of Asia. The highest peak is Mount Aconcagua, which rises to an elevation of 6,962 m above sea level. Numerous glaciers are found along the Andes Mountains and range in size from very small snow patches and glacierets in northern Chile to the 13,000-km^2 Southern Patagonian Ice Field in Argentina and Chile (Lliboutry 1998; Skvarca et al. 2003; Stuefer et al. 2007).

Patagonia region belongs to the southern part of the American continent and includes two neighboring countries, Chile and Argentina. To the west, east and south, Patagonia is limited by the Pacific and Atlantic oceans. Its northern limit corresponds to the northern borders of Neuquén and Río Negro provinces of Argentina and the Chilean region of Maule (Correa 1998). Islands in the proximities of the "continental Patagonia", on the submarine platform, are also included in the more accepted definition of Patagonia (Barthelemy et al. 2008).

The Patagonian Andes possess unique physical and environmental characteristics (Villarosa et al. 2008). The climate in Patagonia is described as temperate or cold-temperate. The prevalence of humid westerly winds has a great influence on the distribution of precipitations in this region, occurring predominantly during the colder period of the year. Toward the south, the influence of polar winds increases, so that the summer is cooler and wetter than further north, and snowfall is a common phenomenon in this region (Correa 1998). Toward the eastern part of Patagonia, temperature fluctuations are moderated by the wind from the Atlantic Ocean (Barthelemy et al. 2008).

Andean Patagonia in Argentina offers a great variety of glacially formed water bodies, which are still fed by glacier meltwater. They cover an ultra-mesotrophic to mesotrophic range of small and large lakes including small high elevation lakes, sometimes surrounded by dense native forest of *Nothofagus* spp. and *Austrocedrus chilensis* trees (Quirós and Drago 1985; Díaz et al. 2000). These lakes are oligotrophic to ultra-oligotrophic temperate water bodies of glacial origin and have been classified as warm monomictic with a period of summer stratification (Díaz et al. 2000).

In the Nahuel Huapi National Park, Argentina, Mount Tronador has 10 glaciers. Four of these are located in Argentina (Río Manso, Castaño Overo, Alerce, Frías), the remaining six are in Chile (Rabassa et al. 1978). Los Glaciares National Park (Argentina) and Patagonian Icefields (Hielos Patagónicos) are the largest temperate ice masses in the Southern Hemisphere, which account for more than 60 % of the Southern Hemisphere's glacial area outside Antarctica. Perito Moreno Glacier is located within these icefields (Skvarca et al. 2003; Stuefer et al. 2007).

Currently, most glaciers of the world are retreating rapidly as a result of global warming (Delgado Granados et al. 2007). Glaciers of Mount Tronador Patagonia Argentina are a clear example of this situation, having experienced a constant retreat for the past 100 years (Villarosa et al. 2008). Perito Moreno Glacier, however, is one of the few glaciers that are in a stable situation, i.e., it is neither advancing nor retreating (Stuefer et al. 2007).

Glacier/climate interactions in the Patagonian region are of relevance to understand the global climate change pattern. In addition, icefields and periglacial areas hold valuable information for Quaternary paleoenvironments (Stuefer et al. 2007). There is also a dense forest cover, associated with recent glacial deposits. Lakes, rivers and peatlands are the main landforms that create conditions for the study of past and present environmental variations in the region (Villarosa et al. 2008).

The relevance of studies on the effects of climatic and environmental changes on continental glaciers has been shown elsewhere, especially in the Northern Hemisphere (Villarosa et al. 2008; Branda et al. 2010). Microorganisms inhabiting these withdrawing glaciers may be released into soil, rivers and oceans, possibly complementing or changing the existing microbial communities (Butinar et al. 2007; de Garcia et al. 2012a).

From a forestry point of view, Patagonian habitats are dominated principally by endemic tree species of *Nothofagus* spp. (southern beech) and by the coniferous tree species *Austrocedrus chilensis*, *Araucaria araucana*, *Fitzroya cupressoides* and *Pilgerodendrum uviferum*, which are among the most important species in Argentina and Chile (Donoso Zegers 2006). The genus *Nothofagus* (Nothofagaceae) has about 35 species, which form endemic forest in the Southern Hemisphere (Manos 1997). In Argentina, large areas of this forest, which include the species *Nothofagus pumilio, Nothofagus nervosa, Nothofagus dombeyi, Nothofagus antarctica* and *Nothofagus betuloides* (Laclau 1997), are included in protected areas characterized by low anthropogenic impact and minimal atmospheric pollution (Satti et al. 2007).

Nothofagus pumilio is a deciduous species and represents the main species of the high-altitude tree-line forests of the Andes (Souza et al. 2000). This species inhabits a wide altitude and latitude range, and *N. pumilio* forests account for 25 %

of the native forest surface in Chile and 4 % of it in Argentina, constituting one of the most important types of forest in both countries (Gonzalez et al. 2006).

Nothofagus nervosa is one of the most economically important tree species in Patagonia, given it yields a highly valuable wood, it was overexploited in the past, and natural populations were drastically reduced. Currently, a program is in place for the conservation and domestication of this species (Marchelli and Gallo 1999; Gallo et al. 2004). All *Nothofagus* species have indehiscent dry fruits containing only one seed (nuts), this pericarp does not split open, so seedlings are cultivated directly from these fruits (Fernández et al. 2012).

Despite studies of microbial communities in soils from cold to temperate environments worldwide (Vishniac 2006; Gilichinsky et al. 2008; Yurkov et al. 2012), knowledge of biodiversity of cold-adapted yeasts in forest soils in the Southern Hemisphere is at present limited (Spaak 2008). The relevance of yeasts to soil function is not yet fully understood, although it is known that they influence soil aggregation, contribute to nutrient cycles and interact with the vegetation (Botha 2011). South America is, biologically speaking, the most diverse region of the globe and offers great potential in terms of macro- and micro-biodiversity (Carvajal et al. 2011).

6.2 Cold-Adapted Yeast Species Described in Patagonia

Microorganisms inhabiting cold environments can be divided into psychrophiles and psychrotolerants: the first predominate in permanent glacial habitats, whereas the second ones are generally predominant in ecosystems periodically exposed to low temperatures. Both are believed to play key roles in the biodegradation of organic matter and the cycling of essential nutrients at the cryosphere level (Margesin and Miteva 2011; Buzzini et al. 2012). An in-depth overview on psychrophilic and psychrotolerant microorganisms and on their global distribution is reported in Chap. 1.

Extensive studies on the occurrence of psychrophilic and psychrotolerant yeasts from environments of Patagonia were carried out since 1990s in aquatic (freshwater, meltwaters, glacial ice and sea water) and terrestrial (flowers, phylloplane, sap exudates, bark, soil, rotten wood, rhizosphere and *Cyttaria* stromata) habitats. These studies allowed the characterization of native yeast communities, the description of novel species and the description of a broad yeast biodiversity with invaluable scientific and technological value (Brizzio and van Broock 1998; Libkind et al. 2003, 2004a, b, 2006, 2007, 2008a, b, 2009a, b, 2011a; Brizzio et al. 2007; de Garcia et al. 2007, 2012a; Fernández et al. 2012; Mestre et al. 2011b).

It has been found that a considerable percentage (25–40 %) of the yeast species recovered from different substrates belong to undescribed taxa. Twenty-six yeast species have been formally described from Patagonian environments (Table 6.1). A list of the cold-adapted yeasts recovered from these environments is presented in Tables 6.2 and 6.3. Table 6.2 provides a comparison between Patagonian aquatic

Table 6.1 Formally described yeast species from Patagonian environments

Novel species	Substrate	References
Candida boleticola	Rotten trunk of *Laurelia sempervirens* and fallen trunk of *L. philippiana* in Valdivian evergreen rainforest, Chile	Ramírez and González (1984c)
Candida castrensis	Decaying wood of *N. dombeyii* in Valdivian evergreen rainforest, Chile	Ramírez and González (1984a)
Candida llanquihuensis	Fallen trunk of *N. dombeyi* in Valdivian evergreen rainforest, Chile	Ramírez and González (1984c)
Candida petrohuensis	Rotten trunk of *N. dombeyi* in Valdivian evergreen rainforest, Chile	Ramírez and González (1984e)
Candida rancensis	Decayed wood of *L. sempervirens* in Valdivian evergreen rainforest, Chile	Ramírez and González (1984e)
Candida santjacobensis	Fallen trunk of *N. dombeyi* in Valdivian evergreen rainforest, Chile	Ramírez and González (1984f)
Candida sophiae-reginae	Rotten wood of *L. sempervirens* in Valdivian evergreen rainforest, Chile	Ramírez and González (1984d)
Candida tepae	Rotten wood of *L. philippiana, L. sempervirens* and from fallen trunk of *Eucryphia cordifolia* in Valdivian evergreen rainforest, Chile	Ramírez and González (1984a)
Cryptococcus fonsecae	Sea water, Cape Horn Meridian Argentinian Sea	de Garcia et al. (2012b)
Cryptococcus frias	Meltwater river from Frias Glacier, NHNP, Argentina	de Garcia et al. (2012b)
Cryptococcus psychrotolerans	Sea water, Cape Horn Meridian Argentinian Sea	de Garcia et al. (2012b)
Cryptococcus tronadorensis	Meltwater, Rio Manso Glacier (Garganta del Diablo waterfall), NHNP, Argentina	de Garcia et al. (2012b)
Cryptococcus spencermartinsiae	Meltwater river from Frias Glacier, NHNP, Argentina	de Garcia et al. (2010a)
Cystofilobasidium lacus-mascardii	Subsurface water from Lake Mascardi, NHNP, Argentina	Libkind et al. (2009b)
Cystofilobasidium macerans	Water from Lake Mascardi and meltwaters from Black glacier, Mount Tronador, NHNP, Argentina	Libkind et al. (2009b)
Holtermanniella takashimae	*Nothofagus pumilio* phylloplane and water from Negra lake, NHNP, Argentina	Wuczkowski et al. (2010)
Lachancea nothofagi	Bark, ectomycorrhizosphere and sap from different species of *Nothofagus* trees, NHNP, Argentina	Mestre et al. (2011a)
Lindnera rhizosphaerae	Rhizosphere of *N. pumilio* ectomycorrhizal roots, NHNP, Argentina	Mestre et al. (2010)
Rhodotorula meli	Meltwaters from Black glacier, Mount Tronador, NHNP, Argentina	Libkind et al. (2010)

(continued)

Table 6.1 (continued)

Novel species	Substrate	References
Rhodotorula nothofagi	Rotten trunk of *N. obliqua* in Valdivian evergreen rainforest, Chile	Ramírez and González (1985)
Saccharomyces eubayanus	Bark and soil of *Nothofagus* spp., stromata of *Cyttaria* fungus, NHNP, Argentina	Libkind et al. (2011a)
Saprochaete chiloensis (formerly *Schizoblastosporion chiloense*)	Rotten trunk of *Eucryphia cordifolia* in Valdivian evergreen rainforest, Chile	Ramírez and González (1984g), de Hoog and Smith (2011)
Sporidiobolus longiusculus	Surface water of Lake Fonck and Lake Ilon, NHNP, Argentina	Libkind et al. (2005a)
Sporobolomyces patagonicus	Surface water of Lake Fonck of Lake Hess, NHNP, Argentina	Libkind et al. (2005a)
Sugiyamaella chiloensis (teleomorph of *Candida bertae*)	Rotted wood of *N. dombeyii* and *Eucryphia cordifolia* in Valdivian evergreen rainforest, Chile	Ramírez and González (1984b)
Wickerhamomyces patagonicus	Sap exudates of *N. dombeyi* and meltwater river from the Castaño Overo Glacier of Mount Tronador, NHNP, Argentina	de Garcia et al. (2010b)

NHNP Nahuel Huapi National Park

Table 6.2 Yeast species isolated from aquatic and terrestrial environments from Patagonia

Cold-adapted yeasts	Patagonian environments		References
	Aquatic	Terrestrial	
Aureobasidium pullulans	Glacial ice, water from glacier meltwater rivers, lakes and high-altitude lagoons, National Parks in Argentina	Phylloplane, fruits and soil from *Nothofagus* forest	de Garcia et al. (2007, 2012a); Libkind et al. (2009a); Muñoz (2010); Brandao et al. 2011); Fernández et al. (2012); Mestre et al. (2011b)
*Candida coipomoensis**	Water from high-altitude lagoons in NHNP, Argentina	Tree trunks in an advanced state of decomposition in the Valdivian rainforest of Chile	Ramírez and González (1984c); Libkind et al. (2009a)
*Candida marítima**	Water from glacier meltwater river Castaño Overo Glacier of Mount Tronador, NHNP, Argentina	Bark soil and rhizosphere of *N. pumilio* and tree trunks in an advanced state of decomposition in the Valdivian rainforest of Chile	Ramírez and González (1984d); Brandao et al. (2011); Mestre et al. (2011b)

(continued)

Table 6.2 (continued)

Cold-adapted yeasts	Patagonian environments		References
	Aquatic	Terrestrial	
*Candida railenensis**	Water from Nahuel Huapi Lake, NHNP, Argentina	Tree trunks in an advanced state of decomposition in the Valdivian rainforest of Chile	Ramírez and González (1984b); Brandao et al. (2011)
Cryptococcus adeliensis	Water from glacier meltwaters rivers, lakes and high-altitude lagoons in NHNP, Argentina	Phylloplane and fruits from *Nothofagus* trees, Argentina	de Garcia et al. (2007); Brandao et al. (2011); Fernández et al. (2012)
Cryptococcus diffluens	Water from Nahuel Huapi Lake, NHNP, Argentina	Fruits from *Nothofagus nervosa*, Argentina	Brandao et al. (2011); Fernández et al. (2012)
Cryptococcus heveanensis	Water from Nahuel Huapi Lake, NHNP, Argentina	Fruits of *Nothofagus nervosa*, Argentina	Brandao et al. (2011); Fernández et al. (2012)
Cryptococcus terricola	Water from glacier meltwater river of Frias Glacier, Mount Tronador, NHNP Argentina	*Nothofagus pumilio* forest soil, Argentina	Brandao et al. (2011); Mestre et al. (2011b)
Cryptococcus victoriae	Glacial ice, water from glacier meltwaters rivers and Nahuel Huapi Lake, National Parks, Argentina	Philoplane from *Nothofagus* trees, NHNP, Argentina	Muñoz (2010); Brandao et al. (2011); de Garcia et al. (2012a)
Cryptococcus wieringae	Water from glacier meltwaters rivers and Nahuel Huapi Lake, NHNP, Argentina	Fruits of *Nothofagus nervosa*, Argentina	de Garcia et al. (2007); Brandao et al. (2011); Fernández et al. (2012)
Dioszegia fristingensis	Water from glacier meltwaters rivers, lakes and high-altitude lagoons in NHNP, Argentina	Philoplane from *Nothofagus* trees, NHNP, Argentina	Libkind et al. (2003); de Garcia et al. (2007); Muñoz (2010); Brandao et al. (2011)
Guehomyces pullulans	Glacial ice and water from Nahuel Huapi Lake, National Parks, Argentina	Ectomycorrhizosphere, rhizosphere of *Nothofagus pumilio,* National Parks in Patagonia, Argentina	Brandao et al. (2011); Mestre et al. (2011b); de Garcia et al. (2012a)

(continued)

and terrestrial environments, whereas Table 6.3 shows a list of cold-adapted yeast

Table 6.2 (continued)

Cold-adapted yeasts	Patagonian environments		References
	Aquatic	Terrestrial	
Rhodotorula colostri	Water from glacier meltwaters rivers, lakes and high-altitude lagoons in NHNP, Argentina	Philoplane, fruits and bulk soil from *Nothofagus* trees, National Parks in Patagonia, Argentina	Libkind et al. (2003); Muñoz (2010); Brandao et al. (2011); Fernández et al. (2012); Mestre et al. (2011b); de Garcia et al. (2012a)
Rhodotorula laryngis	Water from glacier meltwater rivers, lakes and high-altitude lagoons, NHNP, Argentina	Philoplane from *Nothofagus* trees, NHNP, Argentina	Libkind et al. (2003); de Garcia et al. (2007); Libkind et al. (2009a); Muñoz (2010); de Garcia et al. (2012a)
Rhodotorula mucilaginosa	Glacial ice, water from glacier meltwater rivers, lakes and high-altitude lagoons	Philoplane from *Nothofagus* trees, NHNP, Argentina	Libkind et al. (2003); de Garcia et al. (2007); Libkind et al. (2009a); Muñoz (2010); Brandao et al. (2011); de Garcia et al. (2012a)
Sporobolomyces ruberrimus	Water from glacier meltwater rivers, lakes and high-altitude lagoons in NHNP	Philoplane from *Nothofagus* trees, NHNP, Argentina	Libkind et al. (2003); de Garcia et al. (2007); Libkind et al. (2009a); de Garcia et al. (2012a)

*Novel species described for the first time for decaying wood in Valdivian rainforest of Chile and isolated also from other Patagonian habitats; bold, cosmopolite species; *NHNP* Nahuel Huapi National Park

species originally described from Patagonia.

Table 6.3 Cold-adapted yeasts species isolated from environments from Patagonia

Basidiomycetous yeasts	Environments	References
Bensingtonia yamatoana	Glacial ice from Frias Glacier of Mount Tronador	de Garcia et al. (2012a)
Bullera dendrophila	Water from Nahuel Huapi Lake	Brandao et al. (2011)
Cryptococcus aerius	*Nothofagus pumilio* forest soil	Mestre et al. (2011a, b)
Cryptococcus albidus	Water from high-altitude lagoons in NHNP	Libkind et al. (2009a)
Cryptococcus antarticus	Water from high-altitude lagoons in NHNP	Libkind et al. (2009a)
Cryptococcus carnescens	Water from Nahuel Huapi Lake	Brandao et al. (2011)
Cryptococcus gastricus	Water from high-altitude lagoons in NHNP	Libkind et al. (2009a)
Cryptococcus magnus	Water from Nahuel Huapi Lake	Brandao et al. (2011)
Cryptococcus phenolicus	*Nothofagus pumilio* forest soil	Mestre et al. (2011a, b)
Cryptococcus podzolicus	*Nothofagus pumilio* forest soil	Mestre et al. (2011a, b)
Cryptococcus saitoi	Water from lakes and high-altitude lagoons in NHNP	Libkind et al. (2009a); Brandao et al. (2011)
Cryptococcus tephrensis	Water from Nahuel Huapi Lake	Brandao et al. (2011)
Cryptococcus terreus	*Nothofagus pumilio* forest soil	Libkind et al. (2011b)
Cryptococcus cylindricus	Water from glacier meltwater rivers of Mount Tronador	de Garcia et al. (2012a)
Cryptococcus stepossus	Water from glacier meltwater rivers of Mount Tronador	de Garcia et al. (2007)
Curvibasidium cygneicollum (teleomorph of *Rhodotorula fujisanensis*)	Phylloplane, fruits and soil from *Nothofagus* forest	Muñoz (2010); Fernández et al. (2012); Mestre et al. (2011a, b)
Cystofilobasidium infirmominiatum	Water from lakes and high-altitude lagoons in NHNP	Libkind et al. (2003); Brandao et al. (2011)
Cystofilobasidium capitatum	Water from lakes and high-altitude lagoons in NHNP	Libkind et al. (2003); Brandao et al. (2011)
Dioszegia butyracea	Glacial ice from Mount Tronador	de Garcia et al. (2012a)
Dioszegia hungarica	Water from lakes and high-altitude lagoons in NHNP	Brandao et al. (2011); Libkind et al. (2009a)
Dioszegia crocea	Glacial ice and water from glacier meltwater rivers	de Garcia et al. (2007, 2012a)
Holtermanniella festucosa	Glacial ice, water from glacier meltwater rivers and Nahuel Huapi Lake	de Garcia et al. (2007); Brandao et al. (2011); de Garcia et al. (2012a)

(continued)

Table 6.3 (continued)

Basidiomycetous yeasts	Environments	References
Holtermanniella wattica	Phylloplane and ectomycorrhizosphere from *Nothofagus* trees	Muñoz (2010); Mestre et al. (2011a, b)
Leucosporidiella fragaria	Water from glacier meltwater rivers from glaciers of Mount Tronador	de Garcia et al. (2007)
Leucosporidiella muscorum	Water from glacier meltwater rivers and high-altitude lagoons in NHNP	de Garcia et al. (2007); Libkind et al. (2009a)
Leucosporidiella creatinivora	Water from glacier meltwater rivers and high-altitude lagoons in NHNP	de Garcia et al. (2007); Libkind et al. (2009a)
Mastigobasidium intermedium	Glacial ice from Mount Tronador	de Garcia et al. (2012a)
Mrakia frigida	Meltwater from Garganta del Diablo waterfall of Mount Tronador	de Garcia et al. (2007)
Mrakia robertii	Glacial ice from Frias Glacier of Mount Tronador	de Garcia et al. (2012a)
Mrakiella aquatica	Glacial ice from Frias Glacier of Mount Tronador	de Garcia et al. (2012a)
Rhodosporidium babjevae	Water from lakes and high-altitude lagoons in NHNP	Libkind et al. (2003, 2009a)
Rhodosporidium diobovatum	Water from lakes and high-altitude lagoons in NHNP	Libkind et al. (2003); Brandao et al. (2011)
Rhodosporidium kratochvilovae	Water from lakes and high-altitude lagoons in NHNP	Libkind et al. (2003)
Rhodotorula aurantiaca	Phylloplane from *Nothofagus* trees	Muñoz (2010)
Rhodotorula glacialis	Water from glacier meltwater rivers and Nahuel Huapi Lake	Brandao et al. (2011); de Garcia et al. (2012a)
Rhodotorula minuta	Water from high-altitude lagoons in NHNP	Libkind et al. (2003)
Rhodotorula pinicola	Water from high-altitude lagoons in NHNP	Libkind et al. (2003)
Rhodotorula slooffiae	Water from lakes and high-altitude lagoons in NHNP	Libkind et al. (2003); Brandao et al. (2011)
Sporidiobolus metaroseus	Water from high-altitude lagoons in NHNP	Libkind et al. (2003)
Sporidiobolus salmonicolor	Water from Nahuel Huapi Lake and acidic river	Libkind et al. (2003)
Sporobolomyces salicinus	Phylloplane from *Nothofagus* trees	Muñoz (2010)
Trichosporon dulcitum	Bark, soil and fruit of *Nothofagus trees*	Fernández et al. (2012); Mestre et al. (2011a, b)

(continued)

Table 6.3 (continued)

Basidiomycetous yeasts	Environments	References
Trichosporon porosum	*Nothofagus pumilio* forest soil	Mestre et al. (2011a, b)
Udeniomyces megalosporus	Glacial ice and water from glacier meltwater rivers of Mount Tronador	de Garcia et al. (2012a)
Udeniomyces pannonicus	Glacial ice and water from glacier meltwater rivers of Mount Tronador	de Garcia et al. (2007, 2012a)
Udeniomyces pyricola	Glacial ice and water from glacier meltwater rivers of Mount Tronador	de Garcia et al. (2012a)
Ascomycetous yeasts		
Candida delftensis	Water from Nahuel Huapi Lake	Brandao et al. (2011)
Candida mesenterica	Glacial ice from Perito Moreno Glacier	de Garcia et al. (2012a)
Candida sake	Water from Nahuel Huapi Lake	Brandao et al. (2011)
Debaryomyces hansenii	Glacial ice and water from Nahuel Huapi Lake	Brandao et al. (2011), de Garcia et al. (2012a)
Hanseniaspora uvarum	Water from Nahuel Huapi Lake	Brandao et al. (2011)
Hanseniaspora valbyensis	Bark, soil and rhizosphere of *Nothofagus pumilio*	Muñoz (2010)
Hanseniospora opuntiae	Water from high-altitude lagoons in NHNP	Libkind et al. (2009a)
Saccharomyces uvarum	Soil, bark and *Cyttaria* spp. stromata from *Nothofagus* trees	Libkind et al. (2011a)
Dimorphic fungi		
Citeromyces matritensis	Phylloplane from *Nothofagus* trees	Muñoz (2010)
Coniozima leucospermi	Fruits from *Nothofagus nervosa*	Fernández et al. (2012)
Dothiora cannabinae	Bark soil of *N. pumilio*	Muñoz (2010)
Hormonema dematioides (*Sydowia polyspora*)	Phylloplane from *Nothofagus* trees	Muñoz (2010)
Lalaria carpini	Phylloplane from *Nothofagus* trees	Muñoz (2010)
Lecythophora mutabilis	Fruits from *Nothofagus nervosa*	Fernández et al. (2012)
Phaeomoniella zymoides	Fruits from *Nothofagus nervosa*	Fernández et al. (2012)

6.3 Diversity, Taxonomy and Ecology of Cold-Adapted Yeasts in Patagonia

6.3.1 *Aquatic Environments: Freshwater, Meltwater and Glacial Ice*

Yeasts exhibit excellent survival in water; they occur in a variety of freshwater systems, including eutrophic to ultra-oligotrophic lakes, lagoons, rivers, groundwater, glaciers and glacial meltwater (Starmer and Lachance 2011). The yeast diversity in these ecosystems is highly affected by a variety of abiotic and biotic factors, such as temperature, pressure, UV radiation (UVR), salinity, fauna, flora, soil run-off and anthropogenic effluents (Medeiros et al. 2011).

Yeast studies in aquatic environments of Patagonia have reported densities of these microorganisms with average counts from 10^2 to 10^3 CFU L^{-1} in freshwater mountain lakes (Libkind et al. 2003, 2009a; Brandao et al. 2011), from 1×10^2 to 3×10^2 CFU L^{-1} in meltwater rivers (de Garcia et al. 2007), from 1×10^3 to 5×10^3 CFU L^{-1} for continental glacial ice (de Garcia et al. 2012a) and from 1×10^2 to 3×10^2 CFU L^{-1} in Austral and Antarctic sea waters (de Garcia, unpublished data).

From these surveys, yeast diversity was evaluated statistically using the Shannon–Weaver (H) index with Hutcheson's *t* test (a = 0.05) as described in Moreno (2001). Similar diversity indices were obtained when comparing yeasts from meltwater from Mount Tronador, ice from Frias Glacier (H = 2.23 and H = 2.52, respectively) (de Garcia et al. 2012a) and water samples from Nahuel Huapi Lake (Patagonia Argentina; coast sites H = 2.2 and pelagic sites H = 2.8) (Brandao et al. 2011). These values correspond to a relatively higher richness index of taxa when compared to the values reported for different soil fraction (bulk soil H = 1.07, rhizosphere H = 1.06 and ectomycorrhizosphere H = 0.98) in Patagonian forest (Mestre et al. 2011b).

Basidiomycetous yeasts were the predominant group in aquatic cold environments (Libkind et al. 2003; de Garcia et al. 2007, 2012a; Brandao et al. 2011). Similar results from different cold environments worldwide (Antarctica, Alpine glaciers and Arctic) have been reported (Thomas-Hall et al. 2002; Bergauer et al. 2005; Margesin et al. 2005, 2007b; Butinar et al. 2007; Connell et al. 2008, Turchetti et al. 2008; Shivaji and Prasad 2009; Branda et al. 2010; Connell et al. 2010; Thomas-Hall et al. 2010; Uetake et al. 2011; Vaz et al. 2011; Buzzini et al. 2012). An exhaustive overview of the biodiversity of cold-adapted yeasts in worldwide cold habitats is reported in Chaps. 3, 4, 5, 6, 7, 8. A few authors have suggested that the predominance of basidiomycetous yeasts in these extreme cold environments could be due to a higher nutritional versatility and a higher tolerance to extreme environmental conditions compared to ascomycetous yeasts (Sampaio 2004; Frisvad 2008).

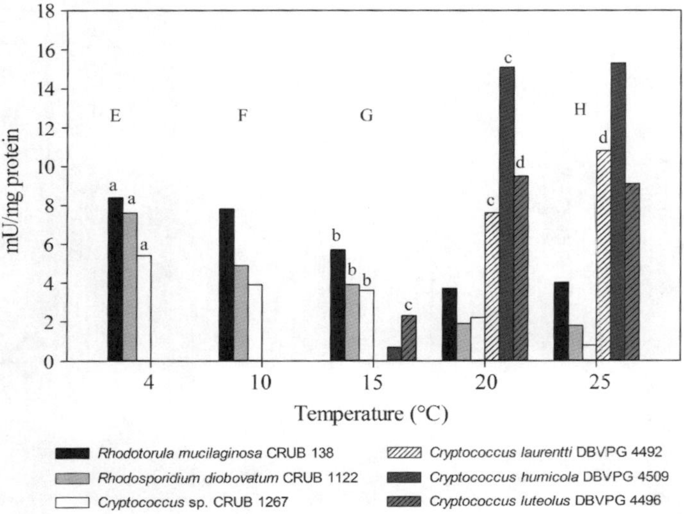

Fig. 6.1 Influence of temperature on extracellular lipase activity of selected cold-adapted yeasts strains from Patagonia (CRUB) and of mesophilic strains from Brazil (DBVPG) *a*, *b*, *c*, *d* denotes significant differences among values for each strain ($P < 0.001$) after ANOVA of triplicates; *E*, *F* and *G* denote significant differences at 4, 10 and 15 °C for cold-adapted yeasts and *H* denotes significant differences at 15, 20 and 25 °C for mesophilic strains after Tukey's HSD test analysis; CRUB, Regional University Center of Bariloche (Centro Regional Universitario Bariloche), Argentina; DBVPG, Industrial Yeasts Collection, University of Perugia, Italy

Studies with laboratory cultures suggested that most yeasts from these cold (extreme) habitats could express some metabolic adaptations to low temperatures (de Garcia, unpublished data). These microorganisms are heterotrophic, and their ability to degrade organic macromolecules by secreting extracellular hydrolytic cold-adapted enzymes suggests that, as proposed by Turchetti et al. (2008), these yeasts may have a significant ecological role in organic matter decomposition and nutrient cycling in glacial environments. This role is also supported by the presence of organic carbon and organic and inorganic nitrogen in glacial meltwater and ice (Skidmore et al. 2000; Foght et al. 2004; Margesin et al. 2007a).

Extracellular protease and lipase activities of selected psychrotolerant yeast strains isolated from aquatic cold environments of Patagonia, Argentina and mesophilic strains from Brazilian rain forests were studied at different temperatures (4–25 °C; Figs. 6.1, 6.2) (Brizzio et al. 2007). Lipase and protease activities of selected psychrophilic strains were significantly higher ($P < 0.05$) at 4 °C than at 20 °C, while opposite results were observed for mesophilic strains where activity was generally limited to temperatures of 20–25 °C and was dramatically affected by a drop of temperature (Brizzio et al. 2007). Cold-active enzymes can be up to ten times more active at low and moderate temperatures than their mesophilic homologues (Huston 2008).

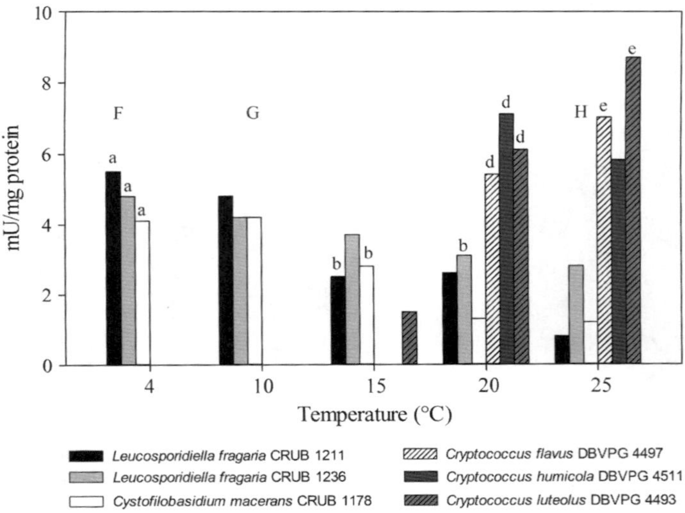

Fig. 6.2 Influence of temperature on protease activity of selected cold-adapted yeasts from Patagonia (CRUB) and mesophilic yeasts from Brazil (DBVPG); *a*, *b*, *c*, *d* and *e* denote significant differences among values for each strain ($P < 0.001$) after ANOVA of triplicates; *F* and *G* denote significant differences at 4 and 10 °C for cold-adapted yeasts; and *H* denotes significant differences at 15, 20 and 25 °C for mesophilic strains after Tukey's HSD test analysis; CRUB, Regional University Center of Bariloche (Centro Regional Universitario Bariloche), Argentina; DBVPG, Industrial Yeasts Collection, University of Perugia, Italy

Semi-quantitative assessment of extracellular enzymatic activity (protease, esterase, amylase, pectinase and cellulase) was studied with over 200 cold-adapted yeasts isolated from meltwater and glacial ice (de Garcia et al. 2007, 2012a). Studies on enzyme activity at 5 and 18 °C showed a higher activity at 5 °C than at 18 °C for all extracellular enzymes studied. A statistically significant relation between basidiomycetous yeast genera and their ability to produce extracellular enzymes was also found, thus confirming their ability to hydrolyze organic macromolecules. This evidence is undoubtedly interesting both from an ecological and from a biotechnological points of view (de Garcia et al. 2012a). Basidiomycetous yeasts are a diverse group of fungi with considerable industrial and medical importance and have undeniable potential for economic exploitation (Frisvad 2008). Cold-adapted enzymes are beneficial for their enhanced selectivity and high catalytic activity at low and moderate temperatures, in addition to their structural lability that can be exploited in multi-step processes requiring rapid and mild inactivation treatments (Huston 2008). An exhaustive discussion of fundamentals of cold-active enzymes is reported in Chap. 15.

Studies of yeasts from aquatic freshwater in Patagonia have not only contributed to suggest a possible ecological role of these microorganisms in natural environments, but have also given a great input to yeast taxonomy: to date, 10 novel species have been formally described from these environments (Table 6.2). Another interesting contribution was the isolation of a few strains from glacier meltwater from Mount Tronador, these strains (CRUB 1261, CRUB 1271 and CRUB 1296) belong possibly to novel *Rhodotorula* species (de Garcia et al. 2012a). They were phylogenetically related to the so-called psychrophilic ecoclade (Gadanho and Sampaio 2009). This "ecoclade" includes some yeast species that are phylogenetically related and have metabolic adaptations associated with physicochemical conditions present in the environment from which they were recovered.

In this regard, the psychrophilic ecoclade includes the species *Rhodotorula glacialis, Rhodotorula psychrophila, Rhodotorula psychrophenolica*, all isolated from alpine glacial environments (Margesin et al. 2007b) and *Rhodotorula himalayensis*, isolated from mountainous areas of the Himalaya (Shivaji et al. 2008). All the species related to this ecoclade are psychrophilic and have been isolated from extreme cold environments. The Patagonian isolates contribute to support the ecoclade proposal (de Garcia et al. 2012a).

The biodiversity of cold-adapted yeasts in sea water from the Austral Argentinian Sea (Beagle channel and meridian of Cape Horn) was also studied. From this survey, two novel *Cryptococcus* species from Tremellales were isolated and formally described: *Cryptococcus psychrotolerans* (related to *Cryptococcus victoriae* group) and *Cryptococcus fonsecae* (belonging to a new clade close to *Cryptococcus laurentii*). Both species were isolated from sea water of Cape Horn Meridian Argentinian Sea and from other cold marine environments of the world (de Garcia et al. 2012b).

6.3.2 Terrestial Environments: Cold-Adapted Yeasts in Phyllosphere

Knowledge of yeast diversity in forest soils in the Southern Hemisphere, until recently, was limited to a few studies performed in New Zealand (di Menna 1955, 1965). In the last years, several studies on biodiversity of cold-adapted yeasts associated with different substrates in *Nothofagus* spp. forest were performed in Patagonia, Argentina. These studies included leaves (phylloplane), seeds, bark, rooting wood, sap, soil, rhizosphere, flowers and the stromata of *Cyttaria* spp. (an exclusive parasite of *Nothofagus* spp. tree).

Terrestrial substrates like flowers, fruits, seeds and leaves can be broadly defined as the phyllosphere. For yeasts, the phyllosphere is a harsh environment, given reduced access to nutrients, high fluctuations in temperature and water availability and exposure to wind and UV radiation (Kowalchuk et al. 2010). In order to study the diversity and abundance of yeasts and dimorphic fungi (i.e., those fungi which can grow in hyphal or yeast form) in leaves (phylloplane) of *N. pumilio*, samples were obtained in exposed (to solar radiation) and non-exposed (covered by vegetation) leaves. Average cell counts varied between 10 and 10^3 CFU cm^{-2} (Muñoz 2010), similar counts (10^1–10^5 CFU cm^{-2}) were reported for other plants worldwide (Fonseca and Inácio 2006). Differences in yeast counts were found between abaxial and adaxial faces of the leaves and between exposed and non-exposed leaves. Ascomycetous fungi were predominant, and the most frequently isolated genus was *Aureobasidium*. These results are in agreement with previous reports of phylloplane. Fonseca and Inácio (2006) mentioned that *Aureobasidium pullulans* is apparently ubiquitous on aerial plant surfaces in temperate or subtropical regions worldwide and has been reported as the dominant fungal epiphyte of intact fruits (apple, pear and grapes) and leaves of trees such as *Acer platanoides* (Norway maple), *Ilex aquifolium* (holly), *Fagus sylvatica* (beech), *Malus domestica* (apple) and *Mangifera indica* (mango), and also from extreme cold and saline environments (Zalar et al. 2008).

Forty percentage of all the identified species may possibly belong to novel species (Muñoz 2010). Also, in order to understand adaptation to extreme conditions (high UV radiation), the ability of these fungi to synthesize photo-protective compounds like carotenoids, melanin and mycosporines was studied. The results obtained suggest a selection in favor of mycosporinogenic yeasts and dimorphic fungi (68 % of total) (Muñoz 2010). This evidence could support the hypothesis that that cell pigments, such as carotenoids and melanin, may play an important role in UV and sunlight protection (Libkind et al. 2004b; Sterflinger 2006; Yurkov et al. 2008; Moliné et al. 2009) and that the so-called red and black yeasts are, therefore, associated with phylloplane, seed surface and other environments with high irradiation rates (Inácio et al. 2005; Fonseca and Inácio 2006; Libkind et al. 2009a). An extensive review on the production of pigments and photo-protective compounds by cold-adapted yeasts is reported in Chap. 9.

Studies on yeasts associated with dry fruits of the economically important tree species *N. nervosa* allowed the isolation of 171 yeast isolates and some strains of dimorphic fungi. Ascomycetous fungi were dominant (78 %), and all were dark pigmented (green, brown and black), melanin-containing yeast-like fungi capable of forming conspicuous mycelium. From all identified species (17), five failed to match known ones. Two novel species (*Dothichiza* sp. 1, *Ascomycetous yeast* sp. 1; Fernández et al. 2012), *A. pullulans* and *Cryptococcus heveanensis,* were the most abundant species; altogether, these four species accounted for 84 % of the total number of isolates (Fernández et al. 2012). The production of UV-protective compounds such as mycosporines was detected in 71 % of the isolates, and most of them were pigmented (melanin). These results are in agreement with those reported for phylloplane of *Nothofagus*, showing the adaptations of these microorganisms to the harsh conditions of phyllosphere in Patagonia. Black yeasts and dimorphic fungi conquer extreme environments characterized by oligotrophic nutrient conditions, cycles of extremely low and high temperatures, UV radiation, osmotic stress and combinations of these factors (Sterflinger 2006; Zalar et al. 2008). The biodiversity and ecology of black yeasts in worldwide cold habitats are reported in Chap. 8.

6.3.3 Occurrence of Cold-Adapted Yeasts in Soil

Diversity, distribution and physiologic properties of yeasts inhabiting bulk soil, rhizosphere and ectomycorrhizosphere were also studied in *N. pumilio* forests. Average yeast counts varied from 9×10^2 to 1×10^4 CFU g^{-1} soil, in close agreement with the range previously reported for other forest soils. Yeast counts in deciduous and coniferous forests range from 1.5×10^3 to 1.1×10^4 CFU g^{-1} soil (Slaviková and Vadkertiová 2000) and from 1×10^2 to 1×10^5 CFU g^{-1} soil in subboreal forests of Russia (Maksimova and Chernov 2004). *Cryptococcus podzolicus*, *Cryptococcus phenolicus* and *Cryptococcus aerius* were the most frequently occurring species, and these yeasts have been defined as pedobiont (soil associated) in *N. pumilio* forest (Mestre et al. 2011b). *Cryptococcus* genus accounted for 58 % of all isolates (3 % were related to Filobasidiales and 23 % to Tremellales), and species of this genus possess some characteristics that could facilitate their development in soils, such as the production of polysaccharide capsules that enable yeasts to sequester and concentrate nutrients (Fonseca and Inácio 2006). Furthermore, their exogenous polysaccharide capsules may play an important role in soil aggregation processes (Mestre et al. 2011b).

Nothofagus spp. trees in Patagonia have a high rate of ectomycorrhizal infection, and yeast–ectomycorrhizal association might play an important role in the successful colonization and development of ectomycorrhizal symbiosis (Mestre et al. 2011b).

Only 20 % of isolates were related to ascomycetous yeasts. Among them, a novel species of *Lindnera* (*Lindnera rhizosphaerae*) was recently isolated and formally described (Mestre et al. 2010). Besides, three novel species related to *Nothofagus* spp. substrates were described: *Wickerhamomyces patagonicus*, *Lachancea nothofagi* and *Holtermanniella takashimae*. The last species was also isolated from other substrates in Europe and Asia. A new order (Holtermanniales) was proposed to include tremellomycetous yeasts of the Holtermannia clade (Wuczkowski et al. 2010) (Table 6.2).

6.4 A Few "Cool Cases" from Cold Habitats of Patagonia

6.4.1 *Phaffia rhodozyma*

The psychrotolerant basidiomycetous yeast *Phaffia rhodozyma* (anamorph of *Xanthophyllomyces dendrorhous*) represents an exceptional fungal case, given that it combines the production of orange-colored colonies with the ability to ferment simple sugars. The main carotenoid pigment synthesized by *Phaffia* is astaxanthin, another exclusive characteristic of this yeast species (Andrewes et al. 1976) and the main reason for which it is currently being exploited biotechnologically as a natural source of astaxanthin in aquaculture feed (Rodríguez-Sáiz et al. 2010).

Fig. 6.3 Photograph of *Cyttaria harioti* stromata growing on trunk tumor of *Nothofagus pumilio*. *Cyttaria* fungus exclusively parasites *Nothofagus* trees and its fructifications (stromata) form annually in spring and can reach the size of golf balls. *Cyttaria* stromata typically have a distinctive yellow to orange color and, when mature, a few species show up to 10 % w/v of simple sugars (Lederkremer and Ranalli 1967). Photo by Diego Libkind

P. rhodozyma was originally found in association with spring sap flows of various broad-leafed trees in Japan, Alaska and Russia (Phaff et al. 1972; Golubev et al. 1977). Afterward, additional strains were isolated from beech trees in Central Europe (Weber et al. 2006) and the USA (Fell et al. 2007). More recently, the range of *P. rhodozyma* was significantly expanded when a South American population associated with *Nothofagus* trees (southern beech), particularly the stromata of its biotrophic fungal parasite *Cyttaria* spp., was discovered (Libkind et al. 2007, 2008b) (Fig. 6.3). Interestingly, the Patagonian population was genetically divergent from its Northern Hemisphere counterparts based on DNA–DNA reassociation experiments, MSP-PCR fingerprinting, as well as ITS and IGS rRNA gene sequencing (Libkind et al. 2007, 2011b). Patagonian strains appeared to be genetically uniform and could be included into a distinct population, supporting the hypothesis that geographic isolation and association with different host species have determined genetically different *P. rhodozyma* populations worldwide. In agreement with this, Weber et al. (2006) described a novel isolate from Chile with marked ITS and LSU sequence differences from the other known populations. This single isolate was obtained from a leaf of the Tasmanian blue gum tree (*Eucalyptus globulus*) in the Mediterranean climate at Concepción, a species originally from Australasia. Interestingly, novel *P. rhodozyma* lineages have been recently collected in Australia and New Zealand (Palma et al. 2013, unpublished data).

Based on these recent results from South America, it has become clear that *P. rhodozyma* possesses a greater genetic variability and geographic distribution than previously suspected, generating the necessity to uncover it and assess its potential for the astaxanthin industry. However, difficulties in the isolation of this yeast hinder extensive environmental surveys. Fortunately, a new and innovative strategy for improving the recovery rate of *P. rhodozyma* and identification from environmental samples has recently been reported (Tognetti et al. 2013).

With the only exception of the Chilean strain from *Eucalyptus*, all the known habitats for *P. rhodozyma* are primarily located in mountainous areas of cold or temperate geographic regions, which is in agreement with its inability to grow above 25 °C and an optimal growth temperature of 17–19 °C. A few physiological adaptations to cold environments were observed in the Patagonian strains: an unusual proportion of polyunsaturated fatty acids (ca. 70 % of total fatty acids) (Libkind et al. 2008c), which largely surpassed that of *P. rhodozyma* from other origins (Sanderson and Jolly 1994) and the ability to produce extracellular enzymes (amylase) preferentially under low temperatures (Libkind et al. 2008b). These results suggest that the Patagonian *P. rhodozyma* population is highly adapted to the cold conditions found in this region.

6.4.2 Wild *Saccharomyces* and the Discovery of *Saccharomyces eubayanus*

The production of the most commercially produced beer, lager beer, was initiated in the fifteenth century and employs the allopolyploid hybrid yeast, *Saccharomyces pastorianus* (formerly *Saccharomyces carlsbergensis*). This is a domesticated microbe created by the fusion of a *Saccharomyces cerevisiae* ale-yeast with a previously unknown psychrotolerant *Saccharomyces* species (Dunn and Sherlock 2008). During yeast biodiversity surveys in Patagonian *Nothofagus* forests, two cold-adapted *Saccharomyces* species were recovered mainly from bark and soil samples but particularly from the stromata of *Cyttaria* fungus (Fig. 6.3) (Libkind et al. 2011a). One of the species was identified as *Saccharomyces uvarum* which had been isolated previously from forests in temperate regions of Europe and North America (Sampaio and Gonçalves 2008). From a taxonomic point of view, *S. uvarum* was previously considered as a variety of *Saccharomyces bayanus* (*S. bayanus* var. *uvarum*) (Vaughan-Martini and Martini 2011). However, Libkind et al. (2011a) recently demonstrated that this species is a biologically meaningful species, while *S. bayanus* is not since it was found to be complex hybrid of *S. uvarum*, *S. cerevisiae* and a third psychrophilic unknown *Saccharomyces* yeast, exclusively known from human-related environments. The later species was discovered in Patagonia, actually was the other *Saccharomyces* yeast found, and was formally described as *S. eubayanus* because of its contribution to the *S. bayanus* hybrid genome. Thus, in Patagonian habitats, *S. eubayanus* and *S. uvarum* (which can be considered two sister species) exist in apparent sympatry in *Nothofagus* (southern beech) forests, but are isolated genetically through intrinsic postzygotic barriers (Libkind et al. 2011a). Previously, it was shown that sympatric

Saccharomyces species tend to have different growth temperature preferences. This is the case of *S. cerevisiae* (thermotolerant) and *Saccharomyces kudriavzevii* (psychrotolerant) co-occurring in Mediterranean regions, as well as *Saccharomyces paradoxus* (thermotolerant) and *S. uvarum* (psychrotolerant) co-inhabiting temperate Europe and North America (Sampaio and Goncalves 2008). Another particular characteristic of Patagonian environments in contrast to North Hemisphere counterparts is the almost complete occupancy of the *Nothofagus* niche by psychrotolerant *Saccharomyces* species. While less than 50 % of the isolates from bark and soil samples from the North Hemisphere belong to *Saccharomyces*, in Patagonia, these values range from 64 to 95 %, for both *S. uvarum* and *S. eubayanus*. The substrate of greatest occupancy is *Cyttaria* stromata (~95 %) (Fig. 6.3), which is in agreement with its high content of simple sugars.

Furthermore, a putative ecological isolation through host preference was detected, given that *S. eubayanus* was found in association with *N. antarctica* and *N. pumilio*, whereas *S. uvarum* was associated with *N. dombeyi*. This finding might explain the co-existence of these two hitherto phenotypically undistinguishable species.

It is worth mentioning that *S. eubayanus* was the first yeast species to be formally described based on a complete genome sequence. Interestingly, it was shown to be 99.5 % identical to the non-*S. cerevisiae* portion of the *S. pastorianus* genome sequence and allowed the detection of specific genetic changes in sugar and sulfite metabolism that were vital for the domestication of the lager yeast in cold brewing environments (Libkind et al. 2011a). Thus, the cold-adapted yeast *S. eubayanus* is the parental species of the lager yeast, and Patagonian populations are so far the only known wild genetic stock of this species. Additional worldwide studies are needed to determine whether there are more *S. eubayanus* populations in the world.

6.5 Conclusions

Biodiversity studies of cold-adapted yeasts from aquatic and terrestrial cold environments of Patagonia have contributed not only to the general knowledge of fungal biodiversity in Patagonia, but also to general yeasts taxonomy and biogeography analysis of a few biotechnologically important species. Studies of cold-adapted yeasts from glacial meltwater and ice mentioned here were the first surveys from continental glaciers in the Southern Hemisphere (outside Antarctica). Future studies are needed to elucidate the role of yeasts in extreme cold environments, as well as their potential biotechnological applications. It has been clearly established that Patagonia's unexplored environments are true cold-adapted yeast reservoirs.

References

Andrewes AG, Phaff HJ, Starr MP (1976) Carotenoids of *Phaffia rhodozyma*, a red-pigmented fermenting yeast. Phytochem 15:1003–1007

Barthelemy D, Brion C, Puntieri J (eds) (2008) Plantas–Plants of Patagonia. Vazques Mazzini, Buenos Aires

Bergauer P, Fonteyne PA, Nolard N, Schinner F, Margesin R (2005) Biodegradation of phenol and phenol-related compounds by psychrophilic and cold-tolerant alpine yeasts. Chemosphere 59:909–918

Branda E, Turchetti B, Diolaiuti G, Pecci M, Smiraglia C, Buzzini P (2010) Yeast and yeast-like diversity in the southernmost glacier of Europe (Calderone Glacier, Apennines, Italy). FEMS Microbiol Ecol 72:354–369

Brandao RL, Libkind D, Vaz BMA, Espirito Santo LC, Moliné M, de Garcia V, van Broock M, Rosa C (2011) Yeasts from an oligotrophic lake in Patagonia (Argentina): diversity, distribution and synthesis of photoprotective compounds and extracelular enzymes. FEMS Microbiol Ecol 76:1–13

Brizzio S, van Broock M (1998) Characteristics of wild yeast killer from Nahuel Huapi National Park (Patagonia, Argentina). Food Biotechnol 36:273–278

Brizzio S, Turchetti B, de Garcia V, Libkind D, Buzzini P, Gasparetti C, van Broock M (2007) Extracellular enzymatic activities (EEA) in basidiomycetous yeasts isolated from glacial and subglacial waters of northwest Patagonia (Argentina). Can J Microbiol 53:519–525

Butinar L, Spencer-Martins I, Gunde-Cimerman N (2007) Yeasts in high Arctic glaciers: the discovery of a new habitat for eukaryotic microorganisms. A van Leeuwenhoek 91:277–289

Buzzini P, Branda E, Goretti M, Turchetti B (2012) Psychrophilic yeasts from worldwide glacial habitats: diversity, adaptation strategies and biotechnological potential. FEMS Microbiol Ecol 82:217–241

Botha A (2011) The importance and ecology of yeasts in soil. Soil Biol Biochem 43:1–8

Carvajal J, Libkind D, Briones A, Úbeda J, Roberts I, James S, Portero P, Alejo J, Rosa C (2011) Yeasts biodiversity significance: case studies in natural and human-related environments, *ex situ* preservation, applications and challenges. In: Grillo O, Venora G (eds) Changing diversity in changing environment. InTech, Croatia, pp 55–86

Connell LB, Redman R, Craig S, Scorzetti G, Iszaard M, Rodriguez R (2008) Diversity of soil yeasts isolated from South Victoria Land, Antarctica. Microb Ecol 56:448–459

Connell LB, Redman R, Rodriguez R, Barrett A, Iszard M, Fonseca A (2010) *Dioszegia antarctica* sp. nov. and *Dioszegia cryoxerica* sp. nov., psychrophilic Basidiomycetous yeasts from polar desert soils in Antarctica. Int J Syst Evol Microbiol 60:1466–1472

Correa MN (ed) (1998) Flora Patagónica (Republica Argentina) Tomo VIII. Colección científica del INTA. Instituto Nacional de Tecnología Agropecuaria, Buenos Aires, pp 31–38

Delgado Granados H, Miranda PJ, Huggel C, Ortega del Valle S, Alatorre Ibarguengoitia MA (2007) Chronicle of a death foretold: Extinction of the small-size tropical glaciers of Popocatépetl volcano (Mexico). Glob Planet Change 56:13–22

Donoso Zegers C (2006) Las especies arbóreas de los bosques templados de Chile y Argentina. In: Cuneo M (Ed) Autoecología. Valdivia, Chile. p. 678

de Garcia V, Brizzio S, Libkind D, Buzzini P, van Broock M (2007) Biodiversity of cold-adapted yeasts from glacial meltwater rivers in Patagonia, Argentina. FEMS Microbiol Ecol 59:331–341

de Garcia V, Brizzio S, Russo G, Rosa CA, Boekhout T, Theelen B, Libkind D, van Broock M (2010a) *Cryptococcus spencermartinsiae* sp. nov., a basidiomycetous yeast isolated from glacial waters and apple fruits. Int J Syst Evol Microbiol 60:707–711

de Garcia V, Brizzio S, Libkind D, Rosa CA, van Broock M (2010b) *Wickerhamomyces patagonicus,* a novel yeasts isolated from Patagonia, Argentina. Int J Syst Evol Microbiol 60:1693–1696 de Garcia V, Brizzio S, van Broock M (2012a) Yeasts from Glacial Ice of Patagonian Andes, Argentina. FEMS Microbiol Ecol 82:540–550

de Garcia V, Zalar P, Brizzio S, Gunde-Cimerman N, van Broock M (2012b) *Cryptococcus* species (Tremellales) from glacial biomes in the southern (Patagonia) and northern (Svalbard) hemispheres. FEMS Microbiol Ecol 82:523–539

de Hoog GS, Smith ThM (2011) Saprochaete Coker & Shanor ex D.T.S Wagner and Dawes (1970). In: Kurtzman CP, Fell JW, Boekhout T (eds) The Yeasts: a taxonomic study, 5th edn. Elsevier, Amsterdam, pp 1327–1328

di Menna M (1955) A search for pathogenic species of yeasts in New Zealand soils. J Gen Microbiol 12:54–62

di Menna M (1965) Yeast in New Zealand soils. New Zealand J Bot 1:194–203

Díaz M, Pedrozo F, Baccala N (2000) Summer classification of Southern Hemisphere temperate lakes (Patagonia, Argentina). Lakes Reserv 5:213–229

Dunn B, Sherlock G (2008) Reconstruction of the genome origins and evolution of the hybrid lager yeast *Saccharomyces pastorianus*. Genome Res 18:1610–1623

Fell JW, Scorzetti G, Statzell-Tallman A, Boundy-Mills K (2007) Molecular diversity and intragenomic variability in the yeast genus *Xanthophyllomyces:* the origin of *Phaffia rhodozyma*? FEMS Yeast Res 7:1399–1408

Fernández N, Mestre MC, Marchelli P, Fontenla S (2012) Yeast and yeast-like fungi associated with dry indehiscent fruits of a tree species in Patagonia, Argentina. FEMS Microbiol Ecol 80:179–192

Foght J, Aislabie J, Turner S, Brown CE, Ryburn J, Saul DJ, Lawson W (2004) Culturable bacteria in subglacial sediments and ice from two southern hemisphere glaciers. Microb Ecol 47:329–340

Fonseca A, Inácio J (2006) Phylloplane Yeast. In: Rosa C, Gabor P (eds) Biodiversity and ecophysiology of yeasts. Springer, Berlin, pp 263–302

Frisvad JC (2008) Fungi in cold ecosystems. In: Margesin R, Schinner F, Marx JC, Gerday C (eds) Psychrophiles: from biodiversity to biotechnology. Springer, Berlin, pp 137–156

Gadanho M, Sampaio JP (2009) *Cryptococcus ibericus* sp. nov., *Cryptococcus aciditolerans* sp. nov. and *Cryptococcus metallitolerans* sp., nov., a new ecoclade of anamorphic basidiomycetous yeasts species from an extreme environment associated with acid rock drainage in Sao Domingos pyrite mine. Portugal. Int J Syst Evol Microbiol 59:2375–2379

Gallo L, Donoso C, Marchelli P, Donoso P (2004) Variación en *Nothofagus nervosa* (Phil.) Dim. et Mil (*N. alpina, N. procera*). In: Donoso C, Premoli A, Gallo L, Ipinza R (eds) Variación intraespecífica en especies arbóreas de los bosques templados de Chile y Argentina. Editorial Universitaria, Santiago, pp 115–144

Gilichinsky D, Vishnivetskaya T, Petrova M, Spirina E, Mamykin V, Rivkina E (2008) Bacteria in permafrost. In: Margesin R, Schinner F, Marx JC, Gerday C (eds) Psychrophiles: from biodiversity to biotechnology. Springer, Berlin, pp 83–102

Golubev VI, Babéva IP, Blagodatskaya VM, Reshetova IS (1977) Taxonomic study of yeasts isolated from spring sap-flows. Microbiol 46:461–466

Gonzalez ME, Donoso Zegers C, Ovalle P, Martinez-Pastur G (2006) *Nothofagus Pumilio* (Poep. et Endl) Krasser. Lenga, roble blanco, leñar, roble de Tierra del Fuego. In: Donoso Zegers C (ed) Las especies arbóreas de los bosques templados de Chile y Argentina. Autoecología. Marisa Cuneo Ediciones, Valdivia, pp 486–500

Huston AL (2008) Biotechnological aspects of cold-adapted enzymes. In: Margesin R, Schinner F, Marx JC, Gerday C (eds) Psychrophiles: from biodiversity to biotechnology. Springer, Berlin, pp 347–363

Inácio J, Portugal L, Spencer-Martins I, Fonseca A (2005) Phylloplane yeasts from Portugal: seven novel anamorphic species in the Tremellales lineage of the Hymenomycetes (Basidiomycota) producing orange-coloured colonies. FEMS Yeast Res 5:1167–1183

Kowalchuk GA, Yergeau E, Leveau JHJ, Sessitsch A, Bailey M (2010) Plant-associated microbial communities. In: Lui WT, Jansson JK (eds) Environmental molecular microbiology. Caister Academic Press, New York, pp 131–148

Laclau P (1997) Los Ecosistemas Forestales y el Hombre en el Sur de Chile y Argentina. Fundación Vida Silvestre Argentina. Bol Técn 34:1–120

Lederkremer RM and Ranalli ME (1967) Hidratos de carbono en hongos de la Patagonia Argentina. Azúcares simples y polioles en Cyttaria hariotii Fischer. An Asoc Quím Arg 55:199–203

Libkind D, Brizzio S, Ruffini A, Gadanho M, van Broock MR, Sampaio JP (2003) Molecular characterization of carotenogenic yeasts from aquatic environments in Patagonia, Argentina. A van Leeuwenhoek 84:313–322

Libkind D, Brizzio S, van Broock M (2004a) *Rhodotorula mucilaginosa* a carotenoid producing yeast strain from a Patagonian high-altitude lake. Folia Microbiol 49:19–25

Libkind D, Diéguez M, Moliné M, Pérez P, Zagarese H, van Broock M (2006) Occurrence of photoprotective compounds in yeasts from freshwater ecosystems of Northwestern Patagonia (Argentina). Photochem Photobiol 82:972–980

Libkind D, Pérez P, Sommaruga R, Diéguez MC, Ferraro M, Brizzio S, Zagarese H, Broock MR (2004b) Constitutive and UV-inducible synthesis of photoprotective compounds (carotenoids and mycosporines) by freshwater yeasts. Photochem Photobiol Sci 3:281–286

Libkind D, Gadanho M, van Broock MR, Sampaio JP (2005a) *Sporidiobolus longiusculus* sp. nov. and *Sporobolomyces patagonicus* sp. nov., two new members of the Sporidiobolales isolated from aquatic environments in Patagonia. Int J Syst Evol Microbiol 55:503–509

Libkind D, Sommaruga R, Zagarese H, van Broock M (2005b) Mycosporines in carotenogenic yeasts. Syst Appl Microbiol 28:749–754

Libkind D, Diéguez M, Moliné M, Pérez P, Zagarese H, van Broock M (2006) Occurrence of photoprotective compounds in yeasts from freshwater ecosystems of Northwestern Patagonia (Argentina).Photochem Photobiol 82:972–980

Libkind D, Ruffini A, van Broock M, Alves L, Sampaio JP (2007) Biogeography, host specifity, and molecular phylogeny of the basidiomycetous yeast *Phaffia rhodozyma* and its sexual form, *Xantophyllomyces dendrorhous*. Appl Environ Microbiol 73:1120–1125

Libkind D, Gadanho M, van Broock MR, Sampaio JP (2008a) Studies on the heterogeneity of the carotenogenic yeast *Rhodotorula mucilaginosa* from Patagonia, Argentina. J Basic Microbiol 48:93–98

Libkind D, Moliné M, de García V, Fontenla S, van Broock M (2008b) Characterization of a novel South American population of the astaxanthin producing yeast *Xanthophyllomyces dendrorhous* (*Phaffia rhodozyma*). J Ind Microbiol Biotechnol 35:151–158

Libkind D, Arts M, van Broock MR (2008c) Fatty acid composition of cold-adapted carotenogenic basidiomycetous yeasts. Rev Argent Microbiol 40:193–197

Libkind D, Moliné M, Sampaio J, van Broock M (2009a) Yeasts from high altitude lakes: influence of UV radiation. FEMS Microbiol Ecol 69:353–362

Libkind D, Gadanho M, van Broock M, Sampaio JP (2009b) *Cystofilobasidium lacus-mascardii* sp. nov., a new basidiomycetous yeast species isolated from aquatic environments of the Patagonian Andes and *Cystofilobasidium macerans* sp. nov., the sexual stage of *Cryptococcus macerans*. Int J Syst Evol Microbiol 59:622–630

Libkind D, Sampaio JP, van Broock M (2010) Cystobasidiomycetes yeasts from Patagonia (Argentina): description of *Rhodotorula meli* sp. nov. from glacial meltwater. J Ind Microbiol Biotechnol 60:2251–2256

Libkind D, Hittinger CT, Valério E, Gonçalves C, Dover J, Johnston M, Gonçalves P, Sampaio JP (2011a) Microbe domestication and the identification of the wild genetic stock of lager-brewing yeast. Proc Natl Acad Sci USA 35:14539–14544

Libkind D, Tognetti C, Ruffini A, Sampaio JP, van Brook M (2011b) *Xantophyllomyces dendrorhous* (*Phaffia rhodozyma*) on stromata of *Cyttaria hariotii* in Northwestern Patagonian *Nothofagus* forest. Rev Argent Microbiol 43:198–202

Lliboutry L (1998) Glaciers of South America: glaciers of Chile and Argentina. In: Williams RS, Ferrigno JG (eds) Satellite image atlas of glaciers of the world. US Geological Survey, Denver, pp 1109–1119

Maksimova IA, Chernov IY (2004) Community structure of yeast fungi in forest biogeocenoses. Microbiology 73:474–481

Manos PS (1997) Systematics of *Nothofagus* (Nothofagaceae) based on rDNA spacer sequences (ITS): taxonomic congruence with morphology and plastid sequences. Am J Bot 84:1137–1155

Marchelli P, Gallo LA (1999) Annual and geographic variation in seed traits of Argentinean populations of southern beech *Nothofagus nervosa* (Phil.) Dim. et Mil. For Ecol Manag 121:239–250

Margesin R, Fauster V, Fonteyne PA (2005) Characterization of cold-active pectate lyases from psychrophilic *Mrakia frigida*. Lett Appl Microbiol 40:453–459

Margesin R, Neuner G, Storey KB (2007a) Cold-loving microbes, plants, and animals-fundamental and applied aspects. Naturwissenschaften 94:77–99

Margesin R, Fonteyne PA, Schinner F, Sampaio JP (2007b) Novel psychrophilic basidiomycetous yeasts from Alpine environments: *Rhodotorula psychrophila* sp. nov., *Rhodotorula psychrophenolica* sp. nov. and *Rhodotorula glacialis* sp. nov. Int J Syst Evol Microbiol 57:2179–2184

Margesin R, Miteva V (2011) Diversity and ecology of psychrophilic microorganisms. Res Microbiol 162:346–361

Medeiros AO, Rosa CA, Brandao RL, Giani A, Gomes ANL, Libkind D (2011) Microbial quality of freshwater ecosystems of South America. In: Ertuo K, Mirza I (eds) Water quality: physical, chemical and biological characteristics. Nova Science Publishers, Inc, New York, pp 43–69

Mestre MC, Ulloa JR, Rosa CA, Lachance MA, Fontenla S (2010) *Lachancea nothofagi* sp. nov., a yeast associated with *Nothofagus* species in Patagonia, Argentina. Int J Syst Evol Microbiol 60:2247–2250

Mestre MC, Rosa CA, Fontenla SB (2011a) *Lindnera rhizosphaerae* sp. nov., a yeast species isolated from rhizospheric soil. Int J Syst Evol Microbiol 61:985–988

Mestre MC, Rosa C, Safar SC, Libkind D, Fontenla S (2011b) Yeast communities associated with bulk-soil, rhizosphere and ectomycorrhizosphere of a *Nothofagus pumilio* forest in Northwestern Patagonia, Argentina. FEMS Microbiol Ecol 78:531–541

Moliné M, Libkind D, Diéguez MC, van Broock M (2009) Photoprotective role of carotenoids in yeasts: response to UV-B of pigmented and naturally-occurring albino strains. J Photochem Photobiol, B 95:156–161

Moreno CE (2001) Métodos para medir la biodiversidad. M & T–Manuales y Tesis SEA, vol 1. Zaragoza, España

Muñoz MI (2010) Levaduras y hongos dimórficos del filoplano de *Nothofagus pumilio* y el papel de la exposición solar en su distribución y producción de metabolitos fotoprotectores. Tesis para optar al grado de licenciado en Ciencias Biológicas, Centro Regional Universitario Bariloche, Universidad Nacional del Comahue. Bariloche, Argentina

Phaff HJ, Miller MW, Yoneyama M, Soneda MA (1972) Comparative study of the yeast florae associated with trees on the Japanese Islands and on the West Coasts of North America. In: Terui G (ed) Proceedings of the 4th IFS: fermentation technology today meeting. Society of Fermentation Technology, Osaka, pp 759–774

Quirós R, Drago E (1985) Relaciones entre variables físicas, morfométricas y climáticas en lagos patagónicos. Rev Asoc Cs Nat Litoral 16:181–199

Rabassa J, Rubulis S, Suárez J (1978) Los Glaciares del Monte Tronador Parque Nacional Nahuel Huapi (Río Negro, Argentina). Anales Parques Nacionales 14:259–318

Ramírez C, González A (1984a) Three new, non-fermenting, and nitrate-negative *Candida* species isolated the evergreen rainy Valdivian forest of southern Chile. Mycopathologia 87:175–180

Ramírez C, González A (1984b) Two new species and one variety of nitrate-utilizing mycelial *Candida* isolated from decayed wood in the evergreen rainy Valdivian forest of southern Chile. Mycopathologia 88:55–60

Ramírez C, González A (1984c) Five new filamentous, glucose-fermenting *Candida* isolated from decayed wood in the evergreen rainy Valdivian forest of southern Chile. Mycopathologia 88:83–92

Ramírez C, González A (1984d) Three new filamentous, fermenting *Candida* isolated from decayed wood in the evergreen rainy Valdivian forest of southern Chile. Mycopathologia 88:93–98

Ramírez C, González A (1984e) Two new amycelial *Candida* isolated from decayed wood in the evergreen rainy Valdivian forest of southern Chile. Mycopathologia 88:99–103

Ramírez C, González A (1984f) *Candida sant-jacobensis* and *Candida ancudensis*, two new species of yeasts isolated from decayed wood in the evergreen rainy Valdivian forest of southern Chile. Mycopathologia 88:105–109

Ramírez C, González A (1984g) Two new filamen-tous, non fermenting *Candida* and a new *Schizoblas-tosporion* isolated from decayed wood in the evergreen rainy Valdivian forest of southern Chile. Mycopathologia 88:165–171

Ramírez C, González A (1985) *Rhodotorula nothofagi* sp. nov., isolated from decayed wood in the evergreen rainy Valdivian forest of southern Chile. Mycopathologia 93:171–173

Rodríguez-Sáiz M, de la Fuente JL, Barredo JL (2010) *Xanthophyllomyces dendrorhous* for the industrial production of astaxanthin. Appl Microbiol Biotechnol 88:645–658

Sampaio JP (2004) Diversity, phylogeny and classification of basidiomycetous yeasts. In: Agerer R, Blanz P, Piepenbring M (eds) Frontiers in basidiomycote mycology. IHWVerlag, Eching, pp 49–80

Sampaio JP, Gonçalves P (2008) Natural populations of *Saccharomyces kudriavzevii* in Portugal are associated with oak bark and are sympatric with *S. cerevisiae* and *S. paradoxus*. Appl Environ Microbiol 74:2144–2152

Sanderson GW, Jolly SO (1994) The value of *Phaffia* yeast as a feed ingredient for salmonid fish. Aquaculture 124:193–200

Satti P, Mazzarino MJ, Roselli L, Crego P (2007) Factors affecting soil P dynamics in temperate volcanic soils of southern Argentina. Geoderma 139:229–240

Shivaji S, Prasad GS (2009) Antarctic Yeasts: biodiversity and potential applications. In: Satyanarayana T, Kunze G (eds) Yeast biotechnology: diversity and application. Springer, Berlin, pp 3–18

Shivaji S, Bhadra B, Rao RS, Pradhan S (2008) *Rhodotorula himalayensis* sp. nov., a novel psychrophilic yeast isolated from Roopkund Lake of the Himalayan mountain ranges. India. Extremophiles 12:375–381

Skidmore ML, Foght JM, Sharp MJ (2000) Microbial life beneath a high Arctic Glacier. App Environ Microbiol 66:3214–3220

Skvarca L, Raup B, De Angelis H (2003) Recent behaviour of Glaciar Upsala, a fast-flowing calving glacier in Lago Argentino, southern Patagonia. Ann Glaciol 36:184–188

Slaviková E, Vadkertiová R (2000) The occurrence of yeast in the forest soils. J Basic Microb 40:207–212

Souza MS, Puntieri JG, Barthélémy D, Brion C (2000) Bud content and its relation to shoot size and structure in *Nothofagus pumilio* (Poepp. et Endl.) Krasser (Nothofagaceae). Ann Bot 85:547–555

Spaak JM (2008) Studies on yeast diversity from New Zealand's primary forests. Bachelor's Thesis. Faculty of biology and biotechnology, Ruhr University, Bochum, Germany

Starmer WT, Lachance M (2011) Yeast Ecology. In: Kurtzman CP, Fell JW, Boekhout T (eds) The yeasts: a taxonomy study, 5th edn. Elsevier, London, pp 65–83

Sterflinger K (2006) Black yeasts and meristematic fungi: ecology, diversity and identification. In: Rosa CA, Peter G (eds) Biodiversity and ecophysiology of yeasts. Springer, Berlin, pp 501–514

Stuefer M, Rott H, Skvarca P (2007) Glaciar Perito Moreno, Patagonia: climate sensitivities and characteristics preceding 2003/04 and 2005/06 damming events. J Glaciology 53:3–16

Thomas-Hall S, Watson K, Scorzetti G (2002) *Cryptococcus statzelliae* sp. nov. and three novel strains of *Cryptococcus victoriae*, yeasts isolated from Antarctic soils. Int J Syst Evol Microbiol 52:2303–2308

Thomas-Hall S, Turchetti B, Buzzini P, Branda E, Boekhout T, Theelen B, Watson K (2010) Cold adapted yeasts from Antarctica and the Italian Alps. Description of three novel species: *Mrakia robertii* sp. nov., *Mrakia blollopis* sp. nov. and *Mrakiella niccombsii* sp. nov. Extremophiles 14:47–59

Tognetti C, Moliné M, van Broock M, Libkind D (2013) Favored isolation and rapid identification of *Xanthophyllomyces dendrorhous* (*Phaffia rhodozyma*) from environmental samples. J Basic Microbiol 53:1–7

Turchetti B, Buzzini P, Goretti M, Branda E, Diolaiuti G, D'Agata C, Smiraglia C, Vaughan-Martini A (2008) Psychrophilic yeasts in glacial environments of Alpine glaciers. FEMS Microbiol Ecol 63:73–83

Uetake J, Kohshima S, Takeuchi N, Fujita K, Miyake T, Narita T, Aizen V, Nakawo M (2011) Evidence for propagation of cold-adapted yeasts in an ice core from a siberian Altai glacier. J Geophys Res 116:G01021. doi:10.1029/2010JG001337

Vaughan-Martini A, Martini A (2011) Saccharomyces Meyen ex Reess (1870) In: Kurtzman CP, Fell JW, Boekhout T (eds) The Yeasts: a Taxonomic Study, 5th edn. Elsevier, Amsterdam, pp 733–746

Vaz ABM, Rosa LH, Vieira M, de Garcia V, Brandão RL, Teixeira L, Moliné M, Libkind D, van Broock M, Rosa CA (2011) The diversity, extracellular enzymatic activities and photoprotective compounds of yeasts isolated in Antarctica. Braz J Microbiol 42:937–947

Villarosa G, Outes V, Masiokas M, Villalba R, Rivas S (2008) El Monte Tronador: historias de hielo y fuego. In: Rastelli D (ed) Sitios de Interés Geológico de la República Argentina. SEGEMAR (Servicio de Geología y Minería de la República Argentina), Buenos Aires, pp 627–641

Vishniac HS (2006) A multivariate analysis of soil yeasts Isolated from a latitudinal gradient. Microb Ecol 52:90–103

Weber RWS, Davoli P, Anke H (2006) A microbial consortium involving the astaxanthin producer *Xanthophyllomyces dendrorhous* on freshly cut birch stumps in Germany. Mycol 20:57–61

Wuczkowski M, Passoth V, Turchetti B, Andersson AC, Olstorpe M, Laitila A, Theelen B, van Broock M, Buzzini P, Prillinger H, Sterflinger K, Schnurer J, Boekhout T, Libkind D (2010) Description of *Holtermanniella takashimae* sp. nov., *Holtermanniella* gen. nov. and proposal of the order Holtermanniales to accommodate Tremellomycetous yeasts of the Holtermannia clade. Int J Syst Evol Microbiol 61:680–689

Yurkov AM, Vustin MM, Tyaglov BV, Maksimova IA, Sineokiy SP (2008) Pigmented Basidiomycetous yeasts are a promising source of carotenoids and ubiquinone Q10. Microbiology 77:1–6

Yurkov AM, Kemler M, Begerow D (2012) Assessment of yeast diversity in soils under different management regimes. Fungal Ecol 5:24–35

Zalar P, Gostinear C, de Hoog GS, Ursie V, Sudhadham M, Gunde-Cimerman N (2008) Redefinition of *Aurebasidium pullulans* and its varieties. Stud Mycol 61:21–38

Chapter 7
Cold-Adapted Yeasts in Deep-Sea Environments

Yuriko Nagano, Takahiko Nagahama and Fumiyoshi Abe

Abstract Deep sea, the world's largest cold environment, is an environment of extreme conditions, such as high hydrostatic pressure and low nutrient availability, and has an average water temperature between −1 and 4 °C in most areas of deep sea. Living organisms in deep sea are considered to be adapted to cold environments. Yeast diversity commonly found in deep sea is represented by *Rhodosporidium* spp., *Rhodotorula* spp., *Candida* spp., *Cryptococcus* spp., *Pichia* spp., *Sporobolomyces* spp., and *Trichosporon* spp. This representation of yeasts is similar to yeasts found in other cold environments. Only psychrotolerant yeasts have been reported from deep-sea environments to date. However, the majority of yeasts isolated from deep-sea environments show better growth in deep-sea simulated conditions, such as 3 °C/40 MPa, than yeasts isolated from terrestrial environments. In comparison with prokaryotic microorganisms, yeasts in deep-sea environments are relatively underexplored, with few studies carried out on their physiology. Although the true yeast diversity and their ecology in deep-sea environments remains unclear, the intention of this chapter is to discuss current knowledge on deep-sea yeast diversity and their physiological characteristics and adaptation mechanisms to cold and high pressure in the model yeast *Saccharomyces cerevisiae*.

Keywords Cold adaptation · Deep-sea · High pressure · Saccharomyces cerevisiae

Y. Nagano (✉) · T. Nagahama · F. Abe
Institute of Biogeosciences, Japan Agency for Marine-Earth Science and Technology (JAMSTEC), 2-15 Natsushima-cho, Yokosuka, Kanagawa 237-0061, Japan
e-mail: y.nagano@jamstec.go.jp

T. Nagahama
Department of Foods and Human Nutrition, The Faculty of Human Life Sciences, Notre Dame Seishin University, 2-16-9, Ifuku-cho, Kita-ku, Okayama 700-8516, Japan

F. Abe
Department of Chemistry and Biological Science, College of Science and Engineering, Aoyama Gakuin University, 5-10-1 Fuchinobe, Chuo-ku, Sagamihara 252-5258, Japan

P. Buzzini and R. Margesin (eds.), *Cold-Adapted Yeasts*,
DOI: 10.1007/978-3-642-39681-6_7,

Contents

7.1 Deep-Sea Environments 150
7.2 Sampling and Isolation of Deep-Sea Yeasts 151
7.3 Yeast Diversity in Deep-Sea Environments 152
7.4 Physiology of Yeasts Isolated from Deep-Sea Environments 160
7.5 Cold and High-Pressure Adaption in the Model Yeast *Saccharomyces cerevisiae* 163
7.6 Conclusions 168
References 169

7.1 Deep-Sea Environments

Nearly three quarters of the Earth's surface area is covered by ocean, and the average depth of ocean is 3,800 m. Deep sea normally refers to ocean depths greater than 200 m. This means that the vast majority of our planet is comprised of deep sea. It is likely that we are only scratching the surface of the largest habitat on Earth, with the greater parts of deep sea remaining unexplored. Nevertheless, extensive deep-sea research in the last few decades has led to new discoveries and contributions to science.

As our land environments differ completely from location to location, deep-sea environments also vary dramatically depending on ocean depth, location, geographic conditions, land factors, effects from the Earth's interior, and many other influences. The absence of sunlight irradiation, temperature, and hydrostatic pressure is a significant influence in deep-sea environments, which varies according to water depth. Ocean is often divided into five zones. From the shallowest zone to the deepest, there is the epipelagic zone (less than 200 m), the mesopelagic zone (200–1,000 m), the bathypelagic zone (1,000–4,000 m), the abyssopelagic zone (4,000–6,000 m), and the hadopelagic zone (the deep trenches below 6,000 m to about 11,000 m) (Fig. 7.1). There is enough sunlight for photosynthesis to take place within the epipelagic zone but not in the other four zones, which are below 200 m. Pressure increases by 1 atmosphere for each 10 m in water depth. Therefore, high hydrostatic pressure up to 110 megapascals (MPa) is present in deep-sea environments in the deepest ocean. From the base of the epipelagic zone, the temperature drops to 5 or 6 °C at 1,000 m (Fig. 7.1). With the exception of geographic proximities near hydrothermal vents where water temperature can be as high as 400 °C, the average water temperature is between −1 and 4 °C in most parts of the deep sea. Deep-sea water temperature is very stable, with no seasonal or annual temperature changes. Therefore, living organisms in deep-sea environments are considered to be adapted to cold environments.

It is noted that the majority of deep-sea research (including investigations on yeast communities) has been carried out in rather unique deep-sea environments, for example, chemosynthetic environments, such as hydrothermal vents (which form

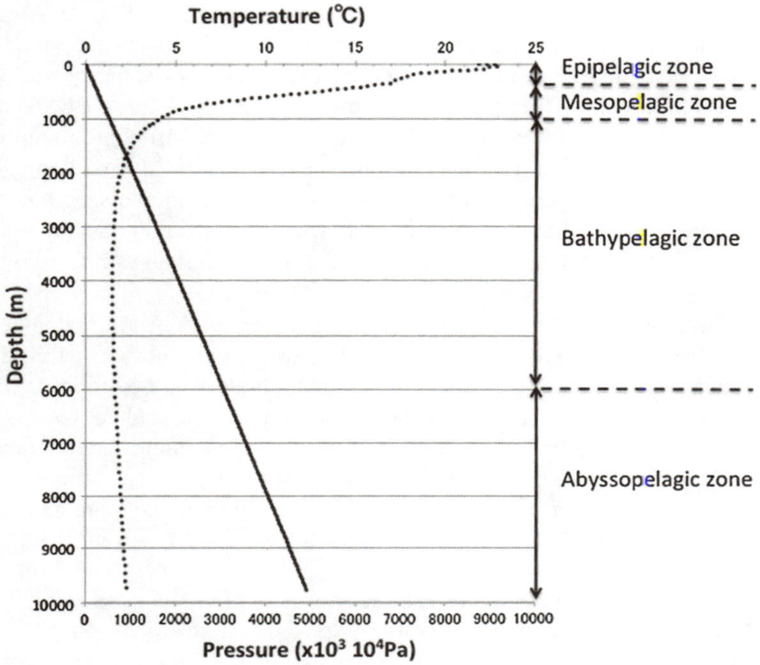

Fig. 7.1 Water temperature and pressure profiles in deep-sea environments with description of different zones. *Straight line* indicates pressure; *dotted line* indicates water temperature

typically along the ocean ridges, where two tectonic plates are diverging, new crust is being formed and hot mineral rich water is spewing) and cold seeps (which form mostly along continental margins and occur over fissures on the seafloor), or geographically interesting environments, such as the deepest part of ocean, the Mariana Trench.

7.2 Sampling and Isolation of Deep-Sea Yeasts

The basic sampling and isolation methods for investigating fungal communities (including yeasts in deep-sea environments) do not differ fundamentally from those used for terrestrial or shallow marine environments. However, great depth and elevated hydrostatic pressure make deep sea a difficult environment to access. Collecting samples from deep sea involves higher costs and more complicated equipment. For example, all deep-sea equipment is required to be high-pressure resistant. There are two basic approaches in reaching deep sea. One is using a human occupied vehicle (HOV), a deep-sea submarine which can carry people inside to the ocean depths. These types of research vessels have some advantages, such as allowing scientists to make direct observations of deep sea. An alternative method is to use a remotely operated vehicle (ROV), which is capable of diving at depths greater than 10,000 m and can collect samples from some of the ocean's deepest environments. Equipment loaded on both HOV and ROVs usually includes a variety of devices, such as cameras (television, still photograph, video), and manipulators, that enable the placement of objects on the seafloor and the collection of various samples. For water sampling, some known samplers are the Nansen bottle, the Niskin sampler, and the van Dorn sampler. Several types of samplers are used to collect deep-sea sediments, such as a grab sampler, box corer, gravity corer, and piston corer.

The isolation procedure for deep-sea yeasts varies depending on the type, volume, and shape of collected samples. Water samples are mostly filtered through membranes and are then used for isolation, due to the low number of yeast cells in the limited volume of water samples. Solid sources, such as sunken wood or deep-sea animals, are normally applied to agar plate media or liquid media after being broken into small pieces or homogenized. Sediment samples are normally directly plated onto medium or mixed with medium agar before they get cold and solidify. Samples can first be diluted with sterile seawater, if they contain high numbers of yeasts. Media often used for the isolation of yeasts from deep-sea environments are not so different from media used to isolate terrestrial yeasts. Several types of media are used to investigate culturable yeasts and filamentous fungal communities in deep-sea environments (Gadanho and Sampaio 2005; Damare et al. 2006; Burgaud et al. 2009; Le Calvez et al. 2009; Jebaraj et al. 2010). Most investigators use organic media, mainly consisting of malt extract, yeast extract, peptone, glucose, and potato starch. Media are usually prepared in artificial seawater, with the addition of antibiotics, such as streptomycin, penicillin, and chloramphenicol, to inhibit bacterial growth. Media are also occasionally used at 1/5 strength of normal condition, in order to simulate the low nutrient conditions found in deep sea. Media are normally incubated between 4 °C and 30 °C. Culturing periods can range from 2 weeks to several months. A long culturing time is sometimes necessary for isolating deep-sea yeasts, as some strains grow extremely slow.

7.3 Yeast Diversity in Deep-Sea Environments

As with other environments, microorganisms are known to play an important role in deep-sea ecosystems. The presence of fungi in deep-sea environments is only recently being recognized, with limited research carried out in comparison with prokaryotic organisms, despite the first isolation of deep-sea fungi has been reported almost 50 years ago (Roth et al. 1964). Since this first report, there have been a growing number of reports noting the isolation and diversity of yeasts and filamentous fungi, including novel species (mostly yeasts) in several deep-sea environments, particularly in the last decade. The reported yeast species isolated by culture-dependent methods from deep-sea environments are shown in Table 7.1.

Table 7.1 Yeast and yeast-like species isolated from deep-sea environments

Yeasts isolated	Location	Source	Latitude	Longitude	Depth (m)	Temp. (°C)	References
Aureobasidium sp.	Central Indian Basin	Sediment	n/a	n/a	4,900–5,390	n/a	Damare et al. (2006)
Candida atlantica	Mid-Atlantic Ridge (Rainbow1)	Water	36° 13.764′ N	33° 54.159′ W	2,295	n/a	Gadanho and Sampaio (2005)
Candida atmosphaerica	Mid-Atlantic Ridge (Rainbow1)	Water	36° 13.764′ N	33° 54.159′ W	2,295	n/a	Gadanho and Sampaio (2005)
Candida carpophila (formerly *Candida fukuyamaensis*)	Mariana Trench	Sediment	11° 21′ N	142° 25′ E	10,897	2.4	Nagahama et al. (2001a)
Candida viswanathii (formerly *Candida lodderae*)	Mid-Atlantic Ridge (Menez Gwen)	Water	37° 50.679′ N	31° 31.138′ W	825	n/a	Gadanho and Sampaio (2005)
Candida oceani	Mid-Atlantic Ridge	Unidentified deep-sea coral	36° 08′ N	34° 00′ W	2,300	n/a	Burgaud et al. (2011)
Candida parapsilosis	Mid-Atlantic Ridge (Menez Gwen)	Water	37° 50.679′ N	31° 31.138′ W	825	n/a	Gadanho and Sampaio (2005)
Candida sake	Yap Trench	Sediment	7° 01′ N	134° 47′ E	6,489	1.7	Nagahama et al. (2001a)
Candida sake	Yap Trench	Sediment	7° 01′ N	134° 45′ E	6,006	1.7	Nagahama et al. (2001a)
Candida sp.	Mid-Atlantic Ridge (Mount Saldanha1)	Water	36° 35.251′ N	33° 26.685′ W	2,116	n/a	Gadanho and Sampaio (2005)
Candida sp.	Mid-Atlantic Ridge (Rainbow1)	Water	36° 13.764′ N	33° 54.159′ W	2,295	n/a	Gadanho and Sampaio (2005)
Candida sp.	Mid-Atlantic Ridge (Rainbow1)	Water	36° 13.764′ N	33° 54.159′ W	2,295	n/a	Gadanho and Sampaio (2005)
Candida sp.	Mid-Atlantic Ridge (Menez Gwen)	Water	37° 50.679′ N	31° 31.138′ W	825	n/a	Gadanho and Sampaio (2005)

(continued)

Table 7.1 (continued)

Yeasts isolated	Location	Source	Latitude	Longitude	Depth (m)	Temp. (°C)	References
Clavispora lusitaniae	Vailulu'u Seamount	Fe-oxide-rich mat	14° 13.056′ S	169° 03.599′ W	881	5	Connell et al. (2009)
Clavispora lusitaniae	Vailulu'u Seamount	Fe-oxide-rich mat	14° 12.889′ S	169° 03.568′ W	707	5.1	Connell et al. (2009)
Coniosporium/*Sarcinomyces* sp.*	Central Indian Basin	Sediment	10–16.5° S	72–77° E	4,000–5,700	n/a	Singh et al. (2010)
Cryptococcus saitoi	Vailulu'u Seamount	Fe-oxide-rich mat	14° 13.056′ S	169° 03.599′ W	881	5	Connell et al. (2009)
Cryptococcus saitoi	Vailulu'u Seamount	Fe-oxide-rich mat	14° 12.889′ S	169° 03.568′ W	707	5.1	Connell et al. (2009)
Cryptococcus surugaensis	Suruga Bay	Sediment	34° 36′ 55 N	138° 34′ 77 E	2,406	3	Nagahama et al. (2003a)
Dioszegia antarctica	Vailulu'u Seamount	Fe-oxide-rich mat	14° 12.889′ S	169° 03.568′ W	707	5.1	Connell et al. (2009)
Dipodascus tetrasporeus	Japan Trench	Sediment	39° 19 · 2359′ N	142° 52 · 5384′ E	1,763	n/a	Nagahama et al. (2008)
Exophiala dermatitidis	Mid-Atlantic Ridge (Menez Gwen)	Water	37° 50.679′ N	31° 31.138′ W	825	n/a	Gadanho and Sampaio (2005)
Graphiola/*Tilletiopsis* sp.*	Central Indian Basin	Sediment	10–16.5° S	72–77° E	4,000–5,700	n/a	Singh et al. (2010)
Kluyveromyces nonfermentans	Suruga or Sagami Bay	Sediment	n/a	n/a	1,200–1,977	n/a	Nagahama et al. (1999)
Kluyveromyces nonfermentans	Suruga Bay	Sediment	n/a	n/a	1,143	2.8	Nagahama et al. (1999)
Kluyveromyces nonfermentans	Suruga Bay	Crab	n/a	n/a	1,182	2.8	Nagahama et al. (1999)
Kluyveromyces nonfermentans	Suruga Bay	Clam *Calyptogena* sp.	n/a	n/a	1,156	3.1	Nagahama et al. (1999)
Meyerozyma guilliermondii (formerly *Pichia guilliermondii*)	Mid-Atlantic Ridge (Mount Saldanha1)	Water	36° 35.251′ N	33° 26.685′ W	2,116	n/a	Gadanho and Sampaio (2005)

(continued)

Table 7.1 (continued)

Yeasts isolated	Location	Source	Latitude	Longitude	Depth (m)	Temp. (°C)	References
Meyerozyma guilliermondii (formerly *Pichia guilliermondii*)	Vailulu'u Seamount	Fe-oxide-rich mat	14° 13.056′ S	169° 03.599′ W	881	5	Connell et al. (2009)
Meyerozyma guilliermondii (formerly *Pichia guilliermondii*)	Vailulu'u Seamount	Fe-oxide-rich mat	14° 12.889′ S	169° 03.568′ W	707	5.1	Connell et al. (2009)
Pseudozyma sp.	Sagami Bay	Clam *Calyptogena* sp.	35° 00′ N	139° 14′ E	1,156	3.1	Nagahama et al. (2001a)
Rhodosporidium diobovatum	Sagami Bay	Clam *Calyptogena* sp.	35° 00′ N	139° 14′ E	1,156	3.1	Nagahama et al. (2001a)
Rhodosporidium diobovatum	Yap Trench	Sediment	6° 20′ N	133° 56′ E	5,561	1.7	Nagahama et al. (2001a)
Rhodosporidium diobovatum	Northern Yap Trench	Sediment	11° 46 · 303′ N	139° 07 · 300′ E	3,702	1.7	Nagahama et al. (2001a)
Rhodosporidium diobovatum	Mid-Atlantic Ridge (Rainbow1)	Water	36° 13.764′ N	33° 54.159′ W	2,295	n/a	Gadanho and Sampaio (2005)
Rhodosporidium diobovatum	Mid-Atlantic Ridge (Mount Saldanha1)	Water	36° 35.251′ N	33° 26.685′ W	2,116	n/a	Gadanho and Sampaio (2005)
Rhodosporidium sp.	Central Indian Basin	Sediment	10–16.5° S	72–77° E	4,000–5,700	n/a	Singh et al. (2010)
Rhodosporidium sphaerocarpum	Yap Trench	Sediment	11° 08′ N	134° 45′ E	6,500	1.7	Nagahama et al. (2001a)
Rhodosporidium sphaerocarpum	Mid-Atlantic Ridge (Menez Hom)	Water	37° 08.222′ N	32° 26.609′ W	1,802	n/a	Gadanho and Sampaio (2005)

(continued)

Table 7.1 (continued)

Yeasts isolated	Location	Source	Latitude	Longitude	Depth (m)	Temp. (°C)	References
Rhodosporidium toruloides	Mid-Atlantic Ridge (Mount Saldanha1)	Water	36° 35.251′ N	33° 26.685′ W	2,116	n/a	Gadanho and Sampaio (2005)
Rhodosporidium toruloides	Vailulu'u Seamount	Fe-oxide-rich mat	14° 13.056′ S	169° 03.599′ W	881	5	Connell et al. (2009)
Rhodotorula/Cryptococcus sp.*	Central Indian Basin	Sediment	10–16.5° S	72–77° E	4,000–5,700	n/a	Singh et al. (2010)
Rhodotorula aurantiaca	Suruga Bay	Sediment	34° 43–57′ N	138° 36–39′ E	1,487–1,977	n/a	Nagahama et al. (2001a)
Rhodotorula benthica	Sagami Bay	Tubeworm	35° 00′ N	139° 14′ E	1,156	3.1	Nagahama et al. (2003b)
Rhodotorula calyptogenae	Sagami Bay	Clam *Calyptogena* sp.	35° 00′ N	139° 14′ E	1,156	3.1	Nagahama et al. (2003b)
Rhodotorula glutinis	Suruga Bay	Sediment	34° 43–57′ N	138° 36–39′ E	1,487–1,977	n/a	Nagahama et al. (2001a)
Rhodotorula glutinis	Sagami Bay	Sediment	34° 59′ N	139° 14′ E	1,241	2.6	Nagahama et al. (2001a)
Rhodotorula glutinis	Japan Trench	Sediment	40° 07′ N	144° 11′ E	6,455	1.7	Nagahama et al. (2001a)
Rhodotorula lamellibranchii	Sagami Bay	Tubeworm *Lamellibrachia* sp.	35° 00′ N	139° 14′ E	1,156	3.1	Nagahama et al. (2001b)
Rhodotorula minuta	Sagami Bay	Tubeworm *Lamellibrachia* sp.	35° 00′ N	139° 14′ E	1,156	3.1	Nagahama et al. (2001a)
Rhodotorula mucilaginosa	Sagami Bay	Tubeworm *Lamellibrachia* sp.	35° 00′ N	139° 14′ E	1,156	3.1	Nagahama et al. (2001a)
Rhodotorula mucilaginosa	Sagami Bay	Clam *Calyptogena* sp.	35° 00′ N	139° 14′ E	1,156	3.1	Nagahama et al. (2001a)
Rhodotorula mucilaginosa	Yap Trench	Sediment	6° 20′ N	133° 56′ E	5,561	1.7	Nagahama et al. (2001a)

(continued)

Table 7.1 (continued)

Yeasts isolated	Location	Source	Latitude	Longitude	Depth (m)	Temp. (°C)	References
Rhodotorula mucilaginosa	Yap Trench	Sediment	7° 18′ N	136° 30′ E	6,455	1.7	Nagahama et al. (2001a)
Rhodotorula mucilaginosa	Yap Trench	Sediment	11° 46′ N	139° 07′ E	3,702	1.7	Nagahama et al. (2001a)
Rhodotorula mucilaginosa	Mariana Trench	Sediment	11° 21′ N	142° 25′ E	10,897	2.4	Nagahama et al. (2001a)
Rhodotorula mucilaginosa	Mid-Atlantic Ridge (Rainbow1)	Water	36° 13.764′ N	33° 54.159′ W	2,295	n/a	Gadanho and Sampaio (2005)
Rhodotorula mucilaginosa	Mid-Atlantic Ridge (Rainbow2)	Water	36° 13.773′ N	33° 54.207′ W	2,316	n/a	Gadanho and Sampaio (2005)
Rhodotorula pacifica	Northern Yap Trench	Sediment	11° 46 · 303′ N	139° 07 · 300′ E	3,702	1.7	Nagahama et al. (2006)
Rhodotorula pacifica	Iheya Ridge	Sediment	27° 27 · 240′ N	126° 53 · 892′ E	991	4.7	Nagahama et al. (2006)
Rhodotorula sp.	Central Indian Basin	Sediment	10–16.5° S	72–77° E	4,000–5,700	n/a	Singh et al. (2010)
Rhodotorula sp.	Mid-Atlantic Ridge (Mount Saldanha2)	Water	36° 33.916′ N	33° 25.887′ W	2,198	n/a	Gadanho and Sampaio (2005)
Rhodotorula sp.	Mid-Atlantic Ridge (Mount Saldanha1)	Water	36° 35.251′ N	33° 26.685′ W	2,116	n/a	Gadanho and Sampaio (2005)
Sarcinomyces petricola	Yap Trench	Sediment	7° 17′ N	136° 30′ E	6,198	1.7	Nagahama et al. (2001a)
Sarcinomyces petricola	Northern Yap Trench	Sediment	11° 46 · 303′ N	139° 07 · 300′ E	3,702	1.7	Nagahama et al. (2001a)
Sporidiobolus salmonicolor	Vailulu'u Seamount	Fe-oxide-rich mat	14° 12.889′ S	169° 03.568′ W	707	5.1	Connell et al. (2009)

(continued)

Table 7.1 (continued)

Yeasts isolated	Location	Source	Latitude	Longitude	Depth (m)	Temp. (°C)	References
Sporidiobolus sp.	Central Indian Basin	Sediment	10–16.5° S	72–77° E	4,000–5,700	n/a	Singh et al. (2010)
Sporidiobolus pararoseus (formerly *Sporobolomyces pararoseus*)	Sagami Bay	Clam *Calyptogena* sp.	35° 00′ N	139° 14′ E	1,156	3.1	Nagahama et al. (2001a)
Sporobolomyces salmonicolor (anamorph of *Sporidiobolus salmonicolor*)	Sagami Bay	Clam *Calyptogena* sp.	35° 00′ N	139° 14′ E	1,156	3.1	Nagahama et al. (2001a)
Sporobolomyces salmonicolor (anamorph of *Sporidiobolus salmonicolor*)	Iheya Ridge	Mussel *Bathymodiolus* sp.	27° 47′ N	126° 54′ E	1,050	n/a	Nagahama et al. (2001a)
Sporobolomyces shibatanus (anamorph of *Sporidiobolus pararoseus*)	Suruga Bay	Sediment	n/a	n/a	1,487–1,977	n/a	Nagahama et al. (2001a)
Trichosporon dermatis	Mid-Atlantic Ridge (Rainbow1)	Water	36° 13.764′ N	33° 54.159′ W	2,295	n/a	Gadanho and Sampaio (2005)
Unidentified yeasts	Central Indian Basin	Sediment	n/a	n/a	4,900–5,390	n/a	Damare et al. (2006)

* Identification by 18S rRNA gene/ITS regions

The yeast species reported from hydrothermal vent environments are also included in this chapter. This is because intense heat from hydrothermal vent openings is limited to a small area and the water temperature drops to cold as the ambient temperature of deep-sea water only a few centimeters away from the vents. Thus, yeasts present in deep-sea hydrothermal environments are also considered to be adapted to cold environments as well as other deep-sea yeasts.

Although sampling sites are often selected with a bias toward interesting geographic locations, such as hydrothermal vents, and may not fully represent the diversity of yeast present in general deep-sea environments, the most frequently isolated yeast species from deep-sea environments belong to the genera *Rhodosporidium*, *Rhodotorula*, and *Candida* (Table 7.1). Some novel species of the genera *Rhodotorula* and *Candida* have been found in deep-sea environments (Nagahama et al. 2001b, 2003a, b, 2006; Burgaud et al. 2011). Yeast species of the genera *Clavispora*, *Cryptococcus*, *Dioszegia*, *Dipodascus*, *Kluyveromyces*, *Pichia*, *Pseudozyma*, *Sporidiobolus*, *Sporobolomyces*, and *Trichosporon* have also been isolated from deep-sea environments. Yeast-like fungi belonging to the genera *Aureobasidium*, *Sarcinomyces*, and *Exophiala* have also been found (Table 7.1). Nagahama et al. (2001a) isolated 99 yeast strains from benthic animals and sediments collected from deep-sea floor in various areas in the northwest Pacific Ocean. Of these 99 yeast strains, 40 strains were pink-pigmented yeasts belonging to the genera *Rhodotorula* and *Sporobolomyces*. Nagahama et al. (2001a) showed that pink-pigmented yeasts are most common among yeasts isolated from deep sea, which is also commonly observed in shallow marine systems (Hagler and Ahearn 1987). However, in deep-sea hydrothermal environments, this disproportion was not observed, with non-pigmented yeasts being found much more abundantly than pink-pigmented yeasts (Gadanho and Sampaio 2005). The same authors have also suggested that this may be due to the unique conditions of hydrothermal vents. However, further study is needed.

With recent advances in technology, molecular-based methods are more often used for studying fungal diversity in deep-sea environments. The advantages of using molecular methods are mainly the requirement of only a small amount of samples and the detection of difficult-to-culture or unculturable species, including rare species. Yeast species which have not been reported by culture-dependent methods have been detected by culture-independent methods from deep-sea environments (Bass et al. 2007; Lai et al. 2007; Le Calvez et al. 2009; Nagano et al. 2010; Singh et al. 2010, 2011; Nagahama et al. 2011). Strains belonging to the genera *Rhodosporidium*, *Rhodotorula*, and *Candida* have been frequently found in deep-sea environments, using culture-independent methods. On the contrary, species of the genera *Cryptococcus*, *Pichia*, and *Trichosporon* seem to be detected more frequently using culture-independent methods than by using culture-dependent methods (Bass et al. 2007; Lai et al. 2007; Nagano et al. 2010; Singh et al. 2010). The basidiomycetous yeast *Cryptococcus curvatus* was found to be the dominant eukaryotic microorganisms in oxygen-depleted sediments from deep-sea methane seeps at the Kuroshima Knoll (Takishita et al. 2006). However, this phenomenon has not been reported in other deep-sea methane cold seeps.

It would be interesting to undertake a physiological study using a cultured isolate and to investigate correlations between the yeast and environmental conditions. Strains of the genus *Malassezia* have never been isolated by culture-dependent methods but are frequently reported from culture-independent methods (Bass et al. 2007; Lai et al. 2007; Singh et al. 2010, 2011; Nagahama et al. 2011). As *Malassezia* species are known to be difficult to grow under laboratory conditions, a diversity study based on culture-dependent methods may often miss these organisms.

Overall, the yeast diversity found in deep-sea environments is relatively limited and similar to those found in other cold environments, such as glacial habitats and polar regions (Buzzini et al. 2012; Singh et al. 2013). However, typical psychrophilic yeasts often found in many cold climates, such as the genera *Mrakia* and *Mrakiella* (de Garcia et al. 2007; Turchetti et al. 2008; Thomas-Hall et al. 2010), have not been found in deep-sea environments to date. Furthermore, culture-independent methods have detected many unknown yeast phylotypes from deep-sea environments. Bass et al. (2007) discovered three novel Ustilaginomycetes phylotypes which were most closely related to the genus *Malassezia*. They also discovered highly novel phylotypes closely related to species belonging to the genera *Candida* and *Metschnikowia*. These phylotypes have been reported frequently from several deep-sea environments, but no isolates have been reported to date (Bass et al. 2007; Lai et al. 2007; Nagano et al. 2010; Nagahama et al. 2011; Thaler et al. 2012). Several other novel phylotypes have been detected by culture-independent methods from deep-sea environments and are comprehensively reviewed in Richards et al. (2012), along with novel phylotypes detected from shallow marine environments.

The abundance of yeasts in deep-sea environments remains unclear. Gadanho and Sampaio (2005) reported that the number of colony-forming units in deep-sea water ranged from 0 to 5.9×10^3 L^{-1}. However, further study needs to be carried out.

7.4 Physiology of Yeasts Isolated from Deep-Sea Environments

Although the physiology of yeasts isolated from deep-sea environments has not been as thoroughly studied as that of yeasts from other cold environments, or prokaryotic organisms from deep-sea environments, the growth characteristics of deep-sea yeast isolates under simulated deep-sea conditions have been examined in several studies (Damare et al. 2006; Singh et al. 2010). Deep-sea conditions are mainly characterized by high hydrostatic pressure, and low temperature and salt content (3.5 % w/v). Many deep-sea yeasts can grow at 20 MPa, with some yeasts showing better biomass production at 20 MPa/5 °C than at 20 MPa/30 °C (Singh et al. 2010). However, even though most deep-sea yeasts are piezotolerant, they usually show better growth at low pressures. No piezophilic fungi have been reported, in contrast to many reported piezophilic prokaryotic microorganisms from deep-sea environments (Nogi et al. 2004; Wang et al. 2004; Takai et al. 2009; Birrien et al. 2011). Similarly, most yeasts recovered from deep-sea environments are psychrotolerant, but isolates of yeasts from deep-sea sediments grew more rapidly at 30 °C than 5 °C (Singh et al. 2010).

The majority of yeasts isolated from deep-sea environments are halotolerant (Burgaud et al. 2009; Le Calvez et al. 2009). Halophilic yeasts which showed an optimal growth at 3 % (w/v) sea salts have been reported from deep-sea hydrothermal vents (Burgaud et al. 2010). However, Damare et al. (2006) reported that isolates of yeasts from deep-sea sediments did not have an absolute requirement for seawater for growth. In this study, we also compared the growth characteristics of a few *Rhodotorula* spp. originating from deep-sea environments and terrestrial environments. There was no clear difference between terrestrial strains and deep-sea strains in terms of growth measurements at different temperatures, i.e., at 3 °C and 10 °C. For all the strains, better growth was observed at 10 °C (Fig. 7.2). In the experiments investigating the effect of hydrostatic pressure on cell growth of deep-sea and terrestrial yeast strains, the majority of deep-sea strains showed better growth under pressures of 20 MPa and 40 MPa than terrestrial strains grown at 3 °C and 10 °C. Interestingly, hydrostatic pressure more strongly affects the growth of some deep-sea yeast strains at 10 °C than at 3 °C (Fig. 7.3). This result compliments results shown by Singh et al. (2010). These results suggest that deep-sea yeasts may have mechanisms to adapt to low temperatures when high pressure

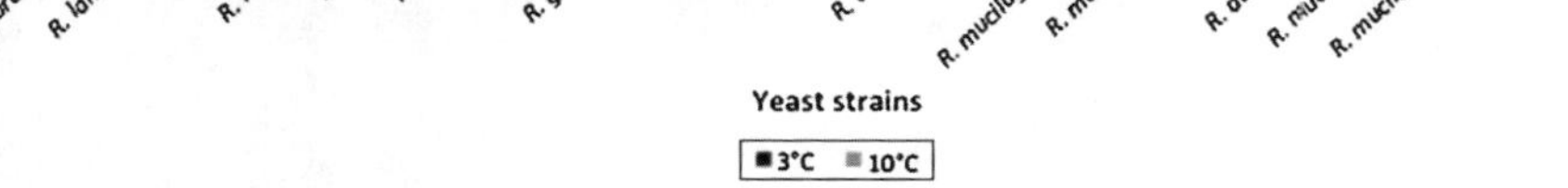

Fig. 7.2 Cell growth of deep-sea and terrestrial yeast strains at 3 °C and 10 °C. Cultures were cultivated with static liquid medium of YM (Difco) for 3 weeks. OD660 was measured from a 10-fold diluted solution of final growth media. Strains used in this study: Erythrobasidiales (*Rhodotorula aurantiaca* and *Rhodotorula lamellibrachii*), Naohideales (*Rhodotorula benthica, Rhodotorula minuta*, and *Rhodotorula slooffiae*), and Sporidiobolales (*Rhodosporidium toruloides, Rhodotorula dairenensis, Rhodotorula glutinis, Rhodotorula mucilaginosa*, and *Rhodotorula pacifica*). Three strains JCM 3777, JCM 3771, and JCM 8208 were isolated from terrestrial origins. The origin of JCM 8115 is unknown. Deep-sea strains were collected from three kinds of sources: sediment (SY-58, SY-96, SY-98, SY-100, SY-102, SY-103), giant white clams *Calyptogena* spp. (SY-77, SY-78, and SY-84), and a tubeworm *Lamellibrachia* sp. (SY-91). Locations of sources are described by Nagahama et al. (2001a, b)

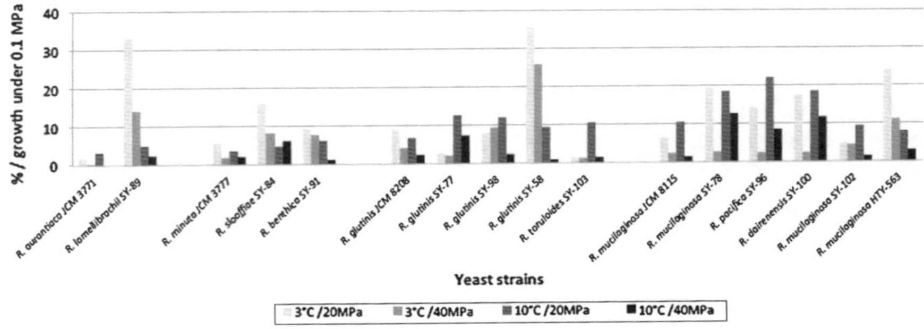

Fig. 7.3 Effect of hydrostatic pressure on cell growth of deep-sea and terrestrial yeast strains. Growth suppression under 20 or 40 MPa hydrostatic pressure is represented by percentage of growth (OD660) per cell growth under atmospheric pressure (0.1 MPa). Strains used in this study: Erythrobasidiales (*Rhodotorula aurantiaca* and *Rhodotorula lamellibrachii*), Naohideales (*Rhodotorula benthica, Rhodotorula minuta*, and *Rhodotorula slooffiae*), and Sporidiobolales (*Rhodosporidium toruloides, Rhodotorula dairenensis, Rhodotorula glutinis, Rhodotorula mucilaginosa*, and *Rhodotorula pacifica*). Cultures were grown with static liquid medium of YM (Difco) for 3 weeks at 3 or 10 °C. OD660 was measured from a 10-fold diluted solution of final growth media

is present. For growth measurement in the presence of different NaCl concentrations, all strains showed better growth with normal yeast mold (YM) broth (yeast extract 0.3 %, malt extract 0.3 %, peptone 0.5 %, dextrose 1 %) without adding extra NaCl (Fig. 7.4); 11 deep-sea strains out of 12 (92 %) and 2 terrestrial strains out of 4 (50 %) showed growth in media in the presence of 10 % (w/v) NaCl

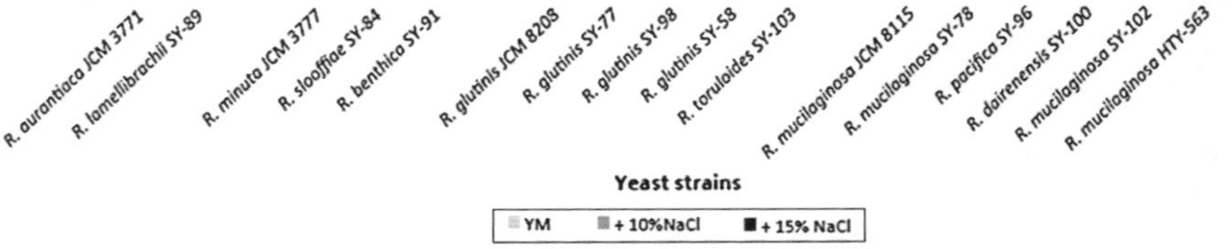

Fig. 7.4 Effect of NaCl concentration in medium on cell growth of red yeast strains from deep-sea and terrestrial origins. Strains used in this study: Erythrobasidiales (*Rhodotorula aurantiaca* and *Rhodotorula lamellibrachii*), Naohideales (*Rhodotorula benthica, Rhodotorula minuta*, and *Rhodotorula slooffiae*), and Sporidiobolales (*Rhodosporidium toruloides, Rhodotorula dairenensis, Rhodotorula glutinis, Rhodotorula mucilaginosa*, and *Rhodotorula pacifica*). Cultures were grown with each static liquid medium 3 weeks at 20 °C. OD660 was measured using a 100-fold diluted solution of final growth media

concentration. However, only deep-sea strains (3 out of 12) showed growth in media containing 15 % (w/v) NaCl. All over, yeast strains isolated from deep-sea environments are generally more adapted to deep-sea environments, such as high pressure and high salinity, but yeasts strains isolated from terrestrial environments can also survive at 3 °C/40 MPa, which is similar to conditions prevailing in deep-sea environments (Fig. 7.3). It is noted that although no piezophilic or psychrophilic yeasts have been reported from deep-sea environments to date, it may be possible to isolate piezophilic fungi and psychrophilic fungi endemic to deep-sea environments by using unique culturing systems, such as selective culturing at high pressure, or long-term culturing in bioreactors with continuous flow.

7.5 Cold and High-Pressure Adaption in the Model Yeast *Saccharomyces cerevisiae*

Marine organisms have developed their physiological functions to survive and reproduce themselves under extreme conditions, such as low temperature and high hydrostatic pressure. Although the effects of low temperature have been widely investigated in a variety of biological systems, the effects of high hydrostatic pressure in organisms have not been thoroughly analyzed. There is growing interest in the potential application of microbes which inhabit in deep-sea extreme environments (Yayanos 1995; Abe and Horikoshi 2001; Abe 2007a). Hydrostatic pressures in the range of several dozen MPa do not readily kill microbial cells but exert adverse effects on the growth of organisms that have adapted to atmospheric pressure (Abe et al. 1999; Bartlett 2002; Aertsen et al. 2009). One of the practical difficulties in cultivating microorganisms at high pressure over extended periods for more than a few days is having a poor oxygen supply in closed hydrostatic chambers. This limits investigations of the physiology of deep-sea yeasts, because most yeasts require sufficient oxygen in order to grow. To solve this problem, a new cultivation system was developed to enable a continuous oxygen supply for aerobes by circulating fresh medium under high pressure (Hiraki et al. 2012). The molecular understandings of the effects of high pressure in deep-sea yeasts are also hampered by difficulties in performing genetic manipulations, e.g., gene disruption, overexpression, or site-directed mutagenesis on microbes occurring in natural environments.

The budding yeast *Saccharomyces cerevisiae* is the first eukaryote of which the complete genome sequence was publicly available (*Saccharomyces* Genome database, http://www.yeastgenome.org/). A number of sophisticated techniques in molecular genetics can be used in any laboratory. *S. cerevisiae* is a facultative anaerobe, and thus, a useful model organism in high-pressure experiments. Although this organism is a mesophile, the systematic analyses have illuminated a substantial amount of cell responses to increasing hydrostatic pressure (Abe 2007b). The yeast genome encodes approximately 6,600 genes, of which 5,000 genes have been verified. Recent large-scale phenotypic screening of the *S. cerevisiae* gene deletion library has revealed numerous unexpected genes and metabolic pathways that are involved in maintaining life under high pressure (Abe and Minegishi 2008).

Amino acids are imported by specific permeases/transporters in the plasma membrane. The uptake of tryptophan in *S. cerevisiae* is mediated by tryptophan permeases Tat1 and Tat2 and is one of the processes most sensitive to high pressure and cold in the physiology of *S. cerevisiae*. Consequently, the availability of tryptophan limits cell growth under high pressure and low temperatures. Experimental wild-type strains generally carry several nutrient auxotrophic markers, such as *ura3* (uracil), *ade2* (adenine), *his3* (histidine), *leu2* (leucine), *lys2* (lysine), or *trp1* (tryptophan) for plasmid selection. Regardless of other nutrient auxotrophy, *trp1* confers growth deficiency at a pressure of 15–25 MPa or at low temperatures of 10–15 °C, whereas tryptophan-prototrophic strains are capable of growth under these pressure and temperature ranges (Abe and Horikoshi 2000). Any factors that lead to increased tryptophan availability enable *trp1* cells to grow under these conditions: The addition of excess tryptophan to the medium, introduction of the *TRP1* gene to confer tryptophan biosynthesis, or overexpression of the genes encoding Tat1 or Tat2 enable *trp1* cells to grow at high pressure or low temperatures (Abe and Horikoshi 2000). Additionally, Tat1 and Tat2 undergo vacuolar degradation in a manner dependent on ubiquitination and the endocytic pathway, when cells are cultured at high pressure or low temperature (Abe and Iida 2003). The similarity between high pressure and low temperature is related to the physicochemical properties of the lipid membrane. Primarily, increasing pressure and decreasing the temperature increase the order of lipid bilayers and restrict the rotational acyl chain motion. With increasing pressure, the gel-to-liquid crystalline coexistence region is shifted toward higher temperatures by approximately 22 °C/100 MPa (Winter and Dzwolak 2005). For example, a pressure increase of 100 MPa increases the main transition (L_b/L_a) temperature of the stearoyloleylphosphatidylcholine (SOPC) and dioleylphosphatidylcholine (DOPC) membrane by 18.1 °C and 23.3 °C, respectively (Kaneshina et al. 1998). In this sense, tryptophan uptake by yeast cells is sensitive to decreases in membrane fluidity caused by either high pressure or low temperature. Interestingly, further investigations have highlighted unambiguous correlations between high-pressure effects and low-temperature effects over a variety of cellular processes (see below).

Global transcriptional responses to high pressure have been analyzed in *S. cerevisiae* in some laboratories using DNA microarray hybridization, but results have varied with the experimental conditions, such as yeast strains, pressure (25–200 MPa), temperature (4–30 °C), and duration of pressurization (10 min to overnight). Iwahashi et al. (2005) reported that high-pressure treatment at 30 MPa and 25 °C upregulated 366 genes more than twofold and downregulated 253 genes more than twofold in a tryptophan-prototrophic strain. Many of the upregulated genes are categorized into stress response and metabolism of carbohydrates, lipids and amino acids including *INO1*, *OPI3*, *PST1*, *RTA1*, *SED1*, *PRM5*, and *POX1* genes, which are involved in biosynthesis and the formation of membrane structure (Iwahashi et al. 2005). Fernandes et al. (2004) reported that exposure of the cells to a lethal pressure of 200 MPa for 30 min upregulated genes involved in stress defense and carbohydrate metabolism and downregulated genes involved in cell cycle progression and protein synthesis.

In our study, we compared the global transcriptional profiles in cells grown under normal growth conditions (0.1 MPa, 24 °C), high pressure (25 MPa, 24 °C), and low temperature (0.1 MPa, 15 °C) (Abe 2007b). There was a greater similarity in transcriptional profiles between the cells grown under high pressure and low temperature. Of the 6,337 genes tested, 561 were concurrently upregulated by high pressure and low temperature, while 161 were downregulated. In particular, the *DAN/TIR* cell wall mannoprotein genes were remarkably upregulated by high pressure and low temperature allowing the cells to establish a physical defensive system on the cell wall structure (Abe 2007b). It is known that expression of the *DAN/TIR* genes is stimulated by hypoxia (Abramova et al. 2001). Accordingly, there is a fascinating cross talk in the regulatory networks of *DAN/TIR* transcription by hypoxia, high pressure, and low temperature. How seemingly unrelated environmental conditions induce *DAN/TIR* gene expression is unknown. Under hypoxia, yeast cells fail to introduce unsaturated bonds in acyl chains of the membrane lipids. This might lead to a decrease in membrane fluidity, which is analogously caused by high pressure and low temperature. In this sense, a membrane sensor(s) is likely to exist in the membrane and may transduce the changes in membrane fluidity into intracellular signals that stimulate transcription of the *DAN/TIR* genes.

A PCR-generated deletion strategy was used to replace each yeast open-reading frame from its start-to-stop codon systematically in a tryptophan-prototrophic strain (http://www-sequence.stanford.edu/group/yeast_deletion_project/project_desc.html#intro). Of the 6,600 genes encoded by the yeast genome, single deletions for approximately 4,800 genes do not cause lethality of the cells under normal culture conditions. The use of the library consisting of the 4,800 deletion strains allowed us to obtain mutants defective in growth under high pressure (25 MPa and 24 °C) or low temperature (0.1 MPa and 15 °C). The global functional screening revealed 80 genes in which 71 genes were required for high-pressure growth and 56 genes were required for low-temperature growth with a significant overlap of 47 genes (Table 7.2) (Abe and Minegishi 2008). These genes are involved in a broad range of cellular functions, such as amino acid biosynthesis, mitochondrial function, the actin cytoskeleton, membrane trafficking, transcription, ribosome biogenesis, chromatin structure, and unknown roles. This in turn means that various cellular functions support cell growth of the yeast under high pressure and low temperature. Consistent with high-pressure sensitivity of tryptophan auxotrophic strains, the losses of genes involved in tryptophan biosynthesis such as *ARO1*, *ARO2*, *TRP1*, *TRP2*, *TRP4*, and *TRP5* resulted in growth defects under high pressure and low temperature. Deletion of amino acid biosynthetic genes, such as *HOM3*, *THR4*, and *SER1*, resulted in auxotrophy for methionine (*hom3*D), threonine (*thr4*D), and serine (*Ser1*D), and thus, the growth of these mutants depends on external amino acids. Because the three mutants also

Table 7.2 Genes required for growth under high-pressure and low-temperature conditions (results from Abe and Minegishi 2008)

Gene name	Description
Amino acid biosynthesis	
TRP1	Phosphoribosylanthranilate isomerase
TRP4	Anthranilate phosphoribosyl transferase
THR4	Threonine synthase
ARO2	Bifunctional chorismate synthase and flavin reductase
ARO1	AROM protein, catalyzes steps 2 through 6 in the biosynthesis of chorismate
TRP5	Tryptophan synthase
HOM3	Aspartate kinase
TRP2	Anthranilate synthase
LEU3	Zinc-finger transcription factor
SER1	3-Phosphoserine aminotransferase
Microautophagy	
GTR2	Cytoplasmic GTP binding protein
GTR1	Cytoplasmic GTP binding protein
EGO1	Component of the EGO/GSE complex
EGO3	Component of the EGO/GSE complex
Mitochondrial function	
MRPL22	Mitochondrial ribosomal protein of the large subunit
MRF1	Mitochondrial polypeptide chain release factor
CAF17	Mitochondrial protein that interacts with Ccr4
ACO1	Aconitase, required for the tricarboxylic acid cycle
MRP51	Mitochondrial ribosomal protein of the large subunit
MRPL38	Mitochondrial ribosomal protein of the large subunit
ATP15	Epsilon subunit of the F1 sector of mitochondrial F_1F_0 ATP synthase
MDJ1	Protein involved in folding of mitochondrially synthesized proteins
MSY1	Mitochondrial tyrosyl-tRNA synthetase
Actin organization/bud formation	
LTE1	Putative GDP/GTP exchange factor
HOF1	Bud neck-localized, SH3 domain-containing protein
SLM3	tRNA-specific 2-thiouridylase
CLA4	Cdc42-activated signal-transducing kinase
CDC50	Endosomal protein that regulates cell polarity
SLM6	Protein with a potential role in actin cytoskeleton organization
Membrane trafficking	
VID24	Peripheral membrane protein located at Vid vesicles
VPS34	Phosphatidylinositol 3-kinase
SEC22	R-SNARE protein
PEP3	Vacuolar peripheral membrane protein
CHC1	Clathrin heavy chain
PEP5	Peripheral vacuolar membrane
VPS45	Protein of the Sec1/Munc-18 family
ERG24	C-14 sterol reductase, acts in ergosterol biosynthesis
VPS54	Component of the Golgi-associated retrograde protein complex
AKR1	Palmitoyl transferase
SAC1	Lipid phosphoinositide phosphatase

(continued)

Table 7.2 (continued)

Gene name	Description
Inositol phosphate metabolism	
PLC1	Phosphoinositide-specific phospholipase C
ARG82	Inositol polyphosphate multikinase
PHO88	Probable membrane protein, involved in phosphate transport
KCS1	Inositol hexaphosphate kinase
Transcriptio/mRNA degradation	
SNF6	Subunit of the SWI/SNF chromatin remodeling complex
MOT2	Component of the CCR4-NOT transcription regulatory complex
POP2	RNase of the DEDD superfamily
SHE3	Protein that acts as an adaptor between Myo4 and the She2-mRNA complex
CDC73	Constituent of Paf1 complex with RNA polymerase II
RPB4	RNA polymerase II subunit B32
HFI1	Adaptor protein required for structural integrity of the SAGA complex
PAF1	RNA polymerase II-associated protein
ELF1	Transcription elongation factor
SNF1	AMP-activated serine/threonine protein kinase
SRB5	Subunit of the RNA polymerase II mediator complex
TAF14	Subunit of TFIID, TFIIF, and SWI/SNF complexes
CCR4	Component of the Ccr4-Not transcriptional complex
SAP155	Protein that forms a complex with the Sit4 protein phosphatase
Ribosome	
RPL1B	*N*-terminally acetylated protein component of the large ribosomal subunit
RPL21A	Protein component of the large ribosomal subunit
RPS30B	Protein component of the small ribosomal subunit
Chromatin maintenance	
NBP2	Protein involved in the HOG pathway
YAF9	Subunit of the NuA4 histone H4 acetyltransferase complex and the SWR1 complex
IES2	Associated protein with the INO80 chromatin remodeling complex
CGI121	Promoting telomere uncapping and elongation and transcription
ARD1	Subunit of the *N*-terminal acetyltransferase NatA
Stress response	
HSP31	Possible chaperone and cysteine protease
YDJ1	Protein chaperone involved in regulation of the Hsp90 and Hsp70 functions
Unknown genes	
AVL9	ND
YDR008C	Complementary to *TRP1*
YKL098 W	ND
DLT1	ND
CSF1	Protein required for fermentation at low temperature
YHR151C	ND
YPR153 W	ND
YBR255 W	ND
YGL218 W	ND
YDL172C	ND
YDL173 W	ND
YDR442 W	ND

ND not determined

exhibit growth defects, high pressure and low temperature impair the ability to import these amino acids, potentially inactivating their permeases. Among 11 deletion strains for mitochondrial proteins, the *aco1*D and *caf17*D mutants require either glutamine or glutamate for growth. Taking the results together, permease-dependent uptake of amino acids generally becomes compromised by high hydrostatic pressure and low temperature.

The establishment of appropriate actin networks is a prerequisite for polarized cell growth in the budding yeast. *CDC50* encodes a protein required for actin cytoskeleton organization and trafficking of proteins between the Golgi complex and the endosome/vacuole. The lack of *CDC50* confers low-temperature sensitivity (Misu et al. 2003) and also high-pressure sensitivity (Abe and Minegishi 2008). Cdc50 is known to associate with Drs2, a P-type ATPase of the aminophospholipid translocase that functions in phospholipids asymmetry (Chen et al. 2006). The lack of *DRS2* causes sensitivity to low temperature (Chen et al. 1999) and also high pressure (Abe and Minegishi 2008). Therefore, the Drs2–Cdc50 complex is likely to play an essential role in actin network formation at high pressure and low temperature by modulating protein trafficking appropriately. Genes involved in membrane trafficking are also responsible for growth at high pressure and low temperature, and hence delivery of newly synthesized proteins to appropriate locations, e.g., the bud neck, cell surface, or cell wall, could be diminished by high pressure and low temperature when one of the membrane trafficking proteins is lost. Ergosterol is a major constituent of the plasma membrane and has a significant role in growth under high pressure and low temperature. Deletion mutants for one of the genes involved in the later steps of ergosterol biosynthesis, such as *ERG24, ERG2, ERG6*, and *ERG3*, are viable but accumulate structurally abnormal sterols in the cells. Such deletion mutants exhibit marked growth defects under high pressure and low temperature, suggesting that the structural motif of ergosterol is required for function and/or trafficking of membrane proteins under the extreme conditions and hence cell growth (Abe and Minegishi 2008).

7.6 Conclusions

The use of genetic databases and applying techniques for molecular biology in studies of the model yeast *S. cerevisiae* have improved our fundamental understanding of the effects of low temperature and high hydrostatic pressure in living cells. Although this understanding may not be readily applicable to natural yeasts occurring in deep sea, they provide clues in searching for survival strategies in high-pressure cold environments. Physiological and biochemical studies on deep-sea yeasts in combination with genetic analysis on the model yeast will shed light on the physiology of mysterious deep-sea animals. Breakthroughs will be based on detailed descriptions of experimental conditions, results, and the cooperative interaction between investigators in high-pressure biology regardless of their objectives or types of target organism.

References

Abe F (2007a) Exploration of the effects of high hydrostatic pressure on microbial growth, physiology and survival: perspectives from piezophysiology. Biosci Biotechnol Biochem 71:2347–2357

Abe F (2007b) Induction of *DAN/TIR* yeast cell wall mannoprotein genes in response to high hydrostatic pressure and low temperature. FEBS Lett 581:4993–4998

Abe F, Horikoshi K (2000) Tryptophan permease gene *TAT2* confers high-pressure growth in *Saccharomyces cerevisiae*. Mol Cell Biol 20:8093–8102

Abe F, Horikoshi K (2001) The biotechnological potential of piezophiles. Trends Biotechnol 19:102–108

Abe F, Iida H (2003) Pressure-induced differential regulation of the two tryptophan permeases Tat1 and Tat2 by ubiquitin ligase Rsp5 and its binding proteins, Bul1 and Bul2. Mol Cell Biol 23:7566–7584

Abe F, Minegishi H (2008) Global screening of genes essential for growth in high-pressure and cold environments: searching for basic adaptive strategies using a yeast deletion library. Genetics 178:851–872

Abe F, Kato C, Horikoshi K (1999) Pressure-regulated metabolism in microorganisms. Trends Microbiol 7:447–453

Abramova N, Sertil O, Mehta S, Lowry CV (2001) Reciprocal regulation of anaerobic and aerobic cell wall mannoprotein gene expression in *Saccharomyces cerevisiae*. J Bacteriol 183:2881–2887

Aertsen A, Meersman F, Hendrickx ME, Vogel RF, Michiels CW (2009) Biotechnology under high pressure: applications and implications. Trends Biotechnol 27:434–441

Bartlett DH (2002) Pressure effects on in vivo microbial processes. Biochim Biophys Acta 1595:367–381

Bass D, Howe A, Brown N, Barton H, Demidova M, Michelle H, Li L, Sanders H, Watkinson SC, Willcock S, Richards TA (2007) Yeast forms dominate fungal diversity in the deep oceans. Proc Biol Sci 274:3069–3077

Birrien J-L, Zeng X, Jebbar M, Cambon-Bonavita M-A, Querellou J, Oger P, Bienvenu N, Xiao X, Prieur D (2011) *Pyrococcus yayanosii* sp nov., an obligate piezophilic hyperthermophilic archaeon isolated from a deep-sea hydrothermal vent. Int J Syst Evol Microbiol 61:2827–2831

Burgaud G, Le Calvez T, Arzur D, Vandenkoornhuyse P, Barbier G (2009) Diversity of culturable marine filamentous fungi from deep-sea hydrothermal vents. Environ Microbiol 11:1588–1600

Burgaud G, Arzur D, Durand L, Cambon-Bonavita MA, Barbier G (2010) Marine culturable yeasts in deep-sea hydrothermal vents: species richness and association with fauna. FEMS Microbiol Ecol 73:121–133

Burgaud G, Arzur D, Sampaio JP, Barbier G (2011) *Candida oceani* sp. nov., a novel yeast isolated from a Mid-Atlantic Ridge hydrothermal vent (−2,300 m). A van Leeuwenhoek 100:75–82

Buzzini P, Branda E, Goretti M, Turchetti B (2012) Psychrophilic yeasts from worldwide glacial habitats: diversity, adaptation strategies and biotechnological potential. FEMS Microbiol Ecol 82:217–241

Chen CY, Ingram MF, Rosal PH, Graham TR (1999) Role for Drs2p, a P-type ATPase and potential aminophospholipid translocase, in yeast late Golgi function. J Cell Biol 147:1223–1236

Chen S, Wang J, Muthusamy BP, Liu K, Zare S, Andersen RJ, Graham TR (2006) Roles for the Drs2p-Cdc50p complex in protein transport and phosphatidylserine asymmetry of the yeast plasma membrane. Traffic 7:1503–1517

Connell L, Barrett A, Templeton A, Staudigel H (2009) Fungal Diversity Associated with an Active Deep Sea Volcano: Vailulu'u Seamount, Samoa. Geomicrobiol J 26:597–605

Damare S, Raghukumar C, Raghukumar S (2006) Fungi in deep-sea sediments of the Central Indian Basin. Deep-Sea Research Part I-Oceanographic Research Papers 53:14–27
de Garcia V, Brizzio S, Libkind D, Buzzini P, van Broock M (2007) Biodiversity of cold-adapted yeasts from glacial meltwater rivers in Patagonia, Argentina. FEMS Microbiol Ecol 59:331–341
Fernandes PM, Domitrovic T, Kao CM, Kurtenbach E (2004) Genomic expression pattern in *Saccharomyces cerevisiae* cells in response to high hydrostatic pressure. FEBS Lett 556:153–160
Gadanho M, Sampaio JP (2005) Occurrence and diversity of yeasts in the mid-Atlantic ridge hydrothermal fields near the Azores Archipelago. Microb Ecol 50:408–417
Hagler AN, Ahearn DG (1987) Ecology of aquatic yeasts. In: Rose AH, Harrison JS (eds) The yeasts, vol 1, 2nd edn. Academic Press, London, pp 181–205
Hiraki T, Sekiguchi T, Kato C, Hatada Y, Maruyama T, Abe F, Konishi M (2012) New type of pressurized cultivation method providing oxygen for piezotolerant yeast. J Biosci Bioeng 113:220–223
Iwahashi H, Odani M, Ishidou E, Kitagawa E (2005) Adaptation of *Saccharomyces cerevisiae* to high hydrostatic pressure causing growth inhibition. FEBS Lett 579:2847–2852
Jebaraj CS, Raghukumar C, Behnke A, Stoeck T (2010) Fungal diversity in oxygen-depleted regions of the Arabian Sea revealed by targeted environmental sequencing combined with cultivation. FEMS Microbiol Ecol 71:399–412
Kaneshina S, Ichimori H, Hata T, Matsuki H (1998) Barotropic phase transitions of dioleoylphosphatidylcholine and stearoyl-oleoylphosphatidylcholine bilayer membranes. Biochim Biophys Acta 1374:1–8
Lai X, Cao L, Tan H, Fang S, Huang Y, Zhou S (2007) Fungal communities from methane hydrate-bearing deep-sea marine sediments in South China Sea. ISME J 1:756–762
Le Calvez T, Burgaud G, Mahe S, Barbier G, Vandenkoornhuyse P (2009) Fungal diversity in deep-sea hydrothermal ecosystems. Appl Environ Microbiol 75:6415–6421
Misu K, Fujimura-Kamada K, Ueda T, Nakano A, Katoh H, Tanaka K (2003) Cdc50p, a conserved endosomal membrane protein, controls polarized growth in *Saccharomyces cerevisiae*. Mol Biol Cell 14:730–747
Nagahama T, Hamamoto M, Nakase T, Horikoshi K (1999) *Kluyveromyces nonfermentans* sp nov., a new yeast species isolated from the deep sea. Int J Syst Bacteriol 49:1899–1905
Nagahama T, Hamamoto M, Nakase T, Takami H, Horikoshi K (2001a) Distribution and identification of red yeasts in deep-sea environments around the northwest Pacific Ocean. A van Leeuwenhoek 80:101–110
Nagahama T, Hamamoto M, Nakase T, Horikoshi K (2001b) *Rhodotorula lamellibrachii* sp. nov., a new yeast species from a tubeworm collected at the deep-sea floor in Sagami bay and its phylogenetic analysis. A van Leeuwenhoek 80:317–323
Nagahama T, Hamamoto M, Nakase T, Takaki Y, Horikoshi K (2003a) *Cryptococcus surugaensis* sp. nov., a novel yeast species from sediment collected on the deep-sea floor of Suruga Bay. Int J Syst Evol Microbiol 53:2095–2098
Nagahama T, Hamamoto M, Nakase T, Horikoshi K (2003b) *Rhodotorula benthica* sp. nov. and *Rhodotorula calyptogenae* sp. nov., novel yeast species from animals collected from the deep-sea floor, and Rhodotorula lysiniphila sp. nov., which is related phylogenetically. Int J Syst Evol Microbiol 53:897–903
Nagahama T, Hamamoto M, Horikoshi K (2006) *Rhodotorula pacifica* sp. nov., a novel yeast species from sediment collected on the deep-sea floor of the north-west Pacific Ocean. Int J Syst Evol Microbiol 56:295–299
Nagahama T, Abdel-Wahab MA, Nogi Y, Miyazaki M, Uematsu K, Hamamoto M, Horikoshi K (2008) *Dipodascus tetrasporeus* sp nov., an ascosporogenous yeast isolated from deep-sea sediments in the Japan Trench. Int J Syst Evol Microbiol 58:1040–1046
Nagahama T, Takahashi E, Nagano Y, Abdel-Wahab MA, Miyazaki M (2011) Molecular evidence that deep-branching fungi are major fungal components in deep-sea methane cold-seep sediments. Environ Microbiol 13:2359–2370

Nagano Y, Nagahama T, Hatada Y, Nunoura T, Takami H, Miyazaki J, Takai K, Horikoshi K (2010) Fungal diversity in deep-sea sediments—the presence of novel fungal groups. Fung Ecol 3:316–325

Nogi Y, Hosoya S, Kato C, Horikoshi K (2004) *Colwellia piezophila* sp nov., a novel piezophilic species from deep-sea sediments of the Japan Trench. Int J Syst Evol Microbiol 54:1627–1631

Richards TA, Jones MDM, Leonard G, Bass D (2012) Marine fungi: their ecology and molecular diversity. Annu Rev Marine Sci 44:495–522

Roth FJ, Orpurt PA, Ahearn DJ (1964) Occurrence and distribution of fungi in a subtropical marine environment. Can J Bot 42:375–383

Singh P, Raghukumar C, Verma P, Shouche Y (2010) Phylogenetic diversity of culturable fungi from the deep-sea sediments of the Central Indian Basin and their growth characteristics. Fung Diversity 40:89–102

Singh P, Raghukumar C, Verma P, Shouche Y (2011) Fungal community analysis in the deep-sea sediments of the central Indian basin by culture-independent approach. Microb Ecol 61:507–517

Singh P, Tsuji M, Singh SM, Roy U, Hoshino T (2013) Taxonomic characterization, adaptation strategies and biotechnological potential of cryophilic yeasts from ice cores of Midre Lovenbreen glacier, Svalbard, Arctic. Cryobiology 66:167–175

Takai K, Miyazaki M, Hirayama H, Nakagawa S, Querellou J, Godfroy A (2009) Isolation and physiological characterization of two novel, piezophilic, thermophilic chemolithoautotrophs from a deep-sea hydrothermal vent chimney. Environ Microbiol 11:1983–1997

Takishita K, Tsuchiya M, Reimer JD, Maruyama T (2006) Molecular evidence demonstrating the basidiomycetous fungus *Cryptococcus curvatus* is the dominant microbial eukaryote in sediment at the Kuroshima Knoll methane seep. Extremophiles 10:165–169

Thaler AD, Van Dover CL, Vilgalys R (2012) Ascomycete phylotypes recovered from a Gulf of Mexico methane seep are identical to an uncultured deep-sea fungal clade from the Pacific. Fung Ecol 5:270–273

Thomas-Hall SR, Turchetti B, Buzzini P, Branda E, Boekhout T, Theelen B, Watson K (2010) Cold-adapted yeasts from Antarctica and the Italian Alps-description of three novel species: *Mrakia robertii* sp nov., *Mrakia blollopis* sp nov and *Mrakiella niccombsii* sp nov. Extremophiles 14:47–59

Turchetti B, Buzzini P, Goretti M, Branda E, Diolaiuti G, D'Agata C, Smiraglia C, Vaughan-Martini A (2008) Psychrophilic yeasts in glacial environments of Alpine glaciers. FEMS Microbiol Ecol 63:73–83

Wang FP, Wang P, Chen MX, Xiao X (2004) Isolation of extremophiles with the detection and retrieval of *Shewanella* strains in deep-sea sediments from the west Pacific. Extremophiles 8:165–168

Winter R, Dzwolak W (2005) Exploring the temperature-pressure configurational landscape of biomolecules: from lipid membranes to proteins. Philos Transact A Math Phys Eng Sci 363:537–563

Yayanos AA (1995) Microbiology to 10,500 meters in the deep sea. Annu Rev Microbiol 49:777–805

Chapter 8
Black Yeasts in Cold Habitats

Laura Selbmann, G. Sybren de Hoog, Laura Zucconi, Daniela Isola and Silvano Onofri

Abstract Black yeasts have already been known since the end of the nineteenth century, but for a number of reasons, only few workers were familiar with them. That was since recently, until the wealth of biodiversity, stunning ecologies and potential applications have become apparent. Some remote and extreme locations, such as mountain tops, glaciers or polar areas, are now being investigated by mycologists. Many rock-colonizing fungi have been interpreted for long time as blackish fly-ash particles or dust on marble monuments or buildings. Black yeasts are easily overlooked in routine studies due to their very slow growth and poor competitive abilities. With the improvement isolation procedures, it has become clear that black yeasts are actually much more common and widespread than previously believed. Identification was hampered by their morphological plasticity, until molecular techniques became a routine approach in fungal systematic. In this chapter, the authors aim to give an overview of all the aspects concerning this unconventional group of fungi, from their peculiar ecology to their wide spectrum of biodiversity. Understanding about their ecological amplitude arose from impressive efforts in sampling remote habitats and concomitant sequencing activity during the last two decades.

Keywords Black yeasts · Dothideomycetes · Evolution · Cold adaptation

L. Selbmann (✉) · L. Zucconi · D. Isola · S. Onofri
Department of Ecological and Biological Sciences, University of Tuscia, Largo dell'Università snc, 01100 Viterbo, VT, Italy
e-mail: selbmann@unitus.it

G. S. de Hoog
CBS Fungal Biodiversity Centre, P.O. Box 85167, 3508 AD, Utrecht, The Netherlands

P. Buzzini and R. Margesin (eds.), *Cold-Adapted Yeasts*,
DOI: 10.1007/978-3-642-39681-6_8, © Springer-Verlag Berlin Heidelberg 2014

Contents

8.1 Introduction 174
8.2 Biodiversity, Locations and Substrate 175
8.3 Phylogenetic Assignment and Evolution of Black Yeasts 177
8.4 Stress Tolerance 178
8.5 Adaptations to Stress: The Importance of Being Simple 179
8.6 Nutrition 182
8.7 Conclusions 183
References 185

8.1 Introduction

The term "black yeasts" comprises melanized fungi able to reproduce by unicellular growth, at least for a part of their life cycle. The group is phylogenetically quite heterogeneous and includes mainly fungi belonging to divergent orders of *Ascomycota* (de Hoog and McGinnis 1987), sharing a number of basic features as melanized cell wall and formation of daughter cells by yeast-like or multilateral polar budding. Yet, since most black yeasts exhibit mycelial growth or may shift to meristematic development, and are phylogenetically diverse, there is no common descriptive term for these fungi. They have variously been indicated as black yeasts and relatives, meristematic fungi or microcolonial fungi. The term meristematic, originally introduced for black fungi by de Hoog and Hermanides-Nijhof (1977), describes non-disintegrating aggregations of thick-walled, melanized cells. Meristematic growth is infrequent in the fungal kingdom and is a specific response to stress: conversion towards isodiametric expansion can be induced in hyphal or yeast-like thalli of *Herpotrichiellaceae* by environmental factors such as acidification of the culture medium (Mendoza et al. 1993; Xi et al. 2009); similar observations have been made in *Dothideaceae* (de Hoog et al. 1999). In some cases, these may be stable mutants (Matsumoto et al. 1986) as for those black yeasts permanently living under extreme conditions (Selbmann et al. 2005, 2008).

This morphological plasticity, coupled in many cases with high polymorphism of synanamorphs and very poor differentiation, hampered morphological identification. Molecular phylogeny supplied an impressive body of data during the last decade enabling to establish the position of black yeast in the fungal kingdom with much more accuracy. These studies revealed that, despite poor morphological differentiation, the biodiversity of black fungi is quite impressive and still far from being elucidated.

One of the characters shared in the whole group is the ability to tolerate or even adapt to extreme conditions. Black yeasts are commonly isolated from saltpans (Plemenitaš and Gunde-Cimerman 2005), acidic and hydrocarbon-contaminated sites (Baker et al. 2004; Seyedmousavi et al. 2011; Selbmann et al. 2012; Isola et al. 2013), exposed natural rocks (Ruibal et al. 2005) and monuments (Sert et al.

2007a, b, c; Marvasi et al. 2012), hot deserts (Staley et al. 1982) and very cold icy habitats (Selbmann et al. 2005, 2008; Turchetti et al. 2008; Zalar et al. 2008; Branda et al. 2010; Brunner et al. 2011). The human body is also an extreme environment, and meristematic growth and melanins represent virulence factors for some human opportunists (Matsumoto et al. 1984; van de Sande et al. 2006; Liu and Nizet 2009; Xi et al. 2009; de Hoog et al. 2011). Apparently, a basic set of traits, among which are melanization and meristematic development, are primarily suited to cope with and adapt to highly diverse environmental conditions.

8.2 Biodiversity, Locations and Substrate

Black fungi are worldwide distributed in cold habitats, thriving from subglacial ice to cold sea water, soils and rocks, one of the most frequently investigated substrata. Some species have a broad ecological amplitude and are found in alternative environments, while others occur specifically in cold habitats.

Aureobasidium pullulans is an osmotolerant black yeast that may thrive in a number of osmotically fluctuating environments including the phyllosphere (Andrews et al. 2002), polluted water (Vadkertiova and Slavikova 1995) and solar salterns (Gunde-Cimerman et al. 2000). Given the known adaptive ability of *A. pullulans* to low water activity (A_w) and oligotrophic conditions, it appeared likely that ice from cryocarstic formations and subglacial ice in polythermal glaciers may constitute a potential natural habitat.

Aureobasidium-like fungi were found among the dominant ascomycetous mycota in ice originating from glacial and subglacial environments as polythermal Arctic glaciers in Svalbard (Spitsbergen, Norway; Butinar et al. 2007, 2011; Sonjak et al. 2006). *A. pullulans* var. *pullulans* and *A. pullulans* var. *melanogenum*, both species with a global distribution, were found in Arctic glaciers using a multilocus approach (Zalar et al. 2008); moreover, an additional genotype, *A. pullulans* var. *subglaciale*, was found exclusively in Kongsfjorden glacial and subglacial ice and sea water (Zalar et al. 2008). The psychrotolerant nature of the last one allows active metabolism under conditions of permanently cold in Arctic glaciers. *A. pullulans* was isolated from other cold locations, such as ice-free Alaskan soils, sediments and melt water from glaciers of the Apennines (Turchetti et al. 2008; Branda et al. 2010; Brunner et al. 2011) and from soils and mosses from different locations of Antarctica, including different locations of the McMurdo Dry Valleys (Onofri et al. 2007).

The dimorphic genus *Exophiala* comprises numerous thermotolerant potential human opportunists or pathogens such as *E. dermatitidis* (Sudhadham et al. 2008), *E. spinifera* (Li et al. 2008) and *E. asiatica* (Li et al. 2009). In contrast, many *Exophiala* spp. lacking thermotolerance were found as pathogens in cold-blooded animals such as *E. salmonis*, *E. pisciphila* and *E. psychrophila* (Richards et al. 1978; Pedersen and Langvad 1989; de Hoog et al. 2009). Their remarkable association with monoaromatic hydrocarbons (Prenafeta-Boldú et al. 2006), which are structural analogues of neurotransmitters, allows these fungi to easily shift to animals, including humans. Species belonging to this genus are normally difficult to recover from the environment as they grow very slowly and are frequently overlooked and require dedicated isolation methods (Prenafeta-Boldú et al. 2006; Sudhadham et al. 2008; Vicente et al. 2008; Zhao et al. 2010). Nonetheless, the genus *Exophiala* was also found in cold habitats. The species *E. nigra* was isolated from Antarctic soils (Lyakh and Ruban 1970). The thermophilic species *E. dermatitidis* was reported from the Antarctic continent associated with mummified seals and from air samples in McMurdo Station (Sun et al. 1978) and in Lake Fryxell (Greenfield 1981), as well as in sediments and meltwater of Calderone glaciers in the Apennines, Italy (Branda et al. 2010), but not all these strains have been confirmed by molecular data.

A number of black yeasts have been found associated with rocks in different cold environments, the majority of which still lack a formal description (Ruibal et al. 2009). Apparently, environmental pressure and geographical isolation promote speciation processes and unknown species are frequently encountered even during routine sampling. Black fungi were observed in the cryptoendolithic communities of the Antarctic (Friedmann 1982), living in spatial association with other extremotolerant microbes such as cyanobacteria and lichens. Several of these black yeasts have been described as endemic genera and species for the Antarctic: *F. endolithicus* (Onofri et al. 1999) and *F. simplex* (Selbmann et al. 2005) have hitherto been recorded exclusively in the Antarctic Victoria Land. The genus *Cryomyces*, reported to be restricted to the McMurdo Dry Valleys (Selbmann et al. 2005), was recently found in the Alps with the novel species *Cryomyces montanus* and *C. funiculosus* (Selbmann et al. 2013a). The genus *Saxomyces*, with still unresolved ancestry and endemic to alpine mountain tops, contains strictly psychrophilic species only (Selbmann et al. 2013a). *Recurvomyces mirabilis* and *Elasticomyces elasticus* were found associated with Antarctic endolithic communities (Selbmann et al. 2008) and reported on exposed rock in the Mediterranean or different cold sites as mountain tops in the Andes and Alps (Isola 2010). *E. elasticus* was also found associated with different species of epilithic psychrophilic Antarctic lichens including the endemic species *Lecanora fuscobrunnea* (Selbmann et al. 2013b). In the same study, further strains closely related to or conspecific with black fungi previously thought to be restricted to Antarctic endolithic microbial communities, such as *F. endolithicus*, were found associated with cosmopolitan lichen species. This ability to make association with different microbes according to location may give further advantage in adaptation and survival of the whole community. As heterotrophs, black fungi take advantage from the association with autotrophic organisms under oligotrophic conditions, but it is still unclear whether or not black fungi may supply benefits to epi- or endolithic lichens as well. It was suggested that black fungi may play a role in hydration or in protection of photobionts by dissipating excessive sunlight (Harutyunyan et al. 2008). This would particularly involve the black barrier just above the photobiont stratification in cryptoendolithic lichens (Selbmann et al. 2005). The presence of black fungi may therefore play a crucial role to allow survival under these highly stressful conditions.

8.3 Phylogenetic Assignment and Evolution of Black Yeasts

The abilities to shift to meristematic development and consistently produce melanin, typical of black yeasts, are not widespread in the fungal kingdom. Molecular phylogenetics revealed that black yeasts are mainly distributed in two classes: Dothideomycetes, where the order Capnodiales is particularly overrepresented, and to a lesser extent in the Eurotiomycetes, order Chaetothyriales. In the latter, with few exceptions, black fungi nearly exclusively thrive in hot, semi-arid climates. Black fungi in the order Chaetothyriales are known for their ability to metabolize aromatic compounds (Prenafeta-Boldú et al. 2006); this may explain their higher presence at sites much more influenced by human activities and rich in pollutants, such as coasts and urban environments rather than mountain tops or the Antarctic (Onofri et al. 2011). Moreover, the class Dothideomycetes shows a significant class-wide tendency to extreme survival with a number of lineages purely constituted of black-specialized extremophiles. To that class, for instance, belongs the halophilic species *Hortaea werneckii*, or the highly acidophilic fungus *Acidomyces acidophilus* (Selbmann et al. 2008) as well as most of the psychrophilic black yeasts mentioned above as the genera *Cryomyces*, *Friedmanniomyces* and *Saxomyces*. Chaetothyriales have instead evolved a larger spectrum of assimilative abilities including toxic organic compounds; this tendency, rather than extremotolerance, has led mainly to diversification and promoted shifting to opportunism (Gueidan et al. 2008) which is the main tendency within the order.

The reason why cold-adapted black yeasts are almost exclusively distributed in the class Dothideomycetes may also be found in the early evolutionary history of both fungal groups. It has been recently demonstrated that the class Dothideomycetes has evolved much earlier than chaetothyrialean lineages (Gueidan et al. 2011). The period of diversification of Dothideomycetes was estimated in the Silurian–Devonian, about 430 million years ago when temperatures were much cooler than today. In contrast, the origin of chaetothyrialean lineages was during a period of recovery after the Permian–Triassic mass extinction and an expansion of arid land masses, about 250 million years ago, when global temperatures were relatively high. Adaptations of extant Dothideomycetes and Eurotiomycetes clearly reflect this evolutionary process.

Both Dothideomycetes and chaetothyrialean fungi evolved in different times but have presumably a common ancestry on oligotrophic organisms living on rock surface or subsurface (Gueidan et al. 2011). At that time, exposed rocks were the most abundant natural substrate and tolerance to radiations could have helped black yeasts to survive and proliferate during historic periods of increased cosmic radiation, e.g. due to weakened or absent magnetic field of the Earth (Dadachova and Casadevall 2008).

There are evidences that symbiotic lifestyle may have played a pivotal role in the evolution of free-living rock ancestor for black yeasts. Molecular phylogeny revealed a certain affinity of some lichenized fungal lineages and different lineages of black fungi since some black yeasts from rocks resulted basal to the large lichenized lineages of Arthoniomycetes and Verrucariales (Gueidan et al. 2008; Ruibal et al. 2009). Moreover, some black yeasts have been observed to develop into lichenoid structures within months when co-cultured with lichen algae (Gorbushina et al. 2005; Brunauer et al. 2007). These peculiar interactions with autotrophic organisms, which help black fungi to improve their carbon supply, may be interpreted as vestiges of the ancestral situation. Black yeasts maintained somehow their connection with this primitive life form and may be easily found associated with rocks or are commonly visible as lichen colonizers (Harutyunyan et al. 2008; Selbmann et al. 2013b).

8.4 Stress Tolerance

Black yeasts may actually be seen as the excellence in the extremes. Along with cold, they may cope with a number of different stresses and also a combination of them (Ruisi et al. 2007), ranging from wide temperature fluctuations to irradiation and osmotic stress (Palmer et al. 1990; Sterflinger 1998; Gorbushina et al. 2003, 2008; Onofri et al. 2007; Marvasi et al. 2012). Black yeasts from cold habitats, such as Antarctic and alpine species of the genera *Friedmanniomyces* and *Cryomyces* spp., have typical psychrophilic profiles (Van Uden 1984). Black yeasts from the Antarctic desert live in almost permanently frozen conditions, but during austral summer, temperature of rock surfaces may fluctuate across the freezing point over 14 times within 40 min, causing a repeated freeze–thawing stress to lithobionts. Antarctic black fungi may actually easily tolerate this stress: repeated freeze–thaw cycles (–20 °C/+20 °C) did not affect growth ability of these fungi (Onofri et al. 2007, 2008). Remarkably, psychrophilic black yeasts were proved to tolerate even very high temperatures. Germination ability of *Cryomyces* spp. is not affected after exposition at 90 °C for 1 h (Onofri et al. 2008). Yet, some black yeasts have an eurithermic behaviour with optimal growth temperature well above the normal ambient temperature of their natural environment. *C. funiculosus*, for instance, a species living in the Alps above 3,000 m a.s.l., grows in the range 0–35 °C with optimal growth at 25 °C. This apparent incongruence may be interpreted as an adaptation to very fluctuating environmental conditions; even in very cold environments, exposed rock surfaces may reach temperatures 20 °C above the air temperatures (Nienow and Friedmann 1993; Selbmann et al. 2013a).

Cold often implies a number of additional stressors such as osmotic stress since the formation of ice crystals leads to lack of water for active life. Black fungi are actually osmotolerant rather than halophiles but, since in some cold location such as the Antarctic desert evaporation may be incredibly high and salt may accumulate conspicuously on rock surface, some of them evolved specific adaptation and may tolerate even considerably high salt concentration. Even if salt tolerance is not comparable with that of a real halophilic fungus, *Cryomyces* spp. still maintain visible growth at NaCl concentration of 25 % (w/v) (Onofri et al. 2007).

Life on exposed rocks in cold environments, such as mountain tops or Antarctic deserts, implies resistance to strong solar radiations since exposition may be even more intense than under hot conditions due to thin atmosphere at mountain tops, or the ozone hole persisting particularly at the South Pole. Resistance to radiation has been largely documented in black yeasts. Antarctic species maintain their ability to germinate after high UV exposition (Onofri et al. 2007) and even space radiations (Onofri et al. 2012) by resisting, rather than repairing potential DNA damages (Selbmann et al. 2011).

8.5 Adaptations to Stress: The Importance of Being Simple

Black yeasts have evolved a number of morphophysiological characteristics and even adopted a lifestyle, allowing them to deal with a wide selection of external pressures. Their adaptations are so effective that their tolerance may push them well beyond the natural conditions they normally experience in their natural environments (Onofri et al. 2008).

Fungi in general have remarkable plasticity and may promptly shift from one growth form to another according to variation in physical–chemical conditions. They adopt unicellular growth when immersed in rapidly fermentable sugar or when they are in their infective phase, as most human pathogens or in vascular plant diseases; alternatively, they may grow as filamentous form, the "search for food" growth form, allowing them to inspect the surroundings. Yet, shifting to meristematic development is less frequent and is the answer of a group of fungi, particularly black yeasts, to stressing conditions. Many black yeasts may shift alternatively from filamentous to yeast-like organization or to meristematic growth according to the external conditions, but, for species living under permanent stress, the last may became a stable character. Examples of morphological organization of "cold" black yeasts are reported in Fig. 8.1. The advantage of meristematic development lies in optimizing the volume/surface ratio since it minimizes exposition to external stressors (Wollenzien et al. 1995).

The dimorphic fungus *A. pullulans* may switch promptly from small colourless yeast cells to thick-walled, heavily melanized, meristematic forms (Bermejo et al. 1981) to mitigate environmental stress. This process has been described at the molecular level (Kogej et al. 2005, 2006; Turk et al. 2007; Gostinčar et al. 2008).

Exophiala dermatitidis can shift into a meristematic form at low pH (de Hoog et al. 1994). However, several stress-tolerant strains of this species are known to irreversibly maintain their slowly expanding, meristematic morphology. The same has been noted even more frequently for the related species *E. phaeomuriformis* (Matsumoto et al. 1986). Meristematic and yeast-like strains do not differ in variable loci of their rDNA (Matos et al. 2003) in these cases, while in other fungal orders, such as *Capnodiales*, some sequence divergence has been seen indicating possible subsequent sympatric speciation (Selbmann, unpublished data).

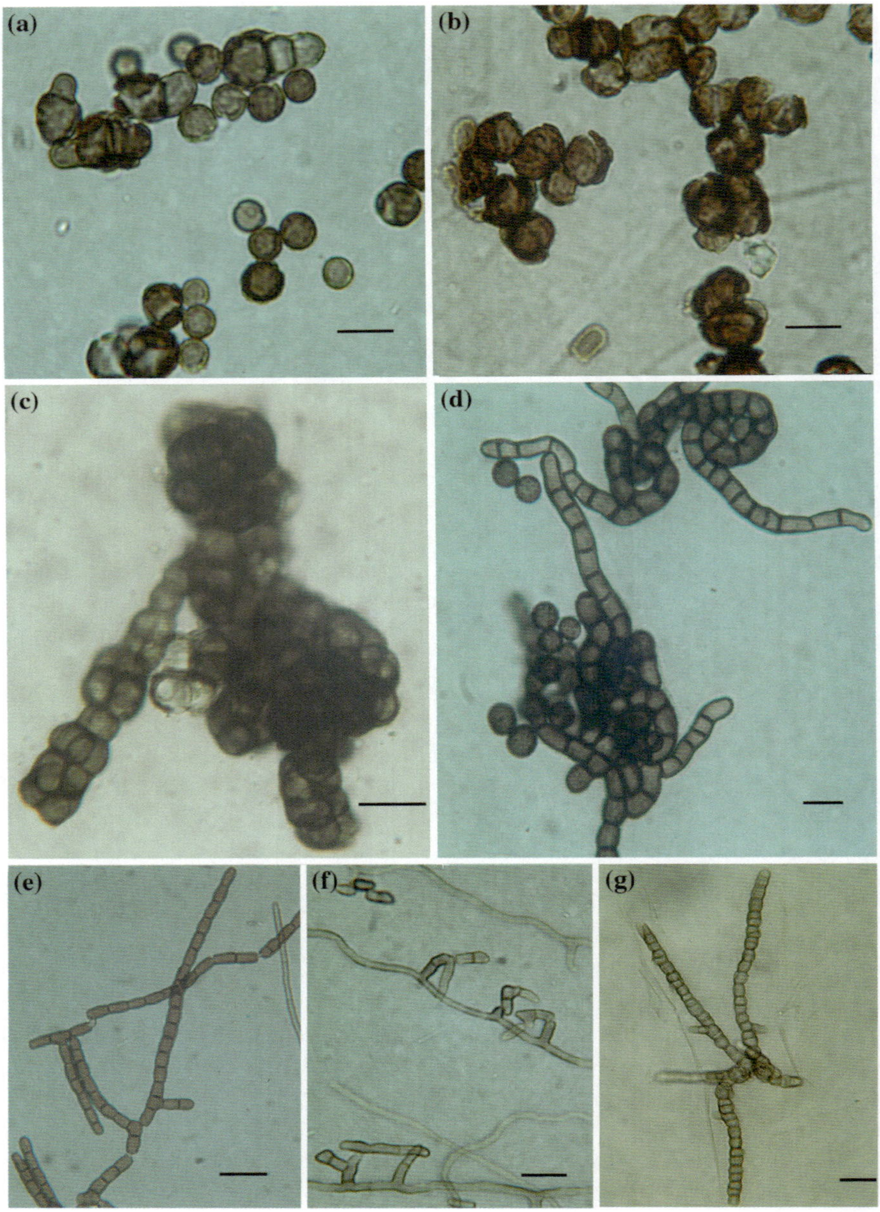

Fig. 8.1 *Cryomyces antarcticus* CCFEE 534 (**a**); *Cryomyces minteri* CCFEE 5187 (**b**); *Cryomyces funiculosus* CCFEE 5554 (**c**); *Saxomyces alpinus* CCFEE 5470 (**d**); *Elasticomyces elasticus* CCFEE 5319 (**e**); *Recurvomyces mirabilis* CCFEE 5264 (**f**); *Friedmanniomyces endolithicus* CCFEE 5208 (**g**). Scale bars = 10 μm

Most of cold-adapted rock black yeasts reproduce by isodiametric enlargement with subdividing cells by production of both longitudinal and horizontal septa. Some, such as *Cryomyces* spp., have mostly a yeast-like organization. To gain success, the extremes black yeasts have learned to save energy: simplification is essential for these fungi. Differentiation is minimal and they show very short life cycles, comprising just some black clumps of cells, to be completed during short periods of time when favourable conditions prevail (Selbmann et al. 2005). The life cycle in these fungi usually concerns just a few cells that subdivide and fall apart for passive dispersal. At most, some species exceptionally show some morphological differentiation, with tiny conidiophores as in *Recurvomyces mirabilis* (Selbmann et al. 2008). The absence in most black yeasts of active and abundant conidia formation hampered the efficiency of dispersal, and some of them have adopted surprising means to spread. Rock black yeasts, for instance, actively penetrate and break up the substratum and are actively dispersed with dust transported by winds. Current biogeographical data suggest that long-distance dispersal occurs in rock-inhabiting fungi (Gorbushina 2007; Selbmann et al. 2008, 2013a). However, in few cases, some rock-inhabiting taxa seem to be only present in unique localities or geographical areas (Selbmann et al. 2005).

Black yeasts, particularly in the class Dothideomycetes, are invariably asexual; the chaethothyrialean genus *Exophiala*, for which *Capronia* as teleomorph is known, represent an exception. Sex is costly, implying not only the trouble of finding a mate but also the cost to keep the genetic machinery for recombination running. Fungal teleomorphs tend to be elaborate, with fruit bodies and specialized cells in which recombination takes place. In the extreme, as in almost permanent frozen environments, there is no place and time for all this. Sexuality is a known driving force in evolution: recombination aids the spread of advantageous traits over populations and purges the genome of deleterious mutations. If this mechanism is absent, as is the case in clonally reproducing organisms, Muller's ratchet applies, harmful mutations accumulate in the course of time and gradually but inevitably lead to extinction (Muller 1964). Black yeasts seem to be a fundamental exception to this rule and do not show to miss their sexuality. Apparently, beyond one certain threshold of stress, asexuality, with perpetuation of few super-adapted genotypes, seems to be advantageous for survival.

Oligotrophy is another important adaptation of black yeasts, enabling these fungi to rely only on sparse, airborne nutrients available such as dust; even if some of them, particularly in the chaetothyrialean lineages, show remarkable degradative abilities and may digest pollutants such as aromatic compounds (Isola et al. 2013), dothidealean black yeasts are metabolically scarcely competent. They produce only few metabolites, which are restricted to substances essential for survival, such as extracellular polymeric substances (EPS), polyols and melanins (Selbmann et al. 2005; Sterflinger 2006). Considering oligotrophy of typical environments for black yeasts, the high metabolic costs for synthesizing all these compounds significantly affect growth velocity of these fungi which typically show a very slow growth rate. EPS play a fundamental role increasing resistance to cold conditions by protecting cells from freeze–thawing damages (Selbmann et al.

2002); moreover, they contribute in creating buffered physicochemical conditions around the cells and their hygroscopic nature increases success under dry conditions (de los Rios et al. 2003). Cold often implies an osmotic stress due to the lack of liquid water and salt accumulation due to strong evaporation. A number of intracellular osmoregulators are accumulated at high concentrations without interfering with enzyme activity; among these compatible solutes, the disaccharide trehalose is obligate in several black yeasts (Sterflinger 1998). It is very efficient for its cryoprotective effects during freezing or desiccation (Weinstein et al. 2000) acting as stabilizer of enzyme conformation and phospholipid bilayers of membranes, allowing these surprising organisms to survive complete dehydration (Onofri et al. 2012).

Melanin, together with a thick cell wall with conspicuous incrustations, is for certain the most important factor to stress resistance in black yeasts. Melanins of black fungi are different types of high molecular weight pigments produced by enzymatic coupling of phenolic units; in black yeasts, these are reported as 1,8-dihydroxynaphthalene (Kogej et al. 2004). They are responsible for the typical dark green to brown or totally black colour of these fungi and confer them the ability to survive a number of different external pressures, such as excessive heat or cold, extreme pH or osmotic conditions, polychromatic UV radiation, and melanins also seem to mediate tolerance toward metals (Gadd and de Rome 1988; Gunde-Cimerman et al. 2000; Onofri et al. 2008; Selbmann et al. 2011; Sterflinger et al. 2012).

Melanin confers also tolerance to ionizing radiation: some melanized fungal species have been found in nuclear reactors and their cooling water systems (Zhdanova et al. 2000) which even feed on radiation by using melanins to convert ionizing gamma radiation into chemical energy by still unknown mechanisms (Dadachova et al. 2007).

Melanin has also a role in osmoadaptation of black yeasts; in *Hortaea werneckii*, for instance, it improves glycerol retention within the cells, which act as osmoregulator, by reducing size of pores in the cell wall (Plemenitaš et al. 2008). This may help to reduce the cost of supplying energy to transmembrane transporters to maintain intracellular concentration of compatible solutes (Gostinčar et al. 2011).

8.6 Nutrition

Cold-adapted black yeasts colonize environments, such as polar or alpine areas, which are strictly oligotrophic and largely unpolluted; therefore, they are adapted to this nutritional constraints relying on airborne nutrients or living in spatial association with autotrophic, similarly stress-resistant organisms. Yet, the environmental constraints exert a high selective pressure; therefore, black fungi do not have the trouble to sharpen their competitive abilities but rather to focus on stress tolerance and survival. The very slow growth rate and ability to shift to dormant state even for long periods make them scarcely exigent from a nutritional point of view; they just need simple sugars for growth or easily degradable polymers such as starch. Some strains living above 3,000 m a.s.l. in the Alps were found to grow with scarce efficiency on potato dextrose agar (PDA) medium, rich in starch, if compared with media containing glucose oligomers. This is probably due to a reduced ability to produce amylases; in those environments, plants are sparse or absent at all and starch is expected to be rare. Therefore, these fungi may benefit from simpler carbon sources such as glucose and other oligomers resulting from photosynthesis of neighbouring lichens and algae (Selbmann et al. 2013b). Most black yeasts have a very slow growth rate with no significant increases even when cultured on rich media; some also maintain visible mycelial growth, even for short periods, when cultivated on media without carbon sources. This behaviour indicates an extreme adaptation to the oligotrophic conditions of their natural environments.

The ubiquity of black fungi in polar and alpine locations indicates that, in addition to their cohabitation with algae, the existence of unconventional pathways of carbon acquisition or energy gain may be supposed (Gostinčar et al. 2012). Atmospheric carbon dioxide fixation could exceptionally take place in fungal metabolism, although it lacks a Calvin cycle comparable with that found in plants. It was observed, using ^{14}C-labelling, that black fungi isolated from Antarctic cryptoendolithic communities of the McMurdo Dry Valleys, the closest terrestrial analogue for Mars, may actually uptake CO_2 (Palmer and Friedman 1988). The authors supposed that Antarctic fungi may actually incorporate CO_2 by carboxylation of pyruvate (Moses et al. 1959) as it was reported earlier for a number of fungi, but a definitive confirmation with a modern approach is still missing. Such mechanisms might be irrelevant for fungi living in nutrient-rich habitats, but could represent a life-sustaining option for slow-growing oligotrophic black yeasts in competitor-free environments.

As already mentioned, black yeasts can also gain energy from unconventional sources such as ionizing radiations which are then transformed into biochemical energy by means of melanins. All these unusual abilities to gain carbon and energy in black yeasts can, in their overall, explain not only the ability to survive, but even to thrive in environments normally neglected or precluded for most life forms, such as oligotrophic, exposed rock surfaces.

8.7 Conclusions

Black yeasts are a group of fungi that have remained rather unexplored since recently due to the difficulties in isolation for the slow growth rate and scarce competitive abilities, and the troubles in identification due to their poor differentiation and morphological plasticity. The "omics" revolution gave new insights into the biology of these unconventional organisms. Rapid progresses in sequencing technologies, for instance, allowed to look deeper into the genomes of black yeasts, and we now have a clearer picture of the amplitude of their biodiversity (Ruibal et al. 2009), even if it still remains largely unknown, and considerable progresses have been done in shedding light on the origin and evolution of these organisms (Gueidan et al. 2008, 2011). Particularly in the class Dothideomycetes novel species of black yeasts, often with unclear phylogenetic relations at order level, are continuously discovered and described (Selbmann et al. 2013a). This suggests that species richness, at least within this class, remains woefully underestimated.

The understanding of the expression, function and regulation of the entire set of genes/proteins encoded by fungal genomes is an intriguing challenge, and proteomics is a recent, promising approach to investigate the presence of stress-associated genes and expressed products and their role in polyextremotolerance and oligotrophism in black yeasts (Isola et al. 2011). Recent studies gave significant evidence that black yeasts have a peculiar response to temperature that differs considerably from the one of common mesophilic hyphomycetes. In particular, it was observed that black yeasts show a significant decrease in protein expression when exposed to temperatures above their growth optimum, indicating a down-regulation of their metabolism, while the mesophilic fungus *Penicillium chrysogenum*, taken as control strain, expressed the highest number of proteins at 40 °C, indicating a temperature-induced reaction (Tesei et al. 2012). It is worth investigating in the future if this is a general stress response in this special group of fungi.

Black yeasts are an excellent material of study in many different fields. Their extraordinary stress tolerance and ability to grow inside the rocks suggested their possible use as model for astrobiological studies, such as searching life on cold planets such as Mars or the under ice oceans of Europa, or to study the possibility of interplanetary transfer of life from one planet to another within meteorites according to the Lithopanspermia theory (Onofri et al. 2008, 2009, 2012).

Among black yeasts, rock inhabitants equally colonize natural rock or monument, and due to their extraordinary capacity to penetrate minerals, they are an attractive subject for applied research in biodeterioration of monuments. Another fascinating field of research concerns the ability of some of them, particularly the chaetothyrialean strains, to metabolize aromatic compounds and their possible exploitation in bioremediation programs (Prenafeta-Boldú et al. 2006); this has encouraged the search of competent strains in new, unconventional habitats such as gasoline tanks or soap dispensers of washing machines (Isola et al. 2013). In the meantime, since assimilation of phenolic compounds and hydrocarbons may represent a virulence factor enabling to infect the central nervous system which has a high content of monoaromatic neurotransmitters, it is of utmost importance to shed light on phylogenetic relations and differences in physio-ecological traits of black yeasts within Chaetothyriales to control and prevent the spreading of biohazardous species.

Since the time when black yeasts were just a subject for few specialists, they have now become an ever-expanding field of study, offering research opportunities in many basic fields, such as microbial ecophysiology, evolution and adaptation to extremes, as well as in applied research, such as human pathogenicity, bioremediation, biodeterioration of monuments and exobiology.

References

Andrews JH, Spear RN, Nordheim EV (2002) Population biology of *Aureobasidium pullulans* on apple leaf surfaces. Can J Microbiol 48:500–513

Baker BJ, Lutz MA, Dawson SC, Bond PL, Banfield JF (2004) Metabolically active eukaryotic communities in extremely acidic mine drainage. Appl Environ Microbiol 70:6264–6271

Bermejo JM, Dominguez JB, Goni FM, Uruburu F (1981) Influence of pH on the transition from yeast-like cells to chlamydospores in *Aureobasidium pullulans*. A van Leeuwenhoek 47:385–392

Branda E, Turchetti B, Diolaiuti G, Pecci M, Smiraglia C, Buzzini P (2010) Yeast and yeast-like diversity in the southernmost Glacier of Europe (Calderone glacier, Apennines, Italy). FEMS Microbiol Ecol 72:354–369

Brunauer G, Blaha J, Hager A, Türk R, Stocker-Wörgötter E, Grube M (2007) Lichenoid structures in vitro of a cultured lichenicolous fungus. Symbiosis 44:127–136

Brunner I, Plotze M, Rieder S, Zumsteg A, Furrer G, Frey B (2011) Pioneering fungi from the Damma glacier fore field in the Swiss Alps can promote granite weathering. Geobiology 9:266–279

Butinar L, Spencer-Martins I, Gunde-Cimerman N (2007) Yeasts in high Arctic glaciers: the discovery of a new habitat for eukaryotic microorganisms. A van Leeuwenhoek 91:277–289

Butinar L, Strmole T, Spencer-Martins I, Gunde-Cimerman N (2011) Relative incidence of ascomycetous yeasts in Arctic coastal environments. Microb Ecol 61:832–843

Dadachova E, Casadevall A (2008) Ionizing radiation: how fungi cope, adapt, and exploit with the help of melanin. Curr Opin Microbiol 11:525–531

Dadachova E, Bryan RA, Huang X, Moadel T, Schweitzer AD, Aisen P, Nosanchuk JD, Casadevall A (2007) Ionizing radiation changes the electronic properties of melanin and enhances the growth of melanized fungi. PLoS ONE 2:e457

de Hoog GS, Hermanides-Nijhof EJ (1977) The black yeasts and allied hyphomycetes. Stud Mycol 15:1–222

de Hoog GS, McGinnis MR (1987) Ascomycetous black yeasts. Stud Mycol 30:187–199

de Hoog GS, Takeo K, Yoshida S, Gottlich E, Nishimura K, Miyaji M (1994) Pleoanamorphic life cycle of *Exophiala* (*Wangiella*) *dermatitidis*. A van Leeuwenhoek 65:143–153

de Hoog GS, Zalar P, Urzì C, de Leo F, Yurlova NA, Sterflinger K (1999) Relationships of dothideaceous black yeasts and meristematic fungi based on 5.8S and ITS2 rDNA sequence comparison. Stud Mycol 43:31–37

de Hoog GS, Guarro J, Figueras MJ, Gené J (2009) Atlas of clinical fungi. 3rd CD-ROM ed. CBS-KNAW Fungal Biodiversity Centre, Utrecht/Universitat Rovira I Virgili, Reus

de Hoog GS, Vicente VA, Harrak MJ, Najafzadeh MJ, Badali H, Seyedmousav S (2011) Waterborne *Exophiala* species causing disease in cold-blooded animals. Persoonia 27:46–72

de los Ríos A, Wierzchos J, Sancho LG, Ascaso C (2003) Acid microenvironments in microbial biofilms of Antarctic endolithic microecosystems. Eviron Microbiol 5:231–237

Friedmann EI (1982) Endolithic microorganisms in the Antarctic cold desert. Science 215:1045–1053

Gadd GM, de Rome L (1988) Biosorption of copper by fungal melanin. Appl Microbiol Biotechnol 29:610–617

Gorbushina AA (2007) Life on the rocks. Environ Microbiol 9:1613–1631

Gorbushina AA, Whitehead K, Dornieden T, Niesse A, Schulte A, Hedges JI (2003) Black fungal colonies as units of survival: hyphal mycosporines synthesized by rock-dwelling microcolonial fungi. Can J Bot 81:131–138

Gorbushina AA, Beck A, Schulte A (2005) Microcolonial rock inhabiting fungi and lichen photobionts: evidence for mutualistic interactions. Mycol Res 109:1288–1296

Gorbushina AA, Kotlova ER, Sherstneva OA (2008) Cellular responses of microcolonial rock fungi to long-term desiccation and subsequent rehydration. Stud Mycol 61:91–97

Gostinčar C, Turk M, Trbuha T, Vaupotič T, Plemenitaš A, Gunde-Cimerman N (2008) Expression of fatty-acid modifying enzymes in halotolerant black yeast *Aureobasidium pullulans* (de Bary) G. Arnaud under salt stress. Stud Mycol 61:1–59

Gostinčar C, Grube M, Gunde-Cimerman N (2011) Evolution of fungal pathogens in domestic environments? Fungal Biol 115:1008–1018

Gostinčar C, Muggia L, Grube M (2012) Polyextremotolerant black fungi: oligotrophism, adaptive potential, and a link to lichen symbioses. Front Microbiol 3:390

Greenfield L (1981) Soil microbiological studies. In: Greenfield L, Wilson G (eds) University of Canterbury, Antarctic Expedition no. 19. Christchurch, pp 4–22

Gueidan C, Ruibal C, de Hoog GS, Gorbushina AA, Untereiner WA, Lutzoni F (2008) A rock-inhabiting ancestor for mutualistic and pathogen-rich fungal lineages. Stud Mycol 61:111–119

Gueidan C, Ruibal C, de Hoog GS, Schneider H (2011) Rock-inhabiting fungi originated during periods of dry climate in the late Devonian and middle Triassic. Fungal Biol 115:987–996

Gunde-Cimerman N, Zalar P, de Hoog S, Plemenitaš A (2000) Hypersaline waters in salterns: natural ecological niches for halophilic black yeasts. FEMS Microbiol Ecol 32:235–240

Harutyunyan S, Muggia L, Grube M (2008) Black fungi in lichens from seasonally arid habitats. Stud Mycol 61:83–90

Isola D (2010) Biodiversity, phylogeny and evolution of rock black fungi. PhD thesis, Università degli Studi della Tuscia, Viterbo

Isola D, Marzban G, Selbmann L, Onofri S, Laimer M, Sterflinger K (2011) Sample preparation and 2-DE procedure for protein expression profiling of black microcolonial fungi. Fungal Biol 115:971–977

Isola D, Selbmann L, de Hoog GS, Fenice M, Onofri S, Prenafeta-Boldú FX, Zucconi L (2013) Isolation and screening of black fungi as degraders of volatile aromatic hydrocarbons. Mycopathologia. doi:10.1007/s11046-013-9635-2

Kogej T, Wheeler MH, Rižner TL, Gunde-Cimerman N (2004) Evidence for 1,8-dihydroxy-naphthalene melanin in three halophilic black yeasts grown under saline and non-saline conditions. FEMS Microbiol Lett 232:203–209

Kogej T, Ramos J, Plemenitaš A, Gunde-Cimerman N (2005) The halophilic fungus *Hortaea werneckii* and the halotolerant fungus *Aureobasidium pullulans* maintain low intracellular cation concentrations in hypersaline environments. Appl Environ Microb 71:6600–6605

Kogej T, Gostinčar C, Volkmann M, Gorbushina AA, Gunde-Cimerman N (2006) Mycosporines in extremophilic fungi—novel complementary osmolytes? Environ Chem 3:105–110

Li DM, de Hoog GS, Lindhardt Saunte DM, Gerrits van den Ende AHG, Chen XR (2008) *Coniosporium epidermidis* sp. nov., a new species from human skin. Stud Mycol 61:131–136

Li DM, Li RY, de Hoog GS, Wang YX, Wang DL (2009) *Exophiala asiatica*, a new species from a fatal case in China. Med Mycol 47:101–109

Liu GY, Nizet V (2009) Color me bad: microbial pigments as virulence factors. TIM 17:406–413

Lyakh SP, Ruban EL (1970) Antarctic "black yeasts" *Nadsoniella nigra* var. *hesuelica* (characteristics and identification of strain 365). Izv Akad Nauk Biol 4:581–592

Marvasi M, Donnarumma F, Frandi A, Mastromei G, Sterflinger K, Tiano P, Perito B (2012) Black microcolonial fungi as deteriogens of two famous marble statues in Florence, Italy. Int Biodet Biodeg 68:36–44

Matos T, Haase G, Gerrits van den Ende AH, de Hoog GS (2003) Molecular diversity of oligotrophic and neurotropic members of the black yeast genus *Exophiala*, with accent on *E. dermatitidis*. A van Leeuwenhoek 83:293–303

Matsumoto T, Matsuda T, McGinnis MR, Ajello L (1984) Clinical and mycological spectra of *Wangiella dermatitidis* infections. Mycoses 36:145–155

Matsumoto T, Padhye AA, Ajello L, McGinnis MR (1986) *Sarcinomyces phaeomuriformis*: a new dematiaceous hyphomycete. J Med Vet Mycol 24:395–400

Mendoza L, Karuppayil SM, Szaniszlo PJ (1993) Calcium regulates in vitro dimorphism in chromoblastomycotic fungi. Mycoses 36:157–164

Moses V, Holm-Hansen O, Calvin M (1959) Non photosynthetic fixation of carbon dioxide by three microorganisms. J Bacteriol 77:70–78

Muller HJ (1964) The relation of recombination to mutational advantage. Mutat Res 1:2–9

Nienow JA, Friedmann EI (1993) Terrestrial litophytic (rock) communities. In: Friedmann EI (ed) Antarctic microbiology. Wiley-Liss, New York, pp 343–412

Onofri S, Pagano S, Zucconi L, Tosi S (1999) *Friedmanniomyces endolithicus* (Fungi, Hyphomycetes) anam. gen. sp. nov., from continental Antarctica. Nova Hedwigia 68:175–181

Onofri S, Selbmann L, de Hoog GS, Grube M, Barreca D, Ruisi S, Zucconi L (2007) Evolution and adaptation of fungi at the boundaries of life. Adv Space Res 40:1657–1664

Onofri S, Barreca D, Selbmann L, Isola D, Rabbow E, Horneck G, de Vera JPP, Hatton J, Zucconi L (2008) Resistance of Antarctic black fungi and cryptoendolithic communities to simulated space and Mars conditions. Stud Mycol 61:99–109

Onofri S, Selbmann L, Barreca D, Isola D, Zucconi L (2009) Fungal survival in space conditions: new insights for lithopanspermia. Plant Biosyst 143:S85–S87

Onofri S, Anastasi A, Del Frate G, Di Piazza S, Garnero N, Guglielminetti M, Isola D, Panno L, Ripa C, Selbmann L, Varese GC, Voyron S, Zotti M, Zucconi L (2011) Biodiversity of rock, beach and water fungi in Italy. Plant Biosyst 145:978–987

Onofri S, de la Torre R, de Vera JP, Ott S, Zucconi L, Selbmann L, Scalzi G, Venkateswaran K, Rabbow E, Horneck G (2012) Survival of rock-colonizing organisms after 1.5 years in outer space. Astrobiology 12:508–516

Palmer RJ, Friedman EI (1988) Incorporation of inorganic carbon by Antarctic cryptoendolithic fungi. Polarforschung 58:189–191

Palmer FE, Staley JT, Ryan B (1990) Ecophysiology of microcolonial fungi and lichens on rocks in Northeastern Oregon. New Phytol 116:613–620

Pedersen OA, Langvad F (1989) *Exophiala psychrophila* sp. nov., a pathogenic species of the black yeasts isolated from Atlantic salmon. Mycol Res 92:153–156

Plemenitaš A, Gunde-Cimerman N (2005) Cellular responses in the halophilic black yeast *Hortaea werneckii* to high environmental salinity. In: Gunde-Cimerman N, Oren A, Plemenitaš A (eds) Adaptation to life at high salt concentrations in Archaea, Bacteria, and Eukarya. Springer, Dordrecht, pp 455–470

Plemenitaš A, Vaupotič T, Lenassi M, Kogej T, Gunde-Cimerman N (2008) Adaptation of extremely halotolerant black yeast *Hortaea werneckii* to increased osmolarity: a molecular perspective at a glance. Stud Mycol 61:67–75

Prenafeta-Boldú FX, Summerbell RC, de Hoog GS (2006) Fungi growing on aromatic hydrocarbons: biotechnology's unexpected encounter with biohazard. FEMS Microbiol Rev 30:109–130

Richards RH, Holliman A, Helgason S (1978) *Exophiala salmonis* infection in Atlantic salmon *Salmo salar* L. J Fish Dis 1:357–368

Ruibal C, Gonzalo P, Bills GF (2005) Isolation and characterization of melanized fungi from limestone formations in Mallorca. Mycol Prog 4:23–38

Ruibal C, Gueidan C, Selbmann L, Gorbushina AA, Crous PW, Groenewald JZ, Muggia L, Grube M, Isola D, Schoch CL, Staley JT, Lutzoni F, de Hoog GS (2009) Phylogeny of rock inhabiting fungi related to Dothideomycetes. Stud Mycol 64:123–133

Ruisi S, Barreca D, Selbmann L, Zucconi L, Onofri S (2007) Fungi in Antarctica. Rev Environ Sci Biotech 6:127–141

Selbmann L, Onofri S, Fenice M, Federici F, Petruccioli M (2002) Production and structural characterization of the exopolysaccharide of the Antarctic fungus *Phoma herbarum* CCFEE 5080. Res Microbiol 153:585–592

Selbmann L, de Hoog GS, Mazzaglia A, Friedmann EI, Onofri S (2005) Fungi at the edge of life: cryptoendolithic black fungi from Antarctic deserts. Stud Mycol 51:1–32

Selbmann L, de Hoog GS, Zucconi L, Isola D, Ruisi S, Gerrits van den Ende AHG, Ruibal C, De Leo F, Urzì C, Onofri S (2008) Drought meets acid: three new genera in a dothidealean clade of extremotolerant fungi. Stud Mycol 61:1–20

Selbmann L, Isola D, Zucconi L, Onofri S (2011) Resistance to UV-B induced DNA damage in extreme-tolerant cryptoendolithic Antarctic fungi: detection by PCR assays. Fungal Biol 115:937–944

Selbmann L, Egidi E, Isola D, Onofri S, Zucconi Z, de Hoog GS, Chinaglia S, Testa L, Tosi S, Balestrazzi A, Lantieri A, Compagno R, Tigini V, Varese G (2012) Biodiversity, evolution and adaptation of fungi in extreme environments. Plant Biosyst 147:237–246

Selbmann L, Isola D, Egidi E, Zucconi L, Gueidan C, de Hoog GS, Onofri S (2013a) Mountain tips as reservoirs for new rock-fungal entities: *Saxomyces* gen. nov. and four new species from the Alps. Fungal Divers doi:10.1007/s13225-013-0234-9

Selbmann L, Grube M, Onofri S, Isola D, Zucconi L (2013b) Antarctic epilithic lichens as niches for black meristematic fungi. Biology 2:784–797

Sert HB, Sümbül H, Sterflinger K (2007a) Microcolonial fungi from antique marbles in Perge/Side/Termessos (Antalya/Turkey). A van Leeuwenhoek 91:217–227

Sert HB, Sümbül H, Sterflinger K (2007b) *Sarcinomyces sideticae*, a new black yeast from historical marble monuments in Side (Antalya, Turkey). Bot J Linn Soc 154:373–380

Sert HB, Sümbül H, Sterflinger K (2007c) A new species of *Capnobotryella* from monument surfaces. Mycol Res 111:1235–1241

Seyedmousavi S, Badali H, Chlebicki A, Zhao J, Prenafeta-Boldu′ FX, de Hoog GS (2011) *Exophiala sideris*, a novel black yeast isolated from environments polluted with toxic alkyl benzene and arsenic. Fungal Biol 115:1030–1037

Sonjak S, Frisvad JC, Gunde-Cimerman N (2006) *Penicillium* mycobiota in Arctic subglacial ice. Microb Ecol 52:207–216

Staley JT, Palmer F, Adams JB (1982) Microcolonial fungi: common inhabitants on desert rocks? Science 215:1093–1095

Sterflinger K (1998) Temperature and NaCl-tolerance of rock inhabiting meristematic fungi. A van Leeuwenhoek 74:271–281

Sterflinger K (2006) Black yeasts and meristematic fungi: ecology, diversity and identification. In: Rosa C, Gabor P (eds) Yeast handbook: biodiversity and ecophysiology of yeasts. Springer, New York, pp 505–518

Sterflinger K, Tesei D, Zakharova K (2012) Fungi in hot and cold deserts with particular reference to microcolonial fungi. Fungal Ecol 5:453–462

Sudhadham M, Prakitsin S, Sivichai S, Chaiwat R, Menken SBJ, Dorrestein GM, de Hoog GS (2008) The neurotropic black yeast *Exophiala dermatitidis* has a possible origin in the tropical rain forest. Stud Mycol 61:145–155

Sun SH, Huppert M, Cameron RE (1978) Identification of some fungi from soil and air of Antarctica. Antarct Res Ser 30:1–26

Tesei D, Marzban G, Zakharova K, Isola D, Selbmann L, Sterflinger K (2012) Alteration of protein patterns in black rock inhibiting fungi as a response to different temperatures. Fungal Biol 116:932–940

Turchetti B, Buzzini P, Goretti M, Branda E, Diolaiuti G, D'Agata C, Smiraglia C, Vaughan-Martini A (2008) Psychrophilic yeasts in glacial environments of Alpine glaciers. FEMS Microbiol Ecol 63:73–83

Turk M, Abramovic Z, Plemenitas A, Gunde-Cimerman N (2007) Salt stress and plasma-membrane fluidity in selected extremophilic yeasts and yeast-like fungi. FEMS Yeast Res 7:550–557

Vadkertiova R, Slavikova E (1995) Killer activity of yeasts isolated from the water environment. Can J Microbiol 41:759–766

Van de Sande W, de Kat J, Ahmed A, Verbrugh H, van Belkum A (2006) Melanin protects *Madurella mycetomatis* against itraconazole and ketoconazole, firstline treatment agents against mycetoma. Ned Tijdschr Med Microbiol 14:520–521

Van Uden N (1984) Temperature profiles of yeasts. Adv Microbiol Physiol 25:195–251

Vicente VA, Attili-Angelis D, Pie MR, Queiroz-Telles F, Cruz LM, Najafzadeh MJ, de Hoog GS, Pizzirani-Kleine AR (2008) Environmental isolation of black yeast-like fungi involved in human infection. Stud Mycol 61:137–144

Weinstein RN, Montiel PO, Johnstone K (2000) Influence of growth temperature on lipid and soluble carbohydrate synthesis by fungi isolated from fellfield soil in the maritime Antarctic. Mycologia 92:222–229

Wollenzien U, de Hoog GS, Krumbein WE, Urzì C (1995) On the isolation of microcolonial fungi occurring on and in marble and other calcareous rocks. Sci Tot Environ 167:287–294

Xi L, Lu C, Sun J, Li X, Liu H, Zhang J, Xie Z, de Hoog GS (2009) Chromoblastomycosis caused by a meristematic mutant of *Fonsecaea monophora*. Med Mycol 47:77–80

Zalar P, Gostincar C, de Hoog GS, Ursic V, Sudhadham M, Gunde-Cimerman N (2008) Redefinition of *Aureobasidium pullulans* and its varieties. Stud Mycol 61:21–38

Zhao J, Zeng J, de Hoog GS, Attili-Angelis D, Prenafeta-Boldú FX (2010) Isolation of black yeasts by enrichment on atmospheres of monoaromatic hydrocarbons. Microb Ecol 60:149–156

Zhdanova NN, Zakharchenko VA, Vember VV, Nakonechnaya LT (2000) Fungi from Chernobyl: mycobiota of the inner regions of the containment structures of the damaged nuclear reactor. Mycol Res 104:1421–1426

Part III
Adaptation Strategies of Cold-Adapted Yeasts to Low Temperatures

Chapter 9
Production of Pigments and Photo-Protective Compounds by Cold-Adapted Yeasts

Martín Moliné, Diego Libkind, Virginia de Garcia and María Rosa Giraudo

Abstract The structure of microbial communities in cold environments is not only shaped by temperature, but also other factors determine which species can survive. Ultraviolet radiation (UVR) is one of these factors and only microorganisms bearing photo-protective defence mechanisms colonize environments exposed to high UVR levels. In yeasts collected from Patagonia, among other cold environments, secondary metabolites such as carotenoids and mycosporines appeared to be produced as the most frequent photo-protective compounds. Production of carotenoids is quite common in cold-adapted yeasts, and some species produce unique carotenoids, like plectaniaxanthin and astaxanthin, which are not found in mesophilic species. On the other hand, the production of mycosporines appears in several yeasts species and has been intensively studied in yeasts isolated from high-altitude Patagonian lakes, revealing that several cold-adapted yeasts, including *Phaffia* and *Dioszegia*, are mycosporine-producing species. Both compounds effectively protect yeast cells against UVR.

Keywords Ultraviolet Radiation · Photo-protection · Carotenoids · Mycosporines

Contents

9.1 Introduction 194
9.2 UV Radiation in Cold Environments 195
    9.2.1 Damages Caused by UVR 196
    9.2.2 Strategies of Yeasts for Surviving to UVR 196
9.3 Carotenoids 197
    9.3.1 Carotenoids in Yeasts 198

M. Moliné (✉) · D. Libkind · V. de Garcia · M. R. Giraudo
Laboratorio de Microbiología Aplicada y Biotecnología, Centro Regional Universitario Bariloche, Universidad Nacional del Comahue, INIBIOMA (CONICET-UNCo), Quintral 1250 C.P. 8400 San Carlos de Bariloche, Río Negro, Argentina
e-mail: mmoline@comahue-conicet.gob.ar

P. Buzzini and R. Margesin (eds.), *Cold-adapted Yeasts*,
DOI: 10.1007/978-3-642-39681-6_9, © Springer-Verlag Berlin Heidelberg 2014

9.3.2 Biosynthesis of Carotenoids in Yeasts ... 199
9.3.3 Bioaccumulation of Carotenoids Among the Lineages and Under Different Factors ... 204
9.3.4 Function of Carotenoids in Yeasts ... 206
9.4 Mycosporines ... 207
9.4.1 Mycosporines in Fungi and Yeasts ... 208
9.4.2 Biosynthesis of Mycosporines in Yeasts ... 210
9.4.3 Bioaccumulation of Mycosporines Among the Lineages and Under Different Factors ... 212
9.4.4 Photo-Protective Role of Mycosporines in Yeasts ... 213
9.5 Ecological Relevance of Photo-Protective Compounds in Yeasts ... 214
9.6 Conclusions ... 215
References ... 216

9.1 Introduction

Cold-adapted microorganisms are capable of developing in low-temperature environments such as glacial ice, sea ice, and snow covered soils. In the last years, the occurrence and abundance of cold-adapted yeasts have been widely described in almost every cold environment, from marine waters, glaciers worldwide, Arctic, Alpine, and Antarctic ecosystems, to glacially formed water bodies, soils, and phylloplane (Margesin and Schinner 1994; Vishniac 1996; Gunasekera et al. 1997; Diaz and Fell 2000; Bergauer et al. 2005; Buzzini et al. 2005, 2012; de Garcia et al. 2007; Turchetti et al. 2008, 2011; Branda et al. 2010; Connell et al. 2010; Fernández et al. 2012). An in-depth overview of the biodiversity of cold-adapted yeasts in worldwide cold habitats is reported in Chaps. 3, 4, 5, 6, 7 and 8. Psychrophiles and psychrotolerant microorganisms are predominant in cold and temperate environments, because they can survive and even reproduce under these conditions. However, the structure of microbial communities in such habitats is not shaped only by temperature, given that additional biotic and abiotic (chemical and physical) factors determine the full ecological niche and thus the survival and reproduction of microorganisms. Solar radiation (in particular ultraviolet radiation, UVR) is one of these factors, and only microorganisms that show mechanisms and/or strategies to cope with such stress can grow and colonize highly UV-irradiated substrates. The impact of UVR on cold-adapted microbial communities has been studied by a few authors (Sommaruga 2001; Fernández Zenoff et al. 2006; Libkind et al. 2009a; Ordoñez et al. 2009). For the specific case of cold-adapted yeasts, several photo-protective compounds and enzymes have been identified in response to UVR stress. In this chapter, we focus on the production of two particular compounds with high biotechnological value (carotenoid pigments and mycosporines) that are synthesized by several cold-adapted yeast species.

9.2 UV Radiation in Cold Environments

Solar radiation reaching the Earth surface is divided into three major bands; the infrared light with wavelengths longer than 800 nm, the visible light or photosynthetically available radiation (PAR) between 400 and 700 nm, and the UVR between 200 and 400 nm. Solar radiation is essential for sustaining life on Earth, but UVR is extremely harmful and produces different kinds of damage on living organisms (Diffey 1991). This damage can be caused by two main mechanisms: (1) when UVR is absorbed by cellular biomolecules (i.e., proteins, DNA, or RNA), chemical transformations and degradation occur leading to the loss of biological function; this phenomenon is referred as direct damage; (2) when UVR is absorbed by an intermediate molecule that generates reactive oxygen species (ROS) that oxidize other vital cell components, the negative effect is referred as indirect damage.

Among the detrimental effects caused by UVR in microorganisms are included: damage in biomolecules like DNA, RNA, proteins, enzymes, lipids, and pigments; inhibition of many metabolic processes, and a detriment in the growth rate (Häder 1991; Cockell and Knowland 1999; Rothschild 1999). The UVR represents only the 5 % of the incident solar radiation and is subdivided into three bands. Only UV-A (between 315 and 400 nm), and UV-B (280 and 315 nm) reach the Earth surface and have significant effects for life. The highly energetic UV-C radiation is completely absorbed by atmospheric ozone and oxygen and does not reach the Earth surface.

The amount of UVR that reaches the Earth's surface differs widely with latitude and through seasons. About 85 % of the biosphere is permanently exposed to temperatures below 5 °C. Most of these environments are concentrated near to the Arctic and the Antarctic circles or in mountain regions (Gounot 1999; Margesin and Miteva 2011). At such latitudes, the incidence of sunlight is much lower than in temperate zones, although during summer the time of sunlight exposition is longer than near the equator. On the other hand, aquatic environments usually have high water transparencies and therefore higher penetration of UV radiation into the water column (Hanelt et al. 2001). Moreover, cold aquatic environments in high-elevation mountains are exposed to increased UV radiation due to high elevation, water transparency, and shallowness (Zagarese et al. 1998). Mountain lakes are extremely transparent, but unlike similarly clear forest lakes, which may be hundreds of meters deep, mountain lakes are typically shallow (<25 m), while shallow forest lakes are less transparent (Laurion et al. 2000). As a result, the biological communities in cold environments are strongly affected by physical factors such as UVR (Libkind et al. 2009a) and microorganisms living in such environments need strategies for coping with radiation.

9.2.1 Damages Caused by UVR

DNA damage is the best known consequence of exposure to UVR (Rothschild 1999) and UV-B is the radiation with more significant effects on this molecule. There are different types of mutagenic lesions caused by UV-B, the most common being the formation of cyclobutane-pyrimidine dimers (CPD), which interfere with basepairing during the DNA replication resulting in the generation of mutations. Another UV-B-induced damage is caused by the formation of (6–4) pyrimidine/pyrimidone dimers and its isomers, called Dewar isomers; these dimers are less frequent than CPD but much more mutagenic. A third type of photo-products is formed by the photoaddition of a water molecule (H_2O) to a pyrimidine. However, this product is dehydrated in hours, having a relatively less cytotoxic effect (Vincent and Neale 2000). Although DNA is not a chromophore for UV-A (except for photons close to 360 nm), UV-A also indirectly exerts cytotoxic and mutagenic effects. The photons of the UV-A spectrum are absorbed by endogenous photosensitizer molecules, which generate complex lesions, including tandem bases damage, cross-linking between DNA strands, and cross-linking between proteins and DNA (Ravanat et al. 2001).

Proteins are also susceptible to UVR damage, even though their absorbance peak is around 280 nm. The mechanism by which proteins are damaged involves a complex series of steps in which the UVR is absorbed by a chromophore located within the protein, causing the rupture and cross-linking of amino acids (Diffey 1991).

Finally, the production of ROS induced by UVR could also have negative effects for life. ROS is a widespread term to refer to any chemically reactive group of molecules containing oxygen in its composition. Among them, we can mention hydrogen peroxide (H_2O_2), anion hypochlorite (OCl^-), radicals such as hydroxyl (OH·), alkoxy (RO·), and peroxyl (ROO·), superoxide anion ($\cdot O_2^-$), and singlet oxygen ($_1O^2$). ROS are generated as an inevitable component of cellular metabolism, particularly during cellular respiration and photosynthesis (Apel and Hirt 2004; Pospíšil 2009). However, sources of stress such as UVR can highly stimulate their production (He and Häder 2002). Some of these molecules are characterized by a layer of unpaired valence electrons (free radicals) that become extremely unstable and rapidly react with other molecules to achieve a stable electronic configuration. ROS can react with macromolecules such as lipids, pigments, proteins and even nucleic acids causing damage that can lead to cell death or apoptosis (Madeo et al. 1999; He et al. 2002; Perrone et al. 2008).

9.2.2 Strategies of Yeasts for Surviving to UVR

Living organisms have different strategies to survive to the damage caused by UVR, which can be summarized in four main types. The simplest strategy is the avoidance of the stress by means of behavioral responses in which the organisms limit the exposure to UVR by moving to sites with lower levels of radiation. Although this strategy is common, in non-motile organisms like yeasts, it is worthless. Other strategies are related with the damage repair, particularly in the DNA, when the damage has already occurred, or with adaptation to the stress by changes in physiological mechanisms. But maybe the most significant and wide-spread strategy for all living beings is the avoidance of the damage by the synthesis of UV-screening compounds. The ability to synthesize photo-protective compounds varies among phylogenetic lineages; however, different UV-screening compounds are produced by almost every organism on Earth with the purpose to reduce the damage before it happens. For example, animals produce melanins but lack the metabolic pathways to *de novo* synthesis of carotenoids and mycosporines. Carotenoid production is exclusive to plants and microbes, while the synthesis of mycosporines is apparently restricted to organisms with the shikimic acid pathway, which is lacking in metazoa. Yeasts are not the exception and production of screening compounds is common.

In cold environments with high levels of UVR, some yeast species predominate and are frequently isolated (Libkind et al. 2009a; Brandão et al. 2011; Vaz et al. 2011). The occurrence of photo-protective compounds is widespread among these yeasts, and their presence has been shown to prevent photodamage. The major groups of photo-protective compounds in cold-adapted yeasts are carotenoids and mycosporines. The presence of such yeast species is a sign of their ability to minimize the damage caused by UVR. Among Patagonian yeasts, carotenogenic species were found in almost all water bodies, with proportions higher than 50 % of total yeast counts (Libkind et al. 2006). Such results are comparable to those found in other aquatic environments with similar characteristics (Libkind et al. 2006). Furthermore, synthesis of mycosporines is frequent in both pigmented and non-pigmented yeasts, and mycosporine-producing yeasts are frequently isolated from environments with high UVR levels, including high-altitude lakes and phylloplane (Libkind et al. 2006, 2009a; Muñoz 2010; Brandão et al. 2011).

9.3 Carotenoids

Carotenoids are natural pigments that occur universally in photosynthetic organism like plants and algae, and sporadically in some bacteria and some fungi, including yeasts. Most carotenoids found in nature are tetraterpenoids formed by a C40 chain (which is considered the backbone of the molecule), produced by the successive condensation of eight isoprene C5 units in such a manner that the arrangement of the units is reversed at the center of the molecule (Britton 1995, 2004). The backbone of carotenoids consists in a series of conjugated double bonds where the π-electrons are highly delocalized conferring a low-energy excited state. For that reason, these compounds absorb effectively the radiation in the violet and near UV sector of the visible spectrum ($\sim$400–500 nm) (Britton 1995). A minimum of seven conjugated double bonds is enough for the absorption of light in the visible spectra. The increase in the conjugated system shifts the color from yellow to dark red in a maximally desaturated molecule with 15 conjugated double bonds. Besides the basic skeleton of carotenoids may be modified in various manners, including hydrogenation and dehydrogenation of the chain, cyclization reactions, isomerization processes, the introduction of oxygen groups, or a combination of these modifications, giving a large number of different carotenes. Despite this, carotenoids have been singled out into two groups, the carotenes that are composed of only carbon and hydrogen, and the xanthophylls that also have oxygen in their chemical composition.

More than 700 types of carotenes and xanthophylls have been described in nature exerting a wide range of functions in living organisms (Britton 2004). The primary role of carotenoids in microorganisms is to quench ROS. The antioxidant capacity of carotenoids has been observed in bacteria, algae, fungi, and plants (Britton 2008). In fungi, different functions have been attributed to carotenoids, i.e., to be precursors of hormones (in the order Mucorales), to protect against ROS and UVR, and to be associated to membrane lipids modifying the membrane permeability and thus giving increased cellular resistance to heat, radiation and oxidation (Lampila et al. 1985; Britton 1995, 2008; Schroeder and Johnson 1995a; Johnson and Schroeder 1996). Despite the biological role that these compounds exert, carotenoids are also interesting because they represent a group of valuable molecules for the pharmaceutical, chemical, food, and feed industries (Ausich 1997).

9.3.1 Carotenoids in Yeasts

Three different groups of Basidiomycota and some Ascomycota species are able to produce carotenoids that are localized in lipid droplets (Davoli and Weber 2002). Despite the large number of possible pigments, red yeasts are thought to synthesize a few carotenoids molecules with typical concentrations ranging from 50 to 350 $\mu g\ g^{-1}$ dry mass (Bhosale 2004; Libkind et al. 2006; Buzzini et al. 2007; Yurkov et al. 2008). First, studies on the production of carotenoid pigments in yeasts began in the 1950s (Mrak et al. 1949; Peterson et al. 1954). Up to date, the synthesis of carotenoids was detected in several yeast genera including *Bannoa, Cryptococcus, Cystofilobasidium, Dioszegia, Erythrobasidium, Lalaria, Occutilfur, Phaffia* (*Xanthophyllomyces*), *Rhodosporidium, Rhodotorula, Saitoella, Sakaguchia, Sporidiobolus, Sporobolomyces* and *Taphrina* (Peterson et al. 1958; Simpson et al. 1964; Margalith and Meydav 1968; Nam et al. 1988; Martelli and da Silva 1993; Perrier et al. 1995; Madhour et al. 2005). Early studies to elucidate the composition of carotenoids in yeasts were carried out by Nakayama et al. (1954), Peterson et al. (1958) and Simpson et al. (1964) . Four pigments were identified in different proportions for most species, torularhodin (3′,4′-Didehydro-β, ψ-caroten-16′-oic acid), torulene (3′,4′-Didehydro-β, ψ-carotene), γ-carotene (β,ψ-carotene), and β-carotene (β,β-carotene). Of these pigments, torulene, β-

carotene, and γ-carotene are common to most yeasts and can be detected in different species and genera of the subphyla Agaricomycotina, Pucciniomycotina, and Ustilaginomycotina. By contrast, torularhodin (a xanthophyll product of the oxidation of torulene) is found only associated with Pucciniomycotina species (Davoli et al. 2004; Buzzini et al. 2007). Just one derivative of torularhodin has been reported, 2-hydroxytorularhodin, a rare pigment among yeasts so far described only in *Sporobolomyces coprosmae* (Weber et al. 2005). Torularhodin is most frequently found in the genus *Rhodotorula* and allies, including several species found in cold environments, such as *Rhodotorula mucilaginosa*, *Rhodotorula colostri, Rhodotorula glutinis, Rhodosporidium babjevae,* among others (Table 9.1).

Regarding other cold-adapted species, it is worth mentioning that the rare xanthophylls 16-hydroxytorulene and torularhodinaldehyde have been found in *Cystofilobasidium capitatum* and *Cystofilobasidium infirmominiatum* (Herz et al. 2007). Both pigments have been also described for some *Rhodotorula* species in minor proportions. Another interesting case is represented by the genus *Dioszegia* in which plectaniaxanthin (3′,4′-didehydro-1′,2′-dihydro-β, ψ-carotene-1′,2′-diol) is almost exclusively synthesized (Madhour et al. 2005; Moliné 2010). The same xanthophyll was described also for the yeast *Cryptococcus flavescens* (Bae et al. 1971), and a derivate, 2-hydroxyplectaniaxanthin, was described for *Rhodotorula aurantiaca* (Shan et al. 1973). However, these results and the correct identification of these two species were never reconfirmed.

Probably, the most unique case of carotenoids produced by cold-adapted yeasts is that of *Phaffia rhodozyma,* given it is the only known species capable of synthesizing the biotechnologically relevant astaxanthin (3,3′-dihydroxy-β,β′-carotene-4,4′-dione) (Andrewes et al. 1976). Astaxanthin, a pigment of high commercial value in aquaculture (Johnson and Schroeder 1995), represents over 80 % of the total carotenoids produced by *P. rhodozyma*. Thus, hyperpigmented mutants of this species are being exploited as a natural source of astaxanthin for aquaculture feed; though, production levels are limited by the low maximum growth temperatures of the yeast.

9.3.2 Biosynthesis of Carotenoids in Yeasts

The biosynthetic pathway of carotenoids is relatively well understood. In yeast, carotenoids are synthetized via the mevalonic acid pathway (Disch and Rohmer 1998). Mevalonic acid is converted to isopentenyl-pyrophosphate through a series of reactions involving phosphorylation by MVA kinase followed by decarboxylation, then a series of condensation of isopentenyl-pyrophosphate units result in molecules of geranyl–geranyl pyrophosphate (GGPP). The carotenoid synthesis begins with the formation of phytoene from the condensation head to head of two molecules of GGPP (Tefft et al. 1970; Liaaen-Jensen and Andrewes 1972). In fungi, a unique enzyme (phytoene desaturase) performs the conversion of phytoene to lycopene by four successive desaturation steps, and for that reason

Table 9.1 Carotenoids and mycosporines among yeasts and dimorphic fungi

Yeast species	Carotenoids	Mycosporines	Observations
Ascomycetous			
Capnodiales			
Trimmatostroma salinum, Cladosporium sp. *Phaeotheca triangularis*	–	myc-gln-glu and myc-glc-glu (Kogej et al. 2006)	Halophilic black yeasts
Chaetothyriales			
Exophiala dermatitidis, Exophiala xenobiotica	Torularhodinaldehyde (Kaiser et al. 2012)	N/D	
Dothideales			
Dothioraceae sp., *Coniozyma leucospermi, Delphinella strobiligena, Hormonema dematioides* (*Sydowia polyspora*)	–	myc-gln-glu (Muñoz 2010; Brandão et al. 2011)	Cold-adapted species from phylloplane and high mountain lakes (Muñoz 2010; Fernández et al. 2012)
Aureobasidium pullulans	–	myc-gln-glu and myc-glc-glu (Kogej et al. 2006; Brandão et al. 2011)	Ubiquitous specie with cold-adapted varieties (Zalar et al. 2008)
Hortaea werneckii	–	myc-gln-glu and myc-glc-glu (Kogej et al. 2006)	Halophilic black yeasts
Leotiomycetes			
Microglossum sp.	Undeterminated pigments	myc-gln-glu (Vaz et al. 2011)	Isolated from Antarctica (Vaz et al. 2011)
Taphrinales			
Protomyces sp.	β-carotene, γ-carotene, α- carotene (Valadon 1976)	N/D	Cold-adapted plant pathogens (Rossi et al. 2006; Schisler et al. 2011)
Taphrina sp., *Lalaria* sp.	β-carotene, γ-carotene, lycopene, β-zeacarotene (Van Eijk and Roeymans 1982)	myc-gln-glu in *Lalaria carpini* (Muñoz 2010)	Cold-adapted plant pathogens. Most species produce low carotenoids concentrations (Van Eijk and Roeymans 1982; Inácio et al. 2005; Rossi et al. 2006; Schisler et al. 2011)
Saitoella sp.	Undeterminated pigments	N/D	

(continued)

Table 9.1 (continued)

Yeast species	Carotenoids	Mycosporines	Observations
Basidiomycetous			
Agaricostilbales—Spiculogloeales			
Agaricostilbum hyphaenes, Agaricostilbum pulcherrimum, Chionosphaera apobasidialis, Mycogloea nipponica, Sporobolomyces coprosmicola, Sporobolomyces linderae, Bensingtonia sp. *Kondoa malvinella, Kurtzmanomyces nectairei, Sporobolomyces clavatus, Sporobolomyces xanthus, Sterigmatomyces halophilus*	–	myc-gln-glu (Libkind et al. 2011b)	
Cystobasidiales—Erythrobasidiales			
Bannoa hahajimensis, Erythrobasidium hasegawianum, Rhodotorula benthica, Rhodotorula lactosa, Rhodotorula lamellibrachii, Rhodotorula laryngis, Rhodotorula marina, Rhodotorula meli, Rhodotorula slooffiae, Occultifur sp., *Sakaguchia dacryoidea, Sporobolomyces bischofiae, Sporobolomyces elongatus, Sporobolomyces gracilis, Sporobolomyces salicinus*	Undeterminated carotenoid pigments	myc-gln-glu (Libkind et al. 2011b; Vaz et al. 2011)	Several species are cold-adapted including species only isolated from cold environments (Libkind et al. 2004a, b, 2006, 2009a, b; Brandão et al. 2011; Fernández et al. 2012)

(continued)

Table 9.1 (continued)

Yeast species	Carotenoids	Mycosporines	Observations
Rhodotorula minuta	Torulene, β-carotene, γ-carotene, other unidentified pigments (Buzzini et al. 2007, 2010; Yurkov et al. 2008)[a]	myc-gln-glu (Libkind et al. 2011b)	Frequently isolated from cold environments (Libkind et al. 2004a, b, 2006, 2009a, b; Brandão et al. 2011)
Sporobolomyces coprosmae	2-hydroxytorularhodine (Weber et al. 2005)	N/D	
Rhodotorula aurantiaca	2-hydroxyplectaniaxanthin (Shan et al. 1973)	N/D	
Cystofilobasidiales			
Cystofilobasidium capitatum, Cystofilobasidium infirmominiatum. Cystofilobasidium lacus-mascardii, Cystofilobasidium macerans	Torularhodinaldehyde, 16′-hydroxytorulene, torulene, β-carotene, γ-carotene (Herz et al. 2007)[a]	–	Cold-adapted yeasts (Birgisson et al. 2003; Libkind et al. 2009b)
Phaffia rhodozyma	Astaxanthin, phoenicoxanthin, echinenone, torulene, β-carotene, γ-carotene (Andrewes et al. 1976; Johnson and Schroeder 1995; Alvarez et al. 2006)	myc-gln-glu (Libkind et al. 2011c)	Cold-adapted yeast (Libkind et al. 2008)
Udeniomyces sp.	–	myc-gln-glu (Libkind et al. 2011c)	
Filobasidiales			
Cryptococcus antarcticus	–	myc-gln-glu (Vaz et al. 2011)	
Microbotryomycetes			
Rhodotorula sonckii	–	myc-gln-glu (Libkind et al. 2011a, b, c)	The only species of this group that produces mycosporines
Mixiomycetes			
Mixia osmundae	Undetermined carotenoid pigments	myc-gln-glu (Libkind et al. 2011a, b, c)	

(continued)

Table 9.1 (continued)

Yeast species	Carotenoids	Mycosporines	Observations
Sporidiobolales			
Rhodotorula colostri, Rhodosporidium diobovatum. Rhodotorula mucilaginosa, Rhodotorula glutinis, Rhodosporidium babjevae, Rhodosporidium sphaerocarpum, Sporidiobolus longiusculus, Sporidiobolus pararoseus, Sporidiobolus salmonicolor, Sporobolomyces japonicus, Sporobolomyces patagonicus Sporidiobolus metaroseus, Sporobolomyces ruberrimus	Torularhodin, torulene, β-carotene, γ-carotene (Nakayama et al. 1954; Davoli and Weber 2002, 2004; Sperstad et al. 2006; Buzzini et al. 2007, 2010; Yurkov et al. 2008)	–	Several species are cold-adapted including species isolated exclusively from cold environments (Libkind et al. 2004a, b, 2006, 2009a, b; Muñoz 2010; Brandão et al. 2011; Fernández et al. 2012)
Septobasidiales			
Septobasidium carestianum	–	myc-gln-glu (Libkind et al. 2011a, b, c)	
Tremellales			
Dioszegia sp., *Cryptococcus flavescens, Cryptococcus* sp.	β-carotene, γ-carotene, plectanixantin (Bae et al. 1971; Madhour et al. (2005)	myc-gln-glu (Libkind et al. 2005)	Cold-adapted yeasts (Inácio et al. 2005; Brandão et al. 2011)
Bullera variabilis	Plectanixantin-like pigment (Kaiser et al. 2012)	N/D	
Bulleromyces albus	β-carotene, γ-carotene, torulene (Fiasson 1972)	N/D	
Cryptococcus peneaus	β-carotene as major pigment (Nakayama et al. 1954)	N/D	

N/D not determinated, *myc-gln-glu* mycosporine-glutaminol-glucoside, *myc-glc-glu* mycosporine-glutamicol-glucoside

[a] Torularhodin has been reported for *Rhodotorula minuta* and *Cystofilobasidium capitatum* by Yurkov et al. (2008), but to our knowledge, this pigment is absent in both species (Moliné 2010; Herz et al. 2007)

intermediates as neurosporene are present only in trace amounts. Finally, lycopene is modified to generate different carotenoids (Frengova and Beshkova 2009). The first transformation common for all pigmented yeasts results in cyclization of one end of lycopene molecule leading to γ-carotene. This pigment represents a precursor and branch point in the synthesis of the other pigments. Cyclization of the acyclic ψ end group of γ-carotene leads to β-carotene, whereas desaturation at 3′,4′ position removes two hydrogen atoms to give torulene.

Torularhodin is then formed by the subsequent oxidation of the methyl group at C-16′ as the last step in the biosynthetic pathway. The biosynthetic details have not yet been completely elucidated; however, it has been proposed that a single enzyme is responsible for this oxidation and for that reason there are not intermediaries (Goodwin 1980). However, using two different strains, *C. infirmominiatum* and *C. capitatum*, Herz et al. (2007) were able to identify three unusual pigments identified as oxidized torulene derivatives (16′-hydroxytorulene, torularhodinaldehyde, and β-apo-2′-carotenal). Since torularhodin was not produced by either species, the authors suggested that these two pigments are intermediates in the torularhodin biosynthesis.

Details on the synthesis of plectaniaxanthin remain unknown; Madhour et al. (2005) suggested that plectaniaxanthin originates from torulene by the addition of water across the C-1′,2′ double bond and subsequent hydroxylation at C-2′. For the synthesis of 2-hydroxyplectaniaxanthin, Shan et al. (1973) suggested an early hydoxylation of phytoene at the C-2 position.

On the other hand, the synthesis of astaxanthin has been extensively studied and most of the enzymes and genes involved in the process have been identified. The biosynthesis of astaxanthin is produced by enzymatic oxidation and hydroxylation of β-carotene. Keto-groups are introduced in positions 4 and 4′, and the hydroxylation occurs at terminals carbons 3 and 3′ of both ionone groups. Originally, it has been suggested that one enzyme, encoded by the gene CRTS, would be responsible for this transformation. But recently it was found that a second gene, which encodes a cytochrome p450 reductase, is necessary for the synthesis of astaxanthin (Alvarez et al. 2006; Ojima et al. 2006; Alcaino et al. 2008). A summary of the synthesis of yeast carotenoids is depicted in Fig. 9.1.

9.3.3 Bioaccumulation of Carotenoids Among the Lineages and Under Different Factors

The proportion and type of carotenoids produced by yeasts is highly variable and depends on the phylogenetic group to which they belong, and the culture media. For example, torulene is generally predominant in *Rhodotorula minuta* and *Sporidiobolus longiusculus*; while in *R. mucilaginosa* and *R. glutinis* torularhodin is the major pigment. In some species, like *R. babjevae*, the produced amounts of torulene and torularhodin are similar. Other yeasts like *R. sphaerocarpum, R. graminis, Sporobolomyces pagatonicus* synthesize predominantly β-carotene (Sperstad et al. 2006; Buzzini et al. 2007, 2010; Yurkov et al. 2008; Moliné et al.

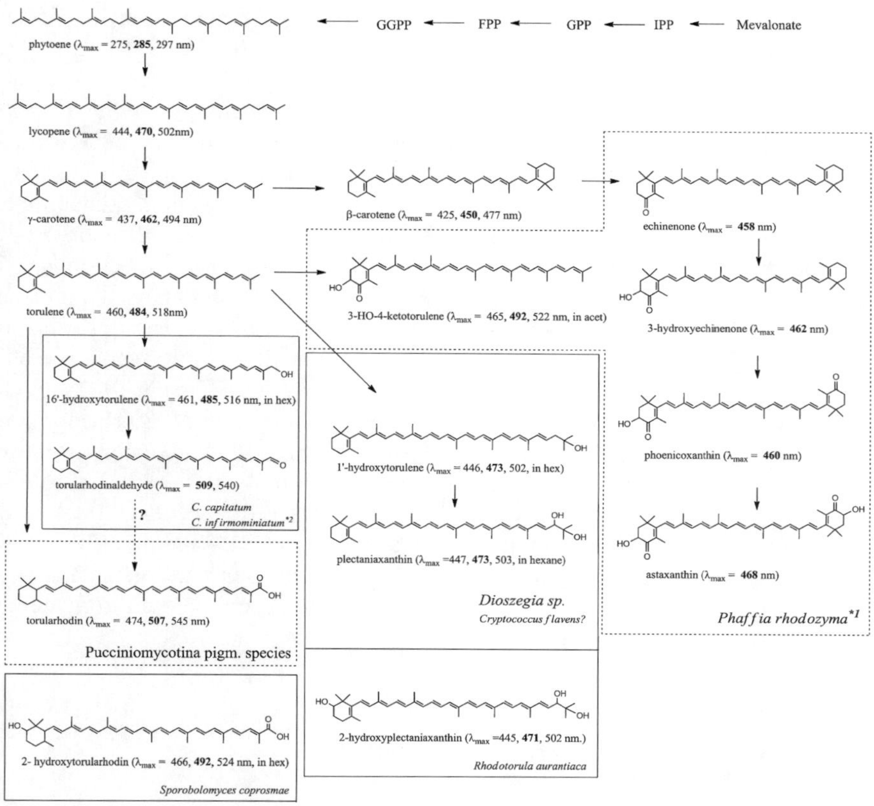

Fig. 9.1 Plausible biosynthetic pathways for the carotenoids of different yeasts species, including cold-adapted species *Phaffia rhodozyma, Dioszegia* sp., *Cystofilobasidium capitatum, Cystofilobasidium infirmominiatum*, among others. Lycopene, torulene, γ-carotene, and β-carotene are common to most pigmented yeasts. Absorption maximum refers to the maxima absorption peaks observed for each pigment in petroleum ether (except were indicated). *IPP* isopentenyl-pyrophosphate, *GPP* geranyl-pyrophosphate, *FPP* farnesyl-pyrophosphate, *GGPP* geranylgeranyl-pyrophosphate; ? is used for non-confirmed data, *1 corresponds to the biosynthetic pathway proposed by Andrewes et al. (1976) and Ojima et al. (2006). Other pathways have been proposed, including a monocyclic carotenoid biosynthetic pathway (An et al. 1999), and other variations (see Rodríguez-Sáiz et al. 2010); *2, 16′-hydroxytorulene has also been described as a minor carotenoid in *Rhodotorula glutinis* and *Rhodotorula mucilaginosa* (former *Rhodotorula rubra*) (Squina and Mercadante 2005). 2-hydroxytorularhodin and 2-hydroxyplectanixantin are probably synthetized from a hydroxylated phytoene

2012). As remarked above, the major pigments in *Dioszegia* species and in *P. rhodozyma* are plectaniaxanthin and astaxanthin, respectively. A summary of different carotenoids produced by pigmented yeasts is shown in Table 9.1.

Many studies pointed out that the concentration and relative proportion of each pigment is affected by physical and chemical factors such as light, carbon/nitrogen relation, temperature, aeration, metals, ions, ROS and solvents (Hayman et al. 1974; Johnson and Lewis 1979; Frengova et al. 1995; Bhosale 2004; Aksu and Eren 2005; Frengova and Beshkova 2009). For some factors, the responses are different depending on the species. For example, in *P. rhodozyma,* a high carbon/nitrogen (C/N) ratio favors the carotenoid production (Yamane et al. 1997; Flores-Cotera et al. 2001; Moliné 2010), but in *R. glutinis,* it is favored by a low C/N ratio (Sakaki et al. 1999), while in *R. mucilaginosa,* the specific carotenoid production is not affected by the C/N ratio (Libkind et al. 2004a; Libkind and van Broock 2006).

Other factors like temperature and pH also affect carotenoid production, and it seems that optimal culture conditions lead to optimal accumulation of pigments (Johnson and Lewis 1979). On the other hand, factors like light, aeration, ethanol, and ROS, which can be detrimental for yeasts growth and survival, generally produce an increase in the carotenoid production (Tada and Shiroishi 1982a, b; Gu et al. 1997; Santopietro et al. 1998; Bhosale 2004; Libkind et al. 2004a; Libkind and van Broock 2006). Moreover, some factors modify the relative concentration of each pigment. In *Rhodotorula* and *Sporobolomyces,* it is common to observe an increased proportion of torularhodin (Sakaki et al. 2001; Simova et al. 2003), in *P. rhodozyma* an increased accumulation of astaxanthin (Santopietro et al. 1998; Schroeder and Johnson 1993), and the same holds for the pigment plectaniaxanthin in *Dioszegia* (Madhour et al. 2005), when these species are exposed to moderate radiation, high levels of aeration, or to different ROS.

9.3.4 Function of Carotenoids in Yeasts

Yeast carotenoids have been historically associated with the antioxidant activity that these compounds exert. Some studies suggest that pigments are capable of compensating the deficiency of certain antioxidant enzymes like superoxide dismutase (SOD) or peroxidase (Schroeder and Johnson 1993). As remarked above, it has been found that is possible to induce production and accumulation of carotenes in fungi by adding ROS to the culture media (Schroeder and Johnson 1995b; Sakaki et al. 2000; Bhosale 2004), and it has been proved that increased production of carotenoids, in particular torularhodin, decreases the susceptibility to damage by ROS in *R. glutinis* and *R. mucilaginosa* species (Moore et al. 1989; Sakaki et al. 2001, 2002). The ability to inactivate free radicals is determined by the presence of distinct functional groups in the molecule (Miller et al. 1996). In the particular case of the torularhodin, quenching activity has not been formally characterized; however, increasing concentration of this pigment under oxidative stress suggests that torularhodin plays an important role in protection against free radicals (Sakaki et al. 2000). Moreover, it has been proved that torularhodin has a higher effect in the scavenging peroxyl radicals and inhibits substrate degradation by singlet oxygen more effectively than β-carotene (Sakaki et al. 2002). On the other hand, aerobic cultures of *P. rhodozyma* (oxidative environment) produce larger amounts of astaxanthin than under microaerophilic growth conditions (Johnson and Lewis 1979), which suggest that this pigment is responsible for protection against ROS.

Over the years, evidence has been accumulated in relation to the potential photo-protective function of carotenoids in yeast. Relation between carotenoid content and survival to high intensity radiation (at 632 nm) was first studied by Maxwell et al. (1966) in *R. glutinis*. Since then, different studies have recorded red cells to be more resistant to UVR than white cells. Using different Antarctic yeasts species Tsimako et al. (2002) found that carotenoid-producing species are more resistant to UV-B than non-pigmented species. By introducing carotenoid biosynthesis genes into *Saccharomyces cerevisiae* Méndez-Álvarez et al. (2000) showed that transformant strains able to produce a torularhodin-like pigment had higher survival to UV-B than the parental strain. On the other hand, we approached the analysis of the photo-protective function of carotenoids using a set of wild cold-adapted Patagonian pigmented and albino strains of *C. capitatum* and *Sporobolomyces ruberrimus.* Albino strains invariably resulted in being less tolerant to UV-B exposure than pigmented strains, and a direct relationship between carotenoid content and survivorship in *C. capitatum* was observed (Libkind et al. 2006; Moliné et al. 2009). Finally, with a numerous set of *R. mucilaginosa* and *P. rhodozyma* strains, a significant positive relationship between intracellular carotenoid concentration and UV-B survival was observed. In *R. mucilaginosa,* the carotenoid content had a correlation higher than 57 % with UV-B survival (Moliné et al. 2010). Even when considering the physiological variability found among strains and the potential presence of alternative photo-protection strategies, carotenoids appear as a significant factor in improving resistance to UV-B. Analysis of carotenoid content pointed out that torularhodin in *R. mucilaginosa* (Moliné et al. 2010) and astaxanthin in *P. rhodozyma* (Moliné 2010) provided effective photo-protection, while other carotenoids like β-carotene showed a lack of correlation with survival to UV-B. The formation of CPDs was independent of cell survival and carotenoid content, thus indicating that the photo-protective activity of carotenoids it is not direct. How these pigments protect yeasts against UV detrimental effects has not yet been elucidated.

9.4 Mycosporines

Mycosporines is a broad term to describe two molecules, mycosporines sensu stricto and mycosporine-like aminoacids (MAAs), but the use of both terms is misleading. In this chapter, mycosporines is used to refer to molecules with a cyclohexenone core, while MAAs refers to molecules with a cyclohexenimine core (Bandaranayake 1998).

Mycosporines and MAAs are water-soluble compounds that contain a cyclic unit, cyclohexenone or cyclohexenimine, respectively, attached to an amino acid or amino alcohol, with the only exception of gadusol and deoxygadusol. These structures possess at least two conjugated double bonds, which can be extended to three and four in some MAAs. For this reason, mycosporines and MAAs absorb radiation in the UV-B, with maximum absorbance at 296 nm for gadusol and deoxygadusol (in water at pH > 7, and at 268 nm in other solvents or at lower pH) (Arbeloa et al. 2010, 2011), at 310 and 320 nm for mycosporines and nor-mycosporines, which possess a OH group in C2, and at 320–360 nm for MAAs (Bandaranayake 1998; Carreto and Carignan 2011). There are more than a dozen of amino acids and amino alcohols described in different mycosporines and MAAs; however, only serine, alanine, α-amino alcohol serinol, pyroglutamic acid and the related pairs glutamine–glutaminol and glutamic acid–glutamicol were described in fungal mycosporines (Arpin et al. 1977; Arpin and Bouillant 1981; Young and Patterson 1982; Favre-Bonvin et al. 1987; Leite and Nicholson 1992; Volkmann et al. 2003; Sommaruga et al. 2004).

More than 30 mycosporines and MAAs have been described in microorganisms, and different functions have been attributed to these compounds (Sinha et al. 2007). Since these compounds absorb light with high efficiency in the UV spectrum, and their synthesis is stimulated in the presence of photosynthetically active radiation and UV radiation (Karsten et al. 1998; Sinha et al. 2001; Libkind et al. 2004b), the primary function assigned was to act as photo-protective UV filters (Garcia-Pichel and Castenholz 1993; Klisch et al. 2001; Shick and Dunlap 2002; Torres et al. 2004).

In recent years, it was suggested that MAAs and mycosporines could play other roles. The antioxidant capacity of mycosporine-glycine, gadusol and deoxygadusol was demonstrated in vitro (Suh et al. 2003; Yakovleva et al. 2004; Arbeloa et al. 2010, 2011) and was suggested for others MAAs (Zhang et al. 2007; Tao et al. 2008); however, its biological significance is yet unknown (Oren and Gunde-Cimerman 2007). These molecules could also play a role in osmoregulation, since the concentration of MAAs and mycosporines in the cell of some microorganisms appears to be related with the salinity of the medium (Oren 1997; Portwich and Garcia-Pichel 1999; Kogej et al. 2006). However, some authors question the biological significance of MAAs in osmoregulation processes (Portwich and Garcia-Pichel 1999). Other functions attributed include resistance to thermal stress, to act as accessory pigments in photosynthesis, to serve as intracellular nitrogen storage (Oren and Gunde-Cimerman 2007), and in sea hares, their role as intraspecific chemical alarm cues has been documented (Kamio et al. 2011; Kicklighter et al. 2011).

9.4.1 Mycosporines in Fungi and Yeasts

Mycosporines were initially discovered in fungal sporulating mycelia, and a series of observations, including that mycosporines were only synthesized during sporulation, led to conclude that they have a reproductive role in fungi (Leach 1965; Leach and Trione 1966; Trione et al. 1966). However, few years later, the sporogenic character of mycosporines has been questioned (Ende and Cornelis 1970), and recently, it has been shown that mycosporines act as inhibitors of spore germination of the filamentous fungus *Colletotrichum graminicola* (Leite and Nicholson 1992). First, mycosporines were isolated from fungi of the subphylum Pezizomycotina. In 1965, Leach suggested the presence of at least four different compounds, which showed an absorbance maximum at 310 nm and different retention times in ion exchange columns. Later, Trione et al. (1966) confirmed the occurrence of three distinct fractions that were subsequently identified, showing that they all had an identical cyclohexenone core attached to a glutaminol, glutamicol, glutamine, or pyroglutamic acid. Moreover, in some cases, these forms were combined with a glucose molecule (Favre-Bonvin et al. 1976; Arpin et al. 1977; Young and Patterson 1982). The existence of different forms of mycosporines was questioned by Pittet et al. (1983a). The presence/absence of the glycosylated form and small amounts (<5 % of total) of the amino acid form (glutamicol) could be the result of the formation of "artifacts" in the process of extraction and purification. Additionally, the pyroglutamic acid form could be due to the dehydration of glutamicol. With these considerations, only a single mycosporine would exist in fungi, the mycosporine-glutaminol-glucoside (MGG). However, Volkmann et al. (2003) observed that the forms glutaminol and glutamicol are present in the fungal genera *Coniosporium* and *Sarcinomyces*, independently of the extraction or purification method. Therefore, their presence cannot always be attributed to artifacts.

It was not until recently that mycosporine synthesis was reported in yeasts by Libkind et al. (2004b) who found that a number of cold-adapted basidiomycetous yeasts synthesized a UV-absorbing compound (maximum absorption at 310 nm) when grown under photosynthetically active radiation. This mycosporine was identified as MGG (Sommaruga et al. 2004), with typical concentrations ranging from less than 1 to 40 mg g^{-1} (Libkind et al. 2004b, 2011b; Moliné et al. 2011). MGG has the general formula $C_{19}H_{32}N_2O_{11}$ and consists of a cyclohexenone attached to a glutaminol and glucose molecule. Its molecular weight is 464.5 g mol^{-1} and corresponds to the major mycosporine fraction found in filamentous fungi. The characteristic absorbance at 310 nm is due to the structure of cyclohexenone with two conjugated double bonds. The degree of resonance delocalization in a molecule affects not only the absorption maximum but also the value of their extinction coefficient (Carreto and Carignan 2011). Mycosporines possess an oxygen in the enone group with higher electronegativity than the nitrogen of MAAs imine affecting the resonance delocalization and therefore the extinction coefficient. In MGG, the molar extinction coefficient is 25,000 M^{-1} which is almost half the one reported for other MAAs. Despite this relative low extinction coefficient, MGG content represents a large proportion of yeast cell dry mass reaching up to 40 mg g^{-1}. This concentration is larger than that observed in other microorganisms producing MAAs, like the cyanobacteria *Nostoc commune* (2.4 mg g^{-1}) or *Gleocapsa* sp. (8–9 mg g^{-1}) (Scherer et al. 1988; Portwich and Garcia-Pichel 1999).

9.4.2 Biosynthesis of Mycosporines in Yeasts

The synthesis of mycosporines and MAAs has not been completely described, but in the last years, great advances were achieved. First studies on the synthesis of mycosporines were performed in the filamentous fungus *Trichothecium roseum* using labeled precursors, and it was concluded that the synthesis of mycosporines was related to the shikimic acid pathway; therefore, the authors proposed that the precursor of the six-carbon ring was 3-dehydroquinate (3-DHQ) (Favre-Bonvin et al. 1987). Further studies showed that glyphosate, an inhibitor of the shikimic acid pathway affecting the enzymes 5-enolpyruvylshikimate-3-phosphate (EPSP) synthase, 3-deoxy-7-phosphoheptulonate synthase, 3-DHQ synthase (Herrmann and Weaver 1999; Knaggs 2003), reduced MAAs synthesis in the corals exposed to UV radiation (Shick et al. 1999). Finally, the discovery of gadusol and deoxygadusol allowed a better understanding of the possible intermediates in the metabolic pathway (Grant et al. 1980), although it was observed that the deoxygadusol in bacteria of the genera *Vibrio* sp. and *Pseudoalternomonas* sp. could be produced by oxidation of MAAs (Masaki et al. 1996). A model in which 3-dehidroquinate is the precursor of deoxygadusol, which was later used for mycosporines and MAAs synthesis, was proposed. However, this model was recently rejected by Balskus and Walsh (2010).

Using a Bioinformatic approach, Singh et al. (2010) identified a set of putative genes involved in the biosynthesis of mycosporines, including a DHQ-like synthase and a O-methyltransferase that were present only in mycosporine-producing microorganisms; however, no experimental evidence accompanied this report. Later, Balskus and Walsh (2010) identified the same MAA genes in the cyanobacteria *Anabaena variabilis* and *Nostoc punctiforme*. These authors cloned and expressed the complete gene cluster in *E. coli* providing indisputable evidence of the role of these genes in mycosporine synthesis. A cluster of four genes was found to be responsible for the conversion of sedoheptulose 7-phosphate precursor into the MAAs shinorine. A dehydroquinate synthase (DHQS) homolog, 2-epi-5-epi-valiolone synthase, and an O-methyltransferase convert the sedoheptulose 7-phosphate into 4-deoxygadusol, after which an ATP-grasp homolog and a non-ribosomal peptide synthase attach glycine and serine to generate mycosporine-glycine and shinorine. The DHQS homolog, O-methyltransferase and ATP-grasp were also identified in nine fungal species (Balskus and Walsh 2010); however, none of these were known to produce mycosporines.

In cold-adapted yeasts, homologous genes were recently reported in *P. rhodozyma*. Genome mining allowed the localization of the hitherto unknown gene cluster responsible for MGG synthesis. This cluster contained three genes (DHQS-like, O-methyltransferase, and ATP-grasp) common to other mycosporine-producing organisms though with significant modifications, in particular in the gene responsible of the ATP-grasp synthesis (Libkind et al. 2011a). All the genes were absent in non-mycosporine-producing strains with available genomes (Libkind 2013, unpublished data). It is highly probable that DHQS-like and

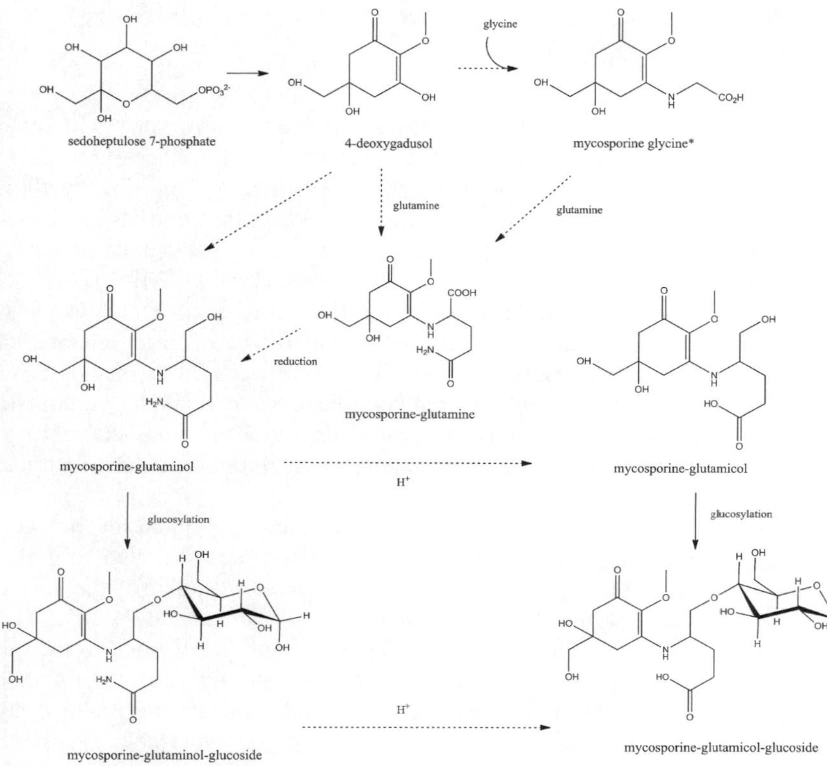

Fig. 9.2 Plausible biosynthetic pathways for mycosporines in yeast, adapted from Balskus and Walsh (2010), Bernillon et al. (1984) and Favre-Bonvin et al. (1987). Genes involved in the synthesis of 4-deoxygadusol have been identified in *Phaffia rhodozyma*. *Asterisk* mycosporine-glycine has never been reported in fungi, but according to the Balskus and Walsh (2010) model, it could be an intermediate in MGG synthesis and rapidly replaced by a glutamine or a glutaminol

O-methyltransferase identified in yeasts are responsible for the synthesis of 4-deoxygadusol. However, the role of ATP-grasp identified in *P. rhodozyma* has not been yet clarified. Two hypotheses can be posed. If we adopt the model proposed by Balskus and Walsh (2010), ATP-grasp adds a glycine which is then replaced by glutaminol (or a glutamine). The second hypothesis posed that ATP-grasp of fungi added the glutaminol directly (Favre-Bonvin et al. 1987), or a glutamine which is subsequently reduced (Bernillon et al. 1984). Finally, the glucose could be transferred from one molecule of uridine diphosphate glucose by a specific glucosyltransferase (Pittet et al. 1983b). The proposed mycosporine pathway for yeasts is depicted in Fig. 9.2.

9.4.3 Bioaccumulation of Mycosporines Among the Lineages and Under Different Factors

Mycosporine production was shown to be a species-specific trait exclusive of certain taxonomic groups. Certain species were able to produce mycosporines after photostimulation, while others did not. The hypothesis that mycosporine synthesis is a plesiomorphic character in fungi seems more parsimonious than a hypothetical appearance of a similar biochemical pathway producing an identical compound in multiple phylogenetically diverse lineages (Libkind et al. 2005, 2011b).

In ascomycetous yeasts and dimorphic fungi (i.e., those fungi which can grow in hyphal or yeast form), significant concentrations of mycosporine-glutaminol-glucoside and mycosporine-glutamicol-glucoside were reported in species of the orders Dothideales and Capnodiales (subphylum Pezizomycotina) and Taphrinales (subphylum Taphrinomycotina) (Gunde-Cimerman and Plemenitaš 2006; Kogej et al. 2006; Muñoz 2010). However, it seems to be missing in the subphylum Saccharomycotina.

In basidiomycetous yeasts, only mycosporine-glutaminol-glucoside has been reported. However, a second mycosporine compound was detected in HPLC chromatograms for a few strains of the genus *Cryptococcus*, and of the species *Rhodotorula pallida*, *Rhodotorula marina*, *Rhodotorula lamellibranchi,* and *Sporobolomyces clavatus*. This compound might probably correspond to the glutamicol form but has not been confirmed yet (Libkind et al. 2011b). Without exception, none of the strains of mycosporine-negative species did produce mycosporines, while strains of mycosporine-positive species were able to produce them. Within the subphylum Pucciniomycotina, mycosporine-producing species were reported in the class Cystobasidiomycetes (with the exception of the Naohideales), in the class Agaricostilbomycetes, in the monotypic class Mixiomycetes, (*Mixia osmundae*), and in Septobasidiales as the only known mycosporine-producing group within the class Pucciniomycetes (Libkind et al. 2011b). In the Agaricomycotina, mycosporines species were reported in the order Tremellales. On the contrary, species of the order Cystofilobasidiales did not produce mycosporines (Libkind et al. 2005) with the only exception of the species *P. rhodozyma* and the genus *Udeniomyces* (Libkind et al. 2011c). The consistent occurrence of mycosporines and carotenoids in some specific phylogenetic groups has potential applications in yeast systematics (Libkind et al. 2005). Moreover, it can be applied for the rapid identification of the cold-adapted yeast *P. rhodozyma* among other pigmented species. *P. rhodozyma* produces MGG and astaxanthin (Libkind et al. 2008, 2011c) both compounds are released when ethanolic extraction at high temperatures is applied. The resulting characteristic UV–visible spectrum allows the unequivocal identification of this species (Tognetti et al. 2013).

The constitutive production of MGG in all these groups is in most cases undetectable, and when detectable, it does not exceed 1 mg g^{-1} (Libkind et al. 2004b, 2011c; Moliné et al. 2011). The highest constitutive MGG accumulation was found in the Patagonian strain *R. minuta* CRUB 76, and in the collection strains *Erythrobasidium hasegawianum* PYCC 4584^T and *R. marina* PYCC 5050^T. All strains exposed to PAR + UV showed more than twofold higher MGG production compared with growth in the dark. Dependence between radiation intensity and wavelength, and accumulation of mycosporines and MAAs has been widely described for most mycosporine-producing organisms (Carreto et al. 1989; Bandaranayake 1998; Hoyer et al. 2002; Klisch and Häder 2002; Kräbs et al. 2004; Singh et al. 2008). In the particular case of yeasts, photoinduction using UVR and PAR was first observed in *Rhodotorula* and *Cryptococcus* species (Libkind et al. 2004b), and this phenomenon is a general response in all mycosporinogenic yeast species (Moliné 2010; Libkind et al. 2011c).

Other factors affecting MAAs accumulation in different organisms have been identified, including osmotic stress, thermal stress, and nitrogen supply; however, studies regarding the influence of these factors on yeasts are scarce. Osmotic stress can induce the synthesis of MAAs in some cyanobacteria (Portwich and Garcia-Pichel 1999; Singh et al. 2008). Using different halophilic and halotolerant strains Kogej et al. (2006) observed that mycosporine production is strongly stimulated when grown in 10 % (w/v) salt, leading to the idea that mycosporines could act as a supplementary compatible solute. However, more recently, they reported that this role is not relevant (Kogej et al. 2007). On the contrary, in *P. rhodozyma* and *Dioszegia* sp, salinity affects negatively mycosporine accumulation (Moliné 2010). Higher nitrogen concentration in the culture media also has significant effects on the synthesis of MAAs and mycosporines. In the cyanobacterium *A. variabilis,* the synthesis of MAAs is clearly favored in the presence of ammonium (Singh et al. 2008). In *P. rhodozyma,* accumulation of MGG is highly favored when peptone is added to culture media, reaching a production of 93 mg g^{-1} dry mass (Moliné 2010). Recently, we identified other factors affecting the production of MGG in this species, including ethanol and different ROS, such as H_2O_2, singlet oxygen and hydroxyl radical. Ethanol is recognized for the stimulation of carotenoid production (Gu et al. 1997); however, it severely affects MGG accumulation. On the contrary, ROS has different effects. While singlet oxygen and hydroxyl radical decrease the accumulation of mycosporines, low concentrations of H_2O_2 (0.01 mM) added to culture media increased the accumulation of mycosporines by 17 % (Moliné 2010).

9.4.4 Photo-Protective Role of Mycosporines in Yeasts

Since their discovery, the function of mycosporines and MAAs has never been fully clarified, and as mentioned above, several authors suggested that they could be multipurpose metabolites. For MAAs, most evidences support their importance as photo-protective and antioxidant compounds. However, these studies refers to indirect evidences like, photoinduction of MAAS synthesis, correlation between MAAs content and irradiance in different locations over the world (Oren and Gunde-Cimerman 2007), and theoretical models of self-shading (Garcia-Pichel 1994).

The photo-protective role of fungal mycosporines was first proposed by Brook (1981), based on the fact that spores produced in light had higher survival and mycosporine concentrations than those of spores produced in the dark. However, evidence was not conclusive since light could also affect other unknown photo-protective mechanisms. When discovered in yeasts, Libkind et al. (2004b) suggested a photo-protective role, and other hypotheses like its role in reproduction and in osmoregulation were instantly discarded given that these microorganisms reproduced asexually and were inhabitants of cold ultraoligotrophic lakes (Libkind et al. 2004a, b, 2005; Sommaruga et al. 2004).

More recently, using a set of *Cryptococcus stepposus* and *P. rhodozyma* strains, a high correlation between survival to UV-B and MGG concentration was observed. This correlation was significant when isolates were cultured under light and a sufficient concentration of mycosporines had accumulated. Moreover, damage of DNA (measured by CPD accumulation) observed after UV-B irradiation was negatively related to MGG accumulation. This is an indication that MGG protects yeasts against the harmful effects of UVR avoiding the direct damage of DNA (Moliné 2010; Moliné et al. 2011).

The mycosporine-glutaminol-glucoside is more stable than other MAAs studied so far. The photostability of MAAs has been described by Conde et al. (2000, 2004, 2007) for porphyra-334, shinorine and palythine, which supported the role of MAAs as potent and stable UV absorbers (Whitehead and Hedges 2005). In this sense, MGG is highly stable and thus could act as a natural sunscreen in these microorganisms. An antioxidant activity of this molecule was also detected. The ability of MGG to act as a quencher of singlet oxygen is comparable with that reported by Suh et al. (2003). The antioxidant activity of mycosporine-glycine has been reported in marine organisms (Dunlap and Yamamoto 1995; Yakovleva et al. 2004) and it is likely that MGG has a similar function in yeasts.

9.5 Ecological Relevance of Photo-Protective Compounds in Yeasts

Little research has been performed to assess the ecophysiological significance of photo-protective compounds (carotenoids and mycosporines) in yeasts. Comparing the proportion of carotenoid-producing yeasts, the synthesis of carotenoids and mycosporine-glutaminol-glucoside and the UV-B tolerance of yeasts isolated from high-altitude lakes (above 1,400 m.a.s.l.) and piedmont lakes (around 700 m.a.s.l.), a greater proportion of carotenogenic yeasts was found in former relative to the latter lakes. However, no differences were detected in the constitutive carotenoid concentration between yeast strains isolated from both groups of lakes. Among pigmented yeasts, mycosporine-synthesizing species were poorly represented in aquatic environments exposed to high UVR (Libkind et al. 2009a). Interestingly, more than 70 % of non-pigmented yeast species found in mountain lakes were able to produce UV-absorbing compounds such as MGG. Strains isolated from high-altitude environments produced significantly higher amounts of mycosporines than those from piedmont lakes (Libkind et al. 2006).

Maybe the most emblematic cold-adapted species isolated from highly UVR-exposed environments is the carotenoid-producing yeast genus *Dioszegia*. Isolates of *Dioszegia* have the highest survival to UV-B radiation recorded so far (Libkind et al. 2009a; Moliné 2010) and the isolates studied showed the highest levels of mycosporines. The high UV-B survival can be explained by the production of both metabolites in *Dioszegia* spp. and helps clarify why it is common in high mountain lakes (Libkind et al. 2009a) and phylloplane (Inácio et al. 2005; Madhour et al. 2005; Wang et al. 2008). Another frequent species isolated in these environments is *R. mucilaginosa*, which produces high concentrations of the photo-protective carotenoid torularhodin (Libkind et al. 2009a; Moliné et al. 2010). A significant intraspecific variability in the UVR tolerance of this species was found. *R. mucilaginosa* isolates obtained from highly transparent mountain lakes were more tolerant to UV-B than those recovered from less transparent water bodies (Libkind et al. 2009a). This is an example of how an ubiquitous species like *R. mucilaginosa* can adapt to high UVR exposure.

9.6 Conclusions

In this chapter, the photo-protective role in yeasts of two families of secondary metabolites has been addressed. Carotenoid pigments, potent antioxidant compounds that mainly afford photo-protection by the intracellular quenching of ROS generated by UVR (indirect damage), are present in a few phylogenetically lineages primarly of basidiomycetous yeasts of which many include cold-adapted species and genera (i.e., *Rhodotorula*, *Rhodosporidium*, *Sporobolomyces*, *Dioszegia*, *Cystofilobasidium*, *Phaffia*, etc.). Yeast carotenoids are chemically diverse and specific pigmented compounds are produced by each phylogenetic group. Among cold-adapted pigmented yeast species, unique and biotechnologically relevant carotenoids are produced, the most relevant case being the xanthophyll astaxanthin (*P. rhodozyma*). The production of carotenoid pigments was shown to be an advantageous trait for yeasts thriving in highly transparent water bodies in Patagonia. On the other hand, mycosporines, in particular MGG, are chemically sunscreen molecules that can absorb UV-B wavelengths and thus avoid direct UVR damage. The phylogenetic distribution of MGG is more widespread compared with carotenoids, although less chemically complex since only MGG has so far been detected for yeasts. This photo-protective compound is not almost only restricted to basidiomycetous as is the case for carotenoids. It is clear that cold environments normally are exposed to high UV irradiances mainly due to high latitude and/or altitude. Production of MGG in yeasts is favored under such conditions as denoted by the large numbers of cold-adapted species that are able to synthesis such metabolite. Given that UVR is an important force in structuring yeast community composition, the ability of yeasts to synthesize carotenoids and mycosporines becomes at least part of the adaptive suite of mechanisms needed for exploiting highly UV-exposed cold habitats.

References

Aksu Z, Eren AT (2005) Carotenoids production by the yeast *Rhodotorula mucilaginosa*: use of agricultural wastes as a carbon source. Process Biochem 40:2985–2991

Alcaino J, Barahona S, Carmona M, Lozano C, Marcoleta A, Niklitschek M, Sepulveda D, Baeza M, Cifuentes V (2008) Cloning of the cytochrome p450 reductase (crtR) gene and its involvement in the astaxanthin biosynthesis of *Xanthophyllomyces dendrorhous*. BMC Microbiol 8:169

Alvarez V, Rodriguez-Saiz M, de la Fuente J, Gudina E, Godio R, Martin J, Barredo J (2006) The crtS gene of *Xanthophyllomyces dendrorhous* encodes a novel cytochrome-P450 hydroxylase involved in the conversion of beta-carotene into astaxanthin and other xanthophylls. Fungal Genet Biol 43:261–272

An G-H, Cho M-H, Johnson EA (1999) Monocyclic carotenoid biosynthetic pathway in the yeast *Phaffia rhodozyma* (*Xanthophyllomyces dendrorhous*). J Biosci Bioeng 88:189–193

Andrewes AG, Phaff HJ, Starr MP (1976) Carotenoids of *Phaffia rhodozyma*, a red-pigmented fermenting yeast. Phytochemistry 15:1003–1007

Apel K, Hirt H (2004) Reactive oxygen species: metabolism, oxidative stress, and signal transduction. Ann Rev Plant Biol 55:373–399

Arbeloa EM, Uez MJ, Bertolotti SG, Churio MS (2010) Antioxidant activity of gadusol and occurrence in fish roes from Argentine Sea. Food Chem 119:586–591

Arbeloa EM, Bertolotti SG, Churio MS (2011) Photophysics and reductive quenching reactivity of gadusol in solution. Photochem Photobiol Sci 10:133–142

Arpin N, Bouillant M (1981) Light and mycosporines. The fungal spore, morphogenetic controls. Academic Press, London, pp 435–454

Arpin N, Favre-Bonvin J, Thivend S (1977) Structure de la mycosporine 2, nouvelle molecule, isolee de *Botrytis cinerea*. Tetrahedron Lett 18:819–820

Ausich RL (1997) Commercial opportunities for carotenoid production by biotechnology. Pure Appl Chem 69:2169–2174

Bae M, Lee TH, Yokoyama H, Boettger HG, Chichester CO (1971) The occurrence of plectaniaxanthin in *Cryptococcus laurentii*. Phytochemistry 10:625–629

Balskus EP, Walsh CT (2010) The genetic and molecular basis for sunscreen biosynthesis in cyanobacteria. Science 329:1653–1656

Bandaranayake W (1998) Mycosporines: are they nature's sunscreens? Nat Prod Rep 15:159–172

Bergauer P, Fonteyne P-A, Nolard N, Schinner F, Margesin R (2005) Biodegradation of phenol and phenol-related compounds by psychrophilic and cold-tolerant alpine yeasts. Chemosphere 59:909–918

Bernillon J, Bouillant M-L, Pittet J-L, Favre-Bonvin J, Arpin N (1984) Mycosporine glutamine and related mycosporines in the fungus *Pyronema omphalodes*. Phytochemistry 23:1083–1087

Bhosale P (2004) Environmental and cultural stimulants in the production of carotenoids from microorganisms. Appl Microbiol Biotechnol 63:351–361

Birgisson H, Delgado O, Arroyo LG, Hatti-Kaul R, Mattiasson B (2003) Cold-adapted yeasts as producers of cold-active polygalacturonases. Extremophiles 7:185–193

Branda E, Turchetti B, Diolaiuti G, Pecci M, Smiraglia C, Buzzini P (2010) Yeast and yeast-like diversity in the southernmost glacier of Europe (Calderone Glacier, Apennines, Italy). FEMS Microbiol Ecol 72:354–369

Brandão LR, Libkind D, Vaz ABM, Espírito Santo LC, Moliné M, de García V, van Broock M, Rosa CA (2011) Yeasts from an oligotrophic lake in Patagonia (Argentina): diversity, distribution and synthesis of photoprotective compounds and extracellular enzymes. FEMS Microbiol Ecol 76:1–13

Britton G (1995) Structure and properties of carotenoids in relation to function. FASEB J 9:1551–1558

Britton G (2004) Carotenoids handbook. Birkhäuser Verlag, Basel

Britton G (2008) Functions of intact carotenoids. In: Britton G, Liaaen-Jensen S, Pfander H (eds) Carotenoids, vol 4. Birkhäuser, Basel, pp 189–212

Brook P (1981) Protective function of an ultraviolet-absorbing compound associated with conidia of *Glomerella cingulata*. N Z J Bot 19:299–304

Buzzini P, Turchetti B, Diolaiuti G, D'Agata C, Martini A, Smiraglia C (2005) Culturable yeasts in meltwaters draining from two glaciers in the Italian Alps. Annals Glaciol 40:119–122

Buzzini P, Innocenti M, Turchetti B, Libkind D, van Broock M, Mulinacci N (2007) Carotenoid profiles of yeasts belonging to the genera *Rhodotorula, Rhodosporidium, Sporobolomyces*, and *Sporidiobolus*. Can J Microbiol 53:1024–1031

Buzzini P, Goretti M, Branda E, Turchetti B (2010) Basidiomycetous yeasts for production of carotenoids. In: Flickinger M (ed) Encyclopedia of industrial biotechnology: bioprocess, bioseparation, and cell technology, vol 1. Wiley, New York, pp 469–481

Buzzini P, Branda E, Goretti M, Turchetti B (2012) Psychrophilic yeasts from worldwide glacial habitats: diversity, adaptation strategies and biotechnological potential. FEMS Microbiol Ecol 82:217–241

Carreto JI, Carignan MO (2011) Mycosporine-like amino acids: relevant secondary metabolites. Chemical and ecological aspects. Mar Drugs 9:387–446

Carreto JI, Marco S, Lutz V (1989) UV-absorbing pigments in the dinoflagellates *Alexandrium excavatum* and *Prorocentrum micans*. Effects of light intensity. In: Okaichi T, Anderson DM, Nemoto T (eds) Red tides; biology, environmental science and toxicology. Elsevier, Amsterdam, pp 333–336

Cockell CS, Knowland J (1999) Ultraviolet radiation screening compounds. Biol Rev 74:311–345

Conde FR, Churio MS, Previtali CM (2000) The photoprotector mechanism of mycosporine-like amino acids. Excited-state properties and photostability of porphyra-334 in aqueous solution. J Photochem Photobiol B: Biology 56:139–144

Conde FR, Churio MS, Previtali CM (2004) The deactivation pathways of the excited-states of the mycosporine-like amino acids shinorine and porphyra-334 in aqueous solution. Photochem Photobiol Sci 3:960–967

Conde FR, Churio MS, Previtali CM (2007) Experimental study of the excited-state properties and photostability of the mycosporine-like amino acid palythine in aqueous solution. Photochem Photobiol Sci 6:669–674

Connell LB, Redman R, Rodriguez R, Barrett A, Iszard M, Fonseca A (2010) *Dioszegia antarctica* sp. nov. and *Dioszegia cryoxerica* sp. nov., psychrophilic basidiomycetous yeasts from polar desert soils in Antarctica. Int J Syst Evol Microbiol 60:1466–1472

Davoli P, Weber RWS (2002) Carotenoid pigments from the red mirror yeast, *Sporobolomyces roseus*. Mycol 16:102–108

Davoli P, Mierau V, Weber RWS (2004) Carotenoids and fatty acids in red yeasts *Sporobolomyces roseus* and *Rhodotorula glutinis*. App Biochem Microbiol 40:392–397

de García V, Brizzio S, Libkind D, Buzzini P, van Broock M (2007) Biodiversity of cold-adapted yeasts from glacial meltwater rivers in Patagonia, Argentina. FEMS Microbiol Ecol 59:331–341

Diaz MR, Fell JW (2000) Molecular analyses of the IGS & ITS regions of rDNA of the psychrophilic yeasts in the genus *Mrakia*. A van Leeuwenhoek 77:7–12

Diffey BL (1991) Solar ultraviolet radiation effects on biological systems. Phys Med Biol 36:299–328

Disch A, Rohmer M (1998) On the absence of the glyceraldehyde 3-phosphate/pyruvate pathway for isoprenoid biosynthesis in fungi and yeasts. FEMS Microbiol Lett 168:201–208

Dunlap WC, Yamamoto Y (1995) Small-molecule antioxidants in marine organisms: antioxidant activity of mycosporine-glycine. Comp Biochem Physiol B: Biochem Mol Biol 112:105–114
Ende G, Cornelis JJ (1970) The induction of sporulation in sclerotinia fructicola and some other fungi and the production of "P310". Netherlands J Plant Pathol 76:183–191
Favre-Bonvin J, Arpin N, Brevard C (1976) Structure de la mycosporine (P 310). Can J Chem 54:1105–1113
Favre-Bonvin J, Bernillon J, Salin N, Arpin N (1987) Biosynthesis of mycosporines: mycosporine glutaminol in *Trichothecium roseum*. Phytochemistry 26:2509–2514
Fernández Zenoff V, Siñeriz F, Farías ME (2006) Diverse responses to UV-B radiation and repair mechanisms of bacteria isolated from high-altitude aquatic environments. Appl Environ Microbiol 72:7857–7863
Fernández NV, Cecilia Mestre M, Marchelli P, Fontenla SB (2012) Yeast and yeast-like fungi associated with dry indehiscent fruits of *Nothofagus nervosa* in Patagonia, Argentina. FEMS Microbiol Ecol 80:179–192
Fiasson JL (1972) Carotenoids in various anascosporous yeasts. C R Acad Sci D Sci Nat 274:3465
Flores-Cotera L, Martín R, Sánchez S (2001) Citrate, a possible precursor of astaxanthin in *Phaffia rhodozyma*: influence of varying levels of ammonium, phosphate and citrate in a chemically defined medium. Appl Microbiol Biotechnol 55:341–347
Frengova G, Beshkova D (2009) Carotenoids from *Rhodotorula* and *Phaffia*: yeasts of biotechnological importance. J Ind Microbiol Biotechnol 36:163–180
Frengova GI, Simova ED, Beshkova DM (1995) Effect of temperature changes on the production of yeast pigments co-cultivated with lacto-acid bacteria in whey ultrafiltrate. Biotechnol Lett 17:1001–1006
Garcia-Pichel F (1994) A model for internal self-shading in planktonic organisms and its implications for the usefulness of ultraviolet sunscreens. Limnol Oceanogr 39:1704–1717
Garcia-Pichel F, Castenholz RW (1993) Occurrence of UV-absorbing, mycosporine-like compounds among cyanobacterial isolates and an estimate of their screening capacity. Appl Environ Microbiol 59:163–169
Goodwin TW (1980) The biochemistry of the carotenoids. Academic Press, London
Gounot AM (1999) Microbial life in permanently cold soils. In: Margesin R, Schinner F (eds) Cold-adapted organisms. Springer, Berlin, pp 3–15
Grant PT, Plack PA, Thomson RH (1980) Gadusol, a metabolite from fish eggs. Tetrahedron Lett 21:4043–4044
Gu WL, An GH, Johnson EA (1997) Ethanol increases carotenoid production in *Phaffia rhodozyma*. J Ind Microbiol Biotech 19:114–117
Gunasekera T, Paul N, Ayres P (1997) Responses of phylloplane yeasts to UV-B (290–320 nm) radiation: interspecific differences in sensitivity. Mycol Res 101:779–785
Gunde-Cimerman N, Plemenitaš A (2006) Ecology and molecular adaptations of the halophilic black yeast *Hortaea werneckii*. Rev Environ Sci Biotechnol 5:323–331
Häder D-P (1991) Effects of enhanced solar ultraviolet radiation on aquatic ecosystems. In: Lenci F, Ghetti F, Colombetti G, Häder DP, Song P-S (eds) Biophysics of photoreceptors and photomovements in microorganisms, vol 211., NATO ASI seriesSpringer, New York, pp 157–172
Hanelt D, Tüg H, Bischof K, Groß C, Lippert H, Sawall T, Wiencke C (2001) Light regime in an Arctic fjord: a study related to stratospheric ozone depletion as a basis for determination of UV effects on algal growth. Mar Biol 138:649–658
Hayman EP, Yokoyama H, Chichester CO, Simpson KL (1974) Carotenoid biosynthesis in *Rhodotorula glutinis*. J Bacteriol 120:1339–1343
He Y-Y, Häder D-P (2002) UV-B-induced formation of reactive oxygen species and oxidative damage of the cyanobacterium *Anabaena* sp.: protective effects of ascorbic acid and N-acetyl-l-cysteine. J Photochem Photobiol B: Biol 66:115–124
He Y-Y, Klisch M, Häder D-P (2002) Adaptation of cyanobacteria to UV-B stress correlated with oxidative stress and oxidative damage. Photochem Photobiol 76:188–196

Herrmann KM, Weaver LM (1999) The shikimate pathway. Ann Rev Plant Biol 50:473–503
Herz S, Weber RWS, Anke H, Mucci A, Davoli P (2007) Intermediates in the oxidative pathway from torulene to torularhodin in the red yeasts *Cystofilobasidium infirmominiatum* and *C. capitatum* (Heterobasidiomycetes, Fungi). Phytochemistry 68:2503–2511
Hoyer K, Karsten U, Wiencke C (2002) Induction of sunscreen compounds in Antarctic macroalgae by different radiation conditions. Mar Biol 141:619–627
Inácio J, Portugal L, Spencer-Martins I, Fonseca A (2005) Phylloplane yeasts from Portugal: seven novel anamorphic species in the Tremellales lineage of the Hymenomycetes (Basidiomycota) producing orange-coloured colonies. FEMS Yeast Res 5:1167–1183
Johnson EA, Lewis MJ (1979) Astaxanthin formation by the yeast *Phaffia rhodozyma*. J Gen Microbiol 115:173–183
Johnson EA, Schroeder WA (1995) Astaxanthin from the yeast Phaffia rhodozyma. Stud Mycol 38:81–90
Johnson E, Schroeder W (1996) Microbial carotenoids. In: Downstream processing biosurfactants carotenoids, vol 53. Adv Biochem Engin/Biotechnol. Springer, Berlin, pp 119–178
Kaiser P, Geyer R, Surmann P, Fuhrmann H (2012) LC–MS method for screening unknown microbial carotenoids and isoprenoid quinones. J Microbiol Methods 88:28–34
Kamio M, Kicklighter CE, Nguyen L, Germann MW, Derby CD (2011) Isolation and structural elucidation of novel mycosporine-like amino acids as alarm cues in the defensive ink secretion of the sea hare *Aplysia californica*. Helv Chim Acta 94:1012–1018
Karsten U, Franklin LA, Lüning K, Wiencke C (1998) Natural ultraviolet radiation and photosynthetically active radiation induce formation of mycosporine-like amino acids in the marine macroalga *Chondrus crispus* (Rhodophyta). Planta 205:257–262
Kicklighter CE, Kamio M, Nguyen L, Germann MW, Derby CD (2011) Mycosporine-like amino acids are multifunctional molecules in sea hares and their marine community. Proc Natl Acad Sci USA 108:11494–11499
Klisch M, Häder D-P (2002) Wavelength dependence of mycosporine-like amino acid synthesis in *Gyrodinium dorsum*. J Photochem Photobiol B: Biol 66:60–66
Klisch M, Sinha RP, Richter PR, Häder D-P (2001) Mycosporine-like amino acids (MAAs) protect against UV-B-induced damage in *Gyrodinium dorsum* Kofoid. J Plant Physiol 158:1449–1454
Knaggs AR (2003) The biosynthesis of shikimate metabolites. Nat Prod Rep 20:119–136
Kogej T, Gostinčar C, Volkmann M, Gorbushina AA, Gunde-Cimerman N (2006) Mycosporines in Extremophilic fungi—novel complementary osmolytes? Environ Chem 3:105–110
Kogej T, Stein M, Volkmann M, Gorbushina AA, Galinski EA, Gunde-Cimerman N (2007) Osmotic adaptation of the halophilic fungus *Hortaea werneckii*: role of osmolytes and melanization. Microbiology 153:4261–4273
Kräbs G, Watanabe M, Wlencke C (2004) A Monochromatic action spectrum for the photoinduction of the UV-absorbing mycosporine-like amino acid shinorine in the red alga *Chondrus crispus*. Photochem Photobiol 79:515–520
Lampila LE, Wallen SE, Bullerman LB (1985) A review of factors affecting biosynthesis of carotenoids by the order Mucorales. Mycopathology 90:65–80
Laurion I, Ventura M, Catalan J, Psenner R, Sommaruga R (2000) Attenuation of ultraviolet radiation in mountain lakes: factors controlling the among-and within-lake variability. Limn Oceanogr 45:1274–1288
Leach CM (1965) Ultraviolet-absorbing substances associated with light-induced sporulation in fungi. Can J Bot 43:185–200
Leach CM, Trione E (1966) Action spectra for light-induced sporulation of the fungi *Pleospora Herbarum* and *Alernaria dauci*. Photochem Photobiol 5:621–630
Leite B, Nicholson RL (1992) Mycosporine-alanine: a self-inhibitor of germination from the conidial mucilage of *Colletotrichum graminicola*. Exp Mycol 16:76–86
Liaaen-Jensen S, Andrewes AG (1972) Microbial carotenoids. Annu Rev Microbiol 26:225–248
Libkind D, van Broock M (2006) Biomass and carotenoid pigment production by patagonian native yeasts. World J Microbiol Biotechnol 22:687–692

Libkind D, Brizzio S, van Broock M (2004a) *Rhodotorula mucilaginosa*, a carotenoid producing yeast strain from a patagonian high-altitude lake. Folia Microbiol 49:19–25

Libkind D, Pérez P, Sommaruga R, del Carmen Diéguez M, Ferraro M, Brizzio S, Zagarese H, van Broock M (2004b) Constitutive and UV-inducible synthesis of photoprotective compounds (carotenoids and mycosporines) by freshwater yeasts. Photochem Photobiol Sci 3:281–286

Libkind D, Sommaruga R, Zagarese H, van Broock M (2005) Mycosporines in carotenogenic yeasts. Syst Appl Microbiol 28:749–754

Libkind D, Diéguez MC, Moliné M, Pérez P, Zagarese HE, Broock M (2006) Occurrence of photoprotective compounds in yeasts from freshwater ecosystems of northwestern Patagonia (Argentina). Photochem Photobiol 82:972–980

Libkind D, Moliné M, de García V, Fontenla S, van Broock M (2008) Characterization of a novel South American population of the astaxanthin producing yeast *Xanthophyllomyces dendrorhous* (*Phaffia rhodozyma*). J Ind Microbiol Biotechnol 35:151–158

Libkind D, Moliné M, Sampaio JP, van Broock M (2009a) Yeasts from high-altitude lakes: influence of UV radiation. FEMS Microbiol Ecol 69:353–362

Libkind D, Gadanho M, van Broock M, Sampaio JP (2009b) *Cystofilobasidium lacus-mascardii* sp. nov., a basidiomycetous yeast species isolated from aquatic environments of the Patagonian Andes, and *Cystofilobasidium macerans* sp. nov., the sexual stage of *Cryptococcus macerans*. Int J Syst Evol Microbiol 59:622–630

Libkind D, Moliné M, Bellora N, Hittinger C, Giraudo M (2011a) Genomic analysis of the astaxanthin producing yeast *Phaffia rhodozyma* (*Xanthophyllomyces dendrorhous*) with special focus on antioxidant-related genes. In: Proceeding meeting on comparative genomics of eukaryotic microorganisms: understanding the complexity of diversity, Saint Feliux de Guixols, España, 15–20 Oct 2011

Libkind D, Moliné M, Sommaruga R, Sampaio JP, van Broock M (2011b) Phylogenetic distribution of fungal mycosporines within the Pucciniomycotina (Basidiomycota). Yeast 28:619–627

Libkind D, Moliné M, van Broock M (2011c) Production of the UVB-absorbing compound mycosporine–glutaminol–glucoside by *Xanthophyllomyces dendrorhous* (*Phaffia rhodozyma*). FEMS Yeast Res 11:52–59

Madeo F, Fröhlich E, Ligr M, Grey M, Sigrist SJ, Wolf DH, Fröhlich K-U (1999) Oxygen stress: a regulator of apoptosis in yeast. J Cell Biol 145:757–767

Madhour A, Anke H, Mucci A, Davoli P, Weber RWS (2005) Biosynthesis of the xanthophyll plectaniaxanthin as a stress response in the red yeast *Dioszegia* (Tremellales, Heterobasidiomycetes, Fungi). Phytochemistry 66:2617–2626

Margalith P, Meydav S (1968) Some observations on the carotenogenesis in the yeast *Rhodotorula mucilaginosa*. Phytochemistry 7:765–768

Margesin R, Miteva V (2011) Diversity and ecology of psychrophilic microorganisms. Res Microbiol 162:346–361

Margesin R, Schinner F (1994) Properties of cold-adapted microorganisms and their potential role in biotechnology. J Biotechnol 33:1–14

Martelli H, da Silva IM (1993) β-carotene synthesis in *Rhodotorula*. In: Lester P (ed) Methods in enzymology, vol 214. Academic Press, New York, pp 386–390

Masaki K, Dunlap W, Yamamoto Y, Karube I, Larsen R, Matsukawa R (1996) A natural antioxidant and its production process. Japanese Patent 9604230

Maxwell WA, Macmillan JD, Chichester C (1966) Function of carotenoids in protection of *Rhodotorula glutinis* against irradiation from a gas laser. Photochem Photobiol 5:567–577

Méndez-Álvarez S, Rüfenacht K, Eggen RI (2000) The oxidative stress-sensitive *yap1* null strain of *Saccharomyces cerevisiae* becomes resistant due to increased carotenoid lLevels upon the introduction of the *Chlamydomonas reinhardtii* cDNA, coding for the 60S ribosomal protein L10a. Biochem Biophys Res Comm 267:953–959

Miller NJ, Sampson J, Candeias LP, Bramley PM, Rice-Evans CA (1996) Antioxidant activities of carotenes and xanthophylls. FEBS Lett 384:240–242

Moliné M (2010) Produccion de compuestos fotoprotectores (carotenoides y micosporinas) por levaduras. PhD thesis, Universidad Nacional de Tucumán, Tucumán

Moliné M, Libkind D, del Carmen Diéguez M, van Broock M (2009) Photoprotective role of carotenoids in yeasts: response to UV-B of pigmented and naturally-occurring albino strains. J Photochem Photobiol B: Biol 95:156–161

Moliné M, Flores MR, Libkind D, del Carmen Diéguez M, Farías ME, van Broock M (2010) Photoprotection by carotenoid pigments in the yeast *Rhodotorula mucilaginosa*: the role of torularhodin. Photochem Photobiol Sci 9:1145–1151

Moliné M, Arbeloa EM, Flores MR, Libkind D, Farías ME, Bertolotti SG, Churio MS, van Broock MR (2011) UVB photoprotective role of mycosporines in yeast: photostability and antioxidant activity of mycosporine-glutaminol-glucoside. Rad Res 175:44–50

Moliné M, Libkind D, van Broock M (2012) Production of torularhodin, torulene, and β-carotene by Rhodotorula yeasts. In: Barredo J-L (ed) Microbial carotenoids from fungi, vol 898. Methods Mol Biol. Humana Press, New York, pp 275–283

Moore MM, Breedveld MW, Autor AP (1989) The role of carotenoids in preventing oxidative damage in the pigmented yeast, *Rhodotorula mucilaginosa*. Arch Biochem Biophys 270:419–431

Mrak EM, Phaff HJ, Mackinney G (1949) A simple test for carotenoid pigments in yeasts. J Bacteriol 57:409–411

Muñoz M (2010) Levaduras y hongos dimórficos del filoplano de Nothofagus pumilio y el papel de la exposición solar en su distribución y producción de metabolitos fotoprotectores. Undergraduate thesis, Universidad Nacional del Comahue, Bariloche

Nakayama T, Mackinney G, Phaff HJ (1954) Carotenoids in asporogenous yeasts. A van Leeuwenhoek 20:217–228

Nam HS, Cho SY, Rhee JS (1988) High-performance liquid chromatographic analysis of major carotenoids from *Rhodotorula glutinis*. J Chromat A:445–447

Ojima K, Breitenbach J, Visser H, Setoguchi Y, Tabata K, Hoshino T, Berg J, Sandmann G (2006) Cloning of the astaxanthin synthase gene from *Xanthophyllomyces dendrorhous* (*Phaffia rhodozyma*) and its assignment as a β-carotene 3-hydroxylase/4-ketolase. Mol Genet Genomics 275:148–158

Ordoñez O, Flores M, Dib J, Paz A, Farías M (2009) Extremophile culture collection from Andean lakes: extreme pristine environments that host a wide diversity of microorganisms with tolerance to UV radiation. Microb Ecol 58:461–473

Oren A (1997) Mycosporine-like amino acids as osmotic solutes in a community of halophilic cyanobacteria. Geomicrobiol J 14:231–240

Oren A, Gunde-Cimerman N (2007) Mycosporines and mycosporine-like amino acids: UV protectants or multipurpose secondary metabolites? FEMS Microbiol Lett 269:1–10

Perrier V, Dubreucq E, Galzy P (1995) Fatty acid and carotenoid composition of *Rhodotorula* strains. Arch Microbiol 164:173–179

Perrone GG, Tan SX, Dawes IW (2008) Reactive oxygen species and yeast apoptosis. Biochim Biophys Acta 1783:1354–1368

Peterson WJ, Bell TA, Etchells JL, Smart WWG (1954) A procedure for demonstrating the presence of carotenoid pigments in yeasts. J Bacteriol 67:708–713

Peterson WJ, Evans WR, Lecce E, Bell TA, Etchells JL (1958) Quantitative determination of the carotenoids in yeasts of the genus *Rhodotorula*. J Bacteriol 75:586–591

Pittet J-L, Bouillant M-L, Bernillon J, Arpin N, Favre-Bonvin J (1983a) Sur la presence de mycosporines-glutamine reduites, nouvelles molecules, chez plusieurs deuteromycetes. Tetrahedron Lett 24:65–68

Pittet J-L, Létoublon R, Frot-Coutaz J, Arpin N (1983b) Soluble uridine diphospho-D-glucose: mycosporin glucosyltransferase from spores of *Ascochyta fabae* Speg. Planta 159:159–164

Portwich A, Garcia-Pichel F (1999) Ultraviolet and osmotic stresses induce and regulate the synthesis of mycosporines in the cyanobacterium *Chlorogloeopsis* PCC 6912. Arch Microbiol 172:187–192

Pospíšil P (2009) Production of reactive oxygen species by photosystem II. Biochim. Biophys. Acta 1787:1151–1160

Ravanat J-L, Douki T, Cadet J (2001) Direct and indirect effects of UV radiation on DNA and its components. J Photochem Photobiol B: Biol 63:88–102

Rodríguez-Sáiz M, de la Fuente JL, Barredo JL (2010) *Xanthophyllomyces dendrorhous* for the industrial production of astaxanthin. Appl Microbiol Biotechnol 88:645–658

Rossi V, Bolognesi M, Languasco L, Giosuè S (2006) Influence of environmental conditions on infection of peach shoots by *Taphrina deformans*. Phytopathology 96:155–163

Rothschild LJ (1999) Microbes and radiation. In: Seckbach (ed) Enigmatic microorganisms and life in extreme environments. Kluwer Academic Publishers, Amsterdam, pp 551–562

Sakaki H, Nochide H, Nakanishi T, Miki W, Fujita T, Komemushi S (1999) Effect of culture condition on the biosynthesis of carotenoids in *Rhodotorula glutinis* No. 21. J Biosci Bioeng 87:400

Sakaki H, Nakanishi T, Satonaka K-Y, Miki W, Fujita T, Komemushi S (2000) Properties of a high-torularhodin-producing mutant of *Rhodotorula glutinis* cultivated under oxidative stress. J Biosci Bioeng 89:203–205

Sakaki H, Nakanishi T, Tada A, Miki W, Komemushi S (2001) Activation of torularhodin production by *Rhodotorula glutinis* using weak white light irradiation. J Biosci Bioeng 92:294–297

Sakaki H, Nochide H, Komemushi S, Miki W (2002) Effect of active oxygen species on the productivity of torularhodin by *Rhodotorula glutinis* No. 21. J Biosci Bioeng 93:338–340

Santopietro LMD, Spencer JFT, Spencer DM, Siñeriz F (1998) Effects of oxidative stress on the production of carotenoid pigments by *Phaffia rhodozyma* (*Xanthophyllomyces dendrorhous*). Folia Microbiol 43:173–176

Scherer S, Chen T, Böger P (1988) A new UV-A/B protecting pigment in the terrestrial cyanobacterium *Nostoc commune*. Plant Physiol 88:1055–1057

Schisler DA, Janisiewicz W, Boekhout T, Kurtzman CP (2011) Agriculturally important yeasts. In: Kurtzman CP, Fell JW, Boekhout T (eds) The yeasts: a taxonomic study, vol 1. Elsevier, London, pp 45–52

Schroeder WA, Johnson EA (1993) Antioxidant role of carotenoids in *Phaffia rhodozyma*. J Gen Microbiol 139:907–912

Schroeder WA, Johnson EA (1995a) Carotenoids protect *Phaffia rhodozyma* against singlet oxygen damage. J Ind Microbiol 14:502–507

Schroeder WA, Johnson EA (1995b) Singlet oxygen and peroxyl radicals regulate carotenoid biosynthesis in *Phaffia rhodozyma*. J Biol Chem 270:18374–18379

Shan L, Terence HL, Yokoyama H, Simpson KL, Chichester CO (1973) Isolation and identification of 2-hydroxyplectaniaxanthin from *Rhodotorula aurantiaca*. Phytochemistry 12:2953–2956

Shick JM, Dunlap WC (2002) Mycosporine-like amino acids and related gadusols: biosynthesis, accumulation, and UV-protective functions in aquatic organisms. Annu Rev Physiol 64:223–262

Shick JM, Romaine-Lioud S, Ferrier-Pages C, Gattuso J (1999) Ultraviolet-B radiation stimulates shikimate pathway-dependent accumulation of mycosporine-like amino acids in the coral *Stylophora pistillata* despite decreases in its population of symbiotic dinoflagellates. Limnol Oceanogr 44:1667–1682

Simova ED, Frengova GI, Beshkova DM (2003) Effect of aeration on the production of carotenoid pigments by *Rhodotorula rubra-Lactobacillus casei* subsp. *casei* co-cultures in whey ultrafiltrate. Z Naturforsch C J Bios 58:225–229

Simpson KL, Nakayama TOM, Chichester CO (1964) Biosynthesis of yeast carotenoids. J Bacteriol 88:1688–1694

Singh SP, Klisch M, Sinha RP, Häder DP (2008) Effects of abiotic stressors on synthesis of the mycosporine-like amino acid shinorine in the cyanobacterium *Anabaena variabilis* PCC 7937. Photochem Photobiol 84:1500–1505

Singh SP, Klisch M, Sinha RP, Häder D-P (2010) Genome mining of mycosporine-like amino acid (MAA) synthesizing and non-synthesizing cyanobacteria: a bioinformatics study. Genomics 95:120–128

Sinha RP, Klisch M, Walter Helbling E, Häder D-P (2001) Induction of mycosporine-like amino acids (MAAs) in cyanobacteria by solar ultraviolet-B radiation. J Photochem Photobiol B: Biol 60:129–135

Sinha RP, Singh SP, Häder D-P (2007) Database on mycosporines and mycosporine-like amino acids (MAAs) in fungi, cyanobacteria, macroalgae, phytoplankton and animals. J Photochem Photobiol B: Biol 89:29–35

Sommaruga R (2001) The role of solar UV radiation in the ecology of alpine lakes J Photochem Photobiol B. Biol 62:35–42

Sommaruga R, Libkind D, van Broock M, Whitehead K (2004) Mycosporine-glutaminol-glucoside, a UV-absorbing compound of two *Rhodotorula* yeast species. Yeast 21:1077–1081

Sperstad S, Lutnæs B, Stormo S, Liaaen-Jensen S, Landfald B (2006) Torularhodin and torulene are the major contributors to the carotenoid pool of marine *Rhodosporidium babjevae* (Golubev). J Ind Microbiol Biotech 33:269–273

Squina FM, Mercadante AZ (2005) Influence of nicotine and diphenylamine on the carotenoid composition of *Rhodotorula* strains. J Food Biochem 29:638–652

Suh HJ, Lee HW, Jung J (2003) Mycosporine glycine protects biological systems against photodynamic damage by quenching singlet oxygen with a high efficiency. Photochem Photobiol 78:109–113

Tada M, Shiroishi M (1982a) Mechanism of Photoregulated Carotenogenesis in *Rhodotorula minuta* I. Photocontrol of carotenoid production. Plant Cell Physiol 23:541–547

Tada M, Shiroishi M (1982b) Mechanism of photoregulated carotenogenesis in *Rhodotorula minuta* II. Aspects of photoregulative reaction. Plant Cell Physiol 23:549–556

Tao C, Sugawara T, Maeda S, Wang X, Hirata T (2008) Antioxidative activities of a mycosporine-like amino acid, porphyra-334. Fish Sci 74:1166–1172

Tefft RE, Goodwin TW, Simpson KL (1970) Aspects of the stereochemistry of torularhodin biosynthesis. Biochem J 117:921–927

Tognetti C, Moliné M, van Broock M, Libkind D (2013) Favored isolation and rapid identificationof the astaxanthin-producing yeast *Xanthophyllomyces dendrorhous* (*Phaffia rhodozyma*) from environmental samples. J Basic Microbiol 53:1–7

Torres A, Hochberg M, Pergament I, Smoum R, Niddam V, Dembitsky VM, Temina M, Dor I, Lev O, Srebnik M (2004) A new UV-B absorbing mycosporine with photo protective activity from the lichenized ascomycete *Collema cristatum*. Eur J Biochem 271:780–784

Trione E, Leach C, Mutch JT (1966) Sporogenic substances isolated from fungi. Nature 212:163

Tsimako M, Guffogg S, Thomas-Hall S, Watson K (2002) Resistance to UVB radiation in Antarctic yeasts. Redox Rep 7:312–314

Turchetti B, Buzzini P, Goretti M, Branda E, Diolaiuti G, D'Agata C, Smiraglia C, Vaughan-Martini A (2008) Psychrophilic yeasts in glacial environments of Alpine glaciers. FEMS Microbiol Ecol 63:73–83

Turchetti B, Thomas Hall SR, Connell LB, Branda E, Buzzini P, Theelen B, Muller WH, Boekhout T (2011) Psychrophilic yeasts from Antarctica and European glaciers: description of *Glaciozyma* gen. nov., *Glaciozyma martinii* sp. nov. and *Glaciozyma watsonii* sp. nov. Extremophiles 15:573–586

Valadon L (1976) Carotenoids as additional taxonomic characters in fungi: a review. Trans British Mycol Soc 67:1–15

Van Eijk G, Roeymans H (1982) Distribution of carotenoids and sterols in relation to the taxonomy of *Taphrina* and *Protomyces*. A van Leeuwenhoek 48:257–264

Vaz ABM, Rosa LH, Vieira MLA, Garcia Vd, Brandão LR, Teixeira LCRS, Moliné M, Libkind D, van Broock M, Rosa CA (2011) The diversity, extracellular enzymatic activities and photoprotective compounds of yeasts isolated in Antarctica. Braz J Microbiol 42:937–947

Vincent WF, Neale PJ (2000) Mechanisms of UV damage to aquatic organisms. In: de Mora S, Demers S, Vernet M (eds) The effects of UV radiation in the marine environment. Cambridge University Press, Cambridge, pp 149–176

Vishniac HS (1996) Biodiversity of yeasts and filamentous microfungi in terrestrial Antarctic ecosystems. Biodivers Conserv 5:1365–1378

Volkmann M, Whitehead K, Rütters H, Rullkötter J, Gorbushina AA (2003) Mycosporine-glutamicol-glucoside: a natural UV-absorbing secondary metabolite of rock-inhabiting microcolonial fungi. Rap Comm Mass Spectrom 17:897–902

Wang Q-M, Jia J-H, Bai F-Y (2008) Diversity of basidiomycetous phylloplane yeasts belonging to the genus *Dioszegia* (Tremellales) and description of *Dioszegia athyri* sp. nov., *Dioszegia butyracea* sp. nov. and *Dioszegia xingshanensis* sp. nov. A van Leeuwenhoek 93:391–399

Weber RWS, Madhour A, Anke H, Mucci A, Davoli P (2005) 2-Hydroxytorularhodin, a new xanthophyll from the red yeast *Sporobolomyces coprosmae*. Helv Chim Acta 88:2960–2966

Whitehead K, Hedges JI (2005) Photodegradation and photosensitization of mycosporine-like amino acids. J Photochem Photobiol B: Biol 80:115–121

Yakovleva I, Bhagooli R, Takemura A, Hidaka M (2004) Differential susceptibility to oxidative stress of two scleractinian corals: antioxidant functioning of mycosporine-glycine. Comp Biochem Physiol Part B: Biochem Mol Biol 139:721–730

Yamane Y, Higashida K, Nakashimada Y, Kakizono T, Nishio N (1997) Astaxanthin production by *Phaffia rhodozyma* enhanced in fed-batch culture with glucose and ethanol feeding. Biotechnol Lett 19:1109–1111

Young H, Patterson VJ (1982) A uv protective compound from *Glomerella cingulata*—a mycosporine. Phytochemistry 21:1075–1077

Yurkov AM, Vustin MM, Tyaglov BV, Maksimova IA, Sineokiy SP (2008) Pigmented basidiomycetous yeasts are a promising source of carotenoids and ubiquinone Q10. Microbiology 77:1–6

Zagarese HE, Tartarotti B, Cravero W, Gonzalez P (1998) UV damage in shallow lakes: the implications of water mixing. J Plankton Res 20:1423–1433

Zalar P, Gostinčar C, De Hoog G, Uršič V, Sudhadham M, Gunde-Cimerman N (2008) Redefinition of *Aureobasidium pullulans* and its varieties. Studies Mycol 61:21–38

Zhang L, Li L, Wu Q (2007) Protective effects of mycosporine-like amino acids of *Synechocystis* sp. PCC 6803 and their partial characterization. J Photochem Photobiol B: Biol 86:240–245

Chapter 10
Changes in Lipids Composition and Fluidity of Yeast Plasma Membrane as Response to Cold

Nina Gunde-Cimerman, Ana Plemenitaš and Pietro Buzzini

Abstract In order to function properly, plasma membranes have to preserve a suitable dynamic state of the bilayer even in changing environments, which alters their fluidity. This is achieved by active restructuring of membrane lipids composition. Perturbations in membrane structure determined by environment may result in significant disruption of its physiological function. Accordingly, its composition is modified as response to environmental changes. Because the fluid state of plasma membrane is mandatory for proper functioning, organisms exposed to changes in the environmental conditions alter their membrane lipid composition, to maintain membrane lipids in a lamellar liquid crystalline phase, and avoid the formation of a lamellar gel phase. One of the main adaptative mechanisms to overcome the negative effect of cold is the ability to finely adjust phospholipid composition and physical properties of their plasma membranes. The minimum growth temperature of a given organism is reached when the degree of fluidity of its plasma membrane becomes too low for allowing the trans-membrane-nutrient transport. Yeasts have long served as an attractive model of a eukaryotic system for studying the regulation of lipid biosynthesis and thermal adaptation of the plasma membrane: the ability of psychrophilic, psychrotolerant and mesophilic yeasts to change the composition of phospholipids may suggest that the extent of

N. Gunde-Cimerman (✉)
Biology Department, Biotechnical Faculty, University of Ljubljana, Vecna pot 111, 1000 Ljubljana, Slovenia
e-mail: nina.gunde-cimerman@bf.uni-lj.si

N. Gunde-Cimerman
Centre of Excellence for Integrated Approaches in Chemistry and Biology of Proteins (CIPKeBiP), Jamova 39, 1000 Ljubljana, Slovenia

A. Plemenitaš
Institute of Biochemistry, Medicine, University of Ljubljana, Vrazov trg 2, 1000 Ljubljana, Slovenia

P. Buzzini
Department of Agricultural, Environmental and Food Sciences and Industrial Yeasts Collection DBVPG, University of Perugia, Borgo XX Giugno 74, 06121 Perugia, Italy

P. Buzzini and R. Margesin (eds.), *Cold-Adapted Yeasts*,
DOI: 10.1007/978-3-642-39681-6_10,

such changes may be an important controlling factor, which determines the yeast growth temperature limits and limits for their distribution in the environment. In this chapter, some essential aspects of the changes in lipids composition and fluidity of yeast plasma membrane as response to cold are reported.

Keywords Plasma membrane · Lipid composition · Fatty acid unsaturation · Fluidity

Contents

10.1 Introduction ... 226
10.2 Fundamentals on Yeast Plasma Membrane ... 227
10.3 Changes of Plasma Membrane Lipid Composition as a Response to Cold ... 230
10.3.1 Changes in Mesophilic Yeasts ... 232
10.3.2 Changes in Psychrophilic and Psychrotolerant Yeasts ... 234
10.3.3 Changes in Yeast Plasma Membrane Fatty Acid Composition During the Growth Phases ... 236
10.4 Yeast Plasma Membrane Fluidity as Determinant of Adaptability to Stressful Conditions ... 237
10.5 Conclusions ... 238
References ... 239

10.1 Introduction

Membranes form a framework, which compartmentalises the different biochemical processes of the cell, controlling movement of substrates and products and providing the anchorage for enzymes and carriers. Besides, they form a continuous dynamic system, the composition of which differs in individual organelles and regions of the cell. To function properly, membranes have to preserve a suitable dynamic state of the bilayer even in changing environments, which alters their fluidity. This is achieved by active restructuring of membrane lipids composition.

The choice of different lipid species is very large. No other class of biological molecules displays a greater diversity of structure over such a narrow range of physical and chemical parameters, so close to physiological conditions. Consequently, environmentally induced perturbations in membrane structure may result in significant disruption of its physiological function. Plasma membrane can be considered the initial sensor of external stress conditions. Its composition is modified as response to environmental changes. For proper functioning, it must be in a fluid state, and therefore, organisms exposed to changes in the environmental conditions alter their membrane lipid composition by synthesising a proper combination of acyl chain and polar head group structures, to maintain membrane lipids in a lamellar liquid crystalline phase and avoid the formation of a lamellar gel phase (Gostincar et al. 2009). In this way, membrane functionality is maintained (Los and Murata 2004; Morgan-Kiss et al. 2006).

Four strategies seem to be utilised in adapting membrane composition according to the prevailing environmental and physiological conditions:

1. changes in the acyl chain structure;
2. changes in the polar head group structure;
3. changes in the amount of sterols; and
4. reshuffling of acyl chains to form new lipid molecular species without changing the average acyl chain composition (Russell 1989; Rilfors and Lindblom 2002).

In spite of biochemical pathways being strongly affected by low temperatures, cold-adapted organisms can grow and metabolise at rates similar to those achieved by closely related species living in temperate habitats (Raspor and Zupan 2006). One of the main adaptative mechanisms to overcome the negative effect of cold is the ability to finely adjust chemical (lipid) composition and physical properties of their plasma membranes (Watson and Rose 1980; Watson 1987; Madigan et al. 1997; Gerday et al. 2000; Chintalapati et al. 2004; Raspor and Zupan 2006; Vishniac 2006; Shivaji et al. 2007). Below the minimum growth temperature of a given organism, the degree of fluidity of its plasma membrane becomes too low for allowing the trans-membrane-nutrient transport (Madigan et al. 1997).

Yeasts are due to their complex membrane organisation, rapid growth reproduction and relative genetic simplicity and, last but not least, due to the availability of a number of mutant strains deficient of some key genes encoding various aspects of lipid metabolism, an attractive model of a eukaryotic system for studying the regulation of lipid biosynthesis and thermal adaptation of the plasma membrane (Arthur and Watson 1976; Daum et al. 1998, 1999; Gaspar et al. 2007). Yeasts exhibit marked qualitative and quantitative variations in composition of their plasma membranes. The ability of psychrophilic, psychrotolerant and mesophilic yeasts to change the composition of phospholipids, such as the unsaturation degree of fatty acids, suggests the extent of such changes may be an important controlling factor, which determines the yeast growth temperature limits and limits for their distribution in the environment (Raspor and Zupan 2006; Shivaji and Prasad 2009).

Some fundamental aspects on the changes in lipids composition and fluidity of yeast plasma membrane as response to cold are treated in this chapter.

10.2 Fundamentals on Yeast Plasma Membrane

The existence of a plasma membrane is essential for the life of yeasts as single cell entities. This structure represents the primary barrier for the passage of hydrophilic molecules and prevents cytoplasm contents mixing freely with external aqueous environment. Like to other eukaryotic homologous structures, the yeast membrane can be described as a lipid bilayer mixed together with a number of globular structural and functional proteins, which form a fluid mixture. Chemical composition, molecular anatomy and functional characters of yeast plasma membranes have been extensively reviewed in the past decades (Henschke and Rose 1991; van der Rest et al. 1995; Höfer 1997; Walker 1998). Phospholipids (glycerophospholipids and sphingolipids) are the major class of membrane lipids, together with glycolipids, sterols and according to some reports other neutral lipids. The amount of individual membrane lipids can vary greatly due to the developmental stage, experimental conditions and methodology used. Most data derive from entire membrane systems or even total lipids, rather than from the separate analysis of individual membrane fractions.

The species *Saccharomyces cerevisiae* is currently used as a model for studying the organisation of plasma membrane in the yeast world. In this species, the thickness of membrane is about 7.5 nm with occasional invaginations protruding into the cytoplasm (Walker 1998). The plasma membrane of S. *cerevisiae* is highly enriched in sterols and sphingolipids, saturated species of phosphatidylserine (PS) and phosphatidylethanolamine (PE) (Schneiter et al. 1999). The lipid components found in the yeast plasma membranes include (Walker 1998):

1. phospholipids, primarily phosphatidylcholine (PC) and phosphatidylethanolamine (PE), with only minor amounts of phosphatidylinositol (PI), phosphatidylserine (PS) and phosphatidylglycerol (PG);
2. sterols, mainly ergosterol and zymosterol, with only minor proportions of fecosterol and lanosterol.

Sterols are essential lipid components of eukaryotic membranes. They are responsible for a number of important physical characteristic of the membranes, such as regulation of membrane permeability, fluidity and stability. Sterol molecules are of a suitable shape to lock across the fatty acid chains of the polar lipids, thus increasing the chain order in liquid crystalline state (condensing effect) and decreasing it in the gel state (liquefying effect). A decrease in the ratio of sterols to phospholipids indicates an increase in membrane fluidity, provided that compensatory changes in other membrane components do not occur.

The arrangement of phospholipid molecules in membranes results from the polar head group being oriented towards the cytoplasm and the hydrophobic fatty acid chains directed inwards, towards the fatty acid chains of the opposite lipid layer. The several types of phospholipids present in membranes differ in their polar groups, and the fatty acids within the individual phospholipid classes vary in both chain length and degree of unsaturation. Through the introduction of one or more double bonds the configuration of the fatty acid chains is altered, affecting the fluidity of the membrane (Losel 1990). Structural roles of the major phospholipids in membranes of yeast cells correspond to those established for other eukaryotes. In fluctuating environmental conditions, the remodelling of the membrane phospholipids by changes in chain length, saturation or positions of fatty acids can facilitate the maintenance of membrane structure and permeability. In short,

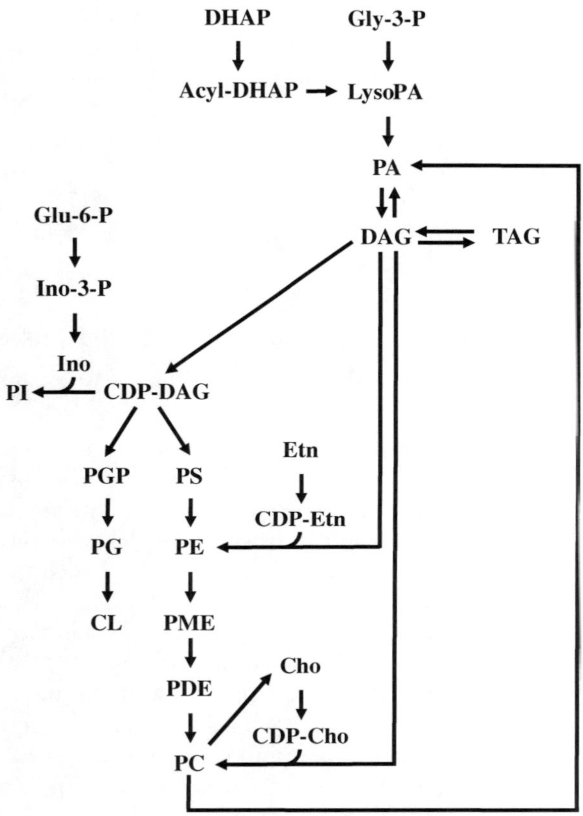

Fig. 10.1 General scheme of phospholipid synthesis pathways in *Saccharomyces cerevisiae*. *Acyl-DHAP* acyl dihydroxyacetone phosphate, *Cho* choline, *CL* cardiolipin, *CTD-Cho* cytidine diphosphate choline, *CTD-DAG* cytidine diphosphate diacylglycerol, *CTD-Etn* cytidine diphosphate ethanolamine, *DAG* diacylglycerol, *DHAP* dihydroxyacetone phosphate, *Etn* ethanolamine, *Glu-6-P* glucose-6-phosphate, *Gly-3-P* glycerol-3-phosphate, *Ino* inositol, *Ino-3-P* inositol-3-phosphate, *LysoPA* lysophosphatidic acid, *PA* phosphatidic acid, *PC* phosphatidylcholine, *PDE* phosphatidyldimethylethanolamine, *PE* phosphatidylethanolamine, *PG* phosphatidylglycerol, *PGP* phosphatidylglycerophosphate, *PI* phosphatidylinositol, *PME* phosphatidylmonomethylethanolamine, *PS* phosphatidylserine, *TAG* triacylglycerol

phospholipids confer fluidity and the sterols confer rigidity to yeast plasma membranes (Walker 1998).

Synthesis of phospholipids in *S. cerevisiae* occurs via pathways that are generally common to those exhibited by higher eukaryotic organisms (Carman and Henry 1999; Gaspar et al. 2007; Carman and Han 2011). The pathways for the synthesis of major phospholipids in *S. cerevisiae* are presented in Fig. 10.1.

The relative amount of the phospholipids in yeast plasma membranes vary with growth conditions (e.g., carbon source, nutrient availability, temperature, growth phase, etc.) and with genetic variations (Rattray et al. 1975; Becker and Lester

1977; Carman and Henry 1989; Paltauf et al. 1992; Chen et al. 2007). Phospholipid synthesis in yeasts is a complex process that is regulated by complex genetic determinants and biochemical mechanisms (Carman and Zeimetz 1996; Carman and Henry 1999, 2007; Carman and Kersting 2004; Carman and Han 2006, 2007; Chen et al. 2007; Chang and Carman 2008) and its regulation is interrelated with the synthesis of other major lipid classes (i.e., fatty acids, triacylglycerols, sterols, and sphingolipids (Carman and Henry 1999; Carman and Han 2006, 2011; Dickson 2008; Gaspar et al. 2008; Malanovic et al. 2008; Rajakumari et al. 2008) (Fig. 10.1).

10.3 Changes of Plasma Membrane Lipid Composition as a Response to Cold

The major response of membrane composition to a change in temperature is an alteration in the fatty acid component of lipids. Microorganisms counteract the propensity for membranes to rigidify at lower temperature by adapting to the conditions in order to maintain a more or less constant degree of membrane fluidity (homoeoviscous adaptation) (Hazel and Williams 1990). Thus, a low-temperature stress initiates a co-ordinated response by fatty acid desaturases and dehydrases induction which in turn increases the ratio of polyunsaturated to saturated fatty acids and/or (but less frequently) actually decreases the length of the fatty acid chains in the membrane (AI-Fageeh and Smales 2006). At low temperatures, an increase in the proportion of unsaturated fatty acids (UFAs) and/or to a lesser extent shortening of fatty acids occurs. The resulting increase in UFAs content causes membrane lipid fluidity to return to its original state, or close to it, with concurrent restoration of normal cellular activity at the lower temperature (Mansilla et al. 2004; Phadtare 2004). Thus, cold tolerant individuals have a higher proportion of unsaturated lipids incorporated into membrane lipids, related to decreased physical order of the resulting bilayer. This effect more or less overcomes the ordered, rigidifying effects of cold and has been invoked as an explanation of both capacity and resistance cold adaptation at the level of the whole organism (Cossins et al. 2002). Composition of phospholipids, almost exclusively concentrated in the plasma membranes, is therefore an excellent parameter for the study of membrane structure and functionality. It can be concluded that the effects of low temperatures on the thickening of membrane bilayers are detrimental for both prokaryotic and eukaryotic organisms (including yeasts). So, the regulation of membrane fluidity and functionality can be realised through above-mentioned changes in phospholipid composition, consisting in incorporation of unsaturated, short-chain, or even branched fatty acids (White et al. 2000; Chintalapati et al. 2004; Feller 2007). Among them, the extent of unsaturation of fatty acids is the most thoroughly investigated mediator of cold adaptation (Russell 1997; Guschina and Harwood 2006; Morgan-Kiss et al. 2006; Rodriguez-Vargas et al. 2007; Rossi et al. 2009).

The precise mechanism whereby yeasts adjust the composition of phospholipids of plasma membranes in dependence to growth temperature is not completely known. The first conventional approaches involved the study of either desaturases or fatty acid synthetases, which act by regulating the unsaturation degree and varying the chain length of fatty acids, and are specific for different yeast species. The adjustment of the activity of those enzymes could be considered the basis of the adaptation mechanisms exhibited by either psychrophilic or mesophilic yeasts, in order to restore membrane bilayer fluidity (via modification of fatty acid unsaturation degree or reduction in chain length) at low temperatures. Pugh and Kates (1975) early found that the desaturase system of *Candida lipolytica* (anamorph of *Yarrowia lipolytica*), which uses oleyl-CoA as substrate, is apparently involved in the modification of membrane fatty acid composition after growth at low temperatures. On the other hand, Okuyama et al. (1979) showed that *S. cerevisiae* responded to changes in growth temperature by varying the activity of a specific fatty acid synthetase, in order to produce more palmitic acid (C16:0) than stearic acid (C18:0) residues and to convert palmitic acid into palmitoleic acid (C16:1) at lower temperatures. Consequently, the ratio between palmitic and stearic acid of *S. cerevisiae* plasma membranes increased from 1.5 to 5.3 when growth temperature decreased at 5 °C.

Kates et al. (1984) found the existence of an inverse relationship between yeast membrane fluidity and phospholipid desaturase activity, which enables self-regulation of membrane fluidity by the cell desaturase system. Interestingly, additional studies reported that *S. cerevisiae* overexpressed the Δ9 desaturase system (which is the key enzyme regulating the synthesis of unsaturated fatty acyl-CoAs) at low temperatures (Stukey et al. 1989; Nakagawa et al. 2002; Martin et al. 2007).

As above reported, phospholipid biosynthesis in yeasts is carried out by a network of highly regulated complex pathways (Fig. 10.1), where the deletion of some key genes could determine an important impact on the phospholipid composition of the plasma membrane. Accordingly, some studies identified a few genes coding the synthesis of phospholipids in *S. cerevisiae* membranes, which are apparently involved in cold adaptation (Blusztajn 1998; McGraw and Henry 1989; Clancey et al. 1993; Trotter et al. 1993; Redón et al. 2012).

Differences in both composition and unsaturation degree of fatty acids between psychrophilic and mesophilic yeasts have been found in the past years (McMurrough and Rose 1973; Arthur and Watson 1976; Watson et al. 1976, 1978; Watson 1978; Rossi et al. 2009). Watson et al. (1976) proposed the use of a suitable Unsaturation Index (UI) in order to compare the unsaturation degree of membrane phospholipids of different yeast species grown at different temperatures:

$$\begin{aligned} \text{UI} = {} & \text{percentage ofmonoenes} + 2(\text{percentages dienes}) \\ & + 3(\text{percentages trienes})/100 \end{aligned}$$

Despite its indubitable utility, a few authors questioned the use of UI for accurate comparison yet, and proposed it just for estimating a tendency, as this index is affected by inter- and intraspecific variability, medium, cell growth phase and additional interfering parameters (Vishniac 2006; Rossi et al. 2009). For example, the concentration of dissolved oxygen may be an additional parameter affecting the

Table 10.1 Fatty acid unsaturation in mesophilic yeasts grown at different temperatures

Species	Original taxonomic designation	Growth temperature (°C)	UI[a]	References
Candida saitoana	*Torulopsis candida*	15	1.42	Watson and Rose (1980)
Candida sake	*T. austromarina*	0	1.87	Watson and Rose (1980)
Candida sake	*T. austromarina*	15	1.44	Watson and Rose (1980)
Kluyveromyces marxianus		4	1.06	Rossi et al. (2009)
Kluyveromyces marxianus		18	0.98	Rossi et al. (2009)
Kluyveromyces marxianus		30	0.75	Rossi et al. (2009)
Saccharomyces cerevisiae		4	0.74–0.77[b]	Rossi et al. (2009)
Saccharomyces cerevisiae		18	0.59–0.62[b]	-
Saccharomyces cerevisiae		30	0.60–0.73[b]	Rossi et al. (2009)
Yarrowia lipolytica	*Candida lipolytica*	10	1.07–1.27[b]	Kates and Baxter (1962)

[a] *UI* Unsaturation index calculated according to Watson et al. (1976)
[b] Range of different strains belonging to the same species

UI detectable at various temperatures, mainly because aerobiosis is mandatory for reactions catalysed by fatty acid desaturase enzymes (Arthur and Watson 1976; Raspor and Zupan 2006). Accordingly, Watson (1987) reported the direct desaturation of palmitic acid to its mono-unsaturated counterpart (palmitoleic acid) in *S. cerevisiae* and *Kazachstania bovina* (formerly *Torulopsis bovina*) grown under anaerobic conditions.

10.3.1 Changes in Mesophilic Yeasts

The UI of some mesophilic yeasts grown at different temperatures is reported in Table 10.1. Overall, with the sole exception of strains belonging to the species *S. cerevisiae*, a general trend of increasing fatty acid unsaturation with decreasing growth temperature was observed.

In parallel with this common tendency, a variable profile of fatty acids as response to cold was also described. Arthur and Watson (1976), who early studied the membrane phospholipid composition of the species *Candida saitoana*, *Clavispora lusitaniae*, *K. bovina*, *Kazachstania slooffiae* and *Kazachstania telluris* (formerly *Torulopsis candida*, *Candida parapsilosis*, *T. bovina*, *Candida slooffiae* and *Saccharomyces telluris*, respectively), revealed a correlation between the growth temperature and both the degree of fatty acid unsaturation and the changes in phospholipid profile. Interestingly, Rossi et al. (2009) found that some mesophilic yeasts (belonging to the species *Kluyveromyces marxianus* and *S. cerevisiae*) synthesised at 4 °C a higher percentage of palmitoleic acid (39.7 %) than those observed in psychrophilic or psychrotolerant species (2.3 and 0.4 %, respectively), in agreement with previous studies (Suutari et al. 1990). Concurrently, both species produced a lower content of oleic (C18:1), linoleic (C18:2) and linolenic (C18:3) acids.

Comparative studies on plasma membrane composition of *Candida utilis* (anamoprh of *Lindnera jadinii*), *Lipomyces starkeyi*, *Rhodosporidium toruloides*, *S. cerevisiae* and *Yarrowia lipolytica* (formerly *Candida oleophila*) revealed that all above four mesophilic species exhibited changes in the absolute amounts of fatty acids and in the fatty acid profile within their growth temperature range (Suutari et al. 1990). Three different patterns of adaptation to low temperatures were observed:

1. *S. cerevisiae* exhibited temperature-induced changes in the form of variation of fatty acid chain length, as also demonstrated by Okuyama et al. (1979). The fatty acid content of this species and the C16/C18 ratio increased with decreasing temperature. On the contrary, the unsaturation degree was virtually unaltered (Suutari et al. 1990). This evidence, firstly suggested by Hunter and Rose (1972), was also confirmed by Watson (1987) and by Rossi et al. (2009). All these authors found that *S. cerevisiae* did not undergo significant changes in the degree of fatty acid unsaturation at low temperatures (Table 10.1).
2. Conversely, *R. toruloides* exhibited adjustment in the degree of fatty acid unsaturation as the preferred way to react to temperature changes: the level of palmitic acid remained unchanged and the concentration of palmitoleic acid was low within the whole temperature range (Suutari et al. 1990).
3. Finally, *C. utilis*, *L. starkeyi* and *Y. lipolytica* were found to use both above responses, by switching from one reaction to another depending on whether the growth temperature was above or below a given threshold (empirically fixed at 20–26 °C). Below 20–26 °C the ratio between C:16/C:18 fatty acids of plasma membranes increased with decreasing temperature, analogously to the response exhibited by *S. cerevisiae*. According to Davis et al. (1981) and Coolbear et al. (1983), the shortening of the chain length and the insertion of the first double bond in a given fatty acid provokes a much greater effect on phospholipid fluidity than that realised by the insertion of subsequent double bond(s). This might explain the increasing content of monoenoic acids as response to low temperature. More recently Suutari et al. (1997) confirmed that, when the temperature decreased from 26–20 to 10 °C, the fatty acid chain length of *C. utilis* also shortened, due to the increase in palmitoleic acid. In parallel, intracellular triacylglycerols accumulated as the temperature decreased from 26–20 to 10 °C. On the other hand, the fatty acid unsaturation of *C. utilis* increased with decreasing growth temperature from 40 to 26–20 °C, due to the increased concentration of linolenic acid in plasma membrane phospholipids. At the same time, an equal phosphatidylcholine and phosphatidylethanolamine or more pronounced phosphatidylinositol and phosphatidylserine decrease

occurred in fatty acids characterised by with lower unsaturation degree and, consequently, the cellular fatty acid content decreased as the temperature was reduced from 40 to 26–20 °C (Suutari et al. 1997).

10.3.2 Changes in Psychrophilic and Psychrotolerant Yeasts

The UI of some psychrophilic and psychrotolerant yeasts grown at different temperatures is reported in Table 10.2. Like to closely related mesophilic species, a common tendency of increasing unsaturation with decreasing growth temperature was generally observed in both psychrophilic and psychrotolerant yeasts.

Arthur and Watson (1976) and Watson (1987) early reported that plasma membranes of psychrophilic yeasts are rich in mono- and poly-UFAs and observed an inverse relationship between growth temperature and the unsaturation degree of membrane lipids: the lower the temperature, the greater the degree of fatty acid unsaturation. The same authors also hypothesised that the high percentage of poly-UFAs (e.g. linoleic and linoleic acid, characterised by low-melting points) found in psychrophilic yeasts would permit plasma membrane to remain sufficiently fluid to allow proper functioning of metabolic processes at low (even sub-zero) temperature. In some psychrophilic yeasts, the UFAs constituted 50–90 % of the total fatty acid composition (Shivaji and Prasad 2009).

The species *Mrakia frigida* (formerly classified as *Leucosporidium frigidum*, *Leucosporidium gelidum* and *Leucosporidium nivalis*) was early considered as a model for the study of physiology of the psychrophilic yeasts (Watson et al. 1976). This species grew well till −1 °C and exhibited a strong correlation between the growth temperature and the unsaturation degree of membrane phospholipids. At sub-zero temperatures, UFAs were about 90 % of the total fatty acids, with linolenic (35–50 %) and linoleic (25–30 %) acids predominating. On the contrary, at temperatures close to the maximum for growth, linolenic acid accounted for less than 20 % of the total fatty acids, whereas oleic (20–40 %) and linoleic (30–50 %) acids were the major components (Watson et al. 1976). Interestingly, Arthur and Watson (1976) postulated the formation of hydrogen bonds between the amino groups of phosphatidylethanolamine (occurring in high amount in *M. frigida* plasma membrane) and the water molecules: this may serve as an additional mechanism to retard ice formation at sub-zero temperatures. Moreover, the same authors suggested a correlation between the degree of fatty acid unsaturation of mitochondrial membranes and the temperature domain of the same species (Arthur and Watson 1976; Watson et al. 1976).

More recently, Thomas-Hall et al. (2002) analysed the fatty acid profile of plasma membrane of *Cryptococcus statzelliae* and *Cryptococcus victoriae* (both isolated from Antarctica) grown at 15 °C. The UFAs predominated: about 50 % were oleic acid, 30 % linoleic acid, 7 % linolenic acid and 10 % palmitic acid. Besides, Thomas-Hall and Watson (2002) found that the psychrophilic species *Cryptococcus nyarrowii* exhibited significantly higher amounts of linolenic acid (3–10 %) and no

Table 10.2 Fatty acid unsaturation in psychrophilic and psychrotolerant yeasts grown at different temperatures

Species	Original taxonomic designation	Growth temperature (°C)	UI[a]	References
Aureobasidium pullulans		4	0.91	Rossi et al. (2009)
Aureobasidium pullulans		18	1.06	Rossi et al. (2009)
Aureobasidium pullulans		30	0.84	Rossi et al. (2009)
Candida psychrophila	*Torulopsis* sp.	~0	1.37–1.78[b]	Watson and Rose (1980)
Cryptococcus antarcticus		14	1.18–1.46[b]	Vishniac and Kurtzmann (1992)
Cryptococcus gilvescens		4	0.82–1.91[b]	Rossi et al. (2009)
Cryptococcus gilvescens		18	0.71–0.82[b]	Rossi et al. (2009)
Cryptococcus gilvescens		30	0.86–1.00[b]	Rossi et al. (2009)
Cryptococcus nyarrowii		6	1.55	Vishniac and Kurtzmann (1992)
Cryptococcus nyarrowii		15	1.38	Vishniac and Kurtzmann (1992)
Cryptococcus statzellieae		15	1.33	Thomas-Hall et al. (2002)
Glaciozyma watsonii	*Leucosporidium* sp.	4	1.02–1.06[b]	Rossi et al. (2009)
Leucosporidiella creatinivora	*Rhodotorula creatinivora*	4	0.95	Rossi et al. (2009)
Leucosporidium scottii	*Candida scottii*	10	1.69–1.72[b]	Kates and Baxter (1962)
Mrakia frigida	*Leucosporidium frigidum, Leucosporidium gelidum, Leucosporidium nivalis*	8	1.50–1.93[b]	Watson and Rose (1980)
Mrakia frigida	*Leucosporidium frigidum, Leucosporidium gelidum, Leucosporidium nivalis*	19	0.91–1.41[b]	Watson and Rose (1980)
Mrakia robertii	*Mrakia* sp.	4	0.92–1.55[b]	Rossi et al. (2009)
Rhodotorula glacialis		4	0.86–1.78[b]	Rossi et al. (2009)
Rhodotorula laryngis		4	0.97–1.03[b]	Rossi et al. (2009)
Rhodotorula laryngis		18	0.90–1.06[b]	Rossi et al. (2009)
Rhodotorula laryngis		30	0.95[c]	Rossi et al. (2009)

[a] *UI* Unsaturation index calculated according to Watson et al. (1976)
[b] Range of different strains belonging to the same species
[c] Average value of different strains belonging to the same species

palmitoleic acid in comparison with *S. cerevisiae*, which synthesised 30–40 % palmitoleic acid, but no linolenic acid. These results were consistent with previous findings reporting the fatty acid profile of some Antarctic yeasts (Watson 1987). Likewise, Rossi et al. (2009) found that some psychrophilic species, namely *Glaciozyma watsonii*, *Leucosporidiella creatinivora*, *Mrakia robertii* (former *Leucosporidium* sp., *Rhodotorula creatinivora* and *Mrakia* sp., respectively) and *Rhodotorula glacialis*, grown at 4 °C in a carbon-rich medium, exhibited a significantly higher unsaturation degree of membrane phospholipids than those observed in some psychrotolerant species (i.e., *Cryptococcus gilvescens* and *Rhodotorula laryngis*), mainly due to their higher amount of linolenic acid (16.8 %). Moreover, the percentage of oleic and linoleic acids in psychrophilic yeasts was significantly higher than that observed in mesophilic ones (Rossi et al. 2009).

10.3.3 Changes in Yeast Plasma Membrane Fatty Acid Composition During the Growth Phases

A few studies revealed that the changes of composition of plasma membrane phospholipids (i.e., profile and unsaturation degree of fatty acids) in both psychrophilic and mesophilic yeasts are also partially dependent by their growth phase (McMurrough and Rose 1973; Watson 1987).

The most manifest changes were observed during the logarithmic phase of growth. In early exponential-growth cultures of *C. utilis*, a decrease in incubation temperature from 20 to 5 °C resulted in a dramatic decrease (0.21–0.09) of the ratio between palmitic and the sum of mono- and poly-UFAs (palmitoleic + oleic + linolenic + linolenic acids oleic acids) and in a parallel increase (0.73–1.12) of the ratio between linolenic and linoleic acids (McMurrogh and Rose 1973). On the contrary, the same variation in incubation temperature gave a less pronounced effect on the fatty acid composition of late exponential-growth cultures: the ratio between palmitic and the sum of mono- and poly-UFAs decreased from 0.21 to 0.13, whereas the ratio between linolenic and linoleic acids increased from 0.44 to 1.12 (McMurrogh and Rose 1973).

More recently, Thomas-Hall and Watson (2002) reported that Antarctic strains of *C. nyarrowii* grown to late exponential phase exhibited no manifest changes in their fatty acid profile across a temperature gradient (2–22 °C). On the contrary, yeast cells grown to stationary phase at 6 °C (14 days) or 15 °C (5 days) showed a higher percentage (50–52 %) of linolenic acid than cells grown to exponential phase, which exhibited a lower content of linoleic (18–22 %) and palmitic (12–15 %) acids.

10.4 Yeast Plasma Membrane Fluidity as Determinant of Adaptability to Stressful Conditions

It has been postulated that a change in membrane fluidity might be the primary signal in the perception of cold stress and, possibly, of osmotic stress. It is very likely that the rigidification of membrane lipids at low temperatures and under hyperosmotic stress is the primary trigger for the corresponding acclimatory responses in cells. The existence of sensors that perceive changes in the physical state of the membrane, disregarding the nature of the stress, is postulated (Swan and Watson 1997; Simons and Toomre 2000; Beney and Gervais 2001; Los and Murata 2004). This would make membranes not just a target of adaptations to environmental conditions, but also a crucial part of the mechanism, involved in triggering the adaptations. Membrane fluidity as an indicator of cell environment would also explain, why many stress responses are common regardless of the kinds of stress (Russell et al. 1995; Russell 2008). As proposed by Panadero et al. (2006), the fluidity of the cell membrane might be a key factor to integrate the sensing mechanism of cold and hyperosmolarity. This phenomenon of cross-protection points on one hand on the existence of a general stress response, while on the other hand, each stress leads to a particular gene expression profile. The fluidity of the membranes may represent a base for the general stress response, which is complemented by additional sensory pathways that fine-tune the adaptations in dependence of the nature of the environmental insult.

Extremophilic organisms are usually determined by the maxima of physico-chemical parameters, such as temperature, defining their growth ranges. Although temperature growth ranges can be very useful in determining the ecological type of the species, they might not be sufficient indicators of their adaptability to extreme conditions. The growth optima determined under in vitro conditions can be very different from those realised in natural habitats. For instance, Antarctic rock-inhabiting meristematic black fungi, also called black yeasts, show optimal growth at significantly higher temperatures than those to which they are usually exposed in their natural environment (Selbmann et al. 2005). Some stenothermal glacial yeasts, such as *Rhodosporidium diobovatum,* can have a very narrow optimal temperature range, while other glacial inhabitants show eurythermal behaviour. In these cases, in vitro determined thermal ranges are insufficient determinants of their psychrophilic nature and of the level of exposure to stress in their natural environments. Since survival in a changing environment depends on the maintenance of an optimal level of membrane fluidity within the lipid matrix, one of the traits that enables a better classification of their ecotype might be membrane fluidity, represented by K value, and its fluctuation in combination with temperature profile might enable classification of microbes according to their specialisation and adaptability. The average membrane K value, calculated from fluidity coefficients of a particular fungus grown and measured at different temperatures, demonstrated clustering of fungi into distinct groups with (1) more fluid membranes (Arctic *R. diobovatum*), (2) with membranes of intermediate fluidity (*Aureobasidium* sp., *Aureobasidium pullulans,*

Cryptococcus liquefaciens, Rhodotorula mucilaginosa, and *S. cerevisiae*), and (3) with more rigid membranes (extremely halotolerant black yeasts of the species *Hortaea werneckii*). Studies on salt adaptation of *H. werneckii* suggested its specialisation to high salinity, which was confirmed by the ability of this species to maintain its membrane fluidity at a certain value over a range of salinities (Turk et al. 2004, 2011). Apparently, this type of specialisation disables adaptation also to low temperatures. Similarly, in *R. diobovatum,* higher average membrane fluidity could prevent
its adaptation to hypersaline conditions. In contrast, the plasma membrane of *Aureobasidium* sp., which was isolated exclusively from the ice of Arctic glaciers, with its intermediate fluidity and high variation under changing temperatures might indicate the specialised nature of this black yeast, since high fluidity variation is negatively correlated with yeast survival under environmental stress (Simonin et al. 2008). This data suggest its adaptation to a very narrow range of conditions, which limits the distribution of this species outside the glaciers. Similar plasma membrane characteristic with slightly lower average membrane fluidity (higher K value) was observed also in mesophilic *S. cerevisiae*, which might prevent its colonisation of extreme environments. On the other hand, different varieties of *A. pullulans*, *C. liquefaciens*, and *Rh. mucilaginosa* with less fluctuating plasma membranes of intermediate fluidity are representatives of globally distributed, phenotypically plastic generalist species with a broad amplitude of ecological tolerance, which can thrive in a variety of extreme environments where competition with other mesophiles is limited. The marine halotolerant yeast *Debaryomyces hansenii* showed similar responses in relation to pH changes. In the range of pH 6–8, its plasma membrane fluidity was hardly affected, while low pH (pH 4) caused a significant decrease in membrane fluidity, indicating that at pH 4 cells are struggling to survive. On the other hand, salt did not have a significant effect on the membrane fluidity and the relative proportions of domains were almost the same in cells growing without salt or at high NaCl concentrations (Turk et al. 2007).

10.5 Conclusions

Yeasts generally display broad temperature optima and can also thrive in the environments that are well outside of their optimum. Temperature growth range is thus not enough to properly describe the ecological type of the species. Membrane fluidity may add valuable information about their preference for certain conditions and can serve as a suitable indicator of psychrotolerance and freeze-thaw resistance, both crucial for successful adaptation to extreme cold environments. In addition to the average fluidity of plasma membrane, the fluctuation of fluidity is an important determinant of stress tolerance: high absolute fluidity fluctuation is tied to decreased survival. The fluidity and its variation thus reflect survival strategy and fitness and are good indicators of the adaptability of microorganisms.

References

Al-Fageeh MB, Smales CM (2006) Control and regulation of the cellular responses to cold shock: the responses in yeast and mammalian systems. Biochem J 397:247–259

Arthur H, Watson K (1976) Thermal adaptation in yeast: growth temperatures, membrane lipid, and cytochrome composition of psychrophilic, mesophilic, and thermophilic yeasts. J Bacteriol 128:56–68

Becker GW, Lester RL (1977) Changes in phospholipids of *Saccharomyces cerevisiae* associated with inositol-less death. J Biol Chem 252:8684–8691

Beney L, Gervais P (2001) Influence of the fluidity of the membrane on the response of microorganisms to environmental stresses. Appl Microbiol Biotechnol 57:34–42

Blusztajn JK (1998) Choline, a vital amine. Science 281:794–795

Carman GM, Han GS (2006) Roles of phosphatidate phosphatase enzymes in lipid metabolism. Trends Biochem Sci 31:694–699

Carman GM, Han GS (2007) Regulation of phospholipid synthesis in *Saccharomyces cerevisiae* by zinc depletion. Biochim Biophys Acta 1771:322–330

Carman GM, Han GS (2011) Regulation of phospholipid synthesis in the yeast *Saccharomyces cerevisiae*. Annu Rev Biochem 80:859–883

Carman GM, Henry SA (1989) Phospholipid biosynthesis in yeast. Annu Rev Biochem 58:635–669

Carman GM, Henry SA (1999) Phospholipid biosynthesis in the yeast *Saccharomyces cerevisiae* and interrelationship with other metabolic processes. Prog Lipid Res 38:361–399

Carman GM, Henry SA (2007) Phosphatidic acid plays a central role in the transcriptional regulation of glycerophospholipid synthesis in *Saccharomyces cerevisiae*. J Biol Chem 282:37293–37297

Carman GM, Kersting MC (2004) Phospholipid synthesis in yeast: regulation by phosphorylation. Biochem Cell Biol 82:62–70

Carman GM, Zeimetz GM (1996) Regulation of phospholipid biosynthesis in the yeast *Saccharomyces cerevisiae*. J Biol Chem 271:13293–13296

Chang YF, Carman GM (2008) CTP synthetase and its role in phospholipid synthesis in the yeast *Saccharomyces cerevisiae*. Prog Lipid Res 47:333–339

Chen M, Hancock LC, Lopes JM (2007) Transcriptional regulation of yeast phospholipid biosynthetic genes. Biochim Biophys Acta 1771:310–321

Chintalapati S, Kiran MD, Shivaji S (2004) Role of membrane lipid fatty acids in cold adaptation. Cell Mol Biol 50:631–642

Clancey CJ, Chang S, Dowhan W (1993) Cloning of a gene (PSD1) encoding phosphatidylserine decarboxylase from *Saccharomyces cerevisiae* by complementation of an *Escherichia coli* mutant. J Biol Chem 268:24580–24590

Coolbear KP, Berde CB, Keough KMW (1983) Gel to liquid-crystalline phase transitions of aqueous dispersions of polyunsaturated mixed-acid phosphatidylcholines. Biochemistry 22:1466–1473

Cossins AR, Murray PA, Gracey AY, Logue J, Polley S, Caddick M, Brooks S, Pestle T, Maclean N (2002) The role of desaturases in cold induced lipid restructuring. Biochem Soc Trans 30:1082–1086

Daum G, Lees ND, Bard M, Dickson R (1998) Biochemistry, cell biology and molecular biology of lipids of *Saccharomyces cerevisiae*. Yeast 14:1471–1510

Daum G, Tuller, G Nemec T, Hrastnik C, Balliano G, Cattel L, Milla P, Rocco F, Conzelmann A, Vionnet C, Kelly DE, Kelly S, Schweizer E, Schüller HJ, Hojad U, Greiner E, Finger K (1999) Systematic analysis of yeast strains with possible defects in lipid metabolism. Yeast 15:601–614

Davis PJ, Fleming BD, Coolbear KP, Keough KMW (1981) Gel to liquid-crystalline transition temperatures of water dispersions of two pairs of positional isomers of unsaturated mixed acid phosphatidyl cholines. Biochemistry 20:3633–3636

Dickson RC (2008) New insights into sphingolipid metabolism and function in budding yeast. J Lipid Res 49:909–921

Feller G (2007) Life at low temperatures: is disorder the driving force? Extremophiles 11:211–216

Gaspar ML, Aregullin MA, Jesch SA, Nunez LR, Villa-Garcia M, Henry SA (2007) The emergence of yeast lipidomics. Biochim Biophys Acta 1771:241–254

Gaspar ML, Jesch SA, Viswanatha R, Antosh AL, Brown WJ, Kohlwein SD, Henry SA (2008) A block in endoplasmic reticulum-to-Golgi trafficking inhibits phospholipid synthesis and induces neutral lipid accumulation. J Biol Chem 283:25735–25751

Gerday C, Aittaleb M, Bentahir M, Chessa JP, Claverie P, Collins T, D'Amico S, Dumont J, Garsoux G, Georlette D, Hoyoux A, Lonhienne T, Meuwis MA, Feller G (2000) Cold-adapted enzymes: from fundamentals to biotechnology. Trend Biotechnol 18:103–107

Gostincar C, Turk M, Gunde–Cimerman N (2009) Environmental impacts on fatty acid composition of fungal mebranes. In: Misra Jk, Desmukh SK (eds) Fungi from different environments. Science Publishers, Enfield, pp 278–327

Guschina IA, Harwood JL (2006) Mechanisms of temperature adaptation in poikilotherms. FEBS Lett 580:5477–5483

Hazel JR, Williams EE (1990) Therole of alterations in membrane lipid composition in enabling physiological adaptation of organisms to their physical environment. Progr Lipid Res 29:167–227

Henschke PA, Rose AH (1991) Plasma membranes. In: Rose AH, Harrison JS (eds) The yeasts, 2nd edn, vol 4, Yeast organelles. Academic Press, London, pp 297–435

Höfer M (1997) Membranes. In: Spencer JFT, Spencer DM (eds) Yeasts in natural and artificial habitats. Springer, Berlin, pp 95–132

Hunter K, Rose AH (1972) Lipid composition of *Saccharomyces cerevisiae* as influenced by growth temperature. Biochim Biophys Acta 260:639–653

Kates M, Baxter RM (1962) Lipid composition of mesophilic and psychrophilic yeasts (*Candida* species) as influenced by environmental temperatures. Can J Biochem Physiol 40:1213–1227

Kates M, Pugh EL, Ferrante G (1984) Regulation of membrane fluidity by lipid desaturases. In: Kates M, Mason LA (eds) Membrane fluidity, vol 12., BiomembranesPlenum Press, New York, pp 379–395

Los DA, Murata N (2004) Membrane fluidity and its roles in the perception of environmental signals. Biochim Biophys Acta 1666:142–157

Losel DM (1990) Lipids in the structure and function of fungal membranes. In: Kuhn PJ, Trinci APJ, Jung MJ, Goosey MW, Copping LG (eds) Biochemistry of the cell walls and membranes in fungi. Springer, Berlin, pp 119–133

Madigan MT, Martinko JM, Parker J (1997) Brock biology of microorganisms, 8th edn. Prentice Hall, Upper Saddle River

Malanovic N, Streith I, Wolinski H, Rechberger G, Kohlwein SD, Tehlivets O (2008) S-adenosyl-L-homocysteine hydrolase, key enzyme of methylation metabolism, regulates phosphatidylcholine synthesis and triacylglycerol homeostasis in yeast: implications for homocysteine as a risk factor of atherosclerosis. J Biol Chem 283:23989–23999

Mansilla MC, Cybulski LE, de Albanesi D, Mendoza D (2004) Control of membrane lipid fluidity by molecular thermosensors. J Bacteriol 186:6681–6688

Martin CE, Oh CS, Jiang Y (2007) Regulation of long chain unsaturated fatty acid synthesis in yeast. Biochim Biophys Acta 1771:271–285

McGraw P, Henry S (1989) Mutations in the *Saccharomyces cerevisiae* Opi3 gene: effects on phospholipid methylation, growth and cross-pathway regulation of inositol synthesis. Genetics 122:317–330

McMurrough I, Rose AH (1973) Effects of temperature variation on the fatty acid composition of a psychrophilic *Candida* species. J Bacteriol 114:451–452

Morgan-Kiss RM, Priscu JC, Pocock T, Gudynaite-Savitch L, Huner NP (2006) Adaptation and acclimation of photosynthetic microorganisms to permanently cold environments. Microbiol Mol Biol R 70:222–252

Nakagawa Y, Sakumoto N, Kaneko Y, Harashima S (2002) Mga2p is a putative sensor for low temperature and oxygen to induce OLE1 transcription in *Saccharomyces cerevisiae*. Biochem Bioph Res Co 291:707–713

Okuyama H, Saito M, Joshi VC, Gunsberg S, Wakil SJ (1979) Regulation by temperature of the chain length of fatty acids in yeast. J Biol Chem 254:12281–12284

Paltauf F, Kohlwein SD, Henry SA (1992) Regulation and compartmentalization of lipid synthesis in yeast. In: Jones EW, Pringle JR, Broach JR (eds) The molecular and cellular biology of the yeast *Saccharomyces*: gene expression. Cold Spring Harbor, New York, pp 415–500

Panadero J, Pallotti C, Rodriguez-Vargas S, Randez-Cil F, Prieto JA (2006) A downshift in temperature activates the high osmolarity glycerol (HOG) pathway, which determines freeze tolerance in *Saccharomyces cerevisiae*. J Biol Chem 281:4638–4645

Phadtare S (2004) Recent developments in bacterial cold-shock response. Curr Issues Mol Biol 6:125–136

Pugh EL, Kates M (1975) Characterization of a membrane-bound phospholipid desaturase system of *Candida lipolytica*. Biochim Biophys Acta 380:442–453

Rajakumari S, Grillitsch K, Daum G (2008) Synthesis and turnover of non polar lipids in yeast. Prog Lipid Res 47:157–171

Raspor P, Zupan J (2006) Yeasts in extreme environments. In: Rosa CA, Peter G (eds) Biodiversity and ecophysiology of yeasts. Springer, Berlin, pp 371–417

Rattray JB, Schibeci A, Kidby DK (1975) Lipids of yeast. Bacteriol Rev 39:197–231

Redón M, Borrull A, López M, Salvadó Z, Cordero R, Mas A, Guillamón JM, Rozès N (2012) Effect of low temperature upon vitality of *Saccharomyces cerevisiae* phospholipid mutants. Yeast 29:443–452

Rilfors L, Lindblom G (2002) Regulation of lipid compositionin biological membranes-biophysical studies of lipids and lipid synthesizing enzymes. Colloids Surf, B 26:112–124

Rodrıguez-Vargas S, Sanchez-Garcıa A, Martınez-Rivas JM, Prieto JA, Randez-Gil F (2007) Fluidization of membrane lipids enhances the tolerance of *Saccharomyces cerevisiae* to freezing and salt stress. Appl Environ Microbiol 73:110–116

Rossi M, Buzzini P, Cordisco L, Amaretti A, Sala M, Raimondi S, Ponzoni C, Pagnoni UM, Matteuzzi D (2009) Growth, lipid accumulation and fatty acid composition in obligate psychrophilic, facultative psychrophilic, and mesophilic yeasts. FEMS Microbiol Ecol 69:363–372

Russell NJ (1989) Structural and functional role of lipids. In: Ratledge C, Wilkinson SG (eds) Microbial lipids, vol 2. Academic Press, New York, pp 249–279

Russell NJ (1997) Psychrophilic bacteria: molecular adaptations of membrane lipids. Comp Biochem Phys A 118:489–493

Russell NJ (2008) Membrane components and cold sensing. In: Margesin R, Schinner F, Marx JC, Gerday C (eds) Psychrophiles: from biodiversity to biotechnology. Springer, Berlin, pp 177–190

Russell NJ, Evans RI, ter Steeg PF, Hellemons J, Verheul A, Abee T (1995) Membranes as a target for stress adaptation. Int J Food Microbiol 28:255–261

Schneiter K, Brugger B, Sandhoff K (1999) Electrospray ionization tandem mass spectrometry (ESI-MS/MS) analysis of the lipid molecular species composition of yeast subcellular membranes reveals acyl chain-based sorting/remodeling of distinct molecular species en route to the plasma membrane. J Cell Biol 146:741–754

Selbmann L, de Hoog GS, Mazzaglia A, Friedmann EI, Onofri S (2005) Fungi at the edge of life: cryptoendolithic black fungi from Antarctic desert. Stud Mycol 51:1–32

Shivaji S, Prasad GS (2009) Antarctic yeasts: biodiversity and potential applications. In: Satyanarayana T, Kunze G (eds) Yeast biotechnology: diversity and applications. Springer, Berlin, pp 3–16

Shivaji S, Kiran MD, Chintalapati S (2007) Perception and transduction of low temperature in bacteria. In: Gerday C, Glansdorff N (eds) Physiology and biochemistry of extremophiles. ASM Press, Washington DC, pp 194–207

Simonin H, Beney L, Gervais P (2008) Controlling the membrane fluidity of yeasts during coupled thermal and osmotic treatments. Biotechnol Bioeng 100:325–333

Simons K, Toomre D (2000) Lipid rafts and signal transduction. Nature Rev Mol Cell Biol 1:31–39

Stukey JE, McDonough VM, Martin CE (1989) Isolation and characterization of OLE1, a gene affecting fatty acid desaturation from *Saccharomyces cerevisiae*. J Biol Chem 264:16537–16544

Suutari M, Liukkonen K, Laakso S (1990) Temperature adaptation in yeasts: the role of fatty acids. J Gen Microbiol 136:1469–1474

Suutari M, Rintamaki A, Laakso S (1997) Membrane phospholipids in temperature adaptation of *Candida utilis*: alterations in fatty acid chain length and unsaturation. J Lipid Res 38:790–794

Swan TM, Watson K (1997) Membrane fatty acid composition and membrane fluidity as parameters of stress tolerance in yeast. Can J Microbiol 43:70–77

Thomas-Hall S, Watson K (2002) *Cryptococcus nyarrowii* sp. nov., a basidiomycetous yeast from Antarctica. Int J Syst Evol Microbiol 52:1033–1038

Thomas-Hall S, Watson K, Scorzetti G (2002) *Cryptococcus statzelliae* sp. nov. and three novel strains of *Cryptococcus victoriae*, yeasts isolated from Antarctic soils. Int J Syst Evol Microbiol 52:2303–2308

Trotter PJ, Pedretti J, Voelker DR (1993) Phosphatidylserine decarboxylase from *Saccharomyces cerevisiae*. Isolation of mutants, cloning of the gene, and creation of a null allele. J Biol Chem 268:21416–21424

Turk M, Mejanelle L, Šentjurc M, Grimalt JO, Gunde-Cimerman N, Plemenitaš A (2004) Salt-induced changes in lipid composition and membrane fluidity of halophilic yeast-like melanized fungi. Extremophiles 8:53–61

Turk M, Montel V, Žigon D, Plemenitas A, Ramos J (2007) Plasma membrane composition of *Debaryomyces hansenii* adapts to changes in pH and external salinity. Microbiology (Soc Gen Microbiol) 153:3586–3592

Turk M, Plemenitaš A, Gunde-Cimerman N (2011) Extremophilic yeasts: plasma-membrane fluidity as determinant of stress tolerance. Fungal Biol 115:950–958

van der Rest ME, Kamminga AH, Nakanoi A, Anraku Y, Poolman B, Konings WN (1995) The plasma membrane of *S. cerevisiae*: structure, function and biogenesis. Microbiol Rev 59:304–322

Vishniac HS (2006) Yeast biodiversity in the Antarctic. In: Rosa CA, Peter G (eds) Biodiversity and ecophysiology of yeasts. Springer, Berlin, pp 420–440

Vishniac HS, Kurtzman CP (1992) *Cryptococcus antarcticus* sp. nov, and *Cryptococcus albidosimilis* sp. nov., basidioblastomycetes from antarctic soils. Int J Syst Bacteriol 42:547–553

Walker GM (1998) Yeast physiology and biotechnology. Wiley, New York

Watson K (1978) Thermal adaptation in yeasts: correlation of substrate transport with membrane lipid composition in psychrophilic and thermotolerant yeasts. Biochem Soc T 6:293–296

Watson K (1987) Temperature relations. In: Rose AH, Harrison JS (eds) The yeasts. Academic Press, London, pp 41–71

Watson K, Rose AH (1980) Fatty-acyl composition of the lipids of *Saccharomyces cerevisiae* grown aerobically or anaerobically in media containing different fatty acids. J Gen Microbiol 117:225–233

Watson K, Arthur H, Shipton WA (1976) *Leucosporidium* yeasts: obligate psychrophiles which alter membrane-lipid and cytochrome composition with temperature. J Gen Microbiol 97:11–18

Watson K, Morton H, Arthur H, Streamer M (1978) Membrane lipid composition: a determinant of anaerobic growth and petite formation in psychrophilic and psychrophobic yeasts. Biochem Soc T 6:380–381

White PL, Wynn-Williams DD, Russell NJ (2000) Diversity of thermal responses of lipid composition in the membranes of the dominant culturable members of an Antarctic fell field soil bacterial community. Antarct Sci 12:386–393

Chapter 11
Cold-Shock Response and Adaptation to Near-Freezing Temperature in Cold-Adapted Yeasts

Masayori Inouye and Sangita Phadtare

Abstract Yeasts, such as *Saccharomyces cerevisiae*, are exposed in their natural habitat to ambient temperatures. They can form colonies at 4 °C and can grow at 10–18 °C. Therefore, the temperature downshift to 10 °C is a moderate cold stress for yeasts. Thus, there is no dramatic induction of cold-shock proteins (CSPs) upon temperature downshift from 30 to 10 °C. On the other hand, the response observed at near-freezing temperatures is more likely to represent the strong cold-shock response for yeasts. The changes in yeast plasma membrane fluidity are the primary signal triggering the cold-shock response. The responses of the yeast cell to temperature downshift to 10 °C can be categorized into three phases. The early and mid-phases are characterized by the initial up-regulation of a number of genes that are associated with the transcriptional machinery, which is then followed by an up-regulation of the translational machinery in the mid-phase. The third phase is characterized by the transcriptional activation of typical stress-marker genes, for example, the heat-shock protein (HSP) genes, and genes that are involved in the cellular processes of metabolism and signal transduction. On the other hand, it has been shown that the cold-induced accumulation of trehalose, glycerol, and HSPs plays a crucial role in protecting the yeast cells against freeze injury. Each of these three cryoprotectants is discussed in detail, along with the relevance of these studies for biotechnological application of yeasts.

Keywords Yeast cold shock · Trehalose · Cryoprotectants

M. Inouye (✉) · S. Phadtare (✉)
Department of Biochemistry and Molecular Biology, Robert Wood Johnson Medical School, UMDNJ, CABM, 679 Hoes Lane, Piscataway, NJ, USA
e-mail: inouye@cabm.rutgers.edu

S. Phadtare
e-mail: phadtasa@rwjms.rutgers.edu

P. Buzzini and R. Margesin (eds.), *Cold-Adapted Yeasts*,
DOI: 10.1007/978-3-642-39681-6_11,

Contents

11.1 Introduction 244
11.2 Cold-Shock Response in Yeasts 245
11.2.1 Yeast Cellular Thermosensors for Cold Shock 246
11.2.2 Cold-Shock Response of Yeasts at 10–18 °C 247
11.2.3 Cold-Shock Response of Yeasts at Near-Freezing Temperatures 249
11.3 Biotechnological Perspectives 253
11.4 Conclusions 254
References 254

11.1 Introduction

Microorganisms that encounter changes in temperatures in their natural habitat are equipped with cellular mechanisms to respond and adapt to these changes. The heat shock causes well-defined damage to the cells, mainly via protein unfolding or denaturation. Cold shock on the other hand may affect the cell at various levels such as stabilization of secondary structures in nucleic acids, hindering transcription and translation, membrane fluidity, ribosome function, and protein folding. However, microorganisms have mechanisms to adapt to cold stress to survive at low temperatures.

Cold-shock adaptation at temperatures above freezing has been quite extensively investigated in bacteria, such as *Escherichia coli*, which encounters cold-shock stress as a result of excretion from animals, as well as in *Bacillus subtilis* (Ermolenko and Makhatadze 2002; Weber and Marahiel 2003; Phadtare 2004; Phadtare and Severinov 2010). Bacteria have evolved several mechanisms to overcome the detrimental effects caused by low temperatures using adaptive strategies for each of the above-mentioned challenges. For example, the decrease in membrane fluidity is counteracted by increasing the proportion of unsaturated fatty acids (UFAs) in the membrane lipids, by shortening the fatty acid chain length or by altering fatty acid branching from *iso* to *anteiso* (Kaneda 1991). An extensive review of changes in lipid composition and fluidity of yeast plasma membrane in response to cold is reported in Chap. 10. The stabilization of secondary structures in DNA and RNA impedes mRNA elongation during transcription, as well as ribosomal movement of mRNAs during translation. In *E. coli*, immediately following a temperature downshift, a major cold-shock protein (CSP), CspA, and its homologs are dramatically induced. These CSPs function as RNA chaperones, which destabilize the secondary structures in RNA and thereby facilitate transcription and translation (Jiang et al. 1997; Bae et al. 2000; Phadtare et al. 2002). During this period, called the "acclimation phase," translation of non-CSPs is reduced. In order to overcome the low-temperature-induced problems in ribosomal assembly, in *E. coli*, two low-temperature-specific ribosome assembly factors, RbfA and CsdA, are produced. These associate with the 30*S* and 50*S* subunits, respectively, and facilitate production of non-CSPs (Xia et al. 2003; Charollais et al. 2004). Cold-shock-induced protein misfolding and denaturation are not as severe as that seen in case of heat shock. Chaperones like trigger factor (Kandror and Goldberg 1997) are produced to aid in proper folding of proteins at low temperature. Production of heat-shock protein (HSP) chaperones, such as GroEL, has also been reported for *E. coli*. In addition, certain sugars, such as maltose, are also implicated for their protective effect during cold-shock adaptation of *E. coli* (Phadtare and Inouye 2004).

11.2 Cold-Shock Response in Yeasts

A comparison of the cold-shock response with adaptation mechanisms between *E. coli* and yeasts belonging to the species *Saccharomyces cerevisiae* immediately throws light on a fundamental difference between these two microorganisms. Yeasts in their natural habitat are exposed to ambient temperatures, whereas enteric *E. coli* grows at warm temperatures inside the animal body and are exposed to ambient temperatures outside of animals. As mentioned above, *E. coli* contains a number of cold-shock-specific proteins that are required only for growth at low temperatures but not at 37 °C. In contrast, in *S. cerevisiae* cells, there is no dramatic induction of CSPs upon temperature downshift from 30 to 10 °C. There is only a modest (twofold–threefold) increase in the concentration of pre-existing proteins.

Yeasts can form colonies at 4 °C and can grow at 10–18 °C. Therefore, the temperature downshift to 10 °C is a moderate cold stress for yeasts. Thus, at this temperature, in yeasts, it is not necessary to produce high amounts of cold-shock-specific proteins except for a moderate increase in the proteins required for ribosomal assembly and translation. On the other hand, the response observed at near-freezing temperatures, which is not observed in *E. coli*, is more likely to represent the strong cold-shock response for yeasts. Also, for these organisms, it is much more critical to survive at or below freezing temperatures than it is for *E. coli*, which has a reservoir in animals.

Among yeasts, the cold-shock response of *S. cerevisiae* has been well studied (Al-Fageeh and Smales 2006; Aguilera et al. 2007). Several approaches such as low-temperature, global transcript profiling using DNA microarray analysis (Sahara et al. 2002; Homma et al. 2003) or differential mRNA display (Rodriguez-Vargas et al. 2002) have been used to explore the genes that are induced by cold shock in this organism. As mentioned above, yeast cells in general exhibit a more moderate cold-shock response than that observed in bacterial systems upon a shift from 30 to 10 °C. The cold-shock response of yeasts can be categorized based on the range of temperature, such as the moderate cold-shock response observed at 10–18 °C and cold-shock response observed at 10 °C or less (near-freezing

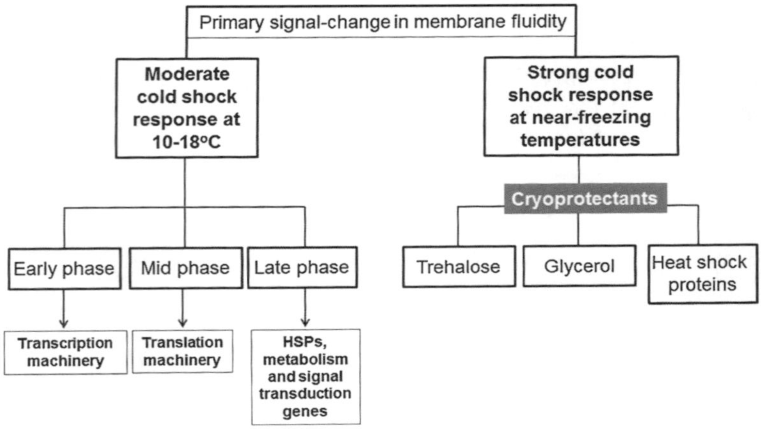

Fig. 11.1 Schematic representation of cold-shock response in yeasts

temperatures) (Al-Fageeh and Smales 2006). The general aspects of the cold-shock response and adaptation in yeasts are schematically represented in Fig. 11.1.

11.2.1 Yeast Cellular Thermosensors for Cold Shock

The cell uses various thermosensors such as cytoplasm membrane, nucleic acids, and ribosomes to sense changes in temperature (Klinkert and Narberhaus 2009). The adaptive changes in the membrane lipid composition represent probably the most highly conserved cold-shock response among bacteria. For example, in *E. coli*, it has been noted that temperature downshift causes the usual liquid crystalline nature of the membrane to change to a gel phase. The temperature downshift leads to an increase in the proportion of UFAs in the membrane phospholipids, which have lower melting points and greater degree of flexibility than phospholipids containing saturated fatty acids. This change, which counteracts the negative effect of low temperature on the physical state of the lipid bilayer, is known as the homeoviscous adaptation (Sinensky 1974). In *E. coli,* the enzyme β-ketoacyl-acyl carrier protein (ACP) synthase II is activated at low temperature and mediates the conversion of palmitoleic acid to *cis*-vaccenic acid, the UFA that is prominent upon cold shock in *E. coli* (Garwin and Cronan 1980; Garwin et al. 1980).

In the cyanobacterium *Synechocystis* sp., it was shown that temperature downshift leads to the reduction in the membrane fluidity, which in turn plays a role in the subsequent signal transduction (Inaba et al. 2003; Los and Murata 2004). Genes encoding fatty acid desaturases played a vital role in this phenomenon by modulating membrane rigidity with respect to temperature. In *Synechocystis* sp., the expression of these desaturase genes upon cold shock is controlled by the histidine kinase Hik33 (Suzuki et al. 2000, 2001). Hik33 is an integral membrane protein that functions as a cold sensor. Similarly, DesK, an integral membrane protein in *B. subtilis* (Aguilar et al. 2001), acts as cold sensor. Overall, these observations suggested that changes in the physical state of the membrane caused by a temperature downshift are recognized by cold sensors anchored in the membrane. In addition to its role in sensing cold temperatures, Hik33 has also been shown to regulate the expression of genes induced by the osmotic stress in *Synechocystis* sp. (Mikami et al. 2002). Thus, one can speculate that both osmotic stress and cold stress can be sensed by common mechanisms.

In yeasts, changes in the physical state of the membrane act as a primary signal of a change in temperature (Vigh et al. 1998). It was demonstrated that a membrane lipid perturbation causes induction of the heat-shock genes in yeasts (Carratu et al. 1996). In addition, Sln1p, a known yeast histidine kinase sensor, has been implicated in its cold-sensing mechanism (Panadero et al. 2006). Similar to Hik33, Sln1P is also implicated in both cold stress and osmotic stress. Sln1p, together with Ypd1p and Ssk1p, forms a phosphorelay system, which transmits the osmotic stress signal through the high-osmolarity glycerol (HOG) pathway, the most important osmotic-stress-responding pathway in *S. cerevisiae* (Hohmann 2002; Westfall et al. 2004). This pathway is also activated by compounds, such as dimethylsulfoxide (DMSO), that modulate the physical state of the membrane (Panadero et al. 2006). DMSO has been reported to mimic the changes in membrane fluidity caused by a significant decrease in temperature and leads to the activation of several cold-induced mitogen-activated protein kinases (MAPKs) (Orvar et al. 2000). Thus, Sln1p may be functioning under both of these stresses, being activated by similar mechanisms (Hayashi and Maeda 2006; Panadero et al. 2006). In spite of the similarities in the mechanism of sensing of temperature changes in bacteria and yeasts, it was observed that instantaneous cold shock leads to an irreversible rigidification of the membrane of exponentially growing cells of *E. coli* and *B. subtilis*, but not of *S. cerevisiae* (Cao-Hoang et al. 2008).

11.2.2 Cold-Shock Response of Yeasts at 10–18 °C

11.2.2.1 Time-Dependent Induction of Genes at 10–18 °C

A DNA microarray-based study investigated the time-dependent response of *S. cerevisiae* upon cold shock at 10 °C. It was observed that transcription of one-fourth of all *S. cerevisiae* genes was affected at this temperature (Sahara et al. 2002); mostly, there is a general reduction in transcription and translation efficiency. The responses can be categorized into three phases; the early and mid-phases are characterized by the initial up-regulation of a number of genes that are associated with the transcriptional machinery, which is then followed by an up-regulation of the translational machinery in the mid-phase. The translational efficiency is decreased upon temperature downshift due to the formation of mRNA secondary structures and an increase in the proportion of inactivated ribosomes in yeasts. Up-regulation of the translational machinery in the mid-phase allows translation to proceed at a rate sufficient to synthesize essential proteins by synthesizing ribosomes *de novo*. The genes, which are prominently induced during early/mid-phase, include *NSR*1, *TIP*1, *TIR*1, and *TIR*2. The products of these genes are involved in pre-rRNA processing, ribosome biogenesis, and maintenance of the cell wall, respectively (Kondo et al. 1992; Schade et al. 2004). Nsr1 protein is required for processing before the ribosomal RNA (rRNA) stage (Kondo and Inouye 1992; Kondo et al. 1992). In the absence of Nsr1, a large rRNA precursor accumulates and cell growth slows. The initial responses (0–2 h) also include the enhanced expression of key genes involved in phospholipid synthesis, such as *INO*1 and *OPI*3, and fatty acid desaturation (*OLE*1). Genes related to transcription (RNA helicases such as *DBP*2, RNA polymerase subunits such as *RPA*49 among others) and a significant number of ribosomal protein genes are also induced in this phase. Increased transcription of ribosome biogenesis genes was also noted in a chemostat-based transcriptome analysis carried out to study acclimation of *S. cerevisiae* (Tai et al. 2007).

Most of these above transcription-related and ribosomal genes are repressed after longer incubation periods (4–24 h) at 10 °C (Sahara et al. 2002; Schade et al. 2004) and also at 4 °C (Homma et al. 2003; Murata et al. 2006). This third phase is characterized by the transcriptional activation of typical stress-marker genes, for example, the HSP genes. In addition, in this third phase of cold-shock response, genes that are involved in the cellular processes of metabolism and signal transduction are up-regulated (Rodriguez-Vargas et al. 2002; Sahara et al. 2002; Homma et al. 2003). The genes induced in the third phase include *HSP*12 and *HSP*26 genes and trehalose-synthesizing enzymes encoded by *TPS*1 and *TPS*2. Certain HSPs as those encoded by *HSP*42, *HSP*104, *SSA*4, *SSE*2, and *YRO*2 are also up-regulated after 4–12 h of transfer to 10 °C as well as at similar times at 4 and 0 °C (Kandror et al. 2004). However, genes encoding other members of the HSPs family encoded by *CIS*3, *HMS*2, *HSC*82, *HSP*30, *HSP*60, *HSP*78, *HSP*82, *HSP*150, *SSA*1, *SSA*2 are repressed at 10 °C and induced by temperatures lower than 10 °C. This suggests that the members of the HSP gene family are differentially regulated. The genes that are induced during mid- to late phase upon cold shock and beyond include those that are involved in the metabolism of glycogen (*GLG*1, *GSY*1, *GLC*3, *GAC*1, *GPH*1, and *GDB*1) and genes for detoxifying reactive oxygen species (*ROS*) and defense against oxidative stress, such as catalase (*CTT*1), glutaredoxin (*TTR*1), thioredoxin (*PRX*1), and glutathione transferase (*GTT*2). Furthermore, it has been suggested that the cAMP/PKA (protein kinase A) pathway plays a central role in the control of gene expression in yeasts upon temperature downshift to 10 °C. It has also been suggested that in the early/mid-phase, the cAMP/PKA pathway is involved in the up-regulation of genes that encode translational machinery, while in the late-phase response, the pathway is involved in the down-regulation of general stress-response genes.

11.2.2.2 Adaptation of Yeast Translation Machinery upon Cold Shock

As discussed above, yeasts adapt to temperature downshift in terms of the duration and the magnitude of the temperature change. At low temperatures at which there is no severe restriction of growth, there is an up-regulation of genes that are essential for cold growth. Most of these genes are expressed minimally or repressed after prolonged periods of stress. Immediately following the temperature downshift, there is a need for restructuring of the translation machinery for efficient translation at the low growth temperature and also destabilization of the secondary structures in mRNAs that may hinder movement of ribosomes on mRNAs. This has been well studied in *E. coli* and also seems to occur in yeasts. Thus, the early and mid-phases of cold adaptation after shift to 10–18 °C are characterized by induction of genes related to translation machinery. Studies with cold-sensitive mutants (Hampsey 1997) revealed that majority of these mutants are affected in ribosomal proteins and translation (Winzeler et al. 1999; Zhang et al. 2001), in proteins involved in pre-rRNA processing for ribosome biogenesis or assembly (Lee and Baserga 1997), protein folding, exocytosis, and nucleus–cytosol exchange (relevant references in Aguilera et al. 2007). Thus, at temperatures up to 10 °C, in *S. cerevisiae,* the specific transcriptional and translational machinery is up-regulated in order to enable cold-adapted growth in the early and mid-phases. These systems are significantly down-regulated in the late phase of cold adaptation, wherein all the transcriptional and translational machinery has already been restructured. This machinery is also down-regulated at severe cold conditions as 4 °C, which are growth restrictive, where the need for *de novo* protein synthesis is significantly diminished. On the other hand, freeze-protective mechanisms, such as fatty acid desaturation and synthesis of osmolytes, are triggered even at permissive temperatures and increased at near-freezing conditions, followed by induction of additional protection systems, such as HSPs as discussed below.

11.2.3 Cold-Shock Response of Yeasts at Near-Freezing Temperatures

In bacteria, it was shown that exposing the cells to cold shock prior to freezing increases the cell viability due to the protective effect of CSPs produced (Willimsky et al. 1992). It was proposed that also in yeasts, exposure to low temperature improves survival at freezing temperature.

Several studies have now shown that the cold-induced accumulation of trehalose, glycerol, and HSPs plays a crucial role in protecting yeast cells against freeze injury. As mentioned above, the changes in membrane fluidity are the primary signal triggering the cold-shock response. Notably, this signal is transduced and regulated through classical stress pathways and transcriptional factors, the HOG MAPK pathway, and Msn2/4p (Aguilera et al. 2007).

11.2.3.1 Trehalose as Cryoprotectant

Trehalose is accumulated at high levels in response to different stress conditions, and several protecting roles, such as membrane and protein stabilization, have been proposed for trehalose (reviewed in Aguilera et al. 2007). High levels of trehalose have also been correlated with freezing resistance. Trehalose is not needed for growth at 10 °C (Schade et al. 2004); however, the observation that the viability of cells incubated at 0 °C for 5–20 days correlated with the intracellular trehalose content suggests that it does protect the cell against very cold temperatures (Kandror et al. 2004).

The biosynthesis of trehalose in *S. cerevisiae* is catalyzed by two enzymes, trehalose-6-phosphate synthase (Tps1) and trehalose-6-phosphate phosphatase (Tps2) (Bell et al. 1998). Near-freezing temperatures lead to the most dramatic induction of the genes encoding these two enzymes, which in turn leads to the production of large amount of trehalose, the chemical chaperone (Kandror et al. 2004). It has been proposed that the accumulation of trehalose corresponds directly to the cell's ability to withstand freezing temperatures and is critical for the maintenance of cell viability at 0 °C; however, the exact mechanism by which trehalose protects the cell is not fully elucidated.

In exponentially growing cells at 30 °C, the mRNA levels of *TPS*1 and *TPS*2 are significantly low; however, upon temperature downshift to 10 °C or less, the mRNA levels increase dramatically. For example, there is a 20-fold increase in the levels of *TPS*1 and *TPS*2 mRNAs 15–20 h after temperature shift (Kandror et al. 2004). Both transcriptional regulation and increased mRNA stabilization are responsible for this increase. Note that *TPS*1 and *TPS*2 mRNAs are extremely stable at 0 °C, but extremely unstable at 30 °C. This is similar to the stabilization of *E. coli cspA* mRNA at low temperature that leads to the robust production of CspA (Fang et al. 1997). The actual mechanism of *TPS*1 and *TPS*2 mRNA stabilization is not known though. On the other hand, two transcription factors, Msn2 and Msn4, are involved in the transcriptional regulation and activation of the *TPS*1 and *TPS*2 genes at 0 °C. These factors bind stress-response elements (STREs) and up-regulate transcription of the *TPS*1 and *TPS*2 genes (Moskvina et al. 1999). This notion is supported by the presence of STREs in the promoter regions of the *TPS*1 and *TPS*2 genes. Upon return to normal growth temperatures, trehalose levels are lowered rapidly. This may be due to the degradation mediated by the Nth1 enzyme whose gene expression is also marginally induced upon cold shock, but is induced further upon a return to 30 °C. In general, there is wide consensus that trehalose confers protection against freeze stress in yeast cells. However, there is no direct correlation observed between trehalose accumulation and freezing resistance, above a certain intracellular concentration.

11.2.3.2 Glycerol as Cryoprotectant

Similar to trehalose, accumulation of glycerol confers protection against damage caused by freezing temperatures. *GPD*1, the gene encoding the main enzyme involved in glycerol synthesis, is activated upon a temperature downshift, and glycerol accumulates in the cells; higher amounts accumulate at 4 °C as compared at 12 °C (Panadero et al. 2006). Glycerol is the only osmoprotectant solute accumulated by yeasts upon hyperosmotic stress and is also a by-product of redox homeostasis. The protective role of glycerol against freezing is suggested to be through the osmotic shrinkage resulting from freezing/thawing processes. The *S. cerevisiae* glycerol active transporter Stl1p plays an important role in the rapid accumulation of glycerol. This gene is expressed under various stress conditions (Tulha et al. 2010).

As mentioned above, upon temperature downshift to 12 or 4 °C, the HOG pathway is activated in yeasts. This activation occurs via the activation and phosphorylation of MAPK Hog1p. It is proposed that the cold-shock-induced changes in the fluidity of the membrane lead to the activation of Hog. In the absence of Hog1p, the induction of those genes required for the biosynthesis of trehalose and glycerol was not observed. The deletion of the *HOG*1 gene had no effect on cells grown at 12 °C, but exhibited decreased tolerance to freezing (Panadero et al. 2006). Cold shock has also been shown to induce a protein-kinase-mediated response in the fission yeast *Schizosaccharomyces pombe* (Soto et al. 2002).

11.2.3.3 Heat-Shock Proteins as Cryoprotectant

Using DNA microarray analysis, the genes induced by the cryopreservation procedure were identified and were shown to be mainly involved in energy metabolism, oxidative stress scavengers, cell rescue system including HSPs, and gene products localized in organelles (Odani et al. 2003). It was suggested that the damage to yeast cells during the cryopreservation treatment is mainly localized to the cell wall and cellular organelle structures, and for the repair of these structures, yeast cells require cell rescue system including HSPs. The induction of heat-shock genes by near-freezing temperatures is particularly intriguing and has been reported by several groups. For example, when yeast cells are cultured at 4 °C for extended period, several HSPs are induced (Homma et al. 2003), suggesting that the induction of these genes might be necessary for adjustment to cold resistance. Under laboratory conditions, it was observed that heat-shock treatment increases cell viability of yeast cells after freeze–thaw treatment (Komatsu et al. 1990). Genome-wide analysis of the yeast transcriptome upon heat and cold shock showed that several HSPs are induced by temperature downshift to 4 °C (Becerra et al. 2003).

The HSPs induced in yeasts by cold shock included proteins encoded by the genes *HSP*12, *HSP*42, *HSP*104, and *SSA*4 (Kandror et al. 2004). The *HSP*12 gene encodes one of the two major small HSPs of *S. cerevisiae* and is induced under different conditions, such as low and high temperatures, osmotic or oxidative stress, and high sugar or ethanol concentrations (Kandror et al. 2004; Murata et al. 2006). It was reported that Hsp12p plays a role in cryoresistance (Pacheco et al. 2009). The study also showed that this protein may compensate for the protective effect of trehalose at freezing temperatures.

The eukaryotic Hsp60 cytoplasmic chaperonin CCT (chaperonin containing the t-complex polypeptide–1) is essential for growth in *S. cerevisiae*, and mutations in CCT subunits affect assembly of tubulin and actin. Unlike most other chaperones, CCT in yeasts does not undergo induction following heat shock. There is a threefold–fourfold increase in mRNA levels of CCTα and CCTβ genes after cold shock at 4 °C (Somer et al. 2002). The role of CCT in cold shock is not known, but may be related to the cytoskeleton. The observation that the mRNA levels of *CCT*a increase threefold–fourfold upon temperature downshift to 4 °C, but an increase in the protein level is observed only upon a shift from 4 to 10 °C, suggests that CCT may be involved in the recovery from cold shock and the transition back to higher temperatures rather than in survival or adaptation to lower temperatures.

11.2.3.4 Additional Genes that Help in Near-Freezing Temperature Adaptation and Growth of Yeasts

At 4 °C, genes involved in glycogen biosynthesis are reportedly induced along with increased levels of phospholipids and mannoproteins (*DAN/TIR* family cell wall mannoproteins). This may be useful in counteracting the changes to membrane maintenance and the permeability of the cell wall. *TIR*-related DNA sequences, seripauperin (Pau) family proteins, which have been shown to display phospholipid-interacting activity, are also up-regulated after cold shock (Al-Fageeh and Smales 2006; Aguilera et al. 2007).

A number of additional genes such as *SEC*11, involved in protein transport, and *BFR*2, involved in protein secretion, have also shown to be involved in the yeast cold-shock response. In *S. cerevisiae*, *BFR*2 is an essential gene. The Bfr2p protein involved in protein trafficking to the Golgi apparatus is thought to be essential for growth and/or cell proliferation in *S. cerevisiae*. The levels of *BFR*2 mRNA increase more than fivefold in response to cold shock (Chabane and Kepes 1998). The mechanism of this induction has not been well elucidated. In addition, the mRNA levels of *CER*1*p* encoding an Hsp70-related protein involved in the translocation of a subset of proteins into the endoplasmic reticulum in *S. cerevisiae* increase upon cold shock due to the possible increased requirement for the chaperone activity of Cer1p at lower temperatures (Hamilton et al. 1999).

11.3 Biotechnological Perspectives

Freezing is an important means of preservation and storage of microbial strains used for many types of industrial applications including food processing. In spite of the recent advances made, the mechanisms of tolerance and sensitivity to freeze or near-freeze stress are not yet fully understood. More and more studies are now being carried out to increase the biotechnological potential of yeasts. Some of these approaches are discussed here.

1. As discussed above, intracellular glycerol is a cryoprotectant that is important for near-freeze temperature adaptation of yeasts and the glycerol active transporter Stl1p plays an important role in the fast accumulation of glycerol. The gene encoding this protein is expressed under gluconeogenic conditions, under osmotic shock and stress, as well as under high temperatures. An extremely high expression of this gene, also concomitant high level of glycerol/H^+ symporter activity level, can be achieved by growing the cells on *STL*1 induction medium (YPGE) (Tulha et al. 2010). Under these conditions, more than 400 mM glycerol was accumulated, whereas the glycerol/H^+ symporter mutant strains accumulated less than 1 mM glycerol. As expected, the strains which accumulated the glycerol showed 25–50 % more survival than the latter strains. These findings are important from biotechnological point of view as existing strains of *S. cerevisiae* strain can be made more resistant to cold/freeze–thaw stress just by adding glycerol to the culture medium. The combination of low temperatures with extracellular glycerol induces the transporter Stl1p. This will also allow the users to avoid transgenic strains, an aspect that is particularly important in the food industry.
2. Production of the heterologous desaturases (from sunflower—*Helianthus annuus*) in yeasts increased the fluidity of the yeast plasma membrane (Rodriguez-Vargas et al. 2007); this in turn increased the freezing tolerance of the yeast cells. Thus, engineering of membrane lipids has the potential to be a useful tool of increasing the freezing tolerance of industrial yeast strains.
3. Yeasts used in bread making are exposed to freeze–thaw stress during frozen-dough baking. A genome-wide screening performed using the complete deletion strain collection of diploid *S. cerevisiae* (Ando et al. 2007) identified 58 gene deletions that conferred freeze–thaw sensitivity. The results suggested the presence of at least two different mechanisms of freeze–thaw injury: oxidative stress generated during the freeze–thaw process and defects in cell wall assembly. The information will help in the further development of freeze–thaw-tolerant yeasts, as overexpression of these genes might increase tolerance to freeze–thaw stress. This in turn will accelerate the development of manufacturing processes for frozen-dough baking.
4. L-proline has cryoprotective activity in *S. cerevisiae*. A freeze-tolerant mutant with L-proline accumulation was shown to carry an allele of gene encoding γ-glutamyl kinase, with a single amino acid substitution (Asp154Asn) (Terao et al. 2003). Interestingly, this mutation enhanced the activities of γ-glutamyl kinase

and γ-glutamyl phosphate reductase, both of which catalyze the first two steps of L-proline synthesis and which together may form a complex in vivo. A high tolerance for freezing correlated with higher levels of L-proline in yeast cells. In addition to its cryoprotective activity, intracellular L-proline may protect yeast cells from damage by oxidative stress. This approach may provide a method for creating novel yeast strains that tolerate the freezing and oxidative stresses better.

11.4 Conclusions

Yeasts such as *S. cerevisiae* are exposed in their natural habitat to ambient temperatures, can form colonies at 4 °C, and can grow at 10–18 °C. Therefore, the temperature downshift to 10 °C is a moderate cold stress for yeasts. On the other hand, the response observed at near-freezing temperatures is more likely to represent the strong cold-shock response for yeasts. The changes in yeast plasma membrane fluidity are the primary signal triggering the cold-shock response. The cold-shock response of the yeast cell at 10 °C occurs in three phases; the early and mid-phases are characterized by the initial up-regulation of a transcription-related genes, which is then followed by an up-regulation of the translational machinery in the mid-phase. In the third phase, the HSP genes, genes that are involved in the cellular processes of metabolism and signal transduction, are induced. Upon temperature downshift to near-freezing temperatures, trehalose, glycerol, and HSPs accumulate and are crucial for protecting yeast cells against freeze injury.

In *S. cerevisiae*, there exists a phenomenon termed as general stress response, as several common genes respond to thermal stress-cold shock or heat shock, high osmolarity, oxidation, or presence of certain poisons. In addition, there also exist shared signal transduction machinery and transcription factors among various stresses. The expansion of the physiological roles of transcription pathways with respect to various stresses is exemplified by membrane perturbation or the HOG pathway. However, there may be certain cold stress-specific systems, such as *TIP/TIR* genes that exist in yeasts. Future studies will shed light on this intriguing aspect of this response. Finally, it should be noted that the majority of studies carried out to explore the changes occurring during the cold-shock response of *S. cerevisiae* were at the mRNA level. There is often a poor correlation between mRNA and protein levels in yeast systems (Al-Fageeh and Smales 2006). Therefore, the response at the protein level may not be fully elucidated.

References

Aguilar PS, Hernandez-Arriaga AM, Cybulski LE, Erazo AC, de Mendoza D (2001) Molecular basis of thermosensing: a two-component signal transduction thermometer in *Bacillus subtilis*. EMBO J 20:1681–1691

Aguilera J, Randez-Gil F, Prieto JA (2007) Cold response in *Saccharomyces cerevisiae*: new functions for old mechanisms. FEMS Microbiol Rev 31:327–341

Al-Fageeh MB, Smales CM (2006) Control and regulation of the cellular responses to cold shock: the responses in yeast and mammalian systems. Biochem J 397:247–259

Ando A, Nakamura T, Murata Y, Takagi H, Shima J (2007) Identification and classification of genes required for tolerance to freeze-thaw stress revealed by genome-wide screening of *Saccharomyces cerevisiae* deletion strains. FEMS Yeast Res 7:244–253

Bae W, Xia B, Inouye M, Severinov K (2000) *Escherichia coli* CspA-family RNA chaperones are transcription antiterminators. Proc Natl Acad Sci USA 97:7784–7789

Becerra M, Lombardia LJ, Gonzalez-Siso MI, Rodriguez-Belmonte E, Hauser NC, Cerdan ME (2003) Genome-wide analysis of the yeast transcriptome upon heat and cold shock. Comp Funct Genomics 4:366–375

Bell W, Sun W, Hohmann S, Wera S, Reinders A, De Virgilio C, Wiemken A, Thevelein JM (1998) Composition and functional analysis of the *Saccharomyces cerevisiae* trehalose synthase complex. J Biol Chem 273:33311–33319

Cao-Hoang L, Dumont F, Marechal PA, Le-Thanh M, Gervais P (2008) Rates of chilling to 0 °C: implications for the survival of microorganisms and relationship with membrane fluidity modifications. Appl Microbiol Biotechnol 77:1379–1387

Carratu L, Franceschelli S, Pardini CL, Kobayashi GS, Horvath I, Vigh L, Maresca B (1996) Membrane lipid perturbation modifies the set point of the temperature of heat shock response in yeast. Proc Natl Acad Sci USA 93:3870–3875

Chabane S, Kepes F (1998) Expression of the yeast BFR2 gene is regulated at the transcriptional level and through degradation of its product. Mol Gen Genet 258:215–221

Charollais J, Dreyfus M, Iost I (2004) CsdA, a cold-shock RNA helicase from *Escherichia coli*, is involved in the biogenesis of 50S ribosomal subunit. Nucleic Acids Res 32:2751–2759

Ermolenko DN, Makhatadze GI (2002) Bacterial cold-shock proteins. Cell Mol Life Sci 59:1902–1913

Fang L, Jiang W, Bae W, Inouye M (1997) Promoter-independent cold-shock induction of *cspA* and its derepression at 37 °C by mRNA stabilization. Mol Microbiol 23:355–364

Garwin JL, Cronan JE Jr (1980) Thermal modulation of fatty acid synthesis in *Escherichia coli* does not involve de novo enzyme synthesis. J Bacteriol 141:1457–1459

Garwin JL, Klages AL, Cronan JE Jr (1980) Beta-ketoacyl-acyl carrier protein synthase II of *Escherichia coli*. Evidence for function in the thermal regulation of fatty acid synthesis. J Biol Chem 255:3263–3265

Hamilton TG, Norris TB, Tsuruda PR, Flynn GC (1999) Cer1p functions as a molecular chaperone in the endoplasmic reticulum of *Saccharomyces cerevisiae*. Mol Cell Biol 19:5298–5307

Hampsey M (1997) A review of phenotypes in *Saccharomyces cerevisiae*. Yeast 13:1099–1133

Hayashi M, Maeda T (2006) Activation of the HOG pathway upon cold stress in *Saccharomyces cerevisiae*. J Biochem 139:797–803

Hohmann S (2002) Osmotic stress signaling and osmoadaptation in yeasts. Microbiol Mol Biol Rev 66:300–372

Homma T, Iwahashi H, Komatsu Y (2003) Yeast gene expression during growth at low temperature. Cryobiology 46:230–237

Inaba M, Suzuki I, Szalontai B, Kanesaki Y, Los DA, Hayashi H, Murata N (2003) Gene-engineered rigidification of membrane lipids enhances the cold inducibility of gene expression in synechocystis. J Biol Chem 278:12191–12198

Jiang W, Hou Y, Inouye M (1997) CspA, the major cold-shock protein of *Escherichia coli*, is an RNA chaperone. J Biol Chem 272:196–202

Kandror O, Goldberg AL (1997) Trigger factor is induced upon cold shock and enhances viability of *Escherichia coli* at low temperatures. Proc Natl Acad Sci USA 94:4978–4981

Kandror O, Bretschneider N, Kreydin E, Cavalieri D, Goldberg AL (2004) Yeast adapt to near-freezing temperatures by STRE/Msn2,4-dependent induction of trehalose synthesis and certain molecular chaperones. Mol Cell 13:771–781

Kaneda T (1991) Iso- and anteiso-fatty acids in bacteria: biosynthesis, function, and taxonomic significance. Microbiol Rev 55:288–302

Klinkert B, Narberhaus F (2009) Microbial thermosensors. Cell Mol Life Sci 66:2661–2676
Komatsu Y, Kaul SC, Iwahashi H, Obuchi K (1990) Do heat shock proteins provide protection against freezing? FEMS Microbiol Lett 60:159–162
Kondo K, Inouye M (1992) Yeast NSR1 protein that has structural similarity to mammalian nucleolin is involved in pre-rRNA processing. J Biol Chem 267:16252–16258
Kondo K, Kowalski LR, Inouye M (1992) Cold shock induction of yeast NSR1 protein and its role in pre-rRNA processing. J Biol Chem 267:16259–16265
Lee SJ, Baserga SJ (1997) Functional separation of pre-rRNA processing steps revealed by truncation of the U3 small nucleolar ribonucleoprotein component, Mpp 10. Proc Natl Acad Sci USA 94:13536–13541
Los DA, Murata N (2004) Membrane fluidity and its roles in the perception of environmental signals. Biochim Biophys Acta 1666:142–157
Mikami K, Kanesaki Y, Suzuki I, Murata N (2002) The histidine kinase Hik33 perceives osmotic stress and cold stress in *Synechocystis* sp. PCC 6803. Mol Microbiol 46:905–915
Moskvina E, Imre EM, Ruis H (1999) Stress factors acting at the level of the plasma membrane induce transcription via the stress response element (STRE) of the yeast *Saccharomyces cerevisiae*. Mol Microbiol 32:1263–1272
Murata Y, Homma T, Kitagawa E, Momose Y, Sato MS, Odani M, Shimizu H, Hasegawa-Mizusawa M, Matsumoto R, Mizukami S, Fujita K, Parveen M, Komatsu Y, Iwahashi H (2006) Genome-wide expression analysis of yeast response during exposure to 4 °C. Extremophiles 10:117–128
Odani M, Komatsu Y, Oka S, Iwahashi H (2003) Screening of genes that respond to cryopreservation stress using yeast DNA microarray. Cryobiology 47:155–164
Orvar BL, Sangwan V, Omann F, Dhindsa RS (2000) Early steps in cold sensing by plant cells: the role of actin cytoskeleton and membrane fluidity. Plant J 23:785–794
Pacheco A, Pereira C, Almeida MJ, Sousa MJ (2009) Small heat-shock protein Hsp12 contributes to yeast tolerance to freezing stress. Microbiology 155:2021–2028
Panadero J, Pallotti C, Rodriguez-Vargas S, Randez-Gil F, Prieto JA (2006) A downshift in temperature activates the high osmolarity glycerol (HOG) pathway, which determines freeze tolerance in *Saccharomyces cerevisiae*. J Biol Chem 281:4638–4645
Phadtare S (2004) Recent developments in bacterial cold-shock response. Curr Issues Mol Biol 6:125–136
Phadtare S, Inouye M (2004) Genome-wide transcriptional analysis of the cold shock response in wild-type and cold-sensitive, quadruple-*csp*-deletion strains of *Escherichia coli*. J Bacteriol 186:7007–7014
Phadtare S, Severinov K (2010) RNA remodeling and gene regulation by cold shock proteins. RNA Biol 7:788–795
Phadtare S, Inouye M, Severinov K (2002) The nucleic acid melting activity of *Escherichia coli* CspE is critical for transcription antitermination and cold acclimation of cells. J Biol Chem 277:7239–7245
Rodriguez-Vargas S, Estruch F, Randez-Gil F (2002) Gene expression analysis of cold and freeze stress in Baker's yeast. Appl Environ Microbiol 68:3024–3030
Rodriguez-Vargas S, Sanchez-Garcia A, Martinez-Rivas JM, Prieto JA, Randez-Gil F (2007) Fluidization of membrane lipids enhances the tolerance of *Saccharomyces cerevisiae* to freezing and salt stress. Appl Environ Microbiol 73:110–116
Sahara T, Goda T, Ohgiya S (2002) Comprehensive expression analysis of time-dependent genetic responses in yeast cells to low temperature. J Biol Chem 277:50015–50021
Schade B, Jansen G, Whiteway M, Entian KD, Thomas DY (2004) Cold adaptation in budding yeast. Mol Biol Cell 15:5492–5502
Sinensky M (1974) Homeoviscous adaptation—a homeostatic process that regulates the viscosity of membrane lipids in *Escherichia coli*. Proc Natl Acad Sci USA 71:522–525
Somer L, Shmulman O, Dror T, Hashmueli S, Kashi Y (2002) The eukaryote chaperonin CCT is a cold shock protein in *Saccharomyces cerevisiae*. Cell Stress Chaperones 7:47–54

Soto T, Beltran FF, Paredes V, Madrid M, Millar JB, Vicente-Soler J, Cansado J, Gacto M (2002) Cold induces stress-activated protein kinase-mediated response in the fission yeast *Schizosaccharomyces pombe*. Eur J Biochem 269:5056–5065

Suzuki I, Los DA, Kanesaki Y, Mikami K, Murata N (2000) The pathway for perception and transduction of low-temperature signals in *Synechocystis*. EMBO J 19:1327–1334

Suzuki I, Kanesaki Y, Mikami K, Kanehisa M, Murata N (2001) Cold-regulated genes under control of the cold sensor Hik33 in *Synechocystis*. Mol Microbiol 40:235–244

Tai SL, Daran-Lapujade P, Walsh MC, Pronk JT, Daran JM (2007) Acclimation of *Saccharomyces cerevisiae* to low temperature: a chemostat-based transcriptome analysis. Mol Biol Cell 18:5100–5112

Terao Y, Nakamori S, Takagi H (2003) Gene dosage effect of L-proline biosynthetic enzymes on L-proline accumulation and freeze tolerance in *Saccharomyces cerevisiae*. Appl Environ Microbiol 69:6527–6532

Tulha J, Lima A, Lucas C, Ferreira C (2010) *Saccharomyces cerevisiae* glycerol/H + symporter Stl1p is essential for cold/near-freeze and freeze stress adaptation. A simple recipe with high biotechnological potential is given. Microb Cell Fact 9:82

Vigh L, Maresca B, Harwood JL (1998) Does the membrane's physical state control the expression of heat shock and other genes? Trends Biochem Sci 23:369–374

Weber MH, Marahiel MA (2003) Bacterial cold shock responses. Sci Prog 86:9–75

Westfall PJ, Ballon DR, Thorner J (2004) When the stress of your environment makes you go HOG wild. Science 306:1511–1512

Willimsky G, Bang H, Fischer G, Marahiel MA (1992) Characterization of *cspB*, a *Bacillus subtilis* inducible cold shock gene affecting cell viability at low temperatures. J Bacteriol 174:6326–6335

Winzeler EA, Shoemaker DD, Astromoff A, Liang H, Anderson K, Andre B, Bangham R, Benito R, Boeke JD, Bussey H, Chu AM, Connelly C, Davis K, Dietrich F, Dow SW, El Bakkoury M, Foury F, Friend SH, Gentalen E, Giaever G, Hegemann JH, Jones T, Laub M, Liao H, Liebundguth N, Lockhart DJ, Lucau-Danila A, Lussier M, M'Rabet N, Menard P, Mittmann M, Pai C, Rebischung C, Revuelta JL, Riles L, Roberts CJ, Ross-MacDonald P, Scherens B, Snyder M, Sookhai-Mahadeo S, Storms RK, Veronneau S, Voet M, Volckaert G, Ward TR, Wysocki R, Yen GS, Yu K, Zimmermann K, Philippsen P, Johnston M, Davis RW (1999) Functional characterization of the *S. cerevisiae* genome by gene deletion and parallel analysis. Science 285:901–906

Xia B, Ke H, Shinde U, Inouye M (2003) The role of RbfA in 16S rRNA processing and cell growth at low temperature in *Escherichia coli*. J Mol Biol 332:575–584

Zhang L, Ohta A, Horiuchi H, Takagi M, Imai R (2001) Multiple mechanisms regulate expression of low temperature responsive (LOT) genes in *Saccharomyces cerevisiae*. Biochem Biophys Res Commun 283:531–535

Chapter 12
Production of Antifreeze Proteins by Cold-Adapted Yeasts

Hak Jun Kim, Jun Hyuck Lee, Hackwon Do and Woongsic Jung

Abstract Antifreeze proteins (AFPs) are proteins that have the ability to bind to ice crystals and inhibit the growth of ice. It is generally accepted that AFPs may protect cell membranes from freezing injury via the inhibition of ice recrystallization, thereby increasing the survival of psychrophilic microorganism at sub-zero temperatures. Screening of antifreeze activity showed that the occurrence of yeast AFPs was limited to basidiomycetous psychrophilic yeasts living under the thermal fluctuations. Currently only three species, *Glaciozyma antarctica*, *Glaciozyma* sp. AY30, and *Rhodotorula glacialis* are known to have AFPs or antifreeze activities. Of ten AFPs whose amino acid sequences are known, two AFPs are relatively well studied: AFP1 from *G. antarctica* and LeIBP from *Glaciozyma* sp. AY30. The amino acid sequence analysis showed that they are closely related with the fungus *Typhula ishikariensis* AFP, which clustered with other fungal, diatom, and bacterial AFPs to form new class of AFPs. The structural investigation of LeIBP revealed that it has an ice-binding motif (T-X-T motif) similar to those of fungal and insect AFPs. Real-time PCR of *Glaciozyma* AFPs exhibited that their expression is regulated by thermal fluctuation.

Keywords Antifreeze protein · Ice-binding protein · *Glaciozyma* · Horizontal gene transfer · Ice recrystallization inhibition

J. H. Lee · H. Do
Department of Polar Sciences, University of Science and Technology, Incheon 406-840, Republic of Korea

H. J. Kim · J. H. Lee
Division of Polar Life Sciences, Korea Polar Research Institute, Incheon 406-840, Republic of Korea

W. Jung
Department of Life Science, College of Natural Sciences, Hanyang University, Seoul 133-791, Republic of Korea

H. J. Kim (✉)
Department of Polar Sciences, Pukyong National University, Incheon 406-840, Republic of Korea
e-mail: hjkim@kopri.re.kr

P. Buzzini and R. Margesin (eds.), *Cold-Adapted Yeasts*,
DOI: 10.1007/978-3-642-39681-6_12, © Springer-Verlag Berlin Heidelberg 2014

Contents

12.1 Introduction 260
12.2 Antifreeze Proteins 261
12.3 AFPs of Psychrophilic Yeasts 263
12.3.1 Antifreeze Activity 263
12.3.2 Antifreeze Genes 265
12.3.3 AFPs 267
12.4 Commercial Application of Yeast AFPs 273
12.5 Conclusions 275
References 276

12.1 Introduction

Yeasts have been frequently spotted in samples from many icy environments of different geographical areas such as samples from Antarctica, the Arctic, Greenland, European Alps, Asian mountain ranges, and Patagonia in South America (Buzzini et al. 2012 and references therein) implying that yeasts are adapted well to these cold environments. Inhabiting the cryosphere demands numerous adjustments at the molecular and cellular levels to the organisms. These changes are increased membrane fluidity, cold-shock and cold-acclimation responses, reduction in the freezing point, protection against reactive oxygen species, production of cold-active enzymes and macromolecular cryoprotectants, etc. (Casanueva et al. 2010 and references therein). A comprehensive discussion of adjustments at the molecular and cellular levels in yeasts as response to cold is reported in the other Chapters of Part III. Interestingly, most of these mechanisms are adopted by most psychrophilic yeasts in order to thrive or survive in their habitats, but antifreeze activity is exhibited by only a few basidiomycetous yeasts. Numerous AFPs have been found in many psychrophilic organisms including bacteria (Raymond et al. 2007, 2008; Garnham et al. 2011; Do et al. 2012), diatoms (Raymond and Knight 2003; Janech et al. 2006; Raymond et al. 2009; Raymond 2011), plants (Worrall et al. 1998; Middleton et al. 2009, 2012), fungi (Hoshino et al. 2003a, b, 2009; Raymond and Janech 2009; Xiao et al. 2010b; Kondo et al. 2012), insect (Doucet et al. 2000a; Graether et al. 2000; Liou et al. 2000b; Leinala et al. 2002b; Graether and Sykes 2004; Mok et al. 2010), crustaceans (Rainer 2010), and fish (DeVries 1969, 1971, 1983, 1986). Antifreeze proteins (AFPs) are a group of proteins that have the ability to bind to ice crystals and inhibit the growth of ice. It is generally accepted that AFPs may protect cell membranes from freezing injury via the inhibition of ice recrystallization, thereby increasing the survival of psychrophilic microorganism at sub-zero temperatures (Raymond et al. 2007, 2008; Garnham et al. 2008). In this Chapter, we describe the distribution, function, structure, and regulation of yeast AFPs.

12.2 Antifreeze Proteins

AFPs possess two distinct activities: thermal hysteresis (TH) and ice recrystallization inhibition (IRI). These activities are all ascribed to their affinity for ice. Because of this, AFPs sometimes can be called ice-binding proteins (IBPs), which is a broad terminology including antifreeze proteins (AFPs), ice recrystallization inhibition proteins, and ice nucleation proteins. Particularly, AFP and IBP are interchangeable in yeast case. In this Chapter, we use the term AFP instead of IBP. TH can be defined as the gap between melting and freezing points. This phenomenon can be explained by an adsorption–inhibition mechanism of AFPs (Raymond and DeVries 1972, 1977). Briefly, AFP binds irreversibly onto the surface of ice, which makes the ice to grow outward in a curved (i.e., convex) manner between the bound AFPs (Fig. 12.1a). The curved growth increases the vapor pressure of the ice to that of the surrounding solution, which halts the further growth of the ice (Wilson 1993; Wilson and Leader 1995; Kristiansen and Zachariassen 2005). This unique protein–ice interaction lowers the freezing point of the solution in non-colligative manner. The temperature separation created by this phenomenon is TH. In addition, TH accompanies changes of ice-crystal morphologies depending on the binding specificity of AFPs (Fig. 12.1b). TH and ice-crystal morphology are examined by a nanoliter osmometer connected to a cold stage with the sample holder made of silver. The disk sample holder has six holes of 0.35 mm diameter. The cold stage is mounted on a light microscope equipped with a digital camera. A droplet of sample to be assayed is layered into a well filled with a mineral oil, eventually sample being enveloped in the oil. The sample holder is then placed on the stage and rapidly frozen down to −40 °C. After freezing, the temperature is raised slowly until a single ice crystal of 10–20 μm remains; this temperature is defined as the melting temperature of the solution. Then, the temperature is lowered again slowly, and the ice-crystal growth and morphology are examined. The temperature at which the rapid or sudden growth of ice crystal occurs is defined as the non-equilibrium freezing temperature.

Ice recrystallization (IR) refers to the growth of larger ice grains at the expense of smaller ones (Knight and Duman 1986; Knight et al. 1988, 1995). The larger ice crystals can burst compartments or cells, leading to cold damage and cell death. Because this process could occur rapidly by thermal fluctuation, it could be detrimental to the overwintering organisms. Extensive studies have shown that AFPs can inhibit this phenomenon (Fig. 12.1c) (Knight et al. 1984; Knight and Duman 1986; Worrall et al. 1998; Smallwood et al. 1999; Sidebottom et al. 2000). Ice recrystallization inhibition (IRI) of AFPs is thought to protect cell membranes from freezing injury (Raymond et al. 2007, 2008, 2009; John et al. 2009; Raymond and Janech 2009; Lauersen et al. 2011). This activity seems to be one of the potential freezing tolerance mechanisms and has been described in a number of freezing-tolerant soil bacteria (Walker et al. 2006; Wilson et al. 2006), Antarctic lake bacteria (Gilbert et al. 2004), Antarctic sea-ice bacteria (Raymond et al.

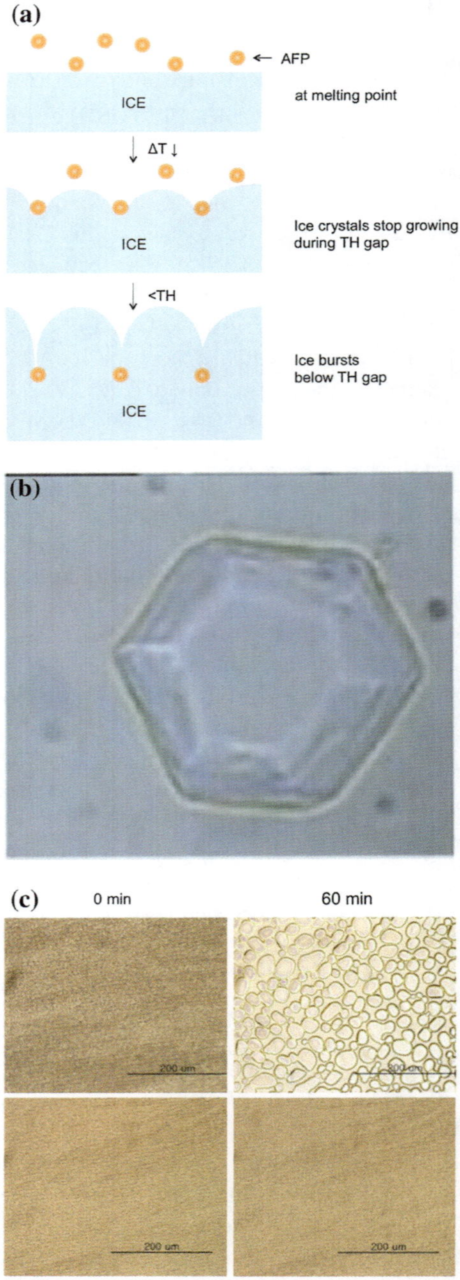

Fig. 12.1 **a** Inhibition of ice-crystal growth by adsorption of AFP (*filled circles*). AFPs bind to the ice front at equilibrium melting point. The ice–water interface during TH gap forms curvatures, leading to ice-growth inhibition by the Kelvin effect. As the temperature drops below TH, ice bursts. **b** Ice-crystal morphology observed in the presence of AFP (designated as LeIBP) from *Glaciozyma* sp. AY30. **c** Ice recrystallization inhibition (IRI) assay of 30 % sucrose solution in the absence (*upper panel*) and presence (*lower panel*) of LeIBP (1 mg ml^{-1}). The sample was flash frozen by lowering the temperature down to −80 °C at a rate of 90 °C min^{-1}, warmed up to −6 °C, and monitored for 1 h. Note that larger crystals were grown in the absence of AFP as a result of the recrystallization process in the *upper panel*, while no discernible changes took place in the *lower panel*

2007), bacteria from Vostok ice core (Raymond et al. 2008), cold-hardy plants (Griffith et al. 1992, 2005; Smallwood et al. 1999; Sidebottom et al. 2000), and snow molds (Snider et al. 2000; Hoshino et al. 2003a, b; Xiao et al. 2010a).

Measurement of ice recrystallization inhibition (RI) activity was first done by Knight et al. (1988) using a splat cooling assay. In this method, an ice disk containing extremely small ice crystals is generated by expelling a small solution droplet from a height of 3 m onto a very cold metal plate. The droplet freezes instantaneously as soon as hitting the metal plate. The ice disk is then transferred to a cold room at high sub-zero temperatures, and the recrystallization process is followed under the microscope over time. Since the limited availability of cold room facility, a modified version of this method was designed. The modified method uses a coverslip sandwich to generate thin ice disk and a cold stage such as a Linkam TMHS600 cold stage (Linkam Scientific Instruments, Surrey, UK) in place of a cold room (Smallwood et al. 1999). In this method, 4 μl of the sample solution, an equal volume mixture of a 60 % sucrose solution and AFP solution, is sandwiched between two circular 16-mm-diameter coverslips. The sandwich is pre-chilled to prevent frost on the surface at −1 °C for a couple of minutes and is transferred to the cooling stage. The temperature is rapidly decreased to −80 °C at a rate of 90 °C min^{-1} and maintained for 1 min. The sandwich is then warmed to −6 °C for 60 min to allow for ice recrystallization to occur.

12.3 AFPs of Psychrophilic Yeasts

12.3.1 Antifreeze Activity

Duman and Olsen (1993) demonstrated antifreeze activity for the first time in fungi. Since then about 200 strains of fungi have been screened for antifreeze activity, but only a handful of strains showed the activity (Snider et al. 2000; Kawahara et al. 2006; Lee et al. 2010). Of these strains tested, only fifteen species were yeasts, all of which belong to basidiomycetes. As summarized in Table 12.1, antifreeze activity was observed with only three different yeast species belonging to the genera *Glaciozyma* and *Rhodotorula* genus: *Glaciozyma antarctica*, *Glaciozyma* sp., and *Rhodotorula glacialis*. In the screening experiments, extracellular antifreeze activity was observed following morphological changes of ice as well as TH.

Lee et al. (2010) reported the first antifreeze activity from the psychrophilic yeast *Glaciozyma* sp. AY30 (formerly classified as *Leucosporidium*), isolated from an ice core sample of a freshwater pond near the Dasan station, Ny-Ålesund, Svalbard archipelago, Norway. Ice-binding ability in culture media was visualized by the ice-pitting method in which binding of AFPs or IBPs to ice leaves pits on the surface of ice plane. Further, the culture medium induced a distinct hexagonal ice-crystal shape with moderate TH of ca. 0.7 °C. The AFP was purified by the

Table 12.1 List of psychrophilic yeasts screened for their antifreeze activity

Basidiomycetous yeast species	Isolation source	Locality	Antifreeze activity	Genes/ proteins	Reference
Cryptococcus adeliensis	GIC[a]	MLG[c]	No	–	Singh et al. (2013)
Cryptococcus albidosimilis	GIC[a]	MLG[c]	No	–	Singh et al. (2013)
Cryptococcus saitoi	GIC[a]	MLG[c]	No	–	Singh et al. (2013)
Cryptococcus sp.	TM[b]	GWS[d]	No	–	Xiao et al. (2010a)
Cryptococcus sp.	GIC[a]	VRS[e]	No	–	Raymond et al. (2008)
Dioszegia sp.	TM[b]	GWS[d]	No	–	Xiao et al. (2010a)
Glaciozyma antarctica	TM[b]	GWS[d]	Yes (7 isolates)	NA	Xiao et al. (2010a)
Glaciozyma antarctica	Sea ice	CRS[f]	Yes	Both	Hashim et al. (2013)
Glaciozyma sp. AY30	Freshwater ice core	DRS[g]	Yes	Both	Lee et al. (2010)
Mrakia sp.	TM[b]	GWS[d]	No	–	Xiao et al. (2010a)
Rhodotorula glacialis	TM[b]	GWS[d]	Yes (2 isolates)	NA	Xiao et al. (2010a)
Rhodosporidium lusitaniae	GIC[a]	MLG[c]	No	–	Singh et al. (2013)
Rhodotoula mucilaginosa	GIC[a]	MLG[c]	No	–	Singh et al. (2013)
Rhodotorula sp.	GIC[a]	VRS[e]	No	–	Raymond et al. (2008)

[a] Glacier ice core; [b] Terrestrial materials (soil, mosses, algal mats); [c] Midre Lovenbreen glacier, Svalbard archipelago, Norway; [d] Great Wall Station, King George Island; Zhongshan Station, Prydz Bay; Soya Coast; [e] Vostok research station; [f] Casey research station; [g] Dasan research station; *NA* data not available

ice-affinity method (Kuiper et al. 2003) and it was confirmed that TH and IRI activities stemmed from itself. As described by Xiao et al. (2010a), ice-crystal morphologies modified by the culture medium of *G. antarctica* had a distorted hexagonal shape, very similar to the ice-crystal shape created by AFPs from the fungus *Typhula ishikariensis*. The culture media displayed weak TH of ca 0.22 °C. Further characterization of AFPs produced by this strain has not yet been reported. Very recently, another *Glaciozyma* yeast demonstrated antifreeze activity in its culture media (Hashim et al. 2013). The culture media of *G. antarctica* PI12 showed both TH and IRI activities. Compared to very closely related *Glaciozyma* sp. AY30, the TH value (0.1 °C) was lower and the ice morphology induced by this species was different. This may mean that *G. antarctica* produces extracellularly different AFPs compared to *Glaciozyma* sp. AY30. Another psychrophilic yeast known to exhibit antifreeze activity is *R. glacialis*. However, the morphological changes due to its culture medium were not described in detail.

Interestingly, the isolates from the glacier ice core did not exhibit any activity, while those from the Arctic freshwater ice, Antarctic sea ice, and Antarctic terrestrial materials did, despite the broad the geographical sampling area and isolation source of these psychrophilic yeasts were widely distributed. This may be related to the environmental fluctuations in temperature. In other words, isolates from the deep ice core may not experience strong temperature fluctuations in spite of the presence of sub-freezing temperatures, while isolates from Antarctic sea ice and terrestrial soil, and from Arctic freshwater ice may (Snider et al. 2000; Raymond et al. 2007, 2008, 2009; Amato et al. 2009; Rainer 2010; Singh et al. 2013). Even in the sea ice, there were significant thermal fluctuations depending on the vertical location and type of sea ice, which also can affect the community of organisms (Rainer 2010). Thermal fluctuations below sub-zero temperature can accelerate ice recrystallization, which is dangerous to the organisms living at sub-zero temperature. Recently, many evidences have shown that bacteria, microalgae, diatoms, and metazoans colonizing at sea ice have evolved IRI by AFPs as a freeze tolerance mechanism (Janech et al. 2006; Raymond et al. 2007, 2009; Lee et al. 2010). It has been suggested that these AFPs, instead of depressing the freezing point, play an important role in preventing cell damage from the recrystallization of extracellular ice because all AFPs from sea-ice organisms reported so far were secreted extracellularly (Raymond et al. 2007, 2009; Raymond and Kim 2012). Psychrophilic *Rhodotorula* species appear to support this hypothesis. Two isolates from severe cold climates with diurnal and seasonal temperature fluctuation possessed the antifreeze activity, but three species isolated from deep ice core did not (Amato et al. 2009; Xiao et al. 2010a). No yeasts living under permanently cold environments (such as prevailing in polar glaciers) had antifreeze activity, but we cannot rule out the possibility that some yeast species (*Cryptococcus* sp. and *Rhodotorula* sp.) may take advantage of commensal relationship with bacteria that had proteins with ice-binding affinity. Further investigations of the distribution of antifreeze activity of isolates from different geographical areas and sources are needed to fully grasp how AFPs have evolved.

Taken together, currently antifreeze activity of psychrophilic yeasts is confined only to basidiomycetous yeasts that inhabit cryosphere with thermal fluctuations.

12.3.2 Antifreeze Genes

To date, there is no psychrophilic yeast whose whole genome is fully sequenced. However, the genome sequencing project of *G. antarctica* PI12 (formerly known as *Leucosporidium antarcticum*) led by a Malaysian group is likely to be done shortly. Pyrosequencing of the *Glaciozyma* sp. AY30 genome was also conducted, producing 10,888 contigs of 1891-nt average length. AFP genes obtained from genome sequencing of *Glaciozyma* species are listed in Table 12.2: nine from *G. antarctica* and one from *Glaciozyma* sp. AY30 (Lee et al. 2010; Boo et al. 2013; Hashim et al. 2013). Since two species belong to the same genus, the Arctic

Table 12.2 List of known AFP sequences from psychrophilic yeasts and their accession numbers

Species	Accession no.	No. of AAs of ORF (aa)	Predicted MW (kDa)
Glaciozyma sp. AY30	GQ336994/ACU30806	261	26.8
G. antarctica PI12	GQ848368/ACX31168	177	18.2
	JF412502/AEG19527	261	26.5
	KC243139/AGE93831	242	24.4
	KC243140/AGE93832	259	26.7
	KC243141/AGE93833	260	27
	KC243142/AGE93834	342	34.6
	KC243143/AGE93835	365	37.2
	KC243144/AGE93836	461	48.2
	KC243145/AGE93837	705	68.1

species probably has more than one AFP in its genome. Therefore, we currently attempt to find AFP or AFP-like sequences from contigs of *Glaciozyma* sp. AY30 searching them against updated AFP sequences. Figure 12.2a shows positions of AFP domains in deduced AFP amino acid sequences. Eight Antarctic AFP genes (AGE19527, AGE93831, AGE93832, AGE93833, AGE93834, AGE93835, AGE93836, and AGE93837) encode the isoforms of well-characterized AFP (ACU30806) from Arctic origin (Lee et al. 2010). The Arctic AFP version is known to form a dimer via the C-terminal tail (see Sect. 12.3 in this Chapter). One (AGE93831) of the Antarctic isoforms encodes the C-terminal tail truncated version of AFP while three isoforms (AGE93831, AGE93832, and AGE93833) have the C-terminal tail. The other four isoforms (AGE93834, AGE93835, AGE93836, and AGE93837) contain one AFP-like domain with varying lengths of the C-terminus in which some harbor a T-X-T ice-binding motif (AGE93835) or glycine-rich region (AGE93837). The AFP1 (ACX31168) gene has a partial AFP domain. All deduced AFPs showed relatively high identities with AFPs of other fungi in the same phylum, and intriguingly with AFPs of sea-ice bacteria and diatoms belonging to other kingdoms (Lee et al. 2010; Hashim et al. 2013). This phylogenetic analysis and the restricted occurrence of yeast AFPs might be a good example of horizontal gene transfer (HGT). Recently, Raymond and Kim (2012) proposed HGT between sea-ice diatoms and bacteria in the sea ice, based on the phylogenetic analysis of many sea-ice AFPs, according to which the closest matches to the IBP genes from sea-ice algae are all bacterial genes, and the algal IBP genes analyzed so far lack introns. Rainer (2010) also suggested HGT of AFPs between diatoms and metazoans in the sea ice. It is not surprising that sea ice can provide a good environment for HGT, in which the donor and recipient dwell very closely, increasing the likelihood of HGT to occur. The growing body of evidence supports HGT between fungi, especially yeasts, and bacteria (Richards 2011). If more information on yeast genome sequence data is available, the question how yeast AFP genes were acquired will be addressed readily in near future.

Sequence Analysis and Expression of AFP Genes: The analysis of about 2-kb genomic DNA sequences from *Glaciozyma* sp. AY30 revealed the presence of 9 exons and 8 introns with an open reading frame (ORF) of 783 bp. The ORF encodes the 261-aa protein including a signal peptide at the N-terminus. The upstream flanking regions presented features slightly different from those frequently found in other yeast genes (Fig. 12.2b). A TATA box (ATATAA) located at position −292 relative to the ATG translation initiation codon is relatively farther upstream compared with other TATA boxes, which locate usually at around −70 (Dobson et al. 1982). A pyrimidine-rich tract, a feature shared by many efficiently expressed *Saccharomyces cerevisiae* genes, is also seen at positions −54 to −75. The promoter region contains a CAAT box (TCAATCT) about 434 bp upstream of the TATA box, which is also quite upstream. The partial AFP1 gene from *G. antarctica* PI12 showed a predicted ORF of 534 bp with no introns. A TATA box was predicted at position −50 and CAAT box at −136 (Fig. 12.2b). However, these elements do not conform to the general features observed in other yeast genes. Hence, we cannot speculate the regulatory effect of these promoter regions. We are currently attempting to delineate *cis*-elements in these regions.

Very recently, the expression of AFP1 and AFP4 (AGE93833) genes at different temperatures was observed quantitatively by RT-PCR experiment (Boo et al. 2013; Hashim et al. 2013). The expression of AFP1 in cells grown at near- and sub-freezing temperatures (0 and −12 °C) as well as at the maximal growth temperature (15 °C) was investigated. Its expression at 0 and −12 °C was two- to three-fold higher than that at 15 °C. The antifreeze activity was not detected or below detectable limit from the culture grown at 15 °C. Boo et al. (2013) also detected a 15-fold and 3-fold increase in expression of AFP4 (AGE93833) at 0 and 5 °C, respectively, compared to the expression level at 12 °C. At heat shock condition (22 °C), its expression was not observed (Boo et al. 2013). Under the cold-shock condition (0 and 5 °C), AFP4 was induced significantly, compared to other genes such as trehalose-6-phosphate synthase, trehalose-phophatase, glutaredoxin, catalase, implying that AFP4 is required for mitigating the cold stress. According to these results, it is clear that AFP genes are regulated by thermal stress.

12.3.3 AFPs

12.3.3.1 AFPs from *Glaciozyma* sp. AY30

Many structure–function studies have been performed in order to understand the antifreeze mechanism. Comparative studies have revealed that different types of AFPs have totally different primary, secondary, and tertiary structure, but they share a common ice-binding function. This diversity of AFPs is a result of evolution, where organisms have evolved an adaptation to cold and the ability to survive sub-zero temperatures. It is thought that the structural shape complementarity with ice-crystal lattice is a general feature of the ice-binding face of AFPs. More

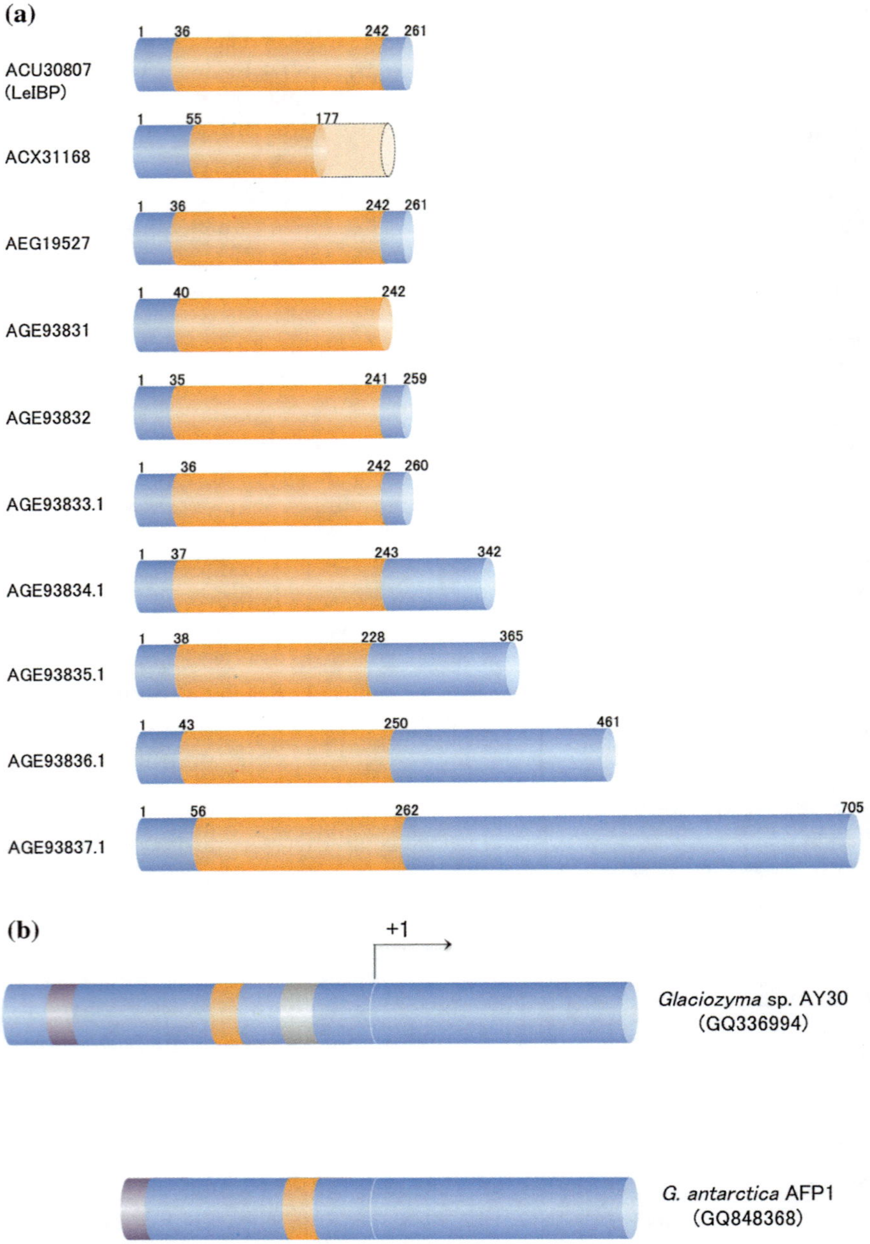

Fig. 12.2 **a** Schematic presentation of deduced AFPs from *Glaciozyma* species. Proteins are drawn to scale. All AFPs contain AFP or IBP domains (in *orange*). *Dotted line* means part of AFP domain is missing in aa sequence. **b** Gene structure of AFPs. Translation start site is denoted as +1. TATA boxes are displayed in *orange*, CAAT in *purple*, and pyrimidine-rich region in *gray*. See the text for details

specifically, the interactions between AFPs and ice are driven by hydrophobic interactions and are stabilized by hydrogen bonds formed between hydrophilic amino acids and ice-crystal lattice (Davies et al. 2002; Jia and Davies 2002).

Recently, a non-canonical AFP (designated as LeIBP, accession no. ACU30806) has been identified from the Arctic yeast *Glaciozyma* sp. AY30 (Lee et al. 2010). The *Glaciozyma* sp. AY30 is a cold-adapted psychrophilic yeast, isolated from an extremely cold environment (average temperature of the isolation source is below −10 °C in late April) of the frozen Tvillingvatnet, a freshwater pond near the Arctic Dasan station in Svalbard archipelago, Norway (Lee et al. 2010). LeIBP is composed of 261 amino acids and has N-terminal signal sequence and one N-glycosylation site. Consistent with the presence of signal sequence, LeIBP band (about 25 kDa) and TH activity were seen in the culture medium (Park et al. 2012). To better understand the ice-binding mechanism and the physiological role of glycosylation, a large amount of recombinant LeIBP was required. For this purpose, two expression systems have been established: one is pColdI-LeIBP, which expresses non-glycosylated bLeIBP in *Escherichia coli*, and the other is pPICZαA-LeIBP, which encodes glycosylated pLeIBP from the eukaryotic expression system of *Komagataella pastoris* (formerly *Pichia pastoris*). Purified recombinant LeIBPs contain apparent TH activity (0.32–0.42 °C at 400 μM) and could effectively inhibit the ice-crystal growth (Lee et al. 2012a; Park et al. 2012). Here, we summarize the optimized experimental procedures for native LeIBP production. Furthermore, a cold-adaptation mechanism model of LeIBP is proposed based on the crystal structure and biochemical studies.

Native LeIBP

Because native LeIBP is secreted outside the cell, it was purified from *Glaciozyma* sp. AY30 culture medium at 4 °C for 7 days in a 4-L flask containing 3 L of nutrient broth (NB). First, native LeIBP was purified using the ice-affinity method from the cell-free supernatant (Kuiper et al. 2003) and the ice fraction was melted and concentrated using tangential flow filtration equipped with a 5,000 MW cut off (Millipore, MA, USA). During the concentration, the protein sample was buffer-exchanged with distilled water. The purified native LeIBP exhibited strong RI and TH activity (1.38 °C at 290 μM). It should be noted that native LeIBP was co-purified with yellowish substances during the ice-affinity purification. These impurities were not completely removed through the iterative rounds of ice-affinity purification and buffer exchange step. This indicated that in addition to LeIBP, the Arctic yeast *Glaciozyma* sp. AY30 may secrete unknown ice-binding materials (Lee et al. 2012a; Park et al. 2012). Currently, we are trying to identify these impurities that enhanced the TH activity of native LeIBP. We will discuss this in detail below.

In the antifreeze activity assay of AFPs, two experimental techniques have been used, i.e., the ice recrystallization inhibition (IRI) assay and the thermal hysteresis assay (TH). Interestingly, ice crystals were needle or hexagonal shaped in the presence of LeIBP; however, in the absence of AFP, ice crystals were irregular round shape (Lee et al. 2010; Park et al. 2012). The antifreeze activities of recombinant LeIBPs were examined and compared to their antifreeze activity with the native LeIBP. Comparative analyses demonstrated that recombinant LeIBP showed similar recrystallization inhibition activity with the wild-type LeIBP although it has lower TH activity than the native LeIBP. There were clear differences in TH activity between native and recombinant LeIBPs. Recombinant LeIBPs show approximately 30 % decreased TH activity compared to the native LeIBP. We initially suspected that these different activities were due to the N-glycosylation, which occurs at Asn185. This result led to the production of glycosylated pLeIBP using the *K. pastoris* expression system. However, the TH activity of glycosylated pLeIBP was similar with non-glycosylated bLeIBP, indicating that glycosylation is not important for antifreeze activity and other factors may induce the increase in TH activity of native LeIBP. Interestingly, we have found that there was a TH activity variation in batch-by-batch sample of native LeIBP. During ice-affinity purification, some materials were carried over to the ice fraction and concentrated together with native LeIBP. These substances could not be removed completely during subsequent purification steps and appeared to enhance the TH activity of native LeIBP (Lee et al. 2012a; Park et al. 2012). Therefore, we proposed that *Glaciozyma* sp. AY30 might secret ice-interacting small molecules and they are capable of enhancing antifreeze activity of LeIBP. Secretion of enhancers, such as salt or polysaccharides, which accumulate at the brine channels in the ice may help to preserve the liquid environment. Moreover, increased antifreeze activity of LeIBP by enhancer molecules has a role in suppressing the growth of ice crystals in brine pockets (Fig. 12.3d). Another possible hypothesis is that *Glaciozyma* sp. AY30 may encode more than one AFP and these isoforms could work in concert. Currently, we do not know the exact molecular mechanism of boosting the antifreeze activity of LeIBP. The mystery of the antifreeze mechanism of *Glaciozyma* sp. AY30 is still under investigation.

Overall Structure of LeIBP

Previous structural studies have shown that different types of AFPs (fish AFP type I-IV, insect AFP and plant AFP) have completely different 3D structures but share a common antifreeze function, which is thought to be a result of convergent evolution (Ewart et al. 1999; Davies et al. 2002; Jia and Davies 2002). Recent structural analysis of the LeIBP revealed a dimeric right-handed β-helix fold consisting of a large coiled domain and a long helix $\alpha3$ (residues 96–115) insertion between $\beta5$ and $\beta6$. The amphipathic nature of helix $\alpha3$ enables hydrophobic interactions with the A face of the β-helical domain. Overall, the β-helical core structure is built from three flat surfaces (A, B and C faces) with dimensions of $\sim$50 Å $\times$ 30 Å $\times$ 20 Å and stabilized by both intra-molecular β-sheet hydrogen bonds and inner hydrophobic interactions (Fig. 12.3a). The β-helical core structure of LeIBP is similar with that of hyperactive AFPs including insect and bacterial AFPs (Graether et al. 2000; Graether and Sykes 2004; Garnham et al. 2011).

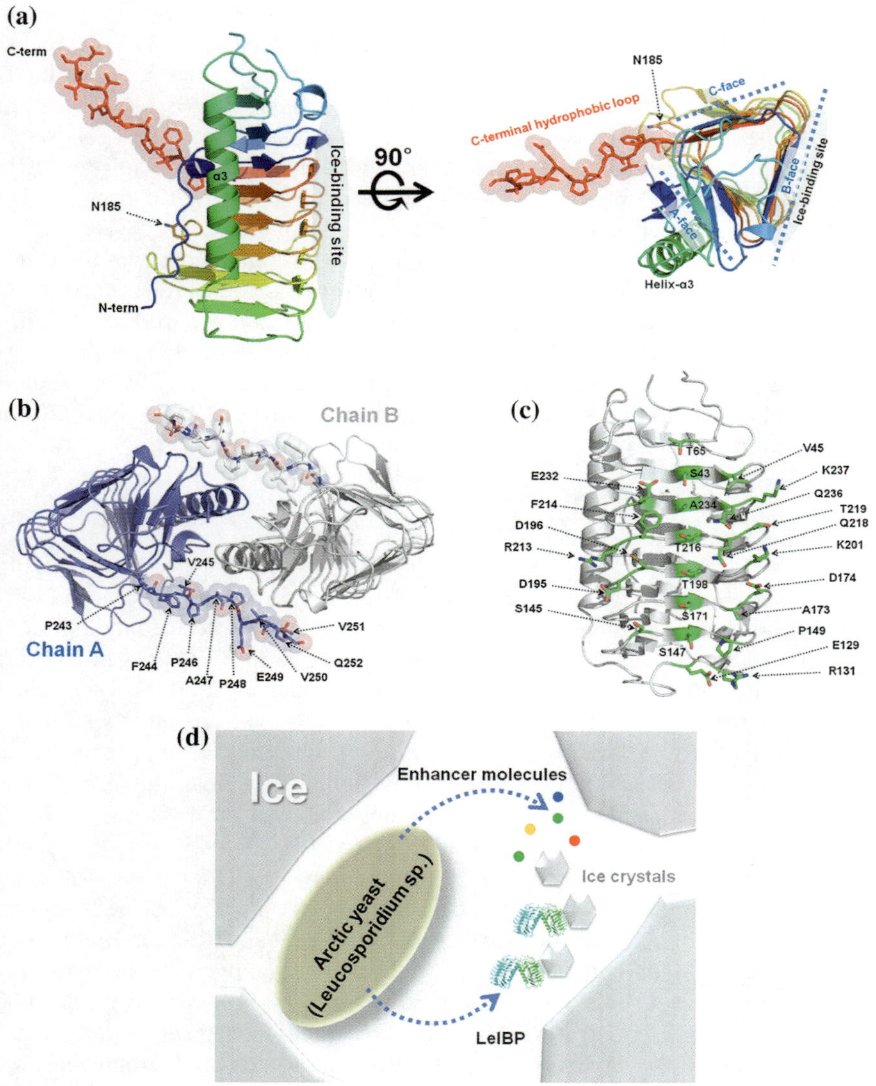

Fig. 12.3 Crystal structure and ice-binding site of LeIBP. **a** Overall structure of LeIBP monomer showing a right-handed β-helical fold. The N-glycosylation residue (Asn185) is marked with stick model. **b** Ribbon diagram of LeIBP dimer interface viewed along a non-crystallographic 2-fold axis. The two sub-units are colored slate *blue* and *gray*. The C-terminal hydrophobic loop residues (243-PFVPAPEVVQ-252) that form hydrophobic dimer interactions are drawn in stick representation. **c** The residues involved in ice binding are drawn in stick representation and labeled. **d** A putative cold-adaptation model of the yeast *Glaciozyma* sp. AY30. The antifreeze activity of LeIBP can be increased by enhancer molecules secreted from *Glaciozyma* sp. AY30 in brine channel

Unlike other known β-helical AFPs, LeIBP can exist and function as dimer. Notably, the C-terminal hydrophobic loop region (243-PFVPAPEVVQ-252) has an extended conformation pointing away from the body of the β-helical coiled domain and forms intertwined dimer interactions (Fig. 12.3b) (Lee et al. 2012a).

In addition, structural analysis of the glycosylated pLeIBP obviously shows the glycosylation at Asn185. The glycosylation site is exposed to the solvent region and is located at the interface between the two protomers but far from the ice-binding site (Lee et al. 2012a). This indicates that the N-linked glycosylation is not involved in the ice binding of LeIBP, which is in good agreement with the results of our antifreeze activity experiment. The glycosylated pLeIBP expressed in *K. pastoris* shows similar antifreeze activity with the non-glycosylated bLeIBP, but these two proteins show different stabilities. In our cross-linking experiments, bLeIBP was more easily oligomerized and aggregated than pLeIBP, which suggest that the glycosylation may contribute to the stability of LeIBP (Lee et al. 2012a; Park et al. 2012).

Ice-Binding Site and Ice-Binding Mechanism of LeIBP

Prior to structural studies, it was difficult to predict the ice-binding sites of LeIBP based on sequence alignment because there was no conserved ice-binding motif, such as T-X-T or T-X-N, which is present in hyperactive AFPs (Liou et al. 2000a, b; Graether et al. 2000; Graether and Sykes 2004; Garnham et al. 2011). Therefore, it was thought that the ice-binding site of LeIBP might be different from those of canonical hyperactive β-helical AFPs. Hyperactive β-helical AFPs have specific ice-crystal plane preferences and specific ice-binding motif, such as T-X-T or T-X-N. Structure-based point mutation studies demonstrated that the B-face residues are essential for antifreeze activity, indicating that B face is the ice-binding site of LeIBP. Comparisons between the LeIBP ice-binding site and the previously determined hyperactive AFP ice-binding site revealed that the ice-binding site residues of LeIBP are more complex and diverse (Fig. 12.3c) (Lee et al. 2012a). LeIBP does not have simple ice-binding motif compared to CfuAFP from the lepidopteron *Choristoneura fumiferana* (T-X-T motif) (Doucet et al. 2000b; Graether et al. 2000; Leinala et al. 2002a, b), TmoAFP from the coleopteron *Tenebrio molitor* (T-X-T motif) (Liou et al. 2000a, b), and MpAFP from the bacterium *Marinomonas primoryensis* (T-X-N motif) (Garnham et al. 2011). In addition, the ice-etching experiment indicates that LeIBP has multiple ice-binding sites and binds preferentially to the basal and primary prism planes. Thus, it is thought that LeIBP serves as a broad ice-binding site for ice-crystal interactions without a strict ice-binding motif arrangement (Lee et al. 2012a; Park et al. 2012).

12.3.3.2 AFP1 from *G. antarctica*

Malaysian scientists recently cloned and characterized a novel antifreeze protein, AFP1 (Hashim et al. 2013). This protein contains 177 amino acid residues with a predicted signal sequence (41 amino acids at the N-terminus) and pI of 5.45 (UniProtKB code D0EKL2). The amino acid sequence of AFP1 showed low similarity with other microbial antifreeze proteins, including 11 % identity with LeIBP from *Glaciozyma* sp. AY30 (UniProtKB code C7F6X3), 10 % identity with FfIBP from *Flavobacterium frigoris* PS1 (UniProtKB code H7FWB6), and 12 % identity with TisAFP from *T. ishikariensis* (UniProtKB code Q76CE2). The expression of recombinant AFP1 in *Escherichia coli* led to the insoluble aggregation called inclusion bodies. Thus, the recombinant AFP1 was refolded from inclusion bodies and purified using nickel affinity chromatography. The purified recombinant AFP1 was still able to exhibit AFP activity, although the activity was lower compared to the *G. antarctica* activity observed from its culture filtrates.

Secondary structure prediction shows that AFP1 consists of four α-helices and three β-strands. Shah et al. (2012) synthesized each helical peptide of AFP1 because solubility and refolding issues hampered obtaining large amount of soluble protein and evaluated the antifreeze activity of each segment using TH and RI assays. The peptides showed apparently RI activity and measurable TH activity. Peptide 1 m showed the highest activity with a TH value of 0.11 °C at 10 mM concentration. At this concentration, peptide 3 and peptide 4 m showed TH values of 0.06 °C and 0.07 °C, respectively. Moreover, the three-dimensional structures of all three peptides (PDB codes 2LQ0, 2LQ1, and 2LQ2) were determined by NMR method and the solution structure of peptides was characterized by a common long α-helical conformation like Type I AFPs (Fig. 12.4). Furthermore, the authors performed molecular dynamics experiment and produced ice-bound peptide models to investigate ice-binding residues.

12.4 Commercial Application of Yeast AFPs

AFPs are a useful material in commercial applications such as food industry (Payne et al. 1994; Griffith and Ewart 1995; Feeney and Yeh 1998), cryosurgery (Muldrew et al. 2001), and cryopreservation of red blood cells (Chao et al. 1996; Lee et al. 2012b), oocytes (Naidenko 1997; Jo et al. 2011, 2012), organs, and tissues (Lee et al. 1992; Tursman and Duman 1995; Koushafar and Rubinsky 1997; Amir et al. 2003). Food industry is a significant field where antifreeze proteins can be applied. AFPs can be added to various food items, such as ice cream to make a better texture and to frozen dough, fruits, and vegetables to keep longer shelf life. However, cost and allergy problems should be considered (Feeney and Yeh 1998; Crevel et al. 2002). To date, fish AFPs (Type I–IV and antifreeze glycopeptide) are used in most applications because they are extensively studied and safe from allergy. A detailed comparison of their structure and

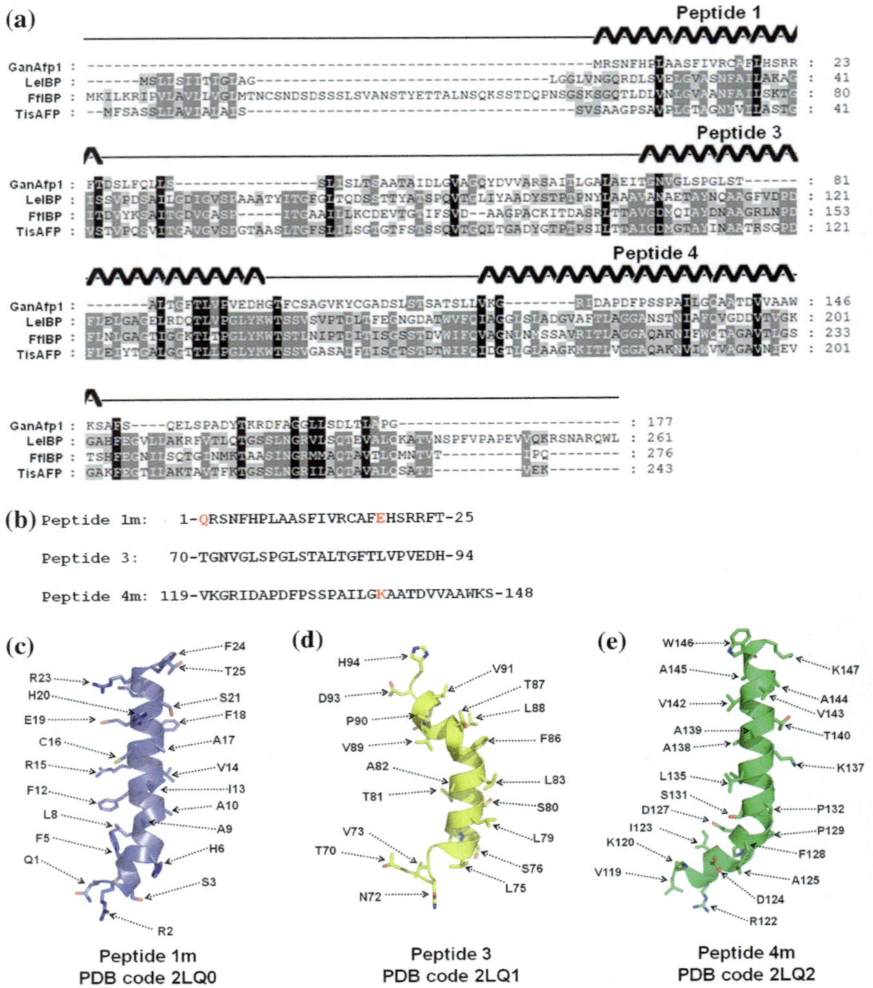

Fig. 12.4 NMR structures of synthetic peptides derived from *G. antarctica* Afp1 **a** The amino acid sequences of Gan Afp1 from *G.a antarctica*, LeIBP from *Glaciozyma* sp. AY30, FfIBP from *F. frigoris PS1* and Tis AFP from *T. ishikariensis* were aligned using ClustalX program (Thompson et al. 1997). Highly conserved residues are shaded *black* and similar residues are shaded *gray*. The presented α-helical regions are proposed by Shah et al. (2012) **b** Designed peptide sequences from the α-helical regions of Gan Afp1. The mutated amino acids in the designed peptide sequence are shown in *red*. **c–e** Ribbon diagrams of Peptide 1 m (PDB code 2LQ0, *blue*), Peptide 3 (PDB code 2LQ1, *yellow*), and Peptide 4 m (PDB code 2LQ2, *green*) structures

functional properties can be found in several reviews (Fletcher et al. 2001; Davies et al. 2002; Harding et al. 2003). Despite the usefulness of fish AFPs, there is still few precedent of large production of recombinant fish AFPs (Tyshenko et al. 2006).

The main reason for its failure is an issue of the low expression level and folding problem of fish AFPs. Because LeIBP does not contain any cysteine residue, it is more suitable for large-scale production without a folding problem. There are also several advantages of using recombinant yeast AFP compared to other types of AFPs for future commercial applications. Large-scale fermentation of recombinant yeast AFP in *K. pastoris* allows to minimize the expenses during the protein purification and also to continuously express AFP into the culture media by facilitating a well-established extracellular secretion system (Lee et al. 2013).

There are two types of cryoprotectant agents (CPAs), cell penetrating cryoprotectants (ethylene glycol, DMSO, and glycerol), and non-penetrating cryoprotectants (glucose, sucrose, trehalose, polyvinylpyrrolidone, and polyvinyl alcohol). In recent years, AFPs have attracted the interest of scientists as a new candidate for less toxic and non-penetrating CPAs, since they effectively inhibit ice growth at much lower concentrations than conventional chemical CPAs. Previous studies also have found that the AFPs have positive cryopreservation effects on carp sperm (Karanova et al. 1997), oyster oocytes (Naidenko 1997), mouse oocytes (Jo et al. 2011, 2012), bovine and porcine oocytes (Rubinsky et al. 1991), red blood cells (Carpenter and Hansen 1992; Chao et al. 1996), cell lines (Tursman and Duman 1995; Koushafar and Rubinsky 1997), and hypothermic storage of heart and liver (Lee et al. 1992; Rubinsky et al. 1994; Amir et al. 2003). Not surprisingly a positive effect has been observed on the cryopreservation of the human red blood cell (RBC) using LeIBP. The study showed that the addition of LeIBP significantly reduced freeze–thaw-induced hemolysis at either rapid (45 °C) or slow warming (22 °C) temperatures (Lee et al. 2012b). Cryopreservation of cells, tissues, sperms, and embryos is one of the most promising areas with commercial applications of AFPs to both medical and biotech industries (Carpenter and Hansen 1992).

12.5 Conclusions

The discovery of AFPs and their structure–function studies expand our knowledge on the mechanism of cold-adaptation in psychrophilic yeast. To our surprise, they were found only in yeasts residing in areas where severe thermal fluctuation exists. It is hypothesized that the AFP genes were acquired by HGT from bacteria or other eukaryotes coexisting in the confined habitats. However, not much is known about genes, regulation, function, and structure of yeast AFPs. Currently, AFPs from two *Glaciozyma* species are deposited in the public domain. To protect themselves from freezing and desiccation, cold-adapted yeasts synthesize and secrete not only polysaccharides and solutes but AFPs (Pavlova et al. 2009, 2011; Poli et al. 2010). Sometimes, these polysaccharides are loosely bound to the cell wall and make a polymer capsule. Recent studies have shown that the biosynthesis of AFPs is regulated by cold stress. In *Glaciozyma* sp. AY30, the co-secretion of LeIBP and enhancer molecules results in an increase in antifreeze activity of LeIBP

(Lee et al. 2012a; Park et al. 2012). Summarizing the structural and biochemical data, we propose a putative "co-secretion strategy" model as one of the cold-adaptation mechanisms of *Glaciozyma* sp. AY30 (Fig. 12.3d). Secreted LeIBP is localized outside the cell and LeIBP may extracellularly protect the cell membrane by preventing the recrystallization of extracellular ice. At the same time, secretion of enhancers such as salt or polysaccharides that accumulate in brine channels of ice help to preserve a liquid environment by increasing salinity or concentration of polysaccharides. In addition, the enhancers increase antifreeze activity of LeIBP, which results in an effective cold-adaptation mechanism allowing the psychrophilic yeast to survive at low temperatures. The continued study of the physiology of psychrophilic yeasts will provide valuable information on how the cells respond and adapt to low temperature stresses.

References

Amato P, Doyle S, Christner BC (2009) Macromolecular synthesis by yeasts under frozen conditions. Environ Microbiol 11:589–596

Amir G, Rubinsky B, Kassif Y, Horowitz L, Smolinsky AK, Lavee J (2003) Preservation of myocyte structure and mitochondrial integrity in subzero cryopreservation of mammalian hearts for transplantation using antifreeze proteins—an electron microscopy study. Eur J Cardiothorac Surg 24:292–297

Boo SY, Wong CMVL, Rodrigues KF, Najimudin N, Murad AMA, Mahadi NM (2013) Thermal stress responses in Antarctic yeast, Glaciozyma antarctica PI12, characterized by real-time quantitative PCR. Polar Biol 36(3):1–9

Buzzini P, Branda E, Goretti M, Turchetti B (2012) Psychrophilic yeasts from worldwide glacial habitats: diversity, adaptation strategies and biotechnological potential. FEMS Microbiol Ecol 82:217–241

Carpenter JF, Hansen TN (1992) Antifreeze protein modulates cell survival during cryopreservation: mediation through influence on ice crystal growth. Proc Natl Acad Sci USA 89:8953–8957

Casanueva A, Tuffin M, Cary C, Cowan DA (2010) Molecular adaptations to psychrophily: the impact of 'omic' technologies. Trends Microbiol 18:374–381

Chao H, Davies PL, Carpenter JF (1996) Effects of antifreeze proteins on red blood cell survival during cryopreservation. J Exp Biol 199(Pt 9):2071–2076

Crevel RW, Fedyk JK, Spurgeon MJ (2002) Antifreeze proteins: characteristics, occurrence and human exposure. Food Chem Toxicol 40:899–903

Davies PL, Baardsnes J, Kuiper MJ, Walker VK (2002) Structure and function of antifreeze proteins. Philos Trans R Soc Lond B Biol Sci 357:927–935

DeVries AL (1969) Freezing resistance in fishes of the Antarctic peninsula. Antarctic J US 4:104–105

Devries AL (1971) Glycoproteins as biological antifreeze agents in Antarctic fishes. Science 172:1152

DeVries AL (1983) Antifreeze peptides and glycopeptides in cold-water fishes. Annu Rev Physiol 45:245–260

DeVries AL (1986) Antifreeze glycopeptides and peptides: interactions with ice and water. Methods Enzymol 127:293–303

Do H, Lee JH, Lee SG, Kim HJ (2012) Crystallization and preliminary X-ray crystallographic analysis of an ice-binding protein (FfIBP) from *Flavobacterium frigoris* PS1. Acta Crystallogr Sect F Struct Biol Cryst Commun 68:806–809

Dobson MJ, Tuite MF, Roberts NA, Kingsman AJ, Kingsman SM, Perkins RE, Conroy SC, Fothergill LA (1982) Conservation of high efficiency promoter sequences in *Saccharomyces cerevisiae*. Nucleic Acids Res 10:2625–2637

Doucet CJ, Byass L, Elias L, Worrall D, Smallwood M, Bowles DJ (2000a) Distribution and characterization of recrystallization inhibitor activity in plant and lichen species from the UK and maritime Antarctic. Cryobiology 40:218–227

Doucet D, Tyshenko MG, Kuiper MJ, Graether SP, Sykes BD, Daugulis AJ, Davies PL, Walker VK (2000b) Structure-function relationships in spruce budworm antifreeze protein revealed by isoform diversity. Eur J Biochem 267:6082–6088

Duman JG, Olsen TM (1993) Thermal hysteresis protein activity in bacteria, fungi, and phylogenetically diverse plants. Cryobiology 30:322–328

Ewart K, Lin Q, Hew C (1999) Structure, function and evolution of antifreeze proteins. Cell Mol Life Sci 55:271–283

Feeney RE, Yeh Y (1998) Antifreeze proteins: current status and possible food uses. Trends Food Sci Technol 9:102–106

Fletcher GL, Hew CL, Davies PL (2001) Antifreeze proteins of teleost fishes. Annu Rev Physiol 63:359–390

Garnham CP, Gilbert JA, Hartman CP, Campbell RL, Laybourn-Parry J, Davies PL (2008) A Ca^{2+}-dependent bacterial antifreeze protein domain has a novel beta-helical ice-binding fold. Biochem J 411:171–180

Garnham CP, Campbell RL, Davies PL (2011) Anchored clathrate waters bind antifreeze proteins to ice. Proc Natl Acad Sci USA 108:7363–7367

Gilbert JA, Hill PJ, Dodd CE, Laybourn-Parry J (2004) Demonstration of antifreeze protein activity in Antarctic lake bacteria. Microbiology 150(Pt 1):171–180

Graether SP, Sykes BD (2004) Cold survival in freeze-intolerant insects: the structure and function of beta-helical antifreeze proteins. Eur J Biochem 271:3285–3296

Graether SP, Kuiper MJ, Gagne SM, Walker VK, Jia Z, Sykes BD, Davies PL (2000) Beta-helix structure and ice-binding properties of a hyperactive antifreeze protein from an insect. Nature 406:325–328

Griffith M, Ewart KV (1995) Antifreeze proteins and their potential use in frozen foods. Biotechnol Adv 13:375–402

Griffith M, Ala P, Yang DS, Hon WC, Moffatt BA (1992) Antifreeze protein produced endogenously in winter rye leaves. Plant Physiol 100:593–596

Griffith M, Lumb C, Wiseman SB, Wisniewski M, Johnson RW, Marangoni AG (2005) Antifreeze proteins modify the freezing process in planta. Plant Physiol 138:330–340

Harding MM, Anderberg PI, Haymet AD (2003) 'Antifreeze' glycoproteins from polar fish. Eur J Biochem 270:1381–1392

Hashim NHF, Bharudin I, Nguong DLS, Higa S, Bakar FDA, Nathan S, Rabu A, Kawahara H, Illias RM, Najimudin N (2013) Characterization of Afp1, an antifreeze protein from the psychrophilic yeast *Glaciozyma antarctica* PI12. Extremophiles 17:63–73

Hoshino T, Kiriaki M, Nakajima T (2003a) Novel thermal hysteresis proteins from low temperature basidiomycete, *Coprinus psychromorbidus*. Cryoletters 24:135–142

Hoshino T, Kiriaki M, Ohgiya S, Fujiwara M, Kondo H, Nishimiya Y, Yumoto I, Tsuda S (2003b) Antifreeze proteins from snow mold fungi. Can J Bot 81:1175–1181

Hoshino T, Xiao N, Tkachenko OB (2009) Cold adaptation in the phytopathogenic fungi causing snow molds. Mycoscience 50:26–38

Janech M, Krell A, Mock T, Kang J-S, Raymond J (2006) Ice-binding proteins from sea ice diatoms (Bacillariophyceae). J Phycol 42:410–416

Jia Z, Davies PL (2002) Antifreeze proteins: an unusual receptor-ligand interaction. Trends Biochem Sci 27:101–106

Jo JW, Jee BC, Lee JR, Suh CS (2011) Effect of antifreeze protein supplementation in vitrification medium on mouse oocyte developmental competence. Fertil Steril 96:1239–1245

Jo JW, Jee BC, Suh CS, Kim SH (2012) The beneficial effects of antifreeze proteins in the vitrification of immature mouse oocytes. PLoS one 7:e37043

John UP, Polotnianka RM, Sivakumaran KA, Chew O, Mackin L, Kuiper MJ, Talbot JP, Nugent GD, Mautord J, Schrauf GE, Spangenberg GC (2009) Ice recrystallization inhibition proteins (IRIPs) and freeze tolerance in the cryophilic Antarctic hair grass *Deschampsia antarctica* E. Desv Plant Cell Environ 32:336–348

Karanova M, Zsvetkova L, Petropavlov N (1997) Effect of antifreeze glycoproteins on quality of cryoconserved carp sperm. Biofizika 42:725–728

Kawahara H, Takemura T, Obata H (2006) Function analysis and screening of antifreeze material from fungi. Cryobiol Cryotechnol 52:151–155

Knight CA, Duman JG (1986) Inhibition of recrystallization of ice by insect thermal hysteresis proteins: a possible cryoprotective role. Cryobiology 23:256–262

Knight CA, DeVries AL, Oolman LD (1984) Fish antifreeze protein and the freezing and recrystallization of ice. Nature 308:295–296

Knight CA, Hallett J, DeVries AL (1988) Solute effects on ice recrystallization: an assessment technique. Cryobiology 25:55–60

Knight CA, Wen D, Laursen RA (1995) Nonequilibrium antifreeze peptides and the recrystallization of ice. Cryobiology 32:23–34

Kondo H, Hanada Y, Sugimoto H, Hoshino T, Garnham CP, Davies PL, Tsuda S (2012) Ice-binding site of snow mold fungus antifreeze protein deviates from structural regularity and high conservation. Proc Natl Acad Sci USA 109:9360–9365

Koushafar H, Rubinsky B (1997) Effect of antifreeze proteins on frozen primary prostatic adenocarcinoma cells. Urology 49:421–425

Kristiansen E, Zachariassen KE (2005) The mechanism by which fish antifreeze proteins cause thermal hysteresis. Cryobiology 51:262–280

Kuiper MJ, Lankin C, Gauthier SY, Walker VK, Davies PL (2003) Purification of antifreeze proteins by adsorption to ice. Biochem Biophys Res Commun 300:645–648

Lauersen KJ, Brown A, Middleton A, Davies PL, Walker VK (2011) Expression and characterization of an antifreeze protein from the perennial rye grass, *Lolium perenne*. Cryobiology 62:194–201

Lee C, Rubinsky B, Fletcher G (1992) Hypothermic preservation of whole mammalian organs with antifreeze proteins. Cryo-Letters 13:59–66

Lee JK, Park KS, Park S, Park H, Song YH, Kang SH, Kim HJ (2010) An extracellular ice-binding glycoprotein from an Arctic psychrophilic yeast. Cryobiology 60:222–228

Lee JH, Park AK, Do H, Park KS, Moh SH, Chi YM, Kim HJ (2012a) Structural basis for antifreeze activity of ice-binding protein from arctic yeast. J Biol Chem 287:11460–11468

Lee SG, Koh HY, Lee JH, Kang SH, Kim HJ (2012b) Cryopreservative effects of the recombinant ice-binding protein from the arctic yeast *Leucosporidium* sp. on red blood cells. Appl Biochem Biotechnol 167:824–834

Lee JH, Lee SG, Do H, Park JC, Kim E, Choe YH, Han SJ, Kim HJ (2013) Optimization of the pilot-scale production of an ice-binding protein by fed-batch culture of *Pichia pastoris*. Appl Microbiol Biotechnol 97:3383–3393

Leinala EK, Davies PL, Doucet D, Tyshenko MG, Walker VK, Jia Z (2002a) A beta-helical antifreeze protein isoform with increased activity. Structural and functional insights. J Biol Chem 277:33349–33352

Leinala EK, Davies PL, Jia Z (2002b) Crystal structure of beta-helical antifreeze protein points to a general ice binding model. Structure 10:619–627

Liou YC, Daley ME, Graham LA, Kay CM, Walker VK, Sykes BD, Davies PL (2000a) Folding and structural characterization of highly disulfide-bonded beetle antifreeze protein produced in bacteria. Protein Expr Purif 19:148–157

Liou YC, Davies PL, Jia Z (2000b) Crystallization and preliminary X-ray analysis of insect antifreeze protein from the beetle Tenebrio molitor. Acta Crystallogr D Biol Crystallogr 56(Pt 3):354–356

Middleton AJ, Brown AM, Davies PL, Walker VK (2009) Identification of the ice-binding face of a plant antifreeze protein. FEBS Lett 583:815–819

Middleton AJ, Marshall CB, Faucher F, Bar-Dolev M, Braslavsky I, Campbell RL, Walker VK, Davies PL (2012) Antifreeze protein from freeze-tolerant grass has a beta-roll fold with an irregularly structured ice-binding site. J Mol Biol 416:713–724

Mok YF, Lin FH, Graham LA, Celik Y, Braslavsky I, Davies PL (2010) Structural basis for the superior activity of the large isoform of snow flea antifreeze protein. Biochemistry 49:2593–2603

Muldrew K, Rewcastle J, Donnelly BJ, Saliken JC, Liang S, Goldie S, Olson M, Baissalov R, Sandison G (2001) Flounder antifreeze peptides increase the efficacy of cryosurgery. Cryobiology 42:182–189

Naidenko T (1997) Cryopreservation of Crassostrea gigas oocytes, embryos and larvae using antioxidant echinochrome A and antifreeze protein AFP1. Cryo-Lett 18:375–382

Park KS, Do H, Lee JH, Park SI, Kim EJ, Kim SJ, Kang SH, Kim HJ (2012) Characterization of the ice-binding protein from Arctic yeast *Leucosporidium* sp. AY30. Cryobiology 64:286–296

Pavlova K, Panchev I, Krachanova M, Gocheva M (2009) Production of an exopolysaccharide by Antarctic yeast. Folia Microbiol 54:343–348

Pavlova K, Rusinova-Videva S, Kuncheva M, Kratchanova M, Gocheva M, Dimitrova S (2011) Synthesis and characterization of an exopolysaccharide by antarctic yeast strain *Cryptococcus laurentii* AL 100. Appl Biochem Biotechnol 163:1038–1052

Payne SR, Oliver JE, Upreti GC (1994) Effect of antifreeze proteins on the motility of ram spermatozoa. Cryobiology 31:180–184

Poli A, Anzelmo G, Tommonaro G, Pavlova K, Casaburi A, Nicolaus B (2010) Production and chemical characterization of an exopolysaccharide synthesized by psychrophilic yeast strain *Sporobolomyces salmonicolor* AL 1 isolated from Livingston Island, Antarctica. Folia Microbiol 55(6):576–581

Rainer K (2010) Acquisition of freeze protection in a sea-ice crustacean through horizontal gene transfer? Polar Biol 33:543–556

Raymond JA (2011) Algal ice-binding proteins change the structure of sea ice. Proc Natl Acad Sci USA 108:E198

Raymond JA, DeVries AL (1972) Freezing behavior of fish blood glycoproteins with antifreeze properties. Cryobiology 9:541–547

Raymond JA, DeVries AL (1977) Adsorption inhibition as a mechanism of freezing resistance in polar fishes. Proc Natl Acad Sci USA 74:2589–2593

Raymond JA, Janech MG (2009) Ice-binding proteins from enoki and shiitake mushrooms. Cryobiology 58:151–156

Raymond JA, Kim HJ (2012) Possible role of horizontal gene transfer in the colonization of sea ice by algae. PLoS one 7:e35968

Raymond JA, Knight CA (2003) Ice binding, recrystallization inhibition, and cryoprotective properties of ice-active substances associated with Antarctic sea ice diatoms. Cryobiology 46:174–181

Raymond JA, Fritsen C, Shen K (2007) An ice-binding protein from an Antarctic sea ice bacterium. FEMS Microbiol Ecol 61:214–221

Raymond JA, Christner BC, Schuster SC (2008) A bacterial ice-binding protein from the Vostok ice core. Extremophiles 12:713–717

Raymond JA, Janech M, Fritsen C (2009) Novel ice-binding proteins from a psychrophilic antarctic alga (Chlamydomonadaceae, Chlorophyceae). J Phycol 45:130–136

Richards TA (2011) Gene transfer into the fungi. Fungal Biol Rev 25:98–110

Rubinsky B, Arav A, Fletcher GL (1991) Hypothermic protection–a fundamental property of "antifreeze" proteins. Biochem Biophys Res Commun 180:566–571

Rubinsky B, Arav A, Hong JS, Lee CY (1994) Freezing of mammalian livers with glycerol and antifreeze proteins. Biochem Biophys Res Commun 200:732–741

Shah SH, Kar RK, Asmawi AA, Rahman MB, Murad AM, Mahadi NM, Basri M, Rahman RN, Salleh AB, Chatterjee S, Tejo BA, Bhunia A (2012) Solution structures, dynamics, and ice growth inhibitory activity of peptide fragments derived from an antarctic yeast protein. PLoS one 7:e49788

Sidebottom C, Buckley S, Pudney P, Twigg S, Jarman C, Holt C, Telford J, McArthur A, Worrall D, Hubbard R, Lillford P (2000) Heat-stable antifreeze protein from grass. Nature 406:256

Singh P, Tsuji M, Singh SM, Roy U, Hoshino T (2013) Taxonomic characterization, adaptation strategies and biotechnological potential of cryophilic yeasts from ice cores of Midre Lovenbreen glacier, Svalbard, Arctic. Cryobiology 66:167–175

Smallwood M, Worrall D, Byass L, Elias L, Ashford D, Doucet CJ, Holt C, Telford J, Lillford P, Bowles DJ (1999) Isolation and characterization of a novel antifreeze protein from carrot (*Daucus carota*). Biochem J 340(Pt 2):385–391

Snider CS, Hsiang T, Zhao G, Griffith M (2000) Role of ice nucleation and antifreeze activities in pathogenesis and growth of snow molds. Phytopathology 90:354–361

Thompson JD, Gibson TJ, Plewniak F, Jeanmougin F, Higgins DG (1997) The CLUSTAL_X windows interface: flexible strategies for multiple sequence alignment aided by quality analysis toos. Nucleic Acids Res 25:4876–4882

Tursman D, Duman JG (1995) Cryoprotective effects of thermal hysteresis protein on survivorship of frozen gut cells from the freezetolerant centipede *Lithobius forficatus*. J Exp Zool 272:249–257

Tyshenko MG, d'Anjou M, Davies PL, Daugulis AJ, Walker VK (2006) Challenges in the expression of disulfide bonded, threonine-rich antifreeze proteins in bacteria and yeast. Protein Expr Purif 47:152–161

Walker VK, Palmer GR, Voordouw G (2006) Freeze-thaw tolerance and clues to the winter survival of a soil community. Appl Environ Microbiol 72:1784–1792

Wilson P (1993) Explaining thermal hysteresis by the Kelvin effect. Cryo-Letters 14:31–36

Wilson PW, Leader JP (1995) Stabilization of supercooled fluids by thermal hysteresis proteins. Biophys J 68:2098–2107

Wilson SL, Kelley DL, Walker VK (2006) Ice-active characteristics of soil bacteria selected by ice-affinity. Environ Microbiol 8:1816–1824

Worrall D, Elias L, Ashford D, Smallwood M, Sidebottom C, Lillford P, Telford J, Holt C, Bowles D (1998) A carrot leucine-rich-repeat protein that inhibits ice recrystallization. Science 282:115

Xiao N, Inaba S, Tojo M, Degawa Y, Fujiu S, Hanada Y, Kudoh S, Hoshino T (2010a) Antifreeze activities of various fungi and *Stramenopila* isolated from Antarctica. North American Fungi 5:215–220

Xiao N, Suzuki K, Nishimiya Y, Kondo H, Miura A, Tsuda S, Hoshino T (2010b) Comparison of functional properties of two fungal antifreeze proteins from *Antarctomyces psychrotrophicus* and *Typhula ishikariensis*. FEBS J 277:394–403

Chapter 13
Role of Sterol Metabolism and Endoplasmic Reticulum-Associated Degradation of Proteins in Cold Adaptation of Yeasts

Kelaine Haas

Abstract The environmental conditions of organisms have significant impact on their ability to grow and thrive. This chapter summarizes recent research related to the understanding of cellular mechanisms necessary for psychrophilic yeasts to overcome molecular challenges present in low-temperature environments. In an effort to identify adaptive mechanisms that allow psychrophilic yeasts to thrive in cold environments, a high number of studies have been done to investigate the low-temperature response of *Saccharomyces cerevisiae*. The focus of this chapter is on evidence that the endoplasmic reticulum-associated degradation (ERAD) pathway and membrane metabolism regulation are connected to the cellular response and adaptation to low temperature in yeasts. Specific consideration is given to three genes in the ERAD pathway, *UBC7*, *CUE1*, and *DOA10*, which play a role in cold adaptation and regulating membrane composition in *S. cerevisiae*.

Keywords Membrane · *Saccharomyces cerevisiae* · Psychrophiles · Sterol · Cold · ERAD · *UBC7*

Contents

13.1 Introduction 282
13.2 Endoplasmic Reticulum-Associated Degradation 283
13.2.1 The ERAD Pathway Regulates HMG-CoA Reductase 284
13.2.2 Mutations in the ERAD Pathway Impact Membrane Biogenesis 285
13.3 Membrane Sterol Composition and Low Temperatures 287
13.4 Conclusions 289
References 290

K. Haas (✉)
Department of Genetics, Cell Biology and Development, University of Minnesota, 420 Washington Ave. SE 3-104 MCB, Minneapolis, MN 55455, USA
e-mail: khaas@umn.edu

P. Buzzini and R. Margesin (eds.), *Cold-Adapted Yeasts*,
DOI: 10.1007/978-3-642-39681-6_13,

13.1 Introduction

As an unavoidable part of their existence, most organisms encounter changing environmental conditions that inflict a variety of stresses. Immediate response and long-term adaptation to these stressors enable organisms to survive both transient (e.g., daily) and longer-term (e.g., seasonal) changes in their environment. These environmental stressors include any condition that interferes with optimal growth like variations in temperature, decreased nutrient availability, hyperosmolarity, and oxygen limitation. The molecular responses to environmental stresses have been studied in organisms ranging from bacteria to mammals (Welch and Suhan 1985; Posas et al. 2000; Gasch and Werner-Washburne 2002; Saito and Posas 2012; Simpson and Ashe 2012). Many of these studies focus on highly conserved molecular responses that organisms utilize due to exposure to high temperatures (reviewed in Richter et al. 2010). Some organisms must activate these conserved molecular mechanisms in response to temporary exposure to elevated temperatures; other thermophilic organisms thrive in environments such as geothermal geysers and deep-sea thermal vents, where temperatures are routinely greater than 100 °C. The ability of thermophilic organisms to grow at high temperatures represents evolutionary adaptations to high temperatures arising through natural selection. Although much is known about transient responses to high temperature, understanding the molecular basis of evolutionary adaptations to extreme temperatures is a topic of ongoing research (Brock 1997; Stetter 2006; Atomi et al. 2011). At the other temperature extreme, a diverse group of microorganisms known as psychrophiles, colonize environments well below freezing temperatures. A discussion on the concept of psychrophiles and psychrotolerance is reported in Chap. 1. Although it is easy to assume that low-temperature environments make up only a marginal part of our biosphere, more than 80 % of the biosphere is exposed to temperatures below 5 °C (Cavicchioli et al. 2000; Margesin and Miteva 2011). These environments occur throughout the planet, including the deep sea, permafrost, and glaciers (Margesin and Miteva 2011). Psychrophiles are found in a broad range of both prokaryotic and eukaryotic phyla, including microorganisms isolated from cold environments around the world. Although these broad distributions suggest that they play a vital role in the evolution and function of planetary ecosystems, we are only recently beginning to understand the cellular mechanisms necessary for cells to overcome challenges presented in low-temperature environments.

A large number of psychrophilic microorganisms isolated from cold environments around the world represent different yeast species from many genera, including *Candida*, *Cryptococcus*, *Leucosporidium*, and many more that were summarized in a recent review of psychrophilic yeasts from Antarctica to South America to Europe (Buzzini et al. 2012). Besides, more in-depth data on the diversity and ecology of cold-adapted (psychrophilic and psychrotolerant) yeasts in worldwide cold habitats are reported in Chaps. 3–8. While these cold-adapted yeast species can be found throughout the globe, we only now are beginning to understand the mechanisms necessary for cells to overcome the challenges presented by low-temperature environments. By studying the response of *Saccharomyces cerevisiae* to short-term and long-term exposure to low temperatures, we may be able to identify adaptive mechanisms that allow psychrophiles to thrive in cold environments. Although *S. cerevisiae* grows best at temperatures around 25 °C and is classified as a mesophile, it can grow at much colder temperatures (Willey et al. 2008). *S. cerevisiae* may encounter these low temperatures during daily or seasonal changes in its natural environment as well as during temporary and long-term storage in the laboratory. As mentioned before, these low temperatures can cause environmental stress that interferes with its optimal growth. Under these conditions, *S. cerevisiae* may activate downstream molecular responses similar or identical to those required for evolutionary adaptation of psychrophilic yeast species to the cold.

13.2 Endoplasmic Reticulum-Associated Degradation

Organisms have evolved molecular responses and cellular mechanisms to overcome the deleterious effects of environmental stressors. For example, when yeasts are exposed to high temperatures, they activate the heat-shock response to combat the protein misfolding, cytoskeletal collapse, and membrane disruption experienced by the cell (Swan and Watson 1998; Richter et al. 2010). Similarly, when yeasts experience stressors that affect protein folding or require the production of endoplasmic reticulum (ER)-targeted proteins at a high rate, accumulation of misfolded proteins activates the unfolded protein response (UPR). The UPR is a highly conserved signaling pathway that monitors the condition of the ER, senses folding stress within the ER lumen, and then communicates the status of the ER to the cell's transcriptional machinery (reviewed in Ron and Walter 2007; Walter and Ron 2011). One of the major responses of the UPR is the transcriptional up-regulation of the Kar2 chaperone protein, which is a member of the Hsp70 family that relieves the folding load of the ER when it is under stress (Normington et al. 1989; Nicholson et al. 1990). If the increased levels of Kar2p are unable to overcome the accumulation of misfolded proteins, the endoplasmic reticulum-associated degradation (ERAD) pathway is activated to target the aberrant proteins for degradation (Travers et al. 2000). Along with misfolded proteins, purposefully short-lived peptides in the ER membrane or lumen are also subject to degradation by the ERAD pathway. ERAD substrates are polyubiquitinated, removed from the ER membrane or lumen, and degraded by the proteasome (Brodsky and McCracken 1999; Hampton 2002; McCracken and Brodsky 2003; Claessen et al. 2012). The ER, which is present in all eukaryotic cells, is a large organelle composed of a membrane and enclosed lumen. Major functions of the ER include lipid and protein biosynthesis. The lipids synthesized in the ER contribute to the plasma membrane and other organelle membranes within the cell, including the Golgi apparatus, the lysosome, peroxisomes, and mitochondria. Significant advances in understanding the function of ER protein quality control mechanisms, and the nuances of the ERAD pathway, have been made using *S. cerevisiae* (Wolf and Fink 1975; Egner et al. 1998; Vashist and Ng 2004; Metzger et al. 2008).

13.2.1 The ERAD Pathway Regulates HMG-CoA Reductase

In the past 30 years, insights into the balance between protein synthesis, functions of the ERAD pathway, and potential ERAD targets have suggested that regulation of the ERAD pathway is complex and impacts many molecular pathways in the cell. For example, increased levels of some ER membrane proteins induce dramatic changes in the synthesis and organization of the ER. A well-characterized example of such protein-induced ER biogenesis occurs when cells express an increased level of the ER-anchored enzyme HMG-CoA reductase (3-hydroxy-3-methyl-glutaryl-CoA reductase or HMGR) (Brown and Goldstein 1980; Anderson et al. 1983; Wright et al. 1988; Goldstein and Brown 1990; Wright et al. 1990). In yeasts, HMGR is encoded by two genes, *HMG1* and *HMG2*, which are the product of a genetic duplication (Basson et al. 1986). HMGR is a highly conserved enzyme with homologs found in eukaryotes, prokaryotes and archaea (Lum and Wright 1995; Loftus et al. 1997) which controls the rate-limiting step in ergosterol biosynthesis in yeasts by converting HMG-CoA to mevalonic acid (Siperstein and Fagan 1966; Nakanishi et al. 1988). Ergosterol is structurally similar to cholesterol, the sterol equivalent found in mammalian cells, which contributes to the function and structure of cellular membranes, and thus plays a role in molecular pathways, protein function, and membrane fluidity. The ergosterol biosynthesis pathway is tightly regulated largely by feedback mechanisms that regulate the amount of HMGR protein available in the cell (Hampton and Rine 1994). A number of these feedback mechanisms involve the regulation of HMGR at levels of transcription, translation, post-translational modification, and degradation (Goldstein and Brown 1990; Goldstein et al. 2006). The targeted degradation of HMGR occurs when members of the ERAD pathway ubiquitinate Hmg1p and Hmg2p to moderate the flux through the ergosterol biosynthetic pathway (Hampton et al. 1996a; Burg and Espenshade 2011). It has also been shown that a 10-fold increase in the level of HMGR induces biogenesis of smooth ER membrane arrays in *S. cerevisiae*; these morphologically unique arrays are called karmellae (Lorenz and Parks 1987; Wright et al. 1988; Lum and Wright 1995; Parrish et al. 1995; Koning et al. 1996). Work done by Wright (1993) took advantage of the HMGR-induced karmellae biogenesis model to explore candidate genes related to ER membrane biogenesis. To better understand which genetic pathways play a role in adaptation to physical changes that affect karmellae biogenesis, Wright et al. (2003) used a competitive growth screen in *S. cerevisiae* to identify genes involved in ER membrane biogenesis. By screening a deletion collection to isolate karmellae-sensitive single mutants, the authors identified genes necessary for cells to grow under karmellae-inducing conditions.

Surprisingly, the results of this screen determined that several genes in the ERAD pathway play a role in adaptation to both chemical and physical changes to the ER membrane (Wright et al. 2003). The results from this work demonstrate a function for ERAD in ER membrane biogenesis; in the process of exploring the function of ERAD in ER membrane biogenesis, the authors serendipitously discovered that some ERAD mutants are cold sensitive. This observation was the first described phenotypic consequence of ERAD mutants and contributed to a hypothesis that ER biogenesis and cold adaptation are functionally linked (Robin Wright, personal communication).

13.2.2 Mutations in the ERAD Pathway Impact Membrane Biogenesis

The karmellae-sensitive screen mentioned above-identified *UBC7* as a strong candidate gene necessary for ER biogenesis. Along with the *ubc7*Δ mutant, two more ERAD mutants were cold sensitive and linked to ER membrane biogenesis, *cue1*Δ and *doa10*Δ. This work indicated that *UBC7*, *CUE1*, and *DOA10*, three genes in the ERAD pathway, are required for *S. cerevisiae* to properly regulate ER membrane biogenesis and to adapt to low-temperature growth (Wright et al. 2003; Loertscher et al. 2006). The protein products of these three genes play a specific role in targeting the substrates of the ERAD pathway for degradation. The proteasomal degradation of ERAD targets is triggered by their polyubiquitination. Conjugation of a single ubiquitin molecule to the target protein requires the enzymatic activity of E1 ubiquitin-activating enzymes, E2 ubiquitin-conjugating enzymes, and E3 ubiquitin ligases (Fig. 13.1a). In *S. cerevisiae*, the E1 enzyme Uba1p activates a single ubiquitin molecule and then passes it to an E2 enzyme. In cooperation with the E1 and E2 enzymes, an E3 enzyme covalently attaches the ubiquitin monomer to its final target substrate (Hershko and Ciechanover 1992; Hershko and Ciechanover 1998; Pickart and Eddins 2004). Although the yeast genome encodes only one E1 enzyme, multiple E2 and E3 enzymes are synthesized in *S. cerevisiae* and unique combinations of these components determine which substrate will be targeted for proteasomal degradation (Haas and Siepmann 1997; Hampton 2002). The ERAD ubiquitination pathway requires two E2 enzymes: Ubc6p, a C-terminally anchored ER membrane protein, and Ubc7p, a peripheral ER membrane protein. Ubc7p is anchored to the ER by the ER membrane protein Cue1p (Hampton 2002; Vashist and Ng 2004) (Fig. 13.1b). Along with the Ubc6 and Ubc7 E2 enzymes, two primary ERAD E3 ubiquitin ligases, Hrd1p and Doa10p, are responsible for the majority of ER-associated proteasomal degradation (Hampton 2002; Vembar and Brodsky 2008; Hirsch et al. 2009). Members of the ERAD pathway are involved in a variety of cellular stress responses ranging from heat tolerance, H_2O_2 stress, and canavanine (Hiraishi et al. 2006). This range of effects implies that proteins of the ERAD pathway may be

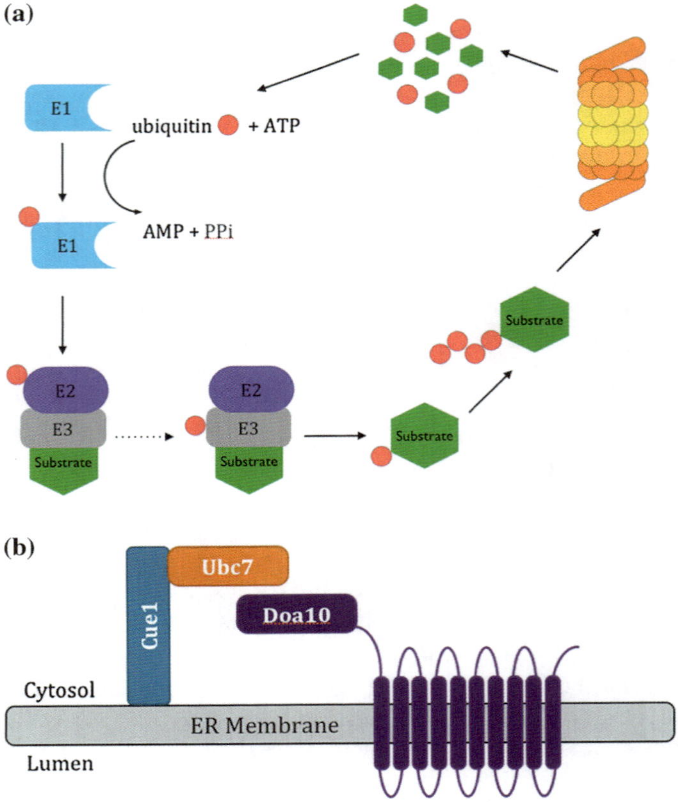

Fig. 13.1 **a** Ubiquitination schematic, adapted from Passmore and Barford (2004). Proteins targeted for degradation by the proteasome are tagged by chains of ubiquitin monomers. The covalent attachment of ubiquitin to a substrate requires three enzymes: an E1 ubiquitin-activating enzyme, an E2 ubiquitin-conjugating enzyme, and an E3 ubiquitin ligase. First, the energy of ATP is used to generate a thioester bond between ubiquitin (Ub) and an E1 enzyme. This Ub molecule is then transferred from the E1 to E2 enzymes. Finally, an E3 ligase transfers the ubiquitin to the specific target substrate. Subsequent ubiquitin molecules are added to the first Ub molecule, and eventually signal the protein to be trafficked to the proteasome for degradation. **b** Schematic of members of the Ubc7-dependent ERAD machinery, adapted from Hampton (2002)

functionally relevant to a similarly large range of environmental stress responses in *S. cerevisiae*.

Since *UBC7*, *CUE1*, and *DOA10* were identified in a screen for genes important to membrane biogenesis, as well as found to be cold sensitive, these results suggest that these members of the ERAD pathway play a role in cold adaptation and membrane regulation (Wright 1993; Wright et al. 2003; Loertscher et al. 2006). The works done by Loertscher et al. (2006) and Wright et al. (2003) are not the only accounts suggesting a relationship between low-temperature growth, membrane composition, and the ERAD pathway. For example, transcription of *OLE1*, an essential gene in the ER, encodes the only Δ9 fatty acid desaturase found in *S. cerevisiae* and is regulated by the transcription factors Mga2p and Spt23p (Zhang et al. 1999). Activation of these two transcription factors is regulated by proteasomal processing initiated by an ERAD complex (Hoppe et al. 2000). More specifically, Nakagawa et al. (2002) concluded that Mga2p serves as a low-temperature sensor in yeasts based on their findings that activation of Mga2p requires proteasomal processing that it is triggered by cold temperatures. This activation of Mga2p induces *OLE1* transcription, which in turn increases the amount of unsaturated fatty acids, thus keeping the membrane more fluid at low temperatures (Nakagawa et al. 2002). One of the most widely accepted mechanisms of cold adaptation in *S. cerevisiae*, psychrophilic yeasts, and most other organisms is the maintenance of membrane function and fluidity despite their disruption at low temperatures. Altering the unsaturated fatty acid and sterol composition of the membrane is a common way for unicellular organisms, including psychrophilic yeasts, to adapt to low temperatures (Finegold 1986). These alterations include an increase in fatty acid unsaturation, decrease in average fatty acid chain length, and a decrease in the sterol:phospholipid ratio (Inniss 1975; Russell 2008; Margesin and Miteva 2011). In addition to regulating Mga2p and Spt23p, ERAD components, including Ubc7p and Cue1p, directly regulate the stability of Ole1p (Braun et al. 2002). Taken together, this information led Loertscher et al. (2006) to suggest that the ultimate cause of cold sensitivity in *ubc7*Δ, *cue1*Δ, and *doa10*Δ cells results from a combination of altered sterol levels and failure to regulate Ole1p levels. Since *OLE1* is the sole desaturase capable of introducing one double bond into the fatty acid chain, the fatty acids present in the phospholipids of yeasts are either unsaturated or monounsaturated (Stukey et al. 1990; Braun et al. 2002). Therefore, the modification of sterols may play a much more important role in the cold adaptation responses of yeasts than in other organisms which produce a wider variety of phospholipids (Daum et al. 1998). In addition to transduction of the low-temperature signal and regulation of *OLE1* transcription, Mga2p has also been linked to transcriptional regulation of enzymes within the ergosterol biosynthetic pathway (Rice et al. 2010). An exhaustive discussion on the changes in lipid composition and fluidity of yeast plasma membranes as response to cold is reported in Chap. 10.

13.3 Membrane Sterol Composition and Low Temperatures

As previously mentioned, the *ubc7*Δ, *cue1*Δ, and *doa10*Δ ERAD mutants were originally identified for their inability to properly regulate membrane biogenesis and karmellae formation in response to elevated expression of HMGR (Wright et al. 2003). Due to the relationship between membrane dynamics and cold adaptation, follow-up experiments were done by Wright et al. (2003) to determine whether growth at low temperatures had an effect on the membrane profiles of ERAD mutant strains. The published data for these experiments explore the membrane profiles of *ubc7Δ* mutant strains. The authors focus on the role of *UBC7* because Ubc7p is the E2 enzyme responsible for specifically targeting HMGR for the ERAD pathway when necessary (Hampton et al. 1996b; Donald et al. 1997; Bays et al. 2001; Ignea et al. 2012). Membrane sterol profiles were generated at 30 °C for wild type and *ubc7Δ* mutant strains. The same sterol analysis was done at 10 °C to determine whether there were changes in sterol composition that occurred at low temperatures in either wild type or *ubc7Δ* mutant strains. At both temperatures, wild type and *ubc7Δ* strains display dramatic differences in membrane sterol composition. Wild-type cells cultured at 30 °C contain approximately 15 % more ergosterol than wild-type cells cultured at 10 °C, and this dramatic change in ergosterol composition is echoed by levels of ergosterol precursors as well. In addition to differences within the same strain at 10 and 30 °C, differences between the overall sterol profiles of wild type and *ubc7Δ* mutants cultured at the same temperature were seen as well (Loertscher et al. 2006). This suggests that low temperature and ergosterol levels are not the only factors impacting sterol concentrations, and that there may be a broad disruption within the ergosterol biosynthetic pathway at low temperatures. Alternatively, although *UBC7* has been shown to directly play a role in HMGR stability in yeasts, it is also possible that other ergosterol biosynthetic enzymes are impacted by the ERAD pathway. Hitchcock et al. (2003) used a large-scale mass spectrometric approach to identify 83 potential membrane-bound Ubc7p-dependent ERAD substrates. Three of these predicted substrates, Erg1p, Erg9p, and Erg27p, are members of the ergosterol biosynthesis pathway; these enzymes are responsible for the synthesis of 2,3-oxidosqualene, squalene, and 4-α-methylzymosterol, respectively (Hitchcock et al. 2003). Disruption in the regulation of the amount of these ergosterol biosynthetic enzymes causes alterations in the flux through the ergosterol pathway and changes the levels of ergosterol pathway intermediates produced (Veen et al. 2003). If low temperatures cause fluctuation of similar ergosterol pathway intermediates, it is possible that proper feedback mechanisms are disturbed at low temperatures, leading to the atypical sterol profiles see in *S. cerevisiae* cultured at 10 °C. The work done by Hitchcock et al. (2003) has identified Ubc7-dependent targets of the ERAD pathway, and similar experiments performed at 10 °C would give additional insight into which ergosterol biosynthesis proteins are targeted by the ERAD pathway at low temperatures. Results from this type of future experiment could be used to identify targets of the ERAD pathway at 10 °C, giving insight into potential mechanisms responsible for the regulation of sterol synthesis in *S. cerevisiae* at low temperature.

It has been shown that genetic and biochemical factors which change the concentration of intermediates within the ergosterol biosynthesis pathway can signal HMGR degradation by the ERAD pathway (Gardner and Hampton 1999). The variation in sterol composition seen in wild-type and *ubc7Δ* cells grown at 10 and 30 °C suggests that either low temperature or ERAD disruption may lead to improper degradation of HMGR. If this is the case, the cold sensitivity of the ERAD mutants suggests that ERAD may be required to compensate for the altered membrane composition at 10 °C. Thus, ERAD mutants grow at wild-type rates at 30 °C, but may be unable to overcome the additional stress of maintaining a functional membrane at low temperatures. The relationship between the ERAD pathway and cold adaptation could be directly due to conformational changes in HMGR protein structure that may target it for ERAD (Gardner and Hampton 1999). If these conformational changes are sufficient to target HMGR for degradation, it is possible that this may lead to its erroneous assignment as a substrate for degradation. Thus, the post-translational regulation and stability of HMGR at low temperatures may be a key piece in understanding the relationship between ERAD, membrane sterol composition, and low-temperature growth.

13.4 Conclusions

If functional HMGR and ERAD machinery are necessary for *S. cerevisiae* to survive in low-temperature environments, it is also possible that organisms that thrive in extreme environments may have evolved to produce members of the ERAD or ergosterol biosynthesis pathways. Therefore, ERAD mechanisms in psychrophilic yeasts may have evolved in response to their natural occurrence in colder climates. The results summarized here only give a glimpse into some of the possible mechanisms that yeasts may use as they adapt and evolve to life in the cold. For example, it is possible that some psychrophiles have adapted molecular mechanisms that control the ergosterol biosynthetic pathway such that their membranes are stabilized at the organism's optimal growth temperature. Work done to determine the effects of temperature on membrane composition in psychrophilic yeasts supports the idea that membrane composition is fundamental to low-temperature adaptation in psychrophilic yeasts (Watson and Arthur 1976). There are still many unanswered questions and much work to be done to better establish possible mechanisms behind the regulation of membrane composition as it relates to survival at low temperatures, as well as how or when the ERAD pathway plays a role in cold adaptation of psychrophilic yeasts.

Areas for further exploration include experiments using psychrophilic yeasts to identify which molecular mechanisms are necessary for their adaptation to cold environments. Based on what has been shown in *S. cerevisiae*, it will be important to determine what role the ERAD pathway plays in psychrophilic yeasts and whether the sequences of *UBC7*, *CUE1*, and *DOA10* are the same in psychrophiles as they are in *S. cerevisiae*. Alterations in gene sequence could potentially highlight functional differences between ERAD proteins, which are necessary for cold adaptation. Additionally, deletion of members of the ERAD pathway in psychrophilic yeasts could also shed light on whether an intact proteasomal degradation system plays a role in adaptation to cold climates. Along with questions about the ERAD pathway in psychrophilic yeasts, understanding the evolution of the HMGR genes in these species may also shed light on evolutionary changes affecting ergosterol biosynthesis at low temperatures. To understand how ergosterol synthesis impacts cold adaptation, it will be important to investigate the sterol profiles of different psychrophilic yeasts at different temperatures and how those profiles impact growth rates, membrane-bound protein function, and adaptation. Not only will these experiments be important for understanding how yeast species adapt to cold environments, but they will provide a better understanding of how yeasts function at low temperatures which may be applicable for industrial use, food preservation, fermentation technology and more.

References

Anderson RG, Orci L, Brown MS, Garcia-Segura LM, Goldstein JL (1983) Ultrastructural analysis of crystalloid endoplasmic reticulum in UT-1 cells and its disappearance in response to cholesterol. J Cell Sci 63:1–20

Atomi H, Sato T, Kanai T (2011) Application of hyperthermophiles and their enzymes. Curr Opin Biotechnol 22:618–626

Basson ME, Thorsness M, Rine J (1986) *Saccharomyces cerevisiae* contains two functional genes encoding 3-hydroxy-3-methylglutaryl-coenzyme A reductase. Proc Natl Acad Sci USA 83:5563–5567

Bays NW, Gardner RG, Seelig LP, Joazeiro CA, Hampton RY (2001) Hrd1p/Der3p is a membrane-anchored ubiquitin ligase required for ER-associated degradation. Nat Cell Biol 3:24–29

Braun S, Matuschewski K, Rape M, Thoms S, Jentsch S (2002) Role of the ubiquitin-selective CDC48 (UFD1/NPL4) chaperone (segregase) in ERAD of OLE1 and other substrates. EMBO J 21:615–621

Brock TD (1997) The value of basic research: discovery of *Thermus aquaticus* and other extreme thermophiles. Genetics 146:1207–1210

Brodsky JL, McCracken AA (1999) ER protein quality control and proteasome-mediated protein degradation. Semin Cell Dev Biol 10:507–513

Brown MS, Goldstein JL (1980) Multivalent feedback regulation of HMG CoA reductase, a control mechanism coordinating isoprenoid synthesis and cell growth. J Lipid Res 21:505–517

Burg JS, Espenshade PJ (2011) Regulation of HMG-CoA reductase in mammals and yeast. Prog Lipid Res 50:403–410

Buzzini P, Branda E, Goretti M, Turchetti B (2012) Psychrophilic yeasts from worldwide glacial habitats: diversity, adaptation strategies and biotechnological potential. FEMS Microbiol Ecol 82:217–241

Cavicchioli R, Thomas T, Curmi PM (2000) Cold stress response in Archaea. Extremophiles 4:321–331

Claessen JH, Kundrat L, Ploegh HL (2012) Protein quality control in the ER: balancing the ubiquitin checkbook. Trends Cell Biol 22:22–32

Daum G, Lees ND, Bard M, Dickson R (1998) Biochemistry, cell biology and molecular biology of lipids of *Saccharomyces cerevisiae*. Yeast 14:1471–1510

Donald KA, Hampton RY, Fritz IB (1997) Effects of overproduction of the catalytic domain of 3-hydroxy-3-methylglutaryl coenzyme A reductase on squalene synthesis in *Saccharomyces cerevisiae*. Appl Environ Microbiol 63:3341–3344

Egner R, Rosenthal FE, Kralli A, Sanglard D, Kuchler K (1998) Genetic separation of FK506 susceptibility and drug transport in the yeast Pdr5 ATP-binding cassette multidrug resistance transporter. Mol Biol Cell 9:523–543

Finegold L (1986) Molecular aspects of adaptation to extreme cold environments. Adv Space Res 6:257–264

Gardner RG, Hampton RY (1999) A highly conserved signal controls degradation of 3-hydroxy-3-methylglutaryl-coenzyme A (HMG-CoA) reductase in eukaryotes. J Biol Chem 274:31671–31678

Gasch A, Werner-Washburne M (2002) The genomics of yeast responses to environmental stress and starvation. Funct Integr Genomics 2:181–192

Goldstein JL, Brown MS (1990) Regulation of the mevalonate pathway. Nature 343:425–430

Goldstein JL, DeBose-Boyd RA, Brown MS (2006) Protein sensors for membrane sterols. Cell 124:35–46

Haas AL, Siepmann TJ (1997) Pathways of ubiquitin conjugation. FASEB J 11:1257–1268

Hampton RY (2002) ER-associated degradation in protein quality control and cellular regulation. Curr Opin Cell Biol 14:476–482

Hampton RY, Rine J (1994) Regulated degradation of HMG-CoA reductase, an integral membrane protein of the endoplasmic reticulum, in yeast. J Cell Biol 125:299–312

Hampton R, Dimster-Denk D, Rine J (1996a) The biology of HMG-CoA reductase: the pros of contra-regulation. Trends Biochem Sci 21:140–145

Hampton RY, Gardner RG, Rine J (1996b) Role of 26S proteasome and HRD genes in the degradation of 3-hydroxy-3-methylglutaryl-CoA reductase, an integral endoplasmic reticulum membrane protein. Mol Biol Cell 7:2029–2044

Hershko A, Ciechanover A (1992) The ubiquitin system for protein degradation. Annu Rev Biochem 61:761–807

Hershko A, Ciechanover A (1998) The ubiquitin system. Annu Rev Biochem 67:425–479

Hiraishi H, Mochizuki M, Takagi H (2006) Enhancement of stress tolerance in *Saccharomyces cerevisiae* by overexpression of ubiquitin ligase Rsp5 and ubiquitin-conjugating enzymes. Biosci Biotechnol Biochem 70:2762–2765

Hirsch C, Gauss R, Horn SC, Neuber O, Sommer T (2009) The ubiquitylation machinery of the endoplasmic reticulum. Nature 458:453–460

Hitchcock AL, Auld K, Gygi SP, Silver PA (2003) A subset of membrane-associated proteins is ubiquitinated in response to mutations in the endoplasmic reticulum degradation machinery. Proc Natl Acad Sci USA 100:12735–12740

Hoppe T, Matuschewski K, Rape M, Schlenker S, Ulrich HD, Jentsch S (2000) Activation of a membrane-bound transcription factor by regulated ubiquitin/proteasome-dependent processing. Cell 102:577–586

Ignea C, Trikka FA, Kourtzelis I, Argiriou A, Kanellis AK, Kampranis SC, Makris AM (2012) Positive genetic interactors of HMG2 identify a new set of genetic perturbations for improving sesquiterpene production in *Saccharomyces cerevisiae*. Microb Cell Fact 11:162

Inniss WE (1975) Interaction of temperature and psychrophilic microorganisms. Annu Rev Microbiol 29:445–465

Koning AJ, Roberts CJ, Wright RL (1996) Different subcellular localization of *Saccharomyces cerevisiae* HMG-CoA reductase isozymes at elevated levels corresponds to distinct endoplasmic reticulum membrane proliferations. Mol Biol Cell 7:769–789

Loertscher J, Larson LL, Matson CK, Parrish ML, Felthauser A, Sturm A, Tachibana C, Bard M, Wright R (2006) Endoplasmic reticulum-associated degradation is required for cold adaptation and regulation of sterol biosynthesis in the yeast *Saccharomyces cerevisiae*. Eukaryot Cell 5:712–722

Loftus SK, Morris JA, Carstea ED, Gu JZ, Cummings C, Brown A, Ellison J, Ohno K, Rosenfeld MA, Tagle DA, Pentchev PG, Pavan WJ (1997) Murine model of Niemann-Pick C disease: mutation in a cholesterol homeostasis gene. Science 277:232–235

Lorenz RT, Parks LW (1987) Regulation of ergosterol biosynthesis and sterol uptake in a sterol-auxotrophic yeast. J Bacteriol 169:3707–3711

Lum PY, Wright R (1995) Degradation of HMG-CoA reductase-induced membranes in the fission yeast, *Schizosaccharomyces pombe*. J Cell Biol 131:81–94

Margesin R, Miteva V (2011) Diversity and ecology of psychrophilic microorganisms. Res Microbiol 162:346–361

McCracken AA, Brodsky JL (2003) Evolving questions and paradigm shifts in endoplasmic-reticulum-associated degradation (ERAD). BioEssays 25:868–877

Metzger MB, Maurer MJ, Dancy BM, Michaelis S (2008) Degradation of a cytosolic protein requires endoplasmic reticulum-associated degradation machinery. J Biol Chem 283:32302–32316

Nakagawa Y, Sakumoto N, Kaneko Y, Harashima S (2002) Mga2p is a putative sensor for low temperature and oxygen to induce *OLE1* transcription in *Saccharomyces cerevisiae*. Biochem Biophys Res Commun 291:707–713

Nakanishi M, Goldstein JL, Brown MS (1988) Multivalent control of 3-hydroxy-3-methylglutaryl coenzyme A reductase. Mevalonate-derived product inhibits translation of mRNA and accelerates degradation of enzyme. J Biol Chem 263:8929–8937

Nicholson RC, Williams DB, Moran LA (1990) An essential member of the HSP70 gene family of *Saccharomyces cerevisiae* is homologous to immunoglobulin heavy chain binding protein. Proc Natl Acad Sci USA 87:1159–1163

Normington K, Kohno K, Kozutsumi Y, Gething MJ, Sambrook J (1989) *S. cerevisiae* encodes an essential protein homologous in sequence and function to mammalian BiP. Cell 57:1223–1236

Parrish ML, Sengstag C, Rine JD, Wright RL (1995) Identification of the sequences in HMG-CoA reductase required for karmellae assembly. Mol Biol Cell 6:1535–1547

Passmore LA, Barford D (2004) Getting into position: the catalytic mechanisms of protein ubiquitylation. Biochem J 379:513–525

Pickart CM, Eddins MJ (2004) Ubiquitin: structures, functions, mechanisms. Biochim Biophys Acta 1695:55–72

Posas F, Chambers JR, Heyman JA, Hoeffler JP, de Nadal E, Ariño J (2000) The transcriptional response of yeast to saline stress. J Biol Chem 275:17249–17255

Rice C, Cooke M, Treloar N, Vollbrecht P, Stukey J, McDonough V (2010) A role for MGA2, but not SPT23, in activation of transcription of ERG1 in *Saccharomyces cerevisiae*. Biochem Biophys Res Commun 403:293–297

Richter K, Haslbeck M, Buchner J (2010) The heat shock response: life on the verge of death. Mol Cell 40:253–266

Ron D, Walter P (2007) Signal integration in the endoplasmic reticulum unfolded protein response. Nat Rev Mol Cell Biol 8:519–529

Russell NJ (2008) Membrane components and cold sensing. In: Margesin R, Schinner F, Marx JC, Gerday C (eds) Psychrophiles: from biodiversity to biotechnology. Springer, Berlin, pp 177–190

Saito H, Posas F (2012) Response to hyperosmotic stress. Genetics 192:289–318

Simpson CE, Ashe MP (2012) Adaptation to stress in yeast: to translate or not? Biochem Soc Trans 40:794–799

Siperstein MD, Fagan VM (1966) Feedback control of mevalonate synthesis by dietary cholesterol. Journal Biol Chem 241:602–609

Stetter KO (2006) Hyperthermophiles in the history of life. Philos Trans R Soc Lond B Biol Sci 361:1837–1844

Stukey JE, McDonough VM, Martin CE (1990) The *OLE1* gene of *Saccharomyces cerevisiae* encodes the Δ9 fatty acid desaturase and can be functionally replaced by the rat stearoyl-CoA desaturase gene. J Biol Chem 265:20144–20149

Swan TM, Watson K (1998) Stress tolerance in a yeast sterol auxotroph: role of ergosterol, heat shock proteins and trehalose. FEMS Microbiol Lett 169:191–197

Travers KJ, Patil CK, Wodicka L, Lockhart DJ, Weissman JS, Walter P (2000) Functional and genomic analyses reveal an essential coordination between the unfolded protein response and ER-associated degradation. Cell 101:249–258

Vashist S, Ng DT (2004) Misfolded proteins are sorted by a sequential checkpoint mechanism of ER quality control. J Cell Biol 165:41–52

Veen M, Stahl U, Lang C (2003) Combined overexpression of genes of the ergosterol biosynthetic pathway leads to accumulation of sterols in *Saccharomyces cerevisiae*. FEMS Yeast Res 4:87–95

Vembar SS, Brodsky JL (2008) One step at a time: endoplasmic reticulum-associated degradation. Nat Rev Mol Cell Biol 9:944–957

Walter P, Ron D (2011) The unfolded protein response: from stress pathway to homeostatic regulation. Science 334:1081–1086

Watson K, Arthur H (1976) *Leucosporidium* yeasts: obligate psychrophiles which alter membrane-lipid and cytochrome composition with temperature. J Gen Microbiol 97:11–18

Welch WJ, Suhan JP (1985) Morphological study of the mammalian stress response: characterization of changes in cytoplasmic organelles, cytoskeleton, and nucleoli, and appearance of intranuclear actin filaments in rat fibroblasts after heat-shock treatment. J Cell Biol 101:1198–1211

Willey JM, Sherwood LM, Prescott LM (2008) Prescott, Harley, and Klein's Microbiology, 7th edn. McGraw-Hill Higher Education, New York

Wolf DH, Fink GR (1975) Proteinase C (carboxypeptidase Y) mutant of yeast. J Bacteriol 123:1150–1156

Wright R (1993) Insights from inducible membranes. Curr Biol 3:870–873

Wright R, Basson M, D'Ari L, Rine J (1988) Increased amounts of HMG-CoA reductase induce "karmellae": a proliferation of stacked membrane pairs surrounding the yeast nucleus. J Cell Biol 107:101–114

Wright R, Keller G, Gould SJ, Subramani S, Rine J (1990) Cell-type control of membrane biogenesis induced by HMG-CoA reductase. New Biol 2:915–921

Wright R, Parrish ML, Cadera E, Larson L, Matson CK, Garrett-Engele P, Armour C, Lum PY, Shoemaker DD (2003) Parallel analysis of tagged deletion mutants efficiently identifies genes involved in endoplasmic reticulum biogenesis. Yeast 20:881–892

Zhang S, Skalsky Y, Garfinkel DJ (1999) MGA2 or SPT23 is required for transcription of the Δ9 fatty acid desaturase gene, OLE1, and nuclear membrane integrity in *Saccharomyces cerevisiae*. Genetics 151:473–483

Chapter 14
Subzero Activity of Cold-Adapted Yeasts

Nicolai S. Panikov

Abstract Cold adaptation is usually interpreted as ability of microorganisms to grow at temperature around zero or recover from deep freezing. This chapter focuses on a recently discovered physiological phenomenon of microbial growth and metabolism in a frozen environment cooled down to −20 °C and deeper. Still, there are numerous controversies about subzero activity (SZA); therefore, the chapter starts from extensive introduction into the SZA phenomenon beyond the yeasts' taxonomic boundary. Critical review of available techniques for SZA detection resulted in a precautionary note that some reports based on account of frozen soil respiration or methane generation could be considerably overestimated. More reliable measurements of SZA are obtained by using methods based on direct microscopy and uptake of labelled gaseous substrates, e.g. $^{14}CO_2$. Two types of below-zero habitats are recognized: homogeneous icy environment (polar sea ice, glaciers, snow) where microbial cells can survive in supercooled liquid veins and permanently frozen soils and sediments containing buried organic matter and allowing slow gas exchange through aeration micropores. Fungi including yeasts are the most successful colonizers of the second type of habitats. Solid-state frozen enrichment culture, phylogenetic survey of soils, round-year seasonal community dynamics and specific inhibitors of protein synthesis - all indicate that fungi have a competitive advantage over prokaryotic organisms in frozen soils. None of the known molecular mechanisms of cold adaptation, e.g. membrane structure, heat- and cold-stress proteins, cold-adapted enzymes, could be uniquely attributed to fungi. Instead, the author discusses wide opportunities given to all fungi by bigger cell size and colonization of frozen heterogeneous environments via mycelial or pseudomycelial growth.

Keywords Frozen · Water · Isotope · Diffusion · Gas exchange

N. S. Panikov (✉)
Department of Immunology and Infectious Deceases, Harvard School of Public Health, Boston, MA 02115, USA
e-mail: npanikov@gmail.com

P. Buzzini and R. Margesin (eds.), *Cold-Adapted Yeasts*,
DOI: 10.1007/978-3-642-39681-6_14,

Contents

14.1 Introduction 296
14.2 History of the SZA Discovery 297
14.3 Measurement of SZA 299
14.4 SZA Variation in Different Cold Habitats and its Dependence on Environmental Factors 303
14.4.1 The Effects of Temperature and Unfrozen Water Content 303
14.4.2 Gas Diffusion as a Limiting Factor 304
14.4.3 Effect of Substrate Amendment 305
14.4.4 Biogenic Heat 305
14.5 Physiological State of Microbial Community in Frozen Soil: Growth or Maintenance? 306
14.6 Microorganisms Responsible for SZA 308
14.6.1 Overall Survey of Permafrost Community 308
14.6.2 Transient Response of Soil Community to Cooling: Seasonal Observations 310
14.6.3 Transient Response of Soil Community to Cooling: Laboratory Incubation Experiments 310
14.6.4 Stable-Isotope Probing 311
14.7 Isolation of Microorganisms by Subzero Enrichment 312
14.8 Why are Fungi Better Adapted to Frozen Environment? 313
14.9 Conclusions 318
References 319

14.1 Introduction

Traditional microbial physiology interprets cold adaptation as ability of microorganisms to grow and metabolize at low temperatures, typically from about 10 °C to zero, assuming that further cooling down below the freezing point prevents any kind of normal physiological activity. Other commonly explored traits are freeze and freeze–thaw resistance expressed as the property of microbial cells to preserve viability after single or multiple cycles of freezing and melting (Chattopadhyay 2000; Zecchinon et al. 2001; Marx et al. 2004; Steven et al. 2006). Relatively new to science and a seriously understudied physiological phenomenon is microbial metabolism *below the freezing point*. Only recently, we have obtained convincing evidences that subzero activity (SZA) of microorganisms is not a laboratory artefact but an important natural phenomenon worth of careful study (Rivkina et al. 2000; Junge et al. 2004; Panikov 2009). This chapter will summarize the current state of SZA studies with special focus on cold-adapted yeasts. It was found that fungi including yeasts are better adapted to cold than other saprotrophic microorganisms having selective advantages in colonizing frozen environments over bacteria. However, contrary to other chapters in this book, this chapter is not restricted to yeasts alone. As SZA is still not yet fully specified and understood, the author refers sometimes to the SZA of the entire microbial community or compares the SZA of eukaryotes versus prokaryotes or, most frequently the chapter is focusing on subzero growth and activity of various fungi, including yeasts as well as dimorphic and mycelial forms of fungi inhabiting cold habitats, such as tundra and permafrost.

14.2 History of the SZA Discovery

The first unambiguous indication of microbial SZA came from Antarctic studies of the cryptoendolithic lichens (Lange and Kappen 1972; Kappen and Friedmann 1983; Kappen 1988, 1993; Schroeter et al. 1992; Friedmann et al. 1993; Kappen et al. 1996, 1998). Lichens are composite organisms consisting of a fungus (the mycobiont) and a photosynthetic partner (the photobiont or phycobiont) growing together in a symbiotic relationship. The photobiont is usually either a green alga or cyanobacterium. The term ‘cryptoendolithic’ refers to organisms colonizing the empty spaces and pores inside rock materials on the surface of polar deserts; they are partly protected by rocks from adverse polar conditions above and below the freezing point. The SZA was reported as *in situ* measured photosynthetic activity of a photobiont under light conditions and total respiratory activity of photo- and mycobionts with possible contribution of free-living microorganisms in the dark day period (Fig. 14.1). The CO_2 balance was positive with net CO_2 uptake of 0.42 mg d^{-1} (g biomass)$^{-1}$ at ambient temperature fluctuating between -10 and

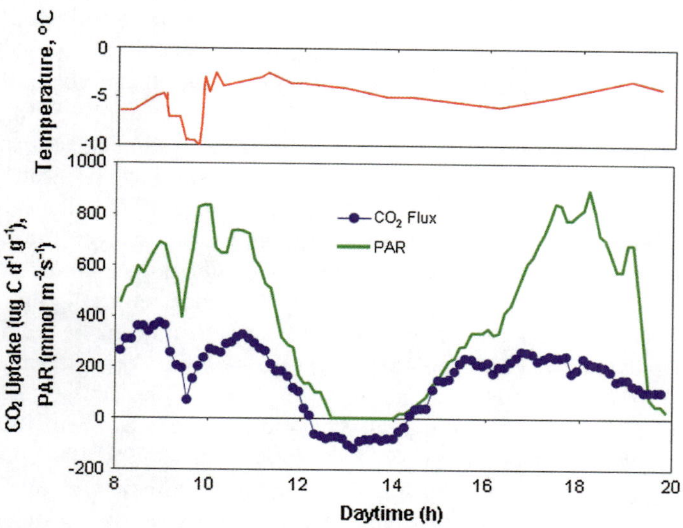

Fig. 14.1 One of the first records of SZA: field measurements of CO_2 exchange (photosynthetic uptake and respiratory release) between frozen lichens and atmosphere at *Bailey Peninsula*, Wilkes Land, continental Antarctica (66°17′S, 100°32′E). Redrawn with permission from (Kappen 1988)

−5 °C, which was equivalent to a specific growth rate (SGR) of the phycobiont of 0.0003 d^{-1} (generation time 6.7 years).

Laboratory studies of SZA with isolation of responsible microorganisms on liquid supercooled media with antifreezes were reported occasionally starting from 1960s (Michener and Elliott 1964; Larkin and Stokes 1968), but these reports did not attract much attention as SZA seemed to be of low environmental significance. A formal definition of psychrophiles was (and is still now) based on a negative condition, such as lack of growth above 20 °C (Morita 1975; Helmke and Weyland 2004), while the low-temperature boundary was either not defined or tacitly assumed to be around zero. A critical review on the significance of the terms psychrophile and psychrotolerance is reported in Chap. 1. Distrust to early publications on SZA was understandable as supercooled media with added antifreeze looked too artificial and rates too low, but most important, the phenomenon of SZA was not compatible with at least two basic principles of modern biology (Kushner 1981). The first principle claims that intracellular metabolic processes should be maintained above a certain minimum level and if cooling brings the metabolism below this level, then cells should die or enter the dormant state. The second important restriction factor was a lack of free liquid water in the frozen environment preventing normal functioning of biopolymers, nucleic acids, enzymes and semi-fluidic membranes.

Remarkable positive impact on the development of subzero microbiology was brought about in 1980–2000 by two emerging research directions: outer space exploration and the International Geosphere-Biosphere Programme (IGBP). In attempts to develop techniques for detection of extraterrestrial life on Mars and other cryogenic planets, moons and comets, astrobiologists initiated microbiological studies of respective terrestrial analogues: polar deserts, permafrost, subsurface polar lakes and other cold habitats (Friedmann and Ocampo-Friedmann 1984; Finegold 1996; Cavicchioli et al. 2002; Jakosky et al. 2003; Marion et al. 2003; Rivkina et al. 2004). Most of these studies clearly indicated that microbial life was not stopped below the freezing point. Another incentive came from the IGBP that studied interactions of the physical, chemical and biological processes within the Earth system responsible for ongoing global change. The main IGBP focus was a greenhouse effect associated with accumulation in atmosphere of CO_2, CH_4 and N_2O. An inventory of their sources demanded setting up the systematic round-year observations on gas emission from soils to atmosphere including cold-season periods. Surprisingly, the winter gas emission was found to account for up to 20 % of the annual CH_4 flux and up to 60 % (!) of the net CO_2 efflux to the atmosphere (Whalen and Reeburgh 1988; Dise 1992; Zimov et al. 1993; Melloh and Crill 1996; Oechel et al. 1997; Fahnestock et al. 1998, 1999; Grogan and Chapin 1999; Panikov and Dedysh 2000). The mechanism behind the cold-season emission was a subject of discussion; it was interpreted not only as manifestation of instant SZA of microorganisms in tundra but also as a delayed release of gas produced by the soil community during the warm season (Coyne and Kelley 1971; Mastepanov et al. 2008). Anyway the 'ice of distrust' was broken, and SZA research acquired their legitimacy to be openly discussed and published in peer-reviewed professional journals. Ironically, the cold-season gas emission eventually turned out to be mostly controlled physical gas transfer at the soil–air interface, but complementary laboratory experiments (isotopes transformation, microbial isolations) unambiguously have proved that SZA occurred in permafrost and frozen soils in the temperature range from −2 to −40 °C (see below).

Simultaneously with terrestrial studies, SZA has been reported for polar sea ice, glaciers and snow (Price 2000, 2007; Christner 2002; Deming 2002; Junge et al. 2002, 2004; Price and Sowers 2004). Presently, SZA research activity is concentrated mostly on the application of molecular culture-independent sequencing techniques to cold environments aimed to embrace diversity of cold-tolerant organisms and to assess their SZA activity *in situ*. Apart from astrobiology and global change studies, the SZA is driven by a wide range of promising biotechnological applications based on stimulation of SZA (bioremediation of polluted tundra) or its suppression (prevention of spoilage of frozen food, vaccines, embryos, etc.). The relevance of cold-adapted yeasts for bioremediation of polluted soils in cold environments and as food-spoiling agents is reported in Chaps. 21 and 23, respectively.

14.3 Measurement of SZA

There are several techniques developed to detect SZA in various natural cold habitats. One set of methods is applied to relatively homogeneous and transparent specimens representing frozen liquid, such as sea ice and glaciers, snow, subsurface polar lakes. Another set of techniques has been developed for heterogeneous non-transparent specimens of seasonally and permanently frozen soils. This material contains a mosaic of solid organic and mineral particles separated by ice fragments and thin films of unfrozen water (UW) as well as tiny air-filled pores. In the first case, transparency of frozen specimens presents a unique opportunity for using microscopic observations that can be arranged in a walk-in cold room; it allows the viewing of bacterial cells (Price 2000, 2007; Campen et al. 2003; Price and Sowers 2004) as well as yeasts cells and microcolonies (Amato et al. 2009) in water-filled veins within ice crystals (Fig. 14.2). Microscopy in combination with differential staining (e.g. CTC, the tetrazolium salt 5-cyano-2,3-ditolyl tetrazolium chloride, revealing respiring cells) was used as evidence of bacterial activity at temperatures down to −20 °C (Junge et al. 2004). Although such observations cannot be used for quantitative estimation of SZA, they provide valuable qualitative information on growth and lysis, survival and transport of microorganisms within frozen space.

Contrary to clear ice, many frozen soils and permafrost material contain air-filled pores and therefore allow very slow but non-zero gas exchange with the ambient atmosphere. Gas diffusion provides a unique methodological opportunity for SZA studies; especially, valuable benefit is that frozen samples can be amended with diffusible growth substrates through gas exchange without their preliminary

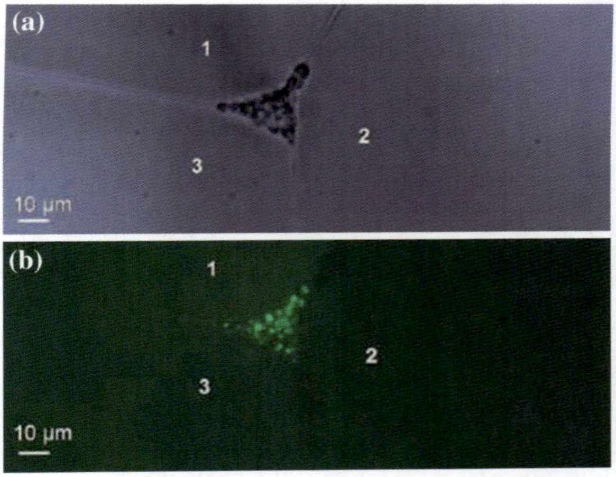

Fig. 14.2 **a** Bright field, **b** epifluorescent microscopic images showing the physical location of cells (*Cryptococcus* sp. 179-4) within a triple junction at −20 °C; numbers *1–3* indicate individual ice crystals. Reproduced with permission from Amato et al. (2009)

thawing. Appropriate mobile substrates include true gases (H_2, CO_2, CH_4, CO) and a long list of volatile organic compounds utilized by the majority of microorganisms: low molecular weight alcohols, aldehydes, ethers, fatty acids, etc. The most common C-substrate is CO_2 utilized by almost every organism, from bacteria to humans. Photo- and chemoautotrophic bacteria assimilate CO_2 via the Calvin cycle as sole C-source, while the heterotrophic organisms perform anapleurotic carboxylation reactions (e.g. carboxylation of pyruvate and propionate) aimed to replenish catabolic intermediates spent for biosynthesis. Heterotrophic growth results in net oxidation of organic substrates to CO_2 which hides anapleurotic CO_2 uptake; therefore, we need to use labelled $^{14}CO_2$ to measure the so-called dark $^{14}CO_2$ fixation (DF) (Fig. 14.3). On average, the DF accounts for about 6 % of the gross cellular synthesis (Panikov 2009) although it varies between phylogenetic groups (Alonso-Saez et al. 2010). DF can be used not only with soil but also with homogeneous icy samples; in this case, ^{14}C-substrate is added as bicarbonate.

Not every gas exchange technique is suitable for measuring SZA. Conventional tests for 'soil respiration' or 'methanogenic activity' as applied to frozen specimens (Rivkina et al. 2002; Panikov et al. 2006; Johnson et al. 2007) could significantly overestimate actual SZA if we assume that measured rates of gas release reflect instant metabolic activity. As a matter of fact, all soils and sediments and especially frozen permafrost layers hold large pools of CH_4, CO_2 and other gases well above the equilibrium with ambient atmosphere. For example, the CO_2 mixing ratio in soil pores varies between 1 and 10 % versus 0.038–0.04 % in the atmosphere. In frozen soil, CO_2 is present as free gas, $CO_2 + HCO_3^-$ dissolved in water and trapped in interstitial ice. In addition to these well-known forms, there is also CO_2 firmly bound to the soil solid phase, most probably 'encapsulated' within soft organomineral micropores. In the course of long-term (up to several months) subzero incubation, the entrapped gas is slowly releasing, giving a wrong impression of steady microbial respiration (Fig. 14.4). Unfortunately, sterile controls do not help in differentiation

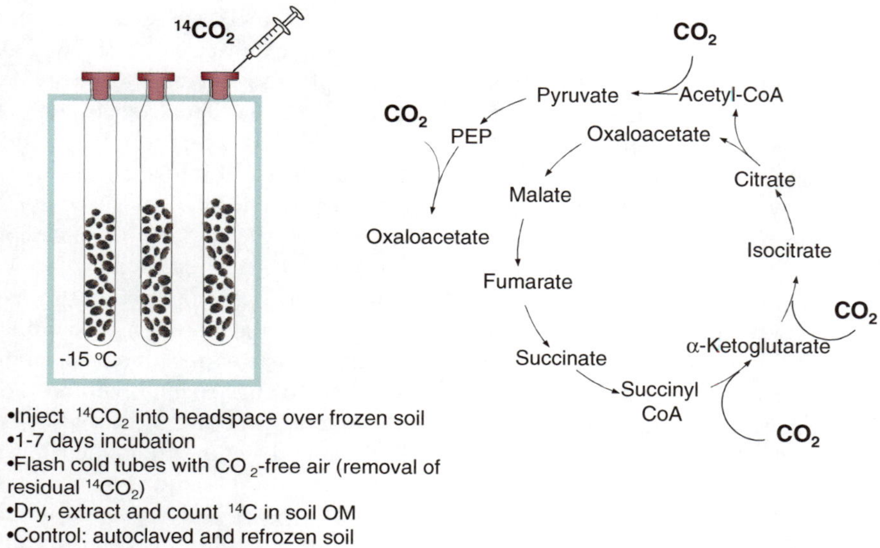

Fig. 14.3 Determination of the SZA in soils and permafrost samples by recording dark CO_2 fixation rate, see other details in Panikov and Sizova (2007)

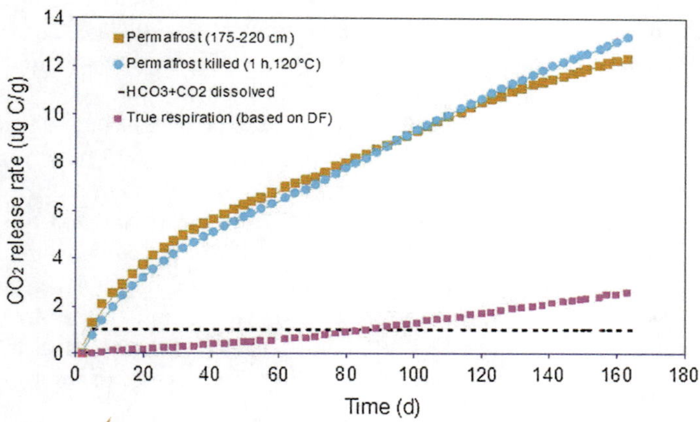

Fig. 14.4 Experimental evidence for the presence of firmly bound CO_2 in permafrost compromising 'soil respiration' data. Both autoclaved and untreated samples of permafrost released similar amounts of CO_2 in the course of subzero (-10 °C) incubation well above the level of inorganic C-pool (sum of dissolved $CO_{2+}HCO_{3-}$ measured as CO_2 evolution after acidification). CO_2 data were obtained by continuous recording of CO_2 by IR gas analyser; true microbial respiration (*TMR*) was calculated from the rates of dark $^{14}CO_2$ fixation (DF) assuming the stoichiometric ratio DF:TMR = 1:10 (Recalculated and redrawn from Panikov 2009)

of true respiration from physical gas release because autoclaving and dry heating cause chemical destruction of soil organic matter and greatly accelerate abiotic CO_2 release (Panikov 2009). Sterilization by γ-radiation or chemical inhibitors is milder but still has some impact, besides the SZA calculated as a difference between two large quantities (CO_2 efflux from untreated soil minus CO_2 efflux from sterile soil) cannot be accurate. Under field conditions, gas fluxes from soil to atmosphere can deviate even more significantly from the instant gas-generation biological activity of microorganisms across the soil profile. Therefore, a cold-season CO_2 emission should be essentially higher than the corresponding instant respiration of cold-adapted microorganisms. In spring and summer time, there could be an inverse trend: lower surface emission than actual cumulative respiration leading to build-up of soil CO_2-pool. Going back to laboratory measurements, a correct implementation of soil respiration as a measure of SZA should be based on account for all forms of CO_2 generated during subzero incubation including the gain of CO_2 in headspace, dissolved inorganic C and firmly bound CO_2. Such protocol is still not developed and is not going to be a fast and rapid procedure. A better way would be to monitor a $^{14}CO_2$ release from ^{14}C-labelled organic substrate added to soil; however, it will characterize a substrate-induced SZA (bigger than true SZA) of some (mostly unknown) fraction of soil community.

Several other techniques are equally applied to ice-like and soil-like frozen specimens. The most popular approach is to measure the intensity of key biosynthetic processes by monitoring the incorporation rate of respective isotope-labelled precursors. Examples include (but are not limited to) the following: incorporation of [^{14}C] acetate into cellular lipids (Rivkina et al. 2000), [3H] thymidine incorporation into DNA and [3H] or [^{14}C] leucine incorporation into microbial proteins (Tibbles and Harris 1996; Carpenter et al. 2000; Sattler et al. 2001; Christner 2002). The procedure requires temporal melting of frozen specimens to add and mix label with the following refreezing and subzero incubation. Several uncertainties compromise the described approach: (1) incorporation rates are sensitive to the concentration of the added intermediates, and too low concentration does not stop *de novo* synthesis of nucleotides which underestimates actual SZA, while too high concentration might induce microbial growth which leads to overestimation of the actual SZA; (2) some microorganisms are not able to transport the labelled intermediates into cells; and (3) a conversion factor from the measured incorporation rate to a more meaningful parameter of microbial growth rate is uncertain.

In summary, there is no single 'ideal' method to measure SZA. The best results so far were produced by methods based on measuring of CO_2 uptake, such as photosynthetic activity of lichens and dark $^{14}CO_2$ fixation. Below the author outlines main results on SZA variation across time and space and on SZA dependence on environmental factors, obtained by the DF method applied to tundra soils and permafrost.

14.4 SZA Variation in Different Cold Habitats and its Dependence on Environmental Factors

In our incubation experiments carried at −10 °C (Panikov 2009), DF varied across all tested seasonally and permanently frozen soils of Alaska from close to zero to 5 nmol CO_2 d^{-1} (g soil)$^{-1}$. The strongest correlation was found between DF and the content of organic C ($r^2 = 0.92$, $n = 60$). Most probable interpretation of this result is that organic matter provides a better microporous environment allowing diffusive gas exchange and thermoinsulation important for keeping the biogenic heat within the microbial microcolony (see below). A role of soil organic matter as microbial nutrient source seems to be less probable because of its recalcitrant nature.

14.4.1 The Effects of Temperature and Unfrozen Water Content

Two factors, temperature and availability of water, are interconnected, and when we reset the incubation temperature, the water status is spontaneously adjusted in a brief transient process. Water availability could be characterized as partial pressure of water vapour over frozen soil by headspace gas analyses (e.g. mass spectrometry or IR gas analyser) or by measuring the content of UW with the TDR method (time-domain reflectometry) (Watanabe and Wake 2009). The temperature *per se* is a measure of kinetic energy of reactants which declines with cooling but does not stop completely because of the exponential character of energy distribution. We were able to detect $^{14}CO_2$ uptake down to −80 °C (Fig. 14.5) in incubation extended to 6 months. The combined effect of temperature and UW on DF obeyed to the Arrhenius equation with added exponential term to account an indirect influence of temperature via the content of available water. SZA normalized per unit of available water displayed linear Arrhenius plot with an apparent activation energy ($E_a = 43.7$ KJ mol^{-1}) typical for enzymatic reactions.

The average water activity in frozen soil is very low, but it could be higher inside microbial cells and around microbial colonies due to self-heating and antifreeze action of exometabolites (see below). Another alleviating factor could be slow equilibration of the subzero system (Wolfe et al. 2002). Most experiments with permafrost samples are accompanied by their mechanical disturbance; after the extraction of the permafrost core and its storage in a freezer, it is split into small aggregates to fit incubation vessels. Such treatment releases certain amounts of available C-substrates and water even if all preparatory steps are done in a −20 °C freezer. During the following subzero incubation, the system goes back to equilibrium but not instantly, and it takes hours and days for equilibration across the entire microporous space when water content remains above the equilibrium level corresponding to the current incubation temperature.

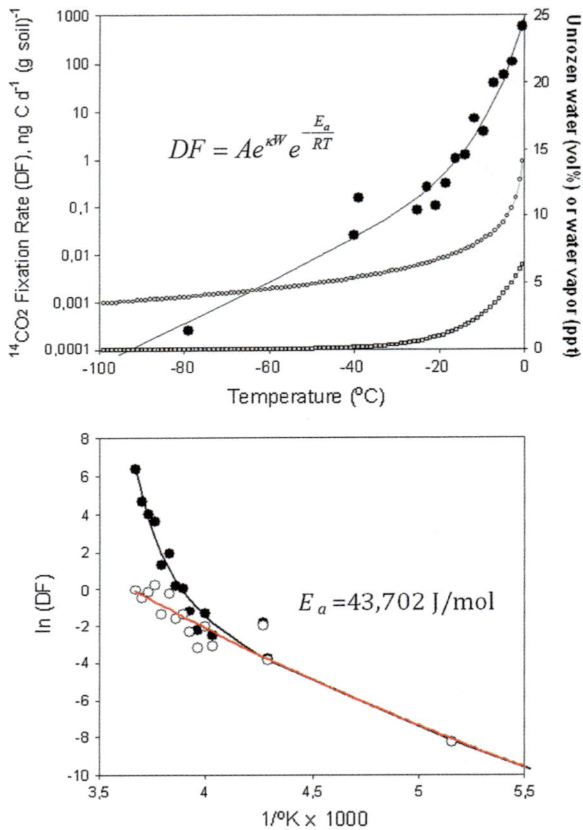

Fig. 14.5 SZA (dark fixation, DF) in permafrost (Sagwon site, AK) as dependent on incubation temperature. *Top* DF (*filled circle*), relative air humidity (open squares) and unfrozen water content (open circles) versus temperature (°C); *Bottom* Arrhenius plot for overall (*filled circle*) and normalized per UW unit (*circle*) rate, DF/exp (κW). *Red line* Arrhenius equation; *black curve* our modification shown on the top panel

14.4.2 Gas Diffusion as a Limiting Factor

Pure frozen water forms perfect ice crystals that do not allow gas diffusion through frozen mass; however, impurities, defects and possibly veins between individual crystals can provide some limited conductance of gases and liquid. As stated above, frozen soils especially rich in coarse mor-type organic matter are more conductive to gases as compared to frozen ice. We measured $^{14}CO_2$ diffusion rate to the inner space of permafrost lumps (Panikov 2009) and found the apparent diffusion coefficient at -20 °C to be as low as 6.9×10^{-9} cm^2 s^{-1} (10^7 times lower than in air). Nevertheless, such gas permeability of organic permafrost is adequate to maintain slow aerobic growth. Mass transfer in the mineral permafrost should be slower (Ostroumov and Siegert 1996).

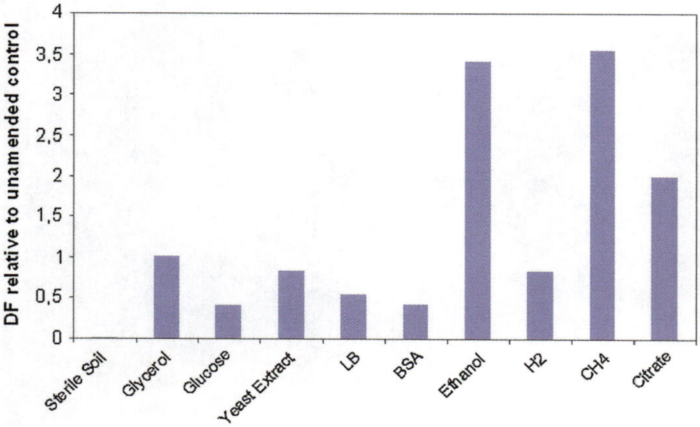

Fig. 14.6 Effect of added substrates on $^{14}CO_2$ uptake by permafrost samples at −11 °C. Note that the complex organic substrates (LB, Luria–Bertani Broth; BSA, bovine serum albumin; YE, yeast extract) inhibited dark fixation (DF), while simple volatile compounds (ethanol, CH_4) produced maximal stimulation

14.4.3 Effect of Substrate Amendment

Complex substrates (yeast and meat extracts, proteins, mixture of sugars) do not stimulate and even suppress DF, while volatile compounds and gases (ethanol, methanol, CH_4 but not H_2) stimulated SZA by 2–3 times (Fig. 14.6). This finding is in a full agreement with data on gas permeability of permafrost, supporting premises that volatile reduced substrates are the best energy sources for microorganisms in a frozen environment. The insoluble polymeric substrates (e.g. lignocellulose, pectin, xylan) are likely to be completely inaccessible. Their utilization requires an extracellular step of hydrolytic breakdown that cannot proceed under fully frozen conditions because neither enzyme nor substrate is able to acquire a flexible tertiary structure needed to form an enzyme–substrate complex. On the other hand, the intracellular space of microbial cells is better fit to severe cold due to production of antifreeze metabolites and self-heating through exothermic catabolic reactions.

14.4.4 Biogenic Heat

Any oxidation reaction generates heat, and microbial respiration is not an exception. Biochemistry typically interprets heat as a non-productive component of respiration, as energy dissipation that reduces growth yield or the capacity of organisms to do mechanical work. Could biogenic heat be a useful part of

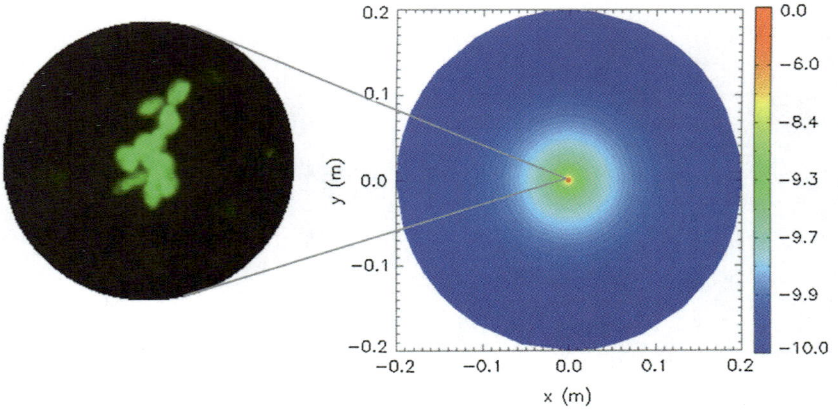

Fig. 14.7 Numerical simulation of the biogenic heat effects on local permafrost warming (Panikov and Zhang unpublished data). Assumptions: size of microbial colony 1 mm^3, respiration rate 0.1 mmol O_2 d^{-1} mm^{-3}, heat quotient 26 kJ (mol O_2)$^{-1}$, ambient temperature −10 °C, permafrost with porosity 0.4. We used a steady-state heat transfer equation solved under spherical coordinates. The *inset* and picture on the left show a real microscopic picture of the yeasts microcolony in Toolik Lake permafrost, cells stained by DTAF

microbial catabolism in frozen habitat? To answer this question, we (Panikov and Zhang, unpublished data) simulated heat generation and transfer for spatially isolated single-yeast colonies usually observed in our experiments with permafrost (Fig. 14.7). Under realistic assumptions of physical and biological properties of the modelled system (thermal conductivity, cell mass, intensity of respiration, etc.), simulation results indicated that microbial self-heating was strong enough to raise the temperature by almost 10 °C inside the colony. The heating effect abruptly declined across the space outside the colony, leaving the permafrost layer unaffected already at the distance of several millimetres. Thus, a heating power of biogenic source is too low to be sensed by traditional geophysical tools aimed at keeping the record of soil temperature. Nevertheless, it is sufficient to sustain steep temperature gradients between metabolizing cells and the surrounding abiotic space as long as the energy source is available. A temperature raise within the microcolony should also result in a respective increase in water availability due to localized partial ice thaw and enhanced pressure of water vapour.

14.5 Physiological State of Microbial Community in Frozen Soil: Growth or Maintenance?

It was hypothesized (Bakermans et al. 2003; Price and Sowers 2004) that in natural frozen habitats, such as permafrost and glaciers, nutrient flux is so low that it could support only the ‘maintenance state’ of microbial cells, which stands for slow catabolic activity without growth. Earlier, the 'maintenance state' was assumed to represent a physiological state for *in situ* microbial populations at any temperature in either terrestrial or oceanic oligotrophic habitats (Chapman and Gray 1986; Morita 1997). The term 'maintenance' means the energy source consumption for physiological functions other than growth, e.g. cell motility, turnover of cellular macromolecules and osmoregulation (keeping a concentration difference between cytoplasm and exterior). In practice, the intensity of *maintenance or basic metabolism* is evaluated indirectly, by recording two variables, the energy source consumption rate (e.g. respiration, R) and SGR, preferentially in a chemostat culture preserving a constancy of cultivation conditions apart from energy flow. The plot of R *versus* SGR is approximated by linear regression, and this intercept formally corresponds to the respiration rate of the non-proliferating cells (SGR = 0) and is called a 'maintenance coefficient' (Pirt 1975). The higher the maintenance goes, the lower is cell yield (Y). In the maintenance state, when ALL consumed substrate is spent for non-growth functions, the yield is zero.

Panikov and Sizova (2007) have tested the hypothesis whether SZA is associated with the non-proliferating maintenance state in a simple incubation experiment: permafrost samples were amended with ^{14}C-ethanol and incubated in a series of subzero temperatures. Three variables were recorded: $^{14}CO_2$ evolution as respiration term (Δp), uptake of ethanol at the end of experiment as a difference between added and residual amount of the water-soluble ^{14}C-compound (Δs) and ^{14}C-incorporation into soil microorganisms as insoluble ^{14}C (Δx). As expected, the consumption was equal to the sum of two production terms $\Delta s = \Delta p + \Delta x$, indicating full recovery of the ^{14}C-balance (Fig. 14.8, left panel). The yield Y was calculated as a ratio $Y = \Delta x/\Delta s$ (Fig. 14.8, right panel), and it varied between 0.33

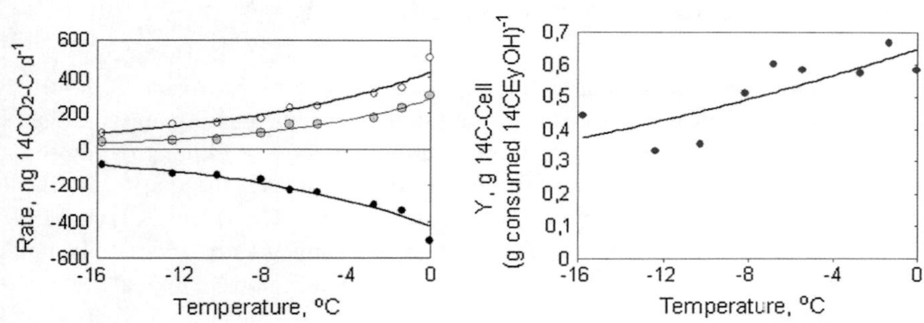

Fig. 14.8 Experimental evidence that permafrost community is growing on added ^{14}C-substrate rather than metabolizes in a maintenance (non-proliferating) state (redrawn with permission from Panikov and Sizova 2007). Permafrost sample was amended with ^{14}C-ethanol and mineral salts and incubated at a series of subzero temperatures. By the end of incubation, we calculated total consumption of added substrate, $^{14}CO_2$ production and ^{14}C incorporation into insoluble fraction presumably into cells of microorganisms grown on ethanol. *Left panel* plot of ^{14}C-uptake data (1), ^{14}C-cells (2) and $^{14}CO_2$ (3) *versus* incubation temperature, uptake is given as negative value and the curve 3 was plotted as a sum ^{14}C-cell + ^{14}C-CO_2 to demonstrate a complete recovery of the ^{14}C-balance. *Right panel* cell yield calculation as Y = ^{14}C-cell/^{14}C-uptake

and 0.67 g C (g C)$^{-1}$ at subzero temperatures similar to above-zero microbial growth (Panikov 1995). Slight decline in *Y* with cooling is expected as a consequence of slowing of microbial growth, but in no way it was a 'maintenance state' as discussed in literature. The most probable reason for the dramatic drop of the yield observed by Bakermans et al. (2003) was the abnormal condition of cultivation of permafrost bacteria in liquid supercooled media. Under natural frozen conditions without antifreeze compounds, microbial metabolism remains productive, and respiring cells sustain normal growth generating new cellular mass. The only difference is that subzero growth is extremely slow with a doubling time about one week at −8 °C and even up to 60 years at −20 °C.

Let us make a preliminary conclusion before moving our story to the main subject, cold-adapted yeasts and other microorganisms. Based on critical analyses of our own and published data, we can with full confidence state that microorganisms in a frozen environment are capable of metabolizing in spite of severe conditions: water deficiency, restricted mass transfer and low level of kinetic energy. We dismissed a number of faulty manifestations of SZA such as 'soil respiration' and left only well-proven evidences based on artefact-free techniques. Contrary to above-zero growth conditions, SZA can be supported only by highly mobile volatile substrates like alcohols, hydrocarbons and volatile fatty acids but not by polymeric and recalcitrant compounds requiring an extracellular step of enzymatic breakdown. The SZA is associated with normal although extremely slow growth generating newborn cells with the yield factor comparable with above-zero growth. Next step is to disclose the identity of organisms able to function in frozen state.

14.6 Microorganisms Responsible for SZA

The majority of available data on microbial diversity in permafrost and seasonally frozen soils were obtained by using culture-dependent (e.g. plating on Petri dishes) or culture-independent techniques through amplification and cloning of ribosomal genes or metagenomic analyses. These data do not answer the question 'who is responsible for SZA', because the colony-forming units (CFU) and extracted DNA could originate from dormant cells resisting freezing but not necessarily being active in frozen ground. With this precautionary note, microbiological studies of permafrost are briefly summarized.

14.6.1 Overall Survey of Permafrost Community

Viable bacteria in permafrost were first documented as a part of investigations of mammoths in Siberia (Becker and Volkmann 1961; Cameron and Morelli 1974). Plating revealed high numbers of viable microorganisms (10^5–10^7 CFU g^{-1}) and high 16S rRNA diversity with 11 established lines of descent of bacteria and some entirely new sequences (Shi et al. 1997; Zhou et al. 1997). Most of the clones belonged to the Alpha- (21 %) and Deltaproteobacteria (26 %), with lesser proportions (<10 %) in the classes of Beta- and Gammaproteobacteria which are typically isolated from soils. The majority of clones (77 %) had sequence similarity less than 95–80 % to those in the database, indicating the predominance of novel not-yet-cultured genera and families. An agreement between plating and 16S rRNA survey was very poor confirming the well-known inadequacy of the plating method to recover the indigenous soil microorganisms. More recent updates on permafrost bacterial diversity were reviewed elsewhere (Steven et al. 2006, 2009; Yergeau et al. 2010). Fungi including yeasts were also surveyed by culture-dependent and culture-independent methods (Gilichinsky et al. 2005; Kochkina et al. 2012) with similar general conclusions about the lack of agreement between them.

Contrary to permafrost, aquatic subzero habitats typically display fairly high agreement between plating and gene-retrieval techniques. Specifically, the analyses of sea ice do not demonstrate a 'great plate anomaly', revealing that up to 65 % of accounted operational taxonomic units (OTUs) are represented by known cultivable species from the two bacterial phyla *Proteobacteria* and *Bacteroidetes* (previously *Cytophaga-Flavobacteria-Bacteroides*) (Gosink and Staley 1995; Irgens et al. 1996; Gosink et al. 1998; Junge et al. 1998; Staley and Gosink 1999; Junge et al. 2002). The reason behind the described difference is that sea ice community is much simpler and younger than permafrost community, every year the sea ice is melted and the process of ice colonization starts from zero in early fall time; even multiyear sea ice in the Arctic Ocean could survive no more than 8–10 melting seasons before circulating out of the Arctic (Deming 2009). The continental subzero water including Lake Vostok accretion ice (Christner et al. 2001), Arctic and Antarctic pack ice (Brinkmeyer et al. 2003), mountainous glaciers (Christner et al. 2003a, b) and cold deep Atlantic sediments (Xu et al. 2003) generally contain more diverse microbial communities, but they are still much less complex than those in permafrost.

Contrary to many other extreme environments, hypersaline, extremely hot, acidic or alkaline, frozen soil communities did not evolve to a specialized state of a strictly selected and narrow list of patient species resistant to particular environmental stress. Instead, permanently frozen soils accommodate a much wider range of microorganisms, including true cold-adapted species displaying SZA and numerous other species preserved in a dormant state since the warmer time in the past when they were active. Therefore, our main task here is to distinguish active permafrost inhabitants versus dormant forms inherited from a warmer geological past. Below, several approaches to solve this problem are considered.

14.6.2 Transient Response of Soil Community to Cooling: Seasonal Observations

An idea behind this approach is very simple: if we follow round-year seasonal dynamics of soil community, then we can identify species involved in subzero processes as those who increase their relative abundance in winter. Dramatic seasonal shifts of communities have been documented for fungi in alpine tundra, boreal forest and temperate grassland, indicating that fungi are not just cold-tolerant organisms but take advantage in a frozen environment (Schadt et al. 2003; Taylor et al. 2010; Dumbrell et al. 2011). Apart from phylogenetic data, this conclusion was supported by the following results:

1. higher winter fungal/bacterial ratios than respective summer ratios as measured by direct microscopy of soil;
2. higher winter fungal/bacterial ratios based on substrate-induced respiration with specific inhibitors of protein synthesis (Lipson et al. 2002);
3. the observation that total tundra soil microbial biomass reached its annual peak under snow, and fungi accounted for most of the biomass (Lipson and Schmidt 2004).

Phylogenetic analysis revealed a high diversity of fungi and three novel clades that constituted the major new groups of fungi (divergent at the subphylum or class level), none of them being represented in culture collections (Schadt et al. 2003). Contrary to fungi, bacterial communities displayed smaller magnitude and direction of changes; in winter, bacterial community had a higher proportion of *Actinobacteria* and members of the *Bacteroidetes* (previously *Cytophaga-Flavobacteria-Bacteroides*, CFB) phylum, indicating that they could be also responsible for SZA (Lipson and Schmidt 2004). In summary, seasonal observation indicates that subzero temperatures do allow the preferential development of fungi. The freezing did not have a stressful effect on their reproduction, although an alternative explanation could be that a winter-induced rise in fungal biomass was the result of reduced grazing pressure caused by winter hibernation of microarthropods, the principal predators of fungal mycelium.

14.6.3 Transient Response of Soil Community to Cooling: Laboratory Incubation Experiments

Under fully controlled laboratory conditions, the competitive advantage of fungi has been confirmed in a most convincing and clear way. We incubated Toolik Lake permafrost samples at -10 °C with ethanol and monitored changes in community structure associated with SZA by PCR-SSCP. In total, three phylogenetic groups were assessed by applying bacterial, archaeal and eukaryotic

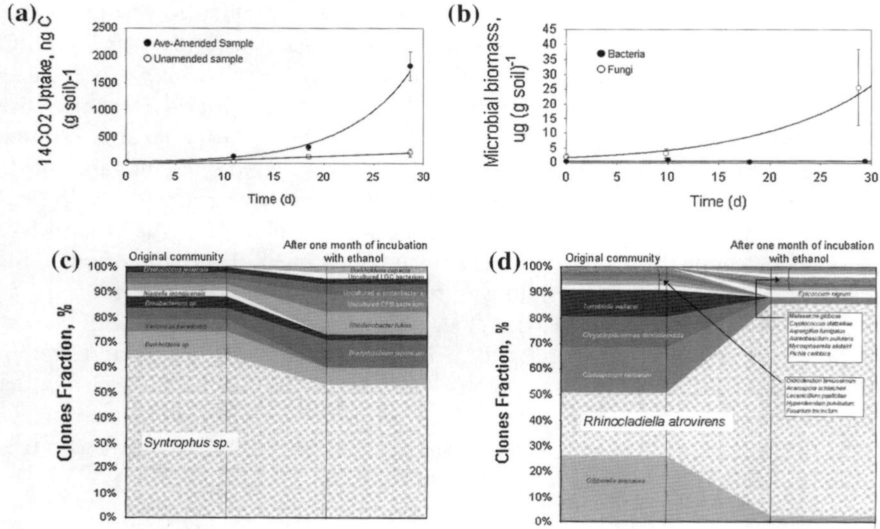

Fig. 14.9 Changes in community structure of permafrost in the course of incubation at −10 °C with ethanol as a C-substrate **a** dynamics of DF indicating exponential growth of microorganisms on added nutrient substrate, **b** example of PCR-SSPC analysis of community structure revealing relative changes in abundance of particular OTUs, **c**, **d** shifts in abundance of bacteria and fungi, respectively, after one month. Note modest or no changes in the bacterial OTUs pattern compared with explosive growth of one single fungal organism (*Rhinocladiella*)

primers (Panikov et al. 2007). Archaea did not contribute significantly to the permafrost community, while bacteria and fungi displayed rich diversity with the following principal difference: the changes in bacterial abundances were minor with some OTUs slightly up or slightly down, while fungi changed dramatically with one single population displaying explosive growth and replacing other species (Fig. 14.9). Another unambiguous indication of the fungal advantage over bacteria was direct microscopy of soil specimens revealing a significant increase in mycelial biomass as compared with stable bacterial counts (Fig. 14.9b).

14.6.4 Stable-Isotope Probing

There are several molecular techniques able to discriminate between active and dormant soil microorganisms. Not all of them could be applied to permafrost, for example, survey based on mRNA still could be misleading because permanent freezing preserves both types of nucleic acids. Some data were obtained by applying stable-isotope probing (SIP). Alaskan permafrost was incubated with ^{13}C-acetate at a range of subzero temperatures from −3 to −20 °C, and then, DNA was extracted and CsCl isopycnic centrifugation was applied to separate heavy ^{13}C-DNA (representing genomes of active microorganisms) from the light ^{12}C-DNA. After PCR amplification of 16S rRNA genes coupled with TRFLP fingerprinting and subsequent cloning, the heavy DNA fraction was used to generate lists of OTUs representing active bacteria. i.e. those organisms who were able to utilize acetate (Tuorto et al. 2011). They belonged mostly to not-yet-cultivable representatives of *Actinobacteria*, Acidobacteria, Betaproteobacteria and Chloroflexi. There was an especially pronounced contribution of Chloroflexi, the group which escaped detection by using molecular techniques other than SIP. Another SIP study of Alaskan permafrost employed bacterial and eukaryotic primers and ^{13}C-cellulose as tested C-source (Taylor et al. 2010) demonstrated the predominance of fungi, e.g. *Sebacina, Geopyxis* and *Geomyces*, in the low-temperature carbon acquisition. Meanwhile, bacterial DNA was less heavily labelled, suggesting that these bacteria (*Cellvibrio* sp., as well as unclassified representatives of Rhizobiales, Sphingobacteriales, and *Actinobacteria*) may be generalists deriving carbon from a variety of sources in addition to cellulose.

14.7 Isolation of Microorganisms by Subzero Enrichment

Most of published subzero isolations were performed with supercooled liquid media containing glycerol as antifreeze (Breezee et al. 2004). The disadvantages of such methods as applied to permafrost are as follows: (1) inhibitory effect of antifreeze; (2) instability of supercooled solutions that only recently received adequate mechanistic explanation in terms of modern physical chemistry (Moore and Molinero 2011); and (3) lack of growth of some fastidious soil organisms requiring the presence of solid phase. We developed a solid-state cultivation system which is free from indicated limitations and closer imitates permafrost as a natural habitat. Liquid nutrient medium is either mixed with microcrystalline cellulose (MCC) or placed inside a polypropylene plastic bag to form a thin liquid film. Growth above or below the freezing point is conveniently recorded as the rates of CO_2 production and/or by direct microscopic observations (Panikov and Sizova 2007). Supercooled and solid media were inoculated with small amounts of permafrost material and incubated at various subzero temperatures until stable enriched cultures were obtained; after that, serial dilutions and plating were used to isolate and purify individual microorganisms.

Interestingly, conventional supercooled liquid enrichment (from 0 to −5 °C) led to the isolation of representatives of the bacterial genera *Polaromonas, Arthrobacter* and *Pseudomonas* from permafrost, similar to those bacteria found in polar sea and Antarctic (Staley and Gosink 1999). The use of solid frozen media consistently ended up with the isolation of microbial eukaryotes; bacteria were detected only as a transient component in primary enrichments. At −8 to −10 °C, several enrichments containing yeasts grew faster than cultures preferentially enriched with mycelial fungi, but both displayed exponential growth with

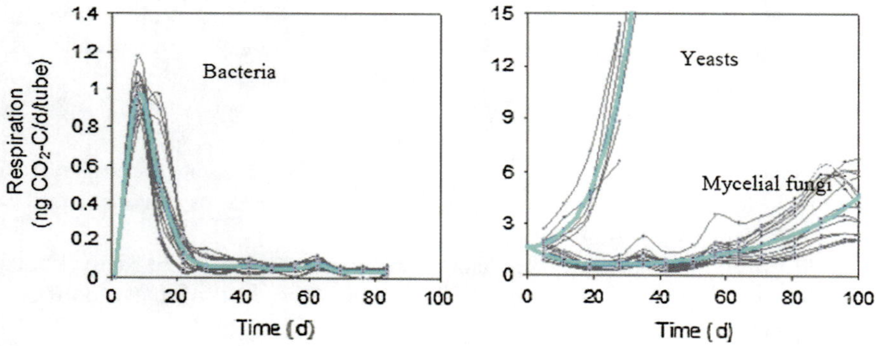

Fig. 14.10 Competitive advantage of fungi over bacteria while growing on solid frozen media (MCC powder + ethanol mineral medium). *Left panel Arthrobacter sp. Right panel* Eukaryotic consortium *Leucosporidium-Geomyces.* Note the difference in y-scale as well as in the growth patterns: declining curves for bacteria and exponential increase for fungi. Reproduced with permission from Panikov and Sizova (2007)

sustained and synchronous oscillations (Fig. 14.10). Bacterial growth progressively decelerated after an initial burst of activity. We never observed the exponential growth pattern in bacteria on solid frozen media and growth always stopped long before the complete consumption of limiting nutrients indicating some other non-nutritional obstacles to culture development.

The isolated yeasts were able to grow exponentially in frozen media without any added antifreeze compounds. Taxonomically, they resembled psychrophilic yeasts already isolated at low positive temperatures belonging to the genera *Leucosporidium, Cryptococcus* and *Mrakia* (Margesin 2009; Rossi et al. 2009; Connell et al. 2010; Kachalkin 2010; Thomas-Hall et al. 2010; Moore and Molinero 2011; Turchetti et al. 2011; Uetake et al. 2011; Buzzini et al. 2012; Maggi et al. 2012; Singh et al. 2013). The mycelial fungi mostly belonged to genus *Geomyces*, and this finding was confirmed by other researchers (Waldrop et al. 2008). Other diverse fungi potentially contributing to SZA were found to be mycorrhiza (Taylor et al. 2010; Bent et al. 2011; Kennedy et al. 2011; Maggi et al. 2012; Timling and Taylor 2012), which require more complex media for their isolation.

14.8 Why are Fungi Better Adapted to Frozen Environment?

Let us summarize a current status of subzero microbiology studies. Culture-independent molecular data and isolations agree in conclusion that fungi are better adapted to frozen habitats than bacteria. This is based on several indirect and direct evidences including

- The observation that fungal biomass (counted by direct microscopy) is increasing during winter season or upon soil subzero incubation with added C-sources;
- The ratio fungi/bacteria derived from microscopy or selective antibiotic treatments also increases at subzero temperature;
- Taxonomic surveys taken as time series consistently indicate associated with freezing the increase in abundance of few selected fungal organisms versus an unclear stochastic pattern of bacterial fluctuations;
- Solid frozen media always give raise to fungal enrichment, while bacteria could be isolated only on supercooled liquid media with added synthetic antifreeze compounds;
- Fungal isolates (yeasts and mycelial forms) display exponential growth in frozen media and completely consume limiting nutrients which is in contrast to progressively degenerating bacterial subzero growth.

On the top of the listed facts related specifically to SZA, we can rely on extensive contemporary and past studies of low-temperature ecosystems referring to yeasts as most frequently isolated organisms, and, what is the most important, able to carry out biogeochemical processes of great biotechnological importance in cold soils (Margesin and Schinner 1994; Margesin et al. 2003, 2007; Bergauer et al. 2005; Turchetti et al. 2008; Margesin 2009; Rossi et al. 2009; Thomas-Hall et al. 2010; Turchetti et al. 2011; Buzzini et al. 2012; Maggi et al. 2012).

The principal physiological difference between fungi and bacteria displaying SZA is that subzero growth is a natural state of fungal populations and is a self-sustained process leading to continuous generation of progeny, while bacteria are just preserving their viability with limited cellular proliferation.

What is behind the indicated difference in terms of known molecular mechanisms? All published reviews list several specific adaptations to cold lifestyle (Chattopadhyay 2000; Deming 2009; Margesin 2009; Rossi et al. 2009; Thomas-Hall et al. 2010; Turchetti et al. 2011; Buzzini et al. 2012; Maggi et al. 2012):

- Changes in the fatty acid profile of the cell membrane to maintain their optimum fluidity at low temperature;
- Synthesis of defensive pigments (e.g. carotenoids) which also optimize membrane functioning under cold conditions;
- Modification of protein synthesis directed to the composition of specific amino acids making the bulk of cellular proteins more cold resistant and functional;
- Changes in kinetic properties of key enzymes (K_m, V^{max}, activation energy E_a) shifting down their optimal temperature;
- Intracellular accumulation of cryoprotecting compounds, from monomeric osmolytes like trehalose, glycine betaine to antifreeze proteins;
- Protective capsular material around cells;
- Stress-response heat-shock and cold-shock proteins;
- Degradosome, a self-digesting system in cells.

An exhaustive overview on the adaptation strategies of cold-adapted yeasts to low temperatures is reported in Chaps. 9, 10, 11, 12, 13, 14 and 15.

By paying well-deserved tributes to all listed mechanisms, we have to admit that none of them gives a clue why both filamentous fungi and yeasts are better adapted to freezing than bacteria. In fact, most of the mentioned qualities were initially discovered in bacteria and then found in fungi. Minor differences in cryoprotecting molecules or membrane fatty acids are not likely to account for the exceptionally deep difference in growth patterns of fungi versus bacteria in frozen media. Microbial ecologists often refer to some other differences between fungi and bacteria, specifically

1. Fungi are mostly moderate acidophiles (pH optimum for growth 4–5) compared to neutrophilic and alkalophilic bacteria, which gives fungi a better chance to colonize acidic habitats such as the majority of soils.
2. Fungi generally display higher non-specific resistance to several stresses, e.g. desiccation, radiation, heavy metals and other toxic compounds, while bacteria tend to show essentially higher resistance to specific stress factors. As a result, extreme resistance to acidity, desiccation or toxins is found among respective prokaryotes.
3. Fungi are better colonizers of heterogeneous coarse particulate environments like soil, peat, plant litter due to their geotropic growth style resulting in formation of a mycelial network, while bacteria are more successful colonizers of homogeneous aquatic habitats (all natural waters from pools to ocean, human/animals bodies, beverages, etc.), they are spread as motile cells and form surface biofilms rather than networks.
4. Fungi and yeasts have a bigger size of cells (smaller surface-to-volume ratio) and a more complex life cycle, and as a consequence, on average they grow and metabolize slower than prokaryotic organisms although exceptions are numerous.

We can capitalize in our discussion these loosely defined qualitative differences, and the most important differences are outlined in points 3 and 4. The first crucial point is mobility within the permafrost body. Motile bacteria are completely helpless; their position is fixed in the frozen space. Fungi are able to move simultaneously with apical growth of their hyphal tips. It is interesting that under poor supply of nutrients, linear hyphae extension is getting faster: fungi use this mechanism to abandon poor microsites as soon as possible by ingrowth into more favourable microloci (Panikov 1995). Note that in permafrost, such style of mobility is highly beneficial in view of the extremely high degree of heterogeneity and mosaic distribution of patches of UW and organic particles. Even with a volatile organic C-substrate (ethanol) growing cells need to absorb mineral nutrients from the nearby frozen outside space. As illustrated in Fig. 14.11, immobile bacterial cells quickly deplete a deficient nutrient around the microcolony, and their growth stops. Fungi have a much higher linear extension rate of their cells, which results in moving through the depletion zone to a new non-

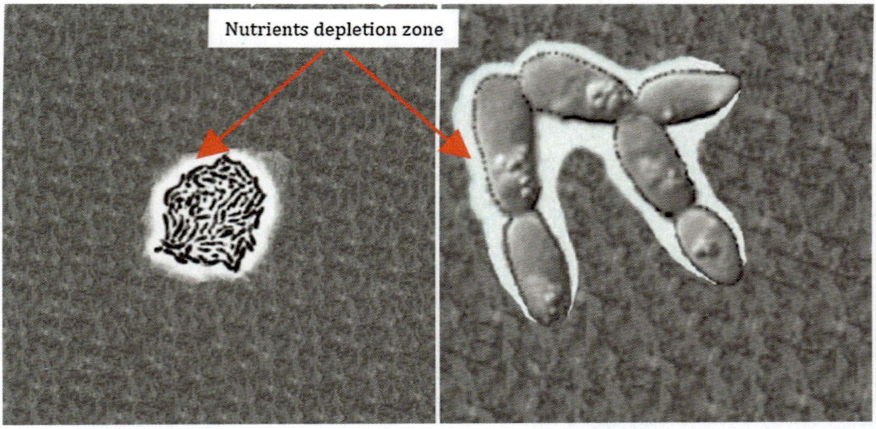

Fig. 14.11 Growth advantage of yeasts (*right panel*) over bacteria (*left panel*) in frozen environment. White halo around microcolonies indicate depletion of deficient nutrient substrates around cells. Freezing does not allow transport of non-volatile nutrients across the microspace, and cells remain immobile because of lack of continuous films of unfrozen water. However, yeasts cells are bigger than bacteria and their linear extension rate is higher, and as a result, the growing tips of yeasts cells or hyphae of mycelial fungi progress outward into non-depleted space. The second advantage is a possibility for exchange of deficient elements across mycelial/pseudomycelial network. The third potential opportunity is essentially larger cell mass of individual fungal cell at smaller surface-to-volume ratio that reduces heat loss and helps to keep the temperature inside cells higher than ambient

exploited space containing nutrient resources. Moreover, the large size of cells allows fungi to transfer deficient nutrient elements from peripheral parts of the mycelium to growing tips.

Finally, the size and low surface-to-volume ratio (what is typically recognized as disadvantage of fungi compared to bacteria) play a highly positive role in permafrost: fungi manage much better to keep their biogenic heat inside the living

Fig. 14.12 Hypothetical population dynamics of microorganisms in tundra taking into account SZA of the cold-adapted yeasts and fungi. *Top panel* C-cycle in tundra. The *arrows* indicate C-fluxes including root exudation, rhizodeposition and spring release of reserved nutrients (*a*), *plant litter* formation (*b*), decomposition into soluble organic matter (*SOM*) (*c*), microbial uptake and oxidation of *SOM* (*d*), CO_2 mass transfer from soil to atmosphere (*e*), plant photosynthesis (*f*) and diffusive flux of volatile organic matter and gases (*g*). Four groups of soil microorganisms are shown in *circles*: mycelial fungi, yeasts, bacteria and oligotrophic microorganisms outside the decomposition zone. *Bottom panel* Soil temperature at 30 cm depth (*1*); *green line* depicts growth of bacteria and fungi (*2*) utilizing root exudates and litter of tundra plants; *blue dotted* line represents subzero growth of yeasts initiated during early spring time (*3*); *grey line* (*4*) reflects near-zero growth of bacteria residing in permafrost scavenging volatile and gaseous substrates. *Black arrows* indicate nutrient fluxes of various intensities including root exudation and rhizodeposition (*a*), *plant litter* formation (*b*), spring release of reserved nutrients (*c*) and slow diffusive flux of volatile organic matter and gases (*d*) ►

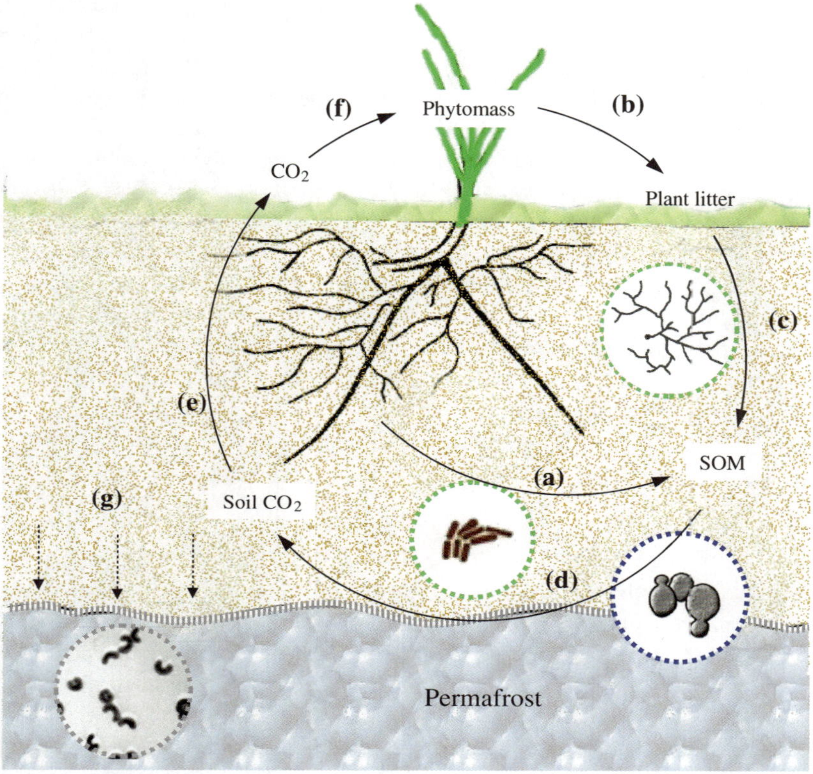
(f)
Phytomass
(b)
CO_2
Plant litter
(c)
(e)
SOM
(a)
(g)
Soil CO_2
(d)
Permafrost

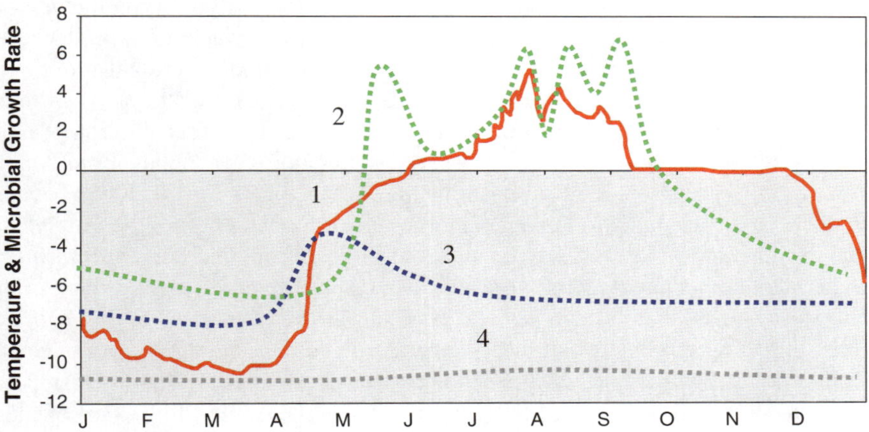
Temperaure & Microbial Growth Rate
8
6
4
2
0
-2
-4
-6
-8
-10
-12
1
2
3
4
J F M A M J J A S O N D

mass. Higher cell mass produces more heat and heat transfer is limited by a smaller relative surface area. Quite possible, there are special biological adaptations to further reduce heat conductivity of fungal cells, e.g. by formation of air-filled vesicles or lipid granules under the cell wall to prevent fast heat exchange through cell boundary and secure biogenic heat.

Yeasts differ from mycelial fungi by their biomorphological structure: budding or dividing cells versus mycelial network. However, yeasts 'prefer' unicellular lifestyle under optimal warm conditions and easily switch to reduced mycelial morphotype under more challenging environmental conditions. Besides even a single-yeast cell is much bigger than a bacterial cell, so all reasoning above is fully applied to all fungi, mycelial or unicellular. It is quite possible that yeasts are better adapted to explosive growth style displaying high SZA under relaxed conditions, e.g. in response to temporal release of nutrients or spring rise of soil temperature approaching zero.

14.9 Conclusions

In conclusion, let us try to outline a 'big picture' concerning the relative contribution of various microbial groups to the functioning of frozen or unfrozen ecosystems. We need here to make a step forward from single factors to a multidimensional ecological space. Figure 14.12 illustrates our approach by giving a simulated population dynamic of several interacting microbial groups functioning in tundra. The permanently frozen subsoil is a domain of near-zero growth of cold-adapted microorganisms (fungi and extremophilic bacteria like the recently discovered *Chloroflexi* group) who receive nutrients as very low intensity diffusive flow of volatile organic compounds generated in surface soil layer (anaerobic decay of soil organic matter with formation of volatile fermentation products) or from deeper lithosphere layers (emanation gases). Another microbial group tentatively represented by psychrophilic yeasts is responding to early spring warming by explosive growth in a still-frozen soil. The actual reason for initiation of their explosive growth could be spring release of organic compounds from tundra plants; their heterotrophic metabolism including root exudation could be activated by increased solar radiation under thinned snow cover and before the start of the photosynthetic uptake of CO_2. Finally, the bulk of soil microorganisms, fungi, bacteria and associated microbivore soil fauna, show up accelerated growth in the late spring, immediately after soil thaw. Then, growth fluctuates during summer time, reflecting changeable surface temperature and nutrients flow from plants as well as a grazing pressure from soil micro- and mesofauna. In the fall, temporal burst of microbial activity is fuelled by plant senescence and formation of plant litter. Soil freezing leads to the decline of microbial growth rate, even of the cold-adapted fungi. However, net growth rates remain positive, which results in accumulation of fungal biomass because of prevention of grazing by winter hibernation of soil animals.

Based on the described simplified scenario, true and continuous SZA in deep permafrost has a relatively small impact on the tundra ecosystem and its C- and N-cycles. However, it could have a significant long-term effect on the integrity and stability of deep subsurface permafrost. Global warming could accelerate SZA and stimulate permafrost degradation long before its physical thaw.

References

Alonso-Saez L, Galand PE, Casamayor EO, Pedros-Alio C, Bertilsson S (2010) High bicarbonate assimilation in the dark by *Arctic bacteria*. ISME J 4:1581–1590

Amato P, Doyle S, Christner BC (2009) Macromolecular synthesis by yeasts under frozen conditions. Environ Microbiol 11:589–596

Bakermans C, Tsapin AI, Souza-Egipsy V, Gilichinsky DA, Nealson KH (2003) Reproduction and metabolism at −10 °C of bacteria isolated from Siberian permafrost. Environ Microbiol 5:321–326

Becker RE, Volkmann CM (1961) A preliminary report on the bacteriology of permafrost in the Fairbanks area. Proc Alaskan Sci Conf 12:188

Bent E, Kiekel P, Brenton R, Taylor DL (2011) Root-associated ectomycorrhizal fungi shared by various boreal forest seedlings naturally regenerating after a fire in interior Alaska and correlation of different fungi with host growth responses. Appl Environ Microbiol 77:3351–3359

Bergauer P, Fonteyne PA, Nolard N, Schinner F, Margesin R (2005) Biodegradation of phenol and phenol-related compounds by psychrophilic and cold-tolerant alpine yeasts. Chemosphere 59:909–918

Breezee J, Cady N, Staley JT (2004) Subfreezing growth of the sea ice bacterium *Psychromonas ingrahamii*. Microbial Ecol 47:300–304

Brinkmeyer R, Knittel K, Jurgens J, Weyland H, Amann R, Helmke E (2003) Diversity and structure of bacterial communities in Arctic versus Antarctic pack ice. Appl Environ Microb 69:6610–6619

Buzzini P, Branda E, Goretti M, Turchetti B (2012) Psychrophilic yeasts from worldwide glacial habitats: diversity, adaptation strategies and biotechnological potential. FEMS Microbiol Ecol 82:217–241

Cameron RE, Morelli FA (1974) Viable microorganisms from ancient Ross Island and Taylor Valley drill core. Antarct J US 9:113–116

Campen RK, Sowers T, Alley RB (2003) Evidence of microbial consortia metabolizing within a low-latitude mountain glacier. Geology 31:231–234

Carpenter EJ, Lin S, Capone DG (2000) Bacterial activity in South Pole snow. Appl Environ Microbiol 66:4514–4517

Cavicchioli R, Siddiqui KS, Andrews D, Sowers KR (2002) Low-temperature extremophiles and their applications. Curr Opin Biotechnol 13:253–261

Chapman SJ, Gray TRG (1986) Importance of cryptic growth, yield factors and maintenance energy in models of microbial growth in soil. Soil Biol Biochem 18:1–4

Chattopadhyay MK (2000) Cold-adaptation of Antarctic microorganisms—possible involvement of viable but nonculturable state. Polar Biol 23:223–224

Christner BC (2002) Incorporation of DNA and protein precursors into macromolecules by bacteria at −15 °C. Appl Environ Microbiol 68:6435–6438

Christner BC, Mosley-Thompson E, Thompson L, Reeve J (2001) Isolation of bacteria and 16S rDNAs from Lake Vostok accretion ice. Environ Microbiol 3:570–577

Christner BC, Mosley-Thompson E, Thompson LG, Reeve JN (2003a) Bacterial recovery from ancient ice. Environ Microbiol 5:433–436

Christner BC, Kvitko BH, Reeve JN (2003b) Molecular identification of bacteria and eukarya inhabiting an Antarctic cryoconite hole. Extremophiles 7:177–183

Connell LB, Redman R, Rodriguez R, Barrett A, Iszard M, Fonseca A (2010) *Dioszegia antarctica* sp. nov. and *Dioszegia cryoxerica* sp. nov., psychrophilic basidiomycetous yeasts from polar desert soils in Antarctica. Int J Syst Evol Microbiol 60:1466–1472

Coyne PI, Kelley JJ (1971) Release of carbon dioxide from frozen soil to the arctic atmosphere. Nature 234:407–408

Deming JW (2002) Psychrophiles and polar regions. Curr Opin Microbiol 5:301–309

Deming JW (2009) Extremophiles: cold environments. In: Schaechter M (ed) Encyclopedia of microbiology, 3rd edn. Academic Press, Oxford, pp 147–158

Dise NB (1992) Winter fluxes of methane from Minnesota peat lands. Biogeochemistry 17:71–83

Dumbrell AJ, Ashton PD, Aziz N, Feng G, Nelson M, Dytham C, Fitter AH, Helgason T (2011) Distinct seasonal assemblages of *arbuscular mycorrhizal fungi* revealed by massively parallel pyrosequencing. New Phytol 190:794–804

Fahnestock JT, Jones MH, Brooks PD, Walker DA, Welker JM (1998) Winter and early spring CO_2 efflux from tundra communities of northern Alaska. J Geophys Res 103:29023–29027

Fahnestock JT, Jones MH, Welker JM (1999) Wintertime CO_2 efflux from arctic soils: implications for annual carbon budgets. Global Biogeochem Cy 13:775–779

Finegold L (1996) Molecular and biophysical aspects of adaptation of life to temperatures below the freezing point. Adv Space Res 18:87–95

Friedmann EI, Ocampo-Friedmann R (1984) The antarctic cryptoendolithic ecosystem: relevance to exobiology. Orig Life Evol B 14:771–776

Friedmann EI, Kappen L, Meyer MA, Nienow JA (1993) Long-term productivity in the cryptoendolithic microbial community of the Ross Desert, Antarctica. Microb Ecol 25:51–69

Gilichinsky D, Rivkina E, Bakermans C, Shcherbakova V, Petrovskaya L, Ozerskaya S, Ivanushkina N, Kochkina G, Laurinavichuis K, Pecheritsina S, Fattakhova R, Tiedje JM (2005) Biodiversity of cryopegs in permafrost. FEMS Microbiol Ecol 53:117–128

Gosink JJ, Staley JT (1995) Biodiversity of gas vacuolate bacteria from Antarctic sea ice and water. Appl Environ Microbiol 61:3486–3489

Gosink JJ, Woese CR, Staley JT (1998) *Polaribacter* gen. nov., with three new species, *P. irgensii* sp. nov., *P. franzmannii* sp. nov. and *P. filamentus* sp. nov., gas vacuolate polar marine bacteria of the Cytophaga-Flavobacterium- Bacteroides group and reclassification of *Flectobacillus glomeratus* as *Polaribacter glomeratus* comb. nov. Int J Syst Evol Microbiol 48:223–235

Grogan P, Chapin IFS (1999) Arctic soil respiration: effects of climate and vegetation depend on season. Ecosystems 2:451–459

Helmke E, Weyland H (2004) Psychrophilic versus psychrotolerant bacteria—occurrence and significance in polar and temperate marine habitats. Cell Mol Biol 50:553–561

Irgens RL, Gosink JJ, Staley JT (1996) *Polaromonas vacuolata* gen. nov., sp. nov., a psychrophilic, marine, gas vacuolate bacterium from Antarctica. Int J Syst Evol Microbiol 46:822–826

Jakosky BM, Nealson KH, Bakermans C, Ley RE, Mellon MT (2003) Subfreezing activity of microorganisms and the potential habitability of Mars' polar regions. Astrobiology 3:343–350

Johnson SS, Hebsgaard MB, Christensen TR, Mastepanov M, Nielsen R, Munch K, Brand T, Gilbert MT, Zuber MT, Bunce M, Ronn R, Gilichinsky D, Froese D, Willerslev E (2007) Ancient bacteria show evidence of DNA repair. Proc Natl Acad Sci USA 104:14401–14405

Junge K, Gosink JJ, Hoppe HG, Staley JT (1998) Arthrobacter, Brachybacterium and Planococcus isolates identified from antarctic sea ice brine. Description of *Planococcus mcmeekinii* sp. nov. Syst Appl Microbiol 21:306–314

Junge K, Imhoff F, Staley T, Deming JW (2002) Phylogenetic diversity of numerically important Arctic sea-ice bacteria cultured at subzero temperature. Microb Ecol 43:315–328

Junge K, Eicken H, Deming JW (2004) Bacterial activity at −2 to −20 °C in Arctic wintertime sea ice. Appl Environ Microbiol 70:550–557

Kachalkin AV (2010) New data on the distribution of some psychrophilic yeasts in the Moscow Region. Microbiology (Moscow) 79:843–847

Kappen L (1988) Field measurements of carbon dioxide exchange of the Antarctic lichen Usnea *sphacekta* in the Frozen state. Antarct Sci 1:31–34

Kappen L (1993) Lichens in the Antarctic region. In: Friedmann EI (ed) Antarctic microbiology. Wiley-Liss, New York, pp 433–490

Kappen L, Friedmann EI (1983) Ecophysiology of lichens in the dry valleys of Southern Victoria Land, Antarctica. Polar Biol 1:227–232

Kappen L, Schroeter B, Scheidegger C, Sommerkorn M, Hestmark G (1996) Cold resistance and metabolic activity of lichens below 0 °C. Proc Symp COSPAR Sci Commission F Adv Space Res 18:119–128

Kappen L, Schroeter B, Green TGA, Seppelt RD (1998) Chlorophyll fluorescence and CO_2 exchange of *Umbilicaria aprina* under extreme light stress in the cold. Oecologia 113:325–331

Kennedy AH, Taylor DL, Watson LE (2011) Mycorrhizal specificity in the fully *mycoheterotrophic Hexalectris* Raf. (Orchidaceae: Epidendroideae). Mol Ecol 20:1303–1316

Kochkina G, Ivanushkina N, Ozerskaya S, Chigineva N, Vasilenko O, Firsov S, Spirina E, Gilichinsky D (2012) Ancient fungi in Antarctic permafrost environments. FEMS Microbiol Ecol 82:501–509

Kushner D (1981) Extreme environments: are there any limits to life? In: Ponnamperuna C (ed) Comets and the origin of life. Reidel, New York, pp 241–248

Lange OL, Kappen K (1972) Photosynthesis of lichens from Antarctica. Antarct Res S 20:83–95

Larkin JM, Stokes JL (1968) Growth of psychrophilic microorganisms at subzero temperatures. Can J Microbiol 14:97–101

Lipson DA, Schmidt SK (2004) Seasonal changes in an alpine soil bacterial community in the Colorado Rocky Mountains. Appl Environ Microbiol 70:2867–2879

Lipson DA, Schadt CW, Schmidt SK (2002) Changes in soil microbial community structure and function in an alpine dry meadow following spring snow melt. Microb Ecol 43:307–314

Maggi O, Tosi S, Angelova M, Lagostina E, Fabbri AA, Pecoraro L, Altobelli E, Picco AM, Savino E, Branda E, Turchetti B, Zotti M, Vizzini A, Buzzini P (2012) Adaptation of fungi, including yeasts, to cold environments. Plant Biosyst 147:247–258

Margesin R (2009) Effect of temperature on growth parameters of psychrophilic bacteria and yeasts. Extremophiles 13:257–262

Margesin R, Schinner F (1994) Properties of cold-adapted microorganisms and their potential role in biotechnology. J Biotechnol 33:1–14

Margesin R, Gander S, Zacke G, Gounot A, Schinner F (2003) Hydrocarbon degradation and enzyme activities of cold-adapted bacteria and yeasts. Extremophiles 7:451–458

Margesin R, Fonteyne PA, Schinner F, Sampaio JP (2007) *Rhodotorula psychrophila* sp. nov., *Rhodotorula psychrophenolica* sp. nov and *Rhodotorula glacialis* sp. nov., novel psychrophilic basidiomycetous yeast species isolated from alpine environments. Int J Syst Evol Microbiol 57:2179–2184

Marion GM, Fritsen CH, Eicken H, Payne MC (2003) The search for life on Europa: limiting environmental factors, potential habitats, and Earth analogues. Astrobiology 3:785–811

Marx JC, Blaise V, Collins T, D'Amico S, Delille D, Gratia E, Hoyoux A, Huston AL, Sonan G, Feller G, Gerday C (2004) A perspective on cold enzymes: current knowledge and frequently asked questions. Cell Mol Biol 50:643–655

Mastepanov M, Sigsgaard C, Dlugokencky EJ, Houweling S, Strom L, Tamstorf MP, Christensen TR (2008) Large tundra methane burst during onset of freezing. Nature 456:628–630

Melloh RA, Crill PM (1996) Winter methane dynamics in a temperate peat land. Global Biogeochem Cy 10:247–254

Michener HD, Elliott RP (1964) Minimum growth temperatures for food-poisoning, fecal-indicator, and psychrophilic microorganisms. Adv Food Res 13:349–396

Moore EB, Molinero V (2011) Structural transformation in supercooled water controls the crystallization rate of ice. Nature 479:506–508

Morita RY (1975) Psychrophilic bacteria. Bacteriol Rev 39:144–167
Morita RY (1997) Bacteria in oligotrophic environments. Chapman and Hall, New York
Oechel WC, Vpurlitis G, Hastings SJ (1997) Cold season CO_2 emission from arctic soils. Global Biogeochem Cy 11:163–172
Ostroumov VE, Siegert C (1996) Exobiological aspects of mass transfer in microzones of permafrost deposits. Adv Space Res 18:79–86
Panikov NS (1995) Microbial growth kinetics. Chapman and Hall, London
Panikov NS (2009) Microbial activity in frozen soils. In: Margesin R (ed) Permafrost soils. Springer, Berlin, pp 119–148
Panikov NS, Dedysh SN (2000) Cold season CH_4 and CO_2 emission from boreal peat bogs (West Siberia): winter fluxes and thaw activation dynamics. Global Biogeochem Cy 14:1071–1080
Panikov NS, Sizova MV (2007) Growth kinetics of microorganisms isolated from Alaskan soil and permafrost in solid media frozen down to −35 °C. FEMS Microbiol Ecol 59:500–512
Panikov NS, Sizova MV, Nechitaylo TJ (2007) Permafrost: a depository or subzero incubator for psychrophilic microroganisms? Small Matters: Microbes and their Role in Conservation Twelfth Annual Spring Symposium. The American Museum of Natural History in New York City. 26–27 April 2007
Panikov NS, Flanagan PW, Oechel WC, Mastepanov MA, Christensen TR (2006) Microbial activity in soils frozen to below −39 °C. Soil Biol Biochem 38:785–794
Pirt SJ (1975) Principles of microbe and cell cultivation. Blackwell Science, Oxford
Price PB (2000) A habitat for psychrophiles in deep Antarctic ice. Proc Natl Acad Sci USA 97:1247–1251
Price PB (2007) Microbial life in glacial ice and implications for a cold origin of life. FEMS Microbiol Ecol 59:217–231
Price PB, Sowers T (2004) Temperature dependence of metabolic rates for microbial growth, maintenance, and survival. Proc Natl Acad Sci USA 101:4631–4636
Rivkina EM, Friedmann EI, McKay CP, Gilichinsky DA (2000) Metabolic activity of permafrost bacteria below the freezing point. Appl Environ Microbiol 66:3230–3233
Rivkina EM, Laurinavichus KS, Gilichinsky DA, Shcherbakova VA (2002) Methane generation in permafrost sediments. Dokl Acad Nauk SSSR 383:179–181
Rivkina E, Laurinavichius K, McGrath J, Tiedje J, Shcherbakova V, Gilichinsky D (2004) Microbial life in permafrost. Adv Space Res 33:1215–1221
Rossi M, Buzzini P, Cordisco L, Amaretti A, Sala M, Raimondi S, Ponzoni C, Pagnoni UM, Matteuzzi D (2009) Growth, lipid accumulation, and fatty acid composition in obligate psychrophilic, facultative psychrophilic, and mesophilic yeasts. FEMS Microbiol Ecol 69:363–372
Sattler B, Puxbaum H, Psenner R (2001) Bacterial growth in supercooled cloud droplets. Geophys Res Letters 28:239–242
Schadt CW, Martin AP, Lipson DA, Schmidt SK (2003) Seasonal dynamics of previously unknown fungal lineages in tundra soils. Science 301:1359–1361
Schroeter B, Green TGA, Seppelt RD, Kappen L (1992) Monitoring photosynthetic activity of crustose lichens using a PAM-2000 fluorescence system. Oecologia 92:457–462
Shi T, Reeves RH, Gilichinsky DA, Friedmann EI (1997) Characterization of viable bacteria from Siberian permafrost by 16S rDNA sequencing. Microb Ecol 33:169–179
Singh P, Tsuji M, Singh SM, Roy U, Hoshino T (2013) Taxonomic characterization, adaptation strategies and biotechnological potential of cryophilic yeasts from ice cores of Midre Lovénbreen glacier, Svalbard, Arctic. Cryobiology 66:167–175
Staley JT, Gosink JJ (1999) Poles apart: biodiversity and biogeography of sea ice bacteria. Annu Rev Microbiol 53:189–215
Steven B, Léveillé R, Pollard WH, Whyte LG (2006) Microbial ecology and biodiversity in permafrost. Extremophiles 10:259–267
Steven B, Niederberger TD, Whyte LG (2009) Bacterial and Archaeal diversity in permafrost. In: Margesin R (ed) Permafrost soils. Springer, Berlin, pp 59–72

Taylor DL, Herriott IC, Stone KE, McFarland JW, Booth MG, Leigh MB (2010) Structure and resilience of fungal communities in Alaskan boreal forest soils. Can J Forest Res 40:1288–1301

Thomas-Hall SR, Turchetti B, Buzzini P, Branda E, Boekhout T, Theelen B, Watson K (2010) Cold-adapted yeasts from Antarctica and the Italian Alps-description of three novel species: *Mrakia robertii* sp. nov., *Mrakia blollopis* sp. nov. and *Mrakiella niccombsii* sp. nov. Extremophiles 14:47–59

Tibbles B, Harris J (1996) Use of radiolabelled thymidine and leucine to estimate bacterial production in soils from continental Antarctica. Appl Environ Microbiol 62:694–701

Timling I, Taylor DL (2012) Peeking through a frosty window: molecular insights into the ecology of Arctic soil fungi. Fungal Ecology 5:419–429

Tuorto SJ, Häggblom M, Panikov N, Zhang T, Kerkhof LJ (2011) Measuring biodiversity and cell division in permafrost. In: Proceedings of gordon research conference applied and environmental microbiology functional interactions from molecules to biomes, Mount Holyoke College, South Hadley, 10–15 July 2011

Turchetti B, Buzzini P, Goretti M, Branda E, Diolaiuti G, D'Agata C, Smiraglia C, Vaughan-Martini A (2008) Psychrophilic yeasts in glacial environments of Alpine glaciers. FEMS Microbiol Ecol 63:73–83

Turchetti B, Thomas Hall SR, Connell LB, Branda E, Buzzini P, Theelen B, Muller WH, Boekhout T (2011) Psychrophilic yeasts from Antarctica and European glaciers: description of *Glaciozyma* gen. nov., *Glaciozyma martinii* sp. nov. and *Glaciozyma watsonii* sp. nov. Extremophiles 15:573–586

Uetake J, Kohshima S, Nakazawa F, Takeuchi N, Fujita K, Miyake T, Narita H, Aizen V, Nakawo M (2011) Evidence for propagation of cold-adapted yeast in an ice core from a Siberian Altai glacier. J Geophys Res-Biogeo 116:G01019

Waldrop MP, White RI, Douglas TA (2008) Isolation and identification of cold-adapted fungi in the Fox Permafrost Tunnel, Alaska. In: Proceedings of the 9th International Conference on Permafrost, pp 1887–1891

Watanabe K, Wake T (2009) Measurement of unfrozen water content and relative permittivity of frozen unsaturated soil using NMR and TDR. Cold Reg Sci Technol 59:34–41

Whalen SC, Reeburgh WS (1988) A CH_4 time series for tundra environments. Global Biogeochem Cy 2:399–409

Wolfe J, Bryant G, Koster KL (2002) What is 'unfreezable water', how unfreezable is it, and how much is there? Cryo Letters 23:157–166

Xu Y, Nogi Y, Kato C, Liang Z, Ruger HJ, De Kegel D, Glansdorff N (2003) *Moritella profunda* sp. nov. and *Moritella abyssi* sp. nov., two psychropiezophilic organisms isolated from deep Atlantic sediments. Int J Syst Evol Microbiol 53:533–538

Yergeau E, Hogues H, Whyte LG, Greer CW (2010) The functional potential of high Arctic permafrost revealed by metagenomic sequencing, qPCR and microarray analyses. ISME J 4:1206–1214

Zecchinon L, Claverie P, Collins T, D'Amico S, Delille D, Feller G, Georlette D, Gratia E, Hoyoux A, Meuwis M-A, Sonan G, Gerday C (2001) Did psychrophilic enzymes really win the challenge? Extremophiles 5:313–321

Zhou J, Davey ME, Figueras JB, Rivkina E, Gilichinsky D, Tiedje JM (1997) Phylogenetic diversity of a bacterial community determined from Siberian tundra soil DNA. Microbiology (UK) 143:3913–3919

Zimov SA, Semiletov IP, Davidov SP, Voropaev YV, Prosiannikov SF, Wong CS, Chan Y-H (1993) Wintertime CO_2 emission from soils of northeastern Siberia. Arctic 46:197–204

Chapter 15
Fundamentals of Cold-Active Enzymes

Charles Gerday

Abstract Cold-active enzymes are produced by organisms adapted to permanently cold habitats. Due to the depressive effect of low temperatures on reaction rates, these enzymes have to be adapted to secure appropriate reaction rates in those organisms that often thrive in environments characterized by temperatures close or below the freezing point of water. They are encountered in all prokaryotic or eukaryotic organisms adapted to cold such as microorganisms, invertebrates, insects and fish originating from the Arctic and Antarctic zones, as well as from alpine regions, glaciers or permafrost zones. They are characterized by a high specific activity at low temperatures, in any case higher than that of their mesophilic and thermophilic counterparts. This higher specific activity is generally accompanied by a decrease in thermal stability illustrated by a shift of the apparent optimum towards low temperatures, and by an important decrease in the thermodynamic stability characterized by a significantly lower stabilization enthalpy. The generally low stability induces an increase in the flexibility of the overall edifice or of crucial zones for activity of the molecular structure. There is apparently a continuum in the adaptation since some enzymes display extreme adaptation illustrated by a severe shift of the activity towards low temperatures whereas others are moderately adapted. This probably depends on their position in a metabolic pathway, on their intracellular or extracellular localization, on the environmental temperature and on the evolutionary history of the organisms.

Keywords Psychrophiles · Cold adaptation · Psychrophilic enzymes · Protein stability

C. Gerday (✉)
University of Liege, Laboratory of Biochemistry, Institute of Chemistry,
B6, Sart-Tilman, B-4000 Liege, Belgium
e-mail: ch.gerday@ulg.ac.be

P. Buzzini and R. Margesin (eds.), *Cold-Adapted Yeasts*,
DOI: 10.1007/978-3-642-39681-6_15,

Contents

15.1 Introduction ... 326
15.2 Catalytic Properties and Substrate Binding ... 327
15.3 The Activity Challenge ... 330
15.4 Conformational and Thermodynamic Stability ... 335
15.5 Determinants of Cold Adaptation and Mutagenesis ... 339
15.6 Conclusions ... 344
References ... 345

15.1 Introduction

The first significant report related to cold-active enzymes dates back to more than a quarter of a century with the publication of the properties of a heat-labile alkaline phosphatase isolated from an Antarctic bacterium (Kobori et al. 1984). This paper can be considered as a seed paper since the two main properties of cold-active catalysts were already correctly described, i.e., a high specific activity at low temperature and a rather high-temperature sensitivity. The biotechnological potential of this enzyme was also underlined since this cold-active enzyme could be, contrary to calf intestinal ATPase, rapidly inactivated by mild heat treatment offering a very convenient method for the 5′-end radio-labelling of DNA fragments. Cold-active enzymes are produced by psychrophiles, a term introduced by Schmidt-Nielsen (1902), but their existence in cold environments was considered as hypothetical due to the fact that their optimum growth temperature was generally above 20 °C. They were considered as cold tolerant or psychrotolerant rather than cold loving or psychrophiles (Ingraham and Stokes 1959). This confusion still persists nowadays since some investigators erroneously consider that the so-called optimum temperature based on the rate of growth is the best possible temperature for microorganisms. It was the merit of Hess (1934) to point out that numerous bacteria, although their maximum growth rate occurred around 20 °C, were true psychrophiles simply because in broth cultures, the largest cell populations were obtained at 5 °C rather than at 20 °C. This was confirmed by more recent works such as that of Margesin (2009) who stated that «cultivation temperatures close to the maximum growth temperature are not appropriate for studying psychrophiles». It was also shown that the production of extracellular enzymes was the highest at temperatures close to that of the environment of the microorganism and far below the so-called optimum temperature (Feller et al. 1994). An overview on the concept of psychrophily and psychrotolerance is reported in Chap. 1.

The adaptative modifications of enzymatic properties as a function of environmental temperature have been discussed long time ago by pioneer scientists such as Peter Hochachka and George Somero (Hochachka and Somero 1973;

Somero 1977). They defined three possible strategies for the adaptation of enzymes to low temperatures: first, the organism can increase the concentration of the enzyme in the cell or secreted; second, the activity can also be modulated by increasing the substrate-binding ability; and third, the organism can modify in a permanent way the catalytic efficiency of the enzyme. The first strategy is, at first sight, improbable due to its energy cost in a cold environment; the second strategy will not be efficient in the case of enzymes working at saturated substrate concentration that probably concerns most of the extracellular enzymes. In the case of intracellular enzymes, it has been shown that the K_m values of phosphoenolpyruvate for pyruvate kinases of fishes adapted to different temperatures were the lowest in the range of temperatures usually experimented by the organisms and that at the respective cell temperatures, the substrate-binding affinities were close to each other (Somero 1977). So, only the third strategy, i.e., the production of enzymes specifically adapted to low temperatures, seemed to be the only one appropriate. These first information, as well as an easier accessibility to Arctic and Antarctic sites, have stimulated the interest for these enzymes and in the 90s several groups were formed, thanks to the support of the European Union, and focused on psychrophiles with the aims to shed some light on the fundamentals aspects and on the biotechnological potential of these extreme organisms. Therefore, significant progresses in the understanding of the molecular adaptation to cold of cold-active enzymes were made. In this context, the characteristics of a cold-active xylanase from the Antarctic yeast, *Cryptococcus adeliensis* (formerly *Cryptococcus adeliae*), was investigated in 1996 (Petrescu et al. 2000). When compared to its mesophilic counterpart, *C. adeliensis,* a maximum cell density was obtained by cultivation at 4 °C and exceeded by sixfold the cell density reached by *C. albidus* at the same temperature. By contrast, at 20 °C, the cell density obtained with the mesophilic strain was 2.5 times as large as that of the psychrophilic organism illustrating the cold-adapted character of *C. adeliensis*. Concomitant differences were observed at the level of the respective catalytic properties of the xylanases with the cold-active enzyme displaying a k_{cat} at 5 °C three times as high as that of the mesophilic counterpart. Interestingly also, the cold-active enzymes displayed an activation energy, ΔG^*, about 5 % lower than that of the mesophilic enzyme. This mainly resulted from a significantly lower activation enthalpy, ΔH^*, partially compensated by a more negative value of the activation entropy ΔS^*. The cold-active enzyme was also rapidly inactivated after exposure at 30 °C contrarily to the mesophilic enzyme.

15.2 Catalytic Properties and Substrate Binding

Few years ago, (Garcia-Viloca et al. 2004), a generalized expression of reaction rate for enzyme-catalysed reactions, based on the transition state theory developed by Eyring (1935), was introduced. This expression, $k_{cat} = \gamma_{(T)} k_B T/h \exp(-\Delta G^*/RT)$, through its coefficient $\gamma_{(T)}$, takes into account the possibility that the activated complex ES* could recross the energy barrier under the influence of various parameters among which the viscosity of the medium should play an important role in inducing a value for $\gamma_{(T)}$ that could differ from 1 in the case of cold-active enzymes. This coefficient, an extended version of the old transmission coefficient κ, has been usually neglected, but the intracellular space represents a very crowded medium with viscosity increasing from 2.5 cP at 20 °C to 5.0 cP at 0 °C. This high viscosity affects reaction rates as demonstrated by studies on lactate dehydrogenases (Demchenko et al. 1989) and alkaline phosphatase (Homchaudhuri et al. 2006). In this latter study, it was also demonstrated that, next to the influence of viscosity, the high fractional cell volume occupancy by the numerous macromolecules could also severely influence reaction rates of enzyme-catalysed reaction in reducing the frequency of enzyme–substrate encounter. Therefore, the catalytic parameters of cold-active enzymes measured in test tubes are probably far from those prevailing in cells. Taking these considerations apart, numerous investigations on cold-active enzymes have clearly demonstrated that these enzymes display a specific activity usually much higher than their mesophilic and *a fortiori* thermophilic counterparts at low and moderate temperatures as shown in Fig. 15.1. This figure concerns, from left to right, the temperature dependence of the activity of ornithine transcarbamylase (OTCase) from psychrophilic (*Moritella abyssi*) and mesophilic bacteria (*Escherichia coli*) and hyperthermophilic archaea (*Pyrococcus furiosus*) (Xu et al. 2003). One can see that the apparent optimum temperature of the OTCases of these microorganisms considerably differs from each other with a difference of nearly 40 °C between the psychrophilic and mesophilic enzymes. The specific activity of the cold-active enzyme is much higher than that of the mesophilic counterpart in the range of 0–30 °C, and the enzyme is also rapidly inactivated by heat since the apparent optimum does not exceed 25 °C. A large majority of cold-active enzymes displays similar properties as shown in Table 15.1. In a limited number of cases, the specific activity of the cold-active enzyme does not exceed that of the mesophilic one as demonstrated by

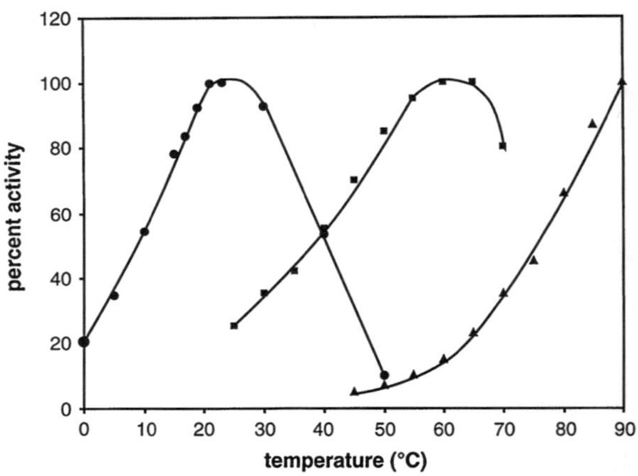

Fig. 15.1 Relative activities of psychrophilic (*Moritella abyssi*; *black circles*) mesophilic (*E. coli*; *black squares*) and thermophilic (*Pyrococcus furiosus*; *black triangles*) OTCases as a function of temperature. Reproduced with permission from Xu et al. (2003)

Table 15.1 K_m and k_{cat} values of some psychrophilic (P) and mesophilic (M) enzymes

Enzyme/organism	T (°C)	K_m	k_{cat} s^{-1}	Reference
Alpha-amylase				D'Amico et al. (2001)
P: *Pseudoalteromonas haloplanktis*	25	234 μM	294 (10 °C)	
M: Pig pancreatic	25	65 μM	97	
Cellulase				Garsoux et al. (2004)
P: *Pseudoalteromonas haloplanktis*	4	600 μM	0.18	
M: *Erwinia chrysanthemi*	4	200 μM	0.01	
DNA ligase				Georlette et al. (2000)
P: *P. haloplanktis*	18	0.30 μM	0.034	
M: *Escherichia coli*	18	0.18 μM	0.004	
Endonuclease I				Altermark et al. (2007)
P: *Vibrio salmonicido*	0.5	246 mM	9.41	
M: *Vibrio cholerae*	0.5	118 mM	1.03	
Isocitrate dehydrogenase				Watanabe et al. (2005)
P: *Colwellia maris*	15	62 mM	70.8	
M: *Escherichia coli*	15	3.3 mM	22.0	
Lactate dehydrogenase				Coquelle et al. (2007)
P: *Champsocephalus gunnarii*	0	0.16 mM	230	
M: *Squalus acanthias*	0	~0.3	72	
Ornithine transcarbamylase				Xu et al. (2003)
P: *Moritella abyssi*	30	45 mM	690	
M: *Saccharomyces cerevisiae*	30	0.9 mM	235	
Subtilisin				Narinx et al. (1997)
P: *Bacillus* sp. (Antarctic)	5	26 μM	32	
M: Subtilisin Carlsberg	5	6 μM	18	

arginine kinase that shows an activity similar to that of the mesophilic enzyme (Suzuki et al. 2012). This type of anomaly has also been mentioned in the case of chitinase (Lonhienne et al. 2001a), and isocitrate reductase (Fedoy et al. 2007), for example. In this latter case, the thermal stability also seems higher than that of the mesophilic counterpart. These data tend to indicate that the enzyme from the psychrophilic bacterium *Desulfotalea psychrophila* is not well adapted to low temperature. It would be interesting to compare the catalytic properties of other enzymes from *D. psychrophila* to those of homologous enzymes from the mesophilic bacterium *Desulfitobacterium hafniense* to see whether this apparently incomplete adaptation to cold can be extended to other enzymes and possibly to the whole organism. In the case of chitinase, the type of substrate is probably important since chitins are structurally very different from each other. As far the arginine kinase from the deep-sea clam *Calyptogena kaikoi* is concerned, the significance of the data is uncertain since, at 25 °C, the activation enthalpy is, as expected, lower than that of the mesophilic enzyme from clams of the genus *Corbicula*. This favourable term is, however, over-compensated by a much more negative activation entropy, the difference leading to a less favourable activation energy. As this enzyme is intracellular, one has in fact to consider the physiological efficiency k_{cat}/K_m rather than the k_{cat} alone because these enzymes are not probably working at saturated substrate concentration. If this is taken into consideration, the cold-active arginine kinase is still, at 10 °C, three times less efficient than the mesophilic enzyme.

Also interesting in Table 15.1 are the values of the Michaelis constant, K_m. With the possible exception of lactate dehydrogenase from the icefish *Champsocephalus gunnarii,* the K_m of cold-active enzymes are systematically higher than the values observed in the case of mesophilic counterparts and the tendency is the same for extracellular and intracellular enzymes. The larger values can be translated into a lower affinity of the cold-active one, provided that one neglects the rate constants not directly involved in the dissociation of the enzyme–substrate complex. In the case of intracellular enzymes, the catalytic efficiency or physiological efficiency is better evaluated by the ratio K_m/k_{cat}. That leads to largely spread situations, favourable for cold-active endonuclease and lactate dehydrogenase but unfavourable for DNA ligase, Isocitrate dehydrogenase and OTCase. So, the usual increase in the rate constant, k_{cat}, of cold-active enzyme is not systematically large enough to compensate for the loss of substrate-binding affinity.

That could indicate that the adaptation of some cold-active enzymes is not complete and depends on the enzyme concerned and probably also on the evolution history of the organism. For extracellular enzymes, the situation appears more favourable since they are probably generally working at saturated substrate concentrations. As already discussed by Somero (1977), the K_m values are rather constant in the range of temperature experimented by the organism but can rapidly increase for temperatures exceeding this zone. This loss in substrate-binding affinity contributes to fix the limit of the thermal tolerance of living organisms. The nearly systematic increase in the K_m values of cold-active enzymes is not fortuitous and constitutes a consequence of the structural modifications needed to improve the catalytic efficiency of these enzymes at low temperatures.

15.3 The Activity Challenge

A quick look to the modern form of the Eyring equation mentioned above indicates that the best way to improve the catalytic efficiency of enzymes at low temperature would be to further decrease the activation energy ΔG^* of the catalysed reaction when compared to the reaction catalysed by the mesophilic counterpart. Simply because the rate constant is exponentially related to the activation energy and that even a small decrease in this value would considerably affect the velocity of the reaction. The activation energy is related to the energy changes necessary to transform the ground state of the enzyme–substrate complex into an activated state. This eventually leads to the transformation of substrate into product. The energy change resulting from the interaction of the enzyme with the substrate has also to be taken into consideration since it is associated with a negative energy change that places the enzyme substrate complex in an unfavourable deep pit of energy as shown in Fig. 15.2. As the K_m values of cold-active enzymes are higher than their mesophilic counterparts, one can consider that the ground state of the ES complex of the cold-active enzyme has a higher free energy level that is favourable in limiting the energy change necessary to reach the activated state. The fact that the ground state of the complex ES occupies a higher level of energy could also be favourable for the activation of ES into ES*. Indeed, if less weak bonds or looser bonds are implicated in the formation of the complex ES, less costly structural modifications will be necessary to reach the activated state. That could lead to the situation represented in Fig. 15.2 showing a significant decrease in the activation energy in the case of the cold-active enzyme. Experimentally, the activation energy of some cold-active enzymes has been measured and compared to that of their mesophilic counterpart and the data are presented in Table 15.2. The activation energy ΔG^* of cold-active enzymes is systematically lower that that of mesophilic homologues to an extent ranging roughly from 1 (amylases) to 10 % (cellulases and bacterial xylanases). These rather moderate figures result from an important decrease in the activation enthalpies ΔH^*, that, for example, amounts to 25 % in the case of the amylases, systematically compensated by a much more negative value of the activation entropy ΔS^*. This indicates that the activated state of the psychrophilic enzymes requires a much more important reordering of the ground state. The activation entropy values of three mesophilic enzymes, i.e., the chitobiase, endonuclease and lysozyme, are positive. That could mean that contrarily to the other enzymes, the activation state ES* is characterized by more disorder than the ground state or, alternatively, that the rearrangement of water molecules, under the form of a release, induces a positive value of the entropy change in the course of the activation. The negative values of the activation entropy of the other cold-active enzymes confirm that they display a

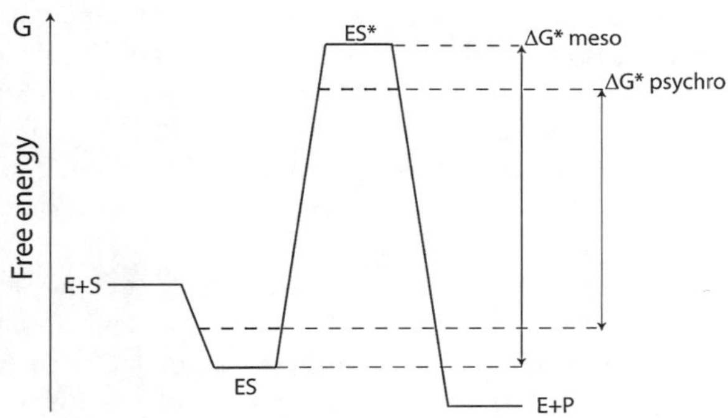

Fig. 15.2 Energetics of the activation of an enzyme-catalysed reaction for mesophilic enzyme (*continuous line*) and psychrophilic counterpart (*dotted line*). The activation in psychrophilic enzymes is rendered easier by a decrease in the affinity of the enzyme for the substrate (higher level of ES) and by a possibly lower energetic level of ES*

Table 15.2 Activation parameters of a few psychrophilic enzymes (P) compared to those of mesophilic counterparts (M)

Enzyme	Type	Temp. (°C)	ΔG* (kJ mol^{-1})	ΔH* (kJ mol^{-1})	TΔS* (kJ mol^{-1})	Reference
Amylase	P	10	57.7	34.7	−23.0	D'Amico et al. (2003)
	M		58.5	46.4	−12.1	
Cellulase	P	4	71.6	46.2	−25.4	Garsoux et al. (2004)
	M		78.2	65.8	−12.4	
Chitobiase	P	15	59.5	44.7	−14.8	Lonhienne et al. (2001b)
	M		63.5	71.5	+8.0	
Endonuclease	P	5	62.8	33.4	−29.4	Altermark et al. (2007)
	M		67.9	74.0	+6.1	
Lysozyme	P	25	45.1	31.9	−13.2	Sotelo-Mundo et al. (2007)
	M		46.2	49.4	+3.2	
Subtilisin	P	15	62.0	36.0	−26.5	Davail et al. (1994)
	M		66.0	46.0	−20.2	
Xylanase (bacteria)	P	10	54.0	21.0	−33.0	Collins et al. (2003)
	M		60.0	58.0	−2.0	
Xylanase (yeast)	P	5	52.3	45.3	−7.0	Petrescu et al. (2000)
	M		54.6	49.9	−4.7	

rather loose structure as suggested by the nearly systematic increase in the K_m values. So the high catalytic efficiency of cold-active enzymes seems to be associated with a rather open structure that has also been associated with a certain loss of specificity of these enzymes. Indeed multi-substrate cold-active enzymes seem to display a broader specificity when compared to their mesophilic counterpart. This has been demonstrated in the case of elastases (Smalas et al. 2000), alcohol dehydrogenase (Tsigos et al. 1998) and α-amylases (D'Amico et al. 2006). These enzymes have the propensity to more easily accept bulky substrates but are less efficient in the case of small-size substrates.

The notion of flexibility in relation with the high specific activity and the generally low thermal stability of cold-active enzymes has been introduced long time ago and notably by Somero (1977). This correlation has been demonstrated in a large majority of cases, but a more refined view of the linkage between the catalytic properties and overall thermal stability of enzymes is probably now emerging. In fact, the catalytic efficiency of a protein is mainly related to the mobility–flexibility of crucial parts of the molecular edifice. An improvement of this flexibility usually leads to an overall decrease in the thermal stability but in a limited number of cases this can occur without significant modifications or visible changes of the thermal stability. Intuitively, one can indeed think that substitutions of amino acids in existing loops, for example, can modify the conformational mobility–flexibility of regions crucial for catalysis without altering the thermal stability. A clear relationship between the activity and local flexibility in psychrophilic and mesophilic enzymes has been recently demonstrated in carbonic anhydrases. It was shown that the cold-active enzyme with a higher specific activity displays a higher flexibility in the region that controls the folding of the protein contrarily to the mesophilic counterpart (Chiuri et al. 2009). EPR spectroscopy has also shown that the catalytic function of a cold-active alkaline phosphatase was closely related to the relative mobility of the helix carrying cysteine 67 and nucleophilic serine 65 (Heidarsson et al. 2009). In subtilisin-like proteases, the relative flexibility of a beta-hairpin close to the active site correlates well with the differential activity of psychrophilic, mesophilic and thermophilic enzymes, but in this example, the homologous enzymes also display a different thermostability (Tiberti and Papaleo 2011).

Protein dynamics obviously play a crucial role in catalytic processes; a systematic investigation using molecular dynamics of 5 pairs of psychrophilic and mesophilic or thermophilic enzymes has shown that amino acid sequences common to both enzymes are generally more flexible in cold-active enzymes, this in relation with differences in the rate or amplitude of opening of the respective active sites. However, in the case of α-amylases, the overall flexibility of the mesophilic enzyme was higher than that of the psychrophilic one due to the insertion of additional loops. It was also shown that the flexibility of orthologous xylanases was approximately the same at the respective apparent optimum temperatures (Spiwok et al. 2007). Experiments using NMR relaxation have also demonstrated that in adenylate kinases, the rate-limiting step was strictly correlated with flexibility in relation with the opening and closing of the active-site lids that shows slower rates in the thermophilic enzyme when compared to the mesophilic counterpart (Wolf-Watz et al. 2004). An apparent increase in the flexibility of a region important for the catalytic efficiency was noticed in the isocitrate dehydrogenase from the cold-adapted *D. psychrophila* and that seems to be induced by the formation of a methionine cluster, but the catalytic efficiency of this enzyme is lower than that of the mesophilic counterpart from *D. hafniense*. Its activation energy ΔG^* is also higher at 5 °C as well as its thermostability. The K_m of the cold-active enzyme for isocitrate is strongly affected by temperatures over 25 °C and is, in any case, higher than that of the mesophilic counterpart (Fedoy et al. 2007). The interpretation of these data seems rather complex since, apart from the K_m value and possibly the local flexibility, this enzyme does not show the usual characteristics of a cold-adapted enzyme, i.e., a higher catalytic efficiency at low temperature and a lower thermal stability. As already mentioned above, it would certainly be interesting to study the biochemical characteristics of other enzymes from this organism. The notion of flexibility is rather complex since the term can refer to structural modifications as a function of time which can be assimilated to a «respiration» of the molecular edifice or to the amplitude of the conformational changes. Both are probably important for the accommodation of the substrates at low temperature and the release of products. The flexibility of proteins has been evaluated by various techniques. Molecular dynamics simulations (MDS) have been applied with a certain success in the case for instance of subtilisin-like proteases from organisms living in different temperature environments (Tindbaek et al. 2004; Tiberti and Papaleo 2011). It was shown that the higher catalytic efficiency of cold-active enzymes was correlated to a higher flexibility of regions important for catalysis, itself correlated to an evolution towards a lower thermal stability from high to low-temperature environments. Recent investigations using MDS have corroborated the idea that the adaptation to cold, in terms of function, is correlated with an improvement of the flexibility of crucial regions in various proteins such as chitinases (Ramli et al. 2012), frataxins (Roman et al. 2013), haemoglobins (Stadler et al. 2012) and beta-tubulins (Chiappori et al. 2012). It has also been shown from studies carried out on homologous 3-isopropylmalate dehydrogenases and thermolysin-like proteases that, at their respective environment temperature, enzymes from mesophilic and thermophilic organisms maintain a balance between their overall rigidity and local flexibility important for their catalytic properties (Radestock and Gohlke 2011) in agreement with the strong correlation between resistance to unfolding and evolutionary adaptation temperature (Hochachka and Somero 2002). Amide hydrogen–deuterium exchanges have provided controversial data. For example, a psychrophilic 3-isopropylmalate dehydrogenase was found to be more rigid than its mesophilic counterpart (Svingor et al. 2001). The «anomalies» observed with the H/D exchange techniques may be due to the use of non-relevant temperatures or time scales. The temperature factor or B-factor has also been used to evaluate the relative flexibility of proteins. It represents the spread of electron densities and can be calculated from known three-dimensional structures. Average B-factors are not systematically correlated with the activity-flexibility relationship (Sun-Yong et al. 1999) but relative B-factors representing the ratio between the B-factors of local parts of the proteins and the average B-factor apparently correlate well with the relative activity and thermostability of psychrophilic and mesophilic enzymes (Russell et al. 1998; Sun-Yong et al. 1999). A recent systematic comparison of the B-factors of twenty pairs of psychrophilic and mesophilic enzymes has demonstrated that psychrophilic enzymes are more flexible in 5-turn and strand secondary structures; they have also larger average cavity sizes and these cavities are lined with an increased frequency of acidic groups (Paredes et al. 2011). Fluorescence quenching of trytophane fluorescence using acrylamide has been used successfully to evaluate the relative flexibility of psychrophilic and mesophilic enzymes. The data have shown that psychrophilic enzymes are much more permeable than their mesophilic counterparts in the case of Ca^{2+}-Zn^{2+} protease (Chessa et al. 2000), α-amylase (D'Amico et al. 2003; Cipolla et al. 2012), DNA ligase (Georlette et al. 2000, 2003), xylanase (Collins et al. 2003), cellulase (Sonan et al. 2007), aminopeptidase (Huston et al. 2008) and thermolysin (Xie et al. 2009). An example of the comparison of tryptophane fluorescence quenching of psychrophilic and mesophilic enzyme is shown in Fig. 15.3. One can see that the cold-active

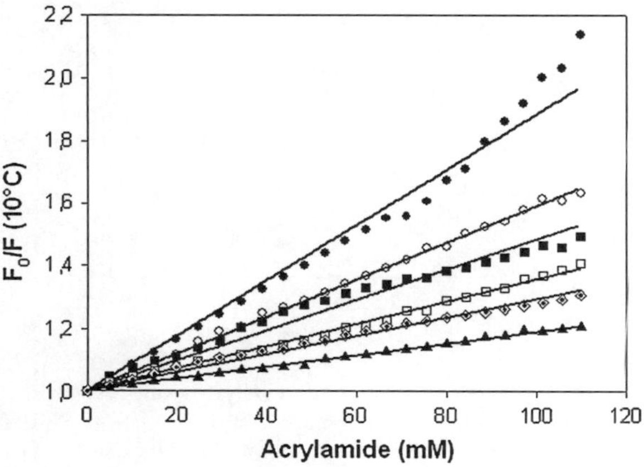

Fig. 15.3 Quenching of tryptophane fluorescence at 10 °C of cold-active cellulase from Antarctic *Pseudoalteromonas haloplanktis* (Cel5G, *black dots*) and variants differing by a shorter length of the linker region (Cel5GΔ1P, *open circles*; Cel5GΔ2P, *black squares*) as well as mutants characterized by higher melting points (Cel5GΔ2PmutCC, *open squares*; $Cel5G_{CM}$, *diamonds*) and mesophilic cellulase from *Erwinia chrysanthemi* (Cel5A, *black triangles*). F_0/F represents the ratio between the fluorescence in the absence (F_0) and presence (F) of acrylamide at various concentrations. Adapted from Sonan et al. (2007)

cellulase presents a much more opened structure when compared to the mesophilic enzyme. Moreover, the mutants that are characterized by a progressive shortening of the linker region (Cel5GΔ1P, Cel5GΔ2P), which induces a progressive increase in the melting point, display a more compact structure, further enhanced in the case of the variant Cel5GΔ2P mutCC carrying an additional disulphide bridge and of the catalytic module $Cel5G_{CM}$. This once again illustrates the good correlation that generally exists between the low stability and flexibility of cold-active enzyme responsible for their higher catalytic efficiency at low temperature (Sonan et al. 2007).

15.4 Conformational and Thermodynamic Stability

The analysis of the thermodependence of the activity of cold-active enzyme, as illustrated in Fig. 15.1, indicates that these enzymes are inactivated at temperatures lower than that of their mesophilic counterparts, and in consequence, they seem to display a lower thermal stability. This thermal stability has been investigated by numerous techniques from the evaluation of the residual activity after exposure for a certain time at a given temperature to biophysical techniques such as circular dichroïsm, fluorescence spectroscopy or microcalorimetry. The method using residual activity is only valid if one demonstrates that the unfolding is irreversible at all temperature tested; an example of the use of this technique is

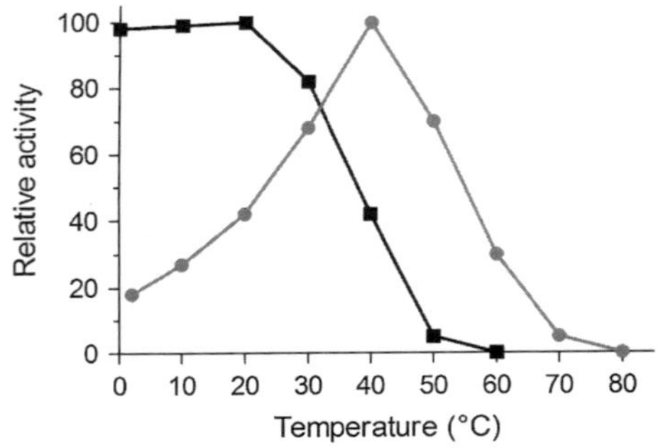

Fig. 15.4 Activity as a function of temperature of polygalacturonase from the Antarctic yeast *Cystofilobasidium capitatum* (formerly *Cystofilobasidium lari-marini*) S3B (*grey circles*) and residual activities of the enzyme after exposure at various temperatures (*black squares*). Note that at the so-called optimum temperature, around 40 °C, the enzyme is already partially unfolded. Adapted from Birgisson et al. (2003)

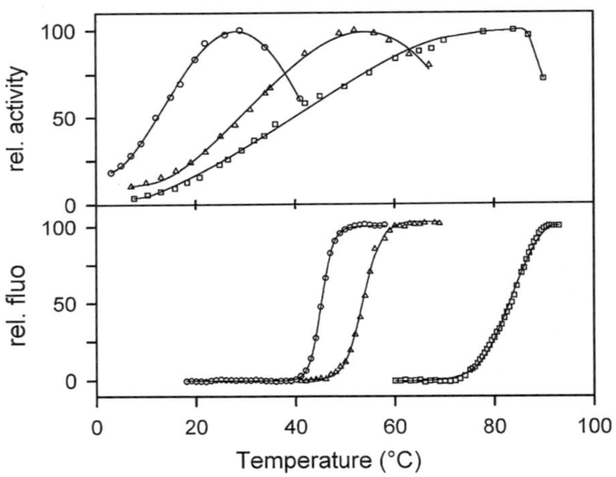

Fig. 15.5 *Upper panel* Relative activity as a function of temperature of psychrophilic (*open circles*), mesophilic (*open triangles*), and thermophilic (*open squares*) α-amylases. *Lower panel* Concomitant thermal unfolding as recorded by fluorescence spectroscopy. Worth noting is the fact that the thermal inactivation of the psychrophilic enzyme occurs without any significant modification of the three-dimensional structure contrary to mesophilic and thermophilic counterparts. Adapted from Georlette et al. (2004)

shown in Fig. 15.4. In this figure, the thermodependence of the activity of a cold-active polygalacturonase from Icelandic yeast (Birgisson et al. 2003) is compared to the thermal stability of the enzyme exposed for one hour at the respective temperatures. One can see that the enzyme starts to unfold at a temperature around 25 °C, far below the so-called optimum temperature which is around 40 °C. This figure clearly demonstrates that the so-called optimum temperature is only an apparent optimum since at this temperature, the half-life of the enzyme is rather short. Another illustration of the thermodependence of the activity and stability of a cold-active enzyme is shown in Fig. 15.5. That concerns a cold-active α-amylase for the Antarctic bacterial strain *Pseudoalteromonas haloplanktis* (AHA) as compared with the mesophilic enzyme from pig pancreas (PPA) and the thermophilic counterpart from *Bacillus amyloliquefaciens* (BAA). The upper panel represents the relative activities and the lower panel the thermal unfolding as followed by fluorescence spectroscopy (D'Amico et al. 2003). A striking feature of these experiments is the fact that the inactivation of the cold-active enzymes precedes the unfolding of the structure. In the mesophilic and thermophilic enzymes, the apparent optimum is strictly correlated to the unfolding transition. That can mean that the active site is very sensitive to temperature or that the ES complex is destabilized before any significant or visible modification of the three-dimensional structure of the enzyme. Such phenomenon has been observed in other cold-active enzymes such as for instance xylanases (Collins et al. 2003, 2007) and DNA ligases (Georlette et al. 2003).The thermal sensitivity of the active site and/or of the ES complex is in agreement with the increase in K_m values reported for cold-active enzymes. In general, the residues lining the catalytic cavities and involved in the interaction of the substrate are strictly conserved. This has been notably demonstrated by comparison of the crystallographic structures of the psychrophilic α-amylase from an Antarctic bacterium with that of the homologous enzyme from pig pancreas. The superimposition of the active site of the two enzymes has shown that the 24 residues forming the catalytic cleft have a position strictly conserved in both enzymes (Aghajari et al. 1998, 2002). Therefore, the flexibility of the catalytic region should originate from other regions of the protein. In other cases, the catalytic cavities have been altered so as to confer a larger opening of the catalytic region of psychrophilic enzymes that favours the accommodation of larger substrates and facilitates the release of products. This has been achieved through the modification, for example, of electric potentials that favours the attraction and orientation of substrates in citrate synthase (Russell et al. 1998), malate dehydrogenase (Kim et al. 1999), uracyl-DNA glycosylase (Leiros et al. 2003) and trypsin (Gorfe et al. 2000). In a cold-active Ca^{2+}-Zn^{2+} protease, an additional Ca^{2+} pulls the backbone lining the entrance of the site and offers in this way a much better accessibility for the substrate (Aghajari et al. 2003). In cold-active iron superoxide dismutases, it has been shown that the residues forming the active site and the active-site channel are more mobile than in mesophilic counterparts (Merlino et al. 2010).

The thermodynamic stability of some cold-adapted proteins has been investigated assuming that the unfolding is a reversible two-state process defined by the equilibrium

$$\Delta G_{N-U} = -RT \ln K_{N-U} \quad \text{with} \quad K_{N-U} = [U]/[N]$$

For a protein mainly in a native state, ΔG_{N-U} is a positive value and can be defined as the energy necessary to unfold the protein. The thermodynamic stability

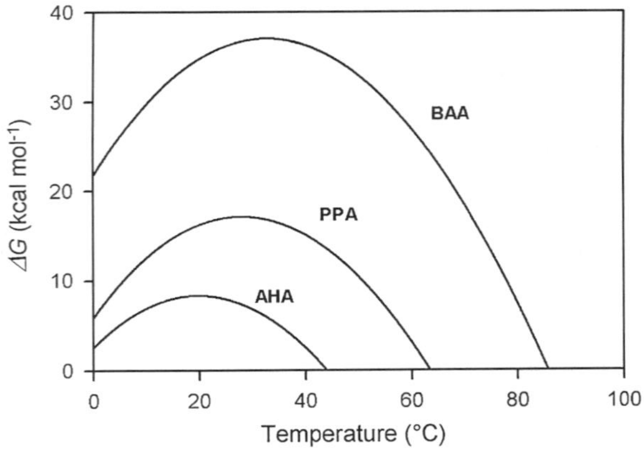

Fig. 15.6 Stabilization energy as a function of temperature of psychrophilic α-amylase (*AHA*) from the Antarctic strain *Pseudoalteromonas haloplanktis*, of mesophilic enzyme from pig pancreas (*PPA*) and of the thermophilic counterpart (*BAA*) from *Bacillus amyloliquefaciens*. Adapted from D'Amico et al. (2003)

of enzymes can be evaluated by non-calorimetric techniques making use of chemical denaturants (Pace and Laurents 1989; Talla-Singh and Stites 2008) or more easily in a microcalorimeter in which the heat absorbed during the unfolding of the protein is measured and defined as ΔH_{cal}. It represents the energy necessary to disrupt the weak bonds that stabilize the three-dimensional structure and appears as a peak often symmetrical in the case of cold-active enzymes but often unsymmetric and flattened in the case of mesophilic and thermophilic counterparts. ΔH_{cal} is equivalent to the area limited by the peak and this area significantly increases from psychrophiles to thermophiles. The symmetry is indicative of a pronounced cooperativity of unfolding and the transition appears in this case sharp contrarily to that of mesophilic and thermophilic proteins that show a flattened profile distributed over a rather broad range of temperatures (Feller 2010). As a function of temperature, the stabilization energy or free energy of unfolding is described by an equivalent of the Gibbs–Helmholtz equation:

$$\Delta G_{(T)} = \Delta H_{cal}(1 - T/T_m) + \Delta C_p(T - T_m) - T\Delta C_p \ln(T/T_m)$$

In this equation, T_m is the so-called melting temperature, located at the top of the peak, and defined as the temperature of half denaturation where [N] = [U], ΔH_{cal} has been defined above and is equivalent to the van't Hoff enthalpy in a two-state model, T is a given temperature, and ΔC_p, the heat capacity change occurring during the transition from the native to the unfolded state. The calculation of the stabilization energy, over a temperature range where the native state is in excess over the unfolded state, gives access to stability curves as shown in Fig. 15.6. These curves, which are parabola-shaped, have been established for cold-active α-amylase AHA, mesophilic α-amylase PPA and thermophilic α-amylase BAA. One can see that over the whole range of temperature, the cold-active enzyme presents a much lower thermodynamic stability than its mesophilic and thermophilic counterparts, even in the low-temperature range. This has been attributed to the lower number or/and strength of the weak bonds that stabilizes the structure. A second striking feature is that the maximum stability for the three orthologous enzymes is found in the rather narrow range of 20–30 °C. This is due to the hydrophobic effect found to be maximum around room temperature (Kumar et al. 2002). Cold denaturation is also more easily achieved in the case of the psychrophilic enzyme with a T_m prediction around −10 °C. On the left side of the curve, the stabilization enthalpy and stabilization entropy are negative. That means that the residual stability of the cold-active enzyme near the environment temperature, around 0 °C (Antarctic sea water), is secured through the negative value of the stabilization entropy resulting from the hydration of the hydrophobic groups of non-polar amino acids that contributes to decrease the entropy of the unfolded state (Dias et al. 2010). Cold denaturation occurs with a release of heat contrarily to heat denaturation. It is also interesting to note that at temperatures corresponding to the most stable conditions, around 20 °C, the cold-active enzyme is marginally stable since the energy required for unfolding is around 30 kJ mol^{-1}. The value drops to around 8 kJ mol^{-1} at 0 °C, the equivalent of about 2 weak bonds.

15.5 Determinants of Cold Adaptation and Mutagenesis

It has been shown that the flexibility of crucial parts of the structure of cold-active enzymes is generally, but not probably always, accompanied by a decrease in the thermal stability of the protein structure. To understand cold adaptation, it is necessary to define the structural changes that are directly or indirectly involved in the adaptation to cold. The problem is rendered more complicated by the high number of neutral mutations resulting from genetic drift that leads to large differences in the amino acid sequences of homologous enzymes produced by related species. The elucidation of the three-dimensional structure of several cold-active enzymes by X-ray crystallography, more than thirty actually, as well as the production of reliable models based on the comparison of the amino acid sequences of mesophilic and psychrophilic proteins, has allowed to detect, in a first approach, the amino acid substitutions that could lead to a decrease in the thermal stability of the molecular structure of cold-active enzymes. Again, only a few of them are intuitively involved in the adaptation to cold since, in low-temperature environments, the absence a high selective pressure on thermal stability can enhance the importance of the genetic drift.

The structural parameters involved in the thermal stability of proteins have been extensively discussed elsewhere (Vieille and Zeikus 2001), and their implications in the adaptation of cold-active enzymes have already been discussed in many reviews (Somero and Low 1976; Feller et al. 1997a; Marshall 1997; Feller and Gerday 1997, 2003; Russell 2000; Smalas et al. 2000; Gerday et al. 2000; Feller 2003, 2010; Hoyoux et al. 2004; Georlette et al. 2004; Siddiqui and Cavicchioli

2006; Marx et al. 2007; Collins et al. 2007, 2008), and there is a consensus to agree that the enhanced flexibility of crucial parts of these enzymes is mainly secured through a weakening of intramolecular weak bonds that stabilize the structure of mesophilic counterparts associated or not to entropic factors that tend to increase the entropy of the folded state. In general, however, the structural changes leading to cold adaptation are rather discrete since the three-dimensional structure of psychrophilic and mesophilic homologous enzymes can be nearly superimposed. Hydrophobic interactions play important roles in driving the folding of proteins in aqueous media and in stabilizing the nascent polypeptide chain. In cold-active enzymes, the number of hydrophobic interactions can be lower and their strength can also be affected by a decrease in the size of the hydrophobic groups and a reduction in the number or/and size of hydrophobic clusters that contribute to decrease the compactness of the molecule in creating cavities that can be occupied by water molecules (Paredes et al. 2011). In other cases, there is an increase in the proportion of hydrophobic groups at the surface of the protein; this induces a destabilization of the molecular edifice through the reduction in the overall entropy of the system through the formation of clathrate-like structures around these groups. The stabilizing effect of salt bridges can be altered through a modification of the orientation of ionized groups and of their reduction in number (Papaleo et al. 2007). The presence of ionized groups carrying similar charge, in general negative, can also modify the stability through repulsion and by increasing the interaction with the solvent (Narinx et al. 1997; Feller et al. 1999; Adekoya et al. 2006). In cold-active uracil-DNA N-glycosylase, however, a more positively charged surface near the active site seems to be important for the adaptation to cold (Moe et al. 2004). In citrate synthases, psychrophilic, mesophilic and thermophilic enzymes have increasingly stronger electrostatic stabilization of the transition state (Bjelic et al. 2008). Some cold-active enzymes are also poorer in arginine residues and display a lower Arg/Arg + Lys ratio (Adekoya et al. 2006). This in relation with the fact that arginine can develop more electrostatic interactions with surrounding amino acids than lysine. α-helices can be assimilated to macroscopic electrostatic dipoles that carry a net positive charge at their N-terminal end and a net negative charge at their C-terminal end (Serrano and Fresht 1989). These secondary structures can be stabilized by various electrostatic interactions in their vicinity that can be weakened in psychrophilic enzymes such as in cold-active trypsins (Leiros et al. 2000). Also the hydrogen bonding network is often weakened in psychrophilic enzymes and found responsible for the increase in flexibility of these enzymes (Xie et al. 2009). The structure of loops connecting secondary structures seems also important in conferring the appropriate flexibility to some cold-active enzymes; their length can be increased, their level of interactions with the internal moiety can be lowered, and their flexibility can be enhanced thanks to the deletion of proline residues (Feller et al. 1997b; Gudmundsdottir 2002; Matsuura et al. 2002). Metal binding can also be involved through the reduction in binding affinity or deletion of site (Narinx et al. 1997; Almog et al. 2003). The deletion of disulphide bridges can also improve the flexibility and activity of cold-active enzymes; this has been demonstrated in cold-active α-amylase and alkaline phosphatase (D'Amico et al. 2003; Asgeirsson et al. 2007; Papaleo et al. 2007).

Following the comparison of the three-dimensional structures of psychrophilic and mesophilic proteins, mutation experiments were carried out in order either to transform psychrophilic enzymes into mesophilic counterparts or to try to force mesophilic enzymes to adopt psychrophilic characteristics.

In a first attempt, random mutagenesis was applied to mesophilic subtilisin BPN' (Taguchi et al. 1998) and a mutant *m63* carrying three mutations, V72I, A92T, G131D, was obtained. It displayed, at 10 °C, an activity, towards the synthetic small-size substrate s-AAPF para nitroanilide, identical to that of the wild-type mesophilic enzyme; only the K_m was twice lower than that of subtilisin BPN'. The activity was not tested on large-size substrates and its thermal stability was slightly lower than that of the original enzyme. More success was obtained later on with the multiple mutant *m 51* also carrying three mutations, A31T, A88 V and A98T (Taguchi et al. 1999). Its specific activity at 10 °C was 1.5 higher than that of subtilisin BPN'; also towards s-AAPF-pNA, the K_m was slightly lower than that of the original enzyme and the thermostability was apparently unchanged. Another cold-active subtilisin (S41) from Antarctic *Bacillus* sp. (Davail et al. 1994) was submitted to random mutagenesis followed by in vitro recombination. Mutant libraries were screened to identify enzymes that displayed greater thermal stability without sacrificing low-temperature activity. After several generations and recombinations, a variant, 3-2G7, carrying 7 mutations was obtained. The specific activity towards a small-size synthetic substrate, s-AAPF-*p*Na, displayed a threefold increase at all temperatures tested with no modification of the K_m values. A higher thermostability was also observed (Miyazaki et al. 2000). This variant was further engineered to give the subtilisin mutant 8-4A9 carrying 13 amino acid substitutions distributed throughout the structure leading to an increase in the affinity for Ca^{2+}. The midpoint of unfolding was about 3 °C higher than the variant 3-2G7 with a specific activity at 10 °C slightly lower than that of 3-2G7 but nearly twice as high as that of the wild-type S41. The K_m fors-AAPF-*p*Na steadily decreased from S41 to 3-2G7 and 8-4A9 (Wintrode et al. 2001). These experiments tended to demonstrate that it was quite possible to increase the thermal stability of an enzyme while keeping its high catalytic activity at low temperature; in other words that the two properties were not necessarily inversely related. It is, however, worth noting that these data have been obtained with non-natural, synthetic and small-size substrates. If it seems possible to modify, by directed evolution, enzymes in such a way to display a higher thermostability associated with a higher specific activity at low temperature towards artificial substrate, the situation is probably much more complicated when natural large-size substrates are concerned. We have indeed learned that the protease mutant 3-2G7 in fact displays a much lower activity than the wild-type S41 cold-active enzymes when natural and large-size substrates such as proteins were used. That does not mean that in vitro it is not possible to increase the thermostability of a cold-active enzyme while preserving or even increasing its catalytic efficiency at low temperature. A cold-active subtilisin, S39, extremely similar to the S41 enzyme, was also submitted to site-directed mutagenesis with the aim to mainly try to increase its stability (Narinx et al. 1997). Several mutations were introduced first to restore a salt bridge found in subtilisin BPN', a hydrophobic interaction found in subtilisin Carlsberg, a disulphide bridge present in aqualysine, a mutation N136S supposed to increase the specific activity towards large-size substrates and a mutation T85D to exchange a poor Ca^{2+} ligand for a strong one. Most of these mutations had a positive effect on the thermostability of the enzyme, especially the mutation T85D, involved in the Ca^{2+} coordination of calcium-binding site 1 that increased the half-life at 50 °C by more than 50 min. As expected, the affinity of this mutant for calcium was drastically improved. At 5 °C, the specific activity of this mutant towards the synthetic substrate s-FAAF-*p*Na was 4 times as high as that of the wild-type enzyme; the K_m value was not affected. The specific activity was, however, only slightly increased when azocasein was used as substrate. In another cold-active subtilisin-like proteinase, Ser-Ala and Xaa-Pro mutations were introduced to produce single, double and triple mutants. Most of these mutations, especially the Xaa-Pro exchanges, led to an increase in the thermostability of the variants with a concomitant decrease in the specific activities (Arnorsdottir et al. 2007). A thermophilic subtilase was also engineered in order to produce cold-active variants (Zhong et al. 2009). All successful single mutations were located within or near the active site and a variant carrying four amino acid substitutions showed a sixfold increase in specific activity towards casein in the temperature range of 15–25 °C, a decrease in thermal stability, and a shift of its apparent optimum towards low temperature by approximately 15 °C. Interestingly, this multiple mutant was less active towards the synthetic substrate s-AAPF-*p*NA due to a large increase in K_m and decrease in k_{cat}. This again underlines the importance, before drawing robust conclusions, of selecting appropriate substrates when studying multi-substrate enzymes by mutagenesis.

Directed evolution was also applied to a thermostable esterase from *B. subtilis* using p-nitrobenzyl acetate as substrate (Giver et al. 1998). After several generations, mutants of higher thermostability were obtained; they show a specific activity slightly higher, about 20 %, than that of the wild-type enzyme. In this case, the natural substrates are unfortunately unknown.

The molecular basis of the adaptation to cold of a psychrophilic α-amylase was also investigated by site-directed mutagenesis. Fourteen amino acid substitutions, including double mutations were introduced with the aim to mimic specific structural characteristics found in mesophilic and thermophilic homologous enzymes (D'Amico et al. 2001, 2002) under the form of hydrogen bonds, salt bridges, helix dipole stabilization, hydrophobic interactions and reinsertion of a disulphide bridge that connects domain A and B in mesophilic counterpart. The highest contribution to stability, in terms of both T_m and ΔH_{cal}, was obtained by the introduction of an electrostatic interaction; a double aromatic interaction increased the stabilization energy by a factor of two at 10 °C whereas the reinforcement of hydrophobic clusters within the hydrophobic core of the enzyme induced the production of multiple calorimetric domains similar to those existing in the mesophilic counterpart from pig pancreas. Most of the introduced mutations aiming to increase the stability were efficient but led to a decrease in k_{cat} and K_m. Any attempts to further increase the catalytic efficiency at low temperature through a decrease in stability gave rise to unfolded forms suggesting that the cold-active α-amylase has reached a limit in stability precluding any improvement of the specific activity via a decrease in thermal stability. The incorporation of a disulphide bridge found in the mesophilic enzyme decreased the specific activity by a factor of two at 5 °C as well as the K_m and showed a microcalorimetric profile similar to that of the mesophilic counterpart with, however, a first T_m transition lower than that of the mesophilic α-amylase. This indicates that the introduction, in the cold-active enzyme, of this disulphide bridge leads to a stabilization of some parts of the cold-adapted protein but also induces an unfavourable structural constraint in another part of this protein. Multiple mutants of this cold-active enzyme were also recently produced to force the cold-active enzyme to display mesophilic characteristics (Cipolla et al. 2012); the specific activity of the variants was lowered whereas the thermal stability was drastically improved with a concomitant decrease in the flexibility of the mutants. It was concluded that these mutants can be considered as structural intermediates between their parents psychrophilic and mesophilic enzymes. The effects of these mutations, single and multiple, on the molecular dynamics of the protein, were recently investigated (Papaleo et al. 2011), and it was shown that the above-mentioned mutants displayed a reduced flexibility in various regions, not only near the active-site and substrate-binding groove, but also at long distance such as in domain C that did not carry any mutation.

Several other cold-active enzymes have been the target of site-directed mutagenesis in order to shed some light on the residues involved in the adaptation. One can mention the work on a cold-active lipase in which the substitutions of three polar residues in the lid region by amino acid residues conserved in homologous mesophilic lipases gave rise to an increase in stability and a modification of the substrate specificity with a preference for C8 rather than C4 fatty acid chains. A substitution, Ser-Gly, increased the specificity for C12 fatty acid chain associated with a decrease in stability this probably in relation with a better accessibility of the active site and improvement of the flexibility in this region (Santarossa et al. 2005).

An interesting study on the adaptation to cold of an alkaline phosphatase from an Antarctic bacterium was carried out by directed evolution to identify amino acid residues involved in the adaptation and stability of this protein (Koutsioulis et al. 2008). Three thermostable and six thermolabile variants were obtained. All the variants showing an increase in stability were also characterized by an increase in the T_m of the first calorimetric transition and persistence, as in the case of wild type, of secondary transitions at higher temperatures; their specific activity at 15 °C was also lower than that of the parent enzyme and this was associated, as expected, with a higher activation energy. By contrast, the two variants, H135E and double mutant H135E/G149D, displaying an activity about 1.5 higher and a much higher K_m than that of the wild-type enzyme had a nearly symmetrical microcalorimetric profile with only one detectable transition shifted towards higher temperature by about 6 °C when compared to the first transition of the parent enzyme. This latter, however, displayed a calorimetric domain with a T_m about 7 °C higher than those of both variants. It has been suggested that the mutation H135E distorts the Mg^{2+} binding site, decreases the affinity for the metal and increases the flexibility of the site presumably through the destabilization of the domain showing the highest thermostability in the wild-type enzyme. In the double mutant, the mutation G149D apparently improves the mobility of residue R148 located in a loop that facilitates the phosphate/substrate coordination during catalysis. Also, after incubation at 60 °C, these two more active variants were also more rapidly inactivated than the wild-type enzyme suggesting an increase in flexibility of the active site.

In a cold-active isocitrate lyase, the replacement of Ala 214 by a Ser residue found in some mesophilic enzyme and close to the Tim barrel involved in substrate binding induced a decrease in the catalytic activity associated with an increase in the thermostability of the mutant presumably due to the two hydrogen bonds formed between Ser and Gln119. It was concluded that Ala 214 was implicated in cold adaptation by providing more flexibility to the active site (Sato et al. 2008).

These experiments of mutagenesis indicate that, in most cases, the high specific activity of cold-active enzymes is related to an improvement of the flexibility of crucial parts of the molecular structure that induces a high thermal sensitivity of the active site and a generally better accessibility for the substrate at low temperatures. The attempts to try to simultaneously increase the thermostability and activity at low temperature have been hardly successful. Directed evolution has produced more stable and more active variants but essentially with multi-substrate enzymes and small-size and artificial substrates, and it remains to be demonstrated that the mutants are also active on large-size substrates at low temperature. One mutated cold-active subtilisin, carrying only one mutation was, however, found more stable and more active on a large-size substrate, azocasein (Narinx et al. 1997), and interestingly, two mutants of a cold-active alkaline phosphatase were found more active than the wild-type enzyme and showed a symmetrical calorimetric profile synonymous of a high cooperativity of unfolding. Although the first transition of the wild-type enzyme had a T_m lower than that of the mutants, these variants had lost the most thermostable calorimetric domain and appear to be more rapidly inactivated than the wild-type alkaline phosphatase. This is correlated to the fact that many cold-active enzymes also show a high cooperativity on unfolding whereas heat inactivation often precedes any significant changes in the three-dimensional structure.

15.6 Conclusions

Psychrophilic organisms living in permanently cold habitats have developed numerous adaptations that allowed them to successfully thrive in low-temperature environments. Their enzymes are a key feature of this adaptation. They are much more active than their mesophilic counterparts at low and moderate temperatures and their high specific activity is undoubtedly due to an improvement of the flexibility of the active site or/and of other regions of the structure indirectly involved. This is induced by rather discrete structural modifications which can be located at long distance from the active site and which generally lead to a higher thermal instability of these enzymes, often associated with an even faster heat inactivation. If the general strategy adopted by nature consists in a weakening of the intramolecular forces that stabilize the structure of their mesophilic and *a fortiori* thermophilic counterparts, this strategy appears to be specific to each enzyme; depends on the position of the enzyme within a metabolic pathway; on its localization, intra-or extra-cellular; on its structural modification capability; on the environment of the organism and on its evolution history. As a consequence, there is a continuum in the adaptation to low temperatures; some cold-active enzymes indeed display high specific activity associated with low T_m while others show rather high T_m and stronger thermal dependence like their mesophilic counterparts.

References

Adekoya OA, Helland R, Willassen NP, Sylte I (2006) Comparative sequence and structure analysis reveal features of cold adaptation of an enzyme in the thermolysin family. Proteins 62:435–439

Aghajari N, Feller G, Gerday C, Haser R (1998) Structures of the psychrophilic *Alteromonas haloplanktis* α-amylase give insights into cold adaptation at a molecular level. Structure 6:1503–1516

Aghajari N, Roth M, Haser R (2002) Crystallographic evidence of a transglycosylation reaction: ternary complexes of a psychrophilic α-amylase. Biochemistry 41:4273–4280

Aghajari N, Van petegem F, Villeret V, Chessa J-P, Gerday C, Haser R, Van Beeumen J (2003) Crystal structures of a psychrophilic metalloprotease reveal new insights into catalysis by cold-adapted proteases. Proteins 50:636–647

Almog O, Gonzales A, Klein D, Greenblatt HM, Braun S, Shoham G (2003) The 0.93Å crystal structure of spericase: a calcium-loaded serine protease from *Bacillus sphaericus*. J Mol Biol 332:1071–1082

Altermark B, Niiranen L, Willassen NP, Smalas AO, Moe E (2007) Comparative studies of endonuclease I from cold-adapted *Vibrio salmonicida* and mesophilic *Vibrio cholerae*. FEBS J 274:252–263

Arnorsdottir J, Helgadottir S, Thorbjarnardottir SH, Eggertsson G, Kristjansson MM (2007) Effect of selected Ser/Ala and Xaa/Pro mutations on the stability and catalytic properties of a cold adapted subtilisin-like serine proteinase. Biochim Biophys Acta 1774:749–755

Asgeirsson B, Adalbjörsson BV, Gylfason GA (2007) Engineered disulfide bonds increase active-site local stability and reduce catalytic activity of a cold-adapted alkaline phosphatase. Biochim Biophys Acta 1774:679–687

Birgisson H, Delgado O, Garcia-Arroyo L, Hatti-Kaul R, Mattiasson B (2003) Cold-adapted yeasts as producers of cold-active polygalacturonases. Extremophiles 7:185–193

Bjelic S, Bransdal BO, Aqvist J (2008) Cold adaptation of enzyme reaction rates. Biochemistry 47:10049–10057

Chessa J-P, Petrescu I, Bentahir M, Van Beeumen J, Gerday C (2000) Purification, physico-chemical characterization and sequence of the heat labile alkaline metalloprotease isolated from a psychrophilic *Pseudomonas* species. Biochim Biophys Acta 1479:265–274

Chiappori F, Pucciarelli S, Merelli I, ballarini P, Miceli C, Milanesi L (2012) Structural thermal adaptation of β-tubulins from the Antarctic psychrophilic protozoan *Euplotes focardii*. Proteins 80:1154–1166

Chiuri R, Majorano G, Rizello A, del Mercato LL, Cingolani R, Rinaldi R, Maffia M, Pompa PP (2009) Exploring local flexibility/rigidity in psychrophilic and mesophilic carbonic anhydrases. Biophys J 96:1586–1596

Cipolla A, Delbrassine F, da Lage JL, Feller G (2012) Temperature adaptations in psychrophilic, mesophilic and thermophilic chloride-dependent alpha-amylases. Biochimie 94:1943–1950

Collins T, Meuwis M-A, Gerday C, Feller G (2003) Activity, stability and flexibility in glycosidases adapted to extreme thermal environments. J Mol Biol 338:419–428

Collins T, D'Amico S, Marx J-C, Feller G, Gerday C (2007) I cold-adapted enzymes. In: Gerday C, Glansdorff N (eds) Physiology and biochemistry of extremophiles. ASM press, Washington DC, pp 165–179

Collins T, Roulling F, Piette F, Marx J-C, Feller G, Gerday C, D'Amico S (2008) Fundamentals of cold-adapted enzymes. In: Margesin R, Schinner F, Marx J-C, Gerday C (eds) Psychrophiles from biodiversity to biotechnology. Springer, Berlin, pp 211–227

Coquelle N, Fioravanti E, Weik M, Vellieux F, Madern D (2007) Activity, stability and structural studies of lactate dehydrogenases adapted to extreme thermal environments. J Mol Biol 374:547–562

D'Amico S, Gerday C, Feller G (2001) Structural determinants of cold adaptation and stability in a large protein. J Biol Chem 276:25791–25796

D'Amico S, Gerday C, Feller G (2002) Dual effects of an extra disulfide bond on the activity and stability of a cold-adapted α-amylase. J Biol Chem 48:46110–46115

D'Amico S, Marx J-C, Gerday C, Feller G (2003) Activity-stability relationship in extremophilic enzymes. J Biol Chem 278:7891–7896

D'Amico S, Sohier J-S, Feller G (2006) Kinetics and energetics of ligand binding determined by microcalorimetry: insight into active site mobility in a psychrophilic alpha-amylase. J Mol Biol 358:1296–1304

Davail S, Feller G, Narinx E, Gerday C (1994) Cold adaptation of proteins. Purification, characterization, and sequence of the heat labile subtilisin from the Antarctic psychrophile *Bacillus* TA 41. J Biol Chem 269:17448–17453

Demchenko AP, Rusyn OI, Saburova EA (1989) Kinetics of the lactate dehydrogenase reaction in high-viscosity media. Biochim Biophys Acta 998:196–203

Dias CL, Ala-Nissila T, Wong-E kkabut J, Vattulainen I, Grant M, Karttunen M (2010) The hydrophobic effect and its role in cold denaturation. Cryobiology 60:91–99

Eyring H (1935) The activated complex in chemical reactions. J Chem Phys 3:107–115

Fedoy A-E, Yang N, Martinez A, Leiros H-K, Stee H (2007) Structural and functional properties of isocitrate dehydrogenase from the psychrophilic bacterium *Desulfotalea psychrophila* reveal a cold-active enzyme with an unusual high thermal stability. J Mol Biol 372:130–149

Feller G (2003) Molecular adaptations to cold in psychrophilic enzymes. Cell Mol Life Sci 60:648–662

Feller G (2010) Protein stability and enzyme activity at extreme biological temperatures. J Phys Condens 22:32101–32118

Feller G, Gerday C (1997) Psychrophilic enzymes: molecular basis of col adaptation. Cell Mol Life Sci 53:830–841

Feller G, Gerday C (2003) Psychrophilic enzymes: hot topics in cold adaptation. Nature Rev Microbiol 1:200–208

Feller G, Narinx E, Arpigny JL, Zekhnini Z, Swings J, Gerday C (1994) Temperature dependence of growth, enzyme secretion and activity of psychrophilic Antarctic bacteria. Appl Microbiol Biotechnol 41:477–479

Feller G, Arpigny J-L, Narinx E, Gerday C (1997a) Molecular adaptations of enzymes from psychrophilic organisms. Comp Biochem Physiol 118:495–499

Feller G, Zekhnini Z, Lamotte-Brasseur J, Gerday C (1997b) Enzymes from cold-adapted microorganisms. The class C beta-lactamase from the Antarctic psychrophile *Psychrobacter immobilis* A5. Eur J Biochem 244:186–191

Feller G, D'Amico S, Gerday C (1999) Thermodynamic stability of a cold-active α-amylase from the Antarctic bacterium *Alteromonas haloplanktis*. Biochemistry 38:4613–4619

Garcia-Viloca M, Gao J, Karplus M, Truhlar DG (2004) How enzymes work: analysis by modern rate theory and computer simulations. Science 303:186–195

Garsoux G, Lamotte-Brasseur J, Gerday C, Feller G (2004) Kinetic and structural optimisation to catalysis at low temperatures in a psychrophilic cellulase from the Antarctic bacterium *Pseudoalteromonas haloplanktis*. Biochem J 384:247–253

Georlette D, Jonsson ZO, van Petegem F, Chessa J-P, van Beeumen J, Hubscher U, Gerday C (2000) A DNA ligase from the psychrophile Pseudoalteromonas haloplanktis gives insights into the adaptation of proteins at low temperatures. Eur J Biochem 267:3502–3512

Georlette D, Damien B, Blaise V, Depiereux E, Uversky VN, Gerday C, Feller G (2003) Structural and functional adaptations to extreme temperatures i psychrophilic, mesophilic, and thermophilic DNA ligases. J Biol Chem 278:37015–37023

Georlette D, Blaise V, Collins T, D'Amico S, Gratia E, Hoyoux A, Marx J-C, Sonan G, Feller G, Gerday C (2004) Some like it cold: biocatalysis at low temperatures. FEMS Microbiol Rev 28:25–42

Gerday C, Aittaleb M, Bentahir M, Chessa J-P, Claverie P, Collins T, D'Amico S, Dumont J, Garsoux G, Georlette D, Hoyoux A, Lonhienne T, Meuwis MA, Feller G (2000) Cold-adapted enzymes: from fundamentals to biotechnology. TIBTECH 18:103–107

Giver L, Gershenson A, Freskgard P-A, Arnold FH (1998) Directed evolution of a thermostable esterase. Proc Natl Acad Sci USA 95(4):12809–12813

Gorfe AA, Bransdal BO, Leiros HK, Helland R, Smalas AO (2000) Electrostatics of mesophilic and psychrophilic trypsin isoenzymes: qualitative evaluation of electrostatic differences at the substrate binding site. Proteins 40:207–217

Gudmundsdottir A (2002) Cold-adapted and mesophilic brachyurins. Biol Chem 383:1125–1131

Heidarsson PO, Sigurdsson ST, Asgeirsson B (2009) Structural features and dynamics of a cold-adapted alkaline phosphatase studied by EPR spectroscopy. FEBS J 276:2725–2735

Hess E (1934) Effects of low temperatures on the growth of marine bacteria. Contribs Can Biol Fisheries Ser C 8:491–505

Hochachka PW, Somero GN (1973) Strategies of biochemical adaptation. WB Saunders, Philadelphia

Hochachka PW, Somero GN (2002) Biochemical adaptation. Oxford University Press, New York

Homchaudhuri L, Sarma N, Swaminathan R (2006) Effect of crowding by dextrans and ficols on the rate of alkaline phosphatase-catalysed hydrolysis: a size-dependent investigation. Biopolymers 83:477–486

Hoyoux A, Blaise V, Collins T, D'Amico S, Gratia E, Huston AL, Marx J-C, Sonan G, Zeng Y, Feller G, Gerday C (2004) Extreme catalysts from low-temperature environments. J Biosci Bioeng 98:317–330

Huston AL, Haeggström JZ, Feller G (2008) Cold adaptation of enzymes: structural, kinetic and microcalorimetric characterizations of an aminopeptidase from the Arctic psychrophile *Colwellia psychrerythraea* and of human leukotriene A (4) hydrolase. Biochim Biophys Acta 1784:1865–1872

Ingraham JL, Stokes JL (1959) Psychrophilic bacteria. Bacteriol Rev 23:97–108

Kim SY, Hwang KY, Kim SH, Sung HC, Han YS, Cho YJ (1999) Structural basis for cold adaptation. Sequence, biochemical properties, and crystal structure of malate dehydrogenase from a psychrophile *Aquaspirillium arcticum*. J Biol Chem 274:11761–11767

Kobori H, Sullivan CW, Shizuya H (1984) Heat-labile alkaline phosphatase from Antarctic bacteria: rapid 5′ end-labelling of nucleic acid. Proc Natl Acad Sci USA 81:6691–6695

Koutsiolis D, Wang E, Tzanodaskalaki M, Nikiforaki D, Deli A, Feller G, Heikinheimo P, Bouriotis V (2008) Directed evolution on the colddapted properties of TAB 5 alkaline phosphatase. Protein Eng 21:319–327

Kumar S, Tsai CJ, Nussinov R (2002) Maximal stabilities of reversible two-state proteins. Biochemistry 41:5359–5374

Leiros HK, Willassen NP, Smalas AO (2000) Structural comparisonof psychrophilic and mesophilic trypsins. Elucidating the molecular basis of cold-adaptation. Eur J Biochem 267:1039–1049

Leiros I, Moe E, Lanes O, Smalas AO, Willassen NP (2003) The structure of uracil-DNA glycosylase from Atlantic cod (*Gadus mohua*) reveals cold-adaptation features. Acta Crystallogr D Biol Crystallogr 59:1357–1365

Lonhienne T, Baise E, Feller G, Bouriotis V, Gerday C (2001a) Enzyme activity determination on macromolecular substrates by isothermal calorimetry: application to mesophilic and psychrophilic chitinases. Biochim Biophys Acta 1545:349–356

Lonhienne T, Zoidakis J, Vorgias E, Feller G, Gerday C, Bouriotis V (2001b) Modular structure, local flexibility and cold-activity of a novel chitobiase from a psychrophilic Antarctic bacterium. J Mol Biol 310:291–297

Margesin R (2009) Effect of temperature on growth parameters of psychrophilic bacteria and yeasts. Extremophiles 13:257–262

Marshall CJ (1997) Cold-adapted enzymes. TIBTECH 15:359–363

Marx J-C, Collins T, D'Amico S, Feller G, Gerday C (2007) Cold-adapted enzymes from marine Antarctic microorganisms. Mar Biotechnol 9:293–304

Matsuura A, Yao M, Aizawa T, Koganesawa N, Masaki K, Miyazawa M, Demura M, Tanaka I, Kawano K, Nitta K (2002) Structural analysis of an insect lysozyme exhibiting catalytic efficiency at low temperature. Biochemistry 41:12086–12092

Merlino A, Russo Kraus I, Castellano I, De venditis E, Rossi B, Conte M, Vergara A, Sica F (2010) Structure and flexibility in cold-adapted iron superoxide dismutases: the case of the enzyme isolated from Pseudoalteromonas haloplanktis. J Struct Biol 172:343–352

Miyazaki K, Wintrode PL, Grayling RA, Rubigh DN, Arnold FH (2000) Directed evolution study of temperature adaptation in a psychrophilic enzyme. J Mol Biol 297:1015–1026

Moe E, Leiros I, Rijse EK, Olufsen M, Lanes O, Smalas A, Willassen NP (2004) Optimisation of the surface electrostatics as a strategy for cold adaptation of uracil-DNA N-glycosylase. J Mol Biol 343:1221–1230

Narinx E, Baise E, Gerday C (1997) Subtilisin from antarctic bacteria: characterisation and site-directed mutagenesis of residues possibly involved in the adaptation to cold. Prot Engineer 10:1271–1279

Pace CN, Laurents DV (1989) A new method for determining the heat capacity change for protein folding. Biochemistry 28:2520–2525

Papaleo E, Olufsen M, de Gioia L, Bransdal BO (2007) Optimization of electrostatics as a strategy for cold-adaptation: a case study of cold- and warm-active elastases. J Mol Graph Model 26:93–103

Papaleo E, Pasi M, Tiberti M, de Gioia L (2011) Molecular dynamics of mesophilic-like mutants of a cold-adapted enzyme: insight into distal effects induced by the mutations. PLoS ONE 6:e24214

Paredes DI, Watters K, Pitman DJ, Bystrff C, Dordick JS (2011) Comparative void-volume analysis of psychrophilic and mesophilic enzymes: Structural bioinformatics of psychrophilic enzymes reveals sources of core flexibility. BMC Struct Biol 11:42

Petrescu I, Lamotte-Brasseur J, Chessa J-P, Ntarima P, Clayessens M, Devreese B, Marino G, Gerday C (2000) Xylanase from the psychrophilic yeast *Cryptococcus adeliae*. Extremophiles 4:137–144

Radestock S, Gohlke H (2011) Protein rigidity and thermophilic adaptation. Proteins 79: 1089–1108

Ramli AN, Mahadi NM, Shamsir MS, Rabu A, Joyce-Tan KH, Murad AM, Illias RM (2012) Structural prediction of a novel chitinase from the psychrophilic *Glaciozyma antarctica* PI12 and an analysis of its structural properties and function. J Comput Aided Mol Des 26:947–961

Roman EA, Faraj SE, Cousido-Siah A, Mitscler A, Podjarny A, Santos J (2013) Frataxin from *Psychromonas ingrahmii* as a model to study stability modulation within the CYaY protein family. Biochim Biophys Acta 1834:1168–1180

Russell NJ (2000) Towards a molecular understanding of cold activity of enzymes from psychrophiles. Extremophiles 4:83–90

Russell RJ, Gerike U, Danson MJ, Hough DW, Taylor GL (1998) Structural adaptations of the cold-active citrate synthase from an Antarctic bacterium. Structure 6:351–361

Santarossa G, Gatti-Lafranconi P, Alquati C, DeGioia L, Alberghina L, Fantucci P, Lotti M (2005) Mutations in the «lid» region affect chain length specificity and thermostability of a *Pseudomonas fragi* lipase. FEBS Lett 579:2383–2386

Sato Y, Watanabe S, Yamaoka N, Takada Y (2008) Gene cloning of cold-adapted isocitrate lyase from a psychrophilic bacterium, *Colwellia psychrerythraea*, and analysis of amino acid residues involved in cold adaptation of this enzyme. Extremophiles 12:107–117

Schmidt-Nielsen S (1902) Ueber einige psychrophile Mikrooganismen und ihr Vorkommen. Centr Bakteriol Parasitenk Abt II 9:145–147

Serrano L, Fersht AR (1989) Capping and alpha-helix stability. Nature 342:296–299

Siddiqui KS, Cavicchioli R (2006) Cold-adapted enzymes. Ann Rev Biochem 75:403–433

Smalas AO, Leiros HK, OS V, Willassen NP (2000) Cold-adapted enzymes. Biotechnol Annu Rev 6:1–57

Somero GN (1977) Temperature as a selective factor in protein evolution: the adaptational strategy of compromise. J Exp Zool 194:175–188

Somero GN, Low PS (1976) Temperature: a «shaping force» in protein evolution. Biochem Soc Symp 41:33–42

Sonan GK, Receveur-Brechot V, Duez C, Aghajari N, Czjzek M, Haser R, Gerday C (2007) The linker region plays a key role in the adaptation to cold of the cellulase from an Antarctic bacterium. Biochem J 407:293–302

Sotelo-Mundo RR, Lopez-Zavala AA, garcia-Orozco KD, Arvizu AA, Velazquez-Contreras EF, Valenzuela-Soto EM, Rojo-Dominguez A, Kanost MR (2007) The lysozyme from insect (manduca sexta) is a cold-adapted enzyme. Protein Pept Lett 14:774–778

Spiwok V, Lipovova P, Skalova T, Duskova J, Dohnalek J, Hasek J, Russell N, Kralova B (2007) Cold-active enzymes studied by comparative molecular dynamics simulation. J Mod Model 13:485–497

Stadler AM, Garvey CJ, Bocahut A, Sacquin-Mora S, Digel I, Schneider GJ, Natali F, Artmann GM, Zaccai G (2012) Thermal fluctuations of haemoglobin from different species: adaptation to temperature via conformational dynamics. J R Soc Interface 7:2845–2855

Sun-Yong K, Kwang-Yeon H, Sung-Hou K, Ha-Chin S, Ye-Sun H, Yunge C (1999) Structural basis for cold adaptation. Sequence, biochemical properties, and crystal structure of malate dehydrogenase from a psychrophile *Aquaspirillium arcticum*. J Biol Chem 274:11761–11767

Suzuki T, Yamamoto K, Tada H, Uda K (2012) Cold-adapted features of arginine kinase from the deep-sea *Calyptogena kaikoi*. Mar Biotechnol 14:294–303

Svingor A, Kardos J, Hajdju I, Nemeth A, Zavodszky P (2001) A better enzyme to cope with cold. Comparative flexibility studies on psychrotrophic, mesophilic, and thermophilic IPMDHs. J Biol Chem 276:28121–28125

Taguchi S, Ozaki A, Momose H (1998) Engineering of a cold-adapted protease by sequential random mutagenesis and a screening system. Appl Environ Microbiol 64:492–495

Taguchi S, Ozaki A, Nonaka T, Momose H (1999) A cold-adapted protease engineered by experimental evolution system. J Biochem 126:689–693

Talla-Singh D, Stites WE (2008) Refinement of noncalorimetric determination of the change in heat capacity, ΔC_p, of protein unfolding and validation across a wide temperature range. Proteins 71:1607–1616

Tiberti M, Papaleo E (2011) Dynamic properties of extremophilic subtilisin-like serine-proteases. J Struct Biol 174:69–83

Tindbaek N, Svendsen A, Oestezrgaard PR, Draborg H (2004) Engineering a substrate-specific cold-adapted subtilisin. Protei Eng Des Sel 17:149–156

Tsigos I, Velonia K, Smonou I, Bouriotis V (1998) Purification and characterization of an alcohol dehydrogenase from the Antarctic psychrophile *Moraxella* sp. TAE123. Eur J Biochem 254:356–362

Vieille C, Zeikus G (2001) Hyperthermophilic enzymes: sources, uses and molecular mechanisms for thermostability. Microbiol Mol Biol Rev 65:1–43

Watanabe S, Yasutake Y, Tanaka I, Takada Y (2005) Elucidation of stability determinants of cold-adapted monomeric isocitrate dehydrogenase from a psychrophilic bacterium, *Colwellia maris*, by construction of chimeric enzymes. Microbiology 151:1083–1094

Wintrode PL, Miyazaki K, Arnold FH (2001) Patterns of adaptation in a laboratory evolved thermophilic enzyme. Biochim Biophys Acta 1549:1–8

Wolf-Watz M, Thai V, Henzler-Wildman K, Hadjipavlou G, Eisemesser EZ, Kern D (2004) Linkage between dynamics and catalysis in a thermophilic-mesophilic enzyme pair. Nat Struct Mol Biol 11:945–949

Xie BB, Bian F, Chen XL, He HL, Guo J, Gao X, Zeng YX, Chen B, Zhou BC, Zhang YZ (2009) Cold adaptation of zinc metalloprotease in the thermolysin family from deep sea and arctic sea ice bacteria revealed by catalytic and structural properties and molecular dynamics: new insights into relationship between conformational flexibility and hydrogen bonding. J Biol Chem 284:9257–9269

Xu Y, Feller G, Gerday C, Glansdorff N (2003) Metabolic enzymes from psychrophilic bacteria: challenge of adaptation to low temperatures in ornithine carbamoyltransferase from *Moritella abyssi*. J Bacteriol 185:2161–2168

Zhong CQ, Song S, Fang N, Liang X, Zhu H, Tang XF, Tang B (2009) Improvement of low-temperature caseinolytic activity of a thermophilic subtilase by directed evolution and site directed mutagenesis. Biotechnol Bioeng 104:862–870

Part IV
Biotechnological Significance of Cold-Adapted Yeasts

Chapter 16
Cold-Active Yeast Lipases: Recent Issues and Future Prospects

Mirosława Szczęsna-Antczak, Janina Kamińska, Tomasz Florczak and Marianna Turkiewicz

Abstract Microbial lipases are, besides proteases, enzymes of the highest biotechnological potential, catalyzing not only hydrolytic reactions but also – in media of low water activity – synthesis reactions. These enzymes are used in pharmaceutical, food, and household chemicals industries and for the treatment of environmental pollution. Production of lipases is constantly increasing and now accounts for more than one-fifth of the global enzyme market. This review is dedicated to cold-active lipases of yeasts, of which lipases A and B of *Pseudozyma* (formerly *Candida*) *antarctica* have been the most thoroughly investigated. This chapter covers distinctive structural features and specificity of these enzymes in comparison with selected mesophilic lipases as well as modifications (together with diverse immobilization techniques on various supports) directed to improving catalytic properties and stability of these proteins. The application potential of cold-active yeast lipases is discussed; the most important applications include enantio- and regioselective biotransformations, production of biofuels, detergents, food additives, structured triacylglycerols, etc. Some lipases from mesophilic yeasts (e.g., non-conventional yeast *Yarrowia lipolytica*) show characteristic features of cold-active enzymes, and examples of their use are also considered.

Keywords Yeast · Cold-active · Lipases · Application · Biotransformation

---

M. Szczęsna-Antczak · T. Florczak · M. Turkiewicz (✉)
Institute of Technical Biochemistry, Faculty of Biotechnology and Food Sciences, Lodz University of Technology, Stefanowskiego Street, 4/10 90-924 Lodz, Poland
e-mail: marianna.turkiewicz@p.lodz.pl

M. Szczęsna-Antczak
e-mail: miroslawa.szczesna-antczak@p.lodz.pl

J. Kamińska
Institute of General Food Chemistry, Faculty of Biotechnology and Food Sciences, Lodz University of Technology, Stefanowskiego Street, 4/10 90-924 Lodz, Poland
e-mail: janina.kaminska@p.lodz.pl

P. Buzzini and R. Margesin (eds.), *Cold-Adapted Yeasts*,
DOI: 10.1007/978-3-642-39681-6_16,

Contents

16.1 Introduction: General Characteristics and Classification of Lipases 354
16.2 Sources and Commercial Preparations of Cold-Active Yeast Lipases 355
16.3 Structural and Kinetic Characteristics and Specificity 356
16.4 Modifications of Cold-Active Lipases 359
16.4.1 Immobilization 360
16.4.2 Examples of Immobilization of Cold-Active Lipases 361
16.4.3 Whole Cell Biocatalysts with Surface-Displayed Lipases 361
16.5 Biotechnological Potential of Cold-Active Yeast Lipases 362
16.5.1 Enantio- and Regioselective Biotransformations 363
16.5.2 Other Applications 364
16.6 Conclusions 368
References 368

16.1 Introduction: General Characteristics and Classification of Lipases

Lipases are believed by many reviewers to be key enzymes in swiftly growing biotechnology and are some of the most important biocatalysts in the food and chemical industries, biomedical sciences, detergent production, environmental management, etc. The great importance of lipases is confirmed by the fact that about 40 % of industrial biotransformations are conducted using these enzymes. The world's output of lipases has been increasing faster than that of other enzymes (by 9.13 % in the years 2001–2010) (Hasan et al. 2010), and in 2012, it accounted for 21 % of a global market worth $2.9 billion. Lipases with the most significant potential, although still not fully exploited in practice, are those obtained from psychrophilic yeasts.

According to the Nomenclature Committee of the International Union of Biochemistry and Molecular Biology (IUBMB), lipases (or, formally, triacylglycerol lipases, EC 3.1.1.3) belong to the first subclass of hydrolases, i.e., hydrolases acting on ester bonds (EC 3.1) and to the sub-subclass of carboxylic ester hydrolases (EC 3.1.1), which also includes carboxylesterases other than lipases (EC 3.1.1.1). Both groups of proteins are serine enzymes. The key difference between lipases and other carboxylesterases is their ability to hydrolyze water-insoluble esters of glycerol and long-chain fatty acids, but it is sometimes difficult to distinguish between these two groups of enzymes. The most typical lipases, also known as true lipases, undergo interfacial activation in contact with insoluble hydrophobic substrates, that is, exhibit significantly increased activity at the lipid–water interface (Verger 1997). However, in some lipases, the interfacial activation is very weak or non-existent. Although formally classified as lipases due to the ability to hydrolyze water-insoluble triacylglycerols, such enzymes display some properties of esterases. Lipases exhibit not only hydrolytic activity; under thermodynamically favorable conditions, e.g., in media with low water activity, they catalyze synthesis reactions—esterification, acidolysis, alcoholysis, and aminolysis of esters, while their chemo-, regio-, and stereoselectivity properties broaden their application potential.

16.2 Sources and Commercial Preparations of Cold-Active Yeast Lipases

The ability to produce lipases is very widespread among yeasts inhabiting cold environments (e.g., the Antarctic, Arctic, Alpine habitats, glaciers, oceans) (Juhl et al. 2010). Over a hundred psychrophilic yeast species have been found to synthesize cell-bound, extracellular, or intracellular lipases (Arenz et al. 2006; Brandão et al. 2011; Buzzini et al. 2012; Carrasco et al. 2012). However, research of these microorganisms is usually limited to selection of lipolytic strains in qualitative plate tests. Quantitative determinations of activity in cell-free extracts or yeast culture supernatants are performed less often, while even fewer studies are devoted to purification and characterization of cold-active yeast lipases. The last group of investigations primarily consists of comprehensive and advanced studies of cold-active lipases A and B (CALA and CALB) (Patkar et al. 1993; Dominguez de Maria et al. 2005) from the psychrophilic yeast *Pseudozyma antarctica* (previously *Candida antarctica*), a strain isolated from sediment from Lake Vanda, Antarctica (Boekhout and Fell 1998). So far, they have been given the greatest attention in the literature dealing with cold-active lipases. Recently, Rashid et al. (2010) have found cold-active lipases in *Glaciozyma antarctica* (previously *Leucosporidium antarcticum*) (Turchetti et al. 2011), while Zaliha et al. (2012) patented a bifunctional lipase from *L. antarcticum* PI12. The optimum temperature of that lipase is 20 °C; at 5 °C, the enzyme exhibits 60 % of its maximum activity. Surprisingly, at temperatures above 30 °C, the enzyme displays considerable proteolytic activity with a maximum at 40 °C, which is activated by zinc ions and repressed by pepstatin (a specific inhibitor of aspartyl proteases). A sequence of 260 amino acids of the lipase, obtained by in silico translation of its gene fragment, reveals a substantial degree of homology with *Pseudomonas* sp. metalloprotease. Importantly, the authors of the patent noted that *L. antarcticum* PI12 cells contain bacterial endosymbionts of the genus *Janthinobacterium*. Since the lipase gene fragment was found in the total pool of DNA isolated from *L. antarcticum* culture, probably including also the symbiont's DNA, it is not certain whether this catalytically promiscuous enzyme is produced by the yeast cells or by the endosymbiotic bacteria. Apart from psychrophilic yeasts, some cold-active lipases have been found in mesophilic yeasts, such as *Yarrowia lipolytica* (previously *Candida lipolytica*) (Alford and Pierce 1961; Yadav et al. 2011), *Wickerhamomyces lynferdii* (formerly *Pichia lynferdii*) (Kim et al. 2010), and *Candida albicans* (Lang et al. 2011). However, it is *Candida antarctica* lipases that biotechnologists have found to be by far the most promising, and they will be discussed in detail in this chapter.

CALB is commercially available in preparations of various form (solid, liquid, immobilized on different carriers) and composition, depending on the producer. The largest manufacturer of enzymes, the Danish company Novozymes, produces CALB in a recombinant strain of *Aspergillus niger* under the commercial names Lipozyme® CALB L (in soluble form) and Novozym®435 (in immobilized form). The liquid preparation is stabilized with sorbitol (25 %) and glycerol (25 %) and is generally recognized as safe (GRAS). The immobilized preparation is obtained by sorption of Lipozyme® CALB L on cross-linked poly-(methyl methacrylate) (Pierre et al. 2006). In turn, Roche Ind. (Germany) offers *C. antarctica* lipase B in lyophilized (Chirazyme L-2 c.-f., lyo) and immobilized (Chirazyme L-2 c.-f., C2) forms, as well as lipase A in the preparation Chirazyme L-5. Some smaller manufacturers also offer *C. antarctica* lipases; for instance, c-LEcta (Leipzig, Germany) sells CALB in the lyophilized and methacrylate-immobilized forms and CALA as lyophilized powder.

16.3 Structural and Kinetic Characteristics and Specificity

The best-studied lipases are those produced by the yeast *C. antarctica*—CALA and CALB (Juhl et al. 2010). In triacylglycerol hydrolysis, lipase A exhibits sn-2 regiospecificity, which is rather rare among microbial lipases, but is not regiospecific in the interesterification reaction. A unique property of this enzyme, which makes it different from most other lipases, including CALB, is its selectivity for *trans* fatty acids. CALA is an ideal catalyst in the bioresolution of sterically hindered compounds (Dominguez de Maria et al. 2005). Lipase B displays sn-1,3 regiospecificity and is one of the most effective catalysts in the resolution of alcohols and amines, allowing the preparation of a variety of secondary alcohols and primary amines of high enantiomeric purity (Martinelle and Hult 1995).

All lipases are α/β hydrolases, that is, they are composed of centrally located β-sheets with α-helices on either side (an α/β sandwich), with the catalytic triad Ser…His…Asp situated in loops; however, they differ in the presence and structure of a characteristic domain called a "lid," which closes or opens access to the active site in contact with the aqueous and hydrophobic phases, respectively. According to Bordes et al. (2010), this process consists of two stages—a partial opening of the lid as a result of the enzyme binding to a hydrophobic surface, and a subsequent full opening as a result of binding to a substrate.

While CALB does not undergo interfacial activation in hydrolysis of water-insoluble substrates, CALA becomes activated by hydrophobic surfaces (Martinelle et al. 1995), albeit not as strongly as the mesophilic lipase HLL from the fungus *Humicola lanuginosa* (now *Thermomyces lanuginosus*) or YlLip2 from *Y. lipolytica* (Fickers et al. 2011). In HLL, as well as in YlLip2 (Bordes et al. 2010), the lid consists of a single helix located in the central area of the hydrophobic

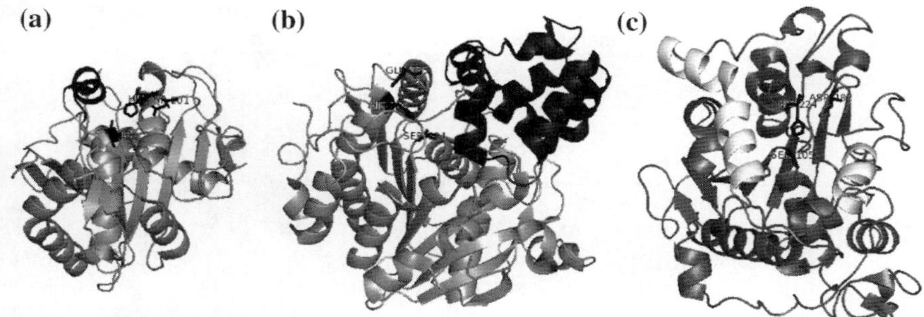

Fig. 16.1 Lipase structures: **a** *H. lanuginosa* lipase, 4DYH; **b** *C. antarctica* lipase A, 3GUU; **c** *C. antarctica* lipase B, 1TCA, The helices forming the lid in HLL (**a**) and in CALA (**b**) are in *black*; helices near the entrance to the active site of CALB (**c**) are in *white*. The catalytic triad is in *black*. The figures were prepared on the basis of crystallographically resolved structures deposited in PDB (codes behind enzyme name)

surface (Fig. 16.1a), while in CALA (Fig. 16.1b) it is composed of six helices located between β-sheets (Svendsen 2000). In CALB, the entrance to the active site is surrounded by four helices (Fig. 16.1c), whose function has not been resolved until recently. Skjøt et al. (2009) suggested that the helices form two lids, but the latest research has contradicted that hypothesis.

Based on molecular dynamics simulations of CALB, Gruber and Pleiss (2012) proposed that while the lids facilitate contact of the lipase with a hydrophobic substrate via the anchor regions Leu147, Leu219, and Val272, they do not completely close access to the active site in water solution. This is corroborated by results of biochemical tests concerning CALB activity toward water-soluble substrates. Similarly to HLL and YlLip2, CALA is a true lipase occurring in two conformations: open (active) and closed (inactive), while CALB, which does not exhibit such a behavior, shows some similarity to esterases.

CALB is a funnel-like lipase (Pleiss et al. 1998) with a binding site that is not easily accessible, due to which the enzyme is less active in triacylglycerol hydrolysis than other lipases (whether from mesophilic or psychrophilic sources) that have large, open, crevice-like binding sites. The binding site of CALB is a hydrophobic pocket dominated by aliphatic amino acids with only one aromatic (Trp104), which is critical for chiral recognition between enantiomers. The W104A mutation leads to reversal of CALB enantiopreference from (R)-1-phenylethanol for the wild enzyme to (S) for the mutant; it also becomes more enantioselective toward sterically hindered α-methyl carboxylates (Naik et al. 2010; Engström et al. 2012). Furthermore, the mutant protein catalyzes kinetic resolution of diarylmethanols, which are not accepted by the wild lipase.

All lipases act due to catalytic triad; however, they may differ in the surroundings of the active serine, which is atypical in CALB, because in contrast to most microbial lipases, including CALA, it contains the sequence TXSXG (TWSQG) rather than GXSXG. The T103G mutation leads to increased stability of CALB at 60 °C from 6 to 25 min, without affecting its activity (Patkar et al. 1997).

Table 16.1 Length and amino acid content of loops in the structure of selected lipases from psychrophilic (CALA, CALB) and mesophilic (HLL, YlLip2) yeasts

Lipase	PDB code	Total loops length/ molecule length	Amino acid composition in loops		
			% Gly	% Pro	Lys/Arg
HLL (*Humicola lanuginosa* lipase)	4DYH	125/269; 46 %	14.4	7.2	4/4 = 1
CALA (*Candida antarctica* lipase A)	3GUU	218/462; 47.18 %	10.1	13.3	10/4 = 2.5
CALB (*Candida antarctica* lipase B)	1TCA	161/317; 50.7 %	11.8	12.4	4/5 = 0.8
YLL lip2 *Yarrowia lipolytica* lipase lip2)	3O0D	156/301; 51.8 %	8.3	9.0	6/4 = 1.5

Calculations based on structures deposited in PDB and ProtParam program (Gasteiger et al. 2005)

The structure of CALA was resolved in 2008 (Ericsson et al. 2008). It was shown that the active site of CALA is significantly different from that of CALB. Naik et al. (2010) found that CALA, as opposed to CALB, accepts esters with alcohol residues larger than acyl groups, due to which these lipases display opposite substrate specificities; that is, substrates hydrolyzed with low yield by CALB are easily hydrolyzed by CALA. Thus, it seems that under physiological conditions, CALA and CALB act synergically.

Elucidation of the structure of psychrophilic enzymes makes it possible to determine, at least partially, the mechanism of their adaptation to cold. The β-sheet which forms the core of the lipase molecule is so rigid that differences in its structure in comparison to homologous mesophilic proteins do not play a major role in cold adaptation, and the same is true for the α-helix structure. Of importance are differences in the length and sequences of the loops connecting the secondary structures. Table 16.1 presents the data that are, according to Cavicchioli et al. (2011), molecular indicators of adaptation to low temperatures. Analysis of these data shows that CALA and CALB do not reveal any particular adaptations to cold, especially as compared to (theoretically) mesophilic enzymes, e.g., the aforementioned HLL or extracellular lipases from the mesophilic yeast *Y. lipolytica* (YLL). However, comparing data from Tables 16.1 and 16.2, one can see certain trends. HLL, which is the most stable, contains the fewest loops, while the most thermosensitive YlLip2 has the greatest number of these structures. CALA, which is characterized by the highest optimum temperature, exhibits the highest proline content in the loops and a higher lysine to arginine ratio than that found in CALB. In turn, analysis of glycine content in the loops is of little explanatory value. While it is assumed that with decreasing optimum temperature of the enzyme, the number of glycines in the loops increases, thus enhancing their flexibility; in light of the data from Table 16.1, this hypothesis is true only for *C. antarctica* lipases A and B. The data from Table 16.2 show that the source of an enzyme does not always determine its adaptations; for instance, lipases from the mesophilic yeast *Y. lipolytica* display properties characteristic of psychrophilic enzymes, while lipases from *C. antarctica* (particularly CALA) exhibit much higher thermal stability, which makes them distinct from the majority of typical cold-active enzymes.

Table 16.2 Selected properties and kinetic constants of lipases from psychrophilic (CALA, CALB) and mesophilic (HLL, YlLip2, NCIM 3639 YLL) yeasts

Lipase	$T_{opt.}$ (°C)	Thermostability T_d (°C)	K_M (mM)	k_{cat} (s^{-1})	k_{cat}/K_M (s^{-1} mM^{-1})	Substrate	References
HLL	38	Up to 100	0.08	ND	ND	α-naphthyl acetate	Omar et al. (1987) Jayawardena (2009)
CALA	50	Up to 70	4.48	1.5×10^6	334×10^3	Tributyrin	Pfeffer et al. (2006) Dominguez de Maria et al. (2005)
	46	≤90	0.123	ND	ND	α-naphthyl acetate	Jayawardena (2009)
CALB	30	Up to 60	0.484	16	3.3×10^1	pNP butyrate	Skjøt et al. (2009)
YLL lip2	37	Up to 45	ND	ND	ND	pNP decanate	Zhao et al. (2011) Fickers et al. (2006)
YLL NCIM 3639[a]	25		20	183	9.1×10^3	pNP palmitate	Grillitsch and Daum (2011) Yadav et al. (2011)

[a] A multimeric lipase, consisting of 20 subunits, produced by *Yarrowia lipolytica* NCIM 3639, exhibiting 46 % homology to the lipase YlLip8

16.4 Modifications of Cold-Active Lipases

Modification of enzymes, including cold-active lipases, is usually conducted using recombinant DNA technology. Such modifications are mostly based on replacement of a single or a group of amino acids in the enzyme molecule (site-directed mutagenesis and directed evolution). Introducing the W104A mutation to CALB, Engström et al. (2012) altered its enantiopreference and specificity, while Juhl et al. (2010) also changed the specificity of this lipase, obtaining considerably increased activity toward ethyl isononanoate by the T138S mutation. In turn, aiming at higher activity and enantioselectivity of CALA toward ibuprofen esters, Sandström et al. (2012) created a library of catalytic pocket variants and selected one with five substitutions in which the active site was almost entirely rearranged, resulting in a nearly 30-fold increase of enantioselectivity as compared to the wild-type lipase. Moreover, the enantioselectivity toward α-methyl carboxylates was also increased.

A very interesting modification of CALB enantioselectivity was proposed by Magnusson et al. (2001), who used substrate-assisted catalysis (SAC). They found that replacement of Thr40 (one of three amino acids stabilizing transition state) with Ala or Val, not containing the hydroxyl group, markedly lowers the activity of the mutant. However, this activity may be partially restored, at the same time modifying the enantioselectivity of the enzyme, if the substrate (e.g., ethyl (S)-2-hydroxypropanoate, not accepted by the wild-type enzyme) contains hydroxyl group in proper configuration. In this way, the synthesis of diol monoesters may be greatly enhanced.

Yet another type of modification facilitates *C. antarctica* lipase expression in a bacterial system (Blank et al. 2006; Seo et al. 2009). This usually involves removal of the signal peptide and the propeptide sequence. It has also been shown that the addition of the 6xHisTag sequence to the N- or C-terminus of the protein does not affect its activity.

Y. lipolytica Lip2 has also been modified, in order to elucidate the mechanism of interfacial activation (Bordes et al. 2010) and facilitate heterologous expression of this protein (the native signal peptide was removed, and HisTag was added; Yu et al. 2007); however, research in this area is less extensive than in the case of *C. antarctica* lipases.

A complementary approach to recombinant DNA technology is chemical modification of cold-active lipases. Most often modified was CALB, which was subjected to PEGylation (using PEG derivatives of various molecular masses; Koops et al. 1999; Forde et al. 2010), glycosylation (with oxidized aldehyde–dextran; Siddiqui and Cavicchioli 2005; Gutarra et al. 2011), amination (with ethylenediamine; Palomo et al. 2007; Barbosa et al. 2012), modification with epoxides (Miletić and Loos 2009) or treated by combined methods (Ruiz et al. 2013). These modifications usually improved its thermostability and sometimes increased the specific activity and enantioselectivity of free or immobilized enzyme. Chemical modification was also used to increase the yield of enzyme immobilization.

16.4.1 Immobilization

Cold-active lipases from yeast, owing to highly flexible molecules and their glycosylation, are particularly suitable for applications in organic solvent media with low water activity, in which their catalytic activity is much higher than that of enzymes derived from mesophilic or thermophilic organisms. Immobilization further improves the stability of these enzymes, additionally expanding their applicability.

While some preparations with immobilized lipases are commercially available (e.g., CALB as Novozym®435), researchers are seeking new methods of immobilization using conventional techniques and supports and genetic methods (such as surface-displayed lipases, see below), with a view to developing biocatalysts with properties optimized for particular applications. Although the prices of commercially available immobilized CALB preparations have declined over the past 15 years, they are still relatively expensive, and thus, they are used for the synthesis of substances whose purity is the overarching priority. Furthermore, Zhao and Song (2010) have found that some low molecular weight compounds (glycerol, benzoic and sorbic acids) were eluted from Novozym®435, contaminating the final product, while Chen et al. (2008) have reported desorption of the enzyme from the carrier. Thus, one can select an immobilization technique more suitable for a given application starting with either purified protein or cheap unpurified liquid lipase preparations (e.g., Chirazyme L-2).

16.4.2 Examples of Immobilization of Cold-Active Lipases

Cold-active lipases have been immobilized by adsorption onto organic and inorganic carriers, covalent attachment, entrapment in various polymers, cross-linking and crystallization (cross-linked dissolved enzymes CLEs, cross-linked enzyme crystals CLECs, and cross-linked enzyme aggregates CLEAs).

The cheapest and the simplest method – adsorption on various supports – is usually performed in an organic solvent medium, which ensures high efficiency. Merck Corporation has used on a large-scale Sepabeads resin Exe120 (octadecyl methacrylate) from Mitsubishi Chemical, which is perfect for CALB adsorption, leading to a biocatalyst that is 15-fold more active than Novozym®435, and also more stable (Sutton et al. 2012).

Recently, sol–gel encapsulation has emerged as a particularly easy and effective way to immobilize lipases. Using this method combined with adsorption on Celite 545 and ionic liquid addition, the soluble CALB preparation from c-LEcta was immobilized and used in large-scale kinetic resolution of racemic secondary alcohols (Ursoiu et al. 2012). Recently, the combination of immobilization, bio-imprinting, and interfacial activation of lipase has been employed to modify Yl-Lip2, CALA, and CALB (Habeych et al. 2011; Kahveci and Xu 2012; Yan et al. 2013). Table 16.3 shows examples of several successful immobilization methods that lead to increased activity or stability of cold-active lipases in selected reactions.

16.4.3 Whole Cell Biocatalysts with Surface-Displayed Lipases

The literature also includes reports of enzymes immobilization by cell surface display. Many attempts have been made to display active *C. antarctica* and *Y. lipolytica* lipases on the surface of the yeasts *S. cerevisiae* and *Komagataella pastoris* (formerly *Pichia pastoris*) and the bacterium *Escherichia coli* (Jiang et al. 2007; Seo et al. 2009; Jin et al. 2012). Narita et al. (2006) obtained CALB-displaying *E. coli* cells catalyzing enantioselective transesterifications. Introducing CALB on the surface of *K. pastoris* cells, Su et al. (2010) developed a catalyst that was more stable than free form of the enzyme. According to many authors, lipases in this form are the most effective catalysts for reactions conducted in non-aqueous media and allow for cost-effective downstream processing (Matsumoto et al. 2004; Gai and Wittrup 2007).

Table 16.3 Some examples of effective methods of cold-active lipases immobilization

Method	Advantages	Reference
Cold-active lipase B (CALB)		
Adsorption onto sepabeads EC-EP, EC-EA, and EC-BU	High stability and reusability in continuous syntheses of polyglycerol esters	Hilterhaus et al. (2008)
Cross-linking with macroporous hydrophilic poly(glycidyl methacrylate-co-ethylene glycol dimethacrylate) particles	Increased hydrolytic and synthetic activity	Miletic et al. (2009)
Adsorption onto organo-modified nanoclays (Laponite, SWy-2 and Kunipia)	Higher activity (3-fold) in α-pinene epoxidation	Tzialla et al. (2010)
Adsorption onto hydrophobic nanoepoxide (polystyrene)	Higher activity	Miletic et al. (2010)
Adsorption onto hydrophobic poly-propylene (Accurel MP1001) from organic solvent	Higher operational stability in oils butanolysis	Séverac et al. (2011)
Adsorption within micron-sized thermosensitive hydrogel particles by solvent exchange (water–isopropanol)	Enhanced specific activity, high stability and reusability	Gawlitza et al. (2012)
Cold-active lipase A (CALA)		
Bioimprinting with oleic acid and adsorption onto hydrophobic carrier Duolite A568	Improved activity and FA selectivity in ethanolysis	Kahveci and Xu (2012)
CLEA obtained using precipitation with 1,2-dimethoxyethane or ammonium sulfate in the presence of Triton X-100.	High activity retention	Lopez-Serrano et al. (2002)
Yarrowia lipolytica lipase (YlLip2)		
Bioimprinting, adsorption on cation-exchange resin and interfacial activation	Better thermal and pH stability, and wider substrate specificity	Yan et al. (2013)
Bound with magnetic iron oxide nanoparticles onto carbon nanotubes	Improved activity for resolution (R,S)-1-phenyl ethanol	Tan et al. (2012)
Adsorption onto octadecyl-sepabeads	10-fold higher operational and storage stability	Cuhna et al. (2008)

16.5 Biotechnological Potential of Cold-Active Yeast Lipases

Biotechnological manufacturing on a multi-ton scale is currently based on two fungal enzymes, *H. lanuginosa* lipase (HLL) and *Rhizopus niveus* lipase (Amano), used in the manufacture of detergents and cocoa butter substitutes, respectively. Among cold-active yeast lipases, *C. antarctica* lipase B has enjoyed a commercial success. It has been the subject of the greatest number of application studies, but relatively little is known about implementations. CALB is used by Unichem International in the manufacture of esters for the cosmetic industry (i.e., isopropyl myristate; Houde et al. 2004), by BASF for the resolution of racemic amines (Breuer et al. 2004), and by Merck for the production of a chiral fluoroleucine intermediate (Sutton et al. 2012). The small number of industrial applications primarily results from the high prices of commercial immobilized CALB preparations. On the other hand, some industrial processes using CALB may remain unknown, as manufacturers are reluctant to disclose their technological know-how.

It should be noted that the last decade has seen a sharp rise in the number of available enantiomerically pure substances used as building blocks in multi-step syntheses of biologically active products, pharmaceuticals, and agrochemicals, which are produced on a much smaller scale than substances obtained by fermentative methods. According to Hauer and Roberts (2004), optically active amino acids (whether proteinogenic or non-natural) whose annual global output does not exceed 10,000 tons are manufactured by biotransformations with the catalysts being isolated enzymes or metabolically inactive cells. Patents from recent years protect not only methods of production (Paal 2011), but also often the biocatalyst form (Hauer 2012; Truppo 2012). Currently, such biocatalysts include numerous variants of CALB (De Maria 2009; Song 2012) or whole cells expressing CALB (Kondo 2008). Emond et al. (2010) obtained CALB mutants with higher activity than the wild type using *Y. lipolytica* as a versatile expression host.

Many studies have also shown considerable usefulness of *C. antarctica* lipase A in industry. Thanks to its good thermal stability, this enzyme has found (patented) applications in the hydrolysis of triacylglycerols in the pulp industry (Fujita et al. 1992; Matsukura et al. 1992) and in the manufacture of wax lubricants (Lund et al. 1997).

The main applications of *Y. lipolytica* lipase involve the enzymes contained in living cells, e.g., in the treatment of wastewater from the oil processing industry (Wu et al. 2009, Goncalves et al. 2009) and cheese ripening (Fickers et al. 2011).

16.5.1 Enantio- and Regioselective Biotransformations

Enantiomerically pure organic compounds are increasingly being used as building blocks, auxiliaries, or advanced intermediates in the pharmaceutical, agrochemical, food, and fine chemical industries. Targeted synthesis of active enantiomers lowers the dosage of the compounds and improves the efficiency and cost-effectiveness of the process while also reducing environmental impact. Alcohols, amines, carboxylic acids, and their derivatives can be produced in enantioenriched forms from racemates or prochiral substrates by enzyme-catalyzed reactions due to the chiral nature of the biocatalysts.

Commercially available since the 1990s, CALA and CALB have been extensively investigated as catalysts for racemate resolution, initially in the hydrolysis of esters (other than triacylglycerols) and subsequently in ester synthesis and acyl transfer reactions performed in non-aqueous solvents. The results have been summarized in recent monographs (Bornscheuer and Kazlauskas 2006; Hogberg 2008) and comprehensive reviews (Breuer et al. 2004; Gotor-Fernandez et al. 2006; Ghanem 2007; Patel 2008). For these applications, CALB has been used more frequently than CALA, because of its exceptionally high activity with respect to a broad range of non-natural substrates and good resistance to non-water solvents. This biocatalyst has been found particularly effective for

(1) kinetic resolution (KR) of racemic secondary alcohols or primary amines in acyl transfer or hydrolytic reactions;
(2) dynamic kinetic resolution (DKR) of alcohols and amines;
(3) desymmetrization of prochiral substrates (1,3-diols, cyclic anhydrides).

In two-step lipase-catalyzed acyl transfer, first the acyl donor reacts with a serine hydroxyl yielding an acyl-enzyme, while in the second step, the acyl group is transferred to the attacking nucleophile. The net effect is acyl group transfer from one nucleophile to another. This mechanism implies chiral recognition in both reaction steps and creates numerous possibilities for enantiomers resolution by lipases, because the stereogenic center could be located in the acyl donor, in the nucleophile, or in both molecules. The commonly used acyl donors include enol esters, acid anhydrides, activated or non-activated esters, thioesters, carboxylic acids, and mixed carboxylic-carbonic anhydrides, while water, alcohol, ammonia, amine, carboxylic ester, or hydrogen peroxide could act as a nucleophile (acyl acceptor). A number of resolved chiral aryl-alkylamines, alkylamines, and amino alcohols are produced by BASF, some on a multi-ton scale (Breuer et al. 2004) by CALB catalyzed N-acylation.

Selected examples of highly enantioselective kinetic resolutions mediated by CALB or CALA reported in the current literature are presented in Table 16.4.

For a kinetic resolution to be effective, the lipase must be sufficiently active and enantioselective (at least $E > 20$). Enantioselectivity (E) is defined as the ratio of the rate constants of the faster to slower reacting enantiomer. Both activity and enantioselectivity are dependent on the biocatalyst, substrate, and solvent used; therefore, different strategies have been applied for improving these parameters: biocatalyst engineering (Lutz 2004; Palomo et al. 2007; Schilke and Kelly 2008; Gutarra et al. 2011; Park et al. 2011), solvent engineering (Dlugy and Wolfson 2007; van Rantwijk and Sheldon 2007; Risso et al. 2012), and substrate modification (Busto et al. 2012; Ni and Lin 2013).

Apart from enantiodifferentiation of numerous chiral substrates, CALB and CALA have been frequently used as regio- and/or chemoselective catalysts toward polyfunctional complex molecules, which enabled selective transformations of sensitive substances under mild conditions without the protection/deprotection steps necessary in chemical synthesis. Examples of regioselective acylation and deacylation of glycosides or nucleosides and chemoselective hydrolysis or alcoholysis of alkyl carboxylates by CALB are also included in Table 16.4.

16.5.2 Other Applications

Thanks to its sn-2 specificity for triacylglycerols, CALA was believed to be useful for the production of symmetrical triacylglycerols (Dominguez de Maria et al.

Table 16.4 Recent applications of CALB, CALA, and YlLip2 in biotransformations

Substrate	Reaction	Lipase	Application	Reference
Ethyl 3-arylalkane-carboxylates	Hydrolysis	CALB	KR	Deasy et al. (2011)
Boron-containing secondary alcohols	O-Acylation with vinyl acetate	CALB	KR	Andrade and Barcellos (2009)
1-Heteroaryl ethanols	O-Acylation with vinyl acetate; Methanolysis of acetates	CALB	KR	Bencze et al. (2010)
2-Aminoalkylpyridines	N-acylation of primary amine with ethyl acetate	CALB	KR	Torre et al. (2007)
1-Aryl and 1-hetero-arylpropan-2-amines	N-Acylation of primary amine with ethyl methoxyacetate	CALB	KR, DKR with Ru catalyst for racemization	Rodriguez-Mata et al. (2011)
1-Heteroarylethanamines	N-Acylation of primary amine with isopropyl butanoate;	CALB	KR	Brem et al. (2012)
	Hydrolysis of N-acylated enantiomer	CALA		
Non-benzylic primary amines	N-acylation of primary amine with ethyl dodecanoate	CALB	DKR	Gastaldi et al. (2007)
Ethyl 2-bromo-phenylacetate	Alcoholysis with 1-octanol	YlLip2p	KR	Guieysse et al. (2004)
6-Azauridine	5′-O-acylation	CALB	Regioselectivity	Wang and Zong (2009)
D-ribono-1,4-lactone	5′-O-acylation	CALB	Regioselectivity	Sebrao et al. (2011)
Aryl alcohols and their acetates	Oxidation of hydroxyl to carbonyl group	CALB	Chemoselectivity	Sharma et al. (2009)
Ethyl carboxylate intermediate for saxagliptin	Ester ammonolysis	CALB	Chemoselectivity	Gill and Patel (2006)

KR, kinetic resolution; DKR, dynamic kinetic resolution

2005), but the low selectivity of the enzyme does not enable the synthesis of sufficiently pure 1,3-diacylglycerols or 2-monoacylglycerols (Kirk and Christensen 2002). Thus, CALA is deemed to be a non-regioselective biocatalyst with regard to industrial interesterification processes. As this potential application of lipase A is very attractive, possibilities of improving its selectivity are being investigated (Pfeffer et al. 2007; Pfeffer 2008). Interestingly, CALA in immobilized form exhibits different selectivity from the free enzyme; for example, in n-hexane, it synthesizes n-butanol esters with oleic acid (*cis*-isomer) several times faster than with elaidic acid (*trans*-isomer) (Borgdorf and Warwel 1999).

CALA is a perfect tool for converting esters (or amino esters) containing branched and bulky alcohol moieties, which are not accepted by most known lipases (Henke et al. 2002). The enzyme exhibits high activity toward sterically hindered (secondary, tertiary) alcohols and α-substituted acids (Kirk and Christensen 2002; Krishna et al. 2002). It has also been used for hydrolysis of amino acids *tert*-butyl esters (Schmidt et al. 2005).

In contrast to other lipases, CALA prefers for hydrolysis substrates containing *trans* fatty acid residues and thus can be used for their removal (e.g., via interesterification) from edible fats. Knowing the structure of the substrate binding site of lipase A, genetic engineers design other lipases with similar specificity (Brundiek et al. 2012).

Lipase B from *C. antarctica* in immobilized form (Novozym®435) has been recognized as the most common biocatalyst for biodiesel production (Hama et al. 2011). This is mostly due to its high specific activity and lack of selectivity in terms of the location of the acyl residue in triacylglycerols, which makes it possible to obtain the highest efficiency for fatty acid alkyl esters, FAAE (Séverac et al. 2011). Many studies dealing with various aspects of catalysis of transesterification were and still are conducted by a number of research institutions (Chang et al. 2005; Hernandez-Martin and Otero 2008; Fedosov et al. 2013).

CALB catalyzes acyl transfer reactions of various oils and acyl acceptors (alcohols or esters, such as methyl and ethyl acetates) showing high stability in organic solvents (particularly in *tert*-butanol) and broad substrate specificity. Using Novozym®435 under optimum conditions, refined plant oils may be transformed into fatty acid methyl esters (FAME) with an efficiency of 80 to >95 % in repeated batch or continuous processes (Severac et al. 2011). An economic analysis of industrial biodiesel production based on *C. antarctica* lipase B has been conducted by Sotoft et al. (2010). Tan et al. (2010) have reported that since 2007 China has been producing biodiesel on an industrial scale using commercially available Novozym®435 (Hainabaichuan Co. Ltd., Hunan Province).

The latest biotechnological research aims at developing super cells that would act as a "factory" processing cheap raw materials into energy and/or biofuels. For instance, Hughes et al. (2012) expressed of the CALB gene in an ethanologenic yeast strain of *S. cerevisiae*. They claim that such a recombinant yeast strain has a potential to increase the profitability of an integrated biorefinery because it can be used to produce two biofuels: ethanol and biodiesel (fatty acid ethyl esters; FAEE).

Apart from producing bioesters, CALB has been tested for many other applications. For instance, Novozym®435 has been found the best candidate for the production of human milk fat substitute (HMFS) in enzymatically catalyzed acidolysis of tripalmitin (Karabulut et al. 2010). In the presence of lipase B, palmitic acid at positions 1 and 3 of tripalmitin is replaced with unsaturated free fatty acids at a rate dependent on the degree of unsaturation (C18:3 > C18:2 > C18:1). This allows for obtaining triacylglycerols with a structure similar to those occurring in human milk, that is, containing 20–25 % palmitic acid, of which ca 70 % are attached to the sn-2 position of the glycerol backbone.

Novozym®435 has also been used for highly efficient (78 % at 53 °C) synthesis of 6-O-lauroyl-erythorbate, recommended as a lipophilic antioxidant in the food industry (Lee et al. 2012). An immobilized CALB preparation has been applied for the ring-opening copolymerization of L-lactide (LLA) and glycolide using various ionic liquids. In an ionic liquid medium, lipase B from a commercially available preparation catalyzes the polymerization of hydroxy acids at temperature of up to 90 °C. Furthermore, Novozym®435 has been proven to be an effective catalyst for the ring-opening polymerization of ε-caprolactone (ε-CL) (Turkan et al. 2011). The resulting macromonomers are based on poly(ε-caprolactone) (PCL) with α,ω-thiophene functional end groups. Macromonomers are being tested for the synthesis of conducting polymer composites. In turn, *Y. lipolytica* Lip2 has been successfully used for the production of polyethers based on ring-opening polymerization (Berrera-Rivera et al. 2008).

A liquid lipase B preparation (fraction B, Chirazyme L-2, cf.-C2, Roche Diagnostics) has been used for the synthesis of sugar esters, which are increasingly appreciated as emulsifiers (Kobayashi et al. 2010). Due to high concentrations of the substrates (palmitic acid and 1,2-O-isopropylidene-α-D-glucofuranose) in acetone/*tert*-butanol mixture, it would be possible to obtain these esters on a larger scale and industrialize their production.

Potential industrial applications of cold-active lipases currently under research also include:

(1) interesterification of fats and plant oils for the production of *trans*-free structured triacylglycerols and special fats (Goli et al. 2008; Pande and Akoh 2012);
(2) synthesis of flavor esters by whole cell lipase (Jin et al. 2012);
(3) epoxidation of the C=C bond by hydrogen peroxide without a metal catalyst (da Silva and Nascimento 2012);
(4) synthesis of biodegradable linear polyesters with pendant ketoprofen (Wang et al. 2010) and mPEG-block-poly (profen amide-co-esters) copolymers (Qian et al. 2011) as carriers for sustained drug release;
(5) synthesis of glycerol-based macroamphiphiles useful as micellar nanotransporters (Gupta et al. 2010).

16.6 Conclusions

Along with proteases and some oxidoreductases, lipases are some of the most useful enzymes, especially in the production of high added value substances, such as fine chemicals, pharmaceuticals, foods, polymers, detergents, biodiesel. Among all known lipases from different environments and organisms, of special importance are cold-active lipases A and B from the yeast *Pseudozyma* (*Candida*) *antarctica*, which have been thoroughly researched at the molecular level and subjected to many genetic and chemical modifications, including immobilization. While the number of papers and patents linked to the use of these enzymes as catalysts for hydrolysis and synthesis reactions on a laboratory scale is impressive, this does not fully translate into industrial implementations, mostly due to the relatively high price of CALB and CALA preparations, both in free and in immobilized forms. Thus, efforts should be made to develop such forms of both enzymes that would be cheap, stable, and reusable. New immobilization carriers and new methods of immobilization (also genetic ones) should also be sought.

In contrast to the wide-ranging and advanced studies on *C. antarctica* lipases, research into lipases from other species of psychrophilic yeasts is in its early stages and undoubtedly should be intensified. First of all, homogeneous preparations of cold-active lipases should be obtained to determine their kinetic and molecular characteristics, followed by their application in a variety of biotransformations, and especially in unconventional biocatalysis. Metatranscriptomics could be used to discover new lipases from uncultured, or rather not yet culturable, yeasts. To date, only a few bacterial lipase genes have been identified in metagenomic DNA from cold environments (Elend et al. 2007; Wei et al. 2009).

References

Alford JA, Pierce DA (1961) Lipolytic activity of microorganisms at low and intermediate temperatures. Activity of microbial lipases at temperatures below 0°C. J Food Sci 26:518–524

Andrade LH, Barcellos T (2009) Lipase-catalyzed highly enantioselective kinetic resolution of boron-containing chiral alcohols. Org Lett 11:3052–3055

Arenz BE, Held BW, Jurgens JA, Farrell RL, Blanchette RA (2006) Fungal diversity in soils and historic wood from the Ross Sea region in Antarctica. Soil Biol Biochem 38:3057–3064

Barbosa O, Ruiz M, Ortiz C, Fernandez M, Torres R, Fernandez-Lafuente R (2012) Modulation of the properties of immobilized CALB by chemical modification with 2,3,4-trinitrobenzenesulfonate or ethylendiamine. Advantages of using adsorbed lipases on hydrophobic supports. Process Biochem 47:867–876

Bencze LC, Paizs C, Tosa MI, Trif M, Irimie FD (2010) Cal-B a highly selective biocatalyst for the kinetic resolution of furylbenzthiazole-2-yl-ethanols and acetates. Tetrahedron: Asymmetry 21:1999–2004

Berrera-Rivera K, Flores-Carreon A, Martinez-Richa A (2008) Enzymatic ring opening polymerization of ε-caprolactone by a new lipase from *Yarrowia lipolytica*. J Appl Pol Sci 109:708–719

Blank K, Morfill J, Gumpp H, Gaub HE (2006) Functional expression of *Candida antarctica* lipase B in *Escherichia coli*. J Biotechnol 125:474–483

Boekhout T, Fell JW (1998) Pseudozyma. In: Kurtzman CP, Fell JW (eds) The yeasts: a taxonomic study. Elsevier Science, Amsterdam, pp 790–797

Bordes F, Barbe F, Escalier P, Mourey L, Andre I, Marty A, Tranier S (2010) Exploring the conformational states and rearrangements of *Yarrowia lipolytica* lipase. Biophys J 99:2225–2234

Borgdorf R, Warwel S (1999) Substrate selectivity of various lipases in the esterification of cis- and trans-9-octadecenoic acid. Appl Microbiol Biotechnol 51:480–485

Bornscheuer UT, Kazlauskas RJ (2006) Hydrolases in organic synthesis. regio- and stereoselective biotransformations. Chapter 5, 2nd ed. Wiley-VCH, Weinheim, pp 61–184

Brandão LR, Libkind D, Vaz AB, Espirito Santo LC, Moline M, de Garcia V, van Broock M, Rosa CA (2011) Yeasts from oligotrophic lake in Patagonia (Argentina): diversity, distribution and synthesis of photoprotective compounds and extracellular enzymes. FEMS Microbiol Ecol 76:1–13

Brem J, Bencze LC, Liljeblad A, Turcu MC, Paizs C, Irimie FD, Kanerva LT (2012) Chemoenzymatic preparation of 1-heteroaryl-ethanamines of low solubility. Eur J Org Chem 2012:3288–3294

Breuer M, Ditrich K, Habicher T, Hauer B, Kesseler M, Stürmer R, Zielinski T (2004) Industrial methods for the production of optically active intermediates. Angew Chem Int Ed 43:788–824

Brundiek HB, Evitt AS, Kourist R, Bornscheuer UT (2012) Creation of a lipase highly selective for trans fatty acid by protein engineering. Angew Chem Int Ed 51:412–414

Busto E, Gotor-Fernandez V, Gotor V (2012) Asymmetric chemoenzymatic synthesis of ramatroban using lipases and oxidoreductases. J Org Chem 77:4842–4848

Buzzini P, Branda E, Goretti M, Turchetti B (2012) Psychrophilic yeasts from worldwide glacial habitats: diversity, adaptation strategies and biotechnological potential. FEMS Microbiol Ecol 82:217–241

Carrasco M, Rozas JM, Barahona S, Alcaino J, Cifuentes V, Baeza M (2012) Diversity and extracellular enzymatic activities of yeast isolated from King George Island, the subantarctic region. BMC Microbiol 12:251–259

Cavicchioli R, Charlton T, Ertan H, Mohd Omar MO, Siddiqui KS, Williams TJ (2011) Biotechnological uses of enzymes from psychrophiles. Microb Biotechnol 4:449–460

Chang HM, Liao HF, Lee CC, Shieh CJ (2005) Optimized synthesis of lipase-catalyzed biodiesel by Novozym 435. J Chem Technol Biotechnol 80:307–312

Chen B, Hu J, Miller EM, Xie W, Cai M, Gross RA (2008) *Candida antarctica* lipase B chemically immobilized on epoxy-activated micro- and nanobeads: catalysts for polyester synthesis. Biomacromolecules 9:463–471

Cuhna AG, Fernandez-Lorente G, Bevilaqua JV, Destain J, Paiva LM, Freire DM, Fernández-Lafuente R, Guisán JM (2008) Immobilization of *Yarrowia lipolytica* lipase—a comparison of stability of physical adsorption and covalent attachment techniques. Appl Biochem Biotechnol 146:49–56

da Silva JMR, Nascimento MG (2012) Chemoenzymatic epoxidation of citronellol catalyzed by lipases. Process Biochem 47:517–522

De Maria L (2009) Lipolytic enzyme variants. EU Patent EP 2099908

Deasy RE, Brossat M, Moody TS, Maguire AR (2011) Lipase catalysed kinetic resolutions of 3-arylalkanoic acids. Tetrahedron: Asymmetry 22:47–61

Dlugy C, Wolfson A (2007) Lipase catalyse glycerolysis for kinetic resolution of racemates. Bioprocess Biosyst Eng 30:327–330

Dominguez de Maria P, Carboni-Oerlemans C, Tuin B, Bergeman G, Meer A, Gemert R (2005) Biotechnological applications of *Candida antarctica* lipase A: state-of-the-art. J Mol Catal B 37:36–46

Elend C, Schmeisser C, Hoebenreich H, Steele HL, Strejt WR (2007) Isolation and characterization of a metagenome-derived and cold-active lipase with high stereoselectivity for (R)-ibuprofen esters. J Biotechnol 130:370–377

Emond S, Montanier C, Nicaud JM, Marty A, Monsan P, Andre I, Remaud-Simeon M (2010) New efficient recombinant expression system to engineer *Candida antarctica* lipase B. Appl Environ Microbiol 76:2684–2687

Engström K, Vallin M, Hult K, Bäckvall J-E (2012) Kinetic resolution of diarylmethanols using a mutated variant of lipase CALB. Tetrahedron 68:7613–7618

Ericsson DJ, Kasrayan A, Johansson P, Berqfors T, Sandstrom AG, Bäckvall JE, Mowbray SL (2008) X-ray structure of *Candida antarctica* lipase B shows a novel lid structure and a likely mode of interfacial activation. J Mol Biol 376:109–119

Fedosov SN, Brask J, Pedersen AK, Nordblad M, Woodley J, Xu X (2013) Kinetic model of biodiesel production using immobilized lipase *Candida antarctica* lipase B. J Mol Cat B 85–86:156–168

Fickers P, Marty A, Nicaud JM (2011) The lipases from *Yarrowia lipolytica*: genetics, biochemical characterization and biotechnological applications. Biotechnol Adv 29:632–644

Fickers P, Ongena M, Destain J, Weekers F, Thonart P (2006) Production and down-stream processing of an extracellular lipase from the yeast *Yarrowia lipolytica*. Enzyme Microb Technol 38:756–759

Forde J, Vakurov A, Gibson TD, Millner P, Whelehan M, Marisin IW, Ó'Fágáin C (2010) Chemical modification and immobilisation of lipase B from Candida antarctica onto mesoporous silicates. J Mol Cat B 66:203–209

Fujita Y, Awaji H, Shimoto H, Sharyou M, Heldt-Hansen HP (1992) Lipase-catalyzed ester hydrolysis. Patent WO/92/18638

Gai SA, Wittrup KD (2007) Yeast surface display for protein engineering and characterization. Curr Opion Struct Biol 17:467–473

Gastaldi S, Escoubet S, Vanthuyne N, Gil G, Bertrand MP (2007) Dynamic kinetic resolution of amines involving biocatalysis and *in situ* free radical mediated racemization. Org Lett 9:837–839

Gasteiger E, Hoogland C, Gattiker A, Duvaud S, Wilkins MR, Appel RD, Bairoch A (2005) Protein identification and analysis tools on the ExPASy server. In: Walker JM (ed) The proteomics protocols handbook. Humana Press, New York, pp 571–607

Gawlitza K, Wu Ch, Georgieva R, Wang D, Ansorge-Schumacher MB, von Klitzing R (2012) Immobilization of lipase B within micron-sized poly-N-isopropylacrylamide hydrogel particles by solvent exchange. Phys Chem Chem Phys 14:9594–9600

Ghanem A (2007) Trends in lipase catalyzed asymmetric access to enantiomerically pure/enriched compounds. Tetrahedron 63:1721–1754

Gill I, Patel R (2006) Biocatalytic ammonolysis of (5S)-4,5-dihydro-1H-pyrrole-1,5- dicarboxylic acid, 1-(1,1-dimethylethyl)-5-ethyl ester: preparation of an intermediate to the dipeptidyl peptidase IV inhibitor saxagliptin. Bioorg Med Chem Lett 16:705–709

Goli SAH, Sari MM, Kadivar M (2008) Enzymatic interesterification of structured lipids containing conjugated linoleic acid with palm stearin for possible margarine production. Eur J Lipid Sci Technol 110:1102–1108

Goncalves C, Lopes M, Ferreira JP, Belo I (2009) Biological treatment of olive mill wastewater by non-conventional yeast. Biores Technol 100:3759–3763

Gotor-Fernandez V, Busto E, Gotor V (2006) *Candida antarctica* lipase B: an ideal biocatalyst for the preparation of nitrogenated organic compounds. Adv Synth Catal 348:797–812

Grillitsch K, Daum G (2011) Triacylglycerols lipases of the yeast. Front Biol 6:219–230

Gruber CC, Pleiss J (2012) Lipase B from *Candida antarctica* binds to hydrophobic substrate–water interfaces via hydrophobic anchors surrounding the active site entrance. J Mol Catal B 84:48–54

Guieysse D, Sandoval G, Faure L, Nicaud JM, Monsan P, Marty A (2004) New efficient lipase from *Yarrowia lipolytica* for the resolution of 2-bromo-arylacetic acid esters. Tetrahedron: Asymmetry 15:3539–3543

Gupta S, Pandey MK, Levon K, Haag R, Watterson AC, Parmar VS, Sharma SK (2010) Biocatalytic approach for the synthesis of glycerol-based macroamphiphiles and their self-assembly to micellar nanotransporters. Macromol Chem Phys 211:239–244

Gutarra MLE, Romero O, Abian O, Torres FAG, Freire DMG, Castro AM, Guisan JM, Palomo JM (2011) Enzyme surface glycosylation in the solid phase: improved activity and selectivity of *Candida antarctica* lipase B. Chem Cat Chem 3:1902–1910

Habeych DI, Juhl PB, Pleiss J, Venegas D, Eggink G, Boeriu CG (2011) Biocatalytic synthesis of polyesters from sugar- based building blocks using immobilized *Candida antarctica* lipase B. J Mol Catal B 71:1–9

Hama S, Tamalampudi S, Yoshida A, Tamadani N, Kuratani N, Noda H, Fukuda H, Kondo A (2011) Enzymatic packed-bed reactor integrated with glycerol-separating system for solvent-free production of biodiesel fuel. Biochem Eng J 55:66–71

Hasan F, Shah AA, Javed S, Hameed H (2010) Enzymes used in detergents: lipases. African J Biotechnol 9:4836–4844

Hauer B (2012) An enzymatically catalyzed method of preparing mono-acylated polyols. EU Patent EP 2478094

Hauer B, Roberts SM (2004) Biocatalysis and biotransformation: Probing the potential usefulness and the mechanisms of action of some novel biocatalysts. Curr Opin Chem Biol 8:103–105

Henke E, Pleiss J, Bornscheuer UT (2002) Activity of lipases and esterases towards tertiary alcohols: insights into structure-function relationships. Angew Chem Int Ed Engl 41:3211–3213

Hernandez-Martin E, Otero C (2008) Different enzyme requirements for the synthesis of biodiesel: Novozym (R) 435 and Lipozyme (R) TL IM. Biores Technol 99:277–286

Hilterhaus L, Minow B, Muller J, Berheide M, Quitmann H, Katzer M, Thum O, Antranikian G, Zeng AP, Liese A (2008) Practical application of different enzymes immobilized on sepabeads. Bioprocess Biosyst Eng 31:163–171

Hogberg HE (2008) Exploiting enantioselectivity of hydrolases in organic solvents. In: Carrea G, Riva S (eds) Organic synthesis with enzymes in non-aqueous media. Wiley-VCH, Weinheim, pp 73–112

Houde A, Kademi A, Leblanc D (2004) Lipases and their industrial applications an overview. Appl Biochem Biotechnol 118:155–170

Hughes SR, Moser BR, Robinson S, Cox EJ, Harmsen AJ, Friesen JA, Bischoff KM, Jones MA, Pinkelman R, Bang SS, Tasaki K, Doll KM, Qureshi N, Liu S, Saha BC, Jackson JS, Cotta MA, Rich JO, Caimi P (2012) Synthetic resin-bound truncated *Candida antarctica* lipase B for production of fatty acid alkyl esters by transesterification of corn and soybean oils with ethanol or butanol. J Biotechnol 159:69–77

Jayawardena MB (2009) Stability and activity of self-cleaning lipase based paints. Dissertation. The University of New South Wales, Sydney

Jiang Z-B, Song H-T, Gupta N, Ma LX, Wu ZB (2007) Cell surface display of functionally active lipases from *Yarrowia lipolytica* in *Pichia pastoris*. Protein Expr Purif 56:35–39

Jin Z, Ntwali J, Han SY, Zheng SP, Lin Y (2012) Production of flavor esters catalyzed by CALB-displaying *Pichia pastoris* whole-cells in a batch reactor. J Biotechnol 159:108–114

Juhl PB, Doderer K, Hollmann F, Thum O, Pleiss J (2010) Engineering of *Candida antarctica* lipase B for hydrolysis of bulky carboxylic acid esters. J Biotechnol 150:474–480

Kahveci D, Xu X (2012) Bioimprinted immobilization of *Candida antarctica* lipase A for concentration of omega-3 polyunsaturated fatty acids. J Am Oil Chem Soc 89:1839–1845

Karabulut I, Durmaz G, Hayaloglu AA (2010) C18 Unsaturated fatty acid selectivity of lipases during the acidolysis reaction between tripalmitin and oleic, linoleic, and linolenic acids. J Am Oil Chem Soc 87:1301–1307

Kim HR, Kim IH, Hou CT, Kwon KI, Shin BS (2010) Production of novel cold-active lipase from *Pichia lynferdii* Y-7723. J Agric Food Chem 58:1322–1326

Kirk O, Christensen MW (2002) Lipases from *Candida antarctica*: unique biocatalysts from a unique origin. Org Process Res Dev 6:446–451

Kobayashi T, Ehara T, Mizuoka T (2010) Efficient synthesis of 6-O-palmitoyl-1,2-O-isopropyl-idene-D-glucofuranose in an organic solvent system by lipase-catalyzed esterification. Biotechnol Lett 32:1679–1684

Kondo A (2008) Highly active arming yeast having highly active lipase B in cell surface layer, method for producing the same and highly active lipase B. JP 2008092849

Koops BC, Papadimou E, Verheij HM, Slotboom AJ, Egmond MR (1999) Activity and stability of chemically modified *Candida antarctica* lipase B adsorbed on solid supports. Appl Microbiol Biotechnol 52:791–796

Krishna H, Persson MM, Bornscheuer UT (2002) Enantioselective transesterification of a tertiary alcohol by lipase A from *Candida antarctica*. Tetrahedron Asymmetry 13:2693–2696

Lang DM, Jang N, Wang WK, Shen YF, Yang B, Wang YH (2011) A novel cold-active lipase from *Candida albicans*: cloning expressions, characterization of recombinant enzyme. Int J Mol Sci 12:3950–3965

Lee DE, Park KM, Choi SJ, Chang PS (2012) Optimal production and structural characterization of erythorbyl laurate obtained through lipase-catalyzed esterification. Food Sci Biotechnol 21:1209–1215

Lopez-Serrano P, Cao L, van Rantwijk F, Sheldon RA (2002) Crosslinked enzyme aggregates with enhanced activity: application to lipases. Biotechnol Lett 24:1379–1383

Lund H, Nilsson T, Pickard T (1997), Treatment of fabrics. Patent WO/97/04160

Lutz S (2004) Engineering lipase B from *Candida antarctica*. Tetrahedron: Asymmetry 15:2743–2748

Magnusson A, Hult K, Holmquist M (2001) Creation of an enantioselective hydrolase by engineered substrate assisted catalysis. J Am Chem Soc 123:4354–4355

Martinelle M, Holmquist M, Hult K (1995) On the interfacial activation of *Candida antarctica* lipase A as compared with *Humicola lanuginosa* lipase. Biochim Biophys Acta 125:272–276

Martinelle M, Hult K (1995) Kinetics and acyl transfer reaction in organic media catalyzed by *Candida antarctica* lipase B. Biochim Biophys Acta 1251:191–197

Matsukura M, Fujita Y, Abo M, Itami R (1992) Methods of avoiding pitch troubles by use of thermostable lipase. Patent WO/92/13130

Matsumoto T, Ito M, Fukuda H, Kondo A (2004) Enantioselective transesterification using lipase-displaying yeast whole-cell biocatalyst. Appl Microbiol Biotechnol 64:481–485

Miletić N, Loos K (2009) Over-stabilization of chemically modified and cross-linked Candida antarctica lipase B using various epoxides and diepoxides. Australian J Chem 62:799–805

Miletic N, Rohandi R, Vukovic Z, Nastasovic A, Loos K (2009) Surface modification of macroporous poly(glycidyl methacrylate-co-ethylene glycol dimethacrylate) resins for improved *Candida antarctica* lipase B immobilization. React Funct Polym 69:68–75

Miletic N, Abetz V, Ebert K, Loos K (2010) Immobilization of *Candida antarctica* lipase B on polystyrene nanoparticles. Macromol Rapid Commun 31:71–74

Naik S, Basu A, Saikia R, Madan B, Paul P, Chaterjee R, Brask J, Svendsen A (2010) Lipases for use in industrial biocatalysis: specificity of selected structural groups of lipases. J Mol Cat B 65:18–23

Narita J, Okano K, Tateno T, Tanino T, Sewaki T, Sung MH, Fukuda H, Kondo A (2006) Display of active enzymes on the cell surface of *Escherichia coli* using PgsA anchor protein and their application to bioconversion. Appl Microbiol Biotechnol 70:564–572

Ni Z, Lin X (2013) Insight into substituent effects in Cal-B catalyzed transesterification by combining experimental and theoretical approaches. J Mol Model 19:349–358

Omar H, Nisho N, Nagai S (1987) Production of a thermostable lipase by *Humicola lanuginosa* grown on sorbitol-corn steep liquor medium. Agr Biol Chem 51:2145–2151

Paal T (2011) A process for the preparation of rosuvastatin intermediate. Patent WO/2011/106546

Palomo JM, Fernandez-Lorente G, Guisan JM, Fernández-Lafuente R (2007) Modulation of immobilized lipase enantioselectivity via chemical amination. Adv Synth Catal 349:1119–1127

Pande G, Akoh C (2012) Enzymatic synthesis of trans-free structured margarine fat analogues using stearidonic acid soybean and high stearate soybean oils. J Am Oil Chem Soc 89:1473–1484

Park CG, Kwon MA, Song JK, Kim DM (2011) Cell-free synthesis and multifold screening of *Candida antarctica* lipase B (CalB) variants after combinatorial mutagenesis of hot spots. Biotech Prog 27:47–53

Patel RN (2008) Synthesis of chiral pharmaceutical intermediates by biocatalysis. Coord Chem Rev 252:659–701

Patkar SA, Björkling F, Zundel M, Schulein M, Svendsen A, Heldt-Hansen HP, Gormsen E (1993) Purification of two lipases from *Candida antarctica* and their inhibition by various inhibitors. Ind J Chem Sect B 32:76–80

Patkar SA, Svendsen A, Kirk O, Clausen IG, Borch K (1997) Effect of mutations in non-consensus Thr-X-Ser- X-Gly in *Candida antarctica* lipase B on lipase specificity, specific activity and thermostability. J Mol Catal B 3:51–54

Pfeffer J (2008) The lipases from *Candida antarctica*: cloning, expression and their application in the synthesis of structured lipids. Dissertation, University of Stuttgart

Pfeffer J, Richter S, Nieveler J, Hansen CE, Rhlid RB, Schmid RD, Rusnak M (2006) High yield expression of lipase A from *Candida antarctica* in the methylotrophic yeast *Pichia pastoris* and its purification and characterisation. Appl Microbiol Biotechnol 72:931–938

Pfeffer J, Rusnak M, Hansen CE, Rhlid RB, Schmid RD, Maurer SC (2007) Functional expression of lipase A from *Candida antarctica* in *Escherichia coli*—a prerequisite for highthroughput screening and directed evolution. J Mol Cat B 45:62–67

Pierre SJ, Thies JC, Dureault A, Cameron NR, van Hest JCM, Carette N, Michon T, Waberskirch R (2006) Covalent enzyme immobilization onto photopolymerized highly porous monoliths. Adv Mat 18:1822–1826

Pleiss J, Fischer M, Schmidt RD (1998) Anatomy of lipases binding sites: the sessile fatty acids binding site. Chem Phys Lipids 93:67–80

Qian X, Wu Q, Xu F, Lin X (2011) Amphiphilic mPEG-block-poly(profen amide-co-esters)copolymers: one pot biocatalytic synthesis, self assembly in water and drug release. Polymer 52:5479–5485

Rashid FAA, Rahim RA, Ibrahim D (2010) Identification of lipase—producing psychrophilic yeast, *Leucosporidium* sp. Internet J Microbiol 9(1), doi: 10.5580/1215

Risso M, Mazzini M, Kroger S, Saenz-Mendez P, Seoane G, Gamenara D (2012) Microwave-assisted solvent-free lipase catalyzed transesterification of β-ketoesters. Green Chem Lett Rev 5:539–543

Rodriguez-Mata M, Gotor-Fernandez V, Gonzalez-Sabin J, Rebolledo F, Gotor V (2011) Straightforward preparation of biologically active 1-aryl and 1-heteroaryl-propan-2-amines in enantioenriched form. Org Biomol Chem 9:2274–2278

Ruiz M, Galvis M, Barbosa O, Ortiz C, Torres R, Fernandez-Lafuente R (2013) Solid-phase modification with succinic polyethyleneglycol of aminated lipase B from *Candida antarctica*: Effect of the immobilization protocol on enzyme catalytic properties. J Mol Cat B 87:75–82

Sandström AG, Wikmark Y, Engström K, Nyhlén J, Bäckvall JE (2012) Combinatorial reshaping of the *Candida antarctica* lipase A substrate pocket for enantioselectivity using an extremely condensed library. Proc Natl Acad Sci USA 109:78–83

Schilke KF, Kelly C (2008) Activation of immobilized lipase in non-aqueous system by hydrophobic poly-DL-tryptophan tethers. Biotechnol Bioeng 101:9–18

Schmidt M, Barbayianni E, Fotakopoulou I (2005) Enzymatic removal of carboxyl protecting groups. 1. Cleavage of the tertbutyl moiety. J Org Chem 70:3737–3740

Sebrao D, Sa MM, Nascimento MG (2011) Regioselective acylation of D-ribono-1,4-lactone catalyzed by lipases. Process Biochem 46:551–556

Seo H-S, Kim S-E, Han K-Y, Park JS, Kim YH, Sim SJ, Lee J (2009) Functional fusion mutant of *Candida antarctica* lipase B (CalB) expressed in *Escherichia coli*. Biochim Biophys Acta 1794:519–525

Séverac E, Galya O, Turond F, Pantel CA, Condoret JS, Monsan P, Marty A (2011) Selection of CalB immobilization method to be used in continuous oil transesterification: analysis of the economical impact. Enzyme Microb Technol 48:61–70

Sharma UK, Sharma N, Kumar R, Kumar R, Shinha AK (2009) Biocatalytic promiscuity of lipase in chemoselective oxidation of aryl alcohols/acetates: a unique synergism of CAL-B and [hmim]Br for metal-free H_2O_2 activation. Org Lett 11:4846–4848

Siddiqui KS, Cavicchioli R (2005) Improved thermal stability of the cold-adapted lipase B from *Candida antarctica* following chemical modification with oxidized polysaccharides. Extremophiles 9:471–476

Skjøt M, De Maria L, Chatterjee R, Svendsen A, Patkar SA, Ostergaard PR, Brask J (2009) Understanding the plasticity of the α/β hydrolase fold: lid swapping on the *Candida antarctica* lipase B results in chimeras with interesting biocatalytic properties. Chem Bio Chem 10:520–527

Song JK (2012) *Candida antarctica*-derived novel lipase variant with enhanced enzyme activity. KR 1020120013462 (Korea Research Institute of Chemical Technology)

Sotoft LF, Rong BG, Christensen KV, Norddahl B (2010) Process simulation and economical evaluation of enzymatic biodiesel production plant. Biores Technol 101:5266–5274

Su G, Zhang X, Lin Y (2010) Surface display of active lipase in *Pichia pastoris* using Sed1 as an anchor protein. Biotechnol Lett 32:1131–1136

Sutton PW, Adams JP, Archer I, Auriol D, Avi M, Branneby C, Collis AJ, Dumas B, Eckrich T, Fotheringham I, ter Halle R, Hanlon S, Hansen M, Holt-Tiffin KE, Howard RM, Huisman GW, Iding H, Kiewel K, Kittelmann M, Kupfer E, Laumen K, Lefèvre F, Luetz S, Mangan DP, Martin VA, Meyer H-P, Moody TS, Osorio-Lozada A, Robins K, Snajdrova R, Truppo MD, Wells A, Wirz B, Wong JW (2012) Biocatalysis in the fine chemical and pharmaceutical industries. In: Whittall J, Sutton PW (eds) Practical methods for biocatalysis and biotransformations 2. Wiley, New York, pp 2–60

Svendsen A (2000) Lipase protein engineering. Biochim Biophys Acta 1543:223–238

Tan T, Lu J, Nie K, Den L, Wang F (2010) Biodiesel production with immobilized lipase: a review. Biotechnol Adv 28:628–634

Tan H, Feng W, Ji P (2012) Lipase immobilized on magnetic multi-walled carbon nanotubes. Biores Technol 115:172–176

Torre O, Busto B, Gotor-Fernandez V, Gotor V (2007) Enzymatic preparation of novel aminoalkylpyridines using lipases in organic solvents. Adv Synth Catal 349:1481–1488

Truppo M (2012) Process for immobilization of *Candida antarctica* lipase B. Patent WO/2012/036973

Turchetti B, Thomas Hall SR, Connell LB, Branda E, Buzzini P, Theelen B, Müller WH, Boekhout T (2011) Psychrophilic yeasts from antarctica and European glaciers: description of *Glaciozyma* gen. nov, *Glaciozyma martini* sp. nov. and *Glaciozyma watsonii* sp. nov. Extremophiles 15:573–586

Turkan A, Yılmaz F, Kücük AÇ, Őzdemir Y (2011) One-pot two-step lipase-catalyzed synthesis of $\alpha\omega$-thiophene-capped poly(ε-caprolactone) macromonomers and their use in electropolymerization. Polym Bull 67:1483–1498

Tzialla AA, Pavlidis IV, Felicissimo MP, Rudolf P, Gournis D, Stamatis H (2010) Lipase immobilization on smectite nanoclays. Characterization and application to the epoxidation of α-pinene. Biores Technol 101:1587–1594

Ursoiu A, Paul C, Kurtán T, Péter F (2012) Sol-gel entrapped *Candida antarctica* lipase B—a biocatalyst with excellent stability for kinetic resolution of secondary alcohols. Molecules 17:13045–13061

Van Rantwijk F, Sheldon RA (2007) Biocatalysis in ionic liquids. Chem Rev 107:2757–2785

Verger R (1997) Interfacial activation of lipases: facts and artifacts. Trends Biotechnol 15:32–38

Wang ZY, Zong MH (2009) Recognition of acyl donor by lipase CAL-B in the acylation of 6-azauridine. Biotechnol Prog 25:784–791

Wang HY, Zhang WW, Wang N, Li C, Li K, Yu XQ (2010) Biocatalytic synthesis and in vitro release of biodegradable linear polyesters with pendant ketoprofen. Biomacromolecules 11:3290–3293

Wei P, Bai L, Song W, Hao G (2009) Characterization of two soil metagenome-derived lipases with high specificity for p-nitrophenyl palmitate. Arch Microbiol 191:233–240

Wu L, Ge G, Wan J (2009) Biodegradation of oil wastewater by free and immobilized *Yarrowia lipolytica* W29. J Environ Sci (China) 21:237–242

Yadav KNS, Adsul MG, Bastawde KB, Jadhav DD, Thulasiram HV, Gokhale DV (2011) Differential induction, purification and characterization of cold-active lipase from *Yarrowia lipolytica* NCIM 3639 Biores Technol 102:10663–10670

Yan Y, Zhang X, Chen D (2013) Enhanced catalysis of *Yarrowia lipolytica* lipase LIP2 immobilized on macroporous resin and its application in enrichment of polyunsaturated fatty acids. Biores Technol 131:179–187

Yu M, Lange S, Richter S, Tan T, Schmid RD (2007) High-level expression of extracellular lipase Lip2 from *Yarrowia lipolytica* in *Pichia pastoris* and its purification and characterization. Protein Expression Purif 53:255–263

Zaliha RN, Salleh AB, Basri M, Mohamad Ali MSB (2012) Cold active enzyme and method thereof. US Patent 2012/0058514 A1

Zhao H, Song Z (2010) Migration of reactive trace compounds from Novozym®435 into organic solvents and ionic liquids. Biochem Eng J 49:113–118

Zhao H, Zheng L, Wang X, Liu Y, Xu L, Yan Y (2011) Cloning, expression and characterization of a new lipase from *Yarrowia lipolytica*. Biotechnol Lett 33:2445–2452

Chapter 17
Miscellaneous Cold-Active Yeast Enzymes of Industrial Importance

Aneta Białkowska and Marianna Turkiewicz

Abstract Cold-adapted organisms, thriving permanently at near-zero temperatures, synthesize cold-active enzymes to sustain their cell cycle. These enzymes are already used in many biotechnological applications requiring high activity at mild temperatures or fast heat-inactivation rate. In this chapter, we describe the main properties of enzymes from cold-adapted yeasts and describe some of their potential biotechnological applications. The useful applications of these enzymes are widespread to a large number of industries like textile industry, food and dairy industry, brewing and wine industry, laundry, etc. Cold-active hydrolytic enzymes like lipases (reviewed in Chap.16), proteases, cellulases, and amylases can be used as an active agent in detergents applied for cold washing. Other potential applications of psychrophilic enzymes, apart from these, are in processes such as the hydrolysis of lactose in milk using β-galactosidases, extraction and clearing fruit juices using pectinases, meat tenderization or taste improvement of refrigerated meat using proteases, betterment of bakery products using glycosidases (e.g., amylases, xylanases).

Keywords Psychrophilic enzymes · Cold-active yeast enzymes · Cold-adapted yeasts · Biotechnological potential

Contents

17.1 Introduction 378
17.2 Biotechnological Potential of Cold-Active Yeast Enzymes 379
17.2.1 Amylases, Xylanases, Chitinases, and Lysozymes 379
17.2.2 Pectinolytic Enzymes 382

A. Białkowska · M. Turkiewicz (✉)
Institute of Technical Biochemistry, Faculty of Biotechnology and Food Sciences, Lodz University of Technology, Stefanowskiego Street, 4/10, 90-924 Lodz, Poland
e-mail: marianna.turkiewicz@p.lodz.pl

P. Buzzini and R. Margesin (eds.), *Cold-Adapted Yeasts*,
DOI: 10.1007/978-3-642-39681-6_17, © Springer-Verlag Berlin Heidelberg 2014

17.2.3 Other Glycosidases ... 385
17.2.4 Proteases ... 387
17.2.5 Phytases and Other Esterases ... 389
17.3 Conclusions and Future Prospects ... 391
References ... 392

17.1 Introduction

The unique properties of the enzymatic apparatus of psychrophilic and psychrotrophic microorganisms, resulting from the special environmental challenges they face, make them an interesting subject of scientific research and lead to their increasingly widespread use in industry (Cavicchioli et al. 2011). The advantages of cold-active enzymes are unquestionable. Being thermolabile, they can be selectively inactivated. Thanks to low-temperature processing, they allow for the reduction of the energy costs of technological processes and prevent infection with mesophilic bacteria thus ensuring a better quality of the final product (Margesin and Schinner 1994).

Interest in psychrophilic microorganisms, which are a source of unique cold-active biocatalysts, has grown considerably over the past twenty years. However, it should be noted that the majority of works on the structure and properties of these proteins, the general mechanism of low-temperature catalysis, and its biological potential, concern enzymes derived from prokaryotic organisms, mostly psychrophilic and psychrotrophic bacteria (Feller and Gerday 1997; Russel 2000; Zecchinon et al. 2001; D'Amico et al. 2002; Feller and Gerday 2003; Feller 2003; Hoyoux et al. 2004; Georlette et al. 2004; Siddiqui and Cavicchioli 2006; Struvay and Feller 2012), while there are few papers dealing with the enzymatic properties of psychrophilic eukaryotes (Shivaji and Prasad 2009; Cavicchioli et al. 2011; Buzzini et al. 2012). Undoubtedly, the least information is available on cold-adapted yeasts, even though they are a versatile group of eukaryotic microorganisms characterized by varied feeding preferences and a surprising ability to survive in extreme environments with diverse physical and geochemical parameters (Buzzini et al. 2012). It has even been suggested that in psychrophilic habitats yeasts are better adapted to low temperatures than bacteria (Margesin et al. 2003). To date, research into these yeasts has mainly focused on their growth conditions, interactions with the abiotic environment, and taxonomical classification. Over the last two decades, there has been increasing interest in psychrophilic yeasts, as it is believed that due to their exceptional ability to grow and maintain metabolism at low temperatures, they produce unique biocatalysts that may be valuable for industrial applications. Examples include lipases A and B synthesized by *Candida antarctica* (now *Pseudozyma antarctica*), which are used in many biotransformations in the food, pharmaceutical, and cosmetic industries (Shivaji and Prasad 2009).

17.2 Biotechnological Potential of Cold-Active Yeast Enzymes

The potential commercial users of cold-active enzymes are primarily the food and household chemical industries. Some of these biocatalysts may also prove a valuable tool for low-temperature biotransformations and in molecular biology (Cavicchioli et al. 2011). Cold-active enzymes may also find applications in the chemical industry for manufacture of extremely volatile organic compounds that can only be modified at low temperatures. Low temperature may facilitate product separation and reduce the process costs (Brenchley 1996).

Among the psychrophilic yeast enzymes explored to date, most attention has been given to hydrolases. However, these enzymes (perhaps except for lipases, discussed at length in Chap. 16) have been studied to varying extent: some authors report characteristics of purified homogeneous protein preparations, while others determine crude enzymatic preparations. Some other works are limited to enzymatic activity screening without in-depth analysis (Brizzio et al. 2007; Khan Pathan et al. 2010; Carrasco et al. 2012; Loperena et al. 2012; Singh and Singh 2012).

17.2.1 Amylases, Xylanases, Chitinases, and Lysozymes

Cold-adapted amylases have emerged as some of the leading biocatalysts with proven potential to be used in a wide array of industrial applications (Kuddus et al. 2011). Most of the cold-adapted amylases studied to date have been derived from prokaryotic organisms. There is only one report concerning these enzymes isolated from psychrophilic yeasts (De Mot and Verachtert 1987), that is, two amylases produced by *C. antarctica* (now *P. antarctica*) CBS 6678, which were purified to homogeneity using various chromatographic techniques and analyzed kinetically. Both of these enzymes, that is, α-amylase and glucoamylase, are monomeric glycoproteins hydrolyzing soluble starches at optimum temperatures of 62 and 57 °C, respectively. The optimum temperature of glucoamylase is in the range of mesophilic yeast glucoamylases (50–60 °C), while the optimum temperature of the other amylolytic enzyme is much higher than that of most of its yeast homologs (below 50 °C). Amylolytic preparations from *C. antarctica* exhibit high specificity for high-molecular-weight substrates, including raw starches. α-Amylase is also active toward cyclodextrins, hydrolyzing α-1,4-glycosidic bonds, while glucoamylase catalyzes the cleavage of α-1,6-glycosidic bonds in dextran, pullulan, and glycogen. Furthermore, De Mot and Verachtert (1987) emphasize the similarity of the amylolytic enzymes from *C. antarctica* CBS 6678 to their mold homologs in terms of glycosylation levels, substrate preferences, and sensitivity to inhibitors, and discuss the potential use of the strain producing these enzymes as a source of genetic material that could be cloned into mesophilic yeasts of the genus *Saccharomyces* to improve the fermentative properties of these microorganisms.

According to Kuddus et al. (2011), cold-active α-amylases have a wide range of potential applications. In the food industry, they may be used for the production of maltotetraose syrups, which are sweeteners and substitutes for sucrose that do not affect the initial taste and aroma of the product. These syrups are characterized by a high ability to bind water, thanks to which they ensure the proper level of humidity and texture in foodstuffs. Cold-active α-amylases may also be applied in the production of maltose and high-molecular-weight branched dextrins, which are used as fillers in the manufacture of bulk foods. In the cotton industry, they may be used for desizing raw fabric and processing denim textiles and in the detergent industry for cold washing. Other potential applications of cold-active α-amylases include the fruit industry (improved juice clarity), the baking industry (delayed bread staling), the paper industry (lower starch viscosity for appropriate coating of paper), the pharmaceutical industry (in dietary supplements to aid digestive processes), and in the treatment of food-processing wastewater (removal of starch waste). The application of psychrophilic enzymes in the above processes brings substantial benefits in terms of reduced energy consumption (and, by the same token, reduced costs) and environmental protection due to the fact that enzymes are biodegradable.

Many companies are interested in financing research on α-amylases despite the high costs and risk that it entails. Cooperation with commercial partners has led to identification of some new amylases from Antarctic microorganisms, which may have potential applications in industry. For instance, *Bacillus licheniformis* amylase was found to be highly active in the temperature range of 10–60 °C in a study by Borchert et al. (2004) sponsored by the company Novozymes. This discovery is the subject of US Patent no. 6673589, although most patents in this field concern processes rather than the substances isolated. Given the above, it should be mentioned that the results of research on amylases have not been commercialized to date.

Other identified cold-active eukaryotic glycosidases are xylanases from the yeasts *Cryptococcus adeliensis* (formerly *Cryptococcus adeliae*) and *Cryptococcus albidus*, which may be applied in low-temperature recycling of farm, industrial, and sewage waste (Shivaji and Prasad 2009). Furthermore, these enzymes are particularly useful in the hydrolysis of xylan, which is the main component of hemicelluloses in plant cell walls. The products of enzymatic degradation of this polysaccharide are used in biotechnological processes for, e.g., manufacture of fuels and organic solvents (Amoresano et al. 2000). The low-temperature activity of xylanases is also taken advantage of in the baking industry. The first tests on cold-adapted bacterial xylanases showed that their high activity at low temperatures and mechanism of action improve the quality of bread by increasing its volume. Some commercial preparations of psychrophilic bacterial xylanases are already available from the company Puratos (Grand-Bigard, Belgium) (Feller 2012).

Gomes et al. (2000) optimized culture conditions for *Cryptococcus adeliensis* with a view to efficient production of xylanase, reaching an activity of 400 nkat ml^{-1} during 168 h of culture at 4 °C. Among 13 tested sources of carbon, xylanolytic activity was induced to the highest degree by xylan, while the most preferred source of nitrogen was yeast extract. In the presence of other sources of carbon in the medium, the level of enzymatic activity was of constitutive nature. An unpurified preparation of xylanase from *C. adeliensis* revealed optimum activity at a pH of 5.0–5.5 and was stable in the pH range of 4–9 (21 h at 4 °C). Despite the fact that this protein exhibited optimum activity at a relatively high temperature (45–50 °C), it was very thermolabile with a half life of 78 min at 35 °C, while temperatures in the range of 40–50 °C led to a loss of 71–95 % of its initial activity within 5 min. The authors suggested that the low thermal stability of *C. adeliensis* xylanase may result from the increased structural flexibility of the molecule. Similar conclusions were advanced by Petrescu et al. (2000), who studied psychrophilic xylanase from the strain *C. adeliensis* ATCC 201412. The mature glycosylated enzyme consists of 338 amino acid residues and is 84 % identical to the amino acid sequence of its mesophilic homolog from *C. albidus*. Similar to the cold-adapted xylanase described above, the enzyme produced by the strain ATCC 201412 is thermolabile with a half life of 60 min at 30 °C, while the activity of its mesophilic homolog is not affected at this temperature. The kinetic adaptation of psychrophilic xylanase has also been confirmed in a study using differential scanning calorimetry (DSC), which revealed a difference between the melting points of these proteins (48 °C and 62 °C for psychro- and mesophilic xylanase, respectively). *C. adeliensis* and *C. albidus* xylanases differed in the degree of their activity in hydrolysis of xylan conducted at 5 °C ($k_{cat} \sim 14.8$ and 4.9 s^{-1} for psychro- and mesophilic xylanase, respectively) as well as in the activation energy of the catalyzed reaction ($E_a = 47.7$ and 52.2 kJ mol^{-1} for C. *adeliensis* and *C. albidus* xylanase, respectively). The suggestion that the considerable thermolability of the cold-adapted xylanase results from increased molecule flexibility was partially confirmed by the analysis of a three-dimensional model of this enzyme. The authors observed some subtle differences between the structures of the psychro- and mesophilic proteins, such as a less tightly packed hydrophobic core, a lack of one salt bridge, and destabilization of α-helix macrodipoles in the psychrophilic enzyme. The spatial structure of the xylanase derived from the Antarctic yeast *Cryptococcus albidosimilis* (*C. albidus* TAE85) was resolved by a more accurate method, using mass spectometry (Amoresano et al. 2000). This enzyme was shown to be a protein composed of 338 amino acids, *N*- or *O*-glycosylated with GalGalNAc only at residue Asn254 in the polypeptide chain. It should be noted that this was the first report concerning the structural characteristics and glycosylation mechanism in a eukaryotic psychrophilic protein.

Furthermore, assay of the enzymatic activity of *C. adeliensis* grown in submerged culture revealed that this yeast produces endoglucanases, β-arabinofuranosidases, β-xylosidases, and β-glucosidases, which are less active than xylanases, as well as β-mannanases (Gomes et al. 2000). The authors suggest that *C. adeliensis*, which produces a wide range of hydrolytic enzymes useful for degradation of plant poly- and oligosaccharides, may be applied as a probiotic ingredient or therapeutic agent in feed.

To produce cold-active enzymes on a larger scale, researchers aim at expression of their genes in mesophilic microorganisms, whose culture is technologically simpler. An example here is the gene of a psychrophilic chitinase from the strain *Glaciozyma antarctica* PI12 with an optimum temperature of 15 °C, which was transferred to a mesophilic host, *Komagataella pastoris* (formerly *Pichia pastoris*). The latter gives a high yield of this enzyme in the presence of 1 % methanol at 25 °C (Ramli et al. 2011). This recombinant yeast enzyme was purified to homogeneity using immobilized metal affinity chromatography and characterized with a view to industrial applications. Based on bioinformatic analysis, it was classified as a class II chitinase, glycosyl hydrolase family 18. The regions forming the active site of *G. antarctica* PI12 chitinase also turned out to be highly conservative revealing strong identity to the catalytic domains of 5 eukaryotic homologous enzymes. The main advantages of this psychrophilic enzyme are its high activity at low temperatures (T_{opt} 15 °C; at 5 °C the enzyme exhibits over 20 % of its maximum activity) and ability to efficiently hydrolyze chitin at acidic pH (pH_{opt} 4.0). Just as the majority of typical cold-adapted enzymes, *G. antarctica* PI12 chitinase is characterized by low thermal stability (following 30 min incubation at 25 °C it lost 30 % of its initial activity). In turn, its activity rose by up to 20 % in the presence of 1 mM K^+, Mn^{2+}, and Co^{2+} ions. Examination of the substrate preferences of this enzyme revealed that its activity was highest in the hydrolysis of colloidal chitin, as compared to swollen chitin, carboxymethyl chitosan, and glycol chitosan. This was confirmed by kinetic experiments which showed that *G. antarctica* PI12 chitinase has higher V_{max} and k_{cat} values in low-temperature degradation of colloidal chitin than for other substrates. The authors proposed that due to its biochemical and kinetic properties, the enzyme should be an attractive tool for the production of chitooligosaccharides and in other biotechnological applications, such as biocontrol of microbial spoilage of refrigerated foods and mycoparasitic activity against phytopathogenic fungi in cold environments (Ramli et al. 2011, 2012).

Another interesting example is a lysozyme from the cold-adapted yeast *Debaryomyces hansenii* (muramidase, EC 3.2.1.17) (Wang et al. 2012). It cleaves the *β*-(1,4)-glycosidic bond between *N*-acetylmuramic acid and *N*-acetylglucosamine in peptidoglycan, the major bacterial cell wall polymer, and acts as a nonspecific innate immunity molecule against the invasion of bacterial pathogens.

Culture conditions for this enzyme were optimized using statistical design methods, the Plackett–Burman design (PBD) and the development and validation of a quadratic model. As a result, the maximum lysozyme yield was about 3.5-fold higher than the baseline value that greatly enhances the application potential of this enzyme. In general, cold-adapted lysozymes are successfully used as food preservatives for fish and meat products as well as fruits and vegetables (Wang et al. 2012).

17.2.2 Pectinolytic Enzymes

Psychrophilic yeasts are a good source of pectinolytic enzymes. Many yeast species produce them, usually several types of pectinases at the same time, often also in the form of isoenzymes. These proteins may be applied in the fruit-processing industry, which would be especially interested in cold-active pectinolytic enzymes for reducing juice viscosity at ambient temperature and for improving the clarity of the final product (Alimardani-Theuil et al. 2011).

All known psychrophilic yeasts with pectinolytic activity (also called PPYs, pectinolytic–psychrophilic yeasts) belong to the genera *Cystofilobasidium*, *Cryptococcus*, and *Mrakia*. In contrast to psychrophilic bacteria, which predominantly synthesize alkaline pectinases, yeasts mostly produce enzymes with optimum activity at acidic pH.

The first yeast strain described in the literature in terms of pectinolytic enzyme production was *Cystofilobasidium capitatum* PPY-1 (Nakagawa et al. 2002). When grown on a pectin or polygalacturonate substrate as a source of carbon at 5 °C and below, this yeast synthesized a wide range of pectinolytic enzymes: pectin methylesterases (PME), polygalacturonases (PG), and pectin lyases (PNL), but it did not produce pectate lyases (PAL). The activity of these enzymes in culture supernatants, determined for hydrolysis of pectins and polygalacturonate at 20 °C, amounted to 85, 205, and 1.7 U mg^{-1} protein, respectively. The authors also tested the ability of *C. capitatum* PPY-1 to degrade pectin substances at 5 °C and obtained activity levels of 267, 101, and 0.01 U mg^{-1} protein for PME, PG, and PNL, respectively. One of these enzymes, that is, PPY-1 PG, was purified to homogeneity using various chromatographic methods and exhibited an activity of 458 U mg^{-1} protein under optimum reaction conditions (T_{opt} 45 °C, pH_{opt} 4.4). At 0 °C, its activity was reduced to about 20 % of its maximum activity, which confirms its adaptation to cold (Nakagawa et al. 2005a). The enzyme was found to be stable in the pH range of 2.5–7.0 and up to about 50 °C. It was shown that the enzyme most effectively hydrolyzes polygalacturonate (degree of esterification (DE) 0 %) and pectins moderately esterified with methanol (DE 60 %). Interestingly, based on identification of polygalacturonate hydrolysis products at the initial stages of enzymatic reaction, the authors suggest that PPY-1 PG exhibits mechanisms of action typical of both endo- and exopolygalacturonases. Nakagawa et al. (2005a) emphasize that the special properties of *C. capitatum* PPY-1 PG make this protein useful for industrial applications and especially for low-temperature pectin degradation in the food industry. PPY-1 PG is quite unique as its N-terminal sequence of 20 amino acids does not show high homology with any sequences of other PG deposited in databases. The highest degree of homology was found for *Saccharomyces cerevisiae* PG (52 %).

Another *C. capitatum* PPY-1 enzyme that has been purified and characterized in terms of cold adaptation was pectin lyase (PNL) (Nakagawa et al. 2005b). An efficient purification procedure was developed for this enzyme using traditional chromatographic techniques. As a result, a homogeneous protein was obtained with a high specific activity (1,230 U mg^{-1} protein). Based on SDS-PAGE, its molecular mass was estimated at 42 kDa and a 25-amino acid N-terminal sequence was determined. PPY-1 PNL exhibits maximum activity in pectin hydrolysis at 40 °C, while in the range of 0–20 °C, it retains about 5–50 % of its maximum activity, similar to its homolog from *Penicillium paxilli*. Moreover, in pectin hydrolysis, the purified enzyme is characterized by high optimum pH (8.0). Thus, it may be considered a cold-adapted alkaline enzyme. Pectin lyase shows a preference for hydrolysis of highly methoxylated pectins and reduces the viscosity of pectin solutions to that of water after 1.5 h reaction at 5 °C.

Nakagawa et al. (2005b) suggest that both enzymes, that is, *C. capitatum* PPY-1 PG and PNL, which depolymerize pectin substances (but differ in substrate specificity and optimum pH) may be used in the form of a mixture in the fruit and vegetable industry to reduce juice and fruit/vegetable pulp viscosity, enhance clarification, and aid depectinization.

Other psychrophilic yeast strains with pectinolytic activity were isolated by Birgisson et al. (2003) from frozen soil, plant leaves, and branches in southwestern Iceland. Based on rRNA sequencing, they classified the yeasts as *C. capitatum*, *Cryptococcus macerans* (anamorph of *Cystofilobasidium macerans*), *Cystofilobasidium lari-marini* (now *C. capitatum*) and *Cryptococcus aquaticus* (now *Mrakiella aquatica*). They showed that the activity of pectinolytic enzymes largely depends on culture conditions, that is, source of carbon, pH, and temperature. Biosynthesis was the most efficient in the presence of pectin as a source of carbon but was repressed on pure glucose. The optimum temperature for the production of polygalacturonases by *Cystofilobasidium* was 14 °C, while in the case of *Cryptococcus,* it was by 5 °C lower. The optimum pH of culture medium also varied among the strains: it was 3.2 for *C. lari-marini* (about 60 U ml^{-1}), 2.6 for *C. macerans* (about 50 U ml^{-1}), and 3.9 for *C. capitatum* (about 37 U ml^{-1}). The authors conducted preliminary analysis of the pectinolytic enzymes contained in the culture supernatant and found that polygalacturonases were produced most efficiently by *Cystofilobasidium* strains, reaching maximum hydrolytic activity at 40 °C and pH 5.0. The same type of enzymes isolated from *Cryptococcus* was characterized by $T_{opt} = 50$ °C and $pH_{opt} = 4.0$. They appear to be cold adapted, as they exhibit relatively high activity in a low-temperature range (0–20 °C): *Cystofilobasidium* PG retains about 20–40 % of its maximum activity, while its *Cryptococcus* homologs retain only 10–30 %. Just as most other psychrophilic enzymes, they are characterized by low thermal stability and undergo partial inactivation at temperatures above 30 °C. Birgisson et al. (2003) emphasize marked kinetic adaptations of yeast polygalacturonases to a cold environment and indicate their usefulness for low-temperature processes as well as ease of selective inactivation after completing the process. Another advantage attributed to the yeast strains is their ability to biosynthesize polygalacturonase isoenzymes, which makes it possible to select proteins with optimum biochemical properties for industrial applications.

Evaluation of the enzymatic activity of psychrophilic yeasts in submerged cultures revealed that they can synthesize other pectinolytic enzymes, albeit not as active as polygalacturonases: *Cystofilobasidium* strains produce pectin lyase, *C. aquaticus* pectin esterase, and *C. macerans* strains produce pectin lyase, pectate lyase, and pectin esterase (Birgisson et al. 2003).

Pectinolytic activity was also observed in yeast strains isolated from forest soil in Abashiri (Hokkaido, Japan) by Nakagawa et al. (2004). Of particular, importance was discovery of a new species of psychrophilic yeast with pectinolytic activity, that is, *Mrakia frigida*. On a substrate with 1 % w/v pectin as a source of carbon, at 5 °C, this strain synthesizes pectin methylesterase and polygalacturonase with a yield of 16.0 and 7.6 mU mg^{-1} protein, respectively.

Margesin et al. (2005) studied two strains of *M. frigida*, one of which was isolated from Alpine glacier cryoconite (A15), while the other originated from a sediment sample containing mud, spring water, and moss from the Gyda peninsula in northern Siberia (AG25). They analyzed the biochemical properties of pectate lyases (PL) and found that their production in both isolates depended on medium composition, pH, and amount of oxygen dissolved in the medium. PL synthesis by both strains was the most efficient in a medium-containing yeast extract at neutral pH, while optimum temperatures were 5 and 1 °C for A15 and AG25, respectively. Both enzymes exhibited highest activity at 30 °C and were highly active in the temperature range of 0–20 °C (about 20–70 % of the maximum activity). Above the optimum temperature, the activity of the lyases was significantly repressed (after 15 min of incubation at 30 °C they lost about 40 % of their maximum activity, while at 50 °C, they underwent complete denaturation). The cold adaptation of these enzymes is also shown by lower values of E_a and other thermodynamic constants (ΔH*, ΔG*, ΔS*) as compared to their mesophilic homologs. Another interesting biochemical property of *M. frigida* PL is their strong alkaline preference (pH 8.5 and 9.0 for A15 and AG25, respectively) and stability (1 h activity in a pH range of 7–10). The enzymes are completely denatured in the presence of EDTA (10 mmol l^{-1}), SDS (1 %), $KMnO_4$ (10 mmol l^{-1}), and L-cysteine (10 mmol l^{-1}); only Ca^{2+} ions enhance the rate of the reaction they catalyze. According to the authors, the functional advantages of pectate lyases from psychrophilic yeasts include their much greater relative activity at low temperatures as compared to the widely used mesophilic enzymes (by a factor of 4–7) and their alkalinity (Margesin et al. 2005).

17.2.3 Other Glycosidases

Considerably less research has been devoted to cold-active yeast enzymes hydrolyzing di- and oligosaccharides than to enzymes degrading polysaccharides, even though such glycosidases would also prove valuable for industrial applications, especially for food processing and synthesis of probiotic oligosaccharides (Sheik Asraf and Gunasekaran 2010).

From the technological standpoint, of greatest interest are cold-active *β*-galactosidases, which might be applied in low-temperature hydrolysis of lactose, which is one of the main components of milk and whey (Loveland et al. 1994; Sheridan and Brenchley 2000). Manufacture of lactose-free milk and milk products is of particular importance for consumers with intolerance of this disaccharide, who often suffer from serious gastrointestinal problems. Cold-active *β*-galactosidases have also been shown to possess transglycosylation activity whereby lactose hydrolysis takes place with simultaneous transfer of monosaccharides to higher oligosaccharides to form tri- and tetrasaccharides (Karasova-Lipovova et al. 2003; Benesova et al. 2005). Such galacto-oligosaccharides (either produced directly in milk or from whey as an additive in dairy products) can be used as ingredients in probiotic foodstuffs to enhance the growth of bifidobacteria in the large intestine, or as low-calorie sweeteners due to their resistance to metabolism in the small intestine (Huston 2008).

The majority of commercial β-galactosidase preparations contain mesophilic enzymes obtained from yeasts and filamentous fungi. The process of removing lactose from milk or sweet whey using mesophilic β-galactosidases from *Kluyveromyces lactis* or *Kluyveromyces fragilis* (now *Kluyveromyces marxianus*) (Ladero et al. 2000) is conducted for a short time (several hours) at pH 6.0–7.0 and 30–40 °C to prevent infection. Such conditions result in a relatively low degree of lactose degradation. Thus, psychrophilic β-galactosidases would be a perfect alternative to these enzymes, because they exhibit sufficiently high activity at below 20 °C and are not inactivated by sodium or calcium ions or by galactose. The application of such β-galactosidases would reduce the risk of infection with microflora and considerably decrease the energy consumption of the process (Białkowska et al. 2009). Only a few cold-active β-galactosidases have been described to date. They are mostly derived from bacteria of the genera *Arthrobacter*, *Flavobacterium*, *Pseudoalteromonas*, *Rahnella*, *Planococcus*, and *Carnobacterium* (Hoyoux et al. 2001; Turkiewicz et al. 2003a; Wierzbicka-Woś et al. 2011), while two are from yeasts of the genus *Guehomyces*. The first yeast β-galactosidase was isolated from the psychrophilic strain *Guehomyces pullulans* R1 by Nakagawa et al. (2006) in selective tests conducted on agar plates containing X-gal (5-bromo-4-chloro-3-indolyl-β-D-galactopyranoside). It was found to be active both in culture supernatants and in biomass. In lactose hydrolysis, the extracellular fraction of the enzyme was characterized by a relatively high optimum temperature for a cold-active enzyme (50 °C), but at 0 °C, it retained as much as 20 % of its maximum activity. Crude enzymatic preparations revealed the highest activity at pH 4.0, with stability in the pH range of 3–7. It is emphasized that the acidic nature of *G. pullulans* R1 β-galactosidase is its unique feature as most of its known psychrophilic homologs hydrolyze lactose most efficiently under near-alkaline pH conditions. Thus, it is suggested that this yeast enzyme may be used for industrial processes conducted in acidic environments. An example of such a process is manufacture of galacto-oligosaccharide syrups from sour whey, where β-galactosidases from filamentous fungi (active in the pH range of 2.5–4.5) are currently applied. β-Galactosidases operating at acidic pH can also reduce the pollution impact and increase the technical usefulness of whey (a product of the cheese industry) by producing glucose- and galactose-rich syrups that can be used as sweeteners in a variety of foodstuffs and substrates that are easily fermented by alcohol-producing microorganisms (Gerday et al. 2005). *G. pullulans* β-galactosidase was found useful for low-temperature hydrolysis of lactose in milk as 10 U ml^{-1} of this enzyme hydrolyzed about 80 % of the disaccharide over 96 h at 10 °C (Nakagawa et al. 2006).

The other yeast β-galactosidase is derived from the psychrotolerant strain *G. pullulans* 17-1, isolated from sea sediment in Antarctica (Song et al. 2010). The biosynthesis of this enzyme, induced by the presence of lactose in the growth medium, is most efficient at pH 4.5. The optimum temperature and pH of the crude protein preparation for hydrolysis of o-nitrophenyl-β-D-galactopyranoside (ONPG) are 50 °C and 4.0, respectively. According to the authors, the cold adaptation of *G. pullulans* 17-1 β-galactosidase is not particularly pronounced if one compares its optimum temperature with those of mesophilic β-galactosidases from the yeasts *K. lactis* ATCC 8583 and *K. fragilis* ($T_{opt} = 37$ and 30 °C, respectively). However, a major advantage of this enzyme is its extracellular secretion and high hydrolytic activity (25.3 U ml^{-1}) in culture supernatants on a medium scale (2 L bioreactor). Song et al. (2010) reported that this represents the highest activity (in hydrolysis of a synthetic substrate) among all β-galactosidases produced by wild yeast strains obtained to date, even as compared to the activity of β-galactosidases from mesophilic yeasts: under optimum conditions, *K. lactis* synthesizes an enzyme characterized by an activity of 0.9 U ml^{-1}, while the enzyme produced by *K. fragilis* 34440 has an activity of 9.2 U ml^{-1}.

Turkiewicz et al. (2005) showed *Leucosporidium antarcticum* (now *Glaciozyma antarctica*; Turchetti et al. 2011) to produce cold-active α-glucosidase and β-fructofuranosidase. The former was found to be active in both culture supernatants and biomass, while the latter was not secreted and was strictly limited to the cell. Unfortunately, following release from the cell, invertase underwent prompt and irreversible inactivation. Although the rapidly progressing inactivation was inhibited in the presence of EDTA (2 mM) and Mn^{2+} (1 mM), the protective action of these compounds declined with time of incubation, due to which it was impossible to obtain even partially purified enzymatic preparations. The high *in vitro* sensitivity of *G. antarctica* to inactivation also led to limited pH stability of the enzyme, which retained maximum activity during 30-min incubation at 4 °C within a very narrow acidic pH range (4.3–4.8). For comparison, α-glucosidase from the same strain, secreted to the growth medium, was stable over the same time within a 3-fold wider neutral pH range (6.0–7.5). These enzymes preferred very different pH of the reaction environment and exhibited different optimum temperatures. Both their low thermal stability and high catalytic activity at 0 °C (over 20 % of the maximum activity for invertase) show that *G. antarctica* glycosidases are indeed cold-active.

17.2.4 Proteases

Other cold-adapted yeast hydrolases with a range of applications as wide as that of glycosidases are two exoproteases: an aspartyl protease isolated from the Antarctic strain *Cryptococcus humicola* (formerly *Candida humicola*) and a serine protease from the endemic yeast *G. antarctica*. There are relatively few data on extracellular yeast proteases, even in respect of mesophilic yeasts, as it is thought that these organisms are not able to synthesize significant amounts of these proteins. It seems that this view largely results from the fact that the predominant yeast species both in research and industrial applications is *S. cerevisiae*. Only a few strains of this yeast exhibit extracellular proteolytic activity and that to a small degree (Sturley and Young 1988).

Psychrophilic proteases have a wide range of industrial and technological applications, e.g., they are used as additives in detergents and foodstuffs, as bio-remediators, in biotransformations, and in molecular biology. These enzymes are also used as an alternative to rennet to reduce production costs by accelerating the maturation of slow-ripening cheeses in a low-temperature and low-moisture environment. In seafood processing, cold-active proteases aid in removing scales and skin from fish as well as in extracting carotenoproteins from shellfish. They are also used to impart tenderness to refrigerated meat products and to improve their flavor. Furthermore, proteases active at room temperature may be used as an ingredient in medicinal gels for wound and scar healing and topical treatment of skin infections (Huston 2008).

The discovery of an extracellular protease produced by *C. humicola* by Ray et al. (1992) confirms the fact that aspartyl (rarely cysteine) enzymes predominate among yeasts inhabiting even very distant biotopes. They are often called acidic proteases as they reveal maximum activity in environments with relatively low-pH levels (3.0–4.5). The Antarctic exoprotease was characterized by an even lower optimum pH in hydrolysis of hemoglobin (pH 1–1.2) and additionally exhibited kinetic adaptations to low temperature: it was active in the temperature range of 0–45 °C with a distinct optimum at 37 °C and retained 12–15 % of its maximum activity at 0 °C. It should be stressed that so far this has been the only aspartyl protease isolated from Antarctic yeasts. The majority of aspartyl protease are derived from mesophilic microorganisms, such as *Rhodotorula glutinis* (Kamada et al. 1972), *Cryptococcus albidus* var. *aerius* (Federici 1982), *Candida albicans* (Rüchel 1981), *Candida parapsilosis* (Rüchel et al. 1986), *Candida tropicalis* (Togni et al. 1991), *Candida olea* (Nelson and Young 1987), *Saccharomycopsis lipolytica* (Yamada and Ogrydziak 1983), *Aureobasidium pullulans* (Chi et al. 2007), *Sporidiobolus ruineniae* (Kim 2009), *Pichia farinosa* CO-2 (Kim 2010), *Metschnikowia pulcherrima* IWBT Y1123, and *Candida apicola* IWBT Y1384 (Reid et al. 2012).

Another interesting proteolytic enzyme isolated from psychrophilic yeasts is *G. antarctica* extracellular serine protease (Turkiewicz et al. 2003b). This class of enzymes, degrading protein and synthetic substrates in neutral or alkaline environments, includes only a few extracellular yeast proteases, mostly produced by mesophilic *Yarrowia lipolytica* (Tobe et al. 1976) and *Aureobasidium pullulans* (Ogrydziak 1993). Interestingly, *G. antarctica* is the first psychrophilic subtilase belonging to the proteinase K subfamily (subtilisin family) to be described. An efficient purification procedure was developed for this enzyme using traditional chromatographic methods. The resulting homogeneous protein had a very high specific activity (85 U mg^{-1} protein). It was shown that the enzyme is stable under conditions of storage and purification without protectants and is not activated by bivalent metal ions. It should be stressed that *G. antarctica* subtilase exhibits a very high specific activity in hydrolysis of protein substrate (hemoglobin denatured with urea), which is much higher than that of many previously isolated microbial or animal psychrophilic proteases. Similar to most other microbial serine proteases, *G. antarctica* 171 subtilase has wide substrate specificity, with the greatest affinity to substrates characteristic of chymotrypsin and chymotrypsin-like enzymes. In this respect, it differs significantly from *Y. lipolytica* AEP, which is of typical trypsin-like nature (Ogrydziak 1993). At 30 °C, the enzyme exhibits an approximately 140-fold higher specificity constant (k_{cat}/K_m) for the hydrolysis of BzTyrOEt (*N*-Benzoyl-L-tyrosine ethyl ester) than for the hydrolysis of BzArgOEt (*N*-Benzoyl-L-arginine ethyl ester) (37.4 and 0.27 s^{-1} mM^{-1}, respectively). Furthermore, *G. antarctica* 171 subtilase displays a significant shift in optimum temperatures toward the low range (25 °C), which is by more than 35 °C lower than the optimum temperature of the best characterized mesophilic subtilisin Carlsberg (T_{opt} = 65 °C; Kulakova et al. 1999). Another unique property of *G. antarctica* 171 subtilase is its high activity at 0 °C (over 20 % of its maximum activity). Interestingly, it remains active at temperatures as low as −10 °C. The enzyme displays high amidase activity (pH_{opt} in the range of 8.0–8.5) against *N*-SucAAPF*p*NA (*N*-Succinyl-AAPF-*p*-nitroanilide) and *N*-SucAAPL*p*NA (*N*-Succinyl-AAPL-*p*-nitroanilide). Similar to most other naturally occurring psychrophilic enzymes, this protease is characterized by high sensitivity to the denaturing action of temperature (half life at 30 °C is 2.5 h). Adaptation to low temperatures is attributed by the authors to the presence of additional loops in the molecule, e.g., a 5-amino acid loop near the N-terminus. Although the gene encoding this 35.5-kDa protein was not isolated, the location of a total of 108 amino acid residues was identified and the N-terminal sequence was determined (58 amino acids; the sequence of 38 was determined by the Edman method, and another 23 by proteomic analysis). Based on the sequences of homologous proteases, the sequences of four peptides defined in proteomic analysis of trypsin-digested *G. antarctica* subtilase were aligned (Turkiewicz, unpublished results), among which are sequences of the catalytic triad of the enzyme Ser224…His69…Asp45 (the two first residues were labeled according to residue numbering in proteinase K), which are highly conservative.

Recently, Zaliha et al. (2012) patented a bifunctional enzyme derived from the strain *L. antarcticum* (now *G. antarctica*) PI12, which exhibited both lipolytic and proteolytic activity (see Chap.16).

17.2.5 Phytases and Other Esterases

Of industrial importance is the cold-active phytase isolated from the Antarctic yeast *Cryptococcus laurentii* AL_{27} (*myo*-inositol hexakisphosphate phosphohydrolase, EC 3.1.3.8) (Pavlova et al. 2008). It is a phosphatase that catalyzes phytate hydrolysis to phosphorylated *myo*-inositol derivatives and inorganic phosphate and the first known yeast-derived Antarctic phytase. This enzyme is used as a dietary ingredient that plays a significant role in feeding monogastric animals. Its main task is to reduce inorganic phosphorus in animal fodder to prevent excessive concentrations of this element in animal waste. Furthermore, this phosphatase increases assimilation of some dietary elements (Ca, Cu, Zn, Mn, and Fe), proteins, and amino acids. Over the past several years, special attention has been given to the products of phytase activity, that is, *myo*-inositol products with varying degrees of phosphorylation, due to the potential application in the pharmaceutical industry (Pavlova et al. 2008).

The strain *C. laurentii* AL_{27} was chosen as a result of a 2-step selection procedure as the best phytase producer among 124 strains of Antarctic yeasts. The first step involved strain cultivation on calcium- and phytate-containing solid medium, and the second step consisted of culturing the isolates selected in the first step in a liquid medium containing glucose as a source of carbon (20 g l^{-1}). *C. laurentii* AL_{27}, grown at 24 °C, was characterized by the highest intracellular phytase activity (23.8 U g^{-1} biomass) on a medium containing sucrose (40 g l^{-1}) and KH_2PO_4 (at a concentration supplying phosphorus at 5 mg l^{-1}). Pavlova et al. (2008) indicate that Antarctic phytase plays a useful role by efficiently degrading phytates directly in the stomachs of monogastric animals (37–40 °C, pH 2–3). The major advantages of the *C. laurentii* enzyme include its relatively low optimum temperature (40 °C as compared to 45–60 °C for most other phytases), an optimum pH of 4.8, and stability over a wide pH range (2.0–5.0; at pH 2.0 the protein preserves about 50 % of its maximum activity).

An interesting cold-active yeast enzyme was described by Zimmer et al. (2006), who isolated an esterase from *Rhodotorula mucilaginosa*. The yeast was grown on *(2R,3R)*-(−)-di-*O*-benzoyl-tartrate, which it stoichiometrically hydrolyzed to monoesters. The monoesters were subsequently broken into benzoates and tartrates and metabolized. The enzyme was obtained from the culture supernatant, purified, and determined to be a monomeric glycosylated 86 kDa benzoyl esterase. Its optimum pH was 7.5, and the optimum temperature was 45 °C. The psychrophilic properties of this esterase were proven by its high activity at 0 °C, which amounted to 20 % of its activity at 30 °C. This shows that the enzyme may be used in low-temperature settings, e.g., in manufacturing volatile aromas and fragrances. *R. mucilaginosa* esterase hydrolyzed short-chain *p*-nitrophenyl-alkyl esters as well as some benzoyl esters, including benzoyl-methyl ester, phenyl-benzoyl ester, ethylene–glycol-dibenzoyl ester, 1,5-anhydro-D-fructose-tribenzoyl ester, and cocaine. The enzyme is potentially useful for hydrolysis of 1,5-anhydro-D-fructose-tribenzoyl ester, which could be used as a reagent for specific insertion or release of protecting groups in sugars (benzoylated carbohydrates are precursors in the synthesis of, e.g., beta-L-nucleoside analogs with antiviral activity). Thus, *R. mucilaginosa* esterase seems to be a useful low-temperature reagent in the production of benzoylated compounds for the pharmaceutical, cosmetic, and fine chemical industries (Zimmer et al. 2006).

Another cold-active esterase is produced by *Y. lipolytica* CL180 (rYli180). It preferentially hydrolyzes the *S*-enantiomer of racemic ofloxacin ester (Kim et al.

2007). The gene of this protein was cloned into *Escherichia coli*, and the product of its expression, with a mass of 53 kDa, was purified to homogeneity by means of immobilized metal affinity chromatography. Based on the determination of its biochemical properties, it was shown that this esterase preferentially hydrolyzes *p*-nitrophenyl esters of fatty acids with short-chain lengths ($\leq$10 carbon atoms) and is a cold-active protein (T_{opt} = 35 °C; 40–60 % of maximum activity in the range of 0–20 °C). According to the authors, these features make it a very attractive enzyme for potential application as a biocatalyst, especially in the manufacture of thermolabile chemicals.

Sabri et al. (2001) described an acyl-CoA thioesterase produced by the psychrophilic Antarctic yeast *Rhodotorula aurantiaca*. Its application potential is not as yet known, albeit it plays an important role in the metabolism of myristoyl-CoA, and, indirectly, also in the regulation of yeast growth at different temperatures.

17.3 Conclusions and Future Prospects

Literature data show cold-active enzymes to have several advantages over their mesophilic and thermophilic homologs. Despite some difficulties linked to their relatively high cost, low activity and stability under typical process conditions, and the low degree of biodiversity of the psychrophilic microbes known to date, cold-active enzymes may successfully replace traditional ones in a range of industrial applications. Some relatively new recombinant DNA technologies, including site-directed mutagenesis and metatranscriptomics, enable the production of an increasing number of recombinant enzymes. Thanks to this, it is possible to obtain cheaper cold-tolerant enzymes with novel or modified activity. It seems advisable to pursue commercially viable production of such enzymes in heterologous hosts and their modification by chemical or protein engineering methods with a view to obtaining higher activity levels and greater robustness. Some strains could be genetically improved to make them more suitable for the expression of cold-adapted enzymes, which would have significant positive implications for their use in various fields of biotechnology and industry.

While research into useful cold-active yeast enzymes is largely limited to hydrolases, it seems that other classes of enzymes could be equally relevant for industrial applications. For instance, oxidoreductases are playing an increasingly important role in biotransformations conducted in organic solvents (microaqueous systems) (Straathof et al. 2002; Burton 2003; Solano et al. 2012). Enzymes that could be particularly suited to such applications include those produced by yeasts from cold biotopes with low water activity, such as, e.g., yeasts inhabiting some Antarctic environments. The only cold-active oxidoreductase from psychrophilic yeasts that has been discovered to date is D-galacturonate reductase from *Cryptococcus diffluens* (Hamada et al. 2011).

References

Alimardani-Theuil P, Gainvors-Claise A, Duchiron F (2011) Yeasts: An attractive source of pectinases—from gene expression to potential applications: a review. Proc Biochem 46:1525–1537

Amoresano A, Andolfo A, Corsaro MM, Zocchi I, Petrescu I, Gerday C, Marino G (2000) Structural characterization of a xylanase from psychrophilic yeast by mass spectrometry. Glycobiology 10:451–458

Benesova E, Markova M, Kralova B (2005) α-glucosidase and β-glucosidase from psychrotrophic strain *Arthrobacter* sp C2–2. Czech J Food Sci 23:116–120

Białkowska AM, Cieśliński H, Nowakowska KM, Kur J, Turkiewicz M (2009) A new β-galactosidase with a low temperature optimum isolated from the Antarctic *Arthrobacter* sp. 20B: gene cloning, purification and characterization. Arch Microbiol 191:825–835

Birgisson H, Delgado O, García Arroyo L, Hatti-Kaul R, Mattiasson B (2003) Cold-adapted yeasts as producers of cold-active polygalacturonases. Extremophiles 7:185–193

Borchert TV, Svendsen A, Andersen C, Nielsen B, Lauesgaard Nissen T, Kjerulff Sorren (2004) Amylase mutants. US Patent 6,673,589

Brenchley JE (1996) Psychrophilic microorganisms and their cold-active enzymes. J Industr Microbiol 17:432–437

Brizzio S, Turchetti B, de Garcia V, Libkind D, Buzzini P, van Broock M (2007) Extracellular enzymatic activities of basidiomycetous yeasts isolated from glacial and subglacial waters of northwest Patagonia (Argentina). Can J Microbiol 53:519–525

Burton SG (2003) Oxidizing enzymes as biocatalyst. Trends Biotechnol 21:543–549

Buzzini P, Branda E, Goretti M, Turchetti B (2012) Psychrophilic yeasts from worldwide glacial habitats: diversity, adaptation strategies and biotechnological potential. FEMS Microbiol Ecol 82:217–241

Carrasco M, Rozas JM, Barahona S, Alcaíno J, Cifuentes V, Baeza M (2012) Diversity and extracellular enzymatic activities of yeasts isolated from King George Island, the sub-antarctic region. BMC Microbiol 12:251–260

Cavicchioli R, Charlton T, Ertan H, Mohd Omar S, Siddiqui KS (2011) Biotechnological uses of enzymes from psychrophiles. Microb Biotechnol 4:449–460

Chi Z, Ma C, Wang P, Li HF (2007) Optimization of medium and cultivation conditions for alkaline protease production by the marine yeast *Aureobasidium pullulans*. Biores Technol 98:534–538

D'Amico S, Claviere P, Collins T (2002) Molecular basis of cold adaptation. Phil Trans R Soc Lond B 357:917–925

De Mot R, Verachtert H (1987) Purification and characterization of extracellular α-amylase and glucoamylase from the yeast *Candida antarctica* CBS 6678. Eur J Biochem 164:643–654

Federici F (1982) A note on milk clotting ability in the yeast genera *Cryptococcus* and *Rhodotorula*. J Appl Bacteriol 52:293–296

Feller G (2003) Molecular adaptations to cold in psychrophilic enzymes. Cell Mol Life Sci 60:648–662

Feller G (2012) Psychrophilic enzymes: from folding to function and biotechnology. Scientifica 2013, ID 512840

Feller G, Gerday C (1997) Psychrophilic enzymes: molecular basis of cold-adaptation. Cell Mol Life Sci 53:830–841

Feller G, Gerday C (2003) Psychrophilic enzymes: hot topics in cold adaptation. Nature Rev Microbiol 1:200–208

Georlette D, Blaise V, Collins T, D'Amico S, Gratia E, Hoyoux A, Marx JC, Sonan G, Feller G, Gerday C (2004) Some like it cold: biocatalysis at low temperatures. FEMS Microbiol Rev 28:25–42

Gerday C, Hoyoux A, Francois J M, Dubois P, Baise E, Jennes I, Genicot S (2005) Cold-active β-galactosidase, the process for its preparation and the use thereof. US Patent 2,005,196,835

Gomes J, Gomes I, Steiner W (2000) Thermolabile xylanase of the Antarctic yeast *Cryptococcus adeliae*: production and properties. Extremophiles 4:227–235

Hamada S, Seike Y, Tanimori S, Sakamoto T, Kishida M (2011) Characterization of *D*-galacturonate reductase purified from the psychrophilic yeast species *Cryptococcus diffluens*. J Biosci Bioeng 111:518–521

Hoyoux A, Jennes I, Dubois P, Genicot S, Dubail F, François JM, Baise E, Feller G, Gerday C (2001) Cold adapted β-galactosidase from the Antarctic psychrophile *Pseudoalteromonas haloplanktis*. Appl Environ Microbiol 67:1529–1535

Hoyoux A, Blaise V, Collins T, D'Amico S, Gratia E, Huston AL, Marx J, Sonan G, Zeng Y, Feller G, Gerday Ch (2004) Extreme catalysts from low-temperature environments. J Biosci Bioeng 98:317–330

Huston AL (2008) Biotechnological aspects of cold-adapted enzymes. In: Margesin R, Schinner F, Marx JC, Gerday C (eds) Psychrophiles: from biodiversity to biotechnology. Springer, Berlin, pp 347–363

Kamada M, Oda K, Murao S (1972) The purification of the extracellular acid protease of *Rhodotorula glutinis* K-24. Agric Biol Chem 36:1103–1108

Karasová-Lipovová P, Strnad H, Spiwok V, Spiwok V, Malá S, Králová B, Russell NJ (2003) The cloning, purification and characterisation of a cold-active β-galactosidase from the psychrotolerant Antarctic bacterium *Arthrobacter* sp. C2–2. Enzyme Microb Technol 33:836–844

Khan Pathan AA, Bhadra B, Begum Z, Shivaji S (2010) Diversity of yeasts from puddles in the vicinity of midre lovenbreen glacier, arctic and bioprospecting for enzymes and fatty acids. Curr Microbiol 60:307–314

Kim JY (2009) Isolation of *Sporidiobolus ruineniae* CO-3 and characterization of its extracellular protease. J Korean Soc Appl Biol Chem 52:1–10

Kim JY (2010) Isolation of protease-producing yeast, *Pichia farinosa* CO-2 and characterization of its extracellular enzyme. J Korean Soc Appl Biol Chem 53:133–141

Kim JT, Kang SG, Woo JH, Lee JH, Jeong BC, Kim SJ (2007) Screening and its potential application of lipolytic activity from a marine environment: characterization of a novel esterase from *Yarrowia lipolytica* CL180. Appl Microbiol Biotechnol 74:820–828

Kuddus M, Roohi Arif JM, Ramteke PW (2011) An overview of cold-active microbial α-amylase: adaptation strategies and biotechnological potentials. Biotechnology 10:246–258

Kulakova L, Galkin A, Kurihara T, Tohru Y, Nobuyoshi E (1999) Cold-active serine protease from psychrophilic *Shewanella* strain Ac10: gene cloning and enzyme purification and characterization. Appl Environ Microbiol 65:611–617

Ladero M, Santos A, Garcia-Ochoa F (2000) Kinetic modeling of lactose hydrolysis with an immobilized β-galactosidase from *Kluyveromyces fragilis*. Enzyme Microb Technol 27:583–592

Loperena L, Soria V, Varela H, Lupo S, Bergalli A, Guigou M, Pellegrino A, Bernardo A, Calvino A, Rivas F, Batista S (2012) Extracellular enzymes produced by microorganisms isolated from maritime Antarctica. World J Microbiol Biotechnol 28:2249–2256

Loveland J, Gutshall K, Kasmir J, Prema P, Brenchley JE (1994) Characterization of psychrotrophic microorganisms producing β-galactosidase activities. Appl Environ Microbiol 60:12–18

Margesin R, Schinner F (1994) Properties of cold-adapted microorganisms and their potential role in biotechnology. J Biotechnol 33:1–14

Margesin R, Gander S, Zacke G (2003) Hydrocarbon degradation and enzyme activities of cold-adapted bacteria and yeasts. Extremophiles 7:451–458

Margesin R, Fauster V, Fonteyne PA (2005) Characterization of cold-active pectate lyases from psychrophilic *Mrakia frigida*. Lett Appl Microbiol 40:453–459

Nakagawa T, Yamada K, Miyaji T, Tomizuka N (2002) Cold-active pectinolytic activity of psychrophilic basidiomycetous yeast *Cystofilobasidium capitatum* strain PPY-1. J Biosci Bioeng 94:175–177

Nakagawa T, Nagaoka T, Taniguchi S, Miyaji T, Tomizuka N (2004) Isolation and characterization of psychrophilic yeasts producing cold-adapted pectinolytic enzymes. Lett Appl Microbiol 38:383–387

Nakagawa T, Nagaoka T, Miyaji T, Tomizuka N (2005a) Cold-active polygalacturonase from psychrophilic basidiomycetous yeast *Cystofilobasidium capitatum* strain PPY-1. Biosci Biotechnol Biochem 69:419–421

Nakagawa T, Nagaoka T, Miyaji T, Tomizuka N (2005b) A cold-active pectin lyase from the psychrophilic and basidiomycetous yeast *Cystofilobasidium capitatum* strain PPY-1. Biotechnol Appl Biochem 42:193–196

Nakagawa T, Ikehata R, Uchino M (2006) Cold-active acid β-galactosidase activity of isolated psychrophilic-basidiomycetous yeast *Guehomyces pullulans*. Microbiol Res 161:75–79

Nelson G, Young TW (1987) Extracellular acid and alkaline proteases from *Candida olea*. J Gen Microbiol 133:1461–1469

Ogrydziak DM (1993) Yeast extracellular proteases. Crit Rev Biotechnol 13:1–55

Pavlova K, Gargova S, Hristozova T, Tankova Z (2008) Phytase from antarctic yeast strain *Cryptococcus laurentii* AL_{27}. Folia Microbiol 53:29–34

Petrescu I, Lamotte-Brasseur J, Chessa JP, Ntarima P, Claeyssens M, Devreese B, Marino G, Gerday Ch (2000) Xylanase from the psychrophilic yeast *Cryptococcus adeliae*. Extremophiles 4:137–144

Ramli AN, Mahadi NM, Rabu A, Murad AM, Bakar FD, Illias RM (2011) Molecular cloning, expression and biochemical characterization of a cold-adapted novel recombinant chitinase from *Glaciozyma antarctica* PI12. Microb Cell Fact 10:94–107

Ramli AN, Mahadi NM, Shamsir MS, Rabu A, Joyce-Tan KH, Murad AM, Illias RM (2012) Structural prediction of a novel chitinase from the psychrophilic *Glaciozyma antarctica* PI12 and an analysis of its structural properties and function. J Comput Aided Mol Des 26:947–961

Ray MK, Uma Devi K, Seshu Kumar G, Shivaji S (1992) Extracellular protease from the antarctic yeast *Candida humicola*. Appl Environ Microbiol 58:1918–1923

Reid VJ, Theron LW, du Toit M, Divol B (2012) Identification and partial characterization of extracellular aspartic protease genes from *Metschnikowia pulcherrima* IWBT Y1123 and *Candida apicola* IWBT Y1384. Appl Environ Microbiol 78:6838–6849

Rüchel R (1981) Properties of a purified proteinase from the yeast *Candida albicans*. Biochem Biophys Acta 659:99–113

Rüchel R, Boning B, Borg M (1986) Characterization of a secretory proteinase of *Candida parapsilosis* and evidence for the absence of the enzyme during infection *in vitro*. Infect Immun 53:411–419

Russell NJ (2000) Toward a molecular understanding of cold activity of enzymes from psychrophiles. Extremophiles 4:83–90

Sabri A, Baré G, Jacques P, Jabrane A, Ongena M, Van Heugen JC, Devreese B, Thonart P (2001) Influence of moderate temperatures on myristoyl-CoA metabolism and acyl-CoA thioesterase activity in the psychrophilic antarctic yeast *Rhodotorula aurantiaca*. J Biol Chem 276:12691–12696

Sheik Asraf S, Gunasekaran P (2010) Current trends of β-galactosidase research and application. In: Mendez-Vilas A (ed) Current research, technology and education topics in applied microbiology and microbial biotechnology. Formatex Research Center, Badajoz, Spain, pp 880–890

Sheridan PP, Brenchley JE (2000) Characterization of a salt-tolerant family 42 β galactosidase from a psychrophilic antarctic *Planococcus isolate*. Appl Environ Microbiol 66:2438–2444

Shivaji S, Prasad GS (2009) Antarctic yeasts: biodiversity and potential applications. In: Satyanarayana T, Kunze G (eds) Yeast biotechnology: diversity and applications. Springer, Berlin, pp 3–16

Siddiqui KS, Cavicchioli R (2006) Cold-adapted enzymes. Annu Rev Biochem 75:403–433

Singh P, Singh SM (2012) Characterization of yeast and filamentous fungi isolated from cryoconite holes of Svalbard, Arctic. Polar Biol 35:575–583

Solano DM, Hoyos P, Hernáiz MJ, Alcántara AR, Sánchez-Montero JM (2012) Industrial biotransformations in the synthesis of building blocks leading to enantiopure drugs. Biores Technol 115:196–207

Song Ch, Chi Z, Li J, Wang X (2010) β-galactosidase production by the psychrotolerant yeast *Guehomyces pullulans* 17–1 isolated from sea sediment in Antarctica and lactose hydrolysis. Bioprocess Biosyst Eng 33:1025–1031

Straathof AJJ, Panke S, Schmidt A (2002) The production of fine chemicals by biotransformations. Curr Opin Biotechnol 13:548–556

Struvay C, Feller G (2012) Optimization to low temperature activity in psychrophilic enzymes. Int J Mol Sci 13:11643–11665

Sturley SL, Young TW (1988) Extracellular protease activity in a strain of *Saccharomyces cerevisiae*. J Inst Brew London 94:23–27

Tobe S, Takami T, Ikeda S, Mitsugi K (1976) Production and some enzymatic properties of alkaline proteinase *Candida lipolytica*. Argic Biol Chem 40:1087–1092

Togni G, Sanglard D, Falchetto R (1991) Isolation and nucleotide sequence of the extracellular acid protease gene (ACP) from yeast *Candida tropicalis*. FEBS Lett 286:181–185

Turchetti B, Hall SRT, Connell LB, Branda E, Buzzini P, Theelen B, Müller WH, Boekhout T (2011) Psychrophilic yeast from Antarctica and European glaciers: description of *Glaciozyma* gen. nov., *Glaciozyma martinii* sp.nov. and *Glaciozyma watsonii* sp. nov. Extremophiles 15:573–586

Turkiewicz M, Kur J, Białkowska A, Cieśliński H, Kalinowska H, Bielecki S (2003a) Antarctic marine bacterium *Pseudoalteromonas* sp. 22b as a source of cold-adapted β-galactosidase. Biomol Eng 20:317–324

Turkiewicz M, Pazgier M, Kalinowska H, Bielecki S (2003b) A cold-adapted extracellular serine proteinase of the yeast *Leucosporidium antarcticum*. Extremophiles 7:435–442

Turkiewicz M, Pazgier M, Donachie SP, Kalinowska H (2005) Invertase and α-glucosidase production by the endemic antarctic marine yeast *Leucosporidium antarcticum*. Pol Polar Res 26:125–136

Wang Q, Hou Y, Yan P (2012) Optimization of cold-adapted lysozyme production from the psychrophilic yeast *Debaryomyces hansenii* using statistical experimental methods. J Food Sci 77:337–342

Wierzbicka-Woś A, Cieśliński H, Wanarska M, Kozłowska-Tylingo K, Hildebrandt P, Kur J (2011) A novel cold-active β-galactosidase from the *Paracoccus* sp. 32d—gene cloning, purification and characterization. Microb Cell Fact 10:108–120

Yamada T, Ogrydziak DM (1983) Extracellular acid proteases produced by *Saccharomycopsis lipolytica*. J Bacteriol 154:23–31

Zaliha RN, Salleh AB, Basri M, Mohamad Ali MSB (2012) Cold active enzyme and method thereof. US Patent 2012/0058514 A1

Zecchinon L, Claverie P, Collins T, D'Amico S, Delille D, Feller G, Georlette D, Gratia E, Hoyoux A, Meuwis MA, Sonan G, Gerday Ch (2001) Did psychrophilic enzymes really win the challenge? Extremophiles 5:313–321

Zimmer Ch, Platz T, Cadez N, Giffhorn F, Kohring GW (2006) A cold-active (2*R*,3*R*)-(-)-di-*O*-benzoyl-tartrate hydrolyzing esterase from *Rhodotorula mucilaginosa*. Appl Microbiol Biotechnol 73:132–140

Chapter 18
Production of Polymers and Other Compounds of Industrial Importance by Cold-Adapted Yeasts

Kostantsa I. Pavlova

Abstract Cold-adapted yeasts from Antarctica, still new and understudied microorganisms, attract considerable interest with their biotechnological potential to synthesise exopolysaccharides (EPS) and other biologically active substances. Yeasts of various genera, species, and strains have been selected as producers of EPS built of residual monosaccharides, the most common ones being mannose and glucose. The physicochemical and rheological properties of the EPS make them suitable for use as emulsifiers and stabilisers in cosmetics. An Antarctic strain produces an EPS with good emulsifying properties; also, its biomass contains biologically active substances that have photoprotective and antioxidant activities. Both products are included in cosmetic compositions designed to enhance the cellular metabolism of the skin and protect it against environmental damage. This chapter reviews the production, characteristics, and application of EPS and metabolites synthesised by cold-adapted Antarctic yeasts.

Keywords Exopolysaccharides · Carotenoids · Coenzyme Q_{10} · Antarctic yeasts

Contents

18.1 Introduction ... 398
18.2 Cold-Adapted Yeasts from Antarctica and Their Biotechnological Potential ... 398
18.3 Production of EPS by Yeasts ... 400
 18.3.1 Production of EPS by Cold-Adapted Yeasts from Antarctica ... 401
 18.3.2 Extraction and Purification of EPS ... 404
 18.3.3 Physicochemical and Rheological Properties of EPS Produced by Cold-Adapted Antarctic Yeasts ... 404
18.4 Biologically Active Compounds in the Biomass of Antarctic Yeasts ... 405

K. I. Pavlova (✉)
Laboratory of Applied Biotechnologies, Institute of Microbiology,
Bulgarian Academy of Sciences, 139 Ruski Blvd, 4000 Plovdiv, Bulgaria
e-mail: konpavlova@yahoo.com

P. Buzzini and R. Margesin (eds.), *Cold-Adapted Yeasts*,
DOI: 10.1007/978-3-642-39681-6_18,

18.5 Application of EPS Produced by Cold-Adapted Yeasts from Antarctica 408
18.6 Conclusions 410
References 410

18.1 Introduction

Polysaccharides produced by microorganisms are natural substances that have specific physicochemical and rheological properties and are widely used in the food, cosmetics, pharmaceutical, and chemical industries. They are incorporated as thickeners in different food products (Duboc and Mollet 2001; Fialho et al. 2008; Rehm 2009, 2010; Freitas et al. 2011; Donot et al. 2012). Others have good emulsifying properties and play an important role in the production and stabilisation of emulsions for the food and cosmetics industries (Williams 2007; Desplanques 2012). In recent years, β-glucans have been subjected to active studies on account of their immune stimulating, anti-tumour, antioxidant, and pre-biotic activities (Chan et al. 2009; Liu et al. 2010; Murphy 2010). All these functional properties exhibited by polysaccharides are dependent on the structure, molecular weight, and concentration of the polysaccharides present.

A large number of microorganisms produce biopolymers with different composition, structure, interesting properties, and functional characteristics (Sutherland 1998). The polysaccharides mannans, glucans, glucomannas, galactomannans, phosphomannans, and glucuronoxylomannans have been synthesised by yeast strains of the *Candida, Cryptococcus, Lipomyces, Pichia, Rhodotorula,* and *Sporobolomyces* genera (De Baets et al. 2002b). Some of these exopolysaccharides (EPS) are very similar to the yeast cell wall polysaccharides, suggesting they are derived from them or at least synthesised by the same mechanism (Van Bogaert et al. 2009). Glucan, mannan, and chitin are known to be structural polysaccharides building the yeast cell wall, which is complex in both composition and structure, but a large number of strains also produce exoglucan and exomannan (Elinov et al. 1998; Tikhomirova et al. 1998; Pavlova and Grigorova 1999; Cho et al. 2001).

18.2 Cold-Adapted Yeasts from Antarctica and Their Biotechnological Potential

The search for new producers for biotechnological production of biologically active substances has led scientists to explore habitats in extreme ecosystems, such as Antarctica. This continent is one of the coldest places on Earth due to its climatic features, i.e., cold, strong winds, humidity, long dark and light periods, high UV radiation, food deficiency and insufficient water amounts for living organisms (Smith et al. 1992; Campbell and Claridge 2000; Raspor and Zupan 2006). Samples of soil, plants (moss, lichen, grass), water, ice, sediments, penguin molten feathers, etc. have been collected from these natural regions and used as sources for the isolation, identification and study of psychrophilic yeasts (Fell et al. 2000; Guffogg et al. 2004; Vishniac 2006; Shivaji and Prasad 2009). The biodiversity of cold-adapted yeasts in the terrestrial Antarctic ecosystem has been the subject of scientific publications and reviews related to taxonomic study, biochemical, and biological characteristics and potential by a large number of researchers (Goto et al. 1969; Abyzov 1993; Vishniac 1996; Wynn-Wiliams 1996; Connell et al. 2008; Thomas-Hall 2010; Buzzini et al. 2012). Antarctic soil and plant samples taken on the territory of the Bulgarian base on Livingston Island, Antarctica, have been used for the isolation of yeasts and evaluation of their biodiversity in that area. Yeasts from the *Candida, Cryptococcus, Debaromyces, Pseudozyma, Rhodotorula, Sporobolomyces* genera have been selected (Pavlova et al. 2001, 2004b). An in-depth discussion of cold-adapted yeasts in Antarctic habitats is reported in Chap. 4.

The interest in cold-adapted Antarctic microorganisms is determined both by the study of their mechanism of adaptation to the environment and by the ability of cells to synthesise significant exocellular and endocellular metabolites. Enzymes constitute the largest group of investigated biologically active substances by cold-adapted yeasts. Protease producers are Antarctic yeasts belonging to different species. Vishniac (1985) reported protease production by the *Cryptococcus friedmannii* strain; the *Cryptococcus humicola* species secreted an extracellular aspartyl protease (Ray et al. 1992) and *Glaciozyma antarctica* (formerly *Leucosporidium antarcticum*) synthesised serine proteinase (Turkiewicz et al. 2003) xylanases were produced by *Cryptococcus albidus* TAE85 (Amoresano et al. 2000); *Cryptococcus adeliensis*, formerly *Cryptococcus adeliae* (Gomes et al. 2000; Petrescu et al. 2000; Scorzetti et al. 2000), the yeast strains were updated according to the latest taxonomic guidelines (Kurtzman et al. 2011). Turkiewicz et al. (2003) reported that *G. antarctica* was a lipase and α-glucosidase producer. The Antarctic yeast strain *Cryptococcus laurentii* isolated from a soil sample produced β-glucosidase (Pavlova et al. 2002), a cold-active pectinolytic enzyme was produced by the psychrophilic-basidiomycetous yeast *Cystofilobasidium capitatum* PPY-1 (Nakagawa et al. 2004), *C. laurentii* AL_{27} synthesised phytase with pH and temperature optimum suitable for its direct use as fodder additive (Pavlova et al. 2008). Additional information on enzymes from cold-adapted yeasts is reported in Chaps. 16 and 17.

Antarctic yeasts exposed to strong UV radiation respond to irradiation by developing their protection through synthesis of liposoluble metabolites, i.e., carotenoids, coenzyme Q_{10}, and ergosterol (Tsimako et al. 2002)—biologically active substances that can be applied to medicine, food, and cosmetics industry. Cold-adapted microorganisms, which can be used for the bioremediation of polluted cold soil and wastewaters, are of great importance for environmental protection (Bergauer et al. 2005; Margesin and Feller 2010).

18.3 Production of EPS by Yeasts

The development of extracellular polymer technologies is essential for polysaccharide production under industrial conditions, and therefore, EPS-synthesising strains are selected. Various mesophilic yeasts are active EPS producers. The use of *C. laurentii* strains resulted in the production of exoglycans (Elinov et al. 1998), heteroglycan (Tikhomirova et al. 1998), and EPS for medical and ecological application (Ananyeva et al. 2002). The *Rhodotorula mucilaginosa* (formerly *Rhodotorula rubra*) strain (Kurtzman et al. 2011) produced linear mannans (Pavlov et al. 1992) and its study by Elinov et al. (1998) showed that mannose units were linked via β-1,3- and β-1,4-linkages. The mannan produced by *Rhodotorula glutinis* KCTC was composed of D-mannose, L-fucose, D-glucose, and D-galactose (Cho et al. 2001). Exocellular mannan synthesised by *R. glutinis* had the following monosaccharide composition: D-mannose, D-glucose, D-galactose, and D-xylose, according to Frengova et al. (1997). The *Rhodotorula acheniorum* MC strain synthesised EPS mannan having 85.6 % carbohydrate content and monosaccharide composition as follows: mannose, glucose, galactose, and fructose in a 1:0.08:0.04:0.01 ratio (Pavlova and Grigorova 1999; Pavlova et al. 2005). There is scarce information on the production, characterisation, and application of polysaccharides synthesised by cold-adapted yeasts. The *Sporobolomyces salmonicolor* (anamorph of *Sporidiobolus salmonicolor*) AL_1 yeast strain (Kurtzmann et al. 2011) isolated from Antarctic soil was selected as the first Antarctic microbial EPS producer (Pavlova et al. 2004a).

Figure 18.1a shows the 1H-NMR spectrum of the EPS of *S. salmonicolor* AL_1. It exhibited two singlets in the anomeric region at 4.65 and 4.77 ppm. These two anomeric signals indicated the presence of two different monosaccharides in the repeating unit, regarding the type or the glycosidic linkage position. The signals between 3.35 and 4.20 ppm represented ring protons highlighting the presence of pyranose. The 13C-NMR spectrum of the EPS of *S. salmonicolor* AL_1. is shown in Fig. 18.1b. It showed two singlets (100.2 and 96.8 ppm) in the anomeric region

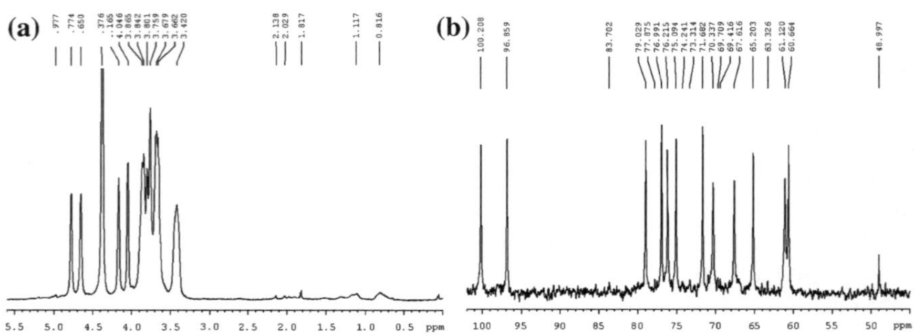

Fig. 18.1 1H-NMR (**a**) and 13C-NMR (**b**) spectra of the EPS of *S. salmonicolor* AL_1 (Poli et al. 2010)

confirming the presence of two units in the repeating group. The presence of a downfield shifted signal at 65.9 ppm confirmed the β backbone structure. The 13C shifts were attributed to a β-mannan configuration by comparison with the shifts of the standard methyl glycoside with a J of 1.5 Hz (Poli et al. 2010).

18.3.1 Production of EPS by Cold-Adapted Yeasts from Antarctica

Heteropolysaccharide synthesis occurs through assimilation of simple sugars and their conversion into nucleotide derivatives, assembly of pentasaccharide subunits attached to a lipid transporter, probably the undecaprenyl phosphate or isoprenoid phosphate, and polymerisation of repeating units of pentasaccharide and secretion into the extracellular environment (Whitfield et al. 1993).

Microbial EPS are related to secondary metabolites. The structure, amount, and properties of microbial EPS depend on genetically determined strains, on the culture medium composition and the growth conditions. The effect of growth-limiting substrates on EPS synthesis clearly demonstrates that the culture medium composition can affect the specific rate of polymer synthesis (Heald and Kristiansen 1985; Lacroix et al. 1995). Media containing a high carbon to limiting-nutrient ratio are favoured for polysaccharide production (Sutherland 1998; Chawla et al. 2009; Rehm 2009). A large number of carbon sources have been used for microbial polysaccharide production: sucrose, glucose, lactose, maltose, mannitol, sorbitol, galactose, whey, starch, hydrolysed corn starch, commercial sugar concentrates, methanol, *n*-alkanes. The type of carbon source influences not only the yield but also the size of the EPS produced (Morin 1998).

Various carbon sources, i.e., pentoses, hexoses, and oligosaccharides, were studied for EPS biosynthesis by the Antarctic strain *C. laurentii* AL_{100}, and raffinose, trehalose, and sucrose were found to undergo the most significant transformation into a polymer. Sucrose proved to be the most suitable and reliable carbon source for EPS production by yeast strains *S. salmonicolor* AL_1, *C. laurentii* $AL_{100,}$ and *C. laurentii* AL_{62} (Pavlova et al. 2004a, 2011; Rusinova-Videva et al. 2011). Sucrose was also used in the biosynthesis of cellulose, dextran, and levan by bacterial producers (Hwang et al.1999; Santos et al. 2000; Bekers et al. 2005) and of pullulan, scleroglucan, and schizophyllan by filamentous fungi (Survase et al. 2007; Kumari et al. 2008; Ravella et al. 2010).

The effect of several nitrogen sources, i.e., ammonium sulphate, sodium nitrate, urea, ammonium chloride, and ammonium phosphate, on EPS production by *R. acheniorum* MC was examined. Comparison of the influence of ammonium salts on yeast growth and metabolic production showed that NH_4^+ played a central role in nitrogen metabolism. All ammonium salts tested ensured active cell growth, only ammonium phosphate was not suitable for EPS synthesis. The study of nitrogen sources for EPS biosynthesis and yeast growth demonstrated the major role of NH_4^+ in nitrogen metabolism and the suitability of ammonium sulphate in low amounts (Grigorova et al. 1999). Ammonium sulphate as a culture medium component affected the composition of *β*-mannan synthesised by *R. mucilaginosa* (formerly *Sporobolomyces albo-rubescens*) (Kurtzmann et al. 2011), as reported by Elinov et al. (1998) and Ananeva and Vitovskaya (1998).

The use of organic nitrogen, peptone, and yeast extract often results in a higher specific growth rate and greater EPS production. A culture medium including peptone and yeast autolysate as an organic nitrogen source stimulated the synthesis of heteroglycan composed of D-mannose, D-glucose, and D-glucuronic acid and trace amounts of D-galactose (Elinov et al. 1998). Ammonium sulphate and yeast extract were the major nitrogen sources in the culture media of Antarctic strains *S. salmonicolor* AL_1, *C. laurentii* AL_{100}, and *C. laurentii* AL_{62} for EPS biosynthesis (Pavlova et al. 2004a, 2011; Rusinova-Videva et al. 2011). Their production was affected by the carbon-to-nitrogen source ratio, dependent on the strain physiology. *S. salmonicolor* AL_1 yeast culture synthesised a maximum polymer amount at a 20:1 carbon/nitrogen ratio, this ratio being 16:1 for the *C. laurentii* AL_{100} and *C. laurentii* AL_{62} strains. The growth medium composition can also indirectly affect the polymer yield by governing the pH change that can occur during fermentation without pH control (Heald and Kristiansen 1985; Lacroix et al. 1995).

Microorganisms usually reach their optimal growth within the initial 24 h of cultivation, and maximal EPS production occurs in the later stages of growth (Vanhooren and Vandamme 1998). The Antarctic strains synthesised EPS in a medium containing sucrose, ammonium sulphate, yeast extract and KH_2PO_4, $MgSO_4.7H_2O$, NaCl, $CaCl_2.2H_2O$, at a temperature of 22 °C. The polymer production started at the 24th h corresponding to the beginning of the exponential growth phase of the producer; the maximum EPS formation was observed at the 96th h, then the polymer content decreased slowly. The culture hydrolysed sucrose to glucose and fructose as early as the beginning of the growth cycle, and at the same time, assimilated glucose intensively and fructose more slowly. During EPS production by *S. salmonicolor* AL_1, the rheology of the fermentation broth changed to 15.37 mPa.s viscosity (Pavlova et al. 2004a). The production of the other two EPS synthesised by cold-adapted Antarctic strains *C. laurentii* AL_{100} and *C. laurentii* AL_{62} occurred at different rates and different polymer compositions depending on their metabolic properties (Pavlova et al. 2011; Rusinova-Videva et al. 2011). A typical feature of EPS formation by yeasts is the significant pH change, which proves to be a regulating factor in their biosynthesis (Adami and Cavazzoni 1990; Elinov et al. 1992). In the course of polysaccharide metabolism, the three cold-adapted strains changed the culture medium from pH 5.3 to pH 1.7–2.0 after 24 h, and these pH values were preserved until fermentation was complete (Pavlova et al. 2004a, 2011; Rusinova-Videva et al. 2011). Hence, Antarctic strains self-adjust the pH of the medium to around pH 2.0, thereby ensuring a high degree of sterility of the biotechnological process, which is of great importance under industrial conditions.

The process of EPS production by *C. laurentii* AL_{100} was transferred from flasks to a laboratory bioreactor with a working volume of 5 L, and the following optimum parameters were established: agitation rate: 400 rpm, air flow rate: 1 L L^{-1} min^{-1} and temperature: 22 °C.

The fermentation duration was reduced by 50 % (from 96 h in flasks to 48 h in a bioreactor) and the accumulated EPS amount was 6.95 g L^{-1}. During the glucomannan biosynthesis by the *S. salmonicolor* AL_1 strain in the bioreactor, at optimum parameters, the hydrodynamic modes were determined under different experimental conditions, i.e., agitator type and agitation rate; the oxygen uptake rate (OUR) and the mass transfer coefficient K_La in the culture medium were also established. The original dual Rushton impeller 2RT at 400 rpm and 1 L L^{-1} min^{-1} were replaced by a dual NS impeller (2NS) at 600 rpm, and the experiments with 2NS agitators showed 30 % higher K_La and 2-fold greater OUR in the course of the fermentation in the logarithmic phase (Vlaev et al. 2012).

Production of biomass and EPS from psychrophilic *S. salmonicolor* AL_1 in a stirred bioreactor was studied. The aspects of production technical-scale parameters, namely bioreactor flow field, biomass, and EPS production rates, oxygen mass transfer per input power, as well as important product properties, such as rheology and stability of EPS mixtures, were considered. The bioprocess was found to proceed in non-Newtonian flow with a consistency coefficient rising typically to 0.03 Pa.s^n and a flow index declining to 0.7. Flow modelling was carried out and showed good homogenization for substrate delivery at agitation rates exceeding 400 rpm. Agitation rates lower than 400 rpm were considered counterproductive due to flow field non-uniformity. The cell density reached 5 g L^{-1}, and EPS production yield reached 5.5 g L^{-1} at a production rate of 0.057 g EPS L^{-1} h^{-1} (0.01 g EPS g^{-1} biomass h^{-1}). OUR and oxygen transfer rate (OTR) were in the range 0.5–1.7 mmol O_2 L^{-1} h^{-1} and 2–4.7 mmol O_2 L^{-1} h^{-1}, respectively. The mass transfer coefficient at reaction conditions was found to be in the range $K\ a\ L \sim 0.004$–$0.01\ s^{-1}$ and was comparable with the reference values for EPS production at similar conditions. The bioprocess biological performance was higher at moderate agitation speed and revealed biomass diminution and cell inactivation by increasing impeller revolutions and shear rate. The product EPS was found to introduce shear-thinning behaviour in water solutions with an apparent viscosity of up to 30 mPa.s. The bioprocess limiting factors and biopolymer properties of a new production process for EPS from *S. salmonicolor* AL_1 in a stirred bioreactor are determined in view of further extrapolation to an elevated production scale. The basic characteristics of the bioprocess fluid dynamics, rheological conditions, flow field velocity structure, and dissolved oxygen transfer were revealed. In order to be able to scale-up the bioprocess hydrodynamics, a computational fluid dynamics (CFD) model was formulated and applied based on the recent modelling experience with commercial codes. In order to be able to determine the mass transfer rate at various scale, it was correlated with specific power. It has been demonstrated that the exoglucomannan produced by the yeast strain *S. salmonicolor* AL_1, selected from the extreme Antarctic environment, is prospective for emulsion preparations. The long-term performance of complex cream mixtures of glucomannan prepared in commercial format is promising and the performance characteristics of the bioprocess can be used for extrapolation (Vlaev et al. 2013).

18.3.2 Extraction and Purification of EPS

EPS preparation from the fermentation liquid is performed by biomass separation, normally through centrifugation, or through separation or filtration in production processes. Polymer precipitation from the supernatant is induced by the addition of an organic solvent (ethanol) in which the polymer is insoluble. The polymer is kept for 24 h at 4 °C, then centrifuged, heat-dried or lyophilised. In order to obtain a higher purity grade polysaccharide, reprecipitation of the polymer from the diluted aqueous solution and chemical deproteinisation can be applied and enzymatic methods and membrane processes, i.e., ultrafiltration and diafiltration, can be used. The crude EPS obtained by the cold-adapted strain *S. salmonicolor* AL_1 was purified on a Sepharose DEAE CL-6B column revealing a first major fraction with the highest carbon content (>90 %) and low presence of protein, nucleic acid and uronic acid (Poli et al. 2010).

18.3.3 Physicochemical and Rheological Properties of EPS Produced by Cold-Adapted Antarctic Yeasts

Microbial isolates from extreme environments offer a great diversity in chemical and physical properties of their EPS compared with anywhere else in the biosphere (Guezennec 2003). The EPS from *S. salmonicolor* AL_1, *C. laurentii* AL_{100} and *C. laurentii* AL_{62} were heteropolysaccharides composed of carbohydrates, protein, and ash (Pavlova et al. 2004a, 2011; Rusinova-Videva et al. 2011). The *S. salmonicolor* AL_1 strain produced glucomannan with the following monosaccharide composition (in %): D-glucose 54.1, mannose 42.6, and fucose 3.3. The recurring monosaccharide units were linked via a β-glucosidic linkage, and the molar mass of the EPSs was >1 MDa (Poli et al. 2010). The arabinomannan EPS produced by *C. laurentii* AL_{100} had the following monosaccharide composition (in %): L-arabinose 61.6, mannose 15, D-glucose 12.6, D-galactose 5.9, L-rhamnose 2.8, and the molar mass was 4200 Da (Pavlova et al. 2011). The xylomannan EPS by the *C. laurentii* AL_{62} strain was built up of D-xylose (45.2 %), mannose (33.6 %), and D-glucose (18.4 %) and had a molar mass of 8,000 Da (Rusinova-Videva et al. 2011). The Antarctic strain *Cryptococcus flavus* was found to produce EPS with the following composition (in %): mannose 55.1, D-glucose 26.1, D-xylose 9.6, and D-galactose 1.9, and the molar mass was 1.01 MDa (Pavlova et al. 2009). The IR-spectra of the glucomannan, arabinomannan and xylomannan showed the characteristic absorption bands observed in natural polysaccharides (Rusinova-Videva et al. 2009; Panchev et al. 2010, 2011). Similar spectral bands were reported by Zhang et al. (2001) for konjac glucomannan. The thermogravimetric analysis indicated a degradation temperature of 280.0 °C for glucomannan (Poli et al. 2010), 175 °C for arabinomannan and 202 °C for xylomannan (Panchev et al. 2011). The EPS were water soluble from 86 to 92 % at temperatures from 20 to 80 °C.

The protein in the structure of the polymer molecule is an important factor in emulsion formation since various intermolecular relations occur between the carbohydrate and the protein part (Dickinson 2009).

Glucomannan demonstrated marked emulsion properties and an emulsion having 100 % stability and fine dispersion characteristics were obtained (Kuncheva et al. 2007). Arabinomannan and xylomannan formed emulsions with 65 and 66 % stability, respectively (Pavlova et al. 2011; Kuncheva et al. 2013). Emulsions with maximum stability were formed using hydrocolloid ingredients, which acted as stabilisers of oil or water drops during homogenisation (Dickinson 2009). In practice, high-quality emulsions are often obtained through the use of combinations of polysaccharides with emulsifying properties and hydrocolloid thickeners due to the synergism occurring between them. The results of the synergistic effect of a combination between 2 % arabinomannan and 0.66 % xanthan showed 100 % stability of the emulsion (Pavlova et al. 2011). Emulsions with similar stability were obtained using 2.0 % xylomannan with 0.5 % xanthan, guar gum and cellulose gum. The physicochemical properties studied showed that the biopolymer synthesised by Antarctic yeasts could be used for the manufacture of emulsion products for the food and cosmetics industries independently, i.e., glucomannan (Kuncheva et al. 2007), or in combination with other hydrocolloids, i.e., arabinomannan and xylomannan (Pavlova et al. 2011; Kuncheva et al. 2013). Desplanques et al. (2012) observed that the effect of xanthan gum and acacia gum mixtures on the stability of oil/water emulsions depended on the impact of the chemical structure of each gum and the occurrence of a synergistic effect.

The mixture of galactomannans with xanthan is an example of synergism that results in the formation of a gel structure (Shobha and Tharanathan 2009). Acacia gum and modified starch showed an emulsifying capacity in a mixture with other hydrocolloids having stabilising properties (Williams 2007; Dickinson 2009).

18.4 Biologically Active Compounds in the Biomass of Antarctic Yeasts

Lipids are products of the living cell, where they perform metabolic and structural functions. The cell membrane is functionally responsible for the passage of different chemical substances and the maintenance of cell viability. An essential factor in the microorganism adaptation to low temperatures is their ability to modulate membrane fluidity by regulation of the fatty acid synthesis, i.e., accumulation of unsaturated fatty acids at a low growth temperature and increase in the saturated fatty acid amount at an increase in temperature (Chintalapati et al. 2004).

In several psychrophilic yeast species, the unsaturated fatty acids constitute 50–90 % of the total fatty acids, which has also been observed in *Mrakia, Candida, Leucosporidium,* and *Cryptococcus* species (Watson 1984; Chintalapati et al. 2004). Vishniac and Kurtzman (1992) observed a temperature effect on the fatty acid composition in yeast cells and found that 87–93 % of the fatty acids of *Cryptococcus antarcticus* strains grown at 13 °C were unsaturated, and 91 % of the *Cryptococcus vishniacii* type were unsaturated. The fatty acid analysis of the biomass from the Antarctic yeasts *R. glutinis, Sporobolomyces roseus* (anamorph of *Sporidiobolus pararoseus*) (Kurtzmann et al. 2011), *C. albidus* and *C. laurentii* cultivated at 15 °C revealed oleic acid as the main and predominant fatty acid (58.6–63.5 %). The values of the polyunsaturated linoleic and linolenic acids were 3.6 and 2.2 %, respectively, for strain *R. glutinis,* and 4.7 and 2.2 %, respectively, for strain *S. roseus*. These levels were higher than the content of polyunsaturated fatty acids in *Cryptococcus* biomasses. The phospholipid content revealed the highest amounts of phosphatidylcholine, phosphatidylinositole and phosphatidylethanolamine, and all classes of tocopherols (α-, β-, and δ-) were identified (Zlatanov et al. 2010). The data obtained on unsaturated fatty acids (66.2–70.5 %) corresponded to the results reported by earlier investigations (Zlatanov et al. 2001). In the Antarctic yeasts obtained by submerged and surface cultivation at 15 °C, the unsaturated fatty acids, mainly oleic acid, were predominant. More in-depth information on changes in lipid composition of cold-adapted yeasts as response to cold is reported in Chap. 10.

Microorganisms that successfully colonised terrestrial habitats in Antarctica are exposed to the coincidental extremes low temperatures and high UV radiation. Yeasts isolated from Antarctic (Tsimako et al. 2002) and freshwater (Libkind 2004) ecosystems have become the focus of attention for researchers who study the biosynthesis of photoprotective substances as a barrier against UV radiation. High UV exposure causes damages to DNA and proteins, including cell membrane lipoproteins and organelles (Karentz 1994). The microorganisms exposed to high UV produce different pigments potentially protecting against UVB damages. UV-screening compounds provide a passive method for the reduction of UV-induced damage, and they are widely distributed across the microbial, plant, and animal kingdoms (Cockel and Knowland 1999). The synthesis of carotenoids that quench oxygen free radicals generated by UV-induced photochemical reactions (Krinsky 1979; Huang et al. 2009) is also an important response in preventing UV-induced damage to a wide diversity of biological macromolecules. Several yeast genera – *Rhodotorula*, *Sporobolomyces*, *Rhodosporidium*, and *Cryptococcus* – produce β-carotene, torulene, torularhodin, which also belong to the carotenoid group (Bhosale and Gadre 2001), coenzyme Q_{10} (CoQ_{10}), (Yajima et al. 2003; Yurkov 2008) and ergosterol (Tuller et al. 1999).

The ability of Antarctic yeasts of the *Sporobolomyces* and *Cryptococcus* genera to synthesise β-carotene and CoQ_{10} when irradiated with Vis, UVA, and mixed light during cultivation has been proved (Dimitrova et al. 2010). The biomass quantities accumulated by Antarctic yeasts are larger under irradiation with low Vis doses (11 J cm^{-2}) than those accumulated under dark cultivation. White light stimulates β-carotene and CoQ_{10} biosynthesis in the pigment-forming *S. salmonicolor* AL_1 strain, whereas the β-carotene quantity in the strains of the *Cryptococcus* genus is not affected by the cultivation conditions. Antarctic yeasts, the highly pigmented species in particular, are essentially resistant to UVA radiation, at least up to 0.5 J cm^{-2}, whereas less than 5 % could survive a similar dose of UVB. It has been suggested that resistance may be associated with pigmentation, possibly carotenoids and related pigments, that may afford important protection (Tsimako et al. 2002). UVA radiation stimulates β-carotene biosynthesis in small doses only and suppresses CoQ_{10} accumulation. The rise in β-carotene in the *Cryptococcus* strains is more pronounced compared with *S. salmonicolor* AL_1, which demonstrates the defence reaction of Antarctic yeasts against UVA irradiation. The same tendency has been observed with the biosynthesis of the investigated substances under the effect of mixed light, when the β-carotene quantity in the biomass of strains from the *Cryptococcus* genera at a dose of 3 + 16 J cm^{-2} (UVA + Vis) exceeded the controls by 100 % (Dimitrova et al. 2010). The highest CoQ10 and β-carotene quantities were obtained during the preliminary irradiation of the *S. salmonicolor* AL_1 strain with UVA, and subsequent deep cultivation at a dose of 11 J cm^{-2} Vis and in the dark. The biomass quantity decreased with all yeast strains studied when they were subjected to UVA radiation. This effect was also observed by Tosi et al. (2005) during experiments with Antarctic soil fungi assemblages. They showed better growth when protected against UV radiation (Tosi et al. 2005). The biomass accumulated during the EPS synthesis by the same *S. salmonicolor* AL_1 strain was suitable for the extraction of biologically active substances. The metabolites ergosterol, torularhodin, torulene, β-carotene, and CoQ_{10} (Fig. 18.2) extracted from the *S. salmonicolor* AL_1 biomass were proved to have antioxidant activity. The emulsion systems, which included all the metabolites produced by the strain, glucomannan as emulsifier and biologically active substances, exhibited UVA protection properties and could therefore be used as components of sunscreen cosmetic products (Dimitrova et al. 2013). The utilisation of the bioactive substances synthesised by *S. salmonicolor* AL_1 would be of biotechnological and industrial significance. Additional information on photo-protective compounds produced by cold-adapted yeasts is reported in Chap. 9.

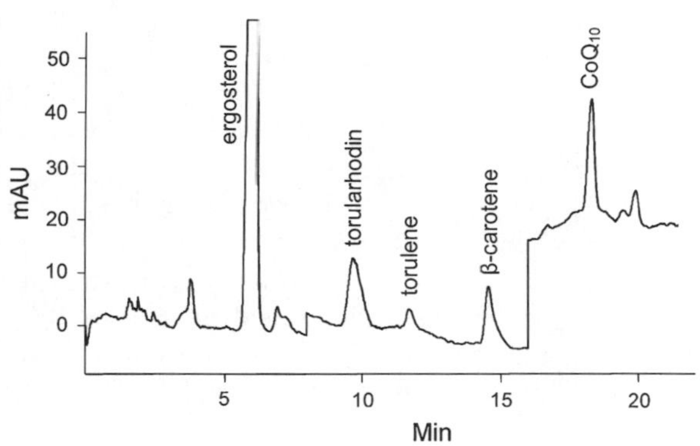

Fig. 18.2 Chromatogram of *S. salmonicolor* AL_1 biomass extract (Dimitrova et al. 2013)

18.5 Application of EPS Produced by Cold-Adapted Yeasts from Antarctica

Water-soluble polymers are widely used as thickeners, emulsifiers, and gelling agents in a broad range of cosmetic products: shampoos and conditioners, hair and body gels, skin creams, liquid and cream foundations, toothpaste, sunscreen lotions and sprays. The natural and synthetic polymers are components of cosmetic products. The modern cosmetics industry demonstrates a tendency for using natural components as substitutes of synthetic ones that are widely used but do not always have positive cosmetic effects. One way of meeting this challenge is the synthesis of naturally active substances via biotechnological methods and their application to different products. Cosmetic compositions have been developed for the creation of novel systems with rheological characteristics and natural ingredients. The system consists of a combination of two naturally occurring polymers, xanthan gum and konjac mannan, which interact synergistically through molecular association leading to the formation of strong thermoreversible gels that can be used in cosmetics (Williams et al. 2003).

An EPS synthesised by a bacterial strain and its cosmetic and dermopharmaceutical compositions can be used for the treatment of the skin, the mucous membranes, hair and nails, especially against the ageing of skin and for the treatment and for the care of disorders and conditions which result from a lack of or decrease in hydration (Lipote and Courtoi 2012). A skin care composition comprising at least one EPS has been derived from microorganisms. The composition is useful in reducing and preventing signs of skin ageing and environmental damage by altering skin cell metabolism and improving hydration (Loing et al. 2011). Crylan produced by *C. laurentii* is a good disperse system stabiliser and can be used for the preparation of cosmetic products, as reported by Tikhomirova et al. (1998). Cosmetic creams have been prepared with natural glucomannan synthesised by *S. salmonicolor* AL_1 and with the synthetic emulsifiers Rofetan N/NS and Arlacel 165 used in cosmetics. The polysaccharide synthesised demonstrated very good interaction with the other standard components of the compositions: they had finely dispersed and smooth consistence, and their stability, studied at different temperature regimes (−10, 22, 45 and a −10 °C/45 °C cycle) for 6 months, confirmed the high emulsion capacity of the EPS which was fully comparable with that of the synthetic emulsifiers and revealed possibilities for their application in the cosmetics industry (Kuncheva 2007).

Day and night creams and cleansing milk have been developed as cosmetic products using arabinomannan, xylomannan, and glucomannan as active components. The rheological characteristic of the cosmetic products were established using the *Ostwald-de Waele* power law $\tau = K\,\gamma^n$. The measurements showed that all compositions were non-Newtonian fluids ($0 \leq n \leq 1$), and among the EPS, the strongest rheological impact was demonstrated by glucomannan, with a viscosity of the day cream of 70.1 $Pa.s^n$. The glucomannan rheological profiles showed that the following were suitable for cosmetics: glucomannan and xylomannan for day and night cream and cleansing milk production, whereas arabinomannan had good characteristics for day cream only (Panchev et al. 2012; Pavlova et al. 2012).

Figure 18.3 shows the scheme for EPS and biomass production by *S. salmonicolor* AL_1, the extraction of biologically active substances from the biomass, and the creation of cosmetic products including all metabolites. Day and night creams have been developed as cosmetic products with active components as follows: glucomannan and biomass containing ergosterol, carotenoides, and coenzyme Q_{10} synthesised by the Antarctic strain *S. salmonicolor* AL_1. The clinical studies of the cosmetic compositions showed compatibility with the skin and lack of objective (erythema, desquamation, oedema) and subjective (irritation, stinging, itch) complaints. The creams nourish, regenerate, and soften the skin, its elasticity and density. Being good antioxidants, the biomass metabolites participate in the neutralisation of free radicals, and in addition, ergosterol plays a protective role

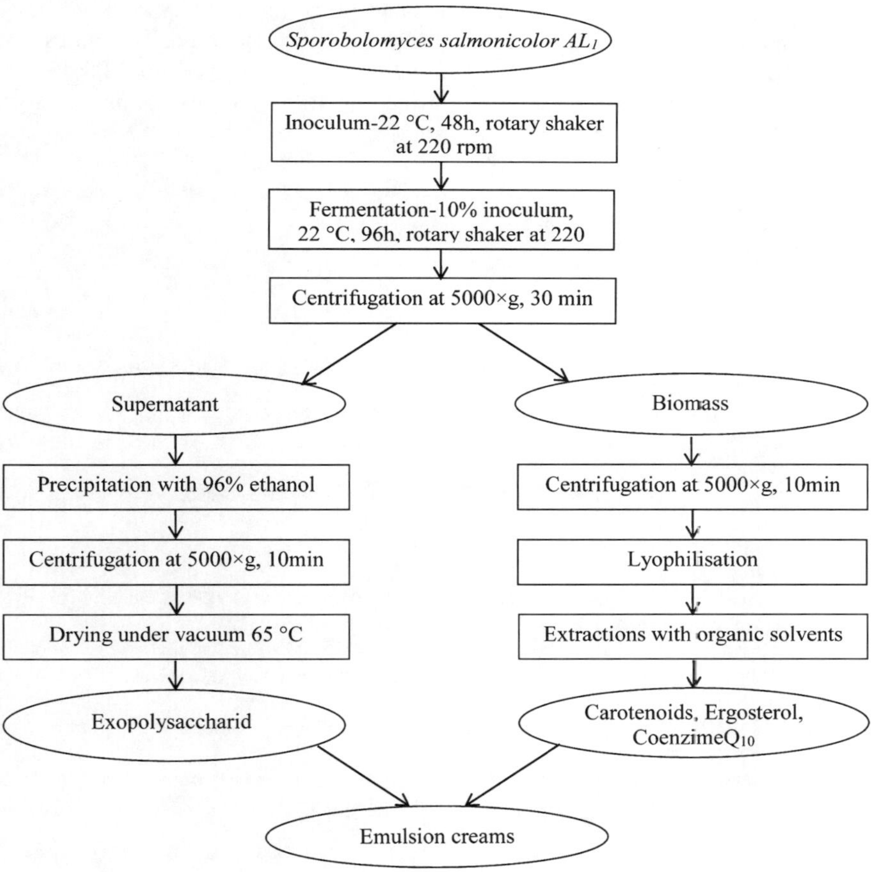

Fig. 18.3 Scheme for the production of biologically active substances by *S. salmonicolor* AL_1

against skin diseases by turning into vitamin D_2 under the influence of sunlight. The role of natural products of microbial origin will gain wider recognition in the future.

18.6 Conclusions

Yeasts constitute a valuable bioresource for the development of microbial biotechnologies. The study of cold-adapted Antarctic yeasts as EPS producers encourages efforts to develop biotechnology for their production and application. The synthesis of EPS with emulsifying properties and metabolites with antioxidant and photoprotective activity by pigment-forming yeasts from the Antarctic ecosystem constitutes a part of the yeasts' potential. Glucomannan, arabinomannan, and xylomannan are good stabilisers and rheological modifiers due to their different chemical composition, physicochemical properties, and specific rheology. These characteristics make them suitable for use in the food, cosmetics and pharmaceutical industries. The optimum parameters established for EPS biosynthesis in a laboratory bioreactor allow for the transfer of the process onto a larger scale, which is of great industrial significance.

References

Abyzov S (1993) Microorganisms in the Antarctic ice. In: Friedman E (ed) Antarctic microbiology. Wiley-Liss and Sons, New York, pp 265–295

Adami A, Cavazzoni V (1990) Exopolysaccharides produced by some yeast strains. Ann Microbiol Enzimol 40:245–253

Amoresano A, Andolfo A, Corsaro M, Zocchi I, Petrescu I, Gerday C, Marino G (2000) Structural characterization of a xylanase from psychrophilic yeast by mass spectrometry. Glycobiology 10:451–458

Ananyeva E, Vitovskaya G (1998) Heteroglycan produced by the yeast *Sporobolomyces albo-rubescens* VKM Y-2450. Microbiology 67:181–183

Ananyeva EP, Vitovskaya GA, Burakova MA, Efimova LS, Karavaeva AV, Ryzhenkov VE, Florova NJ (2002) Russian Patent 2,177,695

Bekers M, Upite D, Kaminska E, Laukevics J, Grube M, Vigants A (2005) Stability of levan produced by *Zymomonas mobilis*. Process Biochem 40:1535–1539

Bergauer P, Fonteyne P, Nolard N, Schinner F, Margesin R (2005) Biodegradation of phenol and phenol-related compounds by psychrophilic and cold-tolerant alpine yeasts. Chemosphere 59:909–918

Bhosale P, Gadre R (2001) Production of β-carotene by a mutant of *Rhodotorula glutinis*. Appl Microbiol Biotechnol 55:423–427

Buzzini P, Branda E, Goretti M, Turchetii B (2012) Psychrophilic yeasts from worldwide glacial habitats: diversity, adaptation strategies and biotechnological potential. FEMS Microbiol Ecol 82:217–241

Campbell I, Claridge G (2000) Soil temperature, moisture and salinity patterns in Transantarctic Mountain cold desert ecosystems. In: Davidson W, Howard-Williams C, Broady P (eds)

Antarctic ecosystems: models for wider understanding. Caxton Press, Christchurch, pp 233–240
Chan G, Chan W, Sze D (2009) The effects of β-glucan on human immune and cancer cells. J Hematol Oncol 2:1–11
Chawla P, Bajaj I, Shrikant A, Singhal R (2009) Microbial cellulose: fermentative production and applications. Food Technol Biotechnol 47:107–124
Chintalapati K, Kiran M, Shivaji S (2004) Role of membrane lipid fatty acids in cold adaptation. Cell Mol Biol 50:631–642
Cho D, Chae H, Kim E (2001) Synthesis and characterization of a novel extracellular polysaccharide by *Rhodotorula glutinis*. Appl Biochem Biotechnol 95:183–193
Cockel C, Knowland J (1999) Ultraviolet radiation screening compounds. Biol Rev 74:311–345
Connell L, Redman R, Craig S, Scorzetti G, Iszard M, Rodriguez R (2008) Diversity of soil yeasts isolated from South Victoria Land, Antarctica. Microb Ecol 56:448–459
De Baets S, Laing S, Vandamme E (2002) In: De Baets S, Vandamme E, Steinbuchel A (eds) Biopolymers: polysaccharides II. Wiley-VCH, Weinheim, pp 93–122
Desplanques S, Renou F, Grisel M, Malhiac C (2012) Impact of chemical composition and acacia gums on the emulsification and stability of oil-in water emulsions. Food Hydrocolloid 27:401–410
Dickinson E (2009) Hydrocolloids as emulsifiers and emulsion stabilizers. Food Hydrocolloid 23:1473–1480
Dimitrova S, Pavlova K, Lukanov L, Zagorchev P (2010) Synthesis of coenzyme Q_{10} and β-caroten by yeasts isolated from Antarctic soil and lichen in response to ultraviolet and visible radiations. Appl Biochem Biotechnol 162:795–804
Dimitrova S, Pavlova K, Lukanov L, Korotkova E, Petrova E, Zagorchev P, Kuncheva M (2013) Production of metabolites with antioxidant and emulsifying properties by Antarctic strain *Sporobolomyces salmonicolor* AL_1. Appl Biochem Biotechnol 169:301–311
Donot F, Fontana A, Baccou J, Schorr-Galindo S (2012) Microbial exopolysaccharides: Main examples of synthesis, excretion, genetics and extraction. Carbohydr Polym 87:951–962
Duboc P, Mollet B (2001) Applications of exopolysaccharides in the dairy industry. Int Dairy J 11:759–768
Elinov N, Ananyeva E, Vitovskaya G (1992) Features of the biosynthesis and characteristics of exoglucans in yeasts of the genus *Sporobolomyces*. Mikrobiologiya 60:466–470
Elinov N, Ananyeva E, Yaskovich G (1998) Activity of exoglycans as sorbents of ions of heavy metals. Appl Biochem Microbiol 35:168–171
Fell J, Boekhout T, Fonseca A, Scorzetti G, Statzell-Tallman A (2000) Biodiversity and systematic of basidiomycetous yeasts as determined by large-subunits rDNA D1/D2 domain sequence analysis. Int J Syst Evol Microbiol 50:1351–1371
Fialho A, Moreira L, Granla A, Popescu A, Sá-Correia I (2008) Occurrence, production and applications of gellan: current state and perspectives. Appl Microbiol Biotechnol 79:889–900
Freitas F, Alves V, Reis M (2011) Advances in bacterial exopolysaccharides: from production to biotechnological applications. Trends Biotechnol 29:388–398
Frengova G, Simova E, Beshkova D (1997) Caroteno-protein and exopolysaccharide production by co-cultures of *Rhodotorula glutinis* and *Lactobacillus helveticus*. J Ind Microbiol Biotechnol 18:272–277
Gomes J, Gomes I, Steiner W (2000) Thermolabile xylanase of the Antarctic yeast *Cryptococcus adeliae*: production and properties. Extremophiles 4:227–235
Goto S, Sugiyama J, Lizuka (1969) Taxonomic study of Antarctic yeasts. Mycologia 61:748–774
Grigorova D, Pavlova K, Panchev I (1999) Preparation and preliminary characterization of exopolysaccharides by yeast *Rhodotorula acheniorum* MC. Appl Biochem Biotechnol 81:181–191
Guezennec J (2003) From extreme environments to biologically active exopolysaccharides. Commun Agric Appl Biol Sci 68:227–234

Guffogg S, Thomas-Hall S, Holloway H, Watson K (2004) A novel psychrotolerant member of the hymenomycetous yeasts from Antarctica: *Cryptococcus watticus* sp.nov. Int J Syst Evol Microbiol 54:275–277

Heald P, Kristiansen B (1985) Synthesis of polysaccharide by yeast-like forms of *Aurebasidium pullulan*. Biotechnol Bioeng 27:1516–1519

Huang Z, Lin Y, Fang J (2009) Biological and pharmacological activities of squalene and related compounds: potential uses in cosmetic dermatology. Molecules 14:540–554

Hwang J, Yang Y, Hwan J, Pyun Y, Kim Y (1999) Effect of pH and dissolved oxygen on cellulose production by *Acetobacter xylinum* BRC5 in agitated culture. J Biosci Bioeng 88:183–189

Karentz D (1994) Ultraviolet tolerance mechanisms in Antarctic marine organisms In: Weler C, Penhale P (eds) UV radiation in Antarctica: measurement and biological effects. American Geophysical Union, Washington DC, pp 92–110

Krinsky N (1979) Carotenoid protection against oxidation. Pure Appl Chem 51:649–660

Kumari M, Survase N, Singhal R (2008) Production of schizophyllan using *Schizophyllum commune* NRCM. Biores Technol 99:1036–1043

Kuncheva M, Pavlova K, Panchev I, Dobreva S (2007) Emulsifying power of mannan and glucomannan produced by yeasts. Int J Cosmet Sci 29:377–384

Kuncheva M, Panchev I, Pavlova K, Rusinova-Videva S, Georgieva K, Dimitrova S et al (2013) Production and characterisation of exopolysaccharide by Antarctic yeast strain C*ryptococcus laurentii* AL_{62}. Biotechnol Biotec Eq. doi:10.5504/bbeq.2013.0009

Kurtzman C, Fell J, Boekhout T (2011) The yeasts: a taxonomy study. Elsevier, Amsterdam

Lacroix C, LeDuy A, Noel G, Choplin L (1995) Effect of pH on the batch fermentation of pullulan from sucrose medium. Biotechnol Bioeng 27:202–207

Libkind D, Pérez P, Sommaruga R, Diéguez M, Ferraro M, Brizzio S (2004) Constitutive and UV-inducible synthesis of photoprotective compounds (carotenoids and mycosporines) by freshwater yeasts. Photochem Photobiol Sci 3:281–286

Lipote C, Courtoi A (2012) Exopolysaccharide for the treatment and/or care of the skin, mucous membranes, hair and/or nails. Patent WO/2012/072245

Liu C, Lu J, Lu L, Liu Y, Wang F, Xiao M (2010) Isolation, structural characterization and immunological activity of an exopolysaccharide produced by Bacillus licheniformis 8-37-0-1. Biores Technol 101:5528–5533

Loing E, Briatte S, Vayssier C, Beaulieu M, Dionne P, Richert L, Moppert X (2011) Cosmetic compositions comprising exopolysaccharides derived from microorganism isolated from microbial mars found in French Polynesia. US Patent 2011/0150795 A1

Margesin R, Feller G (2010) Biotechnological applications of psychrophiles. Environ Technol 31:835–844

Morin A (1998) Screening of polysaccharide-producing microorganisms, factors influencing the production, and recovery of microbial polysaccharides. In: Dumitriu S (ed) Polysaccharides—structural diversity and functional versatility. Marcel Dekker Inc. Publication, New York, pp 275–296

Murphy E, Davis J, Carmichael M (2010) Immune modulating effects of β-glucan. Curr Opin Clin Nutr Metab Care 13:656–661

Nakagawa T, Nagaoka T, Taniguchi S, Miyaj T, Tomizuka N (2004) Isolation and characterization of psychrophilic yeasts producing cold adapted pectinolytic enzymes. Lett Appl Microbiol 38:83–387

Panchev I, Pavlova K, Rusinova-Videva S, Rusinova-Videva S, Georgieva K (2010) Water sorption properties of exopolysaccharides syntjesized by psychrophilic yeasts. Scientific works of University of Food Technologies 57:118–123

Panchev I, Pavlova K, Rusinova-Videva S, Kuncheva M (2011) Thermophysical and IR spectral characteristics of an exopolysaccharide synthesized by the psychrophilic yeast Isolate 62. Sci Res Union Scientists Bulgaria 13:80–85

Panchev I, Dobreva S, Karashanova D, Pavlova K, Kuncheva M, Georgieva K (2012) Physical properties of cosmetic creams containing exopolysaccharides synthesized from Antarctic yeast. Sci Works Univ Food Technol 59:318–323

Pavlov G, Komeeva A, Michailova N, Ananyeva E (1992) Hydrodynamic properties of the fractions of mannan formed by *Rhodotorula rubra* yeast. Carbohydr Polym 19:243–248

Pavlova K, Grigorova D (1999) Production and properties of exopolysaccharide by *Rhodotorula acheniorum* MC. Food Res Int 32:473–477

Pavlova K, Grigorova D, Hristozova T, Angelov A (2001) Yeast strains from Livingston Island, Antarctica. Folia Microbiol 46:397–401

Pavlova K, Angelova G, Savova I, Grigorova D, Kupenov L (2002) Studies of Antarctic yeast for β-glucosidase. World J Microbiol Biotetechnol 18:569–573

Pavlova K, Koleva L, Kratchanova M, Panchev I (2004a) Production and characterization of an exopolysaccharide by yeast. World J Microbiol Biotetechnol 20:435–439

Pavlova K, Gushterova A, Savova I, Nustorova M (2004b) Isolation and taxonomic study of Antarctic yeasts from Livingston Island for exopolysaccharide production. Bulgarian Antarct Res 4:27–34

Pavlova K, Panchev I, Hristozova T (2005) Physico-chemical characterization of exomannan from *Rhodotorula acheniorum* MC. World J Microbiol Biotechnol 21:279–283

Pavlova K, Gargova S, Hristozova T, Tancova Z (2008) Phytase from Antarctic yeast strain *Cryptococcus laurentii* AL_{27}. Folia Microbiol 53:29–34

Pavlova K, Panchev I, Kratchanova M, Gocheva M (2009) Production of an exopolysaccharide by Antarctic yeast. Folia Microbiol 54:343–348

Pavlova K, Rusinova-Videva S, Kuncheva M, Kratchanova M, Gocheva M, Dimitrova S (2011) Synthesis and characterization of an exopolysaccharide by Antarctic yeast strain *Cryptococcus laurentii* AL_{100}. Appl Biochem Biotechnol 163:1038–1052

Pavlova K, Dobreva S, Panchev I, Kuncheva M, Georgieva K, Rusinova-Videva (2012) Cosmetic application of arabinomannan and xylomannan produced by Antarctic yeast strains. Sci Works Univ Food Technol 59:330–333

Petrescu I, Lamotte-Brasseur J, Chessa J, Claeyssens M, Devreese B, Marino G, Gerday C (2000) Xylanase from the psychrophilic yeast *Cryptococcus adeliae*. Extremophiles 4:137–144

Poli A, Anzelmo G, Tommonaro G, Lama L, Gambacorta A, Nicolaus B (2010) Production and chemical characterization of exopolysaccharides synthesized by psychrophilic yeast strain *Sporobolomyces salmonicolor* AL_1 isolated from Livingston Island, Antarctica. Folia Microbiol 55:576–581

Raspor P, Zupan J (2006) Yeasts in extreme environments. In: Rosa C, Peter G (eds) Biodiversity and ecophysiology of yeasts. Springer, Berlin, pp 371–417

Ravella S, Quiñones T, Retter A, Heiermann M, Amon T, Hobbs P (2010) Extracellular polysaccharide (EPS) production by a novel strain of yeast-like fungus *Aureobasidium pullulans*. Carbohydrate Polym 82:728–732

Ray M, Devi K, Kumar G, Shivaji S (1992) Extracellular protease from the Antarctic yeast *Candida humicola*. Appl Environ Microbiol 58:1918–1923

Rehm B (ed) (2009) Microbial production of biopolymers and polymer precursors: applications and perspectives. Caister Academic Press, Norwich

Rehm B (2010) Bacterial polymers: biosynthesis, modifications and applications. Nat Rev Microbiol 8:578–592

Rusinova-Videva S, Pavlova K, Metcheva R (2009) Studies of Antarctic yeast isolates for exopolysaccharide synthesis. Biotechnol Biotec Eq 23:888–891

Rusinova-Videva S, Pavlova K, Georgieva K (2011) Effect of different carbon sources on biosynthesis of exopolysaccharide from Antarctic strain *Cryptococcus*. Biotechnol Biotec Eq 23:888–891

Santos M, Teixeira J, Rodrigues A (2000) Production of dextransucrase, dextran and fructose from sucrose using *Leuconostoc mesenteroides* NRRL B512(f). Biochem Eng J 4:177–188

Scorzetti G, Petrescu I, Yarrow D, Fell J (2000) *Cryptococcus adeliensis* sp.nov., a xylanase producing basidiomycetous yeast from Antarctica. A van Leeuwenhoek 77:153–157

Shivaji S, Prasad G (2009) Antarctic yeasts: biodiversity and potential applications. In: Satyanarayana T, Kunze G (eds) Yeast biotechnology: diversity and applications. Springer Science+Business Media BV, Dordrecht, pp 3–18

Shobha M, Tharanathan R (2009) Rheological behaviour of pullulanase-treated guar galactomannan on co-gelation with xanthan. Food Hydrocolloid 23:749–754

Smith R, Prezelin B, Baker K (1992) Ozone depletion: Ultraviolet radiation and phytoplankton biology in Antarctic waters. Science 255:952–959

Survase S, Saudagar P, Singhal R (2007) Enhanced production of scleroglucan by *Sclerotium rolfsii* MTCC 2156 by use of metabolic precursors. Biores Technol 98:410–415

Sutherland I (1998) Novel and established applications of polysaccharides. Trends Biotechnol 16:41–45

Tikhomirova O, Vitovskaya G, Sinitskaya I (1998) Changes in the composition of *Rhodotorula rubra* (DEMME) Lodder polysaccharides during exomannan synthesis. Microbiology 67:66–70

Thomas-Hall S, Turchetii B, Buzzini P, Branda E, Boekhout T, Theelen B, Watson (2010) Cold adapted yeast from Antarctica and the Italian Alps. Discription of three novel species: *Mrakia robertii sp. nov., Mrakia blollopis sp. nov.* and *Mrakiella niccombsii sp. nov*. Extremophiles 14:47–59

Tosi S, Onofri S, Brusoni M, Zucconi L, Vishniac H (2005) Response of Antarctic soil fungal assemblages to experimental warming and reduction of UV radiation. Polar Biol 28:470–482

Tsimako M, Guffogg S, Thomas-Hall S, Watson K (2002) Resistance to UV radiation in Antarctic yeasts. Redox Rep 7:312–314

Tuller G, Nemec T, Hrastnik C, Daum G (1999) Lipid Analysis-Intermediate Report, *S. cerevisiae*—EUROFAN 2-Node N4

Turkiewicz M, Pazgier M, Kalinowska H, Bielecki S (2003) A cold-adapted extracellular serine proteinase of the yeast *Leucosporidium antarcticum*. Extremophiles 7:435–442

Van Bogaert I, De Maeseneire S, Vandamme E (2009) Extracellular polysaccharides produced by yeasts and yeast-like fungi. In: Satyanarayana T, Kunze G (eds) Yeast biotechnology: diversity and applications. Springer Science + Business Media BV, Dordrecht, pp 651–671

Vanhooren P, Vandamme E (1998) Biosynthesis, physiological role, use and fermentation process characteristics of bacterial exopolysaccharides. Recent Res Dev Ferment Bioeng 1:253–258

Vishniac H (1985) *Cryptococcus friedmannii,* a new species of yeast from the Antarctic. Mycologia 77:149–153

Vishniac H (1996) Biodiversity of yeasts and filamentous microfungi in terrestrial Antarctic ecosystems. Biodivers Conserv 5:1365–1378

Vishniac H (2006) Yeast biodiversity in the Antarctic. In: Rosa C, Peter G (eds) Biodiversity and ecophysiology of yeasts. Springer, Berlin, pp 419–439

Vishniac H, Kurtzman C (1992) *Cryptococcus antarcticus* sp. nov. and *Cryptococcus albidosimilis* sp. nov., basidioblastomycetes from Antarctic soils. Int J Syst Bacteriol 42:547–553

Vlaev S, Martinov M, Pavlova K, Rusinova-Videva S, Georgieva K (2012) Characterization of NS impeller mixing in viscous batches containing exopolysaccharides. In: Proceedings 14th European conference on mixing, Warsaw, Sept 2012, pp 497–502

Vlaev S, Rusinova-Videva S, Pavlova K, Kuncheva M, Panchev I, Dobreva S (2013) Submerged culture process for biomass and exopolysaccharide production by Antarctic yeast: some engineering considerations. Appl Microbiol Biotechnol. 97:5303-5313

Watson K (1984) Membrane lipid adaptation in yeast. In: Kates M, Manson L (eds) Membrane fluidity. Biomembranes, vol 12. Plenum, New York, pp 517–542

Whitfield C, Valvano M, Rose A (1993) Biosynthesis and expression of cell-surface polysaccharides in Gram-negative bacteria. Advances in microbial physiology. Academic Press, London, pp 135–246

Williams P (2007) Handbook of industrial water soluble polymers. Blackwell Publishing LTD, Oxford

Williams P, Hickey M, Mitchell D (2003) Fluid gels based on natural polymers for cosmetic application. Cosmet Toiletries 1187:51–59

Wynn-Williams D (1996) Antarctic microbial diversity: the basis of polar ecosystem process. Biodivers Conserv 5:1271–1293

Yajima K, Kato T, Kanda A, Kitamura S, Veda Y (2003) Process for producing Coenzyme Q_{10}. Japanese Patent WO/056024

Yurkov M, Vustin M, Tyaglov B, Maksimova I, Sineokiy S (2008) Pigmented basidiomycetous yeasts are a promising source of carotenoids and ubiquinone Q_{10}. Microbiology 77:1–6

Zhang H, Yoshimura M, Nishinari K, Williams M (2001) Gelation behaviour of konjac glucomannan with different molecular weights. Biopolymers 59:38–50

Zlatanov M, Pavlova K, Grigorova D (2001) Lipid composition of some yeast strains from Livingston, Antarctica. Biotechnol Biotec Eq 24:402–406

Zlatanov M, Pavlova K, Antova G, Angelova-Romova M, Georgieva K, Rusinova-Videva S (2010) Biomass production by Antarctic yeast strains: an investigation on the lipid composition. Biotechnol Biotec Eq 24:2096–2100

Chapter 19
Low-Temperature Production of Wine, Beer, and Distillates Using Cold-Adapted Yeasts

Maria Kanellaki, Argyro Bekatorou and Athanasios A. Koutinas

Abstract Recently, low-temperature fermentation has been recognized as a valuable tool to improve the flavor of fermented foods such as alcoholic beverages, e.g., in terms of improved ratios of off-flavor compounds to desirable compounds on total volatiles produced during fermentation. Extremely low-temperature fermentation processes can be made feasible using psychrophilic or psychrotolerant yeasts, combined with cell immobilization techniques and suitable bioreactor design. At research level, many studies deal with the optimization of low-temperature alcoholic fermentation (0–15 °C) in order to improve product quality, produce different products from the same raw material, and create added value. Studies dealing with these aspects also discuss ways to maintain bioreactors when production in the factory ceases, as well as the production of low-cost, ready-to-use dried yeast formulations. The application of extremely low-temperature fermentation in winemaking, brewing, distillates, and other fermented food production, as well as cold-adaptation aspects for food yeasts is discussed.

Keywords Cold adaptation · Yeast · Low-temperature fermentation · Alcoholic beverages

Contents

19.1 Introduction 418
19.2 Cold-Adapted Yeasts for Low-Temperature Fermentation 420
 19.2.1 Effect of Growth and Fermentation Temperature on Yeast Performance 420
 19.2.2 Genetically Engineered Yeasts for Improved Fermentation Performance and Stress Resistance 423
19.3 Low-Temperature Winemaking, Brewing, and Ethanol Production 424

M. Kanellaki (✉) · A. Bekatorou · A. A. Koutinas
Department of Chemistry, University of Patras, 26500 Rio, Patras, Greece
e-mail: m.kanellaki@upatras.gr

P. Buzzini and R. Margesin (eds.), *Cold-Adapted Yeasts*,
DOI: 10.1007/978-3-642-39681-6_19, © Springer-Verlag Berlin Heidelberg 2014

19.3.1 Low-Temperature Winemaking 425
19.3.2 Low-Temperature Brewing 428
19.3.3 Low-Temperature Ethanol and Distillates Production 429
19.4 Development of Dried Yeasts for Commercial Low-Temperature Applications 430
19.5 Flavor of Wine, Beers, and Distillates Produced by Low-Temperature Fermentation 431
19.6 Conclusions and Perspectives 433
References 434

19.1 Introduction

Fermentation has been used for thousands of years as a natural process to improve quality, nutritional value, and safety of foods. Fermenting microorganisms, mainly yeasts and lactic acid bacteria, are used traditionally and at industrial scale for the production of bread, meat and dairy products, wine, beer and distillates (Steinkraus 1997; Vlieg et al. 2011; Bourdichon et al. 2012; Navarrete-Bolanos 2012). Recently, low-temperature fermentation (below 15 °C) is recognized as a valuable tool to improve the flavor of fermented foods (Kourkoutas et al. 2004; Beltran et al. 2006). Especially, low-temperature fermentation in winemaking is used to produce white wines with minimized losses of aroma volatile compounds, while higher fermentation temperatures (18–30 °C) are preferred in red winemaking to enhance extraction of pigments (Redon et al. 2011). Brewing on the other hand is traditionally carried out at varying temperatures depending on the type of beer. For example, ale beers are brewed at 16–23 °C, while lager fermentation is carried out at lower temperatures (8–15 °C) (Belloch et al. 2008). Special local products such as the Greek wines *Vinsanto* (sweet dessert wine made from white sun-dried grapes, Santorini Island, Greece) and *Zitsa* (semi-sparkling white wine, Zitsa, Epirus, Greece) or the famous French sparkling Champagnes are produced by low-temperature secondary fermentation during winter. The improved quality of these wines is attributed to the improved ratios of off-flavor compounds (such as higher alcohols) to desirable compounds (such as short-chain fatty acid esters) on total volatiles produced during fermentation (Bakoyianis et al. 1993; Bardi et al. 1997a; Kanellaki and Koutinas 1999; Beltran et al. 2006; Domizio et al. 2007; Parapouli et al. 2010; Torresi et al. 2011).

Industrial winemaking includes fermentation at low temperatures, usually not lower than 15 °C, using selected psychrotolerant wine yeasts. At research level on the other hand, since it is well established that fermentation temperature affects wine flavor, many attempts have been made to optimize winemaking at very low temperatures (0–15 °C) to (1) improve wine quality in terms of aroma volatiles composition; (2) produce different wine products using the same raw material; and (3) create added value. Good operational stability of extremely low-temperature fermentation processes can be facilitated by the use of psychrophilic or psychrotolerant yeasts combined with cell immobilization techniques and suitable bioreactor design (Kanellaki and Koutinas 1999; Kourkoutas et al. 2004). Research

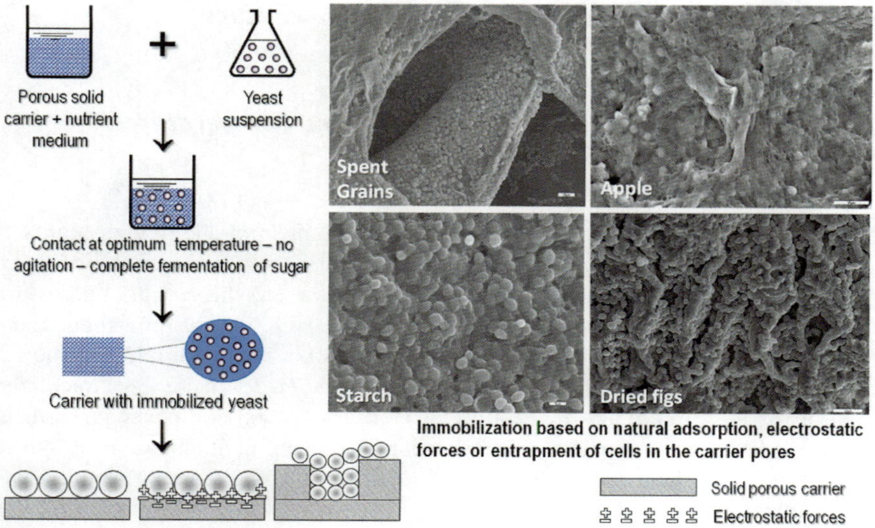

Fig. 19.1 Yeast immobilization technique (*left*) on various porous solid carries and scanning electron micrographs of the immobilized cells (*right*)

efforts regarding optimization of extremely low-temperature fermentations also propose technologies for low-cost production by using cold-adapted yeasts or for ways to maintain the bioreactors when production in the factory ceases. For the production of yeasts ready-to-use and their commercial distribution, various drying techniques have been developed such as freeze-drying or simple thermal drying, focusing on preservation of cell viability during drying and storage (Bekatorou et al. 2001a, b, 2002a; Iconomopoulou et al. 2000, 2002a, b; Tsaousi et al. 2010, 2011). Viable cold-adapted yeasts have been produced in free-cell formulations or immobilized on various carriers, such as porous inorganic (γ-alumina, glass, kissiris etc.), synthetic, or natural organic materials (cellulose, polysaccharides, hydrogels, proteins, etc.), as well as agro-industrial wastes and by-products (cellulosics, pieces of fruit, cereal residues, etc.) (Kourkoutas et al. 2004) (Fig. 19.1). Finally, a number of investigations deal with chemical analysis and sensorial testing to demonstrate the effect of fermentation temperature on the formation of volatile compounds that strongly affect product flavor (Bakoyianis et al. 1993; Bardi et al. 1997a; Mallouchos et al. 2003a, b, 2007; Tsakiris et al. 2004a; Garruti et al. 2006). All these efforts can contribute to the development of novel and integrated low-temperature fermentation technologies for controlled high-quality winemaking and brewing of distillates production.

Recent developments in the application of low-temperature fermentation processes in winemaking, brewing, and distillates production using cold-adapted yeasts are presented and discussed in this Chapter.

19.2 Cold-Adapted Yeasts for Low-Temperature Fermentation

19.2.1 Effect of Growth and Fermentation Temperature on Yeast Performance

Psychrophilic yeasts have a huge biotechnological potential due to their ability to grow and metabolize at low temperatures and can be used in various industrial applications such as (1) production of cold-active enzymes, e.g., lipases for detergents, cellulases for textile and biofuel processing, and in food production; (2) bioremediation of pollutants (e.g. phenolics, hydrocarbons, etc.); (3) production of miscellaneous metabolites of commercial interest (γ-decalactone, extracellular polysaccharides, microbial lipids, etc.); as well as (4) food and beverage fermentations at low temperatures to improve quality, prevent contamination by mesophiles, etc. (Kanellaki and Koutinas 1999; Okuyama et al. 1999; D'Amico et al. 2006; Buzzini et al. 2012).

Metabolic and physiological changes in yeasts are not only induced in psychrophilic species evolved in cold environments, but are also common during growth or fermentation processes at low temperatures. *Saccharomyces cerevisiae* is naturally found in environments, such as the surface of fruit, which can be subjected to low temperatures. In alcoholic fermentation processes, these yeasts can be exposed to temperatures around 10–12 °C, far below their natural physiological temperature (25–30 °C). Also, industrial strains may be stored at very low temperatures (4 °C) at which viability is maintained but growth is restricted (Aguilera et al. 2007). In *S. cerevisiae,* low temperatures induce the expression of genes that display a cold-sensitivity phenotype. Cold-inducible genes upregulated as a response to low temperature, including cold-dependent induction of fatty acid desaturases, a nucleolin-like protein involved in pre-rRNA processing and ribosome biogenesis, serine-, and alanine-rich cell wall proteins, etc., have been identified in *S. cerevisiae* (Kondo and Inouye 1991; Kowalski et al. 1995; Schade et al. 2004). The genetic response to cold stress was found to be time dependent, e.g., genes involved in the metabolism of glycogen and trehalose in detoxifying reactive oxygen species and defending against oxidative stress were induced after 4–6 h of low-temperature incubation (Grant 2001; Sahara et al. 2002; Schade et al. 2004; Murata et al. 2006). The transcriptional response of *S. cerevisiae* to low temperature comprised of two distinct expression patterns: (1) modifications of membrane fluidity and RNA secondary structures for efficient protein translation during the early phase; and (2) environmental stress response due to altered cell physiological state by decreased transport, accumulation of misfolded proteins, and reduced enzyme activities during the late growth phase (Schade et al. 2004). Low temperature was also found to influence wine yeast transcription profile during second fermentation such as that applied in production of sparkling wines by the traditional Champenoise method (Penacho et al. 2012).

Pizarro et al. (2008) showed that temperature differences (15 and 30 °C) most strongly affect nitrogen metabolism and the heat shock response in laboratory and wine *S. cerevisiae* strains. These responses were centered on sugar uptake, nitrogen metabolism, and expression of genes related to organoleptic properties. Salvado et al. (2012) conducted functional analysis to identify genes and proteins involved in adaptation to low-temperature fermentations in commercial wine yeast. They identified nine proteins mainly involved in stress response and in glucose and nitrogen metabolism as the most significant change during the first 24 h of fermentation at 13 °C. Redon et al. (2011) studied the effect of growth and alcoholic fermentation temperature (13 and 30 °C) on yeast viability, vitality, fermentation capacity, and lipid composition of different *Saccharomyces* species with different fermentative origins (wine, beer, baker's and laboratory strains) showing that at low temperature, in spite of specific responses of the different strains/species, the medium-chain fatty acid and the triacylglyceride content increased, whereas the phosphatidic acid content and the phosphatidylcholine/phosphatidylethanolamine ratio decreased. A similar work was conducted by Torija et al. (2003), who determined the production of volatile compounds and the changes in the membrane fatty acids by GC to show the degree of cell adaptation and performance at 13 °C, with 25 °C as reference. They showed that lipid composition changed with the growth temperature, and the optimal membrane fluidity at low temperatures was modulated by changes in the unsaturation degree in *S. cerevisiae* strains. In *S. bayanus*, on the other hand, this change in the unsaturated fatty acid percentage was not observed at different growth temperatures, but the concentration of medium-chain fatty acids at low fermentation temperatures was higher.

Psychrophilic wine yeast strains of the species *S. cerevisiae* were isolated by Kishimoto et al. (1993) and their viability and fermentation ability at low temperatures (7–13 °C) was found superior compared to mesophilic strains (Kanellaki and Koutinas 1999). These strains also produced higher amounts of malic acid, higher alcohols, isoamyl acetate, β-phenylethyl alcohol, β-phenylethyl acetate and lower amounts of acetic acid. Isolation and characterization of psychrotolerant *Saccharomyces* strains from grape musts, able to grow well at 6 °C, was also reported by Castellari et al. (1995). However, publications concerning winemaking at temperatures lower than 10 °C are very few, and in practice, industrial winemaking at temperatures lower than 15 °C is not common due to very low productivity (Kanellaki and Koutinas 1999). To deal with this problem, various *S. cerevisiae* strains were isolated from the Greek agricultural area (the provinces of Achaia and Halkidiki and the Aegean islands Santorini, Samos, and Thasos) and were subjected to cold-adaptation treatments. The psychrotolerant *S. cerevisiae* strains *S. cerevisiae* Visanto and *S. cerevisiae* AXAZ-1 were found suitable for winemaking at low temperatures, by using free or immobilized various solid carriers such as mineral kissiris, delignified cellulosic material (DCM), and gluten pellets (Bakoyianis et al. 1992; Bardi and Koutinas 1994; Argiriou et al. 1996; Bardi et al. 1996a; Kanellaki and Koutinas 1999) (Fig. 19.1). The produced wines had improved flavor and required lower amounts of sulfites for stabilization.

Therefore, regarding alcoholic beverage production, psychrotolerant *S. cerevisiae* strains possess a number of advantages compared to conventional yeasts. In brewing, they can be used for fermentation at low temperatures (6 °C) for the production of beers of superior quality, and in winemaking, they can be used to ferment low acidity musts to produce more malic and succinic acid, glycerol and β-phenylethanol, and less acetic acid (Kishimoto et al. 1993; Giudici et al. 1995; Caridi and Corte 1997; Kanellaki and Koutinas 1999). However, these processes are not commonly used because low temperatures decrease the rates of yeast growth and fermentation and increase the risk of stuck and sluggish fermentations. The competition for nutrients between the selected psychrotolerant yeast and the wild microflora is higher at low temperature, and fermentation can be particularly problematic in musts from southern Europe, which typically have high sugar contents and low nitrogen availability (Llaurado et al. 2005; Salvado et al. 2011). The contribution of indigenous non-*Saccharomyces* species is potentially higher in fermentations carried out below 20 °C, reflecting in the chemical composition and sensory properties of the products. Controlled fermentation with selected yeasts in active dry form may involve rehydration with warm water (35–42 °C) prior to inoculation in a colder must, which may cause physiological stress. Therefore, active dry yeast rehydration at lower temperatures was proposed (Llaurado et al. 2005). On the other hand, Argiriou et al. (1996) pre-adapted the psychrotolerant yeast strain *S. cerevisiae* AXAZ-1 before use to conduct low-temperature fermentation. The yeast was grown in media with gradually increasing amounts of sugar content, while the culture temperature was kept below 20 °C. Improvement of the yeast fermentation performance, faster assimilation of nitrogen, lower acetic acid and fusel alcohol formation and higher glycerol concentrations were obtained (Argiriou et al. 1996; Kanellaki and Koutinas 1999). Fermentation of molasses by *S. cerevisiae* AXAZ-1 was evaluated in an extremely wide temperature range (3–40 °C). Sequence analysis of the 5.8S internal transcribed spacer and the D1/D2 ribosomal DNA (rDNA) regions assigned the isolate to *S. cerevisiae*. Restriction fragment length polymorphism of the mitochondrial DNA showed that the strain is genetically divergent compared to other wild strains of Greek origin or commercial yeast starters. Immobilization of the yeast on brewer's spent grains, the main by-product of the brewing industry, improved the thermo-tolerance of the strain and enabled fermentation at 40 °C. Therefore, the strain *S. cerevisiae* AXAZ-1 is very promising for the production of ethanol from low-cost raw materials, such as molasses, as it is capable to perform fermentations yielding high ethanol concentrations, good-quality products regarding aroma volatile profiles, at both low and high temperatures (Kopsahelis et al. 2009).

19.2.2 Genetically Engineered Yeasts for Improved Fermentation Performance and Stress Resistance

Over the last few years, the wine and brewing industry also showed interest for genetically engineered yeasts with specific functions, with emphasis on the development of *S. cerevisiae* strains with improved fermentation efficiency, biopreservation, and capacities for increased wholesomeness and sensory quality of wine. Genetic modification may involve hybridization, backcrossing, mutagenesis, transformation, and somatic fusion procedures (Kanellaki and Koutinas 1999; Pretorius and Bauer 2002; Pérez-Través et al. 2012; López-Malo et al. 2013). Hybrid strains of *S. cerevisiae*, *Saccharomyces bayanus,* and *Saccharomyces kudriavzevii* are commercially available for wine fermentations (Belloch et al. 2008). Improvement of wine yeasts through genetic engineering has been reported aiming at obtaining certain properties, such as killer activity (Hara et al. 1981), higher yields of glycerol (Eustace and Thornton 1987), enhanced fermentation rate, decreased formation of acetate and improved production of fermentative aroma (Herrero et al. 2008; Cadière et al. 2012), tolerance to SO_2 (Thornton 1982), cell flocculation at the end of fermentation (Watari et al. 1989), and enhanced xylose metabolism for conversion of lignocellulosics into biofuels (Kim et al. 2013), etc.

Improvement of yeast stress responses (including low or high temperature) during alcoholic fermentation processes has also been reported. It is well established that heterologous overexpression of proteins is connected with different stress reactions. Especially, the unfolded protein response in *S. cerevisiae* and other yeasts is well documented. High-cell-density fermentation, the typical process design for recombinant yeasts, exerts growth conditions that deviate far from the natural environment of the cells. Thus, different environmental stresses such as high osmolarity, low pH and low temperature may be exerted on the host. Whereas the molecular pathways of stress responses are well characterized, there is a lack of knowledge concerning the impact of stress responses on industrial production processes. Accordingly, most metabolic engineering approaches conducted so far target at the improvement of protein folding and secretion, and there are only few examples of cell engineering against general stress sensitivity (Mattanovich et al. 2004). Cardona et al. (2007) described a novel genetic manipulation strategy to improve stress resistance in wine yeasts. This strategy involved modification of the expression of the transcription factor *MSN2*, which plays an important role in yeast stress responses. The modified cells were able to carry out vinifications at 15 and 30 °C with higher fermentation rates during the first days of the process compared to the control strain. Jiménez-Martí et al. (2009) introduced several genetic manipulations in two genes induced by several stress conditions: *HSP26*, which encodes a heat shock protein, and *YHR087W*, which encodes a protein of unknown function in two different wine yeasts. Some of these modifications resulted in strains with higher expression of these genes, better resistance to certain stress conditions, and even improved fermentative behavior. Sato et al. (2002) performed hybridization trials between four top-fermenting *S. cerevisiae* strains and a psychrophilic yeast *S. bayanus* in order to improve the fermentability at low-temperature. The hybrids were selected using 5-bromo-4-chloro-3-indolyl-α-D-galactopyranoside and were checked with pulsed-field gel electrophoresis. Their improved brewing performance at 10.5 °C was observed using small-scale fermentation trials.

In brewing, lager beers are fermented with strains collectively known as *S. carlsbergensis*, which are hybrids of *S. cerevisiae* and *S. eubayanus*-like strains, particularly adapted to low-temperature fermentations (Saerens et al. 2010; Sanchez et al. 2012). Selection of new yeast strains with improved properties, or fermentation performance is laborious, due to the allotetraploid nature of lager yeasts. Sanchez et al. (2012) generated new lager yeast and *S. cerevisiae* hybrids by classical genetics, which were improved regarding growth at elevated temperature and resistance against high osmolarity or high ethanol concentrations. Due to the uncertainty of chromosomal make-up of lager yeast spore clones, they introduced molecular markers to analyze mating-type composition and isolated new hybrids between a lager and an ale yeast by micromanipulation. These hybrids were not subject to genetic modification and showed improved stress resistance including improved survival at the end of fermentation and improved fermentation rates at 18–25 °C.

Achievements in genetic engineering of *Saccharomyces* yeast strains applied in food and beverage industry (brewing, winemaking and baking) have been reviewed and discussed (Donalies et al. 2008), yet, it is uncertain whether genetic engineering, although promising for the improvement of yeast properties with regard to product quality and fermentation efficiency, can lead to losses of other natural properties or affect yeast metabolic pathways with unexpected undesirable results (Guerzoni et al. 1985; Lee et al. 1995; Kanellaki and Koutinas 1999).

19.3 Low-Temperature Winemaking, Brewing, and Ethanol Production

A number of studies report the selection or the improvement of cold-adapted yeasts for better fermentation performance and stress tolerance as discussed above, but the largest part of literature reporting the development of low-temperature fermentation processes for alcoholic beverages or ethanol production involves the use of immobilized cell systems. The upsurge of interest in cell immobilization for alcoholic beverages and potable alcohol production is mainly due to the numerous advantages that it offers over free-cell systems, including enhanced fermentation productivity, stress tolerance, feasibility of continuous processing, cell stability and lower costs of recovery and recycling and downstream processing. Nevertheless, industrial use of immobilized cells is still limited and it will depend on the development of processes that can be readily scaled-up (Kourkoutas et al. 2004). Alterations in cell growth, physiology, and metabolic activity may be induced by cell immobilization of both yeast and bacterial species. Comparative studies on immobilized and free cells reported effects on activation of yeast energetic metabolism, increase in storage polysaccharides, altered growth rates, increased substrate uptake and product yield, lower yield of fermentation by-products, higher intracellular pH values, increased tolerance against toxic and inhibitory compounds, and increased invertase activity. Also, the immobilized cells showed increased viability and activity when stored at low temperature for long periods (Kourkoutas et al. 2004).

19.3.1 Low-Temperature Winemaking

Important improvement of fermentation productivity was achieved by *S. cerevisiae* immobilized on inorganic materials such as mineral kissiris (porous volcanic mineral found in Greece; 70 % SiO_2) and γ-alumina, during batch fermentations of glucose and raisin extracts (Tsoutsas et al. 1990; Kana et al. 1989a, b) at ambient temperatures. These systems were further evaluated in batch as well as continuous winemaking at extremely low temperatures (2–10 °C), leading to important productivities (Bakoyianis et al. 1992, 1993, 1997). γ-Alumina promoted the alcoholic fermentation kinetics but showed lower efficiency at low temperatures compared with mineral kissiris (Bakoyianis et al. 1997). Despite the better fermentation performance of these biocatalysts, γ-alumina and kissiris leave mineral residues (e.g. Al) to the produced wine, therefore, studies to remove them were also carried out (Loukatos et al. 2000, 2003). However, these wines can be used as raw material for distillates production (Bakoyianis et al. 1997; Loukatos et al. 2000, 2003).

To avoid the presence of undesirable residues in wine released by inorganic materials, food-grade organic supports such as DCM and gluten pellets were evaluated as yeast immobilization supports for low-temperature fermentation. DCM was prepared after removal of lignin by alkaline (NaOH) treatment of sawdust, which increased the material's porosity, and was used for immobilization of the psychrotolerant and alcohol resistant strain *S. cerevisiae* AXAZ-1. The biocatalyst was found suitable for extremely low-temperature batch fermentation of grape must, with acceptable fermentation rates even at temperatures lower than 5 °C (Bardi and Koutinas 1994). In the same manner, gluten pellets were used for low-temperature winemaking (Bardi et al. 1996a). Both supports presented similar efficiency for grape must fermentation. However, DCM is considered more attractive compared to gluten pellets due to its negligible cost and easier preparation. Wines produced by yeast immobilized on DCM and gluten pellets at extremely low temperature had improved flavor, which was demonstrated by both sensorial tests and chemical analysis of aroma volatiles.

Similarly to DCM, wine fermentations were conducted using free and immobilized psychrotolerant yeast on brewer's spent grains at 25–10 °C. The operational stability of the biocatalyst was good and no decrease of its activity was observed, even at 10 °C. Ethanol and wine productivities were high, showing suitability for low-temperature winemaking.

Most of the research efforts to optimize low-temperature fermentation using immobilized cells concerned white winemaking. In red winemaking, fermentation proceeds in the presence of grape skins for color extraction. Therefore, new bioreactor design is necessary for low-temperature red winemaking to accommodate the immobilized biocatalyst and the grape skins confined in separate compartments (Tsakiris et al. 2004a).

In seek of new supports for yeast immobilization suitable for low-temperature winemaking, starch containing waste materials such as potatoes (Kandylis and Koutinas 2008), wheat (Kandylis et al. 2010a, b), barley (Kandylis et al. 2012a), and starch gels (Kandylis et al. 2008) were evaluated. These biocatalysts were also found to increase productivities of extremely low-temperature grape must fermentations compared to free cells. Comparing starch with alginates, starch is a more abundant natural carbohydrate and of lower cost, which makes it more attractive for commercialization. The biocatalyst produced by immobilization of the psychrotolerant strain *S. cerevisiae* AXAZ-1 on whole wheat grains was used for batch fermentations of glucose and grape must at laboratory and larger scale (80 L) at various temperatures (20–2 °C). The biocatalyst retained its operational stability for a long period, and it was proved capable to produce dry wines of fine clarity even at 5 °C. The scale-up process did not affect the fermentative ability of biocatalyst or the quality of the produced wines (Kandylis et al. 2010b). The same strain was immobilized on whole corn grains and was used for batch fermentations of glucose and grape musts at various temperatures. The biocatalyst retained its operational stability for a long period and it proved able to produce dry wines of fine clarity, even at extremely low temperatures (5 °C). A mathematical model to describe sugar consumption and ethanol formation was developed fitting almost perfectly with the experimental results (Kandylis et al. 2012b).

The idea to use fruit pieces as yeast immobilization supports for use in winemaking was also considered attractive because fruits are the natural habitats of yeasts, they can be easily accepted by consumers and can positively affect the product flavor. Cells immobilized on apple, pear, and quince pieces were successfully used for batch and continuous winemaking at extremely low temperatures (0 °C) (Kourkoutas et al. 2001, 2002a, 2003a; Mallios et al. 2004). These biocatalysts were stored at low temperature for long periods of time without loss of their activity (Kourkoutas et al. 2003b), which is an important observation regarding their potential for commercial application. In a similar manner, *S. cerevisiae* was immobilized on dried raisin berries (Sultanina variety) (Tsakiris et al. 2004b) or grape skins (Mallouchos et al. 2003b) to produce immobilized biocatalyst for use in dry white winemaking 25–15 °C and 25–10 °C, respectively. Raisin berries and grape skins are fully compatible with wine, and consumer acceptance is not an issue. Both biocatalysts were found suitable for good-quality winemaking with good operational stability (more than 4 months in the case of raisin berries). Although the use of fruit in high-volume bioreactors is problematic, such materials could still be employed in low-capacity processes due to the fine quality of the produced wines which can create added value. Table 19.1 shows the major volatiles produced during wine fermentations by cells immobilized on various natural carriers and by free cells

Table 19.1 Major volatiles produced during wine fermentations by cells immobilized on various natural carriers and by free cells of a psychrotolerant yeast strain (*S. cerevisiae* AXAZ-1) at temperatures of 0–30 °C

Carrier	Acetaldehyde			Ethyl acetate			Amyl alcohols			Propanol-1			Isobutyl alcohol		
	$IM_{<15}$	$IM_{>15}$	$FC_{<15}$	$IM_{<15}$	$IM_{>15}$	$FC_{<15}$	$IM_{<15}$	$IM_{>15}$	$FC_{<15}$	$IM_{<15}$	$IM_{>15}$	$FC_{<15}$	$IM_{<15}$	$IM_{>15}$	$FC_{<15}$
DCM (Bardi et al. 1997a)	38–78	18–29	0–30	64–92	98–135	10–36	41–75	118–161	110–173	19–28	32–48	0–22	15–23	22–41	10–25
Gluten pellets (Bardi et al. 1997a)	8–75	16–68	0–30	19–123	53–84	10–36	104–195	118–147	110–173	32–68	23–46	0–22	20–55	38–63	10–25
BSG (Mallouchos et al. 2007)	25–30	21–56	38–41	36–38	16–53	17–25	21–35	98–141	81–116	26–44	22–33	28–30	81–137	98–150	81–116
Apple (Kourkoutas et al. 2002a)	18–90	13–52	–	23–94	37-88	–	96–155	108–176	–	22–40	18–28	–	14–47	15–38	–
Quince (Kourkoutas et al. 2002b)	25–106	30–54	–	14–90	54–80	–	25–99	72–79	–	11–41	22–26	–	18–32	15–22	–
Kissiris (Bakoyianis et al. 1993)	19–48	18–32	34–38	34–82	48–77	7–9	46–126	131–152	108–117	tr	28–45	43–49	5–21	19–28	14–18
Raisins* (Tsakiris et al. 2004b)	20–60	12–25	30–65	21–25	17–44	7–22	31–44	46–95	15–42	22–27	8–23	15–17	8–10	14–57	4–6
Cork* (Tsakiris et al. 2010)	42–121	21–60	45–67	32–55	34–44	24–51	31–96	111–166	28–158	16–35	12–22	11–35	14–26	22–42	16–31

DCM delignified cellulosic materials; *BSG* brewer's spent grains; $IM_{<15}$ fermentation by immobilized cells at <15 °C; $IM_{>15}$ fermentation by immobilized cells at >15 °C; $FC_{<15}$ fermentation by free cells at <15 °C; *commercial *S. cerevisiae* Uvaferme 299 mesophilic wine yeast

of a psychrotolerant yeast (*S. cerevisiae* AXAZ-1) and a mesophilic commercial yeast (*S. cerevisiae* Uvaferme 299) at temperatures of 0–30 °C.

To industrialize a batch or continuous process employing immobilized cells, high operational stability is required. The improvement of biocatalytic stability and productivity of immobilized yeast cells was attempted by successive preservations at 0 °C of a psychrotolerant strain leading to a stability of productivity for at least 2.5 years (Argiriou et al. 1996). This result showed that the biocatalyst in the bioreactor could be maintained without need for emptying, sterilizing, recovering, and refilling, which can reduce the production cost and make handling easier when the factory is out of operation.

19.3.2 Low-Temperature Brewing

In brewing, research during the last 30 years has mainly focused on immobilization techniques in order to apply continuous processes, reduce maturation time, and produce alcohol-free beer (Masschelein et al. 1994; Kourkoutas et al. 2004). Brewing requires long fermentation times (6–7 days), large-scale fermentation, maturation at low temperature and storage capacities and is a highly energy-consuming process. Diacetyl is a key compound in beer maturation as it has an undesired butter flavor. It is produced by an oxidative non-enzymatic reaction from α-acetolactate and converted to acetoin by yeast metabolism during the maturation stage (*diacetyl rest*). These reactions are time-consuming and energy demanding since they are occurring at very low temperatures to avoid product quality degradation. The combination of yeast immobilization and low-temperature fermentation can reduce process times and improve product quality in the brewing industry, leading to significant reduction in production costs (Kourkoutas et al. 2004).

Traditionally, industrial brewing processes involve fermentations at low temperatures (13–15 °C) to obtain beer with quality that mainly depends on the yeast strain. Low-temperature fermentation of wort below common brewing temperatures (0–12 °C) produced beers of improved quality due to alterations in the chemical composition, mainly of aroma volatiles. Likewise, the combination of low-temperature fermentation and cell immobilization was found to lead to reduction in fermentation times and elimination of the maturation stage due to faster reduction in diacetyl (Bardi et al. 1996b; Bekatorou et al. 2001a, 2002a, b). The use of immobilization for the removal of diacetyl and therefore controlling flavor and reducing maturation time and production costs in brewing has been reported as a promising technological application.

As in the case of low-temperature winemaking, DCM was used as support for yeast immobilization for use as biocatalyst in low-temperature fermentation of wort (Bardi et al. 1996b). Beers were produced in batch or continuous processes with important productivities even at very low temperatures. Beers had lower pH, and lower diacetyl and phenolic compounds contents, and reduced bitterness compared with those of free cells. Beers produced by the DCM biocatalyst also had higher ethanol content (about 15 % higher) than those made with free cells, which can be attributed mainly to lower conversion of sugar to cell biomass. The lower pH of beers increases resistance to spoilages, while the lower diacetyl content improves beer flavor and reduces maturation time. The decreased phenolics content reduces the possibility of haze formation in the bottle. Sensorial tests showed important improvement of beers produced by a psychrotolerant yeast strain immobilized on DCM. Gluten was also employed as support for cell immobilization for extremely low-temperature brewing (Bardi et al. 1997b) and the obtained results were similar with those of the DCM biocatalyst.

Extremely low-temperature brewing (7–3 °C) with yeast immobilized on dried pieces of fruit (dried figs) was also investigated (Bekatorou et al. 2002b). Dried figs, which are rich in pectins, cellulose as well as protein, were found suitable for yeast immobilization and brewing, as shown by improved fermentation efficiency and product quality compared to free cells.

For low-temperature brewing, breweries, especially small production capacity units, require suitable isolated psychrophilic strains available for sale in the market. Consequently, the development of dried brewing yeasts was necessary. Freeze-dried *S. cerevisiae* immobilized on DCM and gluten pellets was examined for its efficiency for glucose fermentation (Iconomopoulou et al. 2000; Bekatorou et al. 2001b) and subsequently for extremely low-temperature brewing (Bekatorou et al. 2001a, 2002a), exhibiting all those important attributes, which are necessary for commercial application (increased productivities and improved product quality).

19.3.3 Low-Temperature Ethanol and Distillates Production

Most of the works describing winemaking or brewing using cold-adapted yeasts, free or immobilized on various carries, as described above, include studies with synthetic media (e.g. glucose) or waste effluents such as whey, to evaluate use of these biocatalyst for potable or fuel ethanol production. Distillates are produced through single or double distillation of fermented liquids or by fractional distillation in columns. The quality of distillates depends on the quality of the fermented raw materials, which is improved in the case of low-temperature fermentations by immobilized cells, due to improved ratios of esters to higher alcohols on total volatiles. In distillates production, inorganic materials such as γ-alumina pellets and kissiris can particularly be useful as immobilization supports, since they are cheap, abundant and can be easily reused. Mineral residues in the fermentation broth from such materials are not an issue since they do not distill. Therefore, continuous winemaking using yeast immobilized on γ-alumina, kissiris, DCM and gluten pellets was successfully performed at lower temperatures than the optimal for improved wine distillates production (Loukatos et al. 2000, 2003). Other natural materials have also been used as yeast immobilization carriers for potable alcohol production at ambient and low temperatures such as orange peel (Plessas et al. 2007).

Whey, the main liquid waste of the dairy industry, has also been used as raw material for potable alcohol production by fermentation with lactose fermenting yeasts or mixed cultures such as kefir. Kefir immobilized on DCM was used for batch fermentations at various temperatures in the range of 5–30 °C. The concentration of fusel alcohols was reduced at lower fermentation temperatures. At 5 °C, the content of total volatiles in the product was only 38 % of the volatiles formed during fermentation at 30 °C (Athanasiadis et al. 1999). Kefir immobilized on DCM was also found suitable for continuous whey fermentation supplemented with 1 % raisin extract or molasses. The continuous fermentation system was operated for 39 days, stored for 18 days at 4 °C and operated again for another 15 days without any diminution of the ethanol productivity. Ethanol productivity and residual sugar were at acceptable for the production of potable alcohol and alcoholic drinks at industrial scale (Kourkoutas et al. 2002a). Subsequently, industrial scale-up process of whey alcoholic fermentation, promoted by raisin extracts, using free kefir-yeast cells was successfully developed (Koutinas et al. 2007). The effect of fermentation temperature in the range 30–5 °C, among other parameters, was examined. The scale-up process involved the development of a technology transfer scheme from laboratory-scale experiments to a successive series of increased capacity bioreactors (100, 3000 and 11,000 L).

19.4 Development of Dried Yeasts for Commercial Low-Temperature Applications

Various commercial yeasts for induced fermentations are available usually in the form of dry active formulations. The most common drying technique used for industrial starter culture production is freeze-drying. Therefore, research works proposing new strains for use in food fermentation processes usually involve optimization of drying techniques to produce viable, active ready-to-use cultures. The efficiency of freeze-dried selected yeast strains, free or immobilized on suitable carriers, for extremely low-temperature fermentations has been reported. Freeze-dried psychrotolerant *S. cerevisiae* AXAZ-1 strain immobilized on DCM (Iconomopoulou et al. 2000, 2002a; Bekatorou et al. 2002a) or gluten pellets (Bekatorou et al. 2001a, b; Iconomopoulou et al. 2002b) was successfully used for alcoholic fermentation, winemaking, and brewing. Freeze-drying was carried out without the use of cryoprotectants showing suitability of the technique for commercial immobilized dry yeast production. Yet, freeze-drying has a relatively high investment and production cost compared to other drying methods. Therefore, the development of simple and mild processes for commercial dry yeast production was investigated, such as convective drying at low temperatures (e.g. 30–40 °C, under air stream or vacuum). Such techniques were found suitable for dry yeast production with fermentation efficiency similar to that of freeze-dried and wet-weight cells as evaluated by winemaking experiments (Tsaousi et al. 2010, 2011).

A freeze-dried biocatalyst produced by immobilization of the psychrotolerant strain *S. cerevisiae* AXAZ-1 on gluten pellets and subsequent freeze-drying was used in a multistage fixed-bed tower (MFBT) bioreactor for batch and continuous winemaking. The process resulted in higher alcohol productivity compared to fermentations carried out in a single-bed bioreactor and better operational stability even at low fermentation temperature (5 °C) and after storage for 6 months at 4 °C. The good fermentation performance, operational stability and good quality of the product as shown by both chemical analysis and sensory evaluations, indicated potential industrial application of the freeze-dried biocatalyst and the MFBT bioreactor (Sipsas et al. 2009).

19.5 Flavor of Wine, Beers, and Distillates Produced by Low-Temperature Fermentation

The production and metabolism of compounds such as esters, higher alcohols, aldehydes, vicinal diketones, fatty acids, and sulfur compounds is crucial for the flavor of fermented foods, especially alcoholic beverages. These compounds are mainly associated with the yeast amino acid metabolism (Kourkoutas et al. 2004; Mallouchos et al. 2007). The effect of temperature on fermented food aroma is well established, and it has been shown that low-temperature processes can improve it. The combined effect of low fermentation temperature and use of immobilized cell techniques on wine and beer flavor has also been demonstrated (Bakoyianis et al. 1993; Bardi et al. 1997a; Yajima and Yokotsuka 2001; Mallouchos et al. 2003a, 2007; Kourkoutas et al. 2004, 2005; Kopsahelis et al. 2007). The greatest impact of such technologies is considered to be the improved ratio of esters to alcohols (increased ester and decreased fusel alcohol formation on total volatiles) (Mallouchos et al. 2003a; Kourkoutas et al. 2004, 2005).

Regarding wine, its distinctive aroma is the result of complex associations of varietal and fermentation volatile compounds. Generally, wines produced at low temperatures have aromas with more fruity notes due to increased synthesis or reduced conversion of esters. The use of immobilized cells in very low-temperature fermentation processes (below 10 °C) led to wines of improved aroma due to this effect (Bakoyianis et al. 1993; Jackson 1994; Bardi et al. 1997a; Yajima and Yokotsuka 2001; Mallouchos et al. 2003a, 2007; Kourkoutas et al. 2004, 2005). As described above, the use of natural food-grade supports for cell immobilization such as DCM and gluten pellets proved to be effective for low-temperature winemaking, showing significantly improved fermentation productivity and product quality compared to free cells. To facilitate commercialization of such biocatalyst, freeze-drying techniques were also evaluated for the production of ready-to-use dry formulations (Iconomopoulou et al. 2002a, b).

The production of wine using psychrotolerant yeast immobilized on food-grade materials, fully compatible with wine, such as grape skins and grape berries was also evaluated at 25–10 °C (Mallouchos et al. 2003b, c) and 6–30 °C (Tsakiris et al. 2004a), respectively. Improved fermentation performance of the immobilized biocatalysts was reported compared to free-cell systems and a positive influence on wine aroma, attributed to lower amounts of fusel alcohols which were decreased dramatically at low fermentation temperatures (below 10 °C). Pieces of various fruits like apple, pear, and quince were also proposed as carriers for yeast immobilization. Their use in batch and continuous winemaking processes also had a positive effect on wine flavor (Kourkoutas et al. 2001, 2002a, 2003a; Mallios et al. 2004), with higher ester and lower propanol-1 and isobutanol and amyl alcohol concentrations observed in the range 1–12 °C. Wine fermentations with yeast immobilized on brewer's spent grains carried out at 15–10 °C also resulted in improved productivity and operational stability, and the combination of immobilization and low fermentation temperature affected significantly the flavor of these wines. The immobilized cells produced wines with higher contents of ethyl and acetate esters and volatile fatty acids at 15 and 10 °C, whereas the opposite was observed for free cells at higher temperatures (Mallouchos et al. 2007). Finally, wines produced at low temperature (5 °C) by psychrotolerant yeast immobilized on inorganic materials such as kissiris had an improved aroma with higher ethyl acetate content and lower higher alcohols on total volatiles (Bakoyianis et al. 1992, 1993, 1997; Kourkoutas et al. 2004), which is important for the production of good-quality wine distillates.

Low-temperature fermentation of fresh cashew juice with free *S. bayannus* cells was also reported, producing good-quality wines with cashew, fruity, and sweet characters. Headspace analysis of volatiles by Osme/GC–MS identified volatiles including ethyl esters associated with fruit aromas (Garruti et al. 2006).

Beer, on the other hand, is produced by more complex biochemical and technological processes (macromolecule hydrolysis, malting, kilning, fermentation, hops addition, maturation, etc.), which affect its flavor. Carbon dioxide, ethanol, and glycerol are the major compounds that control the organoleptic effect of other minor constituents such as alcohols, esters, vicinal diketones, aldehydes, sulfur compounds, and fatty acids. Yeast (amino acid) metabolism, a key to the development of beer flavor, is affected by process temperatures and use of cell immobilization techniques; therefore, technologies based on these features as well as process conditions and strain selection have been developed to control beer flavor (Kourkoutas et al. 2004; Branyik et al. 2008). Biochemical changes occurring during the secondary fermentation (maturation) of beer, which is usually conducted at very low temperature (0 °C), also affect its flavor. One of these changes is *diacetyl rest*, i.e., the reduction in diacetyl (buttery flavor) to the flavorless derivatives acetoin and 2,3-butanediol by yeast. The use of immobilized yeast and low-temperature primary fermentation was found to produce beers with low diacetyl amounts, therefore showing potential of low-cost industrial application since maturation is a high energy-consuming process (Kourkoutas et al. 2004). Perpète and Collin (2000) showed that during alcohol-free beer production, the enzymatic reduction in *worty* flavor (caused by Strecker aldehydes) (Fig. 19.2) by brewer's yeast was improved by cold-contact fermentation. Also, as in the case of

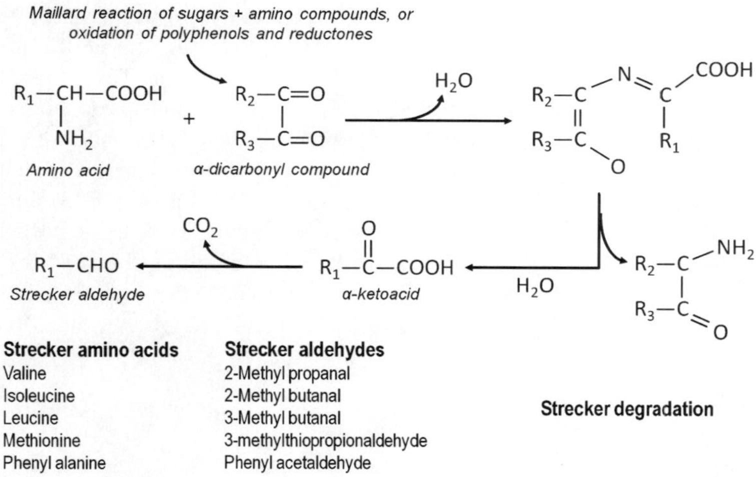

Fig. 19.2 Strecker degradation reaction of amino acids and dicarbonyl compounds responsible for *worty* (*malty*) flavors in beer

winemaking, biocatalysts prepared by immobilization of a psychrotolerant yeast strain on DCM and gluten pellets were found suitable for batch and continuous low-temperature brewing even at 0–5 °C. The produced beers had lower amounts of diacetyl and polyphenols compared to beers made using free cells. The fruity flavor of these beers was also attributed to the improved ratios of esters to alcohols (Bardi et al. 1996b; Bekatorou et al. 2001a, 2002a; Kourkoutas et al. 2004). Similar results were obtained during brewing at low and ambient temperatures (3–20 °C) by yeast immobilized on dried figs (Bekatorou et al. 2002b). The fine clarity of the produced beers and improved flavor after the end of the primary fermentation, in combination with the abundance, food-grade purity, and low cost of the proposed immobilization carriers, encouraged full-scale application for low-temperature brewing.

19.6 Conclusions and Perspectives

The isolation of psychrophilic yeasts, selection of psychrotolerant strains, or preadaptation treatments, in combination with techniques that protect cells from stress during fermentation and the development of drying techniques for commercial culture production, as discussed in this Chapter, describe an integrated approach to optimize low-temperature winemaking, brewing, and alcohol production processes. Low-temperature fermentation can improve quality and reduce maturation time, while cell immobilization can increase productivity, stress tolerance, and facilitate continuous processing of such processes. The proposed technology can also lead to the production of various new types of wines, beers, and distillates, without the need for addition of preservatives (e.g. SO_2) and possibility for direct bottling of the product after the end of the primary fermentation. This is based on the following observations for low-temperature fermentations, especially those involving immobilized cell biocatalysts: (1) the improved flavor of the products, mainly due to improved ratios of esters to alcohols, lower diacetyl content, etc.; (2) increased fermentation performance of cold-adapted yeasts; and (3) fine clarity of the products after the primary low-temperature fermentation due to precipitation of tartrates in winemaking or chill haze in brewing, avoiding the need for clarification before bottling. Application of the described technology is encouraged by more than 50 original publications discussing results of laboratory and scale-up low-temperature fermentation processes.

References

Aguilera J, Randez-Gil F, Prieto JA (2007) Cold response in *Saccharomyces cerevisiae*: new functions for old mechanisms. FEMS Microbiol Rev 31:327–341

Argiriou T, Kanellaki M, Voliotis S, Koutinas AA (1996) Kissiris-supported yeast cells: High biocatalytic stability and productivity improvement of immobilized yeast cells by successive preservations at 0°C. J Agr Food Chem 44:4028–4031

Athanasiadis I, Boskou D, Kanellaki M, Koutinas AA (1999) Low-temperature alcoholic fermentation by delignified cellulosic material supported cells of kefir yeast. J Agr Food Chem 49:658–663

Bakoyianis V, Kanellaki M, Kalliafas A, Koutinas AA (1992) Low temperature wine-making by immobilized cells on mineral kissiris. J Agr Food Chem 40:1293–1296

Bakoyianis V, Kana K, Kalliafas A, Koutinas AA (1993) Low-temperature continuous wine-making by kissiris-supported biocatalyst: volatile by-products. J Agr Food Chem 41:465–468

Bakoyianis V, Koutinas AA, Aggelopoulos K, Kanellaki M (1997) Comparative study of kissiris, γ-alumina and Ca-alginates as supports of cells for batch and continuous wine making at low temperatures. J Agr Food Chem 45:4884–4888

Bardi E, Koutinas AA (1994) Immobilization of yeast on delignified cellulosic material for room temperature and low-temperature wine making. J Agr Food Chem 42:221–226

Bardi E, Bakoyianis V, Koutinas AA, Kanellaki M (1996a) Room temperature and low temperature wine making using yeast immobilized on gluten pellets. Process Biochem 31:425–430

Bardi E, Koutinas AA, Soupioni M, Kanellaki M (1996b) Immobilization of yeast on delignified cellulosic material for low temperature brewing. J Agr Food Chem 44:463–467

Bardi E, Koutinas AA, Psarianos C, Kanellaki M (1997a) Volatile by-products formed in low-temperature wine-making using immobilized yeast cells. Process Biochem 32:579–584

Bardi E, Koutinas AA, Kanellaki M (1997b) Room and low temperature brewing by yeast immobilized on gluten pellets. Process Biochem 32:691–696

Bekatorou A, Koutinas AA, Kaliafas A, Kanellaki M (2001a) Freeze-dried *Saccharomyces cerevisiae* cells immobilized on gluten pellets for glucose fermentation. Process Biochem 36:549–557

Bekatorou A, Koutinas AA, Psarianos C, Kanellaki M (2001b) Low-temperature brewing by freeze-dried immobilized cells on gluten pellets. J Agr Food Chem 49:373–377

Bekatorou A, Soupioni MJ, Koutinas AA, Kanellaki M (2002a) Low-temperature brewing by freeze-dried immobilized cells. Appl Biochem Biotech 97:105–121

Bekatorou A, Sarellas A, Ternan NG, Mallouchos A, Komaitis M, Koutinas AA, Kanellaki M (2002b) Low-temperature brewing using yeast immobilized on dried figs. J Agr Food Chem 50:7249–7257

Belloch C, Orlic S, Barrio E, Querol A (2008) Fermentative stress adaptation of hybrids within the *Saccharomyces* sensu stricto complex. Int J Food Microbiol 122:188–195

Beltran G, Novo M, Leberre V, Sokol S, Labourdette D, Guillamon JM, Mas A, Francois J, Rozes N (2006) Integration of transcriptomic and metabolic analyses for understanding the global responses of low-temperature winemaking fermentations. FEMS Yeast Res 6:1167–1183

Bourdichon F, Casaregola S, Farrokh C, Frisvad JC, Gerds ML, Hammes WP, Harnett J, Huys G, Laulund S, Ouwehand A, Powell IB, Prajapati JB, Seto Y, Ter Schure E, Van Boven A, Vankerckhoven V, Zgoda A, Tuijtelaars S, Hansen EB (2012) Food fermentations: microorganisms with technological beneficial use. Int J Food Microbiol 154:87–97

Branyik T, Vicente AA, Dostalek P, Teixeira JA (2008) A review of flavour formation in continuous beer fermentations. J Inst Brewing 114:3–13

Buzzini P, Branda E, Goretti M, Turchetti B (2012) Psychrophilic yeasts from worldwide glacial habitats: diversity, adaptation strategies and biotechnological potential. FEMS Microbiol Ecol 82:217–241

Cadière A, Aguera E, Caillé S, Ortiz-Julien A, Dequin S (2012) Pilot-scale evaluation the enological traits of a novel, aromatic wine yeast strain obtained by adaptive evolution. Food Microbiol 32:332–337

Cardona F, Carrasco P, Pérez-Ortín JE, del Olmo ML, Aranda A (2007) A novel approach for the improvement of stress resistance in wine yeasts. Int J Food Microbiol 114:83–91

Caridi A, Corte V (1997) Inhibition of malolactic fermentation by cryotolerant yeasts. Biotechnol Lett 19:723–726

Castellari L, Magrini P, Passarelli P, Zambonelli C (1995) Effect of must fermentation temperature on minor products formed by cryo and non-cryotolerant *Saccharomyces cerevisiae* strains. Ital J Food Sci 2:125–132

D'Amico S, Collins T, Marx JC, Feller G, Gerday C (2006) Psychrophilic microorganisms: challenges for life. EMBO Rep 7:385–389

Domizio P, Lencioni L, Ciani M, Di Blasi S, Pontremolesi C, Sabatelli MP (2007) Spontaneous and inoculated yeast populations dynamics and their effect on organoleptic characters of Vinsanto wine under different process conditions. Int J Food Microbiol 115:281–289

Donalies UE, Nguyen HT, Stahl U, Nevoigt E (2008) Improvement of *Saccharomyces* yeast strains used in brewing, wine making and baking. Adv Biochem Eng Biotechnol 111:67–98

Eustace R, Thornton RJ (1987) Selective hybridization of wine yeasts for higher yields of glycerol. Can J Microbiol 33:112–117

Garruti DS, de Abreu FAP, Franco MRB, Aparecida MAP (2006) The influence of fermentation temperature and sulfur dioxide on the volatile composition and flavour profile of cashew wine. Dev Food Sci 43:109–112

Giudici P, Zambonelli C, Passarelli P, Castellari L (1995) Improvement of wine composition with cryotolerant *Saccharomyces* strains. Am J Enol Viticult 46:143–147

Grant CM (2001) Role of the glutathione/glutaredoxin and thioredoxin systems in yeast growth and response to stress conditions. Mol Microbiol 39:533–541

Guerzoni ME, Marchetti R, Giudici P (1985) Modifications of aroma components of wines obtained by fermentation with *Saccharomyces cerevisiae* mutants. B OIV 58:228–234

Hara S, Iimura Y, Oyama H, Kozeki T, Kitano K, Otsuka KI (1981) The breeding of cryophilic killer wine yeasts. Agr Biol Chem 45:1327–1334

Herrero O, Ramón D, Orejas M (2008) Engineering the *Saccharomyces cerevisiae* isoprenoid pathway for de novo production of aromatic monoterpenes in wine. Metab Eng 10:78–86

Iconomopoulou M, Kanellaki M, Psarianos C, Koutinas AA (2000) Delignified cellulosic material supported biocatalyst as freeze-dried product in alcoholic fermentation. J Agr Food Chem 48:958–961

Iconomopoulou M, Kanellaki M, Soupioni M, Koutinas AA (2002a) Freeze dried immobilized cells on delignified cellulosic material in low-temperature and ambient temperature wine making. Appl Biochem Biotechnol 104:23–36

Iconomopoulou M, Psarianos K, Kanellaki M, Koutinas AA (2002b) Low temperature and ambient temperature wine making using freeze dried immobilized cells on gluten pellets. Process Biochem 37:707–717

Jackson RS (1994) Wine science, Principles and applications, 1st edn. Academic Press INC, London

Jiménez-Martí E, Zuzuarregui A, Ridaura I, Lozano N, del Olmo M (2009) Genetic manipulation of HSP26 and YHR087 W stress genes may improve fermentative behaviour in wine yeasts under vinification conditions. Int J Food Microbiol 130:122–130

Kana K, Kanellaki M, Papadimitriou A, Psarianos C, Koutinas AA (1989a) Immobilization of *Saccharomyces cerevisiae* on γ-Alumina pellets and its ethanol production in glucose and raisin extract fermentation. J Ferment Bioeng 68:213–215

Kana K, Kanellaki M, Psarianos C, Koutinas AA (1989b) Ethanol production by *Saccharomyces cerevisiae* immobilized on mineral kissiris. J Ferment Bioeng 68:144–147

Kandylis P, Koutinas AA (2008) Extremely low temperature fermentations of grape must by potatoes supported yeast-strain AXAZ-1. A contribution is performed to catalysis of alcoholic fermentation. J Agr Food Chem 56:3317–3327

Kandylis P, Goula A, Koutinas AA (2008) Corn starch gel for yeast cell entrapment. A view for catalysis of wine fermentation. J Agr Food Chem 56:12037–12045

Kandylis P, Manousi ME, Bekatorou A, Koutinas AA (2010a) Freeze-dried wheat supported biocatalyst for low temperature wine making. LWT-Food Sci Technol 43:1485–1493

Kandylis P, Drouza C, Bekatorou A, Koutinas AA (2010b) Scale-up of extremely low temperature fermentations of grape must by wheat supported yeast cells. Biores Technol 101:7484–7491

Kandylis P, Dimitrellou D, Koutinas AA (2012a) Winemaking by barley supported yeast cells. Food Chem 130:425–431

Kandylis P, Mantzari A, Koutinas AA, Kookos IK (2012b) Modelling of low temperature wine-making, using immobilized cells. Food Chem 133:1341–1348

Kanellaki M, Koutinas AA (1999) Low temperature fermentation by cold-adapted and immobilized yeast cells. In: Margesin R, Schinner F (eds) Biotechnological applications of cold-adapted organisms. Springer, Berlin, pp 118–145

Kim SR, Park YC, Jin YS, Seo JH (2013) Strain engineering of *Saccharomyces cerevisiae* for enhanced xylose metabolism. Biotechnol Adv. doi:10.1016/j.biotechadv.2013.03.004

Kishimoto M, Shinohara T, Soma E, Goto S (1993) Selection and fermentation properties of cryophilic wine yeasts. J Ferment Bioeng 75:451–453

Kondo K, Inouye M (1991) TIP1, a cold shock-inducible gene of *Saccharomyces cerevisiae*. J Biol Chem 266:17537–17544

Kopsahelis N, Kanellaki M, Bekatorou A (2007) Low temperature brewing using cells immobilized on brewer's spent grains. Food Chem 104:480–488

Kopsahelis N, Nisiotou A, Kourkoutas Y, Panas P, Nychas GJE, Kanellaki M (2009) Molecular characterization and molasses fermentation performance of a wild yeast strain operating in an extremely wide temperature range. Biores Technol 100:4854–4862

Kourkoutas Y, Komaitis M, Koutinas AA, Kanellaki M (2001) Wine production using yeast immobilized on apple pieces at low and room temperatures. J Agr Food Chem 49:1417–1425

Kourkoutas Y, Koutinas AA, Kanellaki M, Banat IM, Marchant R (2002a) Continuous wine fermentation using a psychrophilic yeast immobilized on apple cuts at different temperatures. Food Microbiol 19:127–134

Kourkoutas Y, Douma M, Koutinas AA, Kanellaki M, Banat IM, Marchant R (2002b) Continuous winemaking fermentation using quince-immobilized yeast at room and low temperatures. Process Biochem 39:143–148

Kourkoutas Y, Komaitis M, Koutinas AA, Kaliafas A, Kanellaki M, Marchant R, Banat IM (2003a) Wine production using yeast immobilized on quince biocatalyst at temperatures between 30 and 0°C. Food Chem 82:353–360

Kourkoutas Y, Kanellaki M, Koutinas AA, Banat IM, Marchant R (2003b) Storage of immobilized yeast cells for use in wine-making at ambient temperature. J Agr Food Chem 51:654–658

Kourkoutas Y, Bekatorou A, Banat IM, Marchant R, Koutinas AA (2004) Immobilization technologies and support materials suitable in alcohol beverages production: a review. Food Microbiol 21:377–397

Kourkoutas Y, Kanellaki M, Koutinas AA, Tzia C (2005) Effect of fermentation conditions and immobilization supports on the wine-making. J Food Eng 69:115–123

Koutinas AA, Athanasiadis I, Bekatorou A, Psarianos C, Kanellaki M, Agouridis N, Blekas G (2007) Kefir-yeast technology: Industrial scale-up of alcoholic fermentation of whey, promoted by raisin extracts, using kefir-yeast granular biomass. Enzyme Microb Tech 41:576–582

Kowalski LR, Kondo K, Inouye M (1995) Cold-shock induction of a family of TIP1-related proteins associated with the membrane in *Saccharomyces cerevisiae*. Mol Microbiol 15:341–353

Lee S, Villa K, Patino H (1995) Yeast strain development for enhanced production of desirable alcohols/esters in beer. J Am Soc Brew Chem 53:153–156

Llaurado JM, Rozes N, Constanti M, Mas A (2005) Study of some *Saccharomyces cerevisiae* strains for winemaking after preadaptation at low temperatures. J Agr Food Chem 53:1003–1011

López-Malo M, Chiva R, Rozes N, Guillamon JM (2013) Phenotypic analysis of mutant and overexpressing strains of lipid metabolism genes in *Saccharomyces cerevisiae*: Implication in growth at low temperatures. Int J Food Microbiol 162:26–36

Loukatos P, Kiaris M, Ligas I, Bourgos G, Kanellaki M, Komaitis M, Koutinas AA (2000) Continuous wine making by γ-alumina-supported biocatalyst. Appl Biochem Biotechnol 89:1–13

Loukatos P, Kanellaki M, Komaitis M, Athanasiadis I, Koutinas AA (2003) A new technological approach proposed for distillate production using immobilized cells. J Biosci Bioeng 95:35–39

Mallios P, Kourkoutas Y, Iconomopoulou M, Koutinas AA, Psarianos C, Marchant R, Banat IM (2004) Low temperature wine making using yeast immobilized on pear pieces. J Sci Food Agr 84:1615–1623

Mallouchos A, Komaitis M, Koutinas AA, Kanellaki M (2003a) Wine fermentations by immobilized and free cells at different temperatures. Effect of immobilization and temperature on volatile by-products. Food Chem 80:109–113

Mallouchos A, Komaitis M, Koutinas AA, Kanellaki M (2003b) Evolution of volatile by products during wine fermentations using immobilized cells on grape skins. J Agr Food Chem 51:2402–2408

Mallouchos A, Skandamis P, Loukatos P, Komaitis M, Koutinas AA, Kanellaki M (2003c) Volatile compounds of wines produced by cells immobilized on grape skins. J Agr Food Chem 51:3060–3066

Mallouchos A, Loukatos P, Bekatorou A, Koutinas AA, Komaitis M (2007) Ambient and low temperature winemaking by immobilized cells on brewer's spent grains: Effect on volatile composition. Food Chem 104:918–927

Masschelein CA, Ryder DS, Simon JP (1994) Immobilized cell technology in beer production. Crit Rev Biotechnol 14:155–177

Mattanovich D, Gasser B, Hohenblum H, Sauer M (2004) Stress in recombinant protein producing yeasts. J Biotechnol 113:121–135

Murata Y, Homma T, Kitagawa E, Momose Y, Sato MS, Odani M, Shimizu H, Hasegawa-Mizusawa M, Matsumoto R, Mizukami S, Fujita K, Parveen M, Komatsu Y, Iwahashi H

(2006) Genome-wide expression analysis of yeast response during exposure to 4 degrees C. Extremophiles 10:117–128

Navarrete-Bolanos JL (2012) Improving traditional fermented beverages: How to evolve from spontaneous to directed fermentation. Eng Life Sci 12:410–418

Okuyama H, Morita N, Yumoto I (1999) Cold-adapted microorganisms for use in food biotechnology. In: Margesin R, Schinner F (eds) Biotechnological applications of cold-adapted organisms. Springer, Berlin, pp 101–115

Parapouli M, Hatziloukas E, Drainas C, Perisynakis A (2010) The effect of Debina grapevine indigenous yeast strains of *Metschnikowia* and *Saccharomyces* on wine flavour. J Ind Microbiol Biotechnol 37:85–93

Penacho V, Valero E, Gonzalez R (2012) Transcription profiling of sparkling wine second fermentation. Int J Food Microbiol 153:176–182

Pérez-Través L, Lopes CA, Barrio E, Querol A (2012) Evaluation of different genetic procedures for the generation of artificial hybrids in *Saccharomyces* genus for winemaking. Int J Food Microbiol 156:102–111

Perpète P, Collin S (2000) How to improve the enzymatic worty flavour reduction in a cold contact fermentation. Food Chem 70:457–462

Pizarro FJ, Jewett MC, Nielsen J, Agosin E (2008) Growth temperature exerts differential physiological and transcriptional responses in laboratory and wine strains of *Saccharomyces cerevisiae*. Appl Environ Microbiol 74:6358–6368

Plessas S, Bekatorou A, Koutinas AA, Soupioni M, Banat IM, Marchant R (2007) Use of *Saccharomyces cerevisiae* cells immobilized on orange peel as biocatalyst for alcoholic fermentation. Biores Technol 98:860–865

Pretorius IS, Bauer FF (2002) Meeting the consumer challenge through genetically customized wine-yeast strains. Trends Biotechnol 20:426–432

Redon M, Guillamon JM, Mas A, Rozes N (2011) Effect of growth temperature on yeast lipid composition and alcoholic fermentation at low temperature. Eur Food Res Technol 232:517–527

Saerens SMG, Duong CT, Nevoigt E (2010) Genetic improvement of brewer's yeast: current state, perspectives and limits. Appl Microbiol Biotechnol 86:1195–1212

Sahara T, Goda T, Ohgiya S (2002) Comprehensive expression analysis of time-dependent genetic responses in yeast cells to low temperature. J Biol Chem 277:50015–50021

Salvado Z, Arroyo-Lopez FN, Barrio E, Querol A, Guillamon JM (2011) Quantifying the individual effects of ethanol and temperature on the fitness advantage of *Saccharomyces cerevisiae*. Food Microbiol 28:1155–1161

Salvado Z, Chiva R, Roze N, Cordero-Otero R, Guillamo JM (2012) Functional analysis to identify genes in wine yeast adaptation to low-temperature fermentation. J Appl Microbiol 113:76–88

Sanchez RG, Solodovnikova N, Wendland J (2012) Breeding of lager yeast with *Saccharomyces cerevisiae* improves stress resistance and fermentation performance. Yeast 29:343–355

Sato M, Kishimoto M, Watari J, Takashio M (2002) Breeding of brewer's yeast by hybridization between a top-fermenting yeast *Saccharomyces cerevisiae* and a cryophilic yeast *Saccharomyces bayanus*. J Biosci Bioeng 93:509–511

Schade B, Jansen G, Whiteway M, Entian KD, Thomas DY (2004) Cold adaptation in budding yeast. Mol Biol Cell 15:5492–5502

Sipsas V, Kolokythas G, Kourkoutas Y, Plessas S, Nedovic VA, Kanellaki M (2009) Comparative study of batch and continuous Multi-Stage Fixed-Bed Tower (MFBT) bioreactor during wine-making using freeze-dried immobilized cells. J Food Eng 90:495–503

Steinkraus KH (1997) Classification of fermented foods: worldwide review of household fermentation techniques. Food Control 8:311–317

Thornton RJ (1982) Selective hybridization of pure culture wine yeasts II Improvement of fermentation efficiency and inheritance of SO_2 tolerance. Eur J Appl Microbiol Biotechnol 14:159–164

Torija MJ, Beltran G, Novo M, Poblet M, Guillamon JM, Mas A, Rozes N (2003) Effects of fermentation temperature and *Saccharomyces* species on the cell fatty acid composition and presence of volatile compounds in wine. Int J Food Microbiol 85:127–136

Torresi S, Frangipane MT, Anelli G (2011) Biotechnologies in sparkling wine production. Interesting approaches for quality improvement: A review. Food Chem 129:1232–1241

Tsakiris A, Sipsas V, Bekatorou A, Mallouchos A, Koutinas AA (2004a) Red wine making by immobilized cells and influence on volatile composition. J Agr Food Chem 52:1357–1363

Tsakiris A, Bekatorou A, Koutinas AA, Marchant R, Banat IM (2004b) Immobilization of yeast on dried raisin berries for use in dry white wine making. Food Chem 87:11–15

Tsakiris A, Kandylis P, Bekatorou A, Kourkoutas Y, Koutinas AA (2010) Dry red wine making using yeast immobilized on cork pieces. Appl Biochem Biotechnol 162:1316–1326

Tsaousi K, Koutinas AA, Bekatorou A, Loukatos P (2010) Fermentation efficiency of cells immobilized on delignified brewers' spent grains after low and high-temperature thin layer thermal drying. Appl Biochem Biotechnol 162:594–606

Tsaousi K, Velli A, Akarepis F, Bosnea L, Drouza C, Koutinas AA, Bekatorou A (2011) Low-temperature winemaking by thermally dried immobilized yeast on delignified brewer's spent grains. Food Technol Biotechnol 49:379–384

Tsoutsas T, Kanellaki M, Psarianos C, Kaliafas A, Koutinas AA (1990) Kissiris: A mineral support for the promotion of ethanol fermentation by *Saccharomyces cerevisiae*. J Ferment Bioeng 69:93–97

Vlieg JETV, Veiga P, Zhang CH, Derrien M, Zhao LP (2011) Impact of microbial transformation of food on health - from fermented foods to fermentation in the gastro-intestinal tract. Current Opin Biotechnol 22:211–219

Watari J, Takata Y, Ogawa M, Nishikawa N, Kaminura M (1989) Molecular cloning of a flocculation gene in *Saccharomyces cerevisiae* plasmid DNA purification from *E. coli* culture. Agr Biol Chem 53:901–903

Yajima M, Yokotsuka K (2001) Volatile compound formation in white wines fermented using immobilized and free yeast. Am J Enol Viticult 52:210–218

Chapter 20
Cold-Adapted Yeasts as Biocontrol Agents: Biodiversity, Adaptation Strategies and Biocontrol Potential

Marcela P. Sangorrín, Chistian Ariel Lopes, Silvana Vero and Michael Wisniewski

Abstract After harvest, many fruits are kept in cold storage in order to prolong their availability and shelf life. Often, this requires the application of a chemical fungicide to prevent postharvest decay from decay fungi. An alternative approach for preventing postharvest fungal decay during storage could be based on the treatment of the commodity with antagonistic yeasts. In this regard, the use of cold-adapted yeasts may offer a distinct advantage. Numerous cold-adapted yeast species have been isolated from artificial cold environments, as well as cold-stored fruits. Since the method employed to isolate potential antagonists has a major impact on the type and properties of the antagonist to be identified, it is important to evaluate the consequences of the methods that are presently being utilized and to appraise whether or not they can be improved. Although the mechanism(s) by which yeast antagonists suppress postharvest diseases can be quite variable, competition for nutrients and space plays a major role in their antagonistic activity. Additionally, production of antibiotics, direct parasitism and the induction of resistance in the harvested commodity are other modes of action that have been documented and suggested to play a role in how yeasts suppress postharvest pathogens in harvested fruits. While a few yeast-based products have been on the market, this field of study is still in its infancy and it is likely that several new

M. P. Sangorrín (✉) · C. A. Lopes
Grupo de Biodiversidad y Biotecnología de Levaduras, Instituto Multidisciplinario de Investigación y Desarrollo de la Patagonia Norte (IDEPA), (CONICET-UNCo), Buenos Aires 1440, Neuquén, Argentina
e-mail: sangorrinmarcela@conicet.gov.ar

C. A. Lopes
Facultad de Ciencias Agrarias UNCo, Ruta 151, Km 12,5, Cinco Saltos, Argentina

S. Vero
Departamento de Biociencias, Facultad de Química, Cátedra de Microbiología, Gral. Flores 2124, UdelaR, Uruguay

M. Wisniewski
US Department of Agriculture, Agricultural Research Service, 2217 Wiltshire Road, Kearneysville, WV 25430, USA

P. Buzzini and R. Margesin (eds.), *Cold-Adapted Yeasts*,
DOI: 10.1007/978-3-642-39681-6_20,

products will enter the market in near future. Nonetheless, it is necessary to continue to identify new potential microorganisms and to develop a better understanding of the biology of yeast biocontrol systems by involving yeasts, pathogens and host commodity, in order to increase the potential of postharvest biocontrol as a viable alternative to synthetic postharvest fungicides. While the results of this technology are encouraging, we need to continue to explore the potential use of appropriate yeasts worldwide where management practices, types of fruit and decay pathogens can vary considerably.

Keywords Antagonist yeasts · Fruit · Postharvest diseases

Contents

20.1 Introduction 442
20.2 Yeasts for Biological Control of Postharvest Diseases of Fruits 443
20.3 Origin of Biocontrol Yeasts 444
20.4 Basic Strategies for Isolation and Selection of Postharvest BCAs 445
20.5 Improvement in the Strategies Used for Isolation and Selection of Yeasts BCAs ... 447
20.6 Mode of Action of Antagonistic Yeasts 452
20.7 Considerations in Developing a Biocontrol Program 455
20.8 Conclusions 457
References 458

20.1 Introduction

Postharvest losses of stored fruits, vegetables and grains due to decay by fungal pathogens can be very significant, and in fact, addressing this problem is receiving increasing importance as worldwide demand for food increases (Wilson 2013). Postharvest decay of fruits can be reduced by avoiding injury to the fruit during harvest and subsequent handling, stringent sanitation practices, the use of chemical fungicides during cold and/or modified atmosphere storage (Lennox and Spotts 2003; Zhang et al. 2005). These beneficial practices, however, are usually not enough to sufficiently protect harvested fruits from spoilage, caused mainly by several fungal species belonging to *Alternaria, Aspergillus, Botrytis, Fusarium, Geotrichum, Gloeosporium, Mucor, Monilinia, Penicillium, Rhizopus* and other genera. The need for alternatives to the use of chemical fungicides that are safe and effective has become increasingly apparent due to (1) the increasing health and environmental concerns over pesticide disposal and residue levels on fresh commodities; (2) the development of fungicide-resistant strains of postharvest spoilage fungi; and (3) the deregistration of some of the more effective fungicides. Among the alternative approaches that have been explored, the use of microbial antagonists like yeasts, fungi and bacteria has been demonstrated to be quite promising and gained increased attention (Nunes et al. 2001; Scherm et al. 2003; Wisniewski et al. 2007; Droby et al. 2009; Sharma et al. 2009; Teixido et al. 2011). The majority of microbial organisms studied for their potential use as postharvest biocontrol agents belongs to yeasts (Droby et al. 2009; Pimenta et al. 2009; Jijakli 2011; Schisler et al. 2011) due to the fact that many of these microorganisms are considered as generally recognized as safe (GRAS). This chapter reviews the use of yeasts as biological control agents (BCAs) for managing postharvest diseases of fruits during cold storage and discusses the potential of cold-adapted yeasts for this purpose.

20.2 Yeasts for Biological Control of Postharvest Diseases of Fruits

Biological control of postharvest diseases (BCPD), as an aspect of biological control, is relatively young and has been investigated for approximately three decades. Progress in BCPD, especially in the postharvest application of antagonists, may be attributed to the uniqueness and relative simplicity of the postharvest system. Many, if not most, postharvest diseases are caused by necrotrophic pathogens that require a wound site to enter the host, germinate and establish an infection. In the case of biocontrol, antagonists can be applied directly to the targeted area (fruit wounds), and a single application using existing delivery systems (drenches, line sprayers, on-line dips) can significantly reduce fruit decays. Once harvested and processed, fruits are often placed in cold storage for varying periods of time, ranging from a few days to months, depending on the commodity. Environmental conditions, such as temperature and humidity, can be managed to favour antagonist survival. Furthermore, biotic interference is minimal; thus, antagonists encounter little competition from indigenous microorganisms (Janisiewicz and Korsten 2002).

Fruit wounds are rich in nutrients and other germination cues, so the process of germination by fungal spores and the establishment of an infection occur quickly (usually within 24 h). Colonization and growth of a BCA in the wound site must also be very rapid and is thus a key characteristic of a good antagonist (Barkai-Goland 2001). In order to obtain timely yeast colonization of wounds, BCAs can be applied to fruit after harvest in various ways, including drenching entire bins of fruit before placing them in storage, sprays or dips on packing lines, or they can also be applied as part of a wax treatment. However, maintaining good sanitation in packinghouses to reduce fungal inoculum levels is essential in the use of postharvest antagonists or any other control measure.

As recently emphasized by Buzzini et al. (2012), yeasts are a versatile group of eukaryotic microorganisms exhibiting heterogeneous nutritional profiles and a surprising ability to survive in a wide range of natural and artificial habitats (Shivaji and Prasad 2009; Starmer and Lachance 2011). It has also been suggested that yeasts may be better adapted to low temperatures than bacteria (Margesin et al. 2003; Turkiewicz et al. 2003). The ability of yeasts to utilize a plethora of organic compounds as an energy source greatly expands the range of ecological niches that they may colonize well beyond the sugar-rich substrates, such as fermenting fruit juices, with which they are commonly associated. Several studies, as reviewed in Phaff and Starmer (1987) and Fonseca and Inácio (2006), have reported the occurrence of significant shifts in the composition of yeast species present on damaged or fermenting fruits (dominated by ascomycetous and/or fermenting yeasts) compared to those found on intact fruits (generally dominated by basidiomycetous yeasts, exhibiting phenotypes strictly aerobic and generally oligotrophic). Relevant to the use of yeasts as BCAs is the distinctive species composition of epiphytic yeasts isolated from fruit in cold storage, which includes species of yeasts that are also members of the yeast communities found in permanently cold natural environments (e.g. glacial habitats, icy seas, cold deserts and frozen ground) (Buzzini et al. 2012). Overall, approximately 122 yeast species have been identified from worldwide glacial habitats: 13 % ascomycetes and 87 % basidiomycetes. These observations clearly indicate a greater level of adaptation, such as the formation of polysaccharide capsules and changes in the composition of membrane lipids, of basidiomycetous yeasts compared to ascomycetous yeasts to the selective pressure typical of cold ecosystems (Rossi et al. 2009; Shivaji and Prasad 2009). These adaptations would presumably favour the successful use of cold-adapted (psychrophilic and psychrotolerant) yeasts for the BCPD of fruits under cold storage conditions.

20.3 Origin of Biocontrol Yeasts

In general, three different sources have been used in order to obtain potential yeast antagonists: (1) yeasts already present on the commodity of the fruit itself can be specifically supported and managed without isolation; (2) yeasts isolated from fruit that are then evaluated for antagonistic activity, selected and grown in culture and then artificially applied to the fruit in high numbers; and (3) yeasts isolated from an environment other than a fruit surface (Sharma et al. 2009).

In the first approach, significant information on the microbial ecology of the fruit surface and the effect of various management practices on that ecology is needed in order to effectively manipulate the epiphytic microbial community structure (species composition of populations) in a manner that will decrease or prevent decay (Barkai-Goland 2001). Preharvest pesticide applications, as well as various postharvest treatments, such as fungicide and wax sprays, washes and dips, can greatly affect (qualitatively and quantitatively) the resident microbial population. Therefore, this option is fairly complicated because predictive models, as constructed for integrated pest management systems, would be required to guide management practices.

In the second approach, potential antagonistic yeasts are isolated and evaluated from microbial communities obtained from the surfaces of fruits. In general, these isolates are considered to be naturally occurring. Several species of naturally occurring yeast antagonists have been identified and artificially introduced on a variety of harvested commodities for BCPD, including citrus, pome and stone fruits. The number and hence overall representation of microorganisms isolated from fruit surfaces are strongly dependent on the isolation methods used. Various yeast species, such as *Aureobasidum pullulans, Candida sake*, *Candida oleophila* (now *Yarrowia lipolytica*), *Cryptococcus albidus*, *Cryptococcus laurentii*, *Debaryomyces hansenii*, *Wickerhamomyces anomalus* and *Meyerozyma guilliermondii* (formerly *Pichia anomala* and *Pichia guilliermondii*, respectively) and *Rhodotorula glutinis,* have been isolated and evaluated for their potential as BCAs of postharvest fungal diseases of the same fruit from which they have been isolated, as well as from other fruit commodities (Chalutz and Wilson 1990; Viñas et al. 1998; Janisiewicz et al. 2001; Spadaro and Gullino 2004; Zhang et al. 2005; Vero et al. 2009; Jijakli 2011; Robiglio et al. 2011).

In the third approach, yeasts are isolated from a source other than fruit surfaces and then evaluated as BCAs. Keeping in mind the cold environment of fruit storage facilities, Vero et al. (2013) examined the potential of cold-adapted yeasts isolated from Antarctic soils to manage postharvest diseases of fruit. Their isolation and selection criteria were partially based on their ability to grow in apple juice at low temperatures. Among the species recovered, *Leucosporidium scottii* was identified as a good BCA for blue and grey mould of two apple cultivars. Using different criteria, *Rhodosporidium paludigenum,* obtained from cold seawater, was evaluated as a postharvest BCA and demonstrated notable biocontrol activity against *Alternaria* rot on cherry tomatoes and Chinese winter jujube (Wang et al. 2009).

20.4 Basic Strategies for Isolation and Selection of Postharvest BCAs

Isolation and screening are the first crucial steps in the development of a BCA. The approach used in these steps will have a major impact on the species of antagonists isolated and their mode of action; to date, only BCAs from a very narrow range of yeast genera have been isolated (mainly *Aureobasidium, Candida, Cryptococcus, Pichia* and *Rhodotorula*) since the methods used for isolation and screening by most researches are very similar (Chand Goyal and Spotts 1996; Viñas et al. 1998). To overcome this shortcoming, Droby et al. (2009) suggested the use of a variety of new screening procedures in order to increase the range of yeast species examined for their biocontrol efficacy under commercial conditions.

A typical selection strategy for the selection of postharvest biocontrol agents is composed of the following steps: (1) isolation of potential BCAs from washings of fruits collected in the orchard using a general culture medium at room temperature; (2) in vitro screening of potential candidates for antagonistic activity; and (3) an *in situ* assay for biocontrol activity in inoculated fruit wounds using the pathogens to be controlled (Janisiewicz and Roitman 1988; Droby et al. 1999; Bleve et al. 2006; Zhang et al. 2010). A potential problem with this general approach is the fact that both yeast isolation and bioassays are generally performed at room temperature (Viñas et al. 1998; Lima et al. 1998; Droby et al. 1999; Zhang et al. 2005, 2010). Yeasts isolated and selected by using these strategies may not survive and hence efficiently colonize fruit wounds at the low temperature at which fruit is placed during storage.

Baker and Cook (1974) proposed that the best strategy for the isolation of potential antagonists against a particular etiological agent is to look in places where a disease caused by the pathogen can be expected but does not occur. Wilson et al. (1993) incorporated this idea in their approach for both the isolation and screening of yeast antagonists. Their method utilizes the application of fruit washings to fruit wounds, which are then challenged with the fungal spores of a specific pathogen. After a few days of incubation, microorganisms are isolated from the non-infected wounds and streaked on culture plates to obtain pure isolates, which are subsequently screened individually for biocontrol activity. Yeasts have been the predominant microorganisms obtained under these conditions. These organisms may or may not be adapted to fruit packinghouse processing and storage conditions (various postharvest treatments and low temperatures).

The strategy proposed by Wilson et al. (1993) and employed by numerous researchers worldwide enables the rapid selection of potential antagonists with a minimal cost of time and money. The approach particularly favours the selection of fast-growing antagonists capable of efficiently colonizing fruit wounds. Again, however, the limitation of this innovative strategy is that both BCA isolation and bioassays are performed at room temperature (Wisniewski et al. 1995; Chand Goyal and Spotts 1996; Lima et al. 1997, 1998; Viñas et al. 1998; Scherm et al. 2003; Sugar and Basile 2008; Zhang et al. 2011). This may partially explain the lack of correlation generally found between the biocontrol efficacy achieved in laboratory tests, small pilot tests at room temperature, and the performance of the biocontrol agents/products under commercial conditions (Droby et al. 1993; Wisniewski et al. 2001).

Since most harvested fruits are placed in cold storage for varying amounts of time, one of the most important attributes to be considered in the selection of yeasts as BCAs of fruit postharvest diseases is the ability to be effective at low temperatures. In this regard, the ability of yeasts to grow at low temperatures may represent a desirable characteristic for the selection of a potential postharvest BCA. Therefore, inclusion of this characteristic in the first step of the selection process would reduce the number of isolates that need to be tested in time-consuming and labour-intensive biocontrol assays of fruit stored at low temperatures,. Finally, another important potential limitation of most BCA selection programmes is the use of a single pathogen isolate obtained from a personal or public repository. This may not reflect the "real world" in which the potential antagonist may interact with pathogens that exhibit a wide range of genetic diversity, which may impact environmental adaptation and pathogenicity.

20.5 Improvement in the Strategies Used for Isolation and Selection of Yeasts BCAs

Strategies used to obtain and evaluate potential postharvest BCAs should take into account that (1) the temperature used for postharvest storage of fruit is 1 ± 0 °C; (2) different strains belonging to the same or different yeast species exhibit different capacities to grow at low temperatures; and (3) a potential antagonist should be evaluated using regional isolates of postharvest decay fungi isolates under real conditions.

Lutz et al. (2012) proposed an improved strategy that utilizes the following steps (Fig. 20.1):

1. wounded fruits are stored for 6 months at low temperatures;
2. washings from healthy wounds are obtained from the stored fruit, and these washings, along with pathogen spores, are used to inoculate fresh wounds in fruit incubated at low temperatures;
3. use of regional isolates of pathogens selected for their aggressiveness and resistance to fungicides during BCA evaluation;
4. washings exhibiting 50 % reduction in disease incidence after 50 days at low temperature are selected for yeast isolation;
5. yeast isolation is carried out on pear juice agar plates at low temperature;
6. subsequent bioassays of individual yeast isolates are performed under cold storage conditions for several months; and
7. yeasts exhibiting more than 80 % reduction in disease incidence were selected.

During the last 30 years, a number of yeast antagonists have been identified as postharvest BCAs and demonstrated to have commercial potential as an alternative to the use of chemical fungicides for postharvest decay control (Wisniewski et al. 2007; Sharma et al. 2009; Teixido et al. 2011). More than 30 yeast species have been reported as good antagonists for BCPD of fruits stored at low temperatures (Table 20.1). Eighteen of these species have also been isolated from naturally cold environments (species in bold in Table 20.1) along with approximately 100 other psychrophilic species (Brandao et al. 2011; Buzzini et al. 2012). Based on this data, cold-adapted yeasts identified as good postharvest BCAs comprise approximately 13 % of the total species of cold-adapted yeasts that have thus far been identified. The fact that yeast species common to both artificial and natural cold environments were detected as good, potential antagonists was most likely due to the isolation method utilized to obtain the yeasts. Of the 18 cold-adapted species listed, 12 have been extensively reported as BCAs in several reviews (Wisniewski et al. 2007; Sharma et al. 2009; Teixido et al. 2011), while six (*Cryptococcus diffluens, Cryptococcus tephrensis, Cryptococcus victoriae, Cryptococcus wieringae, L. scottii* and *Rhodotorula laryngis*) have been reported after 2011 (Vero et al. 2011; Lutz et al. 2012). The isolation methods used by these authors favour the isolation of higher numbers of yeasts. For example, 81 % out of 11 species reported in Lutz et al. (2012) were identified as cold-adapted species.

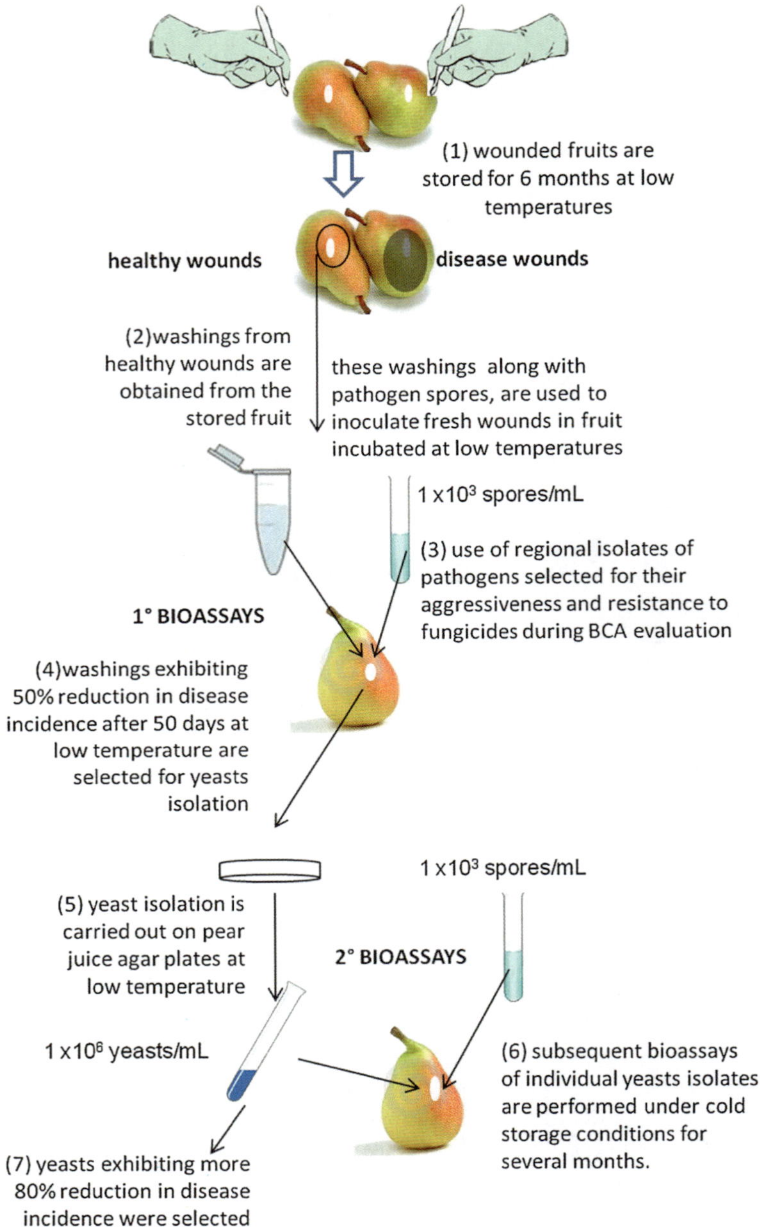

Fig. 20.1 Improved strategy for the selection of cold-adapted antagonist yeasts to control postharvest pear diseases (Lutz et al. 2012)

Table 20.1 Antagonist yeasts used for biocontrol of fruits' postharvest diseases and their relationship to cold natural habitats. Cold-adapted yeasts isolated from cold natural habitats are indicated in bold

Species (original taxonomic designation)	Isolation from fruits	References	Isolation from cold natural sources	References
Aureobasidium pullulans	Apple; citrus; pears; cherry; grapes; strawberry	Castoria et al. (2001), Vero et al. (2009), Lima et al. (1997), Robiglio et al. (2011), Lutz et al. (2012)	Antarctic soil; granitic sediments and water from glacial origin	D'Elia et al. (2009), Turchetti et al. (2008), Branda et al. (2010), Brunner et al. (2011), Brandao et al. (2011)
Candida famata (anamorph of *Debaryomyces hansenii*)	Citrus fruits	Arras et al. (1996)		
Yarrowia lipolytica (*Candida oleophila*)	Apple; banana; citrus; nectarine; papaya; peach; strawberry	Lima et al. (1997), Karabulut et al. (2003), Lahlali et al. (2004), Lassois et al. (2008)		
Candida sake	Apple and pear	Viñas et al. (1998), Torres et al. (2006), Morales et al. (2008)	Penguin-dung sample; water from glacial origin	Goto et al. (1969), Brandao et al. (2011)
Candida saitona (Torulopsis candida)	Pome fruits; citrus	Wilson et al. (1993), El-Ghaouth et al. (1998)	Antarctic soil	di Menna (1966)
Cryptococcus albidus	Apple; pear	Fan and Tian (2001), Roberts (1990), Robiglio et al. (2011), Lutz et al. (2012)	Antarctic and Arctic soil; ice cores; lake water	di Menna (1966), Goto et al. (1969), Butinar et al. (2011), Pavlova et al. (2009)
Cryptococcus diffluens	Pear	Robiglio et al. (2011)	Antarctic and Arctic soil	di Menna (1966)
Cryptococcus flavus	Pear	Roberts (1990)	Moulting feathers of penguins	Pavlova et al. (2009)
Cryptococcus laurentii	Apple; pears; cherry; citrus; jujube; tomatoes; strawberry	Roberts (1990), Vero et al. (2002), Qin et al. (2004), Xi and Tian (2005), Zhang et al. (2007), Sugar and Basile (2008)	Arctic and Antarctic soil; decaying wood; melt water stream from glaciers	Arenz et al. (2006), Pavlova et al. (2009), de García et al. (2007), Butinar et al. (2011)
Cryptococcus tephrensis	Pear	Lutz et al. (2012)	Decaying wood; sediments and water of glacial origin	Arenz et al. (2006), Branda et al. (2010), Brandao et al. (2011)
Cryptococcus victoriae	Pear	Lutz et al. (2012)	Antarctic soil; glacier and basal ice; water of glacial origin	Arenz et al. (2006), Butinar et al. (2011), Branda et al. (2010), Brandao et al. (2011)

(continued)

Table 20.1 (continued)

Species (original taxonomic designation)	Isolation from fruits	References	Isolation from cold natural sources	References
Cryptococcus wieringae	Pear	Lutz et al. (2012)	Decaying wood	Arenz et al. (2006), Branda et al. (2010)
Cystofilobasidium infirmominiatum	Lemon; pear	Chand Goyal and Spotts (1996), Sugar and Basile (2008), Vero et al. (2011), Lutz et al. (2012)	Accretion ice; water from lakes and lagoons of glacial origin	D'Elia et al. (2009), Libkind et al. (2003), Brandao et al. (2011)
Debaryomyces hansenii (teleomorph of *Candida famata*)	Lemon; pear; peach	Chalutz and Wilson (1990), Mandal et al. (2007), Hernández-Montiel et al. (2010)	Antarctic soil; moulting feathers of penguins; seawater; water from lakes of glacial origin	Arenz et al. (2006), Pavlova et al. (2009), Butinar et al. (2011), Brandao et al. (2011), de García et al. (2007)
Kloeckera apiculata	Cherry; citrus	Karabulut et al. (2003), Long et al. (2006)		
Leucosporidium scottii (Candida scotti)	Lemon	Vero et al. (2011)	Arctic and Antarctic soil; algal sample	di Menna (1966), Vero et al. (2013)
Metschnikowia andauensis	Apple; pear	Manso and Nunes (2011)		
Metschnikowia fruticola	Apple; grapes; citrus; peach; strawberry	Kurtzman and Droby (2001), Karabulut et al. (2003)		
Metschnikowia pulcherrima	Apple; grapes	Piano et al. (1997), Spadaro and Gullino (2004), Sipiczki (2006)		
Meyerozyma guilliermondii (Pichia guilliermondii)	Apple; chilli; citrus; tomato; grape	Zhao et al. (2008), Lahlali et al. (2011), Chanchaichaovivat et al. (2007), Scherm et al. (2003)	Cryopegs of permafrost; seawater; subglacial ice	Gilichinsky et al. (2005), Butinar et al. (2011)
Ogataea angusta (Pichia angusta)	Apple	Fiori et al. (2008)		
Pichia membranifaciens	Pear; nectarine; loquat	Qin et al. (2004), Cao et al. (2008), Robiglio et al. (2011), Lutz et al. (2012)		
Rhodotorula glutinis	Apple; pears; jujube; citrus cherry; strawberry	Chand Goyal and Spotts (1996), Vero et al. (2002), Qin et al. (2004), Tian et al. (2005), Lutz et al. (2012)	Antarctic soil; ice cores; lake water	di Menna (1966), Goto et al. (1969)

(continued)

Table 20.1 (continued)

Species (original taxonomic designation)	Isolation from fruits	References	Isolation from cold natural sources	References
Rhodotorula laryngis	Pears	Lutz et al. (2012)	Antarctic soil; sediments and ice from glaciers	Arenz et al. (2006), Turchetti et al. (2008), Branda et al. (2010), Vero et al. (2011)
Rhodotorula mucilaginosa *(Rhodotorula rubra)*	Apple; pear	Robiglio et al. (2011), Lutz et al. (2012), Gholamnejad et al. (2010), Li et al. (2011)	Antarctic soil; water from lakes and lagoons of glacial origin	di Menna (1966), Libkind et al. (2003), Brandao et al. (2011), de García et al. (2007) Vero et al. (2011)
Rhodotorula minuta	Mango	Patiño-Vera et al. (2005)		
Rhodosporidium paludigenum	Jujube	Wang et al. (2009)		
Sporidiobolus metaroseus *(Sporobolomyces roseus)*	Apple; grape; strawberry	Fokkema (1984), Janisiewicz et al. (1994)	Water from lakes and lagoons of glacial origin; Antarctic soils	Libkind et al. (2003), Branda et al. (2010)
Guehomyces pullulans (*Trichosporon pullulans*)	Cherry	Qin et al. (2004)		
Wickerhamomyces anomalus (Pichia anomala)	Banana; citrus	Lahlali et al. (2004), Lassois et al. (2008), Jijakli (2011)		

Recent studies have indicated that glacial habitats harbour a wide diversity of cold-adapted yeasts that evolved strategies to overcome the direct and indirect life-endangering effects of low temperatures (reviewed by Buzzini et al. 2012). Additional research will be required to determine the potential of newly identified cold-adapted yeast species as effective postharvest BCAs on fruit stored at low temperatures. Other environments, such as chilled and frozen food product, have also served as a source for cold-adapted yeasts (Vero et al. 2013).

20.6 Mode of Action of Antagonistic Yeasts

Several mechanisms of action are thought to be involved in the BCPD (Sharma et al. 2009; Jamalizadeh et al. 2011). A significant number of potential adaptive strategies have been identified to be related to the role of cold-adapted yeasts reported as BCAs, most of which are basidiomycetous yeasts (Table 20.2).

Features required for any effective BCA should include a rapid growth rate in fruit wounds, the effective utilization of the nutrients present in the wound, and the capability to survive and develop at the infection site better than the pathogen. And to do so all, this should occur at a low temperature, acidic pH and conditions of osmotic stress (Droby et al. 1996). Mechanisms other than those associated with the ability of an antagonist to colonize wounds have also been reported. These mechanisms, generally based on ecological interactions, fall into the categories of mycoparasitism, antibiosis and the induction of plant defences (Saravanakumar et al. 2009; Vero et al. 2009; Zhang et al. 2011). Attributes associated with these other mechanisms include the ability of BCAs to

1. adhere to host and pathogen cell walls (Wisniewski et al. 2007);
2. secrete lytic enzymes (Jijakli and Lepoivre 1998; Grevesse et al. 2003);
3. induce host resistance mechanisms such as an oxidative burst (Ippolito et al. 2000; Yao and Tian 2005);
4. regulate population density in the infection site (Mc Guire 2000);
5. tolerate accumulation of reactive oxygen species (ROS) (Macarisin et al. 2010; Liu et al. 2011; Xu and Tian 2008); and
6. form biofilms on the inner surface of wounds (Giobbe et al. 2007).

An advantage of the use of yeast antagonists that have multiple modes of action and do not rely on antibiosis for their effectiveness is that it substantially overcomes the problem of pathogen resistance, which is a distinct problem associated with the use of synthetic chemical fungicides. In contrast to synthetic chemical fungicides that have a targeted toxicity, a successful BCA is generally equipped with a combination of mechanisms that work in concert with control disease development. For examples, Vero et al. (2011) demonstrated that *L. scottii* produced siderophores, biofilm, soluble and volatile antifungal substances that were inhibitory to apple pathogens. Similarly, several modes of action have been demonstrated for *Pichia membranifaciens* and *C. victoriae*, including lytic enzyme

Table 20.2 Adaptation strategies of cold-adapted yeast species reported as antagonist yeast used to control postharvest diseases of fruit in cold storage

Species	Adaptation strategies	References
Yeasts-like organism		
Aureobasidium pullulans	Nutritional competition Induction of host resistance Changes in composition of membrane lipids and increased biomass production at low temperature Synthesis of cold-active enzymes and killer proteins	Ippolito et al. (2000), Castoria et al. (2001), Rossi et al. (2009), Lutz et al.(2010)
Ascomycetous yeasts		
Candida sake	Synthesis of cold-active lytic enzymes	Vaz et al. (2011)
Debaryomyes hansenii	Nutritional competition Induction of host resistance Synthesis of cold-active lytic enzymes	Roberts (1990), Chalutz and Wilson (1990), Vaz et al. (2011)
Basidicomycetous yeasts		
Candida satoiana	Synthesis of cold-active lytic enzymes Induction of host resistance	El-Ghaouth et al. (1998)
Cryptococcus albidus	Synthesis of cold-active lytic enzymes and cold-active killer proteins	Brizzio et al. (2007), Lutz et al. (2013)
Cryptococcus diffluens	Synthesis of cold-active lytic enzymes	Brandao et al. (2011)
Cryptococcus laurentii	Nutritional competition Synthesis of cold-active lytic enzymes	Castoria et al. (1997), Brizzio et al. (2007)
Cryptococcus tephrensis	Synthesis of cold-active lytic enzymes	Brandao et al. (2011)
Cryptococcus victoriae	Increased amino acid incorporation at subzero temperature Synthesis of cold-active killer proteins	Amato et al. (2009), Lutz et al. (2013), Vaz et al. (2011)
Cryptococcus wieringae	Synthesis of cold-active lytic enzymes and killer proteins	Lutz et al. (2010)
Cystofilobasidium infirmominiatum	Synthesis of cold-active lytic enzymes Oxidative stress tolerance	Brizzio et al. (2007), Lutz et al. (2010), Liu et al. (2011)

(continued)

Table 20.2 (continued)

Species	Adaptation strategies	References
Leucosporidium scottii (Candida scotti)	Synthesis of cold-active lytic enzymes Nutrient competition	Margesin et al. (2003), Vaz et al. (2011), Vero et al. (2011)
Meyerozyma guilliermondii (Pichia guilliermondii)	Rapid colonization in the cold Nutritional competition Synthesis of cold-active lytic enzymes Induction of host resistance	Wisnieswski et al. (1991), Zhang et al. (2011), Lahl1ali et al. (2011), Vaz et al. (2011)
Rhodotorula glutinis	Nutritional competition Increased amino acid incorporation at subzero temperature	Castoria et al. (1997), Amato et al. (2009)
Rhodotorula laryngis	Changes in composition of membrane lipids and increased biomass production at low temperature Synthesis of cold-active lytic enzymes	Rossi et al. (2009), Lutz et al. (2010), Pathan et al. (2010), Vaz et al. (2011)
Rhodotorula mucilaginosa	Synthesis of cold-active lytic enzymes	Brizzio et al. (2007), Vaz et al. (2011)

production, biofilm formation, killer protein production and germination inhibition (Lutz et al. 2013). Although the ability of yeasts to survive and grow at low temperatures has attracted considerable attention, the mechanisms underlying this phenomenon have not yet been fully described.

Information on the modes of action associated with antagonist activity is essential for several main reasons:

1. it allows the development of more reliable procedures for the effective application of antagonists;
2. it provides a rationale for selecting more effective antagonists (Janisiewicz and Korsten 2002; Jijakli 2011);
3. it facilitates the registration of BCAs by government regulatory agencies;
4. it is useful for optimizing the formulation of preparations composed of mixed yeast cultures showing complementary mechanisms (Chanchaichaovivat et al. 2007).

Finally, identifying genes involved in each particular mechanism and how they are regulated will provide a better understanding of how to select and/or design a superior BCA. For instance, Wisniewski et al. (2005) demonstrated that overexpression of a defensin gene from peach in *Pichia pastoris* (now *Komagataella pastoris*) greatly improved its antagonistic activity against *Botrytis cinerea* in vitro. Studies utilizing gene inactivation or overexpression can provide useful information about the regulation of these genes (Massart and Jijakli 2007).

20.7 Considerations in Developing a Biocontrol Program

Research on the use of yeasts as BCAs has mainly focused on their use for managing postharvest diseases, mainly of fruit; however, this application represents only a small portion of the complete spectrum of plant disease management (Fravel 2005; Nunes 2012). Only six products based on different yeast species have been registered for postharvest use:

1. Aspire, based on *C. oleophila* (Ecogen Inc., Langhorne, USA);
2. Yield Plus, based on *C. albidus* (Lallemand, Montreal, Canada);
3. Shemer™, based on *Metschnikowia fructicola* (Bayer CropScience);
4. Candifruit, based on *Candida sake* (Sipcam-Inaagri, SA Valencia, Spain);
5. Nexy, based on *C. oleophila* (Lesaffre-Bionext, France); and
6. Boni-Protect, based on *A. pullulans* (Biofa, Münsingen, Germany) (Janisiewicz 2010; Jijakli 2011).

The scarcity of commercial products for postharvest use has been discussed by Droby et al. (2009) and Abano and Sam-Amoah (2012). Among the main problems mentioned are the small number of companies involved in the development of biological products, the small size of the postharvest market, the expense and length of time required for selection and registration, the problems associated with scaling-up production at a viable cost while ensuring efficacy, and the scepticism with which the use of biocontrol is still viewed by producers. The legal permits required for the transfer of beneficial organisms and the financial costs involved in their registration have also inhibited the widespread availability of these products in several countries (Sundh and Melin 2011). The growing interest in food free of pesticide residues as well as increased restrictions of pesticide use will increase the growth of niche markets using BCAs. The commercial acceptance of biopesticides and their reliability may also increase if the biopesticide can be complemented or combined with physical treatments, low-risk chemical compounds or natural antimicrobial compounds. Recently, an emphasis has been placed on the further development of the commercial products Shemer™ (Bayer CropScience) and Yield Plus (Lallemand). Both companies intend to develop integrated solutions directed at sustainable production based on their products as a component of the overall approach (Jijakli 2011; Nunes 2012).

It is clear that to broaden the use of biocontrol products, research has to evolve towards integrating the use of BCAs into a production system approach with a greater awareness of industry concerns (Droby et al. 2009). In this regard, the development of new commercial products will be more feasible if the selected microbial antagonists are effective against various pathogens on various commodities. For example, Shemer™ has been reported to be effective against *Rhizopus* on strawberries and sweet potatoes, against *Botrytis cinerea* on grapes and strawberries, and against *Penicillium spp* on citrus (Kurtzman and Droby 2001; Ferrari et al. 2007). Likewise, Aspire™ was registered for use on citrus, pome fruit and strawberries (Hofstein et al. 1994). *M. guilliermondii,* never developed into a commercial product, was effective on citrus, pome fruit, tomatoes and grains (Droby et al. 1996). Additionally, Vero et al. (2011) demonstrated that *Cystofilobasidium infirmominiatum* PL1, selected as a cold-adapted yeast capable of controlling blue and green mould on oranges during cold storage (5 °C), could be also used as an effective antagonist against *Penicillium expansum* and *B. cinerea* on two apple varieties stored at 0 °C. Lutz et al. (2012) used *C. albidus, P. membranifaciens* and *C. victoriae* to control *P. expansum* and *B. cinerea* on pears when the fruits were stored at 0 ± 1 °C for 100 days.

The main factors limiting commercial interest in BCAs include the cost of scaling-up BCA production and the development of a commercial formulation that can be stored for a reasonable length of time without a loss in the viability or the efficacy of the BCA (Fravel 2005; Droby et al. 2009; Rossi et al. 2009; Teixido et al. 2011). Therefore, studies on biomass production, scaling-up and downstream production processes should also be conducted in order to facilitate the development of a commercial product. High substrate costs, low biomass productivity or limited economies of scale are the problems that need to be considered and addressed. The ability to use cost-effective substrates for microbial production, based on industrial by-products or wastes, would help to reduce production costs, provided they do not impact viability and performance of the BCA (Wilson and Wisniewski 1989). In this regard, sugar cane molasses have been used as a source for microbial growth because of its nutritional abundance and low price (Abadias et al. 2003; Wang et al. 2011; Pelinski et al. 2012).

Since the purpose of large-scale production is to obtain the greatest quantity of efficacious microorganisms in the shortest period of time, many variables have to be considered in order to obtain optimal results. Maximum growth rates can be achieved by using optimal growth temperatures. However, because production costs increase if cooling or heating is needed for production, this parameter may need to be balanced in order to address cost versus benefit issues. Another important consideration would be to select psychrotolerant rather than psychrophilic yeasts. An in-depth overview on the concept of psychrophile and psychrotolerance is reported in Chap. 1. The determination of cardinal temperatures (minimum, optimal and maximum) for growth of a specific BCA should be considered in the early stages of the process of identifying and developing a postharvest BCA. Such data are also helpful for determining the biosafety of a BCA since a maximum growth temperature lower than 35 °C can be a significant human health safety factor.

As indicated, for a biocontrol product to have commercial potential, it needs to have a wide scope of application, and perform reliably under a variety of environmental conditions, and adaptable to different packinghouse processing systems. Therefore, studies to test the antagonist on different commodities and cultivars within a commodity are essential. Furthermore, the evaluation of the BCA under different postharvest practices, such as mode of application and storage conditions (Janisiewicz and Korsten 2002; Bastiaanse et al. 2009), is also necessary. These tests also need to be conducted on a large scale under commercial conditions using fresh and formulated product. Pilot or semi-commercial utilization of large amounts of fruit should be carried out in packinghouses using commercial handling practices (Droby et al. 1998; Torres et al. 2007).

Since the method of screening has a major impact on the type and properties of the antagonist to be identified, it is important to evaluate the consequence of the methods for screening that are presently being utilized and to determine whether or not they can be improved. Psychrophilic yeasts are cold-adapted and therefore sufficiently stable at low temperatures. In order for postharvest biocontrol agents to work when fruit is placed in cold storage for long periods of time, screening methods should consider the need to maintain viability and activity at low temperatures in addition to the factors that are presently considered.

20.8 Conclusions

The increasing interest in alternatives to the use of synthetic, chemical fungicides for the control of postharvest diseases has stimulated the development of commercial biocontrol products. The development of new biocontrol products is long and costly (Droby et al. 2009). Therefore, it is of great importance to consider all the criteria needed for a BCA product to be commercially viable in the early stages of the development process. Factors such as fastidious growth and nutritional requirements, as well as questionable association with human diseases or other plant diseases, should immediately eliminate any microorganism from consideration as a BCA. The cooperation with potential industrial partners is also suggested, since the expertise of industry can be greatly beneficial to the successful development of a commercial product (Fravel 2005).

There is growing interest in identifying microorganisms from cold environments and exploring their commercial potential for a wide array of industrial processes. The potential use of cold-adapted yeasts as BCAs for the management of postharvest diseases of commodities placed in cold storage for varying periods of time has been relatively unexplored. While some of the yeast BCAs reported to be good antagonists are in fact cold-adapted, the confirmation of this trait has largely been fortuitous rather than a direct outcome of a selection process. As more and more cold-adapted yeast species are identified, their use as commercial BCAs for the management of postharvest diseases should be explored. Additional areas or exploration should also include the ability to prevent storage diseases of edible tubers, as well as non-edible species such as flower bulbs, and the cold storage of landscape plants such as bare-rooted trees. Other areas of applications include the use of cold-adapted BCAs to preserve cut flowers and the development of retail and consumer products that can be used to extend the shelf life of produce in supermarkets and home refrigerators. Since refrigeration is still the primary method of extending the availability and shelf life of commodities, it is readily apparent that the utilization of cold-adapted yeasts as BCAs has enormous potential.

References

Abadias M, Teixido N, Usall J, Viñas I (2003) Optimization of growth conditions of the postharvest biocontrol agent *Candida sake* CPA-1 in a lab-scale fermenter. J Appl Microbiol 95:301–309

Abano E, Sam-Amoah LK (2012) Application of antagonistic microorganisms for the control of postharvest decays in fruits and vegetables. IJABR 2:1–8

Amato P, Doyle S, Christner BC (2009) Macromolecular synthesis by yeasts under frozen conditions. Environ Microbiol 11:589–596

Arenz BE, Held BW, Jurgens JA, Farrell RL, Blanchette RA (2006) Fungal diversity in soils and historic wood from the Ross Sea Region of Antarctica. Soil Biol Biochem 38:3057–3064

Arras G (1996) Mode of action of an isolate of *Candida famata* in biological control of *Penicillium digitatum* in orage fruits. Postharvest Biol Technol 8:191–198

Baker KJ, Cook RJ (1974) Biological control of plant pathogens. WH Freeman, San Francisco

Barkai-Goland R (2001) Biological control. In: Barkai-Goland R (ed) Postharvest diseases of fruit, vegetables: development and control. Elsevier Sciences, Amsterdam, pp 221–255

Bastiaanse H, Bellaire LL, Lassois L, Misson C, Jijakli MH (2009) Integrated control of crown rot of banana with *Candida oleophila* strain O, calcium chloride and modified atmosphere packaging. Biol Control 53:100–107

Bleve GF, Grieco G, Cozzi A, Logrieco A, Visconti A (2006) Isolation of epiphytic yeasts with potential for biocontrol of *Aspergillus carbonarius* and *A niger* on grape. Int J Food Microbiol 108:204–209

Branda E, Turchetti B, Diolaiuti G, Pecci M, Smiraglia C, Buzzini P (2010) Yeast and yeast-like diversity in the southernmost Glacier of Europe (Calderone glacier, Apennines, Italy). FEMS Microbiol Ecol 72:354–369

Brandao LR, Libkind D, Vaz AB, Espirito Santo LC, Moliné M, de García V, van Broock M, Rosa CA (2011) Yeasts from an oligotrophic lake in Patagonia (Argentina): diversity, distribution and synthesis of photoprotective compounds and extracellular enzymes. FEMS Microbiol Ecol 76:1–13

Brizzio S, Turchetti B, de García V, Libkind D, Buzzini P, Gasparetti C, van Broock M (2007) Extracellular enzymatic activities (EEA) in basidiomycetous yeasts isolated from glacial and subglacial waters of northwest Patagonia (Argentina). Can J Microbiol 53:519–525

Brunner I, Plotze M, Rieder S, Zumsteg A, Furrer G, Frey B (2011) Pioneering fungi from the Damma glacier forefield in the Swiss Alps can promote granite weathering. Geobiology 9:266–279

Butinar L, Strmole T, Gunde-Cimerman N (2011) Relative incidence of ascomycetous yeasts in arctic coastal environments. Microb Ecol 61:832–843

Buzzini P, Branda E, Goretti M, Turchetti B (2012) Psychrophilic yeasts from worldwide glacial habitats: diversity, adaptation strategies and biotechnological potential. FEMS Microbiol Ecol 82:217–241

Cao S, Zheng Y, Tang S, Wang K (2008) Improved control of anthracnose rot in loquat fruit by a combination treatment of *Pichia membranifaciens* with $CaCl_2$. Int J Food Microbiol 126:216–220

Castoria R, de Curtis F, Lima G, de Cicco V (1997) b-1,3-glucanase activity of two saprophytic yeasts and possible mode of action as biocontrol agents against postharvest diseases. Postharvest Biol Technol 12:293–300

Castoria R, de Curtis F, Lima G, Caputo L, Pacifico S, de Cicco V (2001) *Aureobasidium pullulans* (LS-30), antagonist of postharvest pathogens of fruits: study on its mode of action. Postharvest Biol Technol 32:717–724

Chanchaichaovivat A, Ruenwongsa P, Panijpan B (2007) Screening and identification of yeast strains from fruit and vegetables: potential for biological control of postharvest chilli anthracnose (*Colletotrichum capscii*). Biol Control 42:326–335

Chand Goyal T, Spotts RA (1996) Control of postharvest pear diseases using natural saprophytic yeast colonists and their combination with a low dosage of thiabendazole. Postharvest Biol Technol 7:51–64

D'Elia T, Veerapaneni R, Theraisnathan V, Rogers SO (2009) Isolation of fungi from Lake Vostok accretion ice. Mycologia 101:751–763

de García V, Brizzio S, Libkind D, Buzzini P, van Broock M (2007) Biodiversity of cold-adapted yeasts from glacial meltwater rivers in Patagonia, Argentina. FEMS Microbiol Ecol 59:331–341

di Menna ME (1966) Yeasts in Antarctic soil. A van Leeuwenhoek 32:29–38

Droby S, Hofstein R, Wilson CL, Wisniewski M, Fridlender B, Cohen L, Weiss B, Daus A, Timar D, Chalutz E (1993) Pilot testing of *Pichia guilliermondii*: a biocontrol agent for postharvest diseases of citrus fruit. Biol Control 3:47–52

Droby S, Hofstein R, Wilson C, Wisniewski M, Fridlender B, Cohen L, Weiss B, Daus A, Chalutz E (1996) Pilot test of *Pichia guilliermondii*: A biocontrol agent of postharvest diseases of citrus fruit. Biol Control 3:47–52

Droby S, Cohen L, Daus A, Weiss B, Horev B, Chalutz E, Katz H, Keren-Tzur M, Shachnai A (1998) Commercial testing of Aspire: a yeast preparation for the biological control of postharvest decay of citrus. Biol Control 12:97–101

Droby S, Lischinski S, Cohen L, Weiss B, Daus A, Chand-Goyal T, Eckert JW, Manulis S (1999) Characterization of an epiphytic yeast population of grapefruit capable of suppression of green mold decay caused by *Penicillium digitatum*. Biol Control 16:27–34

Droby S, Wisniewski M, Macarisin D, Wilson C (2009) Twenty years of postharvest biocontrol research: Is it time for a new paradigm? Postharvest Biol Technol 52:137–145

El-Ghaouth A, Wilson CL, Wisniewski ME (1998) Ultrastructural and cytochemical aspects of biocontrol activity of *Candida saitona* in apple fruit. Phytopathology 88:282–291

Fan Q, Tian SP (2001) Postharvest biological control of grey mold and blue mold on apple by *Cryptococcus albidus* (Saito) Skinner. Postharvest Biol Technol 21:341–350

Ferrari A, Sicher C, Prodorutti D, Pertot I (2007) Potential new applications of Shemer, a *Metschnikowia fructicola* based product, in post-harvest soft fruit rots control. Bull OILB/SROP 30:43–46

Fiori S, Fadda A, Giobbe S, Berardi E, Migheli Q (2008) *Pichia angusta* is an effective biocontrol yeast against postharvest decay of apple fruit caused by *Botrytis cinerea* and *Monilia fructicola*. FEMS Yeast Res 8:961–963

Fokkema NJ (1984) Competition for endogenous and exogenous nutrients between *Sporobolomyces roseus* and *Cochliobolus sativus*. Can J Bot 162:2463–2468

Fonseca Á, Inacio J (2006) Phylloplane Yeasts. In: Rosa C, Gabor P (eds) Biodiversity and ecophysiology of yeasts. Springer, Berlin, pp 263–301

Fravel DR (2005) Commercialization and implementation of biocontrol. Annu Rev Phytopathol 43:337–359

Gholamnejad J, Etebarian HR, Sahebani N (2010) Biological control of apple blue mold with *Candida membranifaciens* and *Rhodotorula mucilaginosa*. African J Food Sci 4:1–7

Gilichinsky D, Rivkina E, Bakermans C, Shcherbakova V, Petrovskaya L, Ozerskaya S, Ivanushkina N, Kochkina G, Laurinavichius K, Pecheritsina S, Fattakhova R, Tiedje JM (2005) Biodiversity of cryopegs in permafrost. FEMS Microbiol Ecol 53:117–128

Giobbe S, Marceddu S, Scherm B, Zara G, Mazzarello V, Budroni M, Migheli Q (2007) The strange case of a biofilm-forming strain of Pichia fermentans, which controls Monilinia brown rot on apple but is pathogenic on peach fruit. FEMS Yeast Res 7:1389–1398

Goto S, Sugiyama J, Iizuka H (1969) A taxonomic study of Antarctic yeasts. Mycologia 61:748–774

Grevesse C, Lepoivre F, Jijakli MH (2003) Characterization of the exoglucanase-encoding gene PaEXG2 and study of its role in the biocontrol activity of *Pichia anomala* strain K. Phytopathology 93:1145–1152

Hernández-Montiel LG, Ochoa JL, Troyo-Diéguez E, Larralde-Corona CP (2010) Biocontrol of postharvest blue mold (*Penicillium italicum* Wehmer) on Mexican lime by marine and citrus *Debaryomyces hansenii* isolates. Postharvest Biol Technol 56:181–187

Hofstein R, Fridlender B, Chalutz E, Droby S (1994) Large scale production and pilot testing of biocontrol agents of postharvest diseases. In: Wilson CL, Wisniewski ME (eds) Biological control of postharvest diseases of fruits and vegetables - theory and practice. CRC Press, Boca Raton, pp 89–100

Ippolito A, El Ghaouth A, Wilson CL, Wisniewski M (2000) Control of postharvest decay of apple fruit by *Aureobasidium pullulans* and induction of defense responses. Postharvest Biol Technol 19:265–272

Jamalizadeh M, Etebarian HR, Aminian H, Alizadeh A (2011) A review of mechanisms of action of biological control organisms against post-harvest fruit spoilage. Bull OEPP/EPPO 41:65–71

Janisiewicz WJ (2010) Quo vadis of biological control of postharvest diseases In: Prusky D, Gullino ML (eds) Post-harvest pathology. Plant pathology in the 21st century, vol 2. Springer, Berlin, pp 137–148

Janisiewicz WJ, Korsten L (2002) Biological control of postharvest diseases of fruits. Annu Rev Phytopathol 40:411–441

Janisiewicz WJ, Roitman J (1988) Biological control of blue-mold and gray-mold on apple and pear with *Pseudomonas cepacia*. Phytopathology 78:1697–1700

Janisiewicz WJ, Petterson DL, Bors R (1994) Control of storage decay of apples with *Sporobolomyces roseus*. Plant Dis 78:466–470

Janisiewicz WJ, Tworkoski TJ, Kurtzman CP (2001) Biocontrol potential of *Metschnikowia pulcherrima* strains against blue mold of apple. Phytopathology 91:1098–1108

Jijakli HM (2011) *Pichia anomala* in biocontrol for apples: 20 years of fundamental research and practical applications. A van Leeuwenhoek 99:93–105

Jijakli HM, Lepoivre P (1998) Characterization of an exo–1, 3-glucanase produced by *Pichia anomala* strain K, antagonist of *Botrytis cinerea* on apples. Phytopathology 88:335–343

Karabulut OA, Smilanick JL, Gabler FM, Mansour M, Droby S (2003) Near-harvest applications of *Metschnikowia fructicola,* ethanol, and sodium bicarbonate to control postharvest diseases of grape in central California. Plant Dis 87:1384–1389

Kurtzman CP, Droby S (2001) *Metschnikowia fructicola,* a new ascosporic yeast with potential for biocontrol of postharvest fruit rots. Syst Appl Microbiol 24:395–399

Lahlali R, Serrhini MN, Jijakli MH (2004) Efficacy assessment of *Candida oleophila* (strain O) and *Pichia anomala* (strain K) against major postharvest diseases of citrus fruit in Morocco. Commun Agric Appl Biol Sci 69:601–609

Lahlali R, Hamadi Y, El guilli M, Jijakli MH (2011) Efficacy assessment of *Pichia guilliermondii* strain Z1, a new biocontrol agent, against citrus blue mould in Morocco under the influence of temperature and relative humidity. Biol Control 56:217–224

Lassois L, de Bellaire L, Jijakli MH (2008) Biological control of crown rot of bananas with *Pichia anomala* strain K and *Candida oleophila* strain O. Biol Control 45:410–418

Lennox CL, Spotts R (2003) Sensitivity of populations of *Botrytis cinerea* from pear-related sources to benzimidazole and dicarboximide fungicides. Plant Dis 87:645–649

Li R, Zhang H, Liu W, Zheng X (2011) Biocontrol of postharvest gray and blue mold decay of apples with *Rhodotorula mucilaginosa* and possible mechanisms of action. Int J Food Microbiol 146:151–156

Libkind D, Brizzio S, Ruffini A, Gadanho M, van Broock M, Sampaio JP (2003) Molecular characterization of carotenogenic yeasts from aquatic environments in Patagonia, Argentina. A van Leeuwenhoek 84:313–322

Lima G, Ippolito A, Nigro F, Salerno M (1997) Effectiveness of *Aureobasidium pullulans* and *Candida oleophila* against postharvest strawberry rots. Postharvest Biol Technol 10:169–178

Lima G, De Curtis F, Castoria R, De Cicco V (1998) Activity of the yeas *Cryptococcus laurentii* and *Rhodotorula glutinis* against post-harvest rots on different fruits. Biocontrol Sci Technol 8:257–267

Liu J, Michael W, Droby S, Vero S, Tian S, Hershkovitz V (2011) Glycine betaine improves oxidative stress tolerance and biocontrol efficacy of the antagonistic yeast *Cystofilobasidium infirmominiatum.* Int J Food Microbiol 146:76–83

Long CA, Deng BX, Deng XX (2006) Pilot testing of *Kloeckera apiculata* for the biological control of postharvest diseases of citrus. Annals Microbiol 56:13–17

Lutz, MC, Rodríguez ME, Sosa, MC, Lopes CA, Sangorrín MP (2010) Yeast biodiversity on the surface of cold storage pears: potential antagonists of postharvest mold. Asociación Latinoamericana de Genética (ALAG), Viña del Mar, Chile

Lutz MC, Sosa MC, Lopes CA, Sangorrín MP (2012) A new improved strategy for the selection of cold-adapted antagonist yeasts to control postharvest pear diseases. Biocontrol Sci Technol 22:1465–1483

Lutz MC, Sosa MC, Rodriguez ME, Lopes CA, Sangorrín MP (2013) Efficacy and putative mode of action of native and commercial antagonistic yeasts against postharvest pear pathogens. Int J Food Microbiol 164:166–172

Macarisin D, Droby S, Bauchan G, Wisniewski M (2010) Superoxide anion and hydrogen peroxide in the yeast antagonist–fruit interaction: a new role for reactive oxygen species in postharvest biocontrol? Postharvest Biol Technol 58:194–202

Mandal G, Singh D, Sharm RR (2007) Effect of hot water treatment and biocontrol agent (*Debaryomyces hansenii*) on shelf life of peach. Indian J Hortic 64:25–28

Manso T, Nunes C (2011) *Metschnikowia andauensis* as a new biocontrol agent of fruit postharvest diseases. Postharvest Biol Technol 61:64–71

Margesin R, Gander S, Zacke G, Gounot AM, Schinner F (2003) Hydrocarbon degradation and enzyme activities of cold-adapted bacteria and yeasts. Extremophiles 7:451–458

Massart S, Jijakli H (2007) Use of molecular techniques to elucidate the mechanisms of action of fungal biocontrol agents: a review. J Microbiol Meth 69:229–234
Mc Guire RG (2000) Population dynamics of postharvest decay antagonists growing epiphytically and within wounds on grapefruit. Phytopathology 90:1217–1223
Mercier J, Wilson CL (1994) Colonization of apple wounds by naturally-occurring microflora and introduced *Candida oleophila* and their effect on infection by *Botrytis cinerea* during storage. Biol Control 4:138–144
Morales H, Sanchis V, Usall J, Ramos AJ, Marín S (2008) Effect of biocontrol agents *Candida sake* and *Pantoea agglomerans* on *Penicillium expansum* growth and patulin accumulation in apples. Int J Food Microbiol 122:61–67
Nunes CA (2012) Biological control of postharvest diseases of fruit. Eur J Plant Pathol 133:181–196
Nunes CA, Usall J, Teixido N, Viñas I (2001) Biological control of postharvest pear diseases using a bacterium *Pantoea agglomerans* CPA-2. Int J Food Microbiol 70:53–61
Pathan AAK, Bhadra B, Begum Z, Shivaji S (2010) Diversity of yeasts from puddles in the vicinity of Midre Lovénbreen Glacier, Arctic and bioprospecting for enzymes and fatty acids. Curr Microbiol 60:307–314
Patiño-Vera M, Jimenez B, Balderas K, Ortiz M, Allende R, Carrillo A, Galindo E (2005) Pilot-scale production and liquid formulation of *Rhodotorula minuta*, a potential biocontrol agent of mango anthracnose. J Appl Microbiol 99:540–550
Pavlova K, Panchev I, Krachanova M, Gocheva M (2009) Production of an exopolysaccharide by Antarctic yeast. Folia Microbiol 54:343–348
Pelinski R, Cerrutti P, Ponsone ML, Chulze S, Galvagno M (2012) Statistical optimization of simple culture conditions to produce biomass of an ochratoxigenic mould biocontrol yeas. Lett Appl Microbiol 54:377–382
Phaff HJ, Starmer WT (1987) Yeasts associated with plants, insects and soil. In: Rose AH, Harrison JS (eds) The yeasts, biology of yeasts. Academic Press, London, pp 123–180
Piano S, Neyrotti V, Migheli Q, Gullino Ml (1997) Biocontrol capability of *Metschnikowia pulcherrima* against *Botrytis* postharvest rot of apple. Postharvest Biol Technol 11:131–140
Pimenta RS, Morais PB, Rosa CA, Corrêa AJr (2009) Utilization of yeasts in biological control programs. In: Satyanarayana T, Kunze G (eds) Yeast biotechnology: diversity and applications. Springer Science + Business Media BV, New York, pp 199–214
Qin GZ, Tian SP, Xu Y (2004) Biocontrol of postharvest diseases on sweet cherries by four antagonistic yeasts in different storage conditions. Postharvest Technol 31:51–58
Roberts RG (1990) Postharvest biological control of gray mold of apple by *Cryptococcus laurentii*. Phytopathology 80:526–530
Robiglio A, Sosa MC, Lutz MC, Lopes CA, Sangorrín MP (2011) Yeast biocontrol of fungal spoilage of pears stored at low temperature. Int J Food Microbiol 147:211–216
Rossi M, Buzzini M, Cordisco L, Amaretti A, Sala M, Raimondi S, Ponzoni C, Pagnoni UM, Matteuzzi D (2009) Growth, lipid accumulation, and fatty acids composition in obligate psychrophilic, facultative psychrophilic and mesophilic yeasts. FEMS Microbiol Ecol 69:363–372
Saravanakumar D, Spadaro D, Garibaldi A, Gullino LM (2009) Detection of enzymatic activity and partial sequence of a chitinase gene in *Metschnikowia pulcherrima* strain MACH1 used as post-harvest biocontrol agent. Eur J Plant Pathol 123:183–193
Scherm B, Ortu G, Muzzu A, Budroni M, Arras G, Migheli O (2003) Biocontrol activity of antagonistic yeasts against *Penicillium expansum* on apple. J Plant Pathol 85:205–213
Schisler DA, Janisiewicz WJ, Boekhout T, Kurtzman CP (2011) Agriculturally important yeasts: biological control of field and postharvest diseases using yeast antagonists, and yeasts as pathogens of plants. In: Kurtzman CP, Fell JW, Boekhout T (eds) The Yeasts: a taxonomic study. Elsevier, Amsterdam, pp 45–52
Sharma R, Singh D, Singh R (2009) Biological control of postharvest diseases of fruits and vegetables by microbial antagonists: a review. Biol Control 50:205–221

Shivaji S, Prasad GS (2009) Antarctic yeasts: biodiversity and potential applications. In: Satyanarayana T, Kunze G (eds) Yeast biotechnology: diversity and applications. Springer, Berlin, pp 3–16

Sipiczki M (2006) *Metschnikowia* strains isolated from botrytized grapes antagonize fungal and bacterial growth by iron depletion. Appl Environ Microbiol 72:6716–6724

Spadaro D, Gullino ML (2004) State of the art and future prospects of the biological control of postharvest fruit diseases. Int J Food Microbiol 91:185–194

Starmer WT, Lachance MA (2011) Yeast ecology. In: Kurtzman CP, Fell JW, Boekhout T (eds) The yeasts: a taxonomic study. Elsevier, Amsterdam, pp 65–83

Sugar D, Basile SR (2008) Timing and sequence of postharvest fungicide and biocontrol agent applications for control of pear decay. Postharvest Biol Technol 49:107–112

Sundh I, Melin P (2011) Safety and regulation of yeasts used for biocontrol or biopreservation in the food or feed chain. A van Leeuwenhoek 99:113–119

Teixido N, Torres R, Viñas I, Abadias M, Usall J (2011) Biological control of postharvest diseases in fruit and vegetables In: Lacroix C (ed) Protective cultures, antimicrobial metabolites, and bacteriophages for food and beverage bio preservation. Woodhead Publ Ltd, Cambridge, pp 365–402

Tian SP, Qin GZ, Xu Y (2005) Synergistic effects of combining biocontrol agents with silicon against postharvest diseases of jujube fruit. J Food Protect 68:544–550

Torres R, Teixido N, Viñas I, Mari M, Casalini L, Giraud M, Usall J (2006) Efficacy of *Candida sake* CPA-1 formulation for controlling *Penicillium expansum* decay on pome fruit from different Mediterranean regions. J Food Protect 69:2703–2711

Torres R, Nunes C, García JM, Abadias M, Viñas I, Manso T, Olmo M, Usall J (2007) Application of *Pantoea agglomerans* CPA-2 in combination with heated sodium bicarbonate solutions to control the major postharvest diseases affecting citrus fruit at several mediterranean locations. Eur J Plant Pathol 118:73–83

Turchetti B, Buzzini P, Goretti M, Branda E, Diolaiuti G, D'Agata C, Smiraglia C, Vaughan-Martini A (2008) Psychrophilic yeasts in glacial environments of Alpine glaciers. FEMS Microbiol Ecol 63:73–83

Turkiewicz M, Pazgier M, Kalinowska H, Bielecki S (2003) A cold-adapted extracellular serine proteinase of the yeast *Leucosporidium antarcticum*. Extremophiles 7:435–442

Vaz ABM, Rosa LH, Vieira MLA, de Garcia V, Brandão LR, Teixeira LCR, Moliné M, Libkind D, van Broock M, Carlo CA (2011) The diversity, extracellular enzymatic activities and photoprotective compounds of yeasts isolated in Antarctica. Braz J Microbiol 42:937–947

Vero S, Mondino P, Burgueño J, Soubes M, Wisniewski M (2002) Characterization of biocontrol activity of two yeast strains from Uruguay against blue mold of apple. Postharvest Biol Technol 26:91–98

Vero S, Garmendia G, Gonzalez MB, Garat MF, Wisniewski ME (2009) *Aureobasidium pullulans* as a biocontrol agent of postharvest pathogens of apples in Uruguay. Biocontrol Sci Technol 19:1033–1049

Vero S, Garmendia G, Garat MF, Wisniewski ME (2011) *Cystofilobasidium infirmominiatum* as a biocontrol agent of postharvest diseases of apple and citrus. Acta Hortic 905:169–180

Vero S, Garmendia G, González MB, Bentancur O, Wisniewski M (2013) Evaluation of yeasts obtained from Antarctic soil samples as biocontrol agents for the management of postharvest diseases of apple (*Malus x domestica*). FEMS Yeast Res 13:189–199

Viñas I, Usall J, Teixidó N, Sanchis V (1998) Biological control of major postharvest pathogens on apple with *Candida sake*. Int J Food Microbiol 40:9–16

Wang Y, Yu T, Li Y, Cai D, Liu X, Lu H, Zheng XD (2009) Postharvest biocontrol of *Alternaria alternata* in chinese winter jujube by *Rhodosporidium paludigenum*. J Appl Microbiol 107:1492–1498

Wang P, Liu X, Wang Y, Ruan H, Zheng X (2011) Statistical media optimization for the biomass production of postharvest biocontrol yeast *Rhodosporidium paludigenum*. Prep Biochem Biotechnol 41:382–397

Wilson C (1990) Postharvest Biological control of *Penicillium* rots of citrus with antagonistic yeasts and bacteria. Sci Hortic 40:105–112

Wilson C (2013) Establishment of a world food preservation center. Agric Food Secur 2:1–4

Wilson CL, Wisniewski ME (1989) Biological control of postharvest diseases of fruits and vegetables: an emerging technology. Annu Rev Phytopathol 27:425–441

Wilson CL, Wisniewski ME, Droby S, Chalutz E (1993) A selection strategy for microbial antagonists to control postharvest diseases of fruits and vegetables. Sci Hortic 53:183–189

Wisnieswski ME, Biles C, Droby S, McLaughlin R, Wilson C, Chalutz E (1991) Mode of action of the postharvest biocontrol yeast, *Pichia guilliermondii* I. Characterization of attachment to *Botrytis cinerea*. Physiol Mol Plant Pathol 39:245–258

Wisniewski ME, Droby S, Chalutz E, Eilam Y (1995) Effect of Ca^{2+} and Mg^{2+} on *Botrytis cinerea* and *Penicillium expansum* in vitro and on the biocontrol activity of *Candida oleophila*. Plant Pathol 44:1016–1024

Wisniewski ME, Droby S, El-Ghaouth A, Droby S (2001) Non-chemical approaches to postharvest disease control. Acta Hortic 553:407–411

Wisniewski ME, Bassett CL, Artlip TS, Janisiewicz WJ, Norelli JL, Droby S (2005) Overexpression of a peach defensin gene can enhance the activity of post harvest biocontrol agents. Acta Hortic 682:1999–2006

Wisniewski M, Wilson C, Droby S, Chalutz E, El Ghaouth A, Stevens C (2007) Postharvest biocontrol: new concepts and applications. In: Vincent C, Goettel MS, Lazarovits G (eds) Biological control: a global perspective. CAB International, Boca Raton, pp 262–273

Xi L, Tian SP (2005) Control of postharvest diseases of tomato fruit by combining antagonistic yeast with sodium bicarbonate. Sci Agric Sin 38:950–955

Xu XB, Tian SP (2008) Reducing oxidative stress in sweet cherry fruit by *Pichia membranifaciens:* a possible mode of action against *Penicillium expansum*. J Appl Microbiol 105:1170–1177

Yao HJ, Tian SP (2005) Effects of a biocontrol agent and methyl jasmonate on postharvest diseases of peach fruit and the possible mechanisms involved. J Appl Microbiol 98:941–950

Zhang H, Zheng X, Xi Y (2005) Biological control of postharvest blue mold of oranges by *Cryptococcus laurentii* (Kufferath) Skinner. Biol Control 50:331–342

Zhang H, Zheng XD, Yu T (2007) Biological control of postharvest diseases of peach with *Cryptococcus laurentii*. Food Control 18:287–291

Zhang D, Spadaro D, Garibaldi A, Gullino ML (2010) Selection and evaluation of new antagonists for their efficacy against postharvest brown rot of peaches. Postharvest Biol Technol 55:174–181

Zhang D, Spadaro D, Garibaldi A, Gullino ML (2011) Potential biocontrol activity of a strain of *Pichia guilliermondii* against grey mold of apples and its possible modes of action. Biol Control 57:193–201

Zhao Y, Tu K, Shao X, Jing W, Su Z (2008) Effects of the yeast *Pichia guilliermondii* against *Rhizopus nigricans* on tomato fruit. Postharvest Biol Technol 49:113–120

Chapter 21
Bioremediation and Biodegradation of Hydrocarbons by Cold-Adapted Yeasts

Rosa Margesin

Abstract Cold-adapted microorganisms play a significant role in the biodegradation of organic pollutants in cold environments, where ambient temperatures often coincide with their growth temperature range. A number of studies demonstrated the potential of cold-adapted yeasts to degrade a broad range of hydrocarbons, including alkanes, aromatic, and polyaromatic hydrocarbons (PAHs), at low temperatures. The high metabolic versatility and the ability to degrade high amounts of organic pollutants at temperatures down to 1 °C point to the important role of yeasts for biodegradation processes in habitats with permanently low temperatures. The contribution of cold-adapted yeasts in the biodegradation and bioremediation of hydrocarbons in cold environments may be much more important than currently recognized.

Keywords Petroleum hydrocarbons · Phenol · Lipase · Biostimulation

Contents

21.1 Introduction ... 466
21.2 Bioremediation ... 466
21.3 Limiting Factors for Bioremediation in Cold Environments ... 467
21.4 Mycoremediation: The Contribution of Cold-Adapted Yeasts to Bioremediation ... 468
  21.4.1 Soil Biostimulation ... 469
  21.4.2 Soil Bioaugmentation ... 469
  21.4.3 Wastewater Bioaugmentation ... 470
21.5 Lipase Activity: An Indicator of the Hydrocarbon Biodegradation Potential ... 471
21.6 Hydrocarbon Biodegradation by Pure Cultures of Cold-Adapted Yeasts ... 471
  21.6.1 Sources of Isolation ... 472

R. Margesin (✉)
Institute of Microbiology, University of Innsbruck, Technikerstrasse 25, 6020 Innsbruck, Austria
e-mail: rosa.margesin@uibk.ac.at

P. Buzzini and R. Margesin (eds.), *Cold-Adapted Yeasts*,
DOI: 10.1007/978-3-642-39681-6_21,

21.6.2 Growth Temperature Range 472
21.6.3 Low-Temperature Biodegradation of Petroleum Hydrocarbon Fractions: A Comparison of Bacterial and Yeast Strains 473
21.6.4 Low-Temperature Biodegradation of Phenol 474
21.7 Conclusions 477
References 477

21.1 Introduction

The environmental contamination with organic pollutants is a widespread problem in all climates. With increasing attention toward the preservation of the environment, the clean-up of contaminated sites gained increasing interest. A number of methods, including physical, chemical, and biological treatments, are available for the treatment of contaminated soils, ground and surface waters (van Hamme et al. 2003; Singh and Ward 2004; Ma and Jiao 2012).

Microorganisms in contaminated cold environments are subjected to a number of special challenges, nonetheless low-temperature biodegradation of many components of petroleum hydrocarbons at low temperatures has been reported in Arctic, Alpine, and Antarctic environments and can be attributed to cold-adapted hydrocarbon-degrading microorganisms. This chapter is a review of the potential of cold-adapted hydrocarbon-degrading yeasts for the decontamination of pollutants in cold areas.

21.2 Bioremediation

Bioremediation attempts to accelerate the natural biodegradation rates through the optimization of limiting environmental conditions and is an ecologically and economically effective cleanup technology. Bioremediation has been defined as "the use of living organisms to reduce or eliminate environmental hazards resulting from accumulation of toxic chemicals and other hazardous compounds" (Gibson and Saylor 1992). The development of biological remediation methods is based on the capacity of a broad spectrum of microorganisms to utilize organic compounds as the sole source of carbon and energy (biodegradation); microorganisms transform or mineralize organic contaminants into less harmful, non-hazardous substances, which are then integrated into natural biogeochemical cycles. As reported already very early (ZoBell 1946), biodegradative microorganisms are widely distributed in nature and can be enriched in many, if not most, types of environments where contaminants may serve as organic carbon sources. This is not surprising since hydrocarbons are naturally occurring organic compounds, and consequently, microorganisms have evolved the ability to degrade these compounds. Accordingly, hydrocarbon-degrading microorganisms have also been described in cold environments, where they may adapt rapidly to the contamination, as demonstrated by significantly increased numbers of hydrocarbon degraders after a pollution event (Kerry 1990; Aislabie et al. 2001; Margesin and Schinner 2001; Bej et al. 2010; Greer et al. 2010).

Hydrocarbon pollution in cold climates is an area of particular importance since contaminated areas are often remote, and thus, the degradation capacity of indigenous microorganisms is required. Cold environments are increasingly exposed to petroleum exploration, production, and transport, and these activities increase the risk of accidental oil release. Such environments include polar and alpine environments. Also in temperate climatic regions, temperatures of subsoils, groundwater, and industrial wastewater can often decrease to temperature levels around or below 10 °C due to seasonal and/or diurnal fluctuations.

Successful bioremediation of hydrocarbon-contaminated aquatic and terrestrial environments in various cold environments, including Arctic, Antarctic, and Alpine areas, has been reported (Brakstad 2008; Filler et al. 2008; Bej et al. 2010; Greer et al. 2010). Several remediation schemes, such as biopiles, landfarming and engineered bioremediation, have been implemented successfully at petroleum-contaminated cold sites (Filler et al. 2008).

21.3 Limiting Factors for Bioremediation in Cold Environments

The intensity of biodegradation is influenced by a number of environmental factors. The composition, concentration and bioavailability of contaminants as well as the pollution history of the contaminated environment and the chemical and physical characteristics (nutrients and nutrient availability, oxygen, pH value, temperature, water content, salinity) of the contaminated area play a major role.

An important factor governing microbial activity, and thus also biodegradation, is temperature. Due to the Q_{10} effect, reaction rates are reduced in the cold, however, local environmental conditions select for populations with high activities at low temperatures. Cold-adapted microorganisms have evolved a series of adaptation strategies that enable them to compensate for the negative effects of low temperatures on biochemical reactions (Feller 2007; Margesin et al. 2008; Buzzini et al. 2012). Substantial growth and metabolic activity (respiration and biosynthesis) of microorganisms, including yeasts (*Leucosporidum* spp., *Mrakia* spp.) at subzero temperatures down to −20 °C and even −35 °C have been demonstrated (Panikov and Sizova 2007; Bakermans 2008; Amato et al. 2010; see also Chap. 15).

Low temperatures induce the formation of ice crystals, which results in low availability of liquid water, i.e., the creation of low water activity (A_w). Low water activity in habitats like snow, sea ice, and glacier ice influences microbial activity to a significant extent (Gunde-Cimerman et al. 2003).

Temperature affects the rates of biodegradation also by its effects on the physical nature and chemical composition of the contaminants (Atlas and Bartha 1992). Bioavailability and solubility of hydrophobic substances, such as some aliphatic and polyaromatic hydrocarbons (PAHs), are temperature-dependent. At low temperatures, the volatilization of toxic short-chain alkanes is reduced and their water solubility is increased, which results in increased toxicity (Walker and Colwell 1974; Atlas and Bartha 1992; Whyte et al. 1998). A temperature decrease also results in a decrease in diffusion rates of organic compounds and in an increase in viscosity, which affects the degree of distribution (Whyte et al. 1998; Rojo 2009). Under cold conditions, the precipitation of certain alkanes (from crude oil) as waxes greatly diminishes their availability to hydrocarbon-degrading microorganisms (McKenzie and Hughes 1976). All these temperature effects delay the onset of biodegradation under cold conditions.

Until recently, frozen soils were considered a practically impermeable barrier to pollutants. Meanwhile, it is known that hydrocarbons can penetrate into frozen soils. Even ice-saturated soils are not an absolute impermeable barrier for oil penetration (Chuvilin et al. 2001; Barnes and Biggar 2008).

Additional limiting factors for biodegradation, such as low-nutrient availability, dryness, large temperature fluctuations and frequent freeze-thaw events, depend on the local environmental conditions in Arctic, Alpine, and Antarctic regions (Margesin 2004).

21.4 Mycoremediation: The Contribution of Cold-Adapted Yeasts to Bioremediation

Mycoremediation is a new and emerging field in bioremediation and involves the use of fungi to degrade (reduce or eliminate) organic compounds including environmental hazards (Singh 2006; Hughes and Bridge 2010; Harms et al. 2011). The important role of filamentous fungi in the degradation of hydrocarbons or their metabolites (Kerry 1990; Aislabie et al. 2001), and the benefits of their use as bioremediative agents in cold habitats have been recognized (Hughes and Bridge 2010). However, there is little information on the role of cold-adapted yeasts.

According to Hughes and Bridge (2010), yeasts may play an important role in facilitating hydrocarbon degradation for other microbial groups, while Singh (2006) emphasized that yeasts and filamentous fungi may contribute significantly to oil degradation under conditions that select against bacterial growth. Ahearn et al. (1971) recognized already early the important role of yeasts in the *in situ* degradation of surface oil deposits in marine environments and listed a number of advantages of yeasts over bacteria: Vegetative yeast cells are more resistant than those of bacteria to stress conditions, including exposure to UV radiation and alterations of osmotic pressure and salinity. A further advantage is the ability of some yeasts, mainly representatives of the genus *Trichosporon*, to penetrate and develop within oil globules, which offers protection from predators. In contrast, bacterial cells remain attached to the surface of oil globules.

Bacteria and filamentous fungi have been claimed to be the main degraders in soil environments, while bacteria and yeasts appear to be the prevalent hydrocarbon degraders in aquatic ecosystems (Atlas 1981; Singh 2006). A higher occurrence of yeasts was observed in rivers and lakes than in the ocean. Yeast populations increased considerably in oil-contaminated estuarine sediments over a 4-month period but declined after an initial increase in open ocean waters in the presence of petroleum hydrocarbons (Ahearn and Meyers 1976).

Walker and Colwell (1974) compared the utilization of model petroleum in water and sediment samples of marine environments at low temperatures (0–10 °C) by bacteria, yeasts, and filamentous fungi. Hydrocarbon degradation at 0 and 5 °C could be attributed mainly to bacteria, while there was a lack of significant growth of yeasts and filamentous fungi. Yeast populations were higher in April than in February and contributed significantly to petroleum utilization at 10 °C, however, to a significantly lower extent than bacteria. Hydrocarbon degradation by filamentous fungi was not detected at any of the temperatures tested.

Ahearn et al. (1971) established the role of yeasts in the removal of hydrocarbons from oil-contaminated marine environments. Despite the widespread occurrence of hydrocarbonoclastic yeasts, strains able to assimilate high amounts of hydrocarbons (>2 % v/v kerosene or hexadecane) and vapors of aromatic compounds were concentrated in oil-polluted habitats.

21.4.1 Soil Biostimulation

The most widely used bioremediation procedure in cold soils is biostimulation of the indigenous microorganisms by supplementation of appropriate nutrients and optimization of other limiting factors, such as oxygen content, pH, and temperature control. Most commonly hydrocarbon degradation in such studies is attributed to soil bacteria, while unfortunately no studies are available on the contribution of the yeast population to bioremediation processes in soils. There is generally little known about the interactions of soil yeasts *in situ* (Botha 2006). Since ascomycetous and basidiomycetous yeasts constitute a considerable proportion of the indigenous soil population (yeast numbers range from less than 10 to as many as 10^6 culturable cells per gram of soil) and play an important role in mineralization processes (Botha 2006, 2011), it has to be assumed that they contribute to a significant extent to biological decontamination.

21.4.2 Soil Bioaugmentation

Bioaugmentation by inoculating allochthonous hydrocarbon degraders (predominantly bacteria) has been used as a bioremediation option to treat petroleum-contaminated cold and temperate sites. However, this strategy generally underperformed or gave no better results than fertilization (Margesin 2004; Aislabie et al. 2006; Filler et al. 2008). Similar results were obtained when investigating the efficiency of a cold-adapted yeast strain identified as *Yarrowia lipolytica*, which degraded efficiently hydrocarbons (diesel oil) in liquid culture at 10 °C (Margesin and Schinner 1997a), for soil bioaugmentation. Representatives of this yeast species degrade hydrophilic substances very efficiently (Bankar et al. 2009). The inoculation of five diesel oil–contaminated Alpine subsoils resulted only in a small increase (5–7 %) of the total hydrocarbon decontamination at 10 °C. Biostimulation by inorganic nutrients enhanced oil biodegradation to a statistically significantly greater degree than inoculation. In none of the five soils did fertilization plus inoculation result in a higher decontamination than fertilization alone (Margesin and Schinner 1997b).

The observation that this cold-adapted yeast inoculum introduced into soil did not contribute efficiently to the decontamination process led to a further study in order to compare the biodegradation behavior of this cold-adapted yeast in liquid culture and in one of the previously investigated five soils at temperatures between 4 and 30 °C (Margesin and Schinner 1997a). In liquid culture, the inoculum degraded diesel oil over the whole temperature range tested with a maximum activity between 10 and 20 °C (37–41 % biodegradation of the initial oil content); 25–27 % were degraded at 4 and 25 °C, respectively, and still 18 % were utilized at 30 °C. When inoculated into soil, the degradation activity of the inoculum was completely changed: biodegradation ranged from 0 to 3.6 % and was only observed at temperatures $\leq$15 °C, decreased with increasing temperature and time, and was significantly lower in soil than in liquid culture. With increasing incubation time and temperature, hydrocarbon utilization by the inoculum in soil decreased, whereas degradation by the indigenous soil microorganisms increased.

These data clearly showed that the degradation behavior of an inoculum introduced into soil cannot be predicted from liquid culture experiments and that the success of bioaugmentation cannot be predicted from liquid culture experiments. With increasing indigenous biodegradation, the inoculum might have been replaced (Margesin and Schinner 1997a). Introduced microorganisms are subject to various abiotic and biotic stresses. Preconditions of a successful application of bioaugmentation are the expression of the biodegrading activities in the polluted environment and the survival of the inoculated microorganisms at least for the time necessary for remediation (Fritsche et al. 1998). The bioaugmentation of contaminated soils frequently requires a short-term approach since, even under best conditions, the introduced organisms will not survive for extensive periods (Pritchard et al. 1998).

21.4.3 Wastewater Bioaugmentation

The use of cold-adapted microorganisms for low-energy wastewater treatment leads to a significant decrease in operational costs. In cold climates, wastewater temperature often decreases to 10 °C and below, which requires the activity of cold-adapted degraders for an efficient treatment.

The amount of organic pollutants in surface water (e.g., lakes or rivers) or wastewater is commonly determined indirectly by measuring the chemical oxygen demand (COD), which thus is a measure of water quality. High COD removal rates (64–83 %) were obtained in artificial sewage, initially containing 500 mg COD per liter of sewage, after 6 h at 5 °C with six cold-adapted yeast strains that belonged to the genera *Candida*, *Pichia*, *Rhodotorula* and *Saccharomyces* (Ma et al. 2007). COD removal by these yeast strains was better than that obtained with six cold-adapted bacterial strains (45–65 %). All strains were isolated in winter from activated sludge at temperatures below 10 °C. The mixture of bacterial and yeast populations resulted in a COD removal of 87 %, which was only slightly higher that the removal observed with yeasts alone. When this mixed population was inoculated into domestic sewage, COD removal (83 %) was only slightly lower compared to that obtained in artificial sewage (87 %) (Ma et al. 2007). These data demonstrate that cold-adapted yeast strains represent a promising source as inocula for accelerated wastewater treatment in climates, where the activity of mesophilic degraders is limited.

21.5 Lipase Activity: An Indicator of the Hydrocarbon Biodegradation Potential

Several cold-adapted yeasts are known to produce cold-active lipases (Chap. 17). A number of cold-adapted lipase-producing bacterial (Margesin et al. 2013) and yeast strains (Margesin et al. 2003) were able to degrade aliphatic hydrocarbons. A positive relation between lipase activity and hydrocarbon biodegradation in marine environments (Ahearn et al. 1971) and in soils contaminated with petroleum hydrocarbons (Margesin et al. 1999, 2007a) has been recognized. Therefore, cold-adapted yeasts that produce cold-active lipases might represent an interesting source for low-temperature hydrocarbon biodegradation.

21.6 Hydrocarbon Biodegradation by Pure Cultures of Cold-Adapted Yeasts

A wide variety of microorganisms, including bacteria, fungi, and algae, have the ability to metabolize aliphatic and aromatic hydrocarbons (Alexander 1999). Nonetheless, most studies on biodegradation abilities of cold-adapted microorganisms have focused on bacteria. Despite the fact that yeasts are able to degrade a wide range of hydrocarbons (Harms et al. 2011), comparatively little is known on the biodegradation abilities of cold-adapted yeasts.

21.6.1 Sources of Isolation

Hydrocarbon-degrading culturable cold-adapted yeasts have been isolated from oil-contaminated cold environments, such as Antarctic (Atlas et al. 1978; Kerry 1990; Aislabie et al. 2001) and Alpine soils (Margesin and Schinner 1997a; Bergauer et al. 2005; Margesin et al. 2005). Antarctic oil-contaminated soils in the Ross Sea region contained 10^4–10^6 culturable yeasts per gram soil in soil surface; these numbers decreased with soil depth (Aislabie et al. 2001). Similar levels were reported in contaminated Antarctic soils from the former McMurdo Dump site on Ross Island (Atlas et al. 1978). The enrichment of yeasts, especially of fermentative representatives, after pollution events was also evident in marine environments (Ahearn et al. 1971; Kutty and Philip 2008).

Kerry (1990) isolated bacteria and yeasts from a number of Antarctic petroleum contaminated soils and reported the predominance of bacterial isolates. Yeasts could be isolated from 11 to 40 % of the samples investigated, however, only 7–10 % of the samples contained yeasts were able to utilize distillate (the main petroleum product used in Australian Antarctic operations, containing a wide range of hydrocarbon components) as sole carbon source.

Hydrocarbon-degrading yeasts could also be isolated from pristine (uncontaminated) cold environments, such as Alpine glacier cryoconite (Margesin et al. 2003), which indicates the ubiquity of hydrocarbon degraders. The ability of strains from pristine habitats to degrade phenol and related compounds may be linked with a role in the degradation of litter and humification process (Bergauer et al. 2005).

21.6.2 Growth Temperature Range

Cold-adapted yeasts are characterized by a more restricted growth temperature range than cold-adapted bacteria; this could be demonstrated with strains isolated from contaminated and from pristine environments. For example, 60 % of the investigated yeast strains but only 8 % of the investigated bacterial strains were unable to grow above 20 °C (Margesin et al. 2003). Similarly, all representatives of the group of twelve *Microbotryomycetidae* strains (mainly representatives of the genus *Rhodotorula*) could not growth above 20 °C, and four of these strains were even unable to grow above 15 °C (Bergauer et al. 2005). The restricted growth temperature range of yeasts indicates their potential for low-temperature bioremediation processes in permanently cold environments. The application of degraders that are active over a wide temperature range might be advantageous in environments that undergo large temperature fluctuations.

Adaptation of cold-adapted yeasts strains to their natural cold environment is also apparent from optimized growth at low temperatures. Two cold-adapted yeast strains, identified as *Leucosporidiella creatinivora* and *Rhodotorula glacialis* and able to grow at 1–20 °C, produced the highest amount of biomass (as determined by measuring OD600, viable counts and dry mass) at 1 °C, i.e., at 20 °C lower than the maximum temperature for growth (Margesin 2009). This demonstrates that cultivation temperatures close to the maximum growth temperature are not appropriate for studying psychrophiles. Highest yields of cold-adapted cells and their biotechnologically important compounds are generally obtained at cultivation temperatures that correspond to those of the natural environment of the strains, which should be considered for applied aspects.

21.6.3 Low-Temperature Biodegradation of Petroleum Hydrocarbon Fractions: A Comparison of Bacterial and Yeast Strains

Yeasts are known for their metabolic versatility, including the biodegradation of organic compounds (Kutty and Philip 2008). The comparison of culturable cold-adapted bacterial and yeast strains to degrade representative fractions of petroleum hydrocarbons (*n*-alkanes, monoaromatic, and polycyclic aromatic hydrocarbons) demonstrated the efficiency and versatility of yeasts (Margesin et al. 2003). The investigated strains were isolated from uncontaminated Alpine glacier habitats (cryoconite and ice caves) and belonged mainly to the bacterial genera *Pseudomonas* and *Arthrobacter*, and to the yeast genus *Rhodotorula* (Margesin et al. 2007b).

While 79 % of the investigated 28 yeast strains utilized n-hexadecane for growth at 10 °C, only 7 % of the studies 61 bacterial strains were able to degrade *n*-hexadecane and none of them degraded *n*-dodecane. Remarkably, only seven yeast strains but no bacterial strain degraded both *n*-dodecane and *n*-hexadecane. The best *n*-alkane degrader was a representative of the species *Y. lipolytica*.

The efficiency of cold-adapted yeasts compared to bacteria was also shown with regard to the biodegradation of aromatic and PAHs: 13 % of the bacterial strains, but 25 % of the yeast strains degraded 2.5 mM phenol at 10 °C. Similarly, 13 % of the bacterial strains, but 21 % (phenanthrene) or 32 % (anthracene) of the investigated yeast strains utilized three-ring PAHs.

The ability to degrade at least one of the tested hydrocarbons (hexadecane, phenol, phenanthrene or anthracene) at 10 °C was restricted to 26 % of the studied bacterial strains, while 89 % of the yeast strains were degraders. Four yeast strains (*Rhodotorula* spp.) but none of the bacterial strains could grow with both aliphatic and aromatic hydrocarbons. The capability to utilize a wide range of hydrocarbons under cold conditions is advantageous for the low-temperature treatment of mixed pollutions.

21.6.4 Low-Temperature Biodegradation of Phenol

Phenol and phenolic compounds are widely distributed in nature and as environmental pollutants. They are common constituents of many industrial wastewaters, such as those produced from crude oil refineries and coal gasification plants. Due to their toxicity to microorganisms, even low concentrations of phenolic compounds (such as 2 mM) can often cause the breakdown of wastewater treatment plants by inhibition of microbial growth, which can lead to decreased effluent quality (Ren and Frymier 2003). Biological treatment of phenolic compounds is preferable to other methods due to its effectiveness and the production of innocuous end products (Aleksieva et al. 2002).

21.6.4.1 Effect of Phenol Concentration on Biodegradation

Low-temperature phenol biodegradation by cold-adapted Alpine yeasts has been studied using fed-batch cultivation with increasing phenol concentrations as the sole carbon source. This cultivation method has been proven to be efficient for the selection and acclimation of phenol-degrading microorganisms (Guieysse et al. 2001). It could be demonstrated that cold-adapted yeasts tolerate and degrade higher amounts of phenol compared to cold-adapted bacteria.

Amounts of phenol as high as 10–12.5 mM were degraded at 10 °C by cold-adapted *Rhodotorula* species (*Rhodotorula psychrophenolica*, *R. glacialis*) isolated from uncontaminated glacier materials (Margesin et al. 2003, 2007b), while two yeast strains from contaminated soils (*Trichosporon dulcitum* and an urediniomycete later classified as *Glaciozyma watsonii*) even degraded up to 15 mM phenol at 10 °C (Margesin et al. 2005). In comparison, cold-adapted Alpine bacterial strains (*Arthrobacter psychrophenolicus*, *A. sulfureus*, *Rhodococcus* spp.) degraded up to 10–12.5 mM phenol (Margesin et al. 2003, 2005, 2013) at 10 °C. Among mesophilic yeasts, the degradation of amounts up to 27 mM phenol was reported at 30 °C (Krug et al. 1985; dos Santos et al. 2009).

Cold-adapted phenol-degrading yeasts were also shown to degrade phenol faster than bacteria. A concentration of 10 mM phenol was fully degraded after 11–14 days by two rhodococci, but already after 3 days by *G. watsonii*. 12.5 mM phenol was fully degraded after 7 and 10 days by *G. watsonii* and *T. dulcitum*, whereas the rhodococci needed 25 days (Margesin et al. 2005).

21.6.4.2 Effect of Temperature on Phenol Biodegradation

The optimum temperature for phenol degradation has been found to be generally lower for cold-adapted yeasts than for cold-adapted bacteria (Margesin et al. 2003, 2005). The majority of yeast strains preferred a temperature of 10 °C for phenol degradation, and their biodegradation was faster at 1 °C than at 20 or 15 °C. In one case, biodegradation performance at 1 and 10 °C was comparable. Few yeasts had an optimum temperature of 20 °C for phenol degradation.

Similarly, studies on the effect of temperature on growth and biodegradation of 5 mM phenol showed that *G. watsonii* degraded 5 mM phenol at 1 °C faster than two rhodococci at 10 °C, but no growth occurred at 20 °C in the presence of phenol (Margesin et al. 2005).

Microorganisms that degrade high amounts of organic pollutants within a short time at temperatures down to 1 °C represent a promising source of accelerated wastewater treatment.

21.6.4.3 Toxicity of Phenol and Related Compounds

Low-temperature (10 °C) biodegradation of phenol and phenol-related monoaromatic compounds of serious environmental concern (catechol, resorcinol, hydroquinone, benzoate, salicylate, guaiacol, *o*-cresol, *m*-cresol, *p*-cresol, *p*-nitrophenol, *p*-nitrotoluene) was evaluated using 32 basidiomycetous yeast strains isolated from Alpine soils and glacier cryoconite (Bergauer et al. 2005). The strains were representatives of the Hymenomycetes (*Cryptococcus terreus*, *Cryptococcus terricola*) or Urediniomycetes (*Rhodosporidium lusitaniae*, *Rhodotorula ingeniosa*, *R. glacialis*, *Rhodotorula psychrophila, R. psychrophenolica*, *Mastigobasidium intermedium*, *L. creatinivora*, *Sporobolomyces roseus*, which is the anamorph of *Sporidiobolus metaroseus*).

None of the 32 strains utilized any of the highly volatile mono-aromatic compounds (benzene, toluene, ethylbenzene, nitrobenzene, *o*-xylene, *m*-xylene, and *p*-xylene) as the sole carbon source. Non/low volatile aromatic compounds were degraded in the following order: phenol > hydroquinone > resorcinol > benzoate > catechol > salicylate ≫ *p*-cresol > *m*-cresol. *o*-Cresol, guaiacol, *p*-nitrophenol, or *p*-nitrotoluene were not utilized for growth.

The toxicity of the tested 19 monoaromatic compounds was influenced by the chemical structure (functional groups) of the compounds. Methylated compounds were highly toxic, followed by methoxylated and hydroxylated compounds; carboxylated compounds had the lowest toxicity. Biodegradability of phenol-related compounds was influenced by volatility and water solubility of the compounds. Interestingly, the taxonomic affiliation of the strains seemed to influences toxicity and biodegradability. *L. creatinivora* strains were characterized by higher IC_{50} values (50 % growth inhibition in the presence of nutrients) than all other yeast species, whereas *S. roseus* was the most sensitive species. In addition, representatives of *L. creatinivora* were characterized by a higher metabolic versatility (i.e., ability to utilize a wide spectrum of compounds) than representatives of other species (Bergauer et al. 2005).

Strains such as those characterized in this study could be useful, for example, as inocula for the acceleration of low-energy wastewater treatment. Members of *Rhodotorula* and *Leucosporidiella*, esp. *L. creatinivora,* could be especially important, because these strains produced higher amounts of biomass, degraded more non/low volatile mono-aromatic compounds and were able to grow in the presence of higher concentrations than other yeast species.

21.6.4.4 Effect of Immobilization on Phenol Degradation

Immobilization of microorganisms may result in better degradation rates. Due to the promotion of biofilm formation by immobilized cells, they are better protected from damage and can maintain continuous cell growth and biodegradation. The production of exopolymeric substances (EPS) is a characteristic feature of many biofilms. EPS assist in the attachment process and also protect the cells from fluctuating environmental conditions (Chandran and Das 2011).

Immobilized cells of mesophilic *Candida tropicalis* showed better diesel oil (Chandran and Das 2011) and phenol degradation rates and could be exposed without loss of viability to higher amounts of phenol compared to free cells (Juares-Ramirez et al. 2001). The same was reported for *Aureobasidium pullulans* at 30 °C (dos Santos et al. 2009) and for three cold-adapted phenol-degrading yeasts (*L. creatinivora*, *C. terreus*) at 10 °C (Krallish et al. 2006). The *C. terreus* strain was more active in biofilm formation on solid carriers than two *L. creatinivora* strains, however, its phenol degradation performance was probably limited due to the formation of a rich EPS layer, which decreased the diffusion of phenol to the cells surface. All three cold-adapted yeast strains accumulated both trehalose and glycogen during growth on glucose or phenol as sole carbon and energy source (Krallish et al. 2006). Yeasts that accumulate high amounts of intracellular storage compounds (trehalose, glycerol, etc. serve as carbon and energy reserve compounds) are tolerant to adverse environmental conditions (Wiemken 1990). Trehalose plays an important role for the preservation of yeast cells under extreme environmental conditions, such as high temperature, dehydration and high osmotic conditions (Hounsa et al. 1998).

21.6.4.5 Enzymes Involved in Phenol Degradation: Catechol Dioxygenases

Catechol dioxygenases are involved in the second step of phenol degradation and catalyze the ring cleavage of catechol (=*o*-hydroxyphenol) either by the *ortho* (catechol-1,2-dioxygenase; C1,2D) or by the *meta* type (catechol-2,3-dioxygenase; C2,3D). Almost all investigated Alpine cold-adapted phenol-degrading yeast strains oxidize catechol predominantly by the *ortho* type of ring cleavage, while the *meta* type of ring cleavage was only observed in one yeast strain (Margesin et al. 2003, 2005). dos Santos et al. (2009) observed an inhibitory effect on C1,2D activity at phenol amounts above 6 mM phenol.

The presence of C1, 2D activity and the absence of C2,3D activity has also been observed frequently with mesophilic phenol-degrading yeasts. In contrast, bacterial representatives display often both types of ring cleavage.

21.7 Conclusions

Cold-adapted microorganisms play a significant role in the biodegradation of organic pollutants in cold environments, where ambient temperatures often coincide with their growth temperature range. A number of studies demonstrated the potential of cold-adapted yeasts to degrade a broad range of hydrocarbons, including alkanes, aromatic, and PAHs, at low temperatures. The capability to utilize a wide range of hydrocarbons is advantageous for the treatment of mixed pollutions.

The apparently more restricted growth temperature range, substantially lower optimum temperatures for growth and activity as well as the great metabolic versatility of cold-adapted yeasts compared to cold-adapted bacteria point to the important role of yeasts for biodegradation processes in habitats with permanently low temperatures. Yeasts that degrade high amounts of organic compounds within a short time at temperatures down to 1 °C (Margesin et al. 2005) represent an especially useful source for a wide range of applications, such as accelerated wastewater treatment. Due to their ability to tolerate and degrade high amounts of phenol, cold-adapted yeast strains as promising candidates for the biological cleaning of phenol-contaminated environments in climates, where low temperatures can otherwise limit microbial degradation.

Most investigations on low-temperature biodegradation have been concentrated on bacteria, whereas the potential of cold-adapted yeasts may have been underestimated. However, yeasts are a largely unexplored source of cold-adapted hydrocarbon degraders; their contribution in the biodegradation of hydrocarbons in cold environments may be much more important than currently assumed.

References

Ahearn DG, Meyers SP (1976) The role of fungi in the decomposition of hydrocarbons in the marine environment. In: Walters AH, Hueck-va der Plas EH (eds) Biodeterioration of materials. Applied Science, London, pp 12–18

Ahearn DG, Meyers SP, Standard PG (1971) The role of yeasts in the decomposition of oils in marine environments. Dev Ind Microbiol 12:126–134

Aislabie J, Fraser R, Duncan S, Farrell R (2001) Effects of oil spills on microbial heterotrophs in Antarctic soils. Polar Biol 24:308–313

Aislabie J, Saul DJ, Foght JM (2006) Bioremediation of hydrocabon-contaminated polar soils. Extremophiles 10:171–179

Aleksieva Z, Ivanova D, Godjevargova T, Atanasov B (2002) Degradation of some phenol derivates by *Trichosporon cutaneum*. Process Biochem 37:1215–1219

Alexander M (1999) Biodegradation and bioremediation, 2nd edn. Academic Press, London

Amato P, Doyle SM, Battista JR, Christner BC (2010) Implications of subzero metabolic activity on long-term microbial survival in terrestrial and extraterrestrial permafrost. Astrobiology 10:789–798

Atlas RM (1981) Microbial degradation of petroleum hydrocarbons: an environmental perspective. Microbiol Rev 45:180–209

Atlas RM, Bartha R (1992) Hydrocarbon biodegradation and oil spill remediation. Adv Microb Ecol 12:287–338

Atlas RM, Di Menna ME, Cameron RE (1978) Ecological investigations of yeasts in Antarctic soils. Antarct Res Ser 30:27–34

Bakermans C (2008) Limits for microbial life at subzero temperatures. In: Margesin R, Schinner F, Marx JC, Gerday C (eds) Psychrophiles: from biodiversity to biotechnology. Springer, Berlin, pp 17–28

Bankar AV, Kumar AR, Zinjarde SS (2009) Environmental and industrial applications of *Yarrowia lipolytica*. Appl Microbiol Biotechnol 84:847–865

Barnes DL, Biggar K (2008) Movement of petroleum through freezing and frozen soil. In: Filler DM, Snape I, Barnes DL (eds) Bioremediation of petroleum hydrocarbons in cold regions. Cambridge University Press, Cambridge, pp 55–68

Bej AK, Aislabie J, Atlas RM (eds) (2010) Polar microbiology: the ecology, biodiversity and bioremediation potential of microorganisms in extremely cold environments. CRC Press, Boca Raton

Bergauer P, Fonteyne PA, Nolard N, Schinner F, Margesin R (2005) Biodegradation of phenol and phenol-related compounds by psychrophilic and cold-tolerant alpine yeasts. Chemosphere 59:909–918

Botha A (2006) Yeasts in soil. In: Rosa CA, Péter G (eds) The yeast handbook: biodiversity and ecophysiology of yeasts. Springer, Berlin, pp 221–240

Botha A (2011) The importance and ecology of yeasts in soil. Soil Biol Biochem 43:1–8

Brakstad OG (2008) Natural and stimulated biodegradation of petroleum in permafrost-affected regions. In: Margesin R, Schinner F, Marx JC, Gerday C (eds) Psychrophiles: from biodiversity to biotechnology. Springer, Berlin, pp 389–407

Buzzini P, Branda E, Goretti M, Turchetti B (2012) Psychrophilic yeasts from worldwide glacial habitats: diversity, adaptation strategies and biotechnological potential. FEMS Microbiol Ecol 82:217–241

Chandran P, Das N (2011) Degradation of diesel oil by immobilized *Candida tropicalis* and biofilm fomed on gravels. Biodegradation 22:1181–1189

Chuvilin EM, Naletova NS, Miklyaeva EC, Kozlova EV (2001) Factors affecting spreadibility and transportation of oil in regions of frozen ground. Polar Record 37:229–238

dos Santos VL, Monteiro AD, Braga DT, Santoro MM (2009) Phenol degradation by *Aureobasidium pullulans* FE13 isolated from industrial effluents. J Hazard Mat 161: 1413–1420

Feller G (2007) Life at low temperatures: is disorder the driving force? Extremophiles 11:211–216

Filler DM, Snape I, Barnes DL (2008) Bioremediation of petroleum hydrocarbons in cold regions. Cambridge University Press, Cambridge

Fritsche W, Herr A, Kaestner M, Sack U, Scheibner K (1998), Bioremediation of TNT and PAHs by biostimulation and bioaugmentation of fungi. In: Technical University Hamburg-Harburg (eds) Innovative potential of advanced biological systems for remediation. TU Hamburg, Harburg, pp 43–47

Gibson DT, Saylor GS (1992) Scientific foundations of bioremediation: current status and future needs. American Academy of Microbiology, Washington DC

Greer CW, Whyte LG, Niederberger TD (2010) Microbial communities in hydrocarbon-contaminated temperate, tropical, alpine, and polar soils. In: Timmis KN (ed) Handbook of hydrocarbon and lipid microbiology. Springer, Berlin, pp 2313–2328

Guieysse B, Wikström P, Forsman M, Mattiasson B (2001) Biomonitoring of continuous microbial community adaptation towards more efficient phenol degradation in a fed-batch bioreactor. Appl Microbiol Biotechnol 56:780–787

Gunde-Cimerman N, Sonjak S, Zalar P, Frisvad JC, Diderichsen B, Plemenitaš A (2003) Extremophilic fungi in arctic ice: a relationship between adaptation to low temperature and water activity. Phys Chem Earth 28:1273–1278

Harms H, Schlosser D, Wick LY (2011) Untapped potential: exploiting fungi in bioremediation of hazardous chemicals. Nature Rev Microbiol 9:177–192

Hounsa CG, Brandt EV, Thevelein J, Hohmann S, Prior BA (1998) Role of trehalose in survival of *Saccharomyces cerevisiae* under osmotic stress. Microbiology 144:671–680

Hughes KA, Bridge P (2010) Tolerance of Antarctic soil fungi to hydrocarbons and their potential role in soil bioremediation. In: Bej AK, Aislabie J, Atlas RM (eds) Polar microbiology: the ecology, biodiversity and bioremediation potential of microorganisms in extremely cold environments. CRC Press, Boca Raton, pp 277–300

Juares-Ramirez C, Ruiz-Ordaz N, Cristiani-Urbina E, Galindez–Mayer J (2001) Degradation kinetics of phenol by immobilized cells of *Candida tropicalis* in a fluidized bed reactor. World J Microbiol Biotechnol 17:697–705

Kerry E (1990) Microorganisms colonizing plants and soils subjected to different degrees of human activity, including petroleum contamination, in the Vestfold Hills and MacRobertson Land, Antarctica. Polar Biol 10:423–430

Krallish I, Gonta S, Savenkova L, Bergauer P, Margesin R (2006) Phenol degradation by immobilized cold-adapted yeast strains of *Cryptococcus terreus* and *Rhodotorula creatinivora*. Extremophiles 10:441–449

Krug M, Ziegler H, Straube G (1985) Degradation of phenolic compounds by the yeast *Candida tropicalis* HP15. J Basic Microbiol 2:103–110

Kutty SN, Philip R (2008) Marine yeasts—a review. Yeast 25:465–483

Ma W, Han H, Zhong D, Jiang D, Jia Y (2007) Research on domestic sewage treatment my mixed psychrotrophic strains at low temperature. In: Wang Y, Li S, Huang P, Yang Y, An Y, Sun X (eds) Progress in environmental science and technology, vol 1. Science Press USA Inc, Monmouth Junction, pp 871–875

Ma WJ, Jiao BQ (2012) Review of contaminated sites remediation technology. Res J Chem Environ 16:137–139

Margesin R (2004) Bioremediation of petroleum hydrocarbon-polluted soils in extreme temperature environments. In: Singh A, Wards OP (eds) Applied bioremediation and phytoremediation. Springer, Berlin, pp 215–234

Margesin R (2009) Effect of temperature on growth parameters of psychrophilic bacteria and yeasts. Extremophiles 13:257–262

Margesin R, Schinner F (1997a) Effect of temperature on oil degradation by a psychrotrophic yeast in liquid culture and in soil. FEMS Microbiol Ecol 24:243–249

Margesin R, Schinner F (1997b) Efficiency of indigenous and inoculated cold-adapted soil microorganisms for biodegradation of diesel oil in alpine soils. Appl Environ Microbiol 63:2660–2664

Margesin R, Schinner F (2001) Biodegradation and bioremediation of hydrocarbons in extreme environments. Appl Microbiol Biotechnol 56:650–663

Margesin R, Zimmerbauer A, Schinner F (1999) Soil lipase activity—a useful indicator of oil biodegradation. Biotechnol Tech 13:859–863

Margesin R, Gander S, Zacke G, Gounot AM, Schinner F (2003) Hydrocarbon degradation and enzyme activities of cold-adapted bacteria and yeasts. Extremophiles 7:451–458

Margesin R, Fonteyne PA, Redl B (2005) Low-temperature biodegradation of high amounts of phenol by *Rhodococcus* spp. and basidiomycetous yeasts. Res Microbiol 156:68–75

Margesin R, Hämmerle M, Tscherko D (2007a) Microbial activity and community composition during bioremediation of diesel-oil-contaminated soil: effects of hydrocarbon concentration, fertilizers and incubation time. Microb Ecol 53:259–269

Margesin R, Fonteyne PA, Schinner F, Sampaio JP (2007b) *Rhodotorula psychrophila* sp. nov., *Rhodotorula psychrophenolica* sp. nov. and *Rhodotorula glacialis* sp. nov., novel psychrophilic basidiomycetous yeast species from alpine environments. Int J Syst Evol Microbiol 57:2179–2184

Margesin R, Schinner F, Marx JC, Gerday C (eds) (2008) Psychrophiles: from biodiversity to biotechnology. Springer, Berlin

Margesin R, Moertelmaier C, Mair J (2013) Low-temperature biodegradation of petroleum hydrocarbons (n-alkanes, phenol, anthracene, pyrene) by four actinobacterial strains. Int Biodeterior Biodegrad 84:185–191

McKenzie P, Hughes DE (1976) Microbial degradation of oil and petrochemicals in the sea. In: Skinner FA, Carr JG (eds) Microbiology in agriculture, fisheries and food. Society for Applied Bacteriology, Symposium series no 4. Academic Press, London, pp 91–108

Panikov NS, Sizova MV (2007) Growth kinetics of microorganisms isolated from Alaskan soil and permafrost in soild media frozen down to –35°C. FEMS Microbiol Ecol 59:500–512

Pritchard H, Popovic M, Bajpai R, Mueller J G (1998) Metabolic capability and bioavailability; delivery, sustainability, and monitoring. In: Technical University Hamburg-Harburg (eds) Innovative potential of advanced biological systems for remediation. TU Hamburg, Harburg, pp 59–62

Ren S, Frymier PD (2003) Toxicity estimation of phenolic compounds by bioluminescent bacterium. J Environ Engin-ASCE 129:328–335

Rojo F (2009) Degradation of alkanes by bacteria. Environ Microbiol 11:2477–2490

Singh A, Ward OP (eds) (2004) Applied bioremediation and phytoremediation. Springer, Berlin

Singh H (2006) Mycoremediation: fungal bioremediation. Wiley, Hoboken

van Hamme JD, Singh A, Ward OP (2003) Recent advances in petroleum microbiology. Microbiol Mol Biol Rev 67:503–549

Walker JD, Colwell RR (1974) Microbial degradation of model petroleum at low temperatures. Microb Ecol 1:63–95

Whyte LG, Hawari J, Zhou E, Bourbonnière L, Inniss WE, Greer CW (1998) Biodegradation of variable-chain-lenght alkanes at low temperatures by a psychrotrophic *Rhodococcus* sp. Appl Environ Microbiol 64:2578–2584

Wiemken A (1990) Trehalose in yeast, stress protectants rather than reserve carbohydrate. A van Leeuwenhoek 58:209–217

ZoBell CE (1946) Action of microorganisms on hydrocarbons. Bacteriol Rev 10:1–49

Chapter 22
Heterologous Expression of Proteins from Cold-Adapted Yeasts in Suitable Hosts: Methods and Applications

Rosli Md. Illias, Aizi Nor Mazila Ramli, Kheng Oon Low, Nor Muhammad Mahadi, Abdul Munir Abdul Murad and Amir Rabu

Abstract One of the prerequisites for the functional and structural characterisation of proteins is to obtain a sufficiently high amount of the protein sample. Recombinant protein expression systems are usually employed for the overproduction of proteins in cases where isolation from their native host is difficult. For each protein, a suitable expression system must be optimised to obtain the sample in a soluble, correctly folded conformation with high production yield. The recent discovery of psychrophilic yeasts and their potential in industrial applications have sparked interest in overexpression strategies for high-level expression of psychrophilic proteins. Here, we discuss some of the recent developments in the recombinant production of cold-adapted yeast proteins, particularly those from *Glaciozyma antarctica* PI12, in suitable heterologous hosts. Findings from our work, as well as from recent publications, are included.

Keywords Cold-adapted protein · *Escherichia coli* · *Komagataella pastoris* · Cold induction · Psychrophilic yeast

---

R. Md. Illias (✉) · A. N. M. Ramli · K. O. Low
Department of Bioprocess Engineering, Faculty of Chemical Engineering, Universiti Teknologi Malaysia, 81310 Skudai, Johor, Malaysia
e-mail: r-rosli@utm.my

N. Muhammad Mahadi
Malaysia Genome Institute, 43000 Kajang, Selangor, Malaysia

A. M. Abdul Murad · A. Rabu
School of Biosciences and Biotechnology, Faculty of Science and Technology, Universiti Kebangsaan Malaysia, 43600 Bangi, Selangor, Malaysia

P. Buzzini and R. Margesin (eds.), *Cold-Adapted Yeasts*,
DOI: 10.1007/978-3-642-39681-6_22,

Contents

22.1 Introduction 482
22.2 Expression in *Escherichia coli* 483
22.2.1 Cold Induction 484
22.2.2 Coexpression of Chaperone 485
22.2.3 Fusion Technology: Solubility-Enhancing Tag 486
22.2.4 A 'One-Pot' Strategy 487
22.3 Expression in *Komagataella pastoris* 489
22.3.1 Cold Induction 489
22.3.2 Post-translational Modification 490
22.3.3 Multiple Copies of the Target Proteins 491
22.4 Successful Examples 492
22.4.1 Antifreeze Protein from *G. antarctica* PI12 492
22.4.2 Ice-Binding Protein from *Leucosporidium sp* 493
22.5 Conclusions 494
References 495

22.1 Introduction

Obtaining a sufficiently high quantity of target protein is one of the prerequisites for functional and structural characterisation studies. A suitable expression system is important to produce a high yield of soluble and correctly folded target protein. While expression in native host is simple and guarantees the authenticity and 'originality' of the target protein, the yield is often too low and contains significant amounts of other native proteins. Recombinant protein expression in a suitable heterologous host system is a feasible and largely successful approach. Strategies dedicated to enable high-level gene expression and rapid protein purification have been developed and can be customised to meet specific requirements for each protein. Impressive amounts of heterologous proteins have been obtained and well documented using common prokaryotic (i.e. *Escherichia coli*) and eukaryotic (i.e. *Komagataella pastoris*, formerly *Pichia pastoris*) hosts. However, a universal protein expression host is currently not available because not all types of proteins can be produced in a soluble and correctly folded conformation in high yields in a single expression system. The situation can be more challenging particularly when working with a group or class of heterologous proteins for which limited information on successful overexpression is available in the literature.

In recent years, psychrophilic yeasts (and their proteins) have increasingly gained attention in the areas of fundamental and applied studies. To withstand the extreme cold environment, psychrophilic yeasts have developed a series of unique mechanisms, including producing proteins that are functionally stable and active at low temperatures. One of the unique characteristics of psychrophilic proteins is their highly flexible catalytic centres, which are characterised by a high amount of long loops or random coils, thereby allowing the sufficient molecular motions necessary for activity in low-energy environment.

Currently, a majority of the studied psychrophilic yeast proteins have been produced in their respective native hosts. However, attempts at producing proteins from psychrophilic yeasts in heterologous hosts have found limited success. Only two expression hosts, namely *E. coli* and *K. pastoris*, have been shown to be capable of producing sufficient amounts of psychrophilic yeast proteins. Here, we will discuss some recent developments in recombinant production of cold-adapted yeast proteins, particularly those from *Glaciozyma antarctica* PI12, in suitable heterologous hosts. Due to the limited examples, this chapter aims to provide a more general background and strategies to produce cold-adapted proteins. Findings from our work, as well as from other recent publications, are included.

22.2 Expression in *Escherichia coli*

E. coli, the powerhouse for recombinant protein production for decades, is the most popular expression system for recombinant protein production. Although this tiny bacterium is not suitable for proteins that are rich in cysteine residues or those that require extensive post-translational modifications, it offers the easiest, most economical and the quickest route for the expression of heterologous proteins. Other advantages of *E. coli* as a host include cheap and simple cultivation; high product yields; wealth of scientific knowledge about its genetics, molecular biology and fermentation; and easy cloning and expression procedures (Demain and Vaishnav 2009).

Eukaryotic proteins are seldom categorised as 'problematic' or 'difficult to express' when prokaryotic hosts (i.e. *E. coli*) are used for their production. The complexity of the situation is magnified severalfold when one wishes to produce soluble and active eukaryotic psychrophilic proteins (i.e. cold-adapted yeast proteins). The major problem for such an expression is the low solubility and stability of heterologous proteins. Such difficulties arise due to several reasons. First, protein folding may be incomplete in the absence of eukaryotic chaperonins, resulting in inactive protein species, premature degradation or deposition into insoluble inclusion bodies. Second, some eukaryotic proteins are sensitive to heterologous proteases. Third, the difference in codon usage between the eukaryotic genes and the heterologous host may result in premature translation termination, leading to shorter forms of the target protein or very low expression levels. In some cases, the presence of a pseudo Shine-Dalgarno motif in the coding region of the eukaryotic gene may lead to truncated or non-functioning protein species (Tamura et al. 2011).

Despite the seemingly overwhelming problems when expressing eukaryotic proteins, particularly those from psychrophilic yeast, in *E. coli*, promising results have been obtained when suitable strategies have been employed. There are different approaches to improve the solubility of heterologous proteins, including, but not limited to, the following: (1) the use of promoters other than the T7 promoter, in particular, those activated by a temperature downshift; (2) coexpression with molecular chaperones; (3) fusion protein technology; (4) a decrease in the temperature; and/or v) different induction conditions. The most common, and perhaps the most important, strategy is to produce the target protein in a low-temperature environment, a strategy that is known as cold induction.

22.2.1 Cold Induction

The essence of the cold-induction technique is to initiate gene expression (and also protein production) in a cold environment, often at the temperature range of 15–18 °C. The optimum growth temperature for the mesophilic *E. coli* is 37 °C. When grown at a substantially low temperature (i.e. 18 °C), cellular processes in *E. coli*, including protein translation systems, drop to significantly low levels. A lower protein synthesis rate could efficiently reduce protein concentration in the cytoplasm, thus minimising inclusion body formation, and also increase the time available for the protein to fold into a correct/native conformation. In this context, it is interesting to discuss how 'low' the expression temperature should be. Niiranen et al. (2007) suggested that an expression temperature below 18 °C may not be advantageous for protein solubility. While the expression of a psychrophilic protein at low temperature (i.e. 16 °C) is not problematic, the protein is nevertheless less stable and requires stabilising factors from its native host. Overexpression of the target psychrophilic protein may quickly deplete the stabilising factors and lead to protein aggregation. To perform cold induction, recombinant host cells are grown to the exponential phase (A_{600nm} of 0.5–1.0) at 37 °C before immersion in cold water (4 or 18 °C) for approximately 20 min to lower the culture temperature. Then, an appropriate amount of inducer (i.e. 1.0 mM IPTG) is added to the culture, and the incubation is continued at 15–18 °C until a desired amount of heterologous protein is obtained (24 h or longer). Although this simple approach seems logical, it is not without drawbacks. Since *E. coli* is a mesophilic organism, the exposure of the bacteria to a cold environment (i.e. cold induction) would introduce stress or shock (termed cold shock) that may significantly alter gene expression and cell physiology. One of the main consequences of cold shock is the stabilisation of the secondary structures in nucleic acids (Phadtare 2011). The stabilised nucleic acid may prevent transcription, translation and RNA degradation, thus making recombinant gene expression less controllable and predictable. The expression of heterologous proteins following cold induction can be very low and requires a longer cultivation period, often up to 24 h or longer, compared with the 6–8 h required at 37 °C. Therefore, strategies have been devised to address low gene expression levels and slow cell growth even under cold temperatures.

The activity of CspA and its RNA chaperone homologues is essential to adaptation and survival of *E. coli* after cold shock. These proteins act as transcription antiterminators, enabling synthesis of proteins to maintain functional cellular processes, enabling cells to survive. *cpsA* mRNA, expressed at 15 °C,

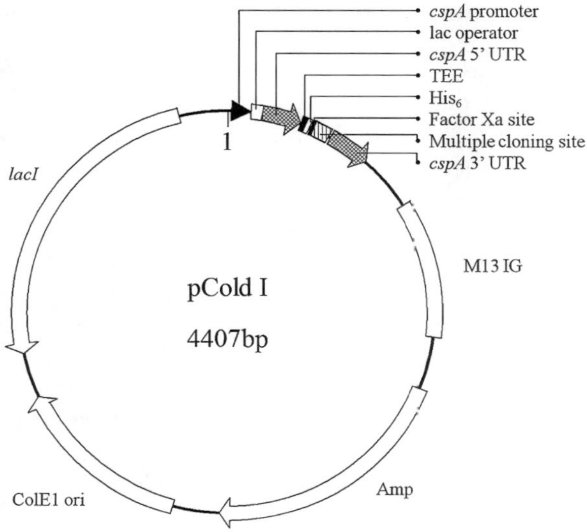

Fig. 22.1 Schematic structure of the pCold I vector (Qing et al. 2004). Important genetic features are labelled. The target gene is inserted using the multiple cloning sites. Gene expression is initiated by simultaneous incubation of the host cell under low temperature (15–18 °C) and the addition of IPTG (see Sect. 22.4 for detailed protocol)

effectively binds to most ribosomes and inhibits the synthesis of other cellular protein (Phadtare 2012). By taking advantage of the highly expressed *cpsA* under cold shock, Qing et al. (2004) constructed a series of cold-induction vectors, named as pCold vectors. Some important features of pCold vectors include *cpsA* promoter, *cpsA* 5′-UTR (untranslated region) with an optimum Shine-Dalgano sequence, *cpsA* five-codon translation-enhancing element (TEE) and *cpsA* 3′ end transcription terminator site (Fig. 22.1). Because the *cpsA* promoter and *lac* operator are utilised in pCold vectors, recombinant gene expression is initiated by the simultaneous addition of IPTG and a temperature shift (15–18 °C). By using a cold-induction protocol, Tamura et al. (2011) successfully expressed human *β*-actin in *E. coli*. Soluble *β*-actin was only obtained when using pCold vector with expression at 15 °C; however, expression using regular plasmids did not yield this product. In another study, the expression of *Gaussia* luciferase using the pCold system in *E. coli* yielded more soluble species at an incubation temperature of 15 °C compared to that at 25 or 37 °C (Rathnayaka et al. 2010).

22.2.2 Coexpression of Chaperone

Another strategy to improve soluble protein production and cell growth during cold induction is by coexpressing chaperones. Chaperones are a class of proteins that aid in maintaining normal cell processes, including protein folding and unfolding. The production of a slow-folding heterologous protein can overwhelm the host chaperones, leading to the accumulation of target proteins as inclusion bodies. Supplementing necessary chaperones, such as the chaperone set DnaK/DnaJ/GrpE

(KJE) or GroEL/GroES (ELS), ClpB, the small heat-shock proteins IbpA and IbpB, and the ribosome-associated trigger factor, can minimise aggregation and improve the solubility of many heterologous proteins (Correa and Oppezzo 2011). Genes encoding for recombinant proteins and chaperone(s) can be cloned into a single plasmid or into two or more separate plasmids. In both cases, it would be beneficial to have a different gene expression system (inducers) for the expression of the chaperone genes (inducer A) and the recombinant proteins (inducer B). Such an expression system offers a higher degree of flexibility when production optimisation is needed, i.e. to achieve a balance between chaperone gene expression and target protein synthesis. Other important considerations for efficient plasmid maintenance in recombinant host cells are the use of different antibiotic resistance genes and origins of replication. Some of the available commercial chaperone plasmid sets can be obtained from Takara Bio Inc. (www.takarabioeurope.com).

The most widely used chaperone coexpression strategy includes the coexpression of two classes of heat-shock proteins, such as GroEL (Hsp60) and DnaK (Hsp70). These proteins are responsible for the majority of the protein folding events in *E. coli*. At a reduced temperature (20 °C), the synthesis of these two chaperones was shown to be significantly downregulated when compared to their synthesis at 37 °C (Kim et al. 2005). Importantly, the overproduction of heat-shock proteins (DnaK, GroEL, GroES) has been shown to reduce cell viability at low temperatures (Kandror and Goldberg 1997). Such an event could minimise the final yield of heterologous proteins.

An interesting chaperone coexpression strategy that addresses the issue of low cell viability and weak chaperone gene expression was demonstrated by Ferrer et al. (2004). By coexpression of two chaperones—Cpn60 and Cpn10—from a psychrophilic bacterium, *Oleispira antarctica* RB8^{T}, Ferrer et al. (2004) successfully engineered a cold-resistant *E. coli* strain that could grow at a higher rate at 4 °C compared to a non-chaperone strain. The application of this strain to express a heterologous, heat-labile protein (esterase) at 4 °C yielded a 180-fold higher specific activity compared to a non-chaperone strain at 37 °C. The strain is commercially available under the trade name ArcticExpress™ (Stratagene, USA). It would be interesting to test the benefits of this strain in expressing psychrophilic yeast proteins.

22.2.3 Fusion Technology: Solubility-Enhancing Tag

Another approach for increasing protein solubility is by fusion with a solubility tag. Table 22.1 lists some of the commonly used fusion tags. One should be careful when considering tags that are designed specifically for rapid purification processes, although most solubility-enhancing tags may also serve purification purposes. The most commonly used purification tags are the histidine (his-tag) and strep tag. These tags do not show solubility-enhancing effects and should be chosen based on their merit in purification processes. On the other hand, fusion tags that consist of complete proteins, such as glutathione-S-transferase (GST),

Table 22.1 Examples of some commonly used fusion tags. Plasmids with specific solubility tags can be obtained from Addgene—a non-profit plasmid repository—and some major molecular biology companies, such as Novagen, New England Biolab and Stratagene

Solubility tag	Size (kDa)	Solubility prediction[a]
MBP	38.5	55 % insoluble
NusA	55	95 % soluble
Thioredoxin (Trx)	11.6	ND
GST	26	56 % soluble
His_6	2.4	ND

[a] According to a modified Wilkinson–Harrison statistical model; *ND* not determined

maltose-binding protein (MBP), and thioredoxin (trx), could improve target protein solubility and stability. While purification tags can be fused to the N- or C-terminus of the recombinant protein, the solubility-enhancing tags are usually placed at the N-terminus of the target protein. In cases where the fusion tag may interfere with subsequent experiments, such as biochemical and biophysical characterisation, a specific endoprotease cleavage site may be placed between the tag and the target protein. Several specific proteases used for this task are commercially available (e.g. thrombin, factor Xa and enterokinase).

So how does a fusion tag improve the solubility and/or stability of its target protein? While there is no concrete evidence to explain this phenomenon, it is believed that the solubility-enhancing tag limits the time that the protein remains in a partially unfolded state and provides another simpler mechanism/route for the protein to fold. The lifetime of partially unfolded intermediates strongly influences the propensity of proteins to aggregate, most likely by exhausting molecular chaperones (Idicula-Thomas and Balaji 2005). Among the many fusion tags (Table 22.1), the Trx tag may be suitable for soluble expression of psychrophilic yeast proteins. The Trx tag is a solubility tag derived from *E. coli* TrxA. This protein is an oxidoreductase that facilitates the reduction in other proteins and aids in catalysing disulphide bonds, thus making it suitable as a fusion partner. When TrxA is expressed in *E. coli*, it can accumulate in a fully soluble state of up to 40 % of the total cellular protein (Lunn et al. 1984). This result suggests that thioredoxin translates very efficiently; thus, if fused at the N-terminus, this property can be conferred to the partner target protein. The high thermal stability (Tm = 85 °C) of thioredoxin is another beneficial property that can contribute to fusion partner stabilisation (LaVallie et al. 1993). We routinely used Trx tag fusion technology to obtain high yield of soluble psychrophilic yeast proteins in *E. coli* host.

22.2.4 A 'One-Pot' Strategy

Different strategies for soluble protein expression can be combined into a more efficient process. For example, fusion with a solubility-enhancing tag and expression under cold-induction conditions can result in higher protein yield compared to the individual approach executed separately. Inouye and Sahara (2008) demonstrated the feasibility of expression of difficult-to-solubilise proteins of eukaryotic origin through fusion with a soluble tag, with the IgG-binding domain of protein A (ZZ-domain) and under cold-induction conditions. The outcome of their experiment was encouraging. When proteins were expressed either under cold-induction conditions (15 °C) or fused with the soluble tag (ZZ-domain), expression levels were low, with recombinant proteins mainly deposited as inclusion bodies. Only when the ZZ-domain-fused proteins were expressed under cold-induction conditions were the proteins soluble with a significant yield. Importantly, most of the expressed proteins retained their biological activities, indicating a correct and native-like folding of eukaryotic proteins in a prokaryotic host. It should be noted that some of the expressed proteins contain a high amount of cysteine residues (up to 31 cysteine residues). This strategy is also feasible for the expression of heat-labile proteins in *E. coli*. A study by Niiranen et al. (2007) showed that MBP and NusA tags are the best fusion tags to increase the solubility of cold-adapted *Vibrio* proteins in *E. coli*. On the other hand, Kataeva et al. (2005) demonstrated that the solubility of heterologous protein from thermophiles increased significantly when fused to a solubility tag (i.e. MBP, NusA and GST). From a total of 66 genes of thermophilic bacterium *Clostridium thermocellum* JW20, 60 genes were found to be expressed in soluble form when attached to an N-terminal fusion tag (MBP) under the induction temperature of 18 °C. In contrast, genes from the mesophilic bacterium *Shewanella oneidensis* MR-1 were largely insoluble (41 of 79 gene products), even in the presence of a fusion tag and low induction temperatures. These findings suggest that a fusion tag coupled with the cold-induction protocol is beneficial for the expression of proteins located at both extreme ends of the heat-sensitivity spectrum (psychrophilic, mesophilic, thermophilic). It should be noted that the solubility of the target protein decreased, even in the presence of a soluble tag, when the total protein size exceeded 60 kDa (Kataeva et al. 2005). It is generally believed that thioredoxin and GST are better suited for mammalian proteins, whereas MBP and NusA work well with bacterial proteins.

The ultimate ‘one-pot’ strategy is to combine cold induction, fusion tag and coexpression of chaperones into a single strategy. This strategy can be realised through the use of the pCold TF vector (www.takara.com). In addition to the beneficial properties of other pCold vectors (promoter of *cspA* and 5’UTR of *cpsA*), the pCold TF vector includes a chaperone gene—trigger factor (TF)—as a solubilisation tag located upstream of the multiple cloning site. It was shown that TF increased cell viability at 4 °C (Kandror and Goldberg 1997) and enhanced the activity of GroEL-dependent protein folding and degradation at 20 °C compared to 37 °C (Kandror et al. 1997). However, the usefulness of this approach for the high yield of soluble and correctly folded conformations of heterologous proteins from psychrophilic yeast requires further examination.

22.3 Expression in *Komagataella pastoris*

Although the *E. coli* system permits rapid and easy recombinant protein production, it lacks many of the post-translational modifications and protein folding mechanisms found in eukaryotes. Indeed, recombinant psychrophilic proteins expressed in a mesophilic host, such as *E. coli*, may contribute to a reduced stability of the protein (Sørensen 2010). Alternatively, the methylotrophic yeast *K. pastoris* may serve as a good candidate for the expression of functional eukaryotic psychrophilic proteins (i.e. cold-adapted yeast proteins). *K. pastoris* can be grown to high cell densities, is characterised by an efficient secretory system and contains many tools for molecular manipulation. Similar to its eukaryotic counterparts, *K. pastoris* possesses many typical eukaryotic post-translational modification systems and complex protein folding pathways. The suitability of this particular expression system for the expression of active eukaryotic psychrophilic proteins from yeast can be attributed to several factors: (1) ability to undergo cold induction; (2) capability to perform many eukaryotic post-translational modifications; (3) incorporation of multiple copies of the target gene; and (4) the presence of strong promoters to drive the expression of foreign gene(s) of interest, thus enabling production of large amounts of the target protein(s).

22.3.1 Cold Induction

Expression of eukaryotic psychrophilic proteins in *K. pastoris* can be advantageous because this species is a psychrotolerant yeast that can grow at temperatures as low as 12 °C (Jahic et al. 2003). Findings from our work have revealed that the expression of cold-adapted chitinase (CHI II) from the psychrophilic yeast *Glaciozyma antarctica* PI12 was optimal at 25 °C; however, its expression decreased with an increase in the cultivation temperature as shown in Fig. 22.2. By lowering the induction temperature to 25 °C, chitinase activity of CHI II exhibited an approximately 3-fold increase compared to the activity obtained at cultivation temperatures of 29 °C or higher. Cold induction in *K. pastoris* was performed as follows. First, the recombinant host cells were grown until an OD_{600nm} value of 2–6 was obtained (approximately 16–18 h). Thereafter, the cells were harvested, and the cell pellet was resuspended in the expression medium using a 1:5 to 1:10 dilution of the original culture volume. The culture growth was continued until the desired heterologous protein was obtained (up to 6 days). An appropriate amount of absolute methanol (100 %) was added to a final concentration of 1–4 % (v/v) every 24 h to induce gene expression.

Similar results have been reported for the expression of endochitinase from the Antarctic bacterium *Sanguibacter antarcticus* KOPRI 21702, where the optimal expression occurred at 25 °C (Lee et al. 2010). Li et al. (2001) also demonstrated that expression of psychrophilic proteins in *K. pastoris* at a lower temperature

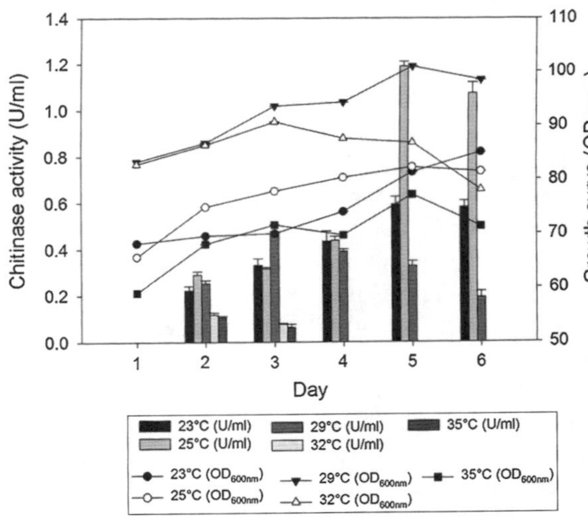

Fig. 22.2 Effect of induction temperature on recombinant CHI II expression and cell growth. Expression of recombinant CHI II in *K. pastoris* was optimum when the induction temperature was maintained at 25 °C. Each bar and point represents the mean ± standard error ($n = 3$)

(23 °C) dramatically increased the yield of the recombinant antifreeze protein, which might be due to the enhanced protein folding pathway as well as the increased cell viability at lower temperatures. Cultivating the cells at lower growth temperatures reduces the rate of protein synthesis and may therefore allow more time for the nascent peptide chains to fold properly.

22.3.2 Post-translational Modification

K. pastoris is a yeast expression system that has the advantage of eukaryotic protein processing and folding. This expression system provides the potential for producing soluble and correctly folded recombinant proteins, especially for eukaryotic proteins that have undergone all the post-translational modifications required for functionality. Therefore, *K. pastoris* has become an alternative expression system for recombinant cold-adapted yeast protein production that requires post-translational modifications. The enhanced stability of proteins in eukaryotes is often achieved by glycosylation. For example, in this species, *N*-glycosylation is a common post-translational modification. As a eukaryotic microbial expression system, *K. pastoris* is a good option for proteins for which expression in a bacterial system leads to the synthesis of improperly folded and inactive protein aggregates or inclusion bodies.

In our previous study, we reported on the expression of recombinant cold-adapted chitinase from the psychrophilic yeast *G. antarctica* PI12, which cannot be expressed in *E. coli* but can be produced as an active cold-adapted protein when expressed in *K. pastoris* (Ramli et al. 2011). In addition, other reports in the literature have described recombinant proteins (other than psychrophilic yeast protein) that do not fold correctly and that are generated as insoluble inclusion bodies when expressed in *E. coli*, but these proteins can be obtained as soluble and correctly folded proteins when expressed in *K. pastoris*, such as herring antifreeze protein (Li et al. 2001) and chitinase from the fungus *Beauveria bassiana* (Fan et al. 2007).

22.3.3 Multiple Copies of the Target Proteins

In addition to the attributes described above, the *K. pastoris* system has also demonstrated the ability to incorporate multiple copies of the expression cassette (with target gene insert) into the host genome through homologous recombination. The expression of recombinant proteins varies depending on the degree of gene recombination (copies of target gene in host genome). Generally, there are two types of vectors used in recombinant protein expression: integrative (integrated into host genome) and replicative (replicate independently from the host genome; has an origin of replication) vectors. An ideal vector is one that can be maintained in most cells in the population (although the selective pressure is absent) and has a high gene copy number for dosage-dependent high-level expression (Domínguez et al. 1998). Transformation in *K. pastoris* is performed by the integration of the expression vector containing the gene of interest into the chromosome at a specific locus to generate genetically stable transformants. This integration will maximise the stability of the expression strain. The employment of chromosomal integration is more desirable than replicative plasmid expression systems (such as in *E. coli*) due to the low copy number of the replicative plasmids that will eventually affect the amount of product expressed (Daly and Hearn 2005). In addition, transformants containing replicative plasmids may require the use of additives, such as antibiotics, to maintain the transformed population of cells. This procedure, in turn, will increase the production costs.

One approach that can be used to construct multi-copy strains involves the use of a vector with the bacterial *Sh ble* gene. The presence of the *Sh ble* gene marker in expression vectors of *K. pastoris* confers resistance to zeocin, which can be used as a tool to detect zeocin-resistant (zeo^R) transformants that harbour multiple copies of the recombinant gene. Strains transformed with the expression vector containing the zeocin marker can be selected directly by growth in the presence of the drug. Consequently, the transformant population can be enriched for multi-copy-expressing vector strains simply by plating on selection plates containing increased concentrations of zeocin (Cereghino and Cregg 2000). Higgins and Cregg (1992) demonstrated the possibility of enhancing the transformant population for colonies with multiple copies of a foreign gene by selecting for hyper-resistance against an appropriate drug.

In our study of the cold-adapted chitinase from *G. antarctica* PI12, the selection effects for hyper-resistance of zeocin were examined using yeast extract peptone dextrose sorbitol (YPDS) plates containing increasing concentrations of zeocin (500, 1,000 and 2,000 $\mu g\ ml^{-1}$). The transformants of the X-33 and GS115 strains, which have the tendency to produce multi-copy transformants, were randomly picked and transferred onto YPDS plates containing zeocin (500, 1,000 and 2,000 μg ml^{-1}). It is desirable to select transformants containing multiple gene integrations because such clones may potentially express significantly higher levels of the recombinant protein due to an increase in gene dosage. The isolation of multi-copy integrants has resulted in dramatically higher yields of the target proteins.

Due to the great potential of *K. pastoris* to perform large-scale production, this yeast system may be used to produce cold-adapted proteins from psychrophilic yeasts. A previous study has reported the use of two different fermentation protocols for expressing lipase A from *Candida antarctica* (now *Pseudozyma antarctica*) in the methylotrophic yeast *K. pastoris* in high yields (Pfeffer et al. 2006). In addition, Lee et al. (2010) reported the use of a fermentation strategy to increase the expression yield of endochitinase from the Antarctic bacterium *Sanguibacter antarcticus* KOPRI 21702 in *K. pastoris*.

22.4 Successful Examples

A literature survey revealed that only a few proteins from *Leucosporidium* sp. and *G. antarctica* have been successfully produced in recombinant hosts. Here, we summarise the methods and important findings from those experiments.

22.4.1 Antifreeze Protein from G. antarctica PI12

Our group has had a long-time interest on studying the structure–function relationships of psychrophilic proteins, particularly the proteins from *G. antarctica* PI12. The first successful attempt was the heterologous expression of chitinase (CHI II) in *K. pastoris*, which has been described in detail in the above Sect. 22.3). A subsequent study from our laboratory has described the structure–function relationship of CHI II using computational modelling (Ramli et al. 2012). However, dealing with the *G. antarctica* PI12 protein was not always a straightforward task. Expression of antifreeze protein, Afp1, resulted in its deposition in inclusion bodies in *E. coli*. Despite extensive efforts at optimisation (including the use of pCold I vector, cold-induction protocol and varied inducer concentrations), soluble expression of the target protein was not detected. Subsequent denaturation and refolding of the inclusion bodies have yielded a sufficient amount of soluble and functional proteins (Hashim et al. 2013). The protocol used is described below:

1. Inclusion bodies denaturation
    (a) Briefly, recombinant cells were grown in LB medium at 37 °C, and protein expression was induced by the addition of 1.0 mM IPTG. Cells were lysed according to the pET manual (Novagen).

(b) Inclusion bodies were collected by centrifugation at 16,000 × g at 4 °C for 20 min and denatured by incubation in Buffer B [8 M urea, 100 mM NaH_2PO_4 and 10 mM Tris–HCl (pH 8.0)] under slow shaking at 4 °C for 30 min.

(c) A soluble fraction (denatured target protein) was obtained by centrifugation at 10,000 × g for 30 min at 4 °C and subjected for purification. It should be noted that the Afp1 protein was expressed with an N-terminal 6xHis fusion tag (pET-28, Novagen).

2. Purification

(a) Four volumes of the protein solution were mixed with one volume of 50 % Ni–NTA resin (Qiagen) and incubated with slow shaking at 4 °C for 1 h.

(b) The mixture was then loaded onto a chromatography column, washed twice with Buffer C [8 M urea, 100 mM NaH_2PO_4 and 10 mM Tris–HCl (pH 6.3)], eluted twice with Buffer D [8 M urea, 100 mM NaH_2PO_4 and 10 mM Tris–HCl (pH 5.9)] and then eluted once with Buffer E [8 M urea, 100 mM NaH_2PO_4 and 10 mM Tris–HCl (pH 4.5)].

3. Refolding

(a) Fractions from the Buffer D eluent were used for refolding via dialysis in 400 volumes of dialysis buffer [10 mM Tris–HCl and 0.1 % Triton X-100 (pH 8.0)].

(b) The buffer was changed after 3 h, and dialysis was continued overnight at 4 °C.

(c) Unfolded aggregates were removed by centrifugation for 1 min at 16,000 × g at 4 °C. The refolded protein species can be analysed by dynamic light scattering, circular dichroism and activity assays.

The work presented here shows that denaturation and refolding of psychrophilic yeast protein is a workable solution for obtaining a sufficient amount of target protein when other available strategies failed to produce a desirable outcome. One main drawback of this approach is that the authenticity and 'native behaviour' of the target protein may be questionable if a comparison with the native protein is not conducted.

22.4.2 Ice-Binding Protein from *Leucosporidium* sp.

Another interesting example is the work of Park et al. (2012) who produced a soluble and functional ice-binding protein (IBP) from the Arctic yeast *Leucosporidium* sp. AY30 in *E. coli* BL21 (DE3) and *K. pastoris*. For expression in *E. coli*, pCold vectors and a cold-induction approach were used. Three expression plasmids were constructed for expression in *E. coli*: bLeIBP (IBP gene only), His-bLeIBP (6xHis tag and factor Xa site fused to IBP gene) and GFP-bLeIBP

(6xHis tag, factor Xa site and green fluorescent protein fused to IBP gene). Gene expression was conducted as per the following protocol:

1. A seed culture was prepared in LB medium supplemented with ampicillin (50 μg ml^{-1}) and incubated at 37 °C overnight with agitation.
2. The seed culture (0.1 % (v/v)) was inoculated into fresh LB medium (with 50 μg ml^{-1} ampicillin) and incubated at 37 °C until the OD_{600} reached 0.5.
3. The culture was cooled down by incubation at 15 °C for 1 h.
4. Gene expression was induced by the addition of IPTG (1.0 mM), followed by incubation at 15 °C for 20 h with agitation.

Most of the recombinant IBP was found in soluble fraction (more than 50 % of total expressed proteins), indicating the effectiveness of the expression system. Among the tested bacterial expression plasmids, His-bLeIBP yielded the most soluble protein at 24.5 mg l^{-1}, compared with 2.1 mg l^{-1} for bLeIBP and 19.3 mg l^{-1} for GFP-LeIBP. However, this yield is significantly less than the corresponding yield in *K. pastoris*, which was 61.2 mg l^{-1}. A codon-optimised mature IBP gene was inserted into a yeast pPICZαA expression vector and transformed into *K. pastoris* X33 strain in a multi-copy manner. Selected *K. pastoris* transformants were cultured in 24 ml of BMGY (buffered glycerol-complex medium) medium containing zeocin (100 μg ml^{-1}). The transformants were grown at 30 °C with agitation until the OD_{600} reached 2.0. Thereafter, the cells were pelleted by centrifugation and resuspended in 1 L of BMMY (buffered methanol-complex medium), and the cultivation was continued at 28 °C. Methanol (0.5 %, v/v), which acts as an inducer, was added to the culture every 24 h to induce recombinant gene expression.

22.5 Conclusions

The choice of host system lies heavily on the nature of the target protein, especially the requirements for post-translational modifications (i.e. glycosylation). Although the *E. coli* system ensures fast, easy and highly customisable and specialised genetic tools (pCold vectors, solubility and purification tag, etc.), the lack of a dedicated modification system limits its usefulness. One of the exciting developments in this field is the engineering of a protein glycosylation pathway in *E. coli* (Fisher et al. 2011). At present, the *K. pastoris* system may serve as a better expression host for proteins from cold-adapted yeasts because its eukaryotic expression system makes it advantageous for the production of near-native-like recombinant psychrophilic yeast proteins.

References

Cereghino JL, Cregg JM (2000) Heterologous protein expression in the methylotrophic yeast *Pichia pastoris*. FEMS Microbiol Rev 24:45–66

Correa A, Oppezzo P (2011) Tuning different expression parameters to achieve soluble recombinant proteins in *E. coli*: advantages of high-throughput screening. Biotechnol J 6:715–730

Daly R, Hearn MTW (2005) Expression of heterologous proteins in *Pichia pastoris*: a useful experimental tool in protein engineering and production. J Mol Recognit 18:119–138

Demain AL, Vaishnav P (2009) Production of recombinant proteins by microbes and higher organisms. Biotechnol Adv 27:297–306

Domínguez A, Fermiñán E, Sánchez M, González FJ, Pérez-Campo FM, García S, Herrero AB, San Vicente A, Cabello J, Prado M, Iglesias FJ, Choupina A, Burguillo FJ, Fernández-Lago L, López MC (1998) Non-conventional yeasts as hosts for heterologous protein production. Int Microbiol 1:131–142

Fan Y, Zhang Y, Yang X, Pei X, Guo S, Pei Y (2007) Expression of a *Beauveria bassiana* chitinase (Bbchit1) in *Escherichia coli* and *Pichia pastoris*. Protein Expr Purif 56:93–99

Ferrer M, Chernikova TN, Timmis KN, Golyshin PN (2004) Expression of a temperature-sensitive esterase in a novel chaperone-based *Escherichia coli* strain. Appl Environ Microbiol 70:4499–4504

Fisher AC, Haitjema CH, Guarino C, Çelik E, Endicott CE, Reading CA, Merritt JH, Ptak AC, Zhang S, Delisa MP (2011) Production of secretory and extracellular N-linked glycoproteins in *Escherichia coli*. Appl Environ Microbiol 77:871–881

Hashim NHF, Bharudin I, Nguong DLS, Higa S, Bakar FDA, Nathan S, Rabu A, Kawahara H, Illias RM, Najimudin N, Mahadi NM, Murad AMA (2013) Characterization of Afp1, an antifreeze protein from the psychrophilic yeast *Glaciozyma antarctica* PI12. Extremophiles 17:63–73

Higgins DR, Cregg JM (1992) *Pichia* protocols: methods in molecular biology. Humana Press, Totowa

Idicula-Thomas S, Balaji PV (2005) Understanding the relationship between the primary structure of proteins and its propensity to be soluble on overexpression in *Escherichia coli*. Protein Sci 14:582–592

Inouye S, Sahara Y (2008) Soluble protein expression in *E. coli* cells using IgG-binding domain of protein A as a solubilizing partner in the cold induced system. Biochem Biophys Res Commun 376:448–453

Jahic M, Wallberg F, Bollok M, Garcia P, Enfors S-O (2003) Temperature limited fed-batch technique for control of proteolysis in *Pichia pastoris* bioreactor cultures. Microb Cell Fact 2:6

Kandror O, Goldberg AL (1997) Trigger factor is induced upon cold shock and enhances viability of *Escherichia coli* at low temperatures. Proc Natl Acad Sci USA 94:4978–4981

Kandror O, Sherman M, Moerschell R, Goldberg AL (1997) Trigger factor associates with GroEL in vivo and promotes its binding to certain polypeptides. J Biol Chem 272:1730–1734

Kataeva I, Chang J, Xu H, Luan C-H, Zhou J, Uversky VN, Lin D, Horanyi P, Liu ZJ, Ljungdahl LG, Rose J, Luo M, Wang B-C (2005) Improving solubility of *Shewanella oneidensis* MR-1 and *Clostridium thermocellum* JW-20 proteins expressed into *Esherichia coli*. J Proteome Res 4:1942–1951

Kim YH, Han KY, Lee K, Lee J (2005) Proteome response of *Escherichia coli* fed-batch culture to temperature downshift. Appl Microbiol Biotechnol 68:786–793

LaVallie ER, Diblasio EA, Kovacic S, Grant KL, Schendel PF, Mccoy JM (1993) A thioredoxin gene fusion expression system that circumvents inclusion body formation in the *E. coli* cytoplasm. Nat Biotechnol 11:187–193

Lee SG, Koh HY, Han SJ, Park H, Na DC, Kim I-C, Lee HK, Yim JH (2010) Expression of recombinant endochitinase from the Antarctic bacterium, *Sanguibacter antarcticus* KOPRI 21702 in *Pichia pastoris* by codon optimization. Protein Expr Purif 71:108–114

Li Z, Xiong F, Lin Q, D'anjou M, Daugulis AJ, Yang DSC, Hew CL (2001) Low-temperature increases the yield of biologically active herring antifreeze protein in *Pichia pastoris*. Protein Expr Purif 21:438–445

Lunn CA, Kathju S, Wallace BJ, Kushner SR, Pigiet V (1984) Amplification and purification of plasmid-encoded thioredoxin from *Escherichia coli* K12. J Biol Chem 259:10469–10474

Niiranen L, Espelid S, Karlsen CR, Mustonen M, Paulsen SM, Heikinheimo P, Willassen NP (2007) Comparative expression study to increase the solubility of cold adapted *Vibrio* proteins in *Escherichia coli*. Protein Expr Purif 52:210–218

Park KS, Do H, Lee JH, Park SI, Kim E, Kim SJ, Kang SH, Kim HJ (2012) Characterization of the ice-binding protein from Arctic yeast *Leucosporidium* sp. AY30. Cryobiology 64:286–296

Pfeffer J, Richter S, Nieveler J, Hansen C-E, Rhlid RB, Schmid RD, Rusnak M (2006) High yield expression of Lipase A from *Candida antarctica* in the methylotrophic yeast *Pichia pastoris* and its purification and characterisation. Appl Microbiol Biotechnol 72:931–938

Phadtare S (2011) Unwinding activity of cold shock proteins and RNA metabolism. RNA Biol 8:394–397

Phadtare S (2012) *Escherichia coli* cold-shock gene profiles in response to over-expression/deletion of CsdA, RNase R and PNPase and relevance to low-temperature RNA metabolism. Genes Cells 17:850–874

Qing G, Ma LC, Khorchid A, Swapna GV, Mal TK, Takayama MM, Xia B, Phadtare S, Ke H, Acton T, Montelione GT, Ikura M, Inouye M (2004) Cold-shock induced high-yield protein production in *Escherichia coli*. Nat Biotechnol 22:877–882

Ramli ANM, Mahadi NM, Rabu A, Murad AMA, Bakar FDA, Illias RM (2011) Molecular cloning, expression and biochemical characterisation of a cold-adapted novel recombinant chitinase from *Glaciozyma antarctica* PI12. Microb Cell Fact 10:1–13

Ramli ANM, Mahadi NM, Shamsir MS, Rabu A, Joyce-Tan KH, Murad AMA, Illias RM (2012) Structural prediction of a novel chitinase from the psychrophilic *Glaciozyma antarctica* PI12 and an analysis of its structural properties and function. J Comput Aided Mol Des 26:947–961

Rathnayaka T, Tawa M, Sohya S, Yohda M, Kuroda Y (2010) Biophysical characterization of highly active recombinant Gaussia luciferase expressed in *Escherichia coli*. BBA—Proteins and Proteomics 1804:1902–1907

Sørensen HP (2010) Towards universal systems for recombinant gene expression. Microb Cell Fact 9:1–4

Tamura M, Ito K, Kunihiro S, Yamasaki C, Haragauchi M (2011) Production of human β-actin and a mutant using a bacterial expression system with a cold shock vector. Protein Expr Purif 78:1–5

Chapter 23
Food Spoilage by Cold-Adapted Yeasts

Anna Maráz and Mónika Kovács

Abstract This Chapter gives an overview of the spoiling potential of cold-adapted yeasts under different food ecological conditions. Special attention is paid to the physiological characterisation of the most important food-spoilage yeasts, illustrating the elevated tolerance of cold-adapted spoilage yeasts to one or more food-related stress factors. Finally, spoilage yeast associations and food deterioration problems caused by cold-adapted spoilage yeasts in different type of foods (fruits, vegetables, beverages, dairy and meat products) are introduced in details.

Keywords Food-spoilage yeasts · Low temperature · Food deterioration

Contents

23.1 Introduction 498
23.2 Cold-Adapted Spoilage Yeasts in the Food Ecosystems 499
  23.2.1 Main Characteristics of Food-Spoilage Yeasts 499
  23.2.2 Symptoms of Food Deterioration Caused by Spoilage Yeasts 501
  23.2.3 Biotic and Abiotic Factors Influencing the Survival and Growth of Yeasts in Food 502
23.3 Ecology and Physiology of Cold-Adapted Spoilage Yeasts in Foods at Low Temperatures 505
  23.3.1 Effect of Low Temperature on the Growth of Yeasts 506
  23.3.2 Tolerance and Adaptation of Spoilage Yeasts to Food-Related Stresses 508
23.4 Problems Caused by Cold-Adapted SpoilageYeasts in Food 510
  23.4.1 Spoilage of Fruits and Vegetables by Cold-Adapted Yeasts 511
  23.4.2 Beverage Spoilage by Cold-Adapted Yeasts 514

A. Maráz (✉) · M. Kovács (✉)
Department of Microbiology and Biotechnology, Faculty of Food Science, Corvinus University of Budapest, Budapest, Somlói ut 14-16 1118, Hungary
e-mail: anna.maraz@uni-corvinus.hu

M. Kovács
e-mail: monika.kovacs@uni-corvinus.hu

P. Buzzini and R. Margesin (eds.), *Cold-Adapted Yeasts*,
DOI: 10.1007/978-3-642-39681-6_23,

23.4.3 Spoilage of Dairy Products by Cold-Adapted Yeasts 516
23.4.4 Spoilage Caused by Cold-Adapted Yeasts of Raw Poultry, Meat, Fish and Derived Products 520
23.5 Strategies to Reduce the Impact of Cold-Adapted Spoilage Yeasts in Foods 523
23.6 Conclusions 525
References 526

23.1 Introduction

Food is generally considered spoiled when it is no longer suitable for human consumption. Spoilage may concern food safety issues in special cases when it causes the illness or even death of consumers; fortunately, in most cases, it merely impacts the quality and not the safety of food.

Spoilage becomes evident by changes in the colour, texture, flavour and taste, depending on the types and components of food. Three main categories of spoilage are distinguished, namely the physical, chemical and microbiological spoilage, although growth of microorganisms frequently leads to changes in the physical and/or the chemical nature of food. All three types of spoilage are influenced by certain ecological factors, the most important being temperature, pH, water activity, exposure to oxygen and light, and chemicals or nutrients available for chemical, enzymatic and microbiological processes (Huis in't Veld 1996; Singh and Anderson 2004).

From an ecological point of view, food is considered a habitat of different microorganisms. Raw material, processing equipments and environment, handling and packaging are the main sources of the contaminating microorganisms. During processing and storage, a limited number of microbial species will grow and cause deterioration. Microorganisms typically associated with food spoilage include a broad range of bacteria, moulds and yeasts; the developing spoiling association is, however, determined by the physical, chemical and biological properties of the products (intrinsic factors), the effects of the surrounding environment (extrinsic factors), the properties and interaction of microorganisms (implicit factors) as well as the modes of food processing and preservation.

Microbial populations will interact directly or indirectly within the food ecosystems including those inoculated as starter cultures for production of specific types of food. In the spoiling associations, yeasts have increased stress tolerance to survive and start growing if the environmental conditions are not favourable to the generally less tolerant bacteria. Yeasts also compete with certain moulds, which can tolerate low water activity, acidic conditions and low temperatures (Fleet 1990a; Huis in't Veld 1996; Pitt and Hocking 1997; Fleet 1999; Deák 2006; Viljoen 2006).

23.2 Cold-Adapted Spoilage Yeasts in the Food Ecosystems

Yeast are ubiquitous microorganisms that are found practically everywhere in the biosphere where organic material is available as a nutrient. They live under all climatic conditions in nature, from the tropical to the polar regions, as has been reviewed in recently published textbooks (Rosa and Peter 2006; Satyanarayana and Kunze 2009; Kurtzman et al. 2011). Beside natural habitats, yeast species are also members of the agri-food ecosystems (Fleet 1990a; Deák 1991; Fleet 1992; Tudor and Board 1993; Stratford 2006; Deák 2008; Fleet 2011). Soil, water, air, plants and animals are all habitats of microbial consortia, comprising the biodiversity of yeasts (Phaff and Starmer 1987; Fleet 1992; Botha 2006; Deák 2006; Fonseca and Inacio 2006; Deák 2008). The majority of yeast species isolated from these habitats belong to the basidiomycetes, which can be attributed to their extended tolerance and adaptation to stressful conditions. The most frequently found species belong to the genera of *Cryptococcus, Rhodotorula, Sporobolomyces, Trichosporon* and *Rhodosporodium.* Stress tolerance is also characteristic to certain ascomycetes living under these circumstances; *Debaryomyces*, *Lipomyces*, *Candida* and *Metschnikowia* are the prevalent ascomycetous yeast genera (van Uden 1984; Nagahama 2006; Choudhary and Johri 2009).

23.2.1 Main Characteristics of Food-Spoilage Yeasts

Pitt and Hocking (1985, 1997) listed the 12–13 most frequent yeast species that cause spoilage in a wide range of foods, which have been processed and packaged under the normal standards of good manufacturing practice (GMP). They assumed that these spoiling species together with the other food-associated yeasts originate from the surrounding environment as contaminating microbes during harvesting the raw material, and handling, processing and packaging food. Tudor and Board (1993) confirmed the list of these most frequent food-spoiling yeasts (called the "top ten") and emphasised that the enrichment of certain yeasts during storage is the consequence of single or combined abiotic stress factors (mainly the low water activity and acidic conditions) and, as a result, the biomass of the most stress-tolerant yeast strain(s) will increase to such an extent that this will cause symptoms of spoilage.

On the other hand, under certain environmental conditions, an additional group of yeasts can cause spoiling, which was described as the "second division" of spoiling yeasts by Tudor and Board (1993). These yeast species were also thought to be selected from the contaminating microbiota and well adapted to one or other environmental extremes. Stratford (2006) made a comprehensive survey of the "top-ten" or "dangerous" spoiling yeast species and pointed out that these species are not generally isolated at a much higher frequency from food than the other frequently isolated food-associated yeasts, nor do they have higher growth rates under normal conditions. Their special physiological features (osmotolerant, acid tolerant, preservative resistant and highly fermentative) enable them to proliferate in food environments that are unfavourable to the great majority of yeast species.

In the new edition of the yeast monograph, Fleet (2011) published the list of yeasts that are frequently associated with the spoilage of foods and beverages. It comprises the lists of "dangerous" spoiling yeasts published by Tudor and Board (1993), Pitt and Hocking (1997) and Stratford (2006), with four other yeast species added, which were the members of the "second-division spoiling yeasts". This list contains some species (e.g. *Saccharomyces cerevisiae, Kluyveromyces marxianus, Schizosaccharomyces pombe, Debaryomyces hansenii*) that are also beneficial in the production or maturation of certain foods, or are even used as starter cultures or adjuncts in different food fermentations, but under special conditions or in an other type of food, these yeasts may cause spoiling. Fleet (2011) also noted that the spoiling potential of different strains belonging to a given species may alter; therefore, certain strains could be selected as beneficial and others could be considered spoiling.

It is worth noting that in the lists of the most frequent spoilage yeast species described by the above-mentioned authors, tolerance to low temperatures (around 8 °C or less) is indicated only in rare cases. *K. marxianus, D. hansenii, Kazachstania exigua, S. cerevisiae, S. pombe, Yarrowia lipolytica, Wickerhamomyces anomalus, Pichia kudriavzevii* and *Rhodotorula* spp. belong to the group of spoiling yeasts that are able to grow at low (refrigeration or freezing) temperatures. The cold tolerance of these species is considered, however, not the primary characteristic, but as an additional one with a role in the spoiling potential. As the primary characters, halotolerance (e.g. *D. hansenii, W. anomalus*), and resistance to high sugar concentrations (e.g. *S. pombe*) are indicated, and these species are sometime considered as extremophiles from these respects. In other cases, high enzymatic (e.g. pectinolytic, proteolytic and/or lipolytic) activities (e.g. *K. marxianus, Y. lipolytica, Rhodotorula* spp.) enable the species to increase their populations and making them able to spoil the product.

Some of the typical spoilage yeasts are not able to grow under cold conditions, even in culture media. Among the food-spoiling *Zygosaccharomyces* species, *Zygosaccharomyces bisporus, Zygosaccharomyces bailii, Zygosaccharomyces mellis* and *Zygosaccharomyces rouxii* were found not to grow at 5 °C or lower, together with *Dekkera anomala, Dekkera bruxellensis, Candida parapsilosis, Candida tropicalis* and *P. kudriavzevii* (Stratford 2006); however, this characteristic seems to be strain-specific in certain cases (Fleet 2011).

Molecular identification has led to the discovery of two new, physiologically related *Zygosaccharomyces* species, namely *Zygosaccharomyces lentus* and *Zygosaccharomyces kombuchaensis*, which are different from all the other currently known *Zygosaccharomyces* species, but resemble *Z. bailii* most closely. Special common characteristics of these two species are good growth at cold temperature (4 °C), failure to grow at 30 °C and sensitivity to oxidative stress. Other main characteristics resembling to *Z. bailii* are tolerance to acidic pH (close to pH 2), osmotolerance (up to 60 % glucose) and resistance to benzoic acid (minimum 2 mM) (Steels et al. 1999a, b, 2002). *Z. kombuchaensis* can be distinguished physiologically from *Z. lentus* only by its sorbic acid sensitivity (Steels et al. 2002). Based on the overall physiological characteristics of *Z. lentus* and *Z. kombuchaensis,* these species could be considered as "dangerous" spoilage organisms of low-pH, high-sugar foods and drinks, especially chilled products. It is highly probable that the formerly identified psychrotolerant strains of *Z. bailii* belong to *Z. lentus* or *Z. kombuchaensis*.

23.2.2 Symptoms of Food Deterioration Caused by Spoilage Yeasts

Based on the extensive study of yeast species isolated from a wide range of raw and processed food, it was found that approximately 10 % (100–150 species) of the accepted yeast species (about 1,500) in the current taxonomic classification (Kurtzman et al. 2011) could be recovered from food (Deák 1991; Fleet 1992; Tudor and Board 1993; Stratford 2006; Deák 2008; Fleet 2011). As mentioned before, only about 20 of these food-associated species are recognised as causing spoilage outbreaks. Spoilage is the consequence of yeast growth, i.e., increases of cell biomass, but it is difficult to find a close correlation between the beginning of spoilage and cell number, because not only the yeast biomass but also their metabolites contribute to spoilage. In general, symptoms of spoilage appear when yeast cell numbers increase to approximately 5 log CFU g^{-1} and become evident if cell numbers increase to about 7 log CFU g^{-1} (Fleet 1992; Deák and Beuchat 1996).

The production of spoilage-related metabolites (e.g. carbon dioxide, hydrogen sulphide, ethanol, acids) depends on the type of food and the metabolic characteristics of the yeast; however, sensory effects (e.g. visible spots or layer on the surface, formation of sediment and off-flavours or off-odours) could be principal in several cases. Therefore, the symptoms of spoilage caused by the same yeast species might be different in cases of different types of food and different yeast species could generate similar symptoms of spoilage. There are very detailed studies available describing the typical spoilage effects in different types of food and beverages and thanks to the routine molecular identification and typing techniques, as well as the development of differential and selective culture media, the causative yeast species and strain(s) of a spoilage outbreak could be easily recognised and exactly described (Fleet 1992; Deák and Beuchat 1996; Loureiro and Querol 1999; Loureiro 2000; Deák 2004, 2008; Fleet 2011). Stratford (2006) made an excellent survey of spoilage symptoms and the causative yeast species and performed a detailed analysis of the physiological background of the characteristic spoiling yeasts together with the role of the physical, chemical and textural characteristics of the food matrix in deterioration.

Spoilage accompanied by excess gas (carbon dioxide) production is the most frequently appreciated problem, which is caused by strongly and weakly fermenting species such as *Z. bailii, S. cerevisiae, D. bruxellensis, Saccharomycodes ludwigii,*

Fig. 23.1 Blown package of fruit juice caused by yeast spoilage at refrigeration temperature (*left*); as a control normal package is also shown (*right*)

or *C. parapsilosis* and *Candida pseudointermedia*, respectively (Fig. 23.1). Visible symptoms of spoilage are very variable and include the sediment and haze/cloudy formation, surface film formation and alterations in the texture and colour of food, from which the surface film formation is characteristic of the majority of spoiling yeast species. Off-taste and off-odour are the consequences of the production of primary and secondary metabolites by yeasts, like ethanol, acetaldehyde, acetic acid, ethyl acetate, hydrogen sulphide and more characteristic volatile compounds such as 4-ethylphenol produced by *D. bruxellensis*.

Table 23.1 gives an overview of the most frequent food-associated cold-adapted spoilage yeasts.

23.2.3 Biotic and Abiotic Factors Influencing the Survival and Growth of Yeasts in Food

In the early dates of food mycology, it was recognised that environmental factors play a crucial role in the growth and selection of food-spoiling yeasts from food-associated microbiota (Mossel and Ingram 1955; Ingram 1958; Walker and Ayres 1970). Since that time, huge progress has been made in the ecology of foodborne yeasts, which is indicated by the publication of several separate book chapters and review articles dealing with yeast ecology (Pitt and Hocking 1985; Fleet 1990a; Deák 1991; Fleet 1992; Tudor and Board 1993; Deák and Beuchat 1996;

Table 23.1 Significant cold-adapted spoilage yeasts

Yeasts		Significant physiological properties	Frequently spoiled food products	Spoilage symptoms
Species name (Teleomorphic)	Anamorphic species name			
Debaryomyces hansenii	*Candida famata*	Salt tolerance Tolerance to high sugar concentrations	Brined vegetables, salad dressing, cheeses, salted meat products	Surface biofilm and layer, Off-flavours
Dekkera anomala	*Brettanomyces anomalus*	Strong fermentation	Alcoholic beverages (especially wine)	Haze, production of organic acids, off-flavours
Dekkera bruxellensis	*Brettanomyces bruxellensis*	Ethanol tolerance		
Hanseniaspora uvarum	*Kloeckera apiculata*	Acid tolerance Ethanol tolerance	Fruits and vegetables, fruit juices, wines	Gas formation, off-flavours
Kazachstania exigua	*Candida holmii*	Moderate salt tolerance Moderate tolerance to acetic acid Moderate resistance to preservatives	Fruits and vegetables, salad dressing, fruit juices, meat and dairy products	Gas formation
Kluyveromyces marxianus	*Candida kefyr*	Lactose fermenting Pectinolytic Proteolytic	Plant products, dairy products	Gas formation, yeasty odour
Kluyveromyces lactis	–	Lactose fermenting Proteolytic	Dairy products	Gas formation, yeasty odour
Pichia kudriavzevii (formerly *Issatchenkia orientalis*)	*Candida krusei*	Tolerance to acetic acid Ethanol tolerance Resistance to preservatives	Fermented vegetables, dairy products	Gas formation, surface layer
Pichia membranifaciens	*Candida valida*	Tolerance to acetic acid Resistance to preservatives	Fruits, fermented vegetables, fruit juices, fermented dairy and meat products	Gas formation, surface biofilm and layer, off-flavours
Rhodosporidium toruloides *Rhodotorula mucilaginosa*		Proteolytic Lipolytic	Fruits, fermented vegetables, beverages, dairy and meat products	Red-pink colonies, softening, haze
Saccharomyces cerevisiae	–	Strong fermentation Ethanol tolerance	Fruit juices, beverages, dairy products	Gas formation, haze, Yeasty odour

(continued)

Table 23.1 (continued)

Yeasts		Significant physiological properties	Frequently spoiled food products	Spoilage symptoms
Species name (Teleomorphic)	Anamorphic species name			
Saccharomycodes ludwigii		Tolerance to acetic acid Strong fermentation Ethanol tolerance	Fruit juices, beverages,	Gas formation, haze
Schizosaccharomyces pombe		Tolerance to high sugar concentrations Strong fermentation Ethanol tolerance Resistance to preservatives	Fruit juices, beverages	Gas formation, haze, off-flavours
Yarrowia lipolytica	*Candida lipolytica*	Proteolytic Lipolytic	Meat and dairy products	Textural changes, off-flavours
Candida alimentaria		Proteolytic Lipolytic	Meat and dairy products	Textural changes, off-flavours
Wickerhamomyces anomalus (formerly *Pichia anomala*)	*Candida colliculosa*	Acid tolerance Salt tolerance Ethanol tolerance	Fruit and dairy products	Surface biofilm and layer, off-flavours
Zygosaccharomyces lentus		Tolerance to acetic acid Tolerance to high sugar concentrations Resistance to preservatives	Fruit juices, beverages	Gas formation, haze
Zygosaccharomyces kombuchaensis		Tolerance to acetic acid Resistance to preservatives	Fruit juices, beverages	Gas formation, haze

Sources Tudor and Board (1993), Stratford M (2006), Fleet (2011)

Pitt and Hocking 1997; Deák 2004, 2006; Stratford 2006; Fleet 2007; Deák 2008; Fleet 2011). Significant progress in the molecular analysis, identification and taxonomy of yeasts, elaboration of the *S. cerevisiae* and several other yeasts' genome projects and construction of genomic databases, as well as achieving new scientific results in the biochemical and genetic analysis of the signal transduction routs, in cell-to-cell communications and in stress responses during the last decades, led to the more precise identification of contamination sources and the yeasts themselves and promoted understanding the behaviour of yeasts under stressful conditions and adaptation to the changing environment.

It has been illustrated in a number of research papers that the increased tolerance of several yeast species against individual stress factors or the combination of them renders these yeasts able to grow to that size of populations which cause food spoilage. This topic was extensively reviewed by Deák (1991, 2004, 2006, 2008); Deák and Beuchat (1996); Fleet (1990a, 1992); Pitt and Hocking (1985); Stratford (2006); Tudor and Board (1993); Walker and Ayres (1970).

Stress factors that favour the growth of yeasts in the food ecosystem are mainly the low water activity (A_w), high solute concentration in the form of sugars and salts, low temperature, starvation, carbon dioxide, low redox potential, as well as antimicrobials added as preservatives for controlling the growth of bacteria, yeasts and moulds or being present as natural antimicrobial compounds of plant and animal tissues. In special cases, limitation in the available oxygen poses another stress factor. Tolerance of different food-associated yeasts against these stress factors and combinations of them are well documented in the above-mentioned literature. Two current references are cited only, which make a comprehensive analysis and overview of this topic (Deák 2008; Fleet 2011). Detailed analysis of these stress factors is not the aim of this Chapter; the effect of low temperature on the survival and growth of food-associated yeasts and interactions of this physical factor with other abiotic environmental factors will be discussed in detail.

23.3 Ecology and Physiology of Cold-Adapted Spoilage Yeasts in Foods at Low Temperatures

Food-associated yeasts frequently originate from the surrounding environment as contaminating microbes during the harvesting and handling of raw material; therefore, the characteristic yeast biota reflects the agricultural environment. All of the further manipulation and processing will repress the less tolerant yeast cells and provide the opportunity for growth of the more stress-adaptive cells. It is also important to consider that the processing environment is another source of contaminating microbes where several well-adapted cells can occur (Ingram 1958; van Uden 1984; Deák 1998; Stratford 2006; Deák 2008).

The general purpose of the application of low temperatures during food production is to increase the storage time and self-life of food raw material, namely harvested fruits and vegetables, milk, poultry, meat and fish, and their processed products, such as salads, heat-treated milk, dairy and other livestock products. It is expected that decreasing the temperature to around 5 °C or below, the growth and metabolic activity of the microbiota will be restricted or even cease, but the difference in the cold stress response of the associated microorganisms can lead to the selection and ongoing multiplication of the more stress-tolerant cells. Yeasts are considered fairly tolerant to cold stress, but growth of the surviving cells is mostly influenced by the adaptation ability to the other stressful conditions typical to the type of food. If the population of the selected cells grows to a size that generates deterioration, food will be spoiled.

In some cheeses and fermented meat products, yeasts contribute to maturation by special enzymatic activities (e.g. proteases, lipases) and the generation of organoleptic metabolites at low temperatures (e.g. *D. hansenii*, *Galactomyces geotrichum*); therefore, they have a beneficial role at this stage. The same yeast species, however, can cause spoiling in other types of foods, even in the same category (e.g. in other types of cheese). It is important to emphasise that quite large differences have been detected among the beneficial and other strains belonging to the same species that support the selection and application of starter and adjunct yeast strains to be used for food fermentation (Fleet 1992; Deák 2008; Tamang and Fleet 2009).

Permanent low temperatures are maintained during processing of certain food, especially dairy, livestock and fish products, which allows the adaptation of psychrotolerant yeasts in the processing line that could be the source of contamination and may lead to spoilage outbreaks.

23.3.1 Effect of Low Temperature on the Growth of Yeasts

Considering the cardinal (minimum, optimum and maximum) temperatures for growth, yeasts have a quite wide range, but this varies between species. Yeasts are generally not able to grow above 50 °C and most species have a maximum temperature around 40 °C. As regards the minimum temperature for growth, a considerable number of yeast species are able to grow around or below 5 °C under optimal nutritional conditions (e.g. in complex yeast culture media) as a pure culture (van Uden 1984; Watson 1987). In terms of the maximum temperature for growth, the majority of yeasts belong to the mesophilic category, but a considerable number of species are psychrophilic or psychrotolerant. Based on the presently accepted terminology suggested by Davenport (1980), the latter category of yeasts is considered as "cold-adapted" species. An extensive overview of the concept of psychrophily and psychrotolerance of cold-adapted yeasts is reported in Chap. 1.

The exposure of yeast cells to temperatures below 10 °C does not support the growth of the majority of the spoilage yeasts; several of them can, however, adapt to refrigeration or even to freezing conditions. It seems a general rule that cells in the exponential phase of growth are more sensitive to the physical and chemical stresses mediated by the food ecosystems than the stationary phase cells. The stationary phase is the consequence of exhausting of nutrients, meaning that cells are starved (Werner-Washburne et al. 1993; Herman 2002; Zakrajsek et al. 2011). Cells remain in this resting state until environmental factors are favourable for starting growth again or die if the environmental conditions do not support the multiplication of cells. Yeasts that are contaminating raw food material of plant and animal origin (e.g. fruits, vegetables, poultry, meat, seafood) originate typically from natural habitats such as soil, water, air and plants and animal themselves. The majority of yeasts cells are in the resting state under these circumstances, because the available nutrients are limited.

Besides nutrient limitation, yeasts are exposed to several other stress factors like solar radiation, highly acidic or alkaline conditions, low water activity and the presence of antimicrobial compounds. Under cold climatic conditions, low temperatures may pose an additional stress factor. Food-associated yeasts are typically exposed to similar stressful conditions and combinations of single stress factors; however, this varies according to the type and processing of food (Deák 2006, 2008). One example when yeasts in food are exposed to low temperature as a single physical stress is the chilled storage of proteinaceous fresh meat, poultry and fish; however, in these cases, yeasts have to compete with fast-growing psychrotolerant and psychrophilic bacteria of high proteolytic and lipolytic activity, thereby acting as an additional biotic stress factor during interaction.

It is not surprising that the biodiversity of yeasts associated with food at low temperatures, including freezing, is very similar to that found in the polar (e.g. Antarctic) regions. Davenport (1980), when studying the general characteristics of psychrotolerant yeasts, concluded that these were mainly non-fermentative and the majority belonged to the basidiomycetes. Several of them (e.g. *D. hansenii*, *Cryptopcoccus albidus* and *Rhodotorula* spp.) were able to grow at subzero temperatures. Among the most frequent psychrotolerant yeast genera isolated from the Antarctic and similar extreme conditions (Gunde-Cimerman et al. 2003; Raspor and Zupan 2006; Shivaji and Prasad 2009), a considerable number of genera, such as *Bullera, Candida, Cryptococcus*, *Cystofilobasidium, Debaryomyces, Leucosporidium, Metschnikowia, Mrakia, Pseudozyma*, *Rhodotorula* and *Trichosporon* comprise species already identified as food-associated psychrotolerant yeasts (Davenport 1980; Dijkman et al. 1980; Schmidt-Lorenz 1982; Banks and Board 1987; Russell 2009).

Guerzoni et al. (1993) investigated 28 commercial chilled food products from seven manufacturers, including fish and meat, mayonnaise salads, cooked vegetables and intermediate moisture vegetables and fish food. The yeast species appeared to be rather uniform irrespective of the type of food; only the frequencies varied. The majority of the isolates belonged to *Y. lipolytica*, *Pichia membranifaciens* and *D. hansenii*. Physiological characterisation of the representative isolates belonging to these three species (minimum growth temperature, tolerance to low water activity and preservatives, protease and lipase activity, cell hydrophobicity) revealed that they have very few common characteristics. The authors came to the conclusion that the spoilage potential of each species is linked to different physiological peculiarities. Diriye et al. (1993) isolated yeasts from the surface of frozen fish, poultry, vegetables and mushroom and identified yeast strains as belonging to the genera *Candida, Pichia, Cryptococcus, Rhodotorula* and *Saccharomyces*; the most frequent species were *Candida zeylanoides, Y. lipolytica, Trichosporon cutaneum* and *Rhodotorula rubra* (now *Rhodotorula mucilaginosa*). Recently published textbooks and reviews analysed the typical yeast biota of different chilled and frozen foods (Fleet 2003; Fröhlich-Wyder 2003; Samelis and Sofos 2003; Stratford 2006; Fleet 2007; Deák 2008; Tamang and Fleet 2009; Fleet 2011) and came into similar conclusions as the previously referred authors, namely, those yeasts which survived the cold stress and prolonged exposure to low temperature (i.e. refrigeration or freezing) and were associated with food raw materials and processed food became cold-adapted; the culturable cells of these populations could be isolated as typical yeast biota associated with these types of food. It is not surprising that these species belong to the same genera of basidiomycetes and ascomycetes yeasts which harbour the most cold-adapted yeast species; several of them are members of the spoiling associations in chilled and frozen food.

The length of the lag phase of a growth curve is a good indication of the adaptation period of the cells to certain food-derived stress factors like high salt and sugar concentrations, low pH, low temperature and preservatives. The application of low temperature (between 3 and 10 °C) as a single stress for *D. hansenii, P. membranifaciens* and *Y. lipolytica* food isolates by Guerzoni et al. (1993) showed that *D. hansenii* had the shortest lag phase among the three yeasts, but *Y. lipolytica* had a comparably short lag phase. It was concluded by Betts et al. (1999, 2000) that the lag phase was the most important factor affecting the spoilage potential of a cocktail of spoilage yeasts and the low temperature had a major effect on the lag phase in high-salt, low-pH food.

23.3.2 Tolerance and Adaptation of Spoilage Yeasts to Food-Related Stresses

What is the effect of temperature decrease to the level of minimum temperature of growth or below, including freezing, and why can different yeast species tolerate it to a different extent? There are quite a few data in answer to these questions, especially in quantitative manner concerning yeast species other than *S. cerevisiae*, although this species is an excellent model in the analysis of different stress responses and tolerance, including food-relevant stresses.

Biochemical, genetic and molecular responses of *S. cerevisiae* to different physical and chemical environmental stress factors have been extensively studied by using chemogenomics, transcriptomics, proteomics and metabolomics approaches (Smits and Brul 2005; Walker and van Dijck 2006; Zakrzewska et al.

2011). A recent publication by Tanghe et al. (2006) provided an excellent overview of the general and specific responses of yeasts to the most important environmental stress factors, where the stress responses of the highly osmotolerant yeasts (*Z. rouxii, D. hansenii*), the thermotolerant yeast *Ogataea polymorpha* (formerly *Hansenula polymorpha*), the human pathogenic *Candida albicans* and the other model yeast *S. pombe* were also analysed at a molecular level beside *S. cerevisiae*. Focused discussion of *D. hansenii* (teleomorph of *Candida famata*), mainly from a biotechnological point of view, displays the adaptive and tolerance mechanisms of cells to high-salt environment, which is highly supported by the available genome sequence data of this yeast (Aggarwal and Mondal 2009; Sibirny and Voronovsky 2009). A recent review article from Teixeira et al. (2011) emphasised the close correlation between the general and specific stress responses and the ability of cells to cope with many environmental insults imposed during food production and preservation processes, which could have an insight into the direction of spoilage. It is important to note that although *S. cerevisiae* is most known from its positive role in the production of food, especially alcoholic beverages, it can cause serious spoilage by the fermentation of fruit juices even at low temperature. Aguilera et al. (2007) described the high-osmolarity glycerol (HOG) signalling pathway as playing an essential role in controlling the adaptive response to cold stress, which is a well-characterised signalling and osmoadaptation pathway in *S. cerevisiae* (Hohmann 2002). Abe and Minegishi (2008) found that cold sensitive mutants exhibited marked high-pressure sensitivity and concluded that tolerance and adaptation to high pressure and cold temperature use very similar signal transduction methods, namely the TOR (target of rapamycin) pathway.

Hydrostatic pressure processing (HPP) is used for inactivation of microorganisms and enzymes in different food products, especially in liquid foods (fruit juices, jams, sauces, milk), fish and fermented milk products (fresh curd cheeses) aiming to extend the self-life and increase the safety of food while retaining the sensory characteristics of fresh products. Vegetative cells of yeasts are similarly or a little more sensitive to pressure at the range of 300-600 MPa than bacteria, although its effectiveness is influenced by a number of extrinsic and intrinsic factors (pH, temperature, A_w, solute concentration, complexity of the food matrix). It has been shown by Patterson (2005) that acidic pH and high hydrostatic pressure (HHP) have synergistic effects in cell inactivation and the cells are more sensitive to HHP treatment at refrigeration or freezing temperatures than ambient temperature. According to Daryaei et al. (2010) *Y. lipolytica* and *C. zeylanoides*—two of the most frequent proteolytic cold-adapted spoilage yeasts—displayed different sensitivity to pressure in fermented milk. *Y. lipolytica* was recovered during chilled storage, while *C. zeylanoides* remained below the detection level during this storage period. Palhano et al. (2004) found that exposure of yeast (*S. cerevisiae*) cells to cold shock (10 °C) prior to HHP resulted the induction of barotolerance.

23.4 Problems Caused by Cold-Adapted Spoilage Yeasts in Food

Processing and/or storage of food at chilled temperatures (between 0 and 8 °C) provide special conditions for the associated microorganisms and ensure the survival of psychrotolerant vegetative cells and spores. Cooling and cold storage of food will inactivate the bacterial cells, while the more tolerant yeast cells survive. As it has been mentioned before, cells in the exponential phase of growth are more sensitive to the environmental stresses than stationary phase cells (Herman 2002; Zakrajsek et al. 2011); therefore, contaminants coming from the environment have a better chance to survive than the food-fermenting yeast biota. The accumulation of protective metabolites like trehalose, glycogen and glycerol in the cells has been proven to be a special response to several food-derived physical and chemical stresses (e.g. nutrient limitation/starvation, osmotic and ethanol stress). Other adaptive responses also aim to allow the species to survive the food preservation processes and will result in the selection of cells with resistance to multiple stress factors (Fleet 1999; Brul et al. 2002, 2003; Querol et al. 2003; Smits and Brul 2005; Walker and Van Dijck 2006; Teixeira et al. 2011).

Food-related factors have a combined effect on the growth of common spoilage yeasts. Available data in most cases only describe the effect of a few factors using kinetic growth models (Betts et al. 1999). To include more factors in mathematical models, the data collection technique based on the onset of growth (growth/no growth models) can be used. This model makes it possible to predict the likelihood of yeast spoilage of fruit-/alcoholic-based foods or drink products and even can give an indication of whether spoilage is likely to be slow or rapid (Evans et al. 2004). Another opportunity is the application of the model for deduction of the region where the combination of different factors inhibits the growth of yeasts, and the determination of efficient preservation techniques is possible (Arroyo López et al. 2007).

In the last decade, a lot of excellent complete books, book chapters and review articles have been published dealing with the biodiversity of yeasts having a role in the environment, biotechnology, food production and spoilage, which discuss both the beneficial and detrimental roles and activities of yeasts in all types of food (Loureiro and Querol 1999; Boekhout and Robert 2003; Querol and Fleet 2006; Rosa and Peter 2006; Fleet 2007; Deák 2008; Satyanarayana and Kunze 2009; Zakrzewska et al. 2011). Here, we make a short overview of the spoiling potential and spoiling activities of yeasts in those types of food which are processed and stored at chilled or freezing temperatures.

23.4.1 Spoilage of Fruits and Vegetables by Cold-Adapted Yeasts

Yeasts are present as part of the surface biota of healthy, undamaged vegetables and fruits. In general, anything that comes into contact with vegetables and fruits has the potential to contaminate it (Lehto et al. 2011). The main sources of yeasts in nature are water, air, soil and insects. Amongst airborne yeasts, basidiomycetes have been reported to predominate, with the airborne biota being largely composed of black and red yeasts together with *Cryptococcus* species. Insects can be regarded as potent carriers of yeasts, especially fruit flies, bees and wasps. They can also carry fermentative or oxidative ascomycetes (*D. hansenii, P. membranifaciens, S. cerevisiae*, *P. kudriavzevii* (formerly *Issatchenkia orientalis*) and *S. ludwigii*), and even osmophilic yeasts belonging to the genus *Zygosaccharomyces* (Boekhout and Phaff 2003; Fleet 2003). Contamination from the soil, especially in the case of vegetables, is facilitated by the close proximity to the soil (Deák 2008). Water as a source of yeasts is not restricted only to nature, but is a potential carrier later when used as a processing aid or as an ingredient in foods.

Although yeasts are often isolated from vegetables and fruits, they typically do not compete well with bacteria and moulds. The occurring yeasts rarely cause spoilage because nutrients are limited in the case of intact skin; however, certain yeasts may contribute to the damage by producing different enzymes like pectinases. As a general rule, vegetables and fruits become more susceptible to infections as they ripen (Deák 2008).

The storage of vegetables and fruits in cold-storage packinghouses is a widely used practice. The low temperature provided during storage slows down the propagation of yeasts; however, this growth at reduced rates can lead to damage, especially when the products are stored for extended periods of time. Moreover, some fruits are sensitive to low temperatures and could suffer chilling injuries, therefore becoming very susceptible to microbial spoilage (Tournas and Katsoudas 2005; Robiglio et al. 2011). Injuries can also be inflicted during harvesting and handling. Once the physical integrity of the surface has been destroyed, the internal tissues are exposed, which provides ideal growth parameters for yeasts (Fleet 2003; Deák 2008; Fleet 2011). Their adaptation to the low temperature used, nutritional conditions prevailing at the wound site and also their ability to colonise wounds makes these yeasts potential spoilage organisms (Robiglio et al. 2011). Differences in the intrinsic properties (A_w, nutrients, pH) of vegetables and fruits are reflected in the general composition of yeast biota. Fruit spoilage occurs because of the ability of yeasts to quickly utilise the simple sugars present in these substrates at the low pH of fruits. Presence of mycocinogenic, proteolytic and pectinolytic yeast species on these substrates could play an important role in the establishment and maintenance of yeast communities. In contrast to fruits, vegetables have near-neutral pH in addition to high water content and adequate nutrient composition. In general, basidiomycetous yeasts (*Cryptococcus*, *Rhodotorula* and *Sporobolomyces* spp.) usually dominate the mycobiota on vegetables, while the spoilage of fresh fruits usually results from the fermentative activity of ascomycetous yeasts (*Saccharomyces*, *Zygosaccharomyces*, *Hanseniaspora*/*Kloeckera*, *Candida* and *Pichia*) (Trindade et al. 2002; Fleet 2003; Deák 2008; Fleet 2011). In addition, air circulation maintained in storehouses contributes to the distribution of potential spoilage microorganisms. Also, the close proximity of healthy vegetables and fruits to damaged or spoiled ones enhances the spoilage potential.

Fresh and healthy characteristics of minimally processed vegetables and fruits support the widespread distribution of ready-to-use or ready-to-eat vegetables and fruits.

Salads are produced incorporating a variety of vegetables or fruits. Many of the ingredients are raw and are therefore sources of microorganisms. Although field contamination is reduced by partial processing, the effective washing and decontamination of fresh-cut vegetables and fruits is difficult to achieve and the reduction of microbial counts is typically 100-fold at best. During the processing stages of fresh-cut products, such as handling, cutting, shredding, slicing and grating, the microbial load can increase. Moreover, some of these contaminant microorganisms can grow during the marketing of fresh-cut vegetables and fruit, especially those that are psychrotolerant, and/or when packaged products are not stored at the temperature recommended by the provider (usually between 1 and 5 °C). Microbial loads of unprocessed (whole) products are, in general, lower than those associated with fresh-cut products. Although great variation in the contamination levels among salad samples can be observed, yeasts are always present with an overall count up to 6–7 log CFU g^{-1}. Ready-to-use products have a shelf life of a week or more if stored at refrigerated temperatures. Spoilage is mainly due to psychrotolerant Gram-negative and lactic acid bacteria (LAB), but a 2–3 log CFU g^{-1} increase in yeast numbers with storage time can be observed for a wide variety of yeast species.

In addition to ready-to-eat salads, various types of sprouts have become very popular in recent years, owing to their nutritional properties; however, the number of contaminating yeasts of these products can be by far the highest, reaching numbers as high as 8 log CFU g^{-1}. It is necessary to mention that the mesophilic and psychrotolerant counts may be of similar magnitude at the time of processing. Storage at refrigeration temperatures generally selects in favour of the growth of psychrotolerant microorganisms. Nevertheless, mesophilic microorganisms can survive and may continue to grow at reduced rates. Selection towards psychrotolerant microorganisms in some cases could take place before packaging (e.g. during storage before processing). Another interesting observation is that the contamination rate of salads shows seasonality, samples of summer and autumn months having, on average, significantly higher counts (Tournas and Katsoudas 2005; Tournas et al. 2006; Ragaert et al. 2007; Abadias et al. 2008; Deák 2008; Caponigro et al. 2010; Lehto et al. 2011).

Yeasts frequently encountered in salads belong to the genera *Saccharomyces*, *Zygosaccharomyces*, *Hanseniaspora*/*Kloeckera*, *Candida*, *Pichia*, *Debaryomyces* and *Metschnikowia* together with the basidiomycetous yeasts *Cryptococcus*, *Rhodotorula*, *Trichosporon* and *Sporobolomyces*. Descriptions of the ability of the species belonging to the mentioned genera to grow at low temperatures differ in the literature, which can be explained by the different physiological properties of strains belonging to the same species (Bonestroo et al. 1993; Trindade et al. 2002; Tornai-Lehoczki et al. 2003; Tournas et al. 2006; Fleet 2011). According to Bonestroo et al. (1993) the growth of spoilage yeasts is only inhibited at storage temperatures below 3.5 °C, while at a temperature of 7 °C many of the spoilage yeasts can reach maximum cell numbers within 7–14 days.

Improved market quality may also be achieved through the use of modified or controlled atmospheres. However, modified atmosphere packaging may encourage the growth of yeasts (Deák 2008). Moreover, elevated concentration of CO_2 could induce physiological disorders independent of microbiological activity (Ragaert et al. 2007).

Some vegetable salads are prepared by the addition of mayonnaise or other salad dressings and may contain even egg, cheese or meat as ingredients. In these cases, the yeast biota characteristic of meat and dairy products could also be present. The used mayonnaise and salad dressings usually contain preservatives; therefore, the role of preservative-resistant yeasts increases, especially when vinegar is applied (Tornai-Lehoczki et al. 2003).

The role of yeasts in the spoilage of vegetables is higher when fermented products are prepared because the lactic acid fermentation decreases the pH, thus providing a favourable environmental condition for the growth of yeasts. In the case of higher spoilage yeast loads than 3 log CFU g^{-1}, the growth of yeasts in fermented salad within 2 weeks cannot generally be prevented even by storage at 7 °C. Some yeast can utilise lactic and acetic acids produced by bacterial fermentation, while the above-mentioned acids act as preservatives for other yeasts, moreover, acetic acid is more inhibitory to yeasts than lactic acid. Besides the fermentative activity of yeasts, sometimes discolouration (as the consequence of *Rhodotorula* growth) and frequent film formation on the surfaces of brines can be observed (e.g. by *D. hansenii*, *Z. rouxii* and *P. kudriavzevii*) (Bonestroo et al. 1993; Trindade et al. 2002; Tornai-Lehoczki et al. 2003; Arroyo-López et al. 2008; Deák 2008; Fleet 2011). Some spoilage yeasts like *D. hansenii*, *W. anomalus* (formerly *P. anomala*), *P. membranifaciens* and *S. cerevisiae* (often associated with table olives) are able to grow at 7 °C and 8 % (w/v) NaCl concentration; however, due to the synergy between pH and NaCl at low temperatures, the spoilage potential can be decreased and selected combinations of pH and NaCl can prevent the growth of strains at 7 °C (Betts et al. 1999; Arroyo-López et al. 2008).

Symptoms of spoilage caused by yeast in case of fruits, vegetables and derived products can be seen as: (1) visual defects like appearance of yeast colonies on the food surface, surface discolouration, slimy appearance and accumulation of fluid inside packages; (2) gas production; and (3) production of off-odours/off-flavours (Stratford 2006; Ragaert et al. 2007).

23.4.2 Beverage Spoilage by Cold-Adapted Yeasts

Another important products of food industry are different fruit juices and concentrates which have high sugar concentrations and low pH (typically between 2.0 and 4.5). The acidic pH and sugar content are ideal for the growth of fermentative yeasts, which can lead to product spoilage by the generation of metabolic by-products (CO_2, acids and off-flavours/off-odours). In some cases (e.g. fermentation by *S. cerevisiae*, *D. bruxellensis*, *S. ludwigii*), the produced CO_2 can lead to exploding of bottles. The most common fermentative spoilage yeasts are similar to those found on fruits and fruit salads (up to 7 log CFU ml^{-1}). In beverages, spoilage yeast growth can be seen as hazes and clouds above 5 log CFU ml^{-1} cell concentration in clear liquids. In case of high sugar concentration, the products are likely to be spoiled by osmotolerant genera such as *Zygosaccharomyces* (Gram et al. 2002; Wareing and Davenport 2005; Stratford 2006; Tournas et al. 2006; Fleet 2011).

Freshly prepared fruit juices require processing to prevent spoilage. Chilling is most commonly used to extend the shelf life of unpasteurised products; however, chemical preservation, pasteurisation, freezing, concentration and/or irradiation are also widely used. Frequently, a single physical or chemical method is not enough to preserve fruit products; therefore, the combined action of pH, A_w, preservatives, refrigeration, and/or pasteurisation is applied (Deák 2008).

In contrast to fruit juices, soft drinks contain less nutrients (e.g. nitrogen sources) and an acidic and low oxygen environment; therefore, these are spoiled by relatively few microorganisms, usually yeasts, notwithstanding that many microorganisms are present from environmental and raw material sources (Tournas et al. 2006; Deák 2008). It has been estimated that 95 % of soft drink spoilage was due to poor factory hygiene (Stratford 2006). In addition, flavourings, water and other chemicals are also potential sources of microbial contamination (Waring and Davenport 2005). When soft drinks are carbonated, the spoilage biota shifts to carbon dioxide tolerant species, like *D. anomala* (Stratford and James 2003). The addition of fruit juices and nutrients to soft drinks enhances the potential for spoilage, and the emerging spoilage biota is similar to that of fruit juices.

Fruit, vegetables and their products can be a primary source of yeast contamination and the spoilage of many other foods, such as dairy products and bakery products if they are used as ingredients. Nowadays, there is a high demand for different frozen unbaked products with different fillings (e.g. pizzas, cakes, pastas). Some of the contaminating yeasts can grow to spoilage levels during the refrigerated storage of these doughs and fillings (Fleet 2011).

Alcoholic beverages like wine and beer have relatively high alcohol content (8–14 and 3–5 %, respectively) and low pH (3.0–4.0). These properties enable yeasts to selectively grow and spoil these products. In addition, wines may contain sulphur dioxide as a preservative, whereas the hop used in beer can have a preservative effect against unwanted microorganisms (Deák 2004; Fleet 2011).

In the case of beer, the contamination and spoilage can also occur at different stages. Any yeast, other than the pitching yeast (usually *Saccharomyces pastorianus* and *S. cerevisiae*), is considered as contaminating wild/foreign yeast during beer production. However, some species other than *Saccharomyces* may be used for fermentation; e.g., *D. bruxellensis* is commonly used in secondary fermentation for Belgian style beers and lambic beer. Re-used pitching yeast in brewery is a potential source of contamination, which can transfer contaminants. Contamination can lead to different defects during fermentation, like

1. producing various off-flavours;
2. producing hazy beer, especially the growth of wild yeast, like psychrotolerant *Rhodotorula* species which sediment very slowly and therefore cannot be removed by fining;
3. the growth of aerobic film-forming yeast, like *P. membranifaciens* and *W. anomalus* which produce high level of esters and also acetic acid from ethanol;
4. contamination with killer toxin-producing *S. cerevisiae* strains which will kill the sensitive brewing yeast strain; and
5. the growth of *S. cerevisiae* strains able to ferment dextrins (formerly classified as *Saccharomyces diastaticus*) may cause attenuation of the beer (Dufour et al. 2003).

During maturation and conditioning, *Pichia*, *Dekkera*, *Debaryomyces*, *Candida* species can cause contamination when hygiene control is too low (Hill 2008; Fleet 2011). Beer usually undergoes filtration and/or pasteurisation as the final step before/after bottling and is therefore not stored refrigerated. After bottling in some cases, carry-over strains of *S. cerevisiae* cause turbidity and/or may form sediment (Deák 2004, 2008; Fleet 2011). There is also a range of new types of beer products containing lower levels of alcohol (or none at all), and/or fruit juices, that are more susceptible to spoilage than traditional ales and lagers (Hill 2008).

In the case of wine, spoilage yeasts can originate from the grapes; however, they can also contaminate the wine during and/or after fermentation. For example, grape berries are the primary source of the dangerous spoilage yeast *D. bruxellensis*, while various off-flavours can be produced by wild yeasts during the alcoholic fermentation of must, especially during sluggish fermentation caused by low-temperature fermentation in the autumn. However, sometimes, the beneficial or spoilage effect of the produced compound is dependent only on its concentration (e.g. in the case of 4-ethylphenol). During the storage of wines, fermentative species like *Z. bailii*, *D. bruxellensis* and oxidative film-forming species such as *P. membranifaciens*, *P. kudriavzevii* can cause problems. Another spoilage type of wine is the re-fermentation of residual sugar in bottles even at low temperature, which can be caused by carry-over strains of wine yeasts (*S. cerevisiae*, *Saccharomyces bayanus*) or *Zygosaccharomyces* spp. The most dangerous spoilage yeasts—aside from *S. cerevisiae*—are *Z. bailii*, *D. bruxellensis* and rarely *S. ludwigii* (Dequin et al. 2003; Loureiro and Malfeito-Ferreira 2003; Deák 2004, 2008; Fleet 2011).

In the case of spoiling by *Z. bailii* strains, it is important to emphasise that strains of *Z. lentus* (a psychrotolerant *Zygosaccharomyces*) have presumably been previously misidentified as *Z. bailii* and, as a result of this misidentification, it is hard to determine how frequently they were participating in spoilage. In addition to tolerance of low temperatures, *Z. lentus* has high resistance to many of the preservatives commonly used in foods (sorbic acid, benzoic acid, acetic acid, dimethyl-dicarbonate, sulphite); moreover, *Z. lentus* strains show resistance to some natural antimicrobials like chitosan and cinnamic acid (Steels et al. 1999b; Martorell et al. 2007; Kisko and Roller 2012). RFLP analysis of the PCR-amplified ITS-5.8S region is a suitable method for detecting misidentified strains (Esteve-Zarzoso et al. 2003).

D. bruxellensis can spoil wines by the production of volatile phenols (e.g. 4-ethylphenol). Spoilage may occur at different stages of the winemaking process; however, it is often observed during ageing in barrels. The repeated use of barrels enhances the contamination risk because *Dekkera* spp. cannot be inactivated by efficient sanitation (Loureiro and Malfeito-Ferreira 2003). According to Couto et al. (2005), a low ageing temperature (10–15 °C) is one of the practices that has been used to limit *Dekkera* activity in wines. However, Barata et al. (2008b) and Dias et al. (2003) showed that *D. bruxellensis* can grow well at 15–16 °C and produce high amounts of 4-ethylphenol. In addition, Echeverrigaray et al. (2013) isolated several *D. bruxellensis* strains from Brazilian wines which could grow even at 10 °C and some of them showed pseudohypha and/or biofilm formation, which can probably contribute to the entrapment in barrel staves. Different techniques have been investigated which could effectively inactivate *D. bruxellensis* strains, like the addition of relatively high doses of sulphur dioxide to prevent proliferation (Barata et al. 2008a), thermal inactivation of living cells (Couto et al. 2005), or the usage of aqueous steam, gaseous or aqueous O_3 to eliminate yeasts in barrels (Guzzon et al. 2011). *D. bruxellensis* is quite an elusive yeast species, being difficult to isolate due to its low growth rate; moreover, upon SO_2 exposure, cells can transform into the VBNC (viable but not culturable) state (Malfeito-Ferreira 2011; Serpaggi et al. 2012). Owing to the potential role in economic losses in the wine industry, various methods have been developed for detection and identification of *D. bruxellensis* strains, like the use of selective/differential media (Rodrigues et al. 2001), long-chain fatty acid analysis (Malfeito-Ferreira et al. 1997; Loureiro 2000), different PCR-based method (Loureiro and Malfeito-Ferreira 2003), real-time PCR analysis (Puig et al. 2011; Tofalo et al. 2012) and PNA FISH (fluorescence *in situ* hybridisation using peptide nucleic acid probes) (Loureiro and Malfeito-Ferreira 2003).

23.4.3 Spoilage of Dairy Products by Cold-Adapted Yeasts

Yeasts could be generally isolated from all types of dairy products including milk, and from the processing and storing environment of dairy products. Yeasts have substantial roles in the production and maturation of certain fermented dairy products like kefyr, cheeses, koumiss and several artisanal products; moreover, they have a positive role in the production of special flavour and aroma components, which is more often the focus of product development (Fleet 1990b; Romano et al. 2006; Tamang and Fleet 2009). As milk and dairy products are typically processed and stored at chilled temperatures, cold-adapted yeasts may have a basic role in spoilage, which are already adapted to one or more of the other intrinsic factors characteristic to the different types of dairy products (low pH values, high salt concentration, preservatives, available nutrients such as lactose, lactic and citric acid, milk proteins and lipids). Besides cold tolerance, one or more of the following inherent properties of the yeasts that are able to survive and multiply in dairy products have been defined: their ability to ferment or assimilate lactose, assimilate lactic or citric acid, produce cold-adapted extracellular proteolytic and lipolytic enzymes, and tolerate low water activity and elevated salt concentrations (Fleet 1990b; Jacobsen and Narvhus 1996; Fröhlich-Wyder 2003).

Biodiversity and characterisation of yeast biota in milk, fermented milk products and cheeses produced with different traditional manufactural and commercial technologies and under different climatic conditions were extensively studied. Molecular biological techniques provide more exact and easier identification and typing of both the culturable and non-culturable (e.g. VBNC) yeasts in comparison to the traditional ones, which resulted in a clearer picture of the biodiversity and survival of yeast biota in dairy products. These new results confirmed that the incidence of different yeast species isolated from a large variety of dairy products shows huge variation, but certain species occur at a higher frequency or even dominate in certain products. These are *Candida catenulata, Candida intermedia, Candida lusitaniae* (now *Clavispora lusitaniae*), *C. parapsilosis, Candida rugosa, C. zeylanoides, Cryptococcus curvatus, D. hansenii, G. geotrichum, P. kudriavzevii, K. marxianus, Klyveromyces lactis, Y. lipolytica, Pichia fermentans, P. membranifaciens, Rhodotorula glutinis, R. mucilaginosa, K. exigua, S. cerevisiae, Torulaspora delbrueckii* and *T. cutaneum* (Fleet and Mian 1987; Fleet 1990b; Fröhlich-Wyder 2003; Vasdinyei and Deak 2003; Lopandic et al. 2006; Latorre-Garcia et al. 2007; Bockelmann et al. 2008; Deák 2008; Tamang and Fleet 2009; Yalcin and Ucar 2009; Chen et al. 2010).

The same non-starter yeasts, which are present in products without influencing the quality or even considered as beneficial, may be detrimental if environmental conditions allow their excess growth. Sources of the indigenous yeast biota have been identified as the environment of milk producing farms, farm workers, raw milk, processing equipments and their environment, and brine or non-dairy ingredients like fruits and flavouring compounds. Therefore, GMP, hygiene and sanitation measures are important in controlling the introduction of potential spoiling yeasts during processing. Dairy yeasts like *D. hansenii*, *C. versatilis* and *T. delbrueckii*, however, have been found to be resistant to commercial sanitisers and cleaning compounds, which could explain the relatively high number of these yeasts, even under careful sanitation of the processing lines (Laubscher and Viljoen 1999).

The introduced spoiling yeasts can cause packages to blow as the result of the production of carbon dioxide, which is typical to the strongly fermenting yeasts *K. marxianus, K. lactis* and *S. cerevisiae*, by the utilisation of milk lactose or the galactose produced by lactic acid bacteria. Fermented milk products and cheeses are more prone to yeast spoilage. Non-fermenting yeasts can form slime on the surface, produce off-flavours and off-odours, may cause softening as the consequence of lipase and protease activity, and cause discolouration or colour spots on the surface (Fleet 1990b; Tudor and Board 1993; Jacobsen and Narvhus 1996; Fröhlich-Wyder 2003; Fleet 2011).

23.4.3.1 Milk

Yeasts are generally not considered to be responsible for milk spoilage, although they constitute the microbiota of fresh cow milk. During the storage of milk at low temperatures, psychrotolerant bacteria (*Pseudomonas, Acinetobacter*) compete with yeasts and quickly overgrow them. Frequently identified milk-associated yeasts belong to the genera of *Candida, Brettanomyces, Dekkera,* and *Geotrichum* (Fleet 1990b; Corbo et al. 2001; de Araújo Viana et al. 2010). PCR-DGGE analysis of raw milk samples for yeasts allowed the detection of twelve different species and showed that *K. marxianus* and *K. lactis* were present in high frequencies beside the dominant *Candida* spp. (Cocolin et al. 2002). de Araújo Viana et al. (2010) isolated *Candida buinensis* with strong extracellular protease activity and it was found that the enzyme exhibited broad stability ranges of pH and temperature. Lavoie et al. (2012) made a comprehensive survey of 19 dairy farms in Quebec (Canada) and found a big biodiversity of yeasts. The isolates belonged to 37 species across 11 genera and each farm had a unique species profile and composition. It has been demonstrated that milk isolates of *I. orientalis* (now *P. kudriavzevii*) could survive the cheese-making process and the same MLST (multilocus sequence typing) genotypes developed in raw milk cheese. Camel milk samples investigated by Njage et al. (2011) contained a similar spectrum of yeast species as the cow milk, but the most frequently isolated yeast was *S. cerevisiae*.

Roostita and Fleet (1996) tested the growth of *C. catenulata, Y. lipolytica, K. marxianus, D. hansenii* and *S. cerevisiae* in milk at 10 and 24 °C and found considerable growth in both temperatures. The addition of NaCl up to 15 % (w/v) inhibited growth, which was more pronounced at low temperature and *S. cerevisiae* died off in the presence of 15 % NaCl at 10 °C. *Y. lipolytica* and *C. catenulata* exhibited both proteolytic and lipolytic activity, but *Y. lipolytica* was stronger in this respect. Protease and lipase activity of *Y. lipolytica* in milk was decreased but not inhibited by either low temperature or the presence of NaCl. Skimmed milk was found to be the best substrate for the production of alkaline extracellular protease and ribonuclease by *Y. lipolytica* cheese isolates secreting also an abundant quantity of acid proteases, lipases and phosphatases (Akpinar et al. 2011).

23.4.3.2 Fermented Milk Products

Typical cold-adapted spoilage yeasts are frequently detected in fermented milk products made from pasteurised milk (butter, yoghurts, sour milk) without any symptoms of spoilage (Fleet and Mian 1987; Fleet 1990b; Fröhlich-Wyder 2003; Vasdinyei and Deak 2003; Lopandic et al. 2006; Tamang and Fleet 2009). During the chilled storage of these products, especially yoghurt, yeasts can start growing because the acidic pH will select the contaminating yeasts in favour of the lactic acid bacteria (Viljoen 2001, 2006). Lactose- and galactose-fermenting yeasts belonging to the genera of *Kluyveromyces, Candida* and *Pichia* are mostly isolated from spoiled plain yoghurt, whereas *Saccharomyces* spp. are frequently the causative agents of fruited and flavoured yoghurts, due to the fermentation of the added fruit and/or sugar. Blowing and the generation of off-flavours and off-odours will be evident around the concentration of 5–6 log CFU g^{-1}. Non-fermenting yeasts such as *Cryptococcus* and *Rhodotorula* were also reported to occur in spoiled plain yoghurt (Fleet 1990b, 1992; Deák and Beuchat 1996; Giudici et al. 1996; Loureiro and Querol 1999; Caggia et al. 2001; Fröhlich-Wyder 2003; Mataragas et al. 2011). Inoculation of yoghurt samples with the most frequent spoiling yeasts in 3 log CFU ml^{-1}, by Fleet and Mian (1987) it was found that *D. hansenii*, *Rhodotorula diffluens* and *C. albidus* were able to reach more than 7 log CFU ml^{-1} during 15 days of incubation at 5 °C, had even better growth than *K. marxianus* at this temperature.

23.4.3.3 Cheese

Cheese microbiota has an essential role in cheese manufacturing; quality and safety are strongly influenced by the activity of the evolving microorganisms during fermentation, maturation and storage. The interaction of yeasts with lactic acid bacteria and other, less acid-tolerant bacteria and with moulds is a very complex phenomenon, which is more commonly understood but it still far from successful. Thanks to the new molecular, analytical and physiological tools, together with the development of the genomics, proteomics and metabolomics of the cheese-related microorganisms, the role and activity of yeasts in the cheese ecosystem have been identified (Viljoen 2001, 2006; Fleet 2007; Irlinger and Mounier 2009).

The role of yeasts in cheese spoilage is well documented from historic to more recent research articles, books and reviews, and it can be concluded that a wide range of yeasts associated with dairy products could be isolated from cheeses, but their spoiling potential is strongly influenced by the type of cheese and ecological factors. Sources of yeasts are the processing environment (brine, water, equipments and workers) if cheese is made from pasteurised milk (Fleet 1992; Tudor and Board 1993; Deák and Beuchat 1996; Tamang and Fleet 2009), while raw milk could be the source of yeasts as well. Clonally identical strains of the stress-resistant yeast species present in raw milk (*P. kudriavzevii*) have been re-isolated from cheese processed from the same raw milk (Lavoie et al. 2012). Fresh, unripe cheeses like cottage cheese are considered to be the most prone to yeast spoilage. Lactose-fermenting (e.g. *K. marxianus*) and non-fermenting yeasts (e.g. *P. membranifaciens*) can grow to populations of 6–7 log CFU g^{-1} during refrigerated storage and cause deterioration with symptoms such as gas formation, the development of visible colonies, layers and slime on the surface of the products or can generate unpleasant odours and flavours (Brocklehurst and Lund 1985; Fleet 1990b; Tudor and Board 1993; Fröhlich-Wyder 2003). A wide range of yeasts, all of which have been considered as cold-adapted species, were reported to be associated with such types of spoilage, namely *C. parapsilosis, C. rugosa, Candida sake, C. zeylanoides, Cryptococcus laurentii, D. anomala, D. hansenii, G. geotrichum, P. kudriavzevii, K. marxianus, Y. lipolytica, P. fermentans, P. membranifaciens, Rhodotorula* spp, *Sporobolomyces roseus* (anamorph of *Sporidiobolus metaroseus*), and *T. delbrueckii*. Over-ripening of hard and semi-hard cheeses by lactose-fermenting yeasts during maturation can lead to increased acidity, gassiness and fruity flavour, or the ongoing hydrolysis of milk proteins and lipids may result in softening, bitter and rancid flavours (Fleet 1990b, 1992; Tudor and Board 1993; Deák 2008).

23.4.4 Spoilage Caused by Cold-Adapted Yeasts of Raw Poultry, Meat, Fish and Derived Products

The application of chill temperature during the storage of meat and meat products or the operation of processing lines is indispensable; therefore, low temperature is the major intrinsic factor that determines the survival and growth of microorganisms under these conditions. There are numerous publications including research articles, review papers and book chapters describing yeast biodiversity of different raw and processed meat, poultry and seafood, which has been recently reviewed together with the analysis of the physiological characteristics of the more frequent meat-associated yeast species by Deák (2008). Yeasts belong to the permanent microbiota of raw meat, but their selection, survival and growth are strongly influenced by the intrinsic properties of the food, environmental factors, their inherent properties and interaction with other members of the indigenous microbiota (mainly bacteria), as well as the processing technologies. Tudor and Board (1993) made a review of the food-associated yeasts and found that more than half of the reported roughly 200 species have been isolated from raw and processed poultry and red meat. The beneficial role of yeast biota is well pronounced only in certain fermented meat products, such as the maturation of dry fermented sausages and raw hams, by the production of specific flavour compounds and a red colour. The hydrolysis of meat proteins and lipids as well as scavenging the oxygen by special strains of *D. hansenii* and *Y. lipolytica* are mainly responsible for these positive effects (Samelis and Sofos 2003; Romano et al. 2006; Tamang and Fleet 2009).

Fluctuating populations of diverse yeast species could be detected in raw meat and under processing conditions if they can successfully compete with the bacterial populations present. Meat-associated yeasts cause spoilage only in rare cases; under situations when the inhibition of bacteria in the spoiling associations is restricted, the elevated resistance or better adaptation to several food-related stresses will allow the yeasts to grow. Another reason for spoiling could be the addition of readily utilisable carbohydrates, which enhances the multiplication of the yeast cells present (Fleet 1992; Tudor and Board 1993). Dillon and Board (1991) considered the main factors governing the biodiversity and activity of yeast biota in meat from the abattoir to the final products: (1) contamination of fresh meat in the abattoir through the hide, hair and fleece of the animals, as well as the local environment, facilities and workers; (2) characteristics of meat as a medium for microbial growth; (3) storage conditions of raw meat (aerobic or anaerobic chilled storage or freezing); and (4) chemical and biochemical processes as a consequence of microbial growth and activity at chilled temperature (excretion of hydrolytic enzymes, production and metabolism of organic acids) v) effect of the processing conditions (A_w, temperature, pH, pO_2, pCO_2, additives and preservatives).

Yeasts associated with the freshly slaughtered livestock (poultry and other animals) originate mainly from the hide, hair, fleece and feather are the sources of primary contamination of meat. As a consequence, the biodiversity of this yeast biota resembles that of the agro-ecosystem typical of livestock farms. Yeasts constitute only a small part (generally less than 1 %) of the residential microbiota; the majority of them, however, belong to the cold-adapted basidiomycota and ascomycota yeasts, and strongly fermentative species are sparse (Dijkman et al. 1980; Dillon and Board 1991; Deák 1998).

Fresh red and poultry meat are the ideal medium for microbial growth because it has a close to neutral pH, high water content ($A_w = 0.99$) and is rich in nutrients and minerals. Under chilled storage conditions, psychrotolerant Gram-negative bacteria (*Pseudomonas, Acinetobacter*, *Aeromonas*, *Hafnia*) will be selected out and grow much faster than the psychrotolerant yeasts, which have also been elected from the original yeast biota.

Ascomycetes will predominate during the storage of poultry meat, but basidiomycetes yeasts are still significant. The most frequently isolated species are *Y. lipolytica* and *C. zeylanoides*; they mostly comprise more than half of the total yeast isolates originating from different raw and processed poultry meat, but other *Candida* and different *Cryptococcus*, *Rhodotorula*, *Metschnikowia* and *Trichosporon* species also occur in considerable quantities. *D. hansenii* isolates were found in a higher ratio in processed poultry than in fresh meat. Most of the species had proteolytic and lipolytic activities, the level of the secreted enzyme activities was, however, strain-dependent (Viljoen et al. 1998; Ismail et al. 2000; Belak et al. 2011). Hsieh and Jay (1984) reported similar yeast diversity in fresh and spoiled ground beef, than in those typical to poultry; *Y. lipolytica* and *C. zeylanoides* were the dominating species in this case, too. Although yeast populations frequently reach 4–6 log CFU g^{-1} in different raw meats during chilled storage, bacteria always grow faster and have cell numbers of two or three orders of magnitude higher than the yeasts; therefore, their putrefactive spoilage can mask any sensory defect caused by yeasts (Nychas et al. 1998; Samelis and Sofos 2003). It should be mentioned that Peter et al. (2004) identified a novel yeast species, *Candida galli* by molecular technique, which have been isolated from poultry and harbours similar physiological characteristics than *Y. lipolytica*.

Spoilage of raw fish by yeasts under chill storage is not important for similar reasons as poultry and red meat. Psychrotolerant bacteria quickly overgrow the yeasts, which originate from the water, but further processing (heat treatment, addition of nitrate or spices) may increase the incidence of spoiling (Tudor and Board 1993; Gram and Huss 1996; Deák 2008).

Diversity of yeast biota in a wide range of cured, dried and fermented meat products has been thoroughly analysed and well documented (Walker and Ayres 1970; Deák and Beuchat 1987; Tudor and Board 1993). Reviews have been published recently by Davies and Board (1998), Dillon and Board (1991), Fleet (1992), Samelis and Sofos (2003) and Tamang and Fleet (2009).

These types of meat products are more prone to spoilage by yeasts than raw meat and it is highly predictable, because bacterial growth is restricted or prevented by the intrinsic factors typical to the cured, dried and fermented meat products; as a consequence, more stress-tolerant yeasts started to grow under chilled storage. The main factors that are conducive to yeast growth are low A_w value (0.83–0.87) and high salt content, the addition of sulphite as a preservative, acidic conditions, nutrient limitation and storage up to -5 °C. Yeasts may have an indirect role in meat spoilage by permitting or even supporting the growth of food-spoiling bacteria. The inhibitory effect of sulphite could be reduced by yeasts via the production of acetaldehyde; moreover, yeasts can utilise organic acids (lactic, citric and acetic acids) produced mainly by lactic acid bacteria; therefore, the pH will increase and the spoiling bacteria have a chance to grow (Tudor and Board 1993). New preservation technologies like γ-irradiation, high hydrostatic pressure (HHP), and modified atmosphere packaging (MAP) also select the generally more resistant yeasts, which can start growing during chilled storage (Deák 2008). Excessive growth of *D. hansenii, Y. lipolytica, C. zeylanoides, C. parapsilosis, Pichia* spp., *Cryptococcus* spp. and *Rhodotorula* spp. has been reported to lead to spoilage by causing yeasty or fruity off-odours, and the development of spots, film layers or slime on the surface of meat products, depending on the type of the products and the spoiling species or strains (Fleet 1992; Tudor and Board 1993). Knutsen et al. (2007) described a novel species, *Candida alimentaria*, which was first isolated from cured meat/ham and resembles *Y. lipolytica* from physiological characteristics. Molecular identification based on the rDNA sequences provides the only way for recognition and exact identification.

In spite of the extensive studies on the biodiversity of yeasts occurring in raw and processed meat products, as the consequence in the differences in the quality of raw materials, processing and storage technologies including the temperatures, it is difficult to draw clear conclusions regarding the positive, neutral or negative effects of the indigenous yeast biota. Great variations in the sampling, cultivation and identification techniques also cause difficulty in the comparison of published results, especially regarding the spoiling potentials of the reported spoiling yeast species. Progress in yeast phylogeny, recognition of the anamorphic and teleomorphic states of certain yeast species as well as the availability of molecular techniques for identification and typing provided an opportunity to make progress in this field. Nielsen et al. (2008) performed a systematic study by quantifying the occurrence of yeasts in five different processed pork meat products (bacon, ham, salami and two different liver patés). Yeasts were isolated, enumerated and identified in raw materials, during processing and in the final products at the beginning of the storage (at 5 °C) and at the end of shelf life. The spoiling potential of the most frequently isolated species has been tested by a challenge test, storing the products under a low oxygen/high carbon dioxide atmosphere at two different temperatures (5 and 8 °C). In general, the number of yeasts was low in the meat and the additives (spices, colourants, etc.), and it decreased considerably to the end of processing, in most cases below the detection limit. Further decrease in the number of yeasts was registered during storage (at 5 °C); only one out of the 25 investigated samples contained yeasts at a detectable level. These results confirm that GMP and satisfactory hygiene standards can reduce the probability of yeast spoilage of processed meat products. At the same time, the identification of the yeast isolates by a polyphasic approach (Jespersen et al. 2005) indicated the presence of a complex yeast biota during the production phases. In general, the prevalence of yeasts was found in the raw materials and during processing, but the smoking step drastically decreased their number below the detectable level. Increases in the ratio of *C. zeylanoides* and *C. alimentaria* during the early processing phase make it probable that these two species are well adapted to the food-related stress at cold temperatures. *D. hansenii* was detected at a much lower frequency, if any, before smoking, its population slightly increased after this step, although it could not be detected at the end of storage. The low incidence of *Y. lipolytica* might indicate that a considerable ratio of *C. alimentaria* isolates were (mis)identified in former investigations as *Y. lipolytica*. Challenge tests with the application of isolates belonging to the most frequent eight different yeast species indicated that *C. zeylanoides, D. hansenii* and *C. sake* were able to grow under MAP conditions to that level, which can potentially cause symptoms of spoilage if the starting (inoculated) cell number is between 2–3 log CFU g^{-1}. In case of *C. alimentaria* and *Cryptococcus victoriae,* a large difference was found in the growth of the different stains tested.

23.5 Strategies to Reduce the Impact of Cold-Adapted Spoilage Yeasts in Foods

Economic consequences of food spoilage include the loss of food products, cost of product recall and disposal, plant clean up and sanitisation, and it may have an adverse impact on company reputation. Estimation of the level of yeast spoilage is difficult, because it is rarely reported and consumer's perception of the symptoms is underestimated; however, the cost must be considerable (Stratford 2006; Fleet 2011). Since majority of spoilage yeasts are cold-adapted, destruction of raw and processed food by spoilage yeasts at chill temperature is highly predictable if growth of the competitive psychrotolerant bacteria is restricted during storage.

It is a general trend in food production that degree of processing and storage under chill conditions are increasing and consumer's attitude turned into the direction of demanding fresh, natural and preservative-free products with satisfactory shelf life. Preparation of minimally processed, ready-to-eat food is also increasing, which also need chilled storage. Ecological factors are generally more favourable for growth of cold-adapted yeasts than psychrotolerant bacteria in these types of products; therefore, food zymologists are challenged by these new trends in product development.

Estimation of shelf life of the products would require not only the exact characterisation and quantification of the intrinsic and extrinsic factors, but also determination of the biodiversity of the associated yeast biota, as well as the detection, identification and quantification of the potential spoiling species. A big progress took place in the detection and identification methodology of yeasts during the last decades, although, techniques for monitoring and controlling the food-associated yeasts at industrial level practically remained unchanged. Some positive examples, however, have to be mentioned like development of selective and differential media for the detection and enumeration of *Y. lipolytica, K. lactis, K. marxianus, D. hansenii* and *Dekkera* spp. Nucleic acid-based typing and identification methods are extremely powerful techniques in research but there is a big gap between the industry and science in the application for routine diagnostic purposes. It would be necessary to prove the efficiency of the new diagnostic tools in detection and controlling of food-spoilage yeasts, which requires well-trained industrial microbiologists and acceptable cost of the tests (Loureiro and Malfeito-Ferreira 2004; Deák 2008).

Predictive microbiology has been established and developed fast during the last decades, which is based on the quantitative evaluation of the microbial responses and growth in different environmental matrices such as the food. Several software packages have been constructed, which allow prediction of microbial responses to the most important food-related environmental factors such as the temperature, A_w, pH, atmosphere composition, preservatives and nutrient concentration. Spoilage yeasts are, however, seldom in these databases. The other drawback of predictive microbiology is that most of the outcomes have not necessarily transferred into the practice of food processing and product development. Present trends in this field include the revision of web-based data-retrieving tools and strengthening the practical use in the food processing settings (Baranyi and Tamplin 2004; Psomas et al. 2012; Koseki 2013).

McMeekin et al. (2013) revisited the "traditional" predictive microbiology and emphasised the crucial role of temperature in the growth rates. Theoretical and actual minimum temperatures for growth have been defined together with eleven regions on the temperature versus growth rate curves, which reflect the physiological responses of the microorganisms (mostly pathogenic and spoilage bacteria). In predictive microbiology, these authors turn from the current empirical to the mechanistic models by the utilisation of the extended knowledge in microbial physiology, "omics" technologies and bioinformatics. Expectedly cold-adapted spoilage and beneficial yeasts will have special attention in the reconstructed models of predictive microbiology. Increasing roles of stress-tolerant yeasts as model organisms in studying the general and specific adaptive responses to food-relevant stresses may contribute to better understanding of the spoiling potential and to reduce the impact of cold-adapted spoilage yeasts in foods.

23.6 Conclusions

Yeasts are permanent constituents of the food-associated microbiota, although they are not able to compete with the faster growing bacteria under favourable conditions. Several yeast species have characteristic tolerance against food-related stress factors like acidic pH, high salt or sugar concentrations and preservatives; therefore, they can overgrow the generally less tolerant bacteria and moulds under these conditions. Spoilage of food by yeasts at low temperature is predictable if growth of psychrophilic and psychrotolerant bacteria is retarded or prevented by other stress factors, thus allowing growth of typical highly tolerant yeast species. Diversity of food-spoilage yeasts is quite narrow, only about 10 % of the food-associated yeast biota (20 out of the 150–200 species) is most frequently reported (Fleet 2011). Majority of the spoilage yeasts are psychrotolerant; therefore, spoilage of chilled food by yeasts could be expected if the environmental conditions are unfavourable for the growth of the psychrophilic and psychrotolerant spoilage bacteria.

Some spoilage yeast species (e.g. *S. cerevisiae, K. marxianus, S. pombe, D. hansenii*) are beneficial in the production or maturation of certain foods, or are even used as starter cultures or adjuncts in fermented dairy and meat products, but under special conditions or in an other types of food, these yeasts may cause spoiling.

Significant progress in the molecular analysis, identification and taxonomy of yeasts, and better understanding of the biochemical, genetic and physiological background of stress tolerance and adaptation mechanisms of potential food-spoilage yeasts allowed more precise identification of contamination sources and the yeasts themselves. Molecular mapping of the signal transduction routs and determination of their regulation promoted understanding the behaviour of yeasts under stressful conditions and adaptation to the changing environment (Smits and Brul 2005; Teixeira et al. 2011). It might be expected that the increasing biotechnological interest of some cold-adapted spoilage yeasts will generate new scientific results and this knowledge will help reduction of the economic losses as the consequence of food spoilage (Rosa and Peter 2006; Satyanarayana and Kunze 2009). Progress in predictive microbiology is an important aid in prediction of spoilage, although more attention should be given to yeasts beside the spoilage bacteria in the future (McMeekin et al. 2013).

References

Abadias M, Usall J, Anguera M, Solsona C, Viñas I (2008) Microbiological quality of fresh, minimally-processed fruit and vegetables, and sprouts from retail establishments. Int J Food Microbiol 123:121–129

Abe F, Minegishi H (2008) Global screening of genes essential for growth in high-pressure and cold environments: searching for basic adaptive strategies using a yeast deletion library. Genetics 178:851–872

Aggarwal A, Mondal AK (2009) *Debaryomyces hansenii*: an osmotolerant and halotolerant yeast. In: Satyanarayana T, Kunze G (eds) Yeast biotechnology: diversity and applications. Springer, Berlin, pp 65–84

Aguilera J, Randez-Gil F, Prieto JA (2007) Cold response in *Saccharomyces cerevisiae*: new functions for old mechanisms. FEMS Microbiol Rev 31:327–341

Akpinar O, Ucar F, Yalcin HT (2011) Screening and regulation of alkaline extracellular protease and ribonuclease production of *Yarrowia lipolytica* strains isolated and identified from different cheeses in Turkey. Ann Microbiol 61:907–915

Arroyo López FN, Durán Quintana MC, Garrido Fernández A (2007) Modelling of the growth-no growth interface of *Issatchenkia occidentalis* an olive spoiling yeast, as a function of the culture media, NaCl, citric and sorbic acid concentrations: Study of its inactivation in the no growth region. Int J Food Microbiol 117:150–159

Arroyo-López FN, Querol A, Bautista-Gallego J, Garrido-Fernández A (2008) Role of yeasts in table olive production. Int J Food Microbiol 128:189–196

Banks JG, Board RG (1987) Some factors influencing the recovery of yeasts and moulds from chilled food. Int J Food Microbiol 4:197–206

Baranyi J, Tamplin ML (2004) ComBase: A common database on microbial responses to food environment. J Food Prot 67:1967–1971

Barata A, Caldeira J, Botelheiro R, Pagliara D, Malfeito-Ferreira M, Loureiro V (2008a) Survival patterns of *Dekkera bruxellensis* in wines and inhibitory effect of sulphur dioxide. Int J Food Microbiol 121:201–207

Barata A, Pagliare D, Piccininno T, Tarantino F, Ciardulli W, Malfeito-Ferreira M, Loureiro V (2008b) The effect of sugar concentration and temperature on growth and volatile phenol production by *Dekkera bruxellensis* in wine. FEMS Yeast Res 8:1097–1102

Belak A, Kovacs M, Hermann Zs, Holczmann A, Marta D, Cenic Stojakovic S, Bajcsi N, Maraz A (2011) Molecular analysis of poultry meat spoiling microbiota and heterogeneity of their proteolytic and lipolytic enzyme activities. Acta Aliment Hung 40:3–22

Betts GD, Linton P, Betteridge RJ (1999) Food spoilage yeasts: effects of pH, NaCl and temperature on growth. Food Control 10:27–33

Betts GD, Linton P, Betteridge RJ (2000) Synergistic effect of sodium chloride, temperature and pH on growth of a cocktail of spoilage yeasts: a research note. Food Microbiol 17:47–52

Bockelmann W, Heller M, Heller KJ (2008) Identification of yeasts of dairy origin by amplified ribosomal DNA restriction analysis (ARDRA). Int Dairy J 18:1066–1071

Boekhout T, Phaff HJ (2003) Yeast biodiversity. In: Boekhout T, Robert V (eds) Yeasts in food: beneficial and detrimental aspects. Behr's Verlag, Hamburg, pp 1–38

Boekhout T, Robert V (eds) (2003) Yeasts in food: beneficial and detrimental aspects. Behr's Verlag, Hamburg

Bonestroo MH, de Wit JC, Kusters BJM, Rombouts FM (1993) Inhibition of the growth of yeasts in fermented salads. Int J Food Microbiol 17:311–320

Botha A (2006) Yeasts in soil. In: Rosa CA, Peter G (eds) Biodiversity and ecophysiology of yeasts. Springer, Berlin, pp 220–240

Brocklehurst TF, Lund B (1985) Microbiological changes in cottage cheese varieties during storage at +7°C. Food Microbiol 2:207–233

Brul S, Klis FM, Oomes SJCM, Montijn RC, Schuren FHJ, Coote P, Hellingwerf KJ (2002) Detailed process design based on genomics of survivors of food preservation processes. Trends Food Sci Tech 13:325–333

Brul S, Klis FM, De Nobel H, Oomes SJCM, Coote P, Hellingwerf KJ (2003) Yeast response to food preservation systems. In: Boekhout T, Robert V (eds) Yeasts in food: beneficial and detrimental aspects. Behr's Verlag, Hamburg, pp 193–2007

Caggia C, Restuccia C, Pulvirenti A, Giudici P (2001) Identification of *Pichia anomala* isolated from yogurt by RFLP of the ITS region. Int J Food Microbiol 71:71–73

Caponigro V, Ventura M, Chiancone I, Amato L, Parente E, Piro F (2010) Variation of microbial load and visual quality of ready-to-eat salads by vegetable type, season, processor and retailer. Food Microbiol 27:1071–1077

Chen L-S, Ma Y, Maubois J-L, Chen L-J, Liu Q-H, Guo J-P (2010) Identification of yeasts from raw milk and selection for some specific antioxidant properties. Intern J Dairy Technol 63:47–54

Choudhary DK, Johri BN (2009) Basidiomycetous yeasts: current status. In: Satyanarayana T, Kunze G (eds) Yeast biotechnology: diversity and applications. Springer, Berlin, pp 21–46

Cocolin L, Aggio D, Manzano M, Cantoni C, Comi G (2002) An application of PCR-DGGE analysis to profile the yeast populations in raw milk. Int Dairy J 12:407–411

Corbo MR, Lanciotti R, Albenzio M, Sinigaglia M (2001) Occurrence and characterization of yeast isolated from milk and dairy products of Apulia region. Int J Food Microbiol 69:147–152

Couto JA, Neves F, Campos F, Hogg T (2005) Thermal inactivation of the wine spoilage yeasts *Dekkera/Brettanomyces*. Int J Food Microbiol 104:337–344

Daryaei H, Coventry J, Versteeg C, Sherkat F (2010) Combined pH and high hydrostatic pressure effects on *Lactococcus* starter cultures and *Candida* spoilage yeasts in a fermented milk test system during cold storage. Food Microbiol 27:1051–1056

Davenport RR (1980) Cold-tolerant yeasts and yeast-like organisms. In: Skinner FA, Passmore SM, Davenport RR (eds) Biology and activities of yeasts. Academic Press, London, pp 215–230

Davies A, Board R (eds) (1998) The microbiology of meat and poultry. Blackie, London

de Araújo Viana D, de Albuquerque Lima C, Neves RP, Mota CS, Moreira KA, de Lima-Filho JL, Cavalcanti MTH, Converti A, Porto ALF (2010) Production and stability of protease from *Candida buinensis*. Appl Biochem Biotech 162:830–842

Deák T (1991) Foodborne yeasts. Adv Appl Microbiol 36:179–278

Deák T (1998) Biodiversity of yeasts in man-made environments. Food Technol Biotech 36:279–283

Deák T (2004) Spoilage yeasts. In: Steele R (ed) Understanding and measuring the self-life of food. Woodhead Publishing Ltd, Cambridge, pp 91–110

Deák T (2006) Environmental factors influencing yeasts. In: Rosa CA, Peter G (eds) Biodiversity and ecophysiology of yeasts. Springer, Berlin, pp 155–174

Deák T (2008) Handbook of food spoilage yeasts, 2nd edn. CRC Press, Boca Raton

Deák T, Beuchat LR (1987) Identification of foodborne yeasts. J Food Protect 50:243–264

Deák T, Beuchat LR (1996) Handbook of food spoilage yeasts. CRC Press, Boca Raton

Dequin S, Salmon JM, Nguqen HV, Blondin B (2003) Wine yeasts. In: Boekhout T, Robert V (eds) Yeast in food: beneficial and detrimental aspects. Behr's Verlag, Hamburg, pp 389–412

Dias L, Pereira-da-Silva S, Tavares M, Malfeito-Ferreira M, Loureiro V (2003) Factors affecting the production of 4-ethylphenol by the yeast *Dekkera bruxellensis* in enological conditions. Food Microbiol 20:377–384

Dijkman KE, Koopmans M, Mossel DAH (1980) The recovery and identification of psychrotrophic yeasts from chilled and frozen comminuted fresh meats. In: Skinner FA, Passmore SM, Davenport RR (eds) Biology and activities of yeasts. Academic Press, London, pp 259–293

Dillon VM, Board RG (1991) Yeasts associated with red meats. J Appl Bacteriol 71:93–108

Diriye FU, Scorzetti G, Martini A (1993) Methods for the separation of yeast cells from the surfaces of processed, frozen foods. Int J Food Microbiol 19:27–37

Dufour JP, Verstrepen K, Derdelinckx G (2003) Brewing yeasts. In: Boekhout T, Robert V (eds) Yeast in food: beneficial and detrimental aspects. Behr's Verlag, Hamburg, pp 347–388

Echeverrigaray S, Randon M, da Silva K, Zacaria J, Longaray Delamare AP (2013) Identification and characterization of non-saccharomyces spoilage yeasts isolated from Brazilian wines. World J Microb Biot. doi:10.1007/s11274-013-1265-9

Esteve-Zarzoso B, Zorman T, Belloch C, Querol A (2003) Molecular characterisation of the species of the genus *Zygosaccharomyces*. Syst Appl Microbiol 26:404–411

Evans DG, Everis LK, Betts GD (2004) Use of survival analysis and Classification and Regression Trees to model the growth/no growth boundary of spoilage yeasts as affected by alcohol, pH, sucrose, sorbate and temperature. Int J Food Microbiol 92:55–67

Fleet GH (1990a) Food spoilage yeasts. In: Spencer JFT, Spencer DM (eds) Yeast technology. Springer, New York, pp 124–166

Fleet GH (1990b) Yeasts in dairy products. J Appl Bacteriol 68:199–211

Fleet GH (1992) Spoilage yeasts. Crit Rev Biotechnol 12:1–44

Fleet GH (1999) Microorganisms in food ecosystems. Int J Food Microbiol 50:101–117

Fleet GH (2003) Yeasts in fruit and fruit products. In: Boekhout T, Robert V (eds) Yeasts in food: beneficial and detrimental aspects. Behr's Verlag, Hamburg, pp 267–287

Fleet GH (2007) Yeasts in food and beverages: impact on product quality and safety. Curr Opin Biotech 18:170–175

Fleet GH (2011) Yeast spoilage of foods and beverages. In: Kurtzman CP, Fell JW, Boekhout T (eds) The yeasts, a taxonomic study, vol 1, 5th edn. Elsevier, Amterdam, pp 53–63

Fleet GH, Mian MA (1987) The occurence and growth of yeasts in dairy products. Int J Food Microbiol 4:145–155

Fonseca A, Inacio J (2006) Phylloplane yeasts. In: Rosa CA, Peter G (eds) Biodiversity and ecophysiology of yeasts. Springer, Berlin, pp 263–301

Fröhlich-Wyder M-T (2003) Yeasts in dairy products. In: Boekhout T, Robert V (eds) Yeasts in food: beneficial and detrimental aspects. Behr's Verlag, Hamburg, pp 209–237

Giudici P, Masini G, Caggia C (1996) The role of galactose fermenting yeasts in plain yoghurt spoilage. Ann Microbiol Enzim 46:11–19

Gram L, Huss HH (1996) Microbiological spoilage of fish and fish products. Int J Food Microbiol 33:121–137

Gram L, Ravn L, Rasch M, Bruhn JB, Christensen AB, Givskov M (2002) Food spoilage—interactions between food spoilage bacteria. Int J Food Microbiol 78:79–97

Guerzoni ME, Lanciotti R, Marchetti R (1993) Survey of the physiological properties of the most frequent yeasts associated with commercial chilled foods. Int J Food Microbiol 17:329–341

Gunde-Cimerman N, Sonjak S, Zalar P, Frisvad JC, Diderichsen B, Plemenitas A (2003) Extremophilic fungi in Arctic ice: a relationship between adaptation to low temperature and water activity. Phys Chem Earth 28:1273–1278

Guzzon R, Widmann G, Malacarne M, Nardin T, Nicolini G, Larcher R (2011) Survey of the yeast population inside wine barrels and the effects of certain techniques in preventing microbiological spoilage. Eur Food Res Technol 23:285–291

Herman PK (2002) Stationary phase in yeast. Curr Opin Microbiol 5:602–607

Hill AE (2008) Microbiological stability of beer. In: Bamforth CW (ed) Beer: a quality perspective. Elsevier, New York, pp 163–183

Hohmann S (2002) Osmotic stress signaling and osmoadaptation in yeasts. Microbiol Mol Biol R 66:300–372

Hsieh DY, Jay JM (1984) Characterization and identification of yeasts from fresh and spoiled ground beef. Int J Food Microbiol 1:141–147

Huis in't Veld JHJ (1996) Microbial and biochemical spoilage of foods: an overview. Int J Food Microbiol 33:1–18

Ingram M (1958) Yeasts in food spoilage. In: Cook AH (ed) The chemistry and biology of food. Academic Press, London, pp 603–633

Irlinger F, Mounier J (2009) Microbial interactions in cheese: implications for cheese quality and safety. Curr Opin Biotech 20:142–148

Ismail SAS, Deak T, Abd-Rahman HA, Zassien MAM, Bezuchat LR (2000) Presence and changes in populations of yeasts associated with raw and processed poultry products stored at refrigeration temperature. Int J Food Microbiol 62:113–121

Jacobsen M, Narvhus J (1996) Yeasts and their possible beneficial and negative effects on the quality of dairy products. Int Dairy J 6:755–768

Jespersen L, Nielsen DS, Hřnholt S, Jakobsen M (2005) Occurence and diversity of yeasts involved in fermentation of West African cocoa beans. FEMS Yeast Res 5:441–453

Kisko G, Roller S (2012) Activity of natural antimicrobials against *Zygosaccharomyces lentus*. Acta Aliment Hung 41:104–108

Knutsen AK, Robert V, Poot GA, Epping W, Figge M, Holst-Jensen A, Skaar I, Smith MT (2007) Polyphasic re-examination of *Yarrowia lipolytica* strains and the description of three novel *Candida* species: *Candida oslonensis* sp. nov., *Candida alimentaria* sp. nov. and *Candida hollandica* sp. nov. Int J Syst Evol Micr 57:2426–2435

Koseki S (2013) Alternative approaches to predicting microbial behaviour: a probabilistic modelling approach for microbial inactivation and a revised web-tool, the microbial responses viewer. Food Control 29:416–421

Kurtzman CP, Fell JW, Boekhout T (eds) (2011) The yeasts, a taxonomic study, 5th edn. Elsevier, Amsterdam

Latorre-Garcia L, del Castillo-Agudo L, Polaina J (2007) Taxonomical classification of yeasts isolated from kefir based on the sequence of their ribosomal RNA genes. World J Microbiol Biotechnol 23:785–791

Laubscher PJ, Viljoen BC (1999) The resistance of dairy yeasts against commercially available cleaning compounds and sanitizers. Food Technol Biotech 37:281–286

Lavoie K, Touchette M, St-Gelais D, Labrie S (2012) Characterization of the fungal microflora in raw milk and specialty cheeses of the province of Quebec. Dairy Sci Technol 92:455–468

Lehto M, Kuisma R, Määttä J, Kymäläinen HR, Mäki M (2011) Hygienic level and surface contamination in fresh-cut vegetable production plants. Food Control 22:469–475

Lopandic K, Zelger S, Banszky LK, Eliskases-Lechner F, Prillinger H (2006) Identification of yeasts associated with milk products using traditional and molecular techniques. Food Microbiol 23:341–350

Loureiro V (2000) Spoilage yeasts in foods and beverages: characterisation and ecology for improved diagnosis and control. Food Res Int 33:247–256

Loureiro V, Malfeito-Ferreira M (2003) Spoilage yeasts in the wine industry. Int J Food Microbiol 86:23–50

Loureiro V, Malfeito-Ferreira M (2004) Detecting spoilage yeasts. In: Steele R (ed) Understanding and measuring the self-life of food. Woodhead Publishing Ltd, Cambridge, pp 233–288

Loureiro V, Querol A (1999) The prevalence and control of spoilage yeasts in foods and beverages. Trends Food Sci Tech 10:356–365

Malfeito-Ferreira M (2011) Yeasts and wine off-flavours: a technological perspective. Ann Microbiol 61:95–102

Malfeito-Ferreira M, Tareco M, Loureiro V (1997) Fatty acid profiling: a feasible typing system to trace yeast contamination in wine bottling plants. Int J Food Microbiol 38:143–155

Martorell P, Stratford M, Steels H, Fernández-Espinar MT, Querol A (2007) Physiological characterization of spoilage strains of *Zygosaccharomyces bailii* and *Zygosaccharomyces rouxii* isolated from high sugar environments. Int J Food Microbiol 114:234–242

Mataragas M, Dimitriou V, Skandamis PN, Drosinos EH (2011) Quantifying the spoilage and shelf-life of yoghurt with fruits. Food Microbiol 28:611–616

McMeekin T, Olley J, Ratkowsky D, Corkrey R, Ross T (2013) Predictive microbiology theory and application: Is it all about rates? Food Control 29:290–299

Mossel DAA, Ingram M (1955) The physiology of microbial spoilage of foods. J Appl Bacteriol 18:232–268

Nagahama T (2006) Yeast biodiversity in freshwater, marine and deep-sea environments. In: Rosa CA, Peter G (eds) Biodiversity and ecophysiology of yeasts. Springer, Berlin, pp 241–262
Nielsen DS, Jacobsen T, Jespersen L, Koch AG, Arneborg N (2008) Occurrence and growth of yeasts in processed meat products—Implications for potential spoilage. Meat Sci 80:919–926
Njage PMK, Dolci S, Jans C, Wangoh J, Lacroix C, Meile L (2011) Characterization of yeasts associated with camel milk using phenotypic and molecular identification techniques. Res J Microbiol 6:678–692
Nychas GJ, Drosinos EH, Board RG (1998) Chemical changes in stored meat. In: Davies A, Board R (eds) The microbiology of meat and poultry. Blackie, London
Palhano FL, Orlando MTD, Fernandes PMB (2004) Induction of baroresistance by hydrogen peroxide, ethanol and cold-shock in *Saccharomyces cerevisiae*. FEMS Microbiol Lett 233:139–145
Patterson MF (2005) Microbiology of pressure-treated food. A review. J Appl Microbiol 98:1400–1409
Peter G, Dlauchy D, Vasdinyei R, Tornai-Lehoczki J, Deak T (2004) *Candida galli* sp. nov., a new yeast from poultry. A van Leeuwenhoek 86:105–110
Phaff HJ, Starmer (1987) Yeasts associated with plants, insects and soil. In: Rose AH, Harrison JS (eds) The yeasts, vol 1, 2nd edn. Academic Press, London, pp 123–180
Pitt JJ, Hocking AD (1985) Fungi and food spoilage. Academic Press, London
Pitt JJ, Hocking AD (1997) Fungi and food spoilage, 2nd edn. Blackie, London
Psomas AN, Nychas G-J, Haroutounian SA, Skandamis P (2012) LabBase: development and validation of an innovative food microbial growth responses database. Comput Electron Agricult 85:99–108
Puig A, Bertran E, Franquet R, García J, Mínguez S (2011) *Brettanomyces bruxellensis* prevalence in wines produced and marketed in Spain. Ann Microbiol 61:145–151
Querol A, Fleet GH (2006) Yeasts in food and beverages. Springer, Berlin
Querol A, Fernandez-Espinar MT, del Olmo M, Barrio E (2003) Adaptive evolution of wine yeast. Int J Food Microbiol 86:3–10
Ragaert P, Devlieghere F, Debevere J (2007) Role of microbiological and physiological spoilage mechanisms during storage of minimally processed vegetables. Postharvest Biol Tec 44:185–194
Raspor P, Zupan J (2006) Yeasts in extreme environments. In: Rosa CA, Peter G (eds) Biodiversity and ecophysiology of yeasts. Springer, Berlin, pp 371–417
Robiglio A, Sosa MC, Lutz MC, Lopes CA, Sangorrin MP (2011) Yeast biocontrol of fungal spoilage of pears stored at low temperature. Int J Food Microbiol 147:211–216
Rodrigues N, Goncalves G, Pereira-da Silva S, Malfeito-Ferreira M, Loureiro V (2001) Development and use of a new medium to detect yeasts of the genera *Dekkera/Brettanomyces*. J Appl Microbiol 90(4):588–599
Romano P, Capece A, Jespersen L (2006) Taxonomic and ecological diversity of food and beverage yeasts. In: Querol A, Fleet GH (eds) The yeast handbook: yeasts in food and beverages. Springer, Berlin, pp 13–53
Roostita R, Fleet GH (1996) Growth of yeasts in milk and associated changes to milk composition. Intern J Food Microbiol 31:205–219
Rosa CA, Peter G (eds) (2006) Biodiversity and ecophysiology of yeasts. Springer, Berlin
Russell N (2009) Psychrophily and resistance to low temperature. In: Gerday C, Glandsdorff N (eds) Extremophyles, vol 2. Eolss Publ Co Ltd, Ontario, pp 1–32
Samelis J, Sofos JN (2003) Yeasts in meat and meat products. In: Boekhout T, Robert V (eds) Yeasts in food: beneficial and detrimental aspects. Behr's Verlag, Hamburg, pp 239–265
Satyanarayana T, Kunze G (eds) (2009) Yeast biotechnology: diversity and applications. Springer, Berlin
Schmidt-Lorenz W (1982) Indicator organisms in frozen food in relation to spoilage. A van Leeuwenhoek 48:625–633

Serpaggi V, Remize F, Recorbet G, Gaudot-Dumas E, Sequeira-Le Grand A, Alexandre H (2012) Characterization of the "viable but nonculturable" (VBNC) state in the wine spoilage yeast *Brettanomyces*. Food Microbiol 30:438–447

Shivaji S, Prasad GS (2009) Antarctic yeasts: biodiversity and potential applications. In: Satyanarayana T, Kunze G (eds) Yeast biotechnology: diversity and applications. Springer, Berlin, pp 3–18

Sibirny AA, Voronovsky AY (2009) Candida famata (Debaryomyces hansenii). In: Satyanarayana T, Kunze G (eds) Yeast biotechnology: diversity and applications. Springer, Berlin, pp 85–112

Singh RP, Anderson BA (2004) The major type of food spoilage: an overwiev. In: Steele R (ed) Understanding and measuring the self-life of food. Woodhead Publishing Ltd, Cambridge, pp 3–23

Smits GJ, Brul S (2005) Stress tolerance in fungi—to kill a spoilage yeast. Curr Opin Biotech 16:225–230

Steels H, Bond CJ, Collins MD, Roberts IN, Stratford M, James SA (1999a) *Zygosaccharomyces lentus* sp. nov., a new member of the yeast genus *Zygosaccharomyces* Barker. Int J Syst Bacteriol 49:319–327

Steels H, James SA, Roberts IN, Stratford M (1999b) *Zygosaccharomyces lentus*: a significant new osmophilic, preservative-resistant spoilage yeast, capable of growth at low temperature. J Appl Microbiol 87:520–527

Steels H, James SA, Bond CJ, Roberts IN, Stratford M (2002) *Zygosaccharomyces kombuchaensis*: the physiology of a new species related to the spoilage yeasts *Zygosaccharomyces lentus* and *Zygosaccharomyces bailii*. FEMS Yeast Res 2:113–121

Stratford M (2006) Food and beverage spoilage yeasts. In: Querol A, Fleet GH (eds) The yeast handbook: yeasts in food and beverages. Springer, Berlin, pp 335–379

Stratford M, James SA (2003) Non-alcoholic beverages and yeasts. In: Boekhout T, Robert V (eds) Yeasts in food: beneficial and detrimental aspects. Behr's Verlag, Hamburg, pp 309–345

Tamang JP, Fleet GH (2009) Yeast diversity in fermented foods and beverages. In: Satyanarayana T, Kunze G (eds) Yeast Biotechnology: diversity and Applications. Springer, Ntherlands, pp 169–198

Tanghe A, Prior B, Thevelein JM (2006) Yeast responses to stresses. In: Rosa CA, Peter G (eds) Biodiversity and ecophysiology of yeasts. Springer, Berlin, pp 175–195

Teixeira MC, Mira NP, Sa′-Correia I (2011) A genome-wide perspective on the response and tolerance to food-relevant stresses in *Saccharomyces cerevisiae*. Curr Opin Biotech 22:150–156

Tofalo R, Schirone M, Corsetti A, Suzzi G (2012) Detection of *Brettanomyces* spp. in red wines using Real-time PCR. J Food Sci 77:545–549

Tornai-Lehoczki J, Peter G, Dlauchy D (2003) CHROMagar Candida medium as a practical tool for the differentiation and presumptive identification of yeast species isolated from salads. Int J Food Microbiol 86:189–200

Tournas VH, Katsoudas E (2005) Mould and yeast flora in fresh berries, grapes and citrus fruits. Int J Food Microbiol 105:11–17

Tournas VH, Heeres J, Burgess L (2006) Moulds and yeasts in fruit salads and fruit juices. Food Microbiol 23:684–688

Trindade RC, Resende MA, Silva CM, Rosa C (2002) Yeast associated with fresh and frozen pulps. Syst Appl Microbiol 25:294–300

Tudor EA, Board RG (1993) Food spoilage yeasts. In: Rose AH, Harrison JS (eds) The yeasts, vol 5, 2nd edn. Academic Press, London, pp 435–516

van Uden N (1984) Temperature profiles of yeasts. Adv Microb Physiol 25:195–251

Vasdinyei R, Deak T (2003) Characterization of yeast isolates originating from Hungarian dairy products using traditional and molecular identification techniques. Int J Food Microbiol 86:123–130

Viljoen BC (2001) The interaction between yeasts and bacteria in dairy environments. Int J Food Microbiol 69:37–44

Viljoen BC (2006) Yeast ecological interactions. Yeast–yeast, yeast–bacteria, yeast–fungi Interactions and yeasts as biocontrol agents. In: Querol A, Fleet GH (eds) Yeasts in food and beverages. Springer, Berlin, pp 83–110
Viljoen BC, Gernaras I, Lamprecht A, Von Holy A (1998) Yeast populations associated with processed poultry. Food Microbiol 15:113–117
Walker GM, Van Dijck P (2006) Physiological and molecular responses of yeasts to the environment. In: Querol A, Fleet GH (eds) Yeasts in food and beverages. Springer, Berlin, pp 111–152
Walker HW, Ayres JC (1970) Yeasts as spoilage organisms. In: Rose AH, Harrison JS (eds) The yeasts, vol 3., Yeast technologyAcademic Press, London, pp 463–527
Wareing P, Davenport RR (2005) Microbiology of soft drinks and fruit juices. In: Ashurst PR (ed) Chemistry and technology of soft drinks and fruit juices, 2nd edn. Blackwell Publishing Ltd, Bodmin, pp 279–299
Watson KG (1987) Temperature relations. In: Rose AH, Harrison JS (eds) The yeasts, vol 2., Yeasts and the environmentAcademic Press, London, pp 41–72
Werner-Washburne M, Braun E, Johnston GC, Singer RA (1993) Stationary phase in the yeast *Sacharomyces cerevisiae*. Microbiol Rev 57:383–401
Yalcin HT, Ucar FB (2009) Isolation and characterization of cheese spoiler yeast isolated from Turkish white cheeses. Annals Microbiol 59:477–483
Zakrajsek T, Raspor P, Jamnik P (2011) *Saccharomyces cerevisiae* in the stationary phase as a model organism—characterization at cellular and proteome level. J Proteomics 74:2837–2845
Zakrzewska A, van Eikenhorst G, Burggraaff JEC, Daniel J, Vis DJ, Huub Hoefsloot H, Delneri D, Oliver SG, Stanley Brul S, Smits GJ (2011) Genome-wide analysis of yeast stress survival and tolerance acquisition to analyze the central trade-off between growth rate and cellular robustness. Mol Biol Cell 22:4435–4446

Subject Index

A

Ablation, 51, 66, 80, 103, 104, 111
Acclimation, 13, 474
  cold, 13, 244, 248, 260
Adaptation
  kinetic, 314, 358, 359, 381, 384
  mechanism, 12, 105, 149, 231, 245, 269, 276, 525
  near-freezing temperature, 14, 243, 250, 252
  osmoadaptation, 182, 509
  strategies, 3, 5, 14, 16, 68, 315, 453, 454
Albedo, 66, 111, 113
Alcohol, 275, 300, 332, 356, 358, 361, 363–366, 418, 421, 427, 431–433
  fermentation, 421, 423, 425, 430, 515
  production, 430
Alcoholic beverage
  ale beer, 418, 515
  beer, 417, 421, 422, 428, 429, 431–433, 514, 515
  beer, alcohol-free , 428, 432
  brewing, 25, 142, 418–434, 419, 422–430, 432–434
  brewing yeast, 40, 515
  distillates, 417, 419, 425, 429–432
  lager beer, 141, 142, 418, 424, 515
  malting, 432
  wine, 26, 40, 417–432, 503, 514
  wine yeast, 36, 418, 421, 423, 427, 515
  winemaking, 16, 418, 420–432
Alps, 9, 26, 61, 65, 66, 100–106, 109–112, 117, 176, 178, 183
  Alpine contaminated soil, 470, 472
  Alpine cryoconite, 12, 33, 49, 113–116, 385, 472, 473, 475
  Alpine glacier, 12, 33, 66, 99–101, 104, 113, 115, 176
Amino acid, 164, 165, 168, 196, 208, 209, 269, 340, 357, 389, 390, 453, 454
  biosynthesis, 165, 166
  composition, 85, 273, 358
  metabolism, 431, 432
  l-proline, 85, 253, 254, 340, 358
  sequence, 85, 259, 266, 273, 274, 333, 339, 355
  substitution, 253, 339, 341–343, 359
  tryptophan uptake, 164–166
Amino alcohol, 208
Andes, 9, 61, 123, 124, 176
  Aconcagua, 124
Antarctica
  Adélie Land, 90
  Antarctic peninsula, 90
  Livingston Island, 399
  McMurdo Dry Valleys, 76–82, 88, 89, 176
  Queen Maud Land, 90
  Ross Island, 77, 90, 472
  South Victoria Land, 26, 77–79, 88
  Vestfold Hills area, 90
  Vostok Lake/Station, 29, 90, 110
  Wilkes Land, 297
Antimicrobial compound, 456, 505, 507, 516
Antioxidant, 85, 198, 206, 208, 213–215, 367, 397, 407
Apennines, 9, 101–106, 110, 112, 117
  Apennine glacier, 99–118, 108, 109, 112, 116, 117, 175, 176
Arctic
  Svalbard archipelago/islands, 33, 50, 51, 63, 65, 67, 111, 112, 175, 263, 264, 269
Argentina, 26, 33, 35, 61, 79, 124–130, 134, 137
Ash, 173, 404
ATP, 286
  ATP-grasp, 210, 211

P. Buzzini and R. Margesin (eds.), *Cold-Adapted Yeasts*,
DOI: 10.1007/978-3-642-39681-6, © Springer-Verlag Berlin Heidelberg 2014

Austria, 33
Auxotrophy, 164, 165

B

Bacteria, 8, 15, 29, 35, 58, 61, 66, 67, 86, 104, 110, 111, 113–116, 197, 244–249, 247, 249, 260, 265, 299, 300, 308–316, 318, 408, 468, 469, 473, 512, 518–522
Bark, 123, 126–128, 132, 133, 137, 141
Beagle channel, 137
Biocatalyst, 41, 179, 360, 361, 363, 364, 366, 378, 391, 425, 426, 428, 429, 431–434
Biocontrol, 382, 441–446, 449, 455–457
Biodegradation, 126, 465–467, 473
  effect of temperature, 468, 474
  immobilization, 476
  potential, 471
  toxicity, 475
Biodeterioration, 184, 497
Biodiesel, 366
Biodiversity, 11, 15–17, 35, 50, 59, 60, 80, 101, 110, 112, 126, 175, 507, 517, 521
Biogenic heat, 303, 305, 306, 316, 318
Biomass, 6, 61, 62, 115, 116, 160, 297, 310, 311, 314, 318, 386, 397, 429, 453, 454, 473, 499, 501
Biopesticide, 456
Bioreactor, 163, 387, 402, 403, 419, 426, 428, 430
  multistage fixed-bed tower bioreactor (MFBT), 431
Bioremediation, 465, 469, 472
  bioaugmentation, 469, 470
  biostimulation, 469
  cold soil, 469, 470
  mycoremediation, 468
  wastewater, 470, 474, 475
  heterologous expression, 7, 15, 360, 423, 481–486, 488, 489, 492
Biotransformation, 354, 363, 365, 378, 379, 388, 391
Brazil, 31, 135, 136

C

Carbohydrate, 68, 164, 165, 390, 400, 404, 405, 521
  arabinose, 404
  glucose, 25, 34, 183, 209, 211, 275, 397, 400–402, 404, 421, 429
  mannose, 397, 400, 402, 404
  trehalose, 86, 182, 243, 248, 250–252, 275, 314, 401, 420, 476, 510
  xylose, 404
Carbon, 53, 59, 104, 113, 116, 380, 401
  acquisition, 183, 312
  CO_2 fixation, 300–305
  dissolved organic carbon (DOC), 29
  fixation, 118, 183
  inorganic, 113
  organic, 79, 104
  respiration, 118
Carotenoid, 40, 139, 193, 197–207, 212–216, 314, 399, 406, 407, 409
  astaxanthin, 40
  carotene, 198, 206
  β-carotene, 198, 200, 202–204, 206, 207, 406, 407
  γ-carotene, 198, 199, 200, 202–205
  16-hydroxytorulene, 199, 202, 204, 205
  2-hydroxyplectaniaxanthin, 199, 202, 204, 205
  2-hydroxytorularhodin, 199, 202, 205
  lycopene, 199, 200, 204, 205
  plectaniaxanthin, 193, 199, 204–206
  torularhodin, 198, 199, 203–207, 215, 406, 407
  torularhodinaldehyde, 199, 200, 202
  torulene, 198, 202–205, 406, 407
Cell
  capsule, 11, 58, 81, 105, 139, 444
  dormant, 58, 182, 298, 308, 309, 311
  endoplasmic reticulum, 252, 281, 283
  plasma membrane, 86, 164, 168, 225–231, 233–234, 236–238, 243, 253, 283
  ribosome, 165, 167, 244, 246, 248, 485, 486
  size, 315, 316
Cellular thermosensor, 246
Chaperone, 167, 244, 245, 250, 252, 484–488
Chemosynthetic environment, 150
Chile, 79, 124–129
Coenzyme Q_{10}, 399, 406, 407, 409
Cold
  induction, 484, 485, 487–489, 493
  shock, 13, 82, 243–254, 260, 267, 314, 484–485, 509
Colonization, 12, 80, 139, 309, 443, 454
Communities
  cryptoendolithic, 9, 61, 80, 176, 183, 297
  endolithic, 60, 61, 80, 176
Conservation/preservation
  culture collection, 17, 23, 36–38, 52
  ex situ, 23, 68, 101
Cryoconite, 9, 12, 31–33, 35, 37, 57, 66, 67, 80, 107, 113–118, 385, 472, 473, 475
Cryopeg, 29, 31, 58, 60, 68

Cryoprotectant agent (CPA), 275
Cycle
  biogeochemical, 68, 88, 99, 101
  freeze–thaw, 58, 79, 82, 178
  iron, 88, 118
  nutrient, 8, 67, 111, 117, 126
  organic matter, 103

D

Debris, 9, 66, 101, 104, 110
  endoglacial, 104
  subglacial, 64, 104
  supraglacial, 117
Deep sea, 6, 41, 62, 149–168, 329
  abyssopelagic zone, 150
  bathypelagic zone, 150
  epipelagic zone, 150
  hadopelagic zone, 150
  mesopelagic zone, 150
Deglaciation, 101
Delignified cellulosic material (DCM), 421, 425–433
Desert
  Antarctic, 58, 75–82, 88, 89, 178
  Arctic, 61
  Polar, 297, 298
Desiccation, 50, 53, 59, 182, 315
  resistance, 77, 81, 86, 87, 89
Dirt cone, 104
DNA/RNA
  16S, 41, 61, 309, 312
  18S, 9, 41, 61, 63, 158
  26S, 9, 105
  5.8S, 9, 422, 516
  D1/D2, 9, 105, 108, 422
  damage, 87, 179, 196, 214, 406
  intergenic spacer (IGS), 140
  internal transcribed spacer (ITS), 41, 67, 105, 140, 158
  (6-4) pyrimidine/pyrimidone dimer, 196
  mRNA, 167, 244, 245, 247, 249, 250, 252, 254, 311, 484
  rRNA, 67, 248, 249, 384, 420
  sequence, 41, 252, 267, 522
Dormant, 58, 182, 298, 308, 309, 311

E

Ecoclade, 137
Ecology, 8, 10, 17, 42, 69, 99, 101, 105, 112, 114, 116, 118, 130, 138, 149, 173, 282, 444, 502, 505
Ecosystem, 3, 4, 8–13, 16, 17, 28, 58, 64, 80, 89, 99, 101, 103, 105, 108, 110, 111, 114, 116–118, 123, 126, 130, 152, 194, 282, 314, 319, 398, 399, 406, 444, 469, 498, 499, 505, 507, 519, 521
Environmental factor, 4, 6, 60, 125, 134, 160, 174, 227, 237, 282, 303, 443, 466, 467, 502, 505, 525
Enzyme
  activation energy, 303, 314, 327, 330–333, 343, 381
  activation parameters, 332
  adenylate kinase, 333
  alkaline phosphatase, 328, 343, 344
  amylase, 15, 136, 141, 183, 329, 331–344, 336–338, 379, 380
  ATP synthase, 166
  ATPase, 168, 326
  carbonic anhydrase, 333
  catalytic efficiency, 327, 330, 331, 333–335, 341, 343
  catechol dioxygenase, 476
  cellulase, 136, 329, 331, 332, 334, 335
  Chirazyme, 356, 360, 367
  chitinase, 329, 334, 379, 382, 489–492
  cold denaturation, 339
  cold-active, 105, 115, 135, 325–328, 330–335, 337–345, 358–362, 377–381, 385–391, 399, 453, 454
  dehydroquinate synthase (DHQS), 210
  desaturase, 199, 230, 231, 246, 253, 287, 420
  enantioselectivity, 359, 360, 361, 364
  esterase, 136, 342, 354, 357, 384, 389–391, 486
  fatty acid dehydrase, 230
  fatty acid desaturase, 230
  fatty acid synthetase, 231
  flexibility, 325, 332–335, 337, 339, 340, 343, 344
  β-galactosidase, 385–387
  geranyl–geranyl pyrophosphate (GGPP), 199
  glucosyl transferase, 211
  glycosidase, 377, 380, 385
  HMG-CoA reductase (HMGR), 284, 288, 289
  hydrolase, 379, 387
  hydrolytic, 65, 68, 99, 105, 116, 135, 305, 353, 364, 381, 384, 387, 521
  kcat, 327, 329, 330, 342, 359, 381, 382, 389
  Km, 314, 327, 329, 330, 333, 337, 341–343, 359, 381, 382, 389

Enzyme (*cont.*)
kinetic resolution, 357, 361, 364, 365
lipase, 15, 135, 343, 353–368, 399, 420, 471, 506, 518
lipase A (CALA), 355–368, 378, 492
lipase B (CALB), 355–368, 378
lipozyme, 356
mesophilic, 327–338, 341–344
O-methyltransferase, 210, 211
molecular dynamics simulation, 357
ornithine transcarbamylase, 328
pectinase, 115, 136, 382, 383
pectinolytic, 382–384, 399, 500, 503
phytase, 389, 390, 399
polygalacturonase, 336, 383, 384
protease, 15, 135, 333, 337, 387–389, 399, 483, 506, 518
psychrophilic, 331–334, 336–341, 358, 381
regioselective, 363, 364, 366
ring opening polymerization, 367
structure, 14, 332, 334–341
substrate binding, 327, 330, 343, 366
surface displayed, 360, 361
thermal stability, 325, 329, 332–335, 339, 341, 358, 363, 381, 382, 384, 487
thermodynamic stability, 335, 337, 339
thermophilic, 333, 340
xylanase, 327, 331–334, 337, 379–381, 399
Ethanol, 26, 59, 206, 213, 252, 305, 307, 311, 313, 315, 366
production, 422, 424, 429, 430, 501, 503
Exopolymeric substances (EPS), 397, 398, 400–405, 407–410, 476
Extrinsic factor, 524

F

Fermentation, 275, 402–404, 492, 503, 509, 513, 515, 519
Food
associated yeasts, 499, 505, 507, 520, 524
baking, 253, 380, 424
bread, 5, 253, 380
cheese, 363, 386, 388, 503, 506, 509, 513, 518–520
dairy, 430, 503–506, 516–519
fish, 388, 520, 522
frozen, 35, 37, 38, 508
fruit, 502–504, 506, 507, 511, 512, 514
meat, 503, 504, 520–523
milk, 516–520
poultry, 520–522
spoilage, 506, 508–525
vegetable, 511, 512
Freezing
point, 53, 58, 115, 261, 296–298, 312
tolerance, 251, 253, 254, 261, 265
Fungi
dimorphic, 90, 133, 138, 175, 179, 200, 212, 297
yeast-like, 52, 60, 61, 68, 90, 107, 112, 138, 153, 159, 179, 181

G

Gas
CH_4, 298, 300, 305
CO_2, 58, 183, 297, 298, 300–303, 312, 316, 432, 501, 505, 513, 514, 518, 523
$^{14}CO_2$, 300–307
diffusion, 299, 304
exchange, 297, 299, 300, 303
N_2O, 298
SO_2/sulphur dioxide, 423, 516
Gene
disruption, 163
expression, 165, 237, 248, 250, 484–486
horizontal transfer, 266
manipulation, 163, 423
mutation, 196, 252, 253, 272, 341–344, 359
overexpression, 163, 164, 253
Gene/genome sequence, 15, 142, 163, 266, 289
Genetic modification
backcrossing, 423
genetic engineering, 423
hybridization, 423
mutagenesis, 341–344, 423
site-directed mutagenesis, 163, 342
somatic fusion, 423
transformation, 423, 491
Glacier
alerce, 125
alpine, 12, 33, 66, 99–117, 176
apennine, 99–117, 175, 176
black, 127
calderone, 101–106, 113, 118, 176
castaño overo, 125
forni, 33, 102
frias, 125, 127, 129, 131, 132, 134
géant, 102, 106, 112
miage, 102, 104, 106, 108
perito moreno, 125, 133
río manso, 125, 127
sforzellina, 102
temperate, 64, 103, 110, 117, 118
tronador, 125, 127–129, 131–134, 137

Global warming, 26, 53, 101, 319
Gluten, 421, 425, 427, 429–431, 433
Glycerol, 34, 182, 243, 249, 251, 253, 275, 312, 354, 422, 423, 432, 510
Glyphosate, 210
Grape must, 421, 425, 426
Grass, 53, 310, 399
Greenhouse effect, 298
Greenland, 30, 33, 63, 65
Growth
  active, 117
  asexual, 7, 61, 181, 214
  fruiting body, 7
  in situ, 109, 111, 307
  kinetic parameters, 5, 6
  lag phase, 508
  meristematic, 61
  sexual, 181
  slow, 25, 61, 81, 173, 181–183, 308
  temperature, 5–7, 13, 23, 36–40, 86, 114, 115, 140, 141, 178, 227, 230–235, 326, 457, 472, 473
Gypsum, 61, 65

H
Habitat
  englacial, 64, 109, 110, 117
  extreme, 4, 14, 50, 52, 53, 58, 60, 68, 77, 80, 103, 105, 109, 112, 116, 117, 135–138, 163, 168, 173, 178, 237, 269, 289, 404, 499
  frozen, 26, 28, 35, 37, 57–59, 113, 178, 295–318, 468, 508
  hypersaline, 29, 31, 238, 309
  icy, 110, 111, 175, 295, 444
  oligotrophic, 6, 32, 65, 81, 82, 88–89, 112, 116, 125, 134, 138, 175–177, 182, 183, 214, 307, 316, 444
  salinity, 8, 59, 62, 163, 208, 213, 238, 468
  subglacial, 26, 28, 31–40, 54–57, 63–65, 101–105, 109–112, 116–118, 175, 450
  temperate, 12, 26, 39, 58, 63, 68, 78, 80, 108–118, 124, 126, 138, 141, 142, 227, 310, 467
Heterologous expression, 7, 15, 360, 423, 481–494
Himalaya, 12, 137
Histidine kinase sensor, 247
Humidity, 87, 304, 380, 398, 443
Hydrocarbon
  n-alkanes, 401, 468, 473
  anthracene, 473
  aromatic compounds, 177
  biodegradation, 465, 470
  diesel oil, 115, 470, 476
  n-dodecane, 473
  n-hexadecane, 473
  monoaromatic, 175, 473, 475
  petroleum hydrocarbons, 466, 469, 471, 473
  phenanthrene, 473
  phenol, 109, 115, 184, 472–476
  pollution in cold climates, 174, 476
  polyaromatic hydrocarbons, 473
  toxicity, 475
Hydrogen, 198
  bond, 234, 269, 270, 340, 342, 344
  peroxide, 196, 364, 367
Hydrolysis, 115, 116, 356, 363–366, 380–390, 432, 520
Hydrothermal vent, 62, 150, 159
Hypochlorite, 196
Hypoxia, 50, 165

I
Ice
  ancient, 33
  bed interfac, 64
  binding mechanism, 269, 272
  binding site, 271, 272
  core, 9, 11, 28–33, 63, 90, 107, 109–112, 263–265, 449, 450
  dust, 113
  englacial, 109, 117
  flow, 101, 118
  mass, 99, 101, 117, 125
  melting, 103, 111
  recrystallization, 260–265
  recrystallization inhibition (IRI), 261–263, 269, 270
  sheet, 65, 66, 80, 111
  surface, 38, 111
Iceland, 39, 50, 384
Immobilization
  carrier/support, 368, 426–433
  cell, 418, 419, 422, 424–433, 476
  enzyme, 360–362, 368
in situ
  activity, 111, 297, 299
  enrichment, 99, 117, 118
  growth, 109, 307
  selection, 23
Inclusion body, 484
Intrinsic factor, 498, 517, 520, 522
Italy, 33, 35, 36, 39, 101, 102, 176

J
Japan, 30, 34, 35, 38, 76, 140, 154, 156, 384

K
Karmellae, 284–286

L
Lagoon, 9, 128–133, 450, 451
Lake
  fonck, 128
  hess, 128
  ilon, 128
  mascardi, 127
  nahuel Huap, 129, 131–133
  negra, 127
  vostok, 90, 110, 309
Lichen, 4, 9, 60, 176–178, 183, 297, 302
Light
  absorbance, 115, 208
  dependence, 111
  exposure, 111, 297, 498
  spore production, 214
  sunlight, 138, 150, 176, 195, 214, 407
  UV light, 79, 138, 195, 407
Lyophilisation, 356, 404
Lipid
  acyl chain, 164, 165, 226, 227
  ergosterol, 168, 228, 284, 288, 399, 406–409
  fatty acid, 13, 68, 85, 116, 141, 226–228, 230–236, 244, 246, 248, 249, 287, 300, 308, 314, 315, 343, 354, 356, 366, 367, 391, 405, 406, 418, 420, 421, 431, 432, 516
  glycerophospholipid, 84
  glycolipid, 228
  phosphatidylethanolamine (PE), 84, 228, 229, 233, 234, 421
  phosphatidylserine (PS), 228, 229, 233
  phospholipid, 84, 168, 182, 227–236, 246, 248, 252, 287, 406
  polar head group, 226–228
  poly-unsaturated, 13, 68, 141, 230, 246, 406
  saturated, 228, 230
  sphingolipid, 228, 230
  sterol, 80, 85, 86, 168, 227–230, 281–289
  unsaturated, 13, 84, 116, 227–236, 244, 287, 367, 406, 421
  unsaturation degree, 231, 233, 234, 421
  unsaturation index (UI), 231, 232, 235
  zymosterol, 228, 288

M
Maintenance state, 306–308
Mangrove, 39
Media
  CMA, 34
  DG18, 33–34
  dichloran, 34
  DRBCA, 33–34
  MEA, 33–34
  MYP, 33–34
  nutrient media, 32–35, 312
  PDA, 33–34
  rose Bengal (RB), 33–34
  SDA, 33–34
  YPD, 33–34
Melanin, 138, 177, 181–183
Melt water, 50, 64–66, 102, 103, 107, 109, 111, 112, 116–118, 175, 449
Melting
  point, 14, 64, 103, 109, 118, 234, 253, 261, 262, 335, 381
  temperature Tm, 261, 338
Membrane
  fluidity, 13, 81, 84, 86, 164, 165, 225–238, 243, 244, 284, 314, 405, 420, 421
  lipid bilayer, 13
  permeability, 198, 228, 252
Metabolism
  microbial, 58, 68, 103, 176, 183, 184, 248, 253, 296, 308, 428
  photosynthetic, 67, 103, 114, 183, 197, 208, 209, 297, 302
Metabolite, 7, 82, 181, 193, 213, 215, 303, 305, 397, 399, 401, 407, 409, 420, 468, 501–510
Metagenomic, 14, 17, 51, 90, 308, 368
Metaproteomic/proteomic, 14, 389, 508, 519
Method
  aseptic, 27, 30
  coagulant, 32
  culture-based, 16, 24, 81, 90, 159, 160, 309, 312
  culture-independent, 16, 24, 40, 42, 81, 90, 159, 160, 299, 308
  DGGE, 40, 518
  dilution, 32, 33, 312
  drill, 29, 30, 90, 110
  drying, 419, 430–433
  enrichment, 59, 99, 117, 118, 295, 312, 314, 472, 499
  enumeration, 24, 524
  extraction, 209, 212, 377, 404, 407, 409, 418, 426
  filtration, 24, 25, 32, 269, 404, 515

FISH, 40, 516
fluorescent microspheres, 29
homogenization, 30, 403
illumina sequencing, 41
isotope labelling, 302
most probable number (MPN), 24
Nansen bottle, 27, 152
next-generation sequencing (NGS), 41
Niskin sampler, 27, 152
nitrocellulose membrane, 32, 33
operational taxonomic unit (OTU), 41, 60, 309
PCR, 40, 41, 63, 140, 165, 267, 310, 312, 516
plate count, 24, 25
precipitation, 25, 362, 404, 468
pyrosequencing, 41, 265
sampling, 27, 29, 151, 152
Secchi disk, 27, 28
separation, 32, 379, 404
Silva reference database, 41
solid-state cultivation, 312
stable-isotope probing (SIP311), 311
surface sterilization, 30
surfactant, 32
TGGE, 40
TRFLP, 41, 312
tube worms, 30
ultraperformance liquid chromatography (UPLC), 82, 83
UNITE database, 41
van Dorn sampler, 27, 152
Microorganism/organism
eukaryotic, 3–5, 8, 16, 116, 159, 230, 325, 378, 443
eurythermal, 7
prokaryotic, 4, 149, 152, 160, 161, 230, 295, 315, 325, 378, 379
psychrophilic, 5, 7, 100, 116, 126, 194, 259, 260, 282, 327, 344, 378, 391
psychrotolerant, 5–7, 26, 100, 116, 126, 194, 512
stenothermal, 7
viable, 4, 30, 32, 63, 109–111, 308
Molasses, 422, 430, 456
Moss, 9, 90, 175, 264, 385, 399
Mycosporine, 138, 193, 197, 200–203, 207–215
cyclobutane-pyrimidine dimers (CPD), 196, 207, 214
3-dehydroquinate (3-DHQ), 210
mycosporine-glutaminol-glucoside (MGG), 203, 209–215
mycosporine-glycine, 208, 214
mycosporine-like aminoacids (MAAs), 207–210, 213, 214
palythine, 214
porphyra-334, 214
sedoheptulose 7-phosphate, 210
shinorine, 210, 214

N

Neurotransmitters, 175, 184
Nitrogen, 11, 25, 67, 104, 105, 113, 116, 205, 206, 208, 209, 213, 381, 401, 402, 421, 422, 514
inorganic, 111, 135
organic, 28, 104, 111, 135, 402
Norway, 33, 34, 36, 40, 50, 138, 175, 263, 264, 269

O

Ocean, 4, 62, 63, 69, 125, 150–152, 184, 315, 355, 469
Atlantic, 124
Arctic, 4, 49, 50, 60, 62, 309
Pacific, 35, 124, 159
Oligotrophy, 117, 181
Organic acid
acetic acid, 421, 422, 502–504, 513, 515, 516, 522
citric acid, 517, 522
lactic acid, 33, 418, 513, 518, 522
malic acid, 421, 422
succinic acid, 422
Oxygen, 159, 163, 195, 196, 198, 209, 231, 282, 385, 403, 406, 467, 469, 471, 498, 505, 514, 520, 523
singlet, 196, 206, 213, 214

P

Packinghouse, 443, 446, 457, 511
Patagonia, 26, 33, 66, 123–127, 129–131, 134–139, 141, 142, 193, 215, 260
Pathway
carotenoid synthesis, 199, 204, 205
ERAD, 281, 283–290
high-osmolarity glycerol (HOG), 247, 250, 251, 254, 509
phospholipid synthesis, 229–231, 248
shikimic acid, 197, 210
mevalonic acid, 199
mycosporine synthesis, 209–211
target of rapamycin (TOR), 509
Pcold vector, 485, 488, 493, 494

Penguin, 9, 399, 449, 450
Peptide, 85, 87, 166, 210, 252, 267, 273, 274, 283, 340, 360, 389, 490, 516
Permafrost, 4, 9, 11–13, 26, 28–32, 34–36, 41, 49, 50, 53–60, 68, 80, 297–312, 315–317, 319, 325, 450
pH, 11, 28, 34, 53, 58, 59, 61, 78, 79, 113, 179, 182, 206, 208, 238, 315, 362, 381–390, 399, 402, 423, 425, 452, 467, 469, 493, 498–501, 508, 509, 511, 513, 514, 517–519, 521, 522, 524, 525
Phenolic compound, 109, 184, 428, 474
Phosphorous, 104
Photoinduction, 213
Photo-protection, 193, 207, 215
Photostability, 214
Phylloplane, 123, 126–129, 131–133, 137, 138, 194, 197, 200, 215
Phyllosphere, 137, 138, 175
Polysaccharide, 11, 81, 139, 270, 275, 276, 380, 385, 398, 400–402, 404, 405, 408, 419, 420, 425, 444
  arabinomannan, 404, 405, 408–410
  cosmetic, 397–399, 405, 408–410
  emulsifier, 397, 407, 408
  emulsion properties, 405
  exopolysaccharide (EPS), 397, 398, 400–405, 407–410
  galactomannans, 398, 405
  gelling agent, 408
  glucans, 398
  glucomannan, 403–405, 407–410
  glucuronoxylomannans, 398
  heteropolysaccharide, 401, 404
  mannans, 398, 400
  phosphomannans, 398
  rheological properties, 397, 398, 404
  thickeners, 398, 405, 408
  xylomannan, 404, 405, 408, 410
Post harvest, 441–444, 446, 447, 455, 457
  control, 16, 442, 443, 445–449, 452, 453, 457
  disease, 441–443, 445, 446, 448, 449, 453, 455, 457, 458
  pathogen, 441–443, 446, 447, 452, 456
Post-translational modification, 284, 483, 489, 490, 494
Pressure, 6, 27, 29, 52, 64, 68, 103, 104, 109, 111, 118, 134, 149–151, 161, 163–166, 168, 176, 179, 182, 261, 303, 306, 310, 318, 319, 444, 491, 509
  hydrostatic, 4, 13, 149–151, 160–163, 168, 509, 522
  osmotic, 468
Protein
  antifreeze (AFP), 53, 59, 259–270, 272–275, 314, 490–492
  cold shock (CSP), 13, 82, 243–245, 249, 251, 314
  endoplasmic reticulum-associated degradation (ERAD), 14, 281, 283–290
  heat shock (HSP), 13, 243, 245, 248, 249, 251, 252, 254, 314, 423, 486
  ice interaction, 261
  ice-binding (IBP), 14, 259, 261, 263, 493, 494
  3D structure, 14, 270

Q

Quencher, 214

R

Radiation/irradiation, 69, 86, 138, 150, 177–179, 182, 195, 197, 198, 206–209, 213, 315, 407, 514
  gamma, 182
  ionizing, 183
  solar, 4, 10, 13, 64, 66, 103, 113, 137, 179, 194, 195, 318, 507
  ultraviolet (UV), 6, 30, 50, 61, 134, 137, 138, 182, 193–197, 208, 210, 215, 398, 399, 406, 407, 468
Radical
  alkoxy, 196
  free, 196, 206, 406, 409
  hydroxyl, 196, 213
  peroxyl, 206
Rain forest, 127–130
  Valdivian, 127–130
Reactive oxygen species (ROS), 195, 196, 198, 206, 207, 213, 215, 248, 260, 420, 452
Refolding, 273, 492, 493
Respiration, 13, 118, 196, 295, 300–302, 305–308, 333, 467
  $^{14}CO_2$ dark fixation, 300–302
  $^{14}CO_2$ uptake, 303, 305
  substrate-induced (SIR), 310
Rhizosphere, 80, 123, 126–129, 133, 134, 137, 139
  ectomycorrhizosphere, 127, 129, 132, 134, 139
Rock, 9, 50, 52, 58, 60, 61, 64, 67, 69, 88, 103, 118, 173–179, 181, 183, 184, 237, 297
  weathering, 88, 104, 118
Rotten wood, 123, 126–128

S

Saltpans, 174
Screening, 17, 164, 165, 197, 253, 259, 263, 284, 379, 406, 445, 446, 457
Seasonal change, 282, 283
Sediment, 9, 26–28, 32, 37, 39, 57–60, 62, 64, 67, 69, 102–105, 107, 111–113, 116–118, 152–158, 175, 176, 295, 300, 309, 385, 387, 399, 449, 451, 469, 501, 502, 515
  deep piping, 28, 102, 104
  deep sea, 23, 27, 34–36, 42, 62, 152, 161
  lake, 37, 76
  subglacial, 28, 31, 33, 37, 99, 101–105, 109, 111, 112, 116–118
  supraglacial, 28, 31, 33, 99, 101–105, 108, 109, 112, 114, 116–118
Seep
  cold, 151, 159
  methane, 159
Selection, 13, 23, 42, 118, 138, 164, 179, 282, 355, 390, 424, 432, 433, 445–448, 455, 456, 458, 474, 491, 502, 506, 510, 512, 520
Sexuality, 181
Siderophores, 87–89, 452
Simplification, 61, 181
Snow, 9, 31, 33, 36–39, 80, 102, 103, 107, 109, 112, 117, 194, 295, 299
  drifting, 101
  fall, 103, 109, 112, 117, 124
  mold, 263
  pack, 53, 111, 112
  patch, 113, 124
Soil
  Alpine, 109, 472, 475
  Antarctic, 11, 31, 49, 52, 77 , 108, 176, 399, 400, 407, 445, 449–451
  Arctic, 53, 59 , 387, 468
  cold, 101, 109, 314, 399, 469
  frozen, 37, 58, 299–301, 303, 304, 306, 308, 309, 384, 468
  organic matter, 302, 303, 318
  polluted/contaminated, 101, 109, 299, 470–474
  population dynamics, 316, 318
  tundra, 35, 58, 297–299, 302, 310, 316, 317, 318
Solubility-enhancing tag, 486, 487
Species
  cosmopolitan, 52, 58, 59, 61, 65, 89, 105
  differentiation, 174, 181
  dominant, 16, 49, 52, 65–67, 80, 81, 105, 109, 112, 113, 134, 137, 138, 159, 175, 388, 518
  endemic, 12, 26, 52, 61, 62, 108, 109, 163, 176, 387
  extremophilic, 53, 86
  novel, 9, 10, 63, 66, 108, 126–130, 136–139, 152, 159, 176, 184, 522
  ubiquitous, 12, 36, 62, 112, 138, 200, 215
Starter cultures, 430, 498, 500
Stress
  adaptation, 85, 179, 197, 237, 505, 508
  factor, 50, 61, 213, 245, 250–252, 315, 499, 503, 507
  osmotic, 13, 138, 178, 182, 213, 237, 247, 251–253, 452, 510
  oxidative, 4, 13, 206, 248, 251–254, 420, 500
  resistance, 10, 182, 208, 315, 423, 424, 519
  response, 13, 164, 167, 174, 184, 230, 276, 282–286, 314, 420–423, 505, 506
  tolerance, 178, 179, 184, 238, 251, 254, 250, 276, 429, 433, 453, 489, 498, 499, 506, 508, 525
Stromata, 123, 128, 133, 137–141
Subzero activity, 295–319
Superoxide, 196, 206, 337
Survival, 11, 12, 86, 87, 116, 134, 176, 177, 181, 182, 207, 214, 215, 238, 249, 260, 299, 424, 443, 470, 502, 520

T

Taxonomy, 8, 17, 52, 134
Temperature
  growth, 5–7, 13, 23, 36–40, 86, 114, 115, 141, 142, 178, 227, 230–235, 457, 472, 473
  optimum, 7, 115, 238, 326, 336, 355, 358, 379, 384–390, 474
Thermal hysteresis (TH), 14, 261, 269
Tolerance, 178–179, 184, 237, 238, 247, 248, 250, 253, 276, 429, 433, 489, 499, 506, 522, 535
Translation machinery, 249
Tree, 39, 53, 68, 123, 125, 128–133, 137–140, 458

U

Ubiquitination, 164, 285, 286

V
Volatile compound, 300, 305, 308, 316, 379, 431, 502, 516
  acetoin, 428, 432
  2,3-butanediol, 432
  aroma, 380, 390, 418, 422–428, 431, 432
  buttery flavour, 432
  diacetyl, 428, 429, 432, 433
  fusel alcohol, 422, 430–432
  isoamyl acetate, 421
  off-flavor, 418, 501–504, 513–519
  β-phenylethyl acetate, 421
  β-phenylethyl alcohol, 421
  Strecker aldehyde, 432, 433
  worty flavor, 432, 433

W
Water
  activity (Aw), 26, 32, 35, 53, 64, 103, 175, 303, 360, 391, 467, 498, 505, 507, 517
  fall, 127, 132
  transparency, 27, 28, 195
Wind, 26, 58, 62, 66, 78, 79, 109, 113–117, 124, 137, 181, 398

Y
Yeast
  antagonistic, 441, 444, 445, 452, 455
  ascomycetous, 5, 7, 14, 31, 35, 40, 41, 54, 59, 62–65, 81, 106, 133–139, 175, 200, 212, 444, 453, 499, 508, 512, 521
  basidiomycetous, 31, 35, 40, 41, 52–54, 59, 60, 62–67, 81, 105–109, 131–139, 159, 201, 209, 212, 259, 263–265, 399, 444, 452, 469, 475, 499, 507, 508, 511
  black, 52, 61, 65, 66, 81, 138, 173–184, 200, 237, 238
  conservation, 68, 101
  cultivation, 24, 25, 32–35, 40, 115, 163, 312, 326, 390, 406, 407, 473, 474, 489
  culturable, 10, 30, 58, 59, 67, 104, 115, 152, 469, 472, 508, 517
  diversity, 11, 23, 25, 89, 101, 102, 112, 134, 137, 149, 152, 160, 521
  dried, 430
  freeze-dried, 429, 430, 431
  halophilic, 60, 161, 177, 178, 200, 201, 213
  halotolerant, 60, 63, 161, 213, 238, 500
  identification, 8, 41, 59, 63, 140, 158, 183, 199, 212, 399, 500, 505, 516, 522–524
  isolation, 9, 11, 12, 23–25, 32, 33, 59, 62–64, 68, 136, 151, 152, 176, 264, 269, 312, 313, 399, 445–447, 449–451, 472
  lipolytic, 355, 359, 362–367, 389, 500, 503–507, 517, 518
  mesophilic, 6, 15, 59, 84, 85, 134–136, 184, 193, 231–238, 355, 358, 359, 379, 381–388, 400, 421, 474, 476
  non-conventional, 5, 353
  non-culturable, 10, 24, 58, 159, 368, 516, 517
  osmophilic, 511
  osmotolerant, 25, 26, 65, 175, 500, 503, 509, 514
  pectinolytic, 382–384, 399, 500, 503, 511
  physiology, 101, 149, 160, 163, 164, 234, 402, 424, 505
  piezophilic, 163
  piezotolerant , 160
  proteolytic, 355, 388, 389, 500–521
  psychrophilic, 3, 66–69, 82, 84, 85, 101, 123, 126, 234–236, 243, 259, 260, 263–266, 269, 275, 276, 281–283, 287, 289, 290, 318, 354, 355, 368, 379, 382–385, 388, 399, 406, 420, 424, 433, 444, 457, 482, 483, 486–490, 494
  psychrotolerant, 3, 23, 39, 40, 68, 69, 101, 123, 126, 134, 149, 234, 235, 282, 417, 422, 425, 427–429, 431–433, 444, 489, 506, 507, 521
  psychrotrophic, 5, 6, 26
  spoilage, 299, 442, 497–525
  thermophilic, 6, 176

Microorganism Index

A

Acidomyces, 61
A. acidophilus, 177
Acinetobacter, 518, 521
Aeromonas, 521
Agaricostilbum
A. hyphaenes, 201
A. pulcherrimum, 201
Alternaria, 442, 445
Anabaena
A. variabilis, 210
Arthrobacter, 313, 473
A. psychrophenolicus, 474
A. sulfureus, 474
Aspergillus, 442
A. niger, 15, 356
Aureobasidium, 37, 107, 138, 153, 159, 175, 237, 238, 445
A. pullulans, 54, 90, 107, 112, 128, 138, 175, 200, 235, 237, 388, 449, 453, 476
A. pullulans var. melanogenum, 65, 175
A. pullulans var. pullulans, 60, 65, 175
A. pullulans var. subglaciale, 65, 175

B

Bacillus, 15, 329, 341
B. amyloliquefaciens, 337, 338
B. licheniformis, 380
B. subtilis, 244, 247, 342
Bacteroides, 110, 309, 310
Bannoa, 198
B. hahajimensis, 201
Barnettozyma, 37
Bensingtonia, 37
B. yamatoana, 54, 131
Botryozyma, 37
Botrytis, 442
B. cinerea, 455, 456
Brettanomyces, 518
B. anomalus, 503
B. bruxellensis, 503
Bullera, 37, 507
B. dendrophila, 131
B. variabilis, 203
Bulleromyces, 37, 38, 49, 60
B. albus, 54, 106, 203

C

Candida, 37, 49, 60, 62, 65, 90, 106, 115, 149, 153, 159, 160, 282, 398, 399, 406, 445, 471, 499, 507, 508, 512, 515, 518, 519, 521
C. albicans, 355, 388, 509
C. alimentaria, 504, 522, 523
C. antarctica, 12, 353, 355–363, 366, 368, 378, 379, 492
C. apicola, 54, 388
C. atlantica, 153
C. atmosphaerica, 153
C. bertae, 128
C. boleticola, 127
C. buinensis, 518
C. carpophila, 153
C. castrensis, 127
C. catenulata, 517, 518
C. coipomoensis, 128
C. delftensis, 133
C. famata, 449, 450, 503, 509
C. fukuyamaensis, 153
C. humicola, 387
C. intermedia, 517
C. krusei, 54, 503
C. lipolytica, 231, 232, 355, 504
C. llanquihuensis, 127
C. lodderae, 153
C. lusitaniae, 517
C. marítima, 128
C. mesenterica, 133

P. Buzzini and R. Margesin (eds.), *Cold-Adapted Yeasts*,
DOI: 10.1007/978-3-642-39681-6,

Candida (*cont.*)
C. molischiana, 54
C. norvegica, 54
C. oceani, 153
C. olea, 388
C. oleophila, 233, 445, 449, 455
C. parapsilosis, 54, 63, 153, 232, 388, 500, 502, 517, 520, 522
C. petrohuensis, 127
C. pseudointermedia, 54, 502
C. psychrophila, 235
C. quercitrusa, 54
C. railenensis, 129
C. rancensis, 127
C. rugosa, 517, 520
C. saitoana, 232
C. sake, 54, 133, 153, 232, 445, 449, 453, 455, 520, 523
C. santamariae, 54, 106
C. santjacobensis, 127
C. scottii, 235, 450, 454
C. slooffiae, 232
C. sophiae-reginae, 127
C. tepae, 127
C. tropicalis, 388, 476, 500
C. utilis, 233, 236
C. valida, 503
C. viswanathii, 153
C. zeylanoides, 508, 509, 517, 520–523
Capronia, 181
Cellvibrio, 312
Chionosphaera
C. apobasidialis, 201
Citeromyces
C. matritensis, 133
Cladosporium, 200
Clavispora, 60, 90, 159
C. lusitaniae, 154, 232, 517
Clostridium
C. thermocellum, 488
Colletotrichum
C. graminicola, 209
Colwellia
C. maris, 329
Coniosporium, 154, 209
Coniozyma
C. leucospermi, 200
Cryomyces, 61, 176–178, 181
C. antarcticus, 180
C. funiculosus, 176, 178, 180
C. minteri, 180
C. montanus, 176
Cryptococcus, 11, 37, 38, 49, 52, 60, 62, 67, 75, 81, 90, 99, 105, 106, 109, 115, 137, 138, 149, 156, 159, 198, 203, 212, 213, 264, 265, 282, 300, 313, 384, 398, 399, 406, 407, 445, 499, 507, 508, 511, 512, 519, 521, 522
C. adeliae, 327, 380, 399
C. adeliensis, 54, 63, 64, 106, 129, 264, 327, 380, 381, 399
C. aerius, 54, 106, 131, 139, 388
C. albidosimilis, 55, 64, 106, 264, 381
C. albidus, 42, 55, 64, 106, 110, 131, 327, 380, 381, 399, 406, 445, 449, 453, 455, 456, 507, 519
C. albidus var. aerius, 388
C. antarcticus, 12, 55, 105, 106, 202, 235, 406
C. aquaticus, 55, 384
C. carnescens, 55, 106, 108, 131
C. chernovii, 106, 112
C. curvatus, 159, 517
C. cylindricus, 131
C. diffluens, 106, 112, 129, 391, 447, 449, 453
C. dimennae, 106
C. flavescens, 199, 203
C. flavus, 55, 136, 404, 449
C. fonsecae, 55, 127, 137
C. frias, 127
C. friedmannii, 12, 61, 399
C. fuscescens, 105, 106
C. gastricus, 55, 65, 105, 106, 112, 117, 118, 131
C. gilvescens, 55, 105, 106, 112, 115, 117, 118, 235, 236
C. heimaeyensis, 55, 108
C. heveanensis, 55, 129, 138
C. humicola, 55, 387, 399
C. hungaricus, 55
C. laurentii, 36, 55, 106, 137, 389, 390, 399–402, 404, 406, 408, 445, 449, 453, 520
C. liquefaciens, 55, 64, 238
C. luteolus, 55
C. macerans, 36, 55, 106, 384
C. magnus, 55, 106, 131
C. nyarrowii, 12, 234–236
C. oeirensis, 55, 106
C. peneaus, 203
C. phenolicus, 131, 139
C. podzolicus, 55, 131, 139
C. psychrotolerans, 55, 127, 137
C. saitoi, 55, 64, 86, 106, 131, 154, 264
C. skinnerii, 55
C. spencermartinsiae, 127
C. statzelliae, 234

C. stepossus, 131
C. surugaensis, 154
C. tephrensis, 55, 106, 108, 131, 447, 449, 453
C. terreus, 106, 109, 131, 475, 476
C. terricola, 55, 105, 106, 109, 129
C. terricolus, 65
C. tronadorensis, 127
C. victoriae, 12, 26, 56, 105, 106, 108, 112, 129, 137, 234, 447, 449, 452, 453, 456, 523
C. vishniacii, 12, 26, 41, 82–85, 87, 406
C. watticus, 56
C. wieringae, 106, 129, 450, 453
Cuniculitrema, 37
Curvibasidium, 37
C. cygneicollum, 56, 131
Cystofilobasidium, 37, 49, 52, 90, 198, 215, 383, 384, 507
C. capitatum, 56, 106, 112, 131, 199, 202–205, 207, 336, 383, 384
C. infirmominiatum, 106, 131, 199, 202, 205, 450, 453, 456
C. lacus-mascardii, 127, 202
C. lari-marini, 56, 336, 384
C. macerans, 36, 106, 127, 202, 384
Cytophaga, 110, 309, 310

D
Debaryomyces, 37, 49, 60, 62, 499, 507, 512, 515
D. hansenii, 54, 62–65, 133, 238, 382, 445, 449, 450, 453, 500, 503, 506–509, 511, 513, 517–521, 523–525
D. maramus, 54, 63
Deinococcus
D. radiodurans, 87
Dekkera, 515, 516, 518, 524
D. anomala, 500, 504, 514, 520
D. bruxellensis, 500–503, 514–516
Delphinella
D. strobiligena, 200
Desulfitobacterium
D. hafniense, 329
Desulfotalea
D. psychrophila, 329
Dioszegia, 37, 52, 60, 107, 159, 193, 198, 199, 203, 205, 206, 213, 215, 264
D. antarctica, 12, 86, 88, 154
D. butyracea, 131
D. crocea, 107, 131
D. cryoxerica, 12, 82–85, 88
D. fristingensis, 107, 129
D. hungarica, 55, 107, 131
D. statzelliae, 12, 234
Dipodascus, 159
D. tetrasporeus, 154
Dothichiza, 138
Dothiora
D. cannabinae, 133

E
Elasticomyces, 61
E. elasticus, 176, 180
Erwinia
E. chrysanthemi, 329, 335
Erythrobasidium, 52, 198
E. hasegawianum, 107, 201, 213
Escherichia
E. coli, 40, 210, 244–247, 249, 250, 269, 273, 328, 329, 361, 391, 481–494
Exophiala, 159, 175, 176, 181
E. asiatica, 175
E. dermatitidis, 66, 107, 154, 175, 179, 200
E. phaeomuriformis, 179
E. pisciphila, 175
E. psychrophila, 175
E. salmonis, 175
E. spinifera, 175
E. xenobiotica, 200

F
Filobasidium
F. albidus, 64
F. globisporum, 41
F. uniguttulatum, 56
Flavobacterium, 110, 386
F. frigoris, 273, 274
Friedmanniomyces, 61, 177, 178
F. endolithicus, 176, 180
F. simplex, 176
Fusarium, 442

G
Galactomyces
G. geotrichum, 506, 517, 520
Geomyces, 312, 313
Geopyxis, 312
Geotrichum, 442, 518
Glaciozyma, 37, 66, 67, 88–90, 259, 262–271, 273–276
G. antarctica, 12, 66, 86, 88, 259, 263–267, 273, 274, 355, 382, 387–389, 399, 481, 483, 489–492

Glaciozyma (*cont.*)
G. martinii, 66, 87, 107, 108
G. watsonii, 66, 88, 107, 108, 235, 236, 474, 475
Gleocapsa, 209
Gloeosporium, 442
Graphiola, 154
Guehomyces, 37, 52, 60, 386
G. pullulans, 57, 107, 129, 386, 387, 451

H
Hafnia, 521
Hannaella, 37
Hanseniaspora, 37, 512
H. opuntiae, 133
H. uvarum, 133, 503
H. valbyensis, 133
Hansenula
H. polymorpha, 509
Holtermanniella, 37
H. festucosa, 107, 131
H. takashimae, 105, 107, 127, 139
H. wattica, 107, 132
Hormonema
H. dematioides, 133, 200
Hortaea
H. werneckii, 177, 200, 238
Humicola
H. lanuginosa, 356–358, 362

I
Issatchenkia
I. orientalis, 503, 511, 518

J
Janthinobacterium, 355

K
Kazachstania, 37
K. bovina, 232
K. exigua, 500, 503, 517
K. slooffiae, 232
K. telluris, 232
Kloeckera, 512
K. apiculata, 450, 503
Kluyveromyces, 37, 62, 159, 519
K. fragilis, 386, 387
K. lactis, 386, 387, 503, 517, 518, 524
K. marxianus, 232, 386, 500, 503, 517–520, 524, 525
K. nonfermentans, 154
Komagataella
K. pastoris, 269, 270, 272, 275, 361, 382, 455, 481–483, 489–494
Kondoa, 49, 52
K. malvinella, 201
Kuraishia
K. molischiana, 54
Kurtzmanomyces
K. nectairei, 201

L
Lachancea, 37
L. meyersii, 39
L. nothofagi, 127, 139
Lalaria, 198, 200
L. carpini, 133, 200
Lecanora
L. fuscobrunnea, 176
Lecythophora
L. mutabilis, 133
Leucosporidiella, 37, 49, 52, 475
L. creatinivora, 56, 107, 109, 132, 235, 236, 473, 475, 476
L. fragaria muscorum, 57, 65, 132
L. yakutica, 57
Leucosporidium, 37, 52, 59, 60, 107, 235, 263, 282, 313, 406, 492, 493
L. antarcticum, 56, 66, 265, 355, 387, 389, 399
L. frigidum, 234, 235
L. gelidum, 234, 235
L. nivalis, 234, 235
L. scottii, 56, 62, 235, 445, 447, 450, 452, 454
Lindnera
L. jadinii, 233
L. rhizosphaerae, 127, 139
Lipomyces, 398, 499
L. starkeyi, 233

M
Malassezia, 49, 52, 60, 81, 89, 90, 160
M. globosa, 81
M. restricta, 81
Mastigobasidium, 49, 52
M. intermedium, 107, 132, 475
Metschnikowia, 25, 37, 49, 65, 160, 499, 507, 512, 521
M. andauensis, 450
M. bicuspidata, 54, 63
M. fructicola, 455

M. pulcherrima, 54, 388, 450
M. reukaufii, 54
M. zobellii, 54, 63
Meyerozyma
M. guilliermondii, 54, 63, 154, 155, 450, 454, 456
Microglossum, 200
Mixia
M. osmundae, 202, 212
Monilinia, 442
Moritella
M. abyssi, 328, 329
Mrakia, 26, 36–38, 49, 52, 59, 60, 65–67, 90, 160, 235, 236, 264, 313, 383, 406, 467, 507
M. blollopis, 66, 107, 108
M. frigida, 36, 38, 56, 115, 132, 234, 235, 385
M. gelida, 36, 56, 65, 107
M. psychrophila, 36, 65, 107
M. robertii, 66, 105, 107, 108, 132, 235, 236
Mrakiella, 26, 36, 37, 49, 52, 60, 65, 160
M. aquatica, 42, 107, 132, 384
M. cryoconiti, 56, 107, 108, 115
M. niccombsii, 66, 107, 108
Mucor, 442
Mycogloea
M. nipponica, 201

N
Nostoc
N. commune, 209
N. punctiforme, 210

O
Occultifur, 201
Ogataea, 37
O. angusta, 450
O. polymorpha, 509
Oleispira
O. antarctica, 486

P
Penicillium, 442, 456
P. expansum, 456
P. paxilli, 383
Phaeomoniella
P. zymoides, 133
Phaeotheca
P. triangularis, 200
Phaffia, 139, 193, 198, 215
P. rhodozyma, 40, 139, 140, 199, 202, 205–207, 211–215
Phytophthora, 37
Pichia, 37, 49, 60, 62, 65, 149, 159, 398, 445, 471, 508, 512, 515, 519, 522
P. angusta, 450
P. anomala, 445, 451, 504, 513
P. farinosa, 388
P. fermentans, 517, 520
P. guilliermondii, 54, 63, 154, 155, 445, 450, 454
P. kudriavzevii, 54, 500, 503, 511, 513, 515, 517–520
P. lynferdii, 355
P. membranifaciens, 450, 452, 456, 503, 507, 508, 511, 513, 515, 520
P. pastoris, 269, 361, 382, 455, 482
P. spartinae, 54
Planococcus, 386
Polaromonas, 312
Protomyces, 49, 200
P. inouyei, 54
Pseudoalteromonas, 386
P. haloplanktis, 329, 335, 337, 338
Pseudomonas, 312, 355, 518, 521
Pseudozyma, 60, 155, 159, 353, 399, 507
P. antarctica, 12, 14, 17, 353, 355, 378, 379, 492
Pyrococcus
P. furiosus, 328

R
Rahnella, 386
Recurvomyces, 61
R. mirabilis, 176, 180, 181
Rhizopus, 442
R. niveus, 362
Rhodococcus, 474
Rhodosporidium, 159, 198, 215, 406
R. babjevae, 132, 199, 203, 204
R. diobovatum, 41, 56, 132, 155, 203, 237, 238
R. kratochvilovae, 56, 132
R. lusitaniae, 64, 107, 109, 264, 475
R. paludigenum, 445, 451
R. sphaerocarpum, 155, 203, 204
R. toruloides, 156, 161, 162, 233, 503
Rhodotorula, 37, 38, 49, 50, 52, 60, 62, 67, 75, 89, 90, 99, 107, 109, 115, 137, 149, 156, 157, 159, 161, 198, 199, 206, 213, 215, 263–265, 399, 406, 445, 471–475, 499, 500, 507, 508, 511–513, 515, 519–522

Rhodotorula (*cont.*)
R. acheniorum, 400, 401
R. arctica, 56
R. aurantiaca, 56, 107, 132, 156, 161, 162, 199, 202, 391
R. bacarum, 107
R. benthica, 156, 161, 162, 201
R. buffonii, 56
R. calyptogenae, 156
R. colostri, 107, 130, 199, 203
R. creatinivora, 56, 109, 235, 236
R. dairenensis, 161, 162
R. diffluens, 519
R. fujianensis, 56
R. glacialis, 56, 65, 107–109, 112, 115, 132, 137, 235, 236, 259, 263, 264, 473–475
R. glutinis, 156, 161, 162, 199, 203–207, 400, 406, 445, 450, 454, 517
R. graminis, 204
R. himalayensis, 137
R. hordea, 105, 107
R. hylophila, 56
R. ingeniosa, 107, 109, 475
R. lactosa, 201
R. lamellibrachii, 156, 161, 162, 201
R. laryngis, 56, 107, 112, 130, 201, 235, 236, 447, 451, 454
R. marina, 201, 212, 213
R. meli, 127, 201
R. minuta, 42, 56, 132, 156, 161, 162, 202–204, 212, 451
R. mucilaginosa, 36, 41, 56, 63, 64, 87, 110, 130, 156, 157, 161, 162, 199, 203–207, 215, 238, 264, 390, 400, 402, 451, 454, 503, 508, 517
R. muscorum, 57, 65
R. nothofagi, 128
R. pacifica, 157, 161, 162
R. pinicola, 132
R. psychrophenolica, 66, 105, 107–109, 112, 115, 137, 474, 475
R. psychrophila, 12, 107, 109, 137, 475
R. rubra, 205, 400, 451, 508
R. slooffiae, 132, 161, 162, 201
R. sonckii, 202
R. sphaerocarpum, 155, 203, 204
R. yakutica, 57

S

Saccharomyces, 5, 25, 37, 60, 62, 90, 141, 163, 379, 421, 422, 424, 471, 508, 512, 515, 519
S. bayanus, 141, 421, 423, 424, 515
S. bayanus var. uvarum, 141
S. carlsbergensis, 141, 424
S. cerevisiae, 5, 40, 83, 84, 86–89, 141, 142, 149, 163, 164, 168, 228, 229, 231–233, 236, 238, 243, 245, 247–254, 267, 281, 283–289, 329, 361, 366, 383, 388, 420–431, 500, 501, 503, 505, 508, 509, 511, 513–515, 517, 518, 525
S. diastaticus, 515
S. eubayanus, 128, 141, 142, 424
S. kudriavzevii, 142
S. pastorianus, 141, 142, 515
S. telluris, 232
S. uvarum, 133, 141
Saccharomycodes
S. ludwigii, 501, 504, 511, 514, 515
Saccharomycopsis
S. lipolytica, 388
Saitoella, 198, 200
Sakaguchia
S. dacryoidea, 201
Sanguibacter
S. antarcticus, 489, 492
Saprochaete
S. chiloensis, 128
Sarcinomyces, 154, 159, 209
S. petricola, 157
Saxomyces, 176, 177
S. alpinus, 180
Scheffersomyces
S. spartinae, 54
Schizoblastosporion
S. chiloense, 128
Schizosaccharomyces
S. pombe, 251, 500, 504, 509, 525
Sebacina, 312
Septobasidium
S. carestianum, 203
Shewanella
S. oneidensis, 488
Sporidiobolus, 37, 49, 52, 60, 158, 159, 198
S. longiusculus, 128, 203, 204
S. metaroseus, 57, 132, 203, 451, 475, 520
S. pararoseus, 158, 203, 406
S. ruineniae, 388
S. salmonicolor, 57, 132, 157, 158, 203, 400–404, 407–409
Sporobolomyces, 37, 49, 52, 60, 62, 90, 149, 159, 198, 206, 215, 398, 399, 406, 511, 512
S. albo-rubescens, 402
S. bischofiae, 201
S. clavatus, 201, 212

S. coprosmae, 199, 202
S. coprosmicola, 201
S. elongatus, 201
S. gracilis, 201
S. japonicus, 203
S. linderae, 201
S. pararoseus, 158
S. patagonicus, 128, 203
S. roseus, 57, 107, 406, 451, 475, 520
S. ruberrimus, 130, 203, 207
S. salicinus, 132, 201
S. salmonicolor, 57, 158, 400–404, 407–409
S. xanthus, 201
Starmerella, 25
Sterigmatomyces
S. halophilus, 201
Sugiyamaella
S. chiloensis, 128
Sydowia
S. polyspora, 133, 200
Synechocystis, 246

T
Taphrina, 198, 200
Thermomyces
T. lanuginosus, 356
Tilletiopsis, 154
Torulaspora, 37, 60
T. delbrueckii, 517, 520
Torulopsis
T. austromarina, 232
T. bovina, 232
T. candida, 232, 449
Trichosporon, 37, 38, 49, 52, 60, 149, 159, 468, 499, 507, 512
T. cutaneum, 57, 508, 517
T. dermatis, 158
T. dulcitum, 107, 109, 132, 474
T. mucoides, 57
T. pullulans, 57
T. porosum, 133
Trichothecium
T. roseum, 210
Trimmatostroma
T. salinum, 200
Typhula
T. ishikariensis, 259, 264, 273, 274

U
Udeniomyces, 37, 202, 212
Udeniomyces megalosporus, 133
Udeniomyces pannonicus, 107, 133
Udeniomyces pyricola, 133
Ustilago, 86

V
Vibrio, 210
V. cholerae, 329
V. salmonicido, 329

W
Wickerhamomyces, 37
W. anomalus, 445, 451, 500, 503, 504, 513, 515
W. lynferdii, 355
W. patagonicus, 128, 139
Williopsis, 37

X
Xanthophyllomyces, 37, 198
X. dendrorhous, 139

Y
Yarrowia
Y. lipolytica, 231–233, 353, 355, 356, 358–363, 367, 388–390, 445, 449, 470, 473, 500, 504, 507–509, 517, 518, 520–524

Z
Zygosaccharomyces, 25, 500, 511, 512, 515, 516
Zygosaccharomyces bailii, 500, 501, 515, 516
Zygosaccharomyces bisporus, 500
Zygosaccharomyces kombuchaensis, 500, 501, 504
Zygosaccharomyces lentus, 500, 501, 504, 516
Zygosaccharomyces mellis, 500
Zygosaccharomyces rouxii, 500, 509, 513

Printed by Printforce, the Netherlands